T0186611

Nanosilicon

Nanosilicon

Properties, Synthesis, Applications, Methods of Analysis and Control

A. A. Ischenko

Professor and Head of the Department of Analytical Chemistry
Moscow Lomonosov State University of Fine Chemical Technologies

G. V. Fetisov

Professor of Chemistry, Moscow Lomonosov State University

L. A. Aslanov

Professor and Head of the Laboratory of Structural Chemistry, Moscow

CISP

CRC Press
Taylor & Francis Group
Boca Raton London New York

CRC Press is an imprint of the
Taylor & Francis Group, an **informa** business

CRC Press
Taylor & Francis Group
6000 Broken Sound Parkway NW, Suite 300
Boca Raton, FL 33487-2742

First issued in paperback 2019

© 2015 by CISP
CRC Press is an imprint of Taylor & Francis Group, an Informa business

No claim to original U.S. Government works

ISBN-13: 978-1-4665-9422-7 (hbk)
ISBN-13: 978-0-367-37851-6 (pbk)

Visit the Taylor & Francis Web site at
http://www.taylorandfrancis.com

and the CRC Press Web site at
http://www.crcpress.com

Contents

Part I

Part II

Part III

Part IV

Foreword

Numerous experimental and theoretical studies show that the transition of particle sizes – the structural elements of matter – to the nanometric range, when the particle contains only a few tens or hundreds of atoms, leads to a qualitative change in the properties of the object. At the same time, the structural elements and material in general (nanomaterial), built on their basis can acquire the physical, physico-chemical and chemical properties differing substantially from the material, which is, for example, a bulk crystal, have identical qualitative composition or a substance constructed from particles of the micron or sub-micron size. On the other hand, the properties of the particles with dimensions of the order of several nanometers (several tens of angstroms) can be also very different from the properties of individual atoms and molecules, their generators. This refers to the nanosized crystals and clusters.

One of the essential, inherent characteristics of the nanocrystal is the distribution function of the particle size. The main parameters of the distribution function of the particle size – the position of the maximum, half-width, asymmetry and excess are essentially manifested in, for example, the optical properties of the nanomaterial and composites based on it. The directional change of these characteristics of the size distribution functions of the silicon nanoparticles leads to the correlated change of their spectral properties. The purposeful change of the particle size distribution function is achieved primarily by modification of the technological conditions for their preparation, fractionation, material, chemical effects in etching and annealing in different environments, monitoring the composition and amount of impurities in the core and the shell of nanocrystalline silicon, nanoparticle surface modification.

A striking example of the effect of the size on the properties of materials is nanocrystalline silicon, which in the form of nanosized objects – films of nanometer thickness – is transparent to visible light, but capable of absorbing ultraviolet light, whereas macroscopic samples of silicon are opaque to visible light. With the change in size the type of conduction of particles can change. For example, changing the diameter and chirality of carbon nanotubes changes the character of the conductivity of a metal to a semiconductor. It is known that gold is diamagnetic. However, direct measurements of spin polarization using the magnetic circular dichroism method at synchrotron radiation show that gold clusters ~ 1.9 nm in size, which contain about 210 atoms exhibit ferromagnetic properties unusual for gold. There are many other such examples, showing changes in the physical and chemical properties of systems with a decrease in their size to the nanometer scale.

The reasons for significant differences in the properties of nanoparticles in comparison with the properties of the bulk material are explained by their having a highly developed specific area surface, as well as quantum effects, manifested by size limitations. These factors can modify properties such as reactivity, strength, electrical and magnetic characteristics. The size constraint of the matter particles by the quantity comparable to the de Broglie wavelength of an electron leads to the situation in which the the quantum effects may dominate its behavior.

Reducing the particle size to a few atomic or molecular diameters leads to a drastic restriction of the range of possible energy states of electrons or excitons (the 'electron–hole' pair) in the particles, resulting in pronounced specific electro-optical and magnetic properties of the particles.

The objects in which these effects are pronounced, are quantum dots and quantum wells, which are used in optoelectronics. When the particle size decreases to a few hundred atoms, the density of states in the conduction band – upper zone containing free electrons – changes radically. The continuous density of states in the band is replaced by the discrete levels, the intervals between which may be greater than thermal energy, which leads to the formation of a slit similar to the bandgap in semiconductors. Among the variety of different quantum dots we can defined several main types that are most frequently used in experimental studies and numerous applications. First of all, it is nanocrystals in liquids, glasses, and wide-band dielectric matrices. If they are grown in glass matrices then they generally have spherical form. It is in such a system, in which CuCl quantum dots were embedded in a silicate glass, in which the study of one-photon absorption revealed for the first time the effect of three-dimensional quantum confinement of excitons described by Ekimov and Onuschenko in 1981 (Ekimov A.I., Onuschenko A.A., Quantum size effect in three-dimensional microscopic semiconductors, Pis'ma ZhETF, (1981), vol. 34, No. 6, pp. 363–366). This work marked the beginning of the rapid development of the physics of quasi-zero-dimensional systems.

The nanosized cluster consisting of atoms is somewhat similar to the molecule – the presence of a discrete set of energy levels, types of orbitals electronic transitions between which determine the spectral properties of the nanoparticles. Eventually, it is possible to reduce the cluster size to the level at which its diameter approaches the wavelength of the electron (de Broglie wavelength). In this case, the energy levels can be calculated by considering the quantum problem of elementary particles in a potential well or a limited periodic potential. The advent of new electronic properties of the nanoparticles in comparison with the material of macroscopic dimensions (bulk materials) can also be explained on the basis of the Heisenberg uncertainty principle: reduction of the region of spatial localization of the electron leads to an increase the range of the values of its momentum and energy. The energy spectrum of the electron will be determined not only by the chemical nature of atoms or molecules forming the particle, but also by the particle size. Interestingly, the quantum size effect is manifested in semiconductors at larger particle sizes than in metals because of the greater de Broglie wavelength of electrons in semiconductors. In semiconductors, the de Broglie wavelength can reach tens of nanometers, while in metals it is ~0.5 nm.

With the particle size decreasing to nanometers silicon exhibits properties unusual for the bulk material, which causes an interest in it, as in an optical material. At the end of 1990 there was a real sensation in the science of semiconductors. Lee Canham, a researcher from the UK, in the article (Canham L.T., Silicon quantum wire array fabrication by electrochemical and chemical dissolution of wafers, Appl. Phys. Lett., 1990. vol. 57, 1046–1048) reported the observation of an effective red-orange photoluminescence of porous silicon at room temperature, which is not characteristic of conventional silicon as the typical representative of indirect-gap semiconductors. This report triggered great interest to researchers around the world. It was shown that the effect observed is due to quantum size constraints of silicon nanostructures. It was established that reducing the size of the Si crystal to several hundred atoms makes it a completely new material with new properties – both electronic or non-electronic, including the ultra-bright and very stable luminescence which could not be observed previously for bulk samples.

Particular interest to the ultra-bright luminescence of silicon nanostructures is due to several reasons:

- First, it is interesting to know why this effect is negligible in bulk samples of Si, how to make it more prominent in the nanoscale state and whether it is possible to create solid-state lasers based on nanocrystalline silicon?
- Second, the synthesis of light-emitting nanoparticles can be very cheap and economically attractive for the production of light sources.
- Third, silicon is available in large quantities and can be a very effective material for microelectronics. Light-emitting silicon nanoparticles could be the basis for a new generation of electronic instruments and extend the functionality of silicon technology from microelectronics to optoelectronics and biophotonics.

The influence of the size effect on the photophysical properties of nanosilicon has been studied intensively and extensively from the beginning of the 90's of the XX century. The results show the influence of the surface of the silicon nanoparticles on the same properties that are influenced by the dimensional effect. nc-Si particles with a diameter of 2 nm are composed of 280 silicon atoms and 120 of them (43%) are on the surface and at a diameter of 1 nm nc-Si consists of 29 atoms of which only five are inside the nanoparticle. The composition and surface structure of nc-Si strong influence the LUMO energy, and it affects the band gap and optical properties of nc-Si. The ratio of the influence of the size effect and the surface on the optical properties of nc-Si is formulated in general terms as follows: the smaller the size of nc-Si, the larger the surface and the stronger the effect of the surface on the properties of nc-Si. Conversely, the larger the nc-Si, the stronger its effect of the internal atoms on its properties and the model of the size effect works better.

Crystal defects are traps of excitons and they are especially abundant on the surface of nc-Si. The excitons, captured by defects in the crystal structure, are called Frenkel excitons or small radius excitons, since the electron and the hole are located in the neighborhood at a distance of several angströms, i.e. much smaller than the size of the nanoparticle. In bulk silicon the excitons, captured by surface defects, can be both inside the forbidden zone and outside it. In the first case, they manifest themselves optically, in the second case – no. With decreasing particle size of nc-Si the width of the bandgap increases, and the surface states which in crystalline silicon were outside the bandgap fall into it, which changes photoluminescence (PL) spectrum. Thus, the size effect and the surface jointly affect the optical properties of nc-Si.

The rate of radiative recombination of electron-hole pairs localized on the surface is greatly increased compared to the free exciton in the volume of nc-Si, since the distance between the hole and the electron in the localized exciton (Frenkel exciton) is very small, and their wave functions will inevitably overlap. Using PL spectra with a time resolution it has been shown that the decay time of the PL emanating from the surface states is 4.1–4.2 ns, and the high rate of radiative recombination leads to the emission of many photons from one nc-Si, which is very important for practical applications. Usually, nc-Si emits red light from the volume with lifetimes in the microsecond and longer time ranges, and photoluminescence from surface states with lifetimes in the nanosecond range is shifted to the blue region of the spectrum.

Recently, it has been possible to get the PL spectra from single isolated nc-Si. It was found that the overall yield of PL varies considerably from one nc-Si to another. Such changes of quantum efficiency are attributed, in particular, to the effect of unexpected termination of the light emission from the nc-Si, which is caused by the presence of traps of charges on the surface of each particle of nc-Si. These traps can capture one of the charges of the electron–hole pair, making the nc-Si particle charged and, therefore, non-radiative. nc-Si particle returns to the radiative state when the charge carrier is released from the trap. Particles of nc-Si 'blink', and each nc-Si particle has its own dark time which depends on the configuration of the traps unique to each particle of nc-Si.

PL depends on the method of synthesis of nc-Si, and also on the surface modification of nc-Si with, for example, silicon dioxide, hydrogen or organic groups compounds. Some authors state that oxygen more efficiently modifies the surface of nc-Si than hydrogen and this results in a much greater increase in the luminescence intensity. Besides, it was found that the presence of (Si = O) and (Si–O–Si) groups on the surface of nc-Si results in an impurity level inside the band gap from which the radiative recombination of the exciton can take place. Thus, the luminescence spectrum should show the Stokes shift relative to the photoexcitation wavelength. This result was obtained by the theoretical method of the density functional for nc-Si with a diameter of less than 2 nm. However, the experimental results show that silica particles with a diameter less than 2.15 nm behave themselves as an amorphous substance, and since the properties of the amorphous and crystalline silicon are different, the possible amorphous state of very small silicon particles should be taken into account.

In nanosilicon there is competition between radiative and non-radiative recombination of electron–hole pairs as a result of excitation. One of the channels of non-radiative recombination centers are P_b-centres – single electrons in the dangling bonds of the surface atoms of silicon. Their elimination is one of the main tasks of modifying the surface of nc-Si. The surface of nc-Si to a large extent determines the quantum efficiency of luminescence of nc-Si.

After Canham's discovery, intensive research started into nanosilicon and the number of publications in the scientific periodical literature has been rapidly increasing from year to year (see the English-language statistics in Fig. F.1).

A large number of scientific publications has also come out in the national languages. For example, the statistics for 2010 show that, if we

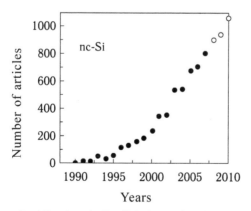

Fig. F.1. Number of publications in English devoted to the nanocrystalline silicon and silicon nanoclusters during the 1990–2010 period. Statistics for the period 1990–2008 is taken from the book (Pavesi & Turan, 2010) – see the following list of monographs. Statistics for 2009–2010 assembled by the same Internet search method using Google Scholar. Keywords: 'low-dimensional silicon' or 'silicon nanocrystal (s)' or 'silicon nanocluster (s)'.

take the number of English-language publications (often the results of international studies) as 1, the relative number of publications in the 'rare' languages (purely national research) is as follows: 1.64 – Japanese, 0.335 – Chinese, 0.038 – in Russian, 0.024 – Korean.

Published research for nanosilicon is regularly analyzed by international teams of experts and published in collective monographs, especially on the problems of synthesis, characterization, surface modification and application of nanosilicon. Only in English in the last 20 years after the discovery of luminescence in porous silicon more than 10 monographs were published, of which the most complete and well-known are:

- Properties of Porous Silicon, Ed. by L. Canham, INSPEC, London, UK., 1997.
- Nanoparticles and Nanostructured Films: Preparation, Characterization and Applications, Ed. by J.H. Fendler, Wiley-VCH Verlag GmbH, 1998.
- Light emitting silicon for microphotonics, Ed. by S. Ossicini, L. Pavesi, F. Priolo, Springer-Verlag, 2003.
- Towards the first silicon laser, Ed. by L. Pavesi, S.V. Gaponenko, L. Dal Negro, Springer, 2003.
- Silicon: evolution and future of a technology, Ed. by P. Siffert, E. F. Krimmel, Springer, 2004.
- Silicon Nanoelectronics, Ed. by S. Oda, D. Ferry, Taylor & Francis Group, LLC, 2006.
- Nanosilicon, Ed. by V. Kumar, Elsevier Ltd., 2007.
- Silicon nanophotonics: Basic principles, present status and perspectives, Ed. by L. Khriachtchev, World Scientific Publishing Co. Pty. Ltd., 2009.
- Device applications of silicon nanocrystals and nanostructures, Ed. by Nobuyoshi Koshida, Springer Science-Business Media, LLC, 2009.
- Silicon nanocrystals: Fundamentals, synthesis and applications, Ed. by L. Pavesi, R. Turan, Wiley-VCH Verlag GmbH & Co. KGaA, Weinheim, 2010.

However, unfortunately, in Russian so far no one book devoted to the systematic exposition of the scientific problems of nanosilicon and the prospects for its practical application. The authors of this monograph tried to fill this gap and present the main achievements in the field of physics and chemistry of nanosilicon. The book describes methods for production nanosilicon, its electronic and optical properties, research methods that allow to characterize the spectral and structural properties of this unique material, and some of its possible applications.

The book consists of 16 chapters, which are divided into four interrelated parts. At the same time, the book is structured so that each chapter can be read as an independent review on the subject indicated by its title. For this reason it was decided to leave the lists of references in each chapter and not constitute a single long list to the entire monograph. In general, the monograph reflects the results of more than 1500 publications.

The first part deals with the basic properties of semiconductors, including causes of the size dependence of these properties, structural and electronic properties and physical characteristics of the various forms of silicon with the theoretical and experimental research results. Examples of porous silicon and the quantum dots are used to analyze the quantum size effects enabling the control of the properties of nanocrystalline silicon. Because the book is intended for a wide audience interested in nanocrystalline silicon, to avoid distracting the reader by referring to the special literature, the beginning of this part contains a chapter that gives a summary of the physics of semiconductors and solid-state theory, which may facilitate the understanding of the subsequent material. Readers who are familiar with these items can skip this chapter.

The second part is central to the monograph and is devoted to the synthesis of nanosilicon, modification of the surface of nanoparticles and the properties of the resulting particles. Much attention is paid to the study of photoluminescence of silicon nanoparticles

The modified nanoparticles obtained to date are soluble in organic solvents and in water. It is shown that the properties of the obtained material depend on the size of the central silicon core and the protective shell. However, only a few studies clearly shows the influence of the shell on the optical properties of nanosilicon. Based on the analysis of the studies presented in this chapter, it is possible to conclude that the numerous results are ambiguous and, in some cases, contradicting. The chemical composition of the surface layer significantly affects the LUMO energy and, respectively, the bandgap value of nc-Si with a diameter less than 5 nm. This helps to understand the mismatch of data across different studies of the photoluminescent properties nanoparticles, which have identical dimensions.

A significant difference between the properties of the material built, for example, from individual the atoms or molecules of the substance having the same composition but comprising nanoparticles is linked not only with the manifestation of the quantum size effect, but also with the absence of the identity of structural elements of the nanomaterial. In contrast to the atoms or molecules, the nanoparticles are not completely identical. Strictly speaking, they differ not only in size but also in composition. It can be assumed that even nanoparticles of the same size are not identical because of possible differences in the quantity, composition and distribution of impurities which each nanoparticle contains. The presence and composition of the impurities are determined by the entire 'history' of nanoparticles, depending on the method of synthesis and subsequent location in the environment. Each nanoparticle is a single compound, and we are dealing with a system of particles for which it is possible to determine the function of the size distribution, but the question of the composition of the individual particles remains open. Appropriate methods of analysis of the nanoparticles have not as yet been developed.

The third part of the book discussed the methods used for the study and controlling the structure and properties of nc-Si. First of all, it is a 'standard' set methods – electron microscopy, spectroscopic and diffraction methods with the features of their application to the study of nanoparticles – which is used to determine the characteristics of nanosilicon. In addition, a separate chapter is devoted to the use of femtosecond spectroscopy, ultrafast electron nanocrystallography and dynamic transmission electron microscopy.

In the past two decades has been possible to observe the motion of the nuclei in the time interval corresponding to the period of oscillation of the nuclei. The observed coherent changes in the nuclear subsystem at such intervals of time determined the fundamental shift from the standard kinetics to the dynamics of the phase trajectories – quantum state tomography.

Therefore, despite considerable progress in the study of nc-Si, with traditional methods of analysis and research, we need to develop new methods for studying the structure and dynamics of nanoparticles. These effective methods that can clarify the relationship of the central core and the nc-Si surface layer are femtosecond spectroscopy and time-resolved diffraction techniques. The application of these methods to the study of nano-objects and nanosilicon generally provides new opportunities for studying the behavior of systems in 4D space-time continuum.

The fourth part describes in detail some of the practical applications of nanocrystalline silicon. One of them is the use of nanoparticles as additives–absorbers of UV radiation and in sunscreens.

It is known that the properties of nanomaterials, as artificial objects, integrable in human activity, are significantly different from the natural properties, in an environment where the humanity has survived for thousands of years. Cosmetics manufacturers often ignore this factor, even neglecting the proper observance of the federal regulations of application of nanomaterials. Numerous examples, given in this monograph, show that reducing the particle size of nanoscale objects changes many of the physical and chemical properties, including, e.g., colour, solubility, material strength, electrical conductivity, magnetic properties, mobility (including the environment and the humans), chemical and biological activity.

Increasing the surface area to volume ratio also increases the chemical activity, increasing the speed of processes, which in the case of bulk materials can be very slow. For example, the effect of the oxygen in the air atmosphere on a nanoparticle can lead to the formation of singlet oxygen and other reactive oxygen-containing species (ROS). ROS have been found in the metal oxide nanoparticles of TiO_2 and ZnO, carbon nanotubes and fullerenes, and is the active component oxidative stress, inflammation and consequent damage to DNA, proteins and membranes. Furthermore, the action of ROS opens the possibility of formation of a series of reaction channels involving free radicals. In the case of small TiO_2 nanoparticles DNA damage was significantly greater than with the large TiO_2 particles. Although TiO_2 microparticles (~500 nm) can also cause damage to DNA, the nanoparticles of medium size (~20 nm) cause complete destruction of the DNA helix at low doses of UV radiation or without its effect.

Because of the potential toxicity of metal nanoparticles, to replace them in sunscreens it is essential to find alternative ingredients for sunscreen materials which are non-toxic and, at the same time, effective in blocking ultraviolet radiation. As shown in a number of works nanosilicon, being a biocompatible and biodegradable material can become one of the most promising materials for use in sunscreens.

The monograph ends with the subject index.

Much of the material presented in this monograph includes the results obtained by the authors or their participation, and leading experts in this field of research were invited to write some chapters that reflect the latest applications and research methods,

This work was supported by RFBR Fund: Grants 11-02-07031-a, 10-02-92000-HHC-a and 11-02-00868-A.

Acknowledgements

One of the authors (A.A. Ischenko) expresses his sincere appreciation to Prof. V.N. Bagratashvili (Institute of Problems of Laser and Information Technologies, Russian Academy of Sciences), Prof. V.R. Flid (Lomonosov Moscow University of Fine Chemical Technology), Corr. Member of the Russian Academy of Sciences, Prof. P.A. Storozenko (State Scientific Research Institute of Chemistry and Technology of Organic Element Compounds, Lomonosov Moscow University of Fine Chemical Technology), Dr. S.G. Dorofeev (Chemistry Department of Moscow State University), Dr. N.N. Kononov (General Physics Institute, Russian Academy of Sciences), Dr. A.A. Ol'khov (Lomonosov Moscow University of Fine Chemical Technology), Dr. V.A. Radtsig (Institute of Chemical Physics, Russian Academy of Sciences), Dr. A.O. Rybaltovsky (Institute of Nuclear Physics, Moscow State University), Dr. V.N. Seminogov (Institute of Problems of Laser and Information Technologies, Russian Academy of Sciences), A.V. Bakhtin (Lomonosov Moscow University of Fine Chemical Technology), G.V. Bulkhova and Ya.A. Ischenko for valuable discussions, comments and invaluable help in writing the monograph.

Abbreviations

AES   – Auger electron spectroscopy
AEY   – Auger electron yield – measuring mode of the Auger electron yield (one of the mode of measurements of XAFS).
AFM   – atomic force microscopy
a-Si   – amorphous silicon
a-Si:H – hydrogenated amorphous silicon
BET   – Brunauer–Emmett–Teller method for describing the adsorption of gases
CARS   – coherent anti-Stokes Raman scattering
CCD   – charge coupled device
CL   – cathodoluminescence (see EL – electroluminescence)
CRN   – continuous random network
c-Si   – crystalline silicon with a particle size >0.1 μm
CVD   – chemical vapour deposition
DAFS   – diffraction absorption fine structure – the fine absorption structure as measured by X-ray diffraction
DAS   – dimer–adatom–stacking fault
DCA   – diffusion cluster approximation
DCS   – dynamic light scattering
DFT   – density functional theory – the method of calculation of the electronic structure of many-particle systems in quantum physics and quantum chemistry
DFTEM   – dark field transmission electron microscopy
DMC   – diffusion Monte Carlo method
DMF   – N, N-dimethylformamide
DMSO   – dimethylsulphoxide
DOS   – density of states – the density of electronic states – the number of energy levels in the energy range per unit volume
DOX   – doxorubicin
DSC   – differential scanning calorimetry
DTEM   – dynamical transmission electron microscopy
ECL   – electrochemical luminescence
EDEXAFS – energy dispersive EXAFS spectroscopy
EDMR   – electric detection of magnetic resonance
EDS   – energy dispersive X-ray spectroscopy
EELS   – electron energy loss spectrometry
EFTEM   – energy-filtered transmission electron microscopy
EL   – electroluminescence (see CL)

EMA   – effective medium approximation
EMF   – electromotive force
EPR   – electron paramagnetic resonance – electron spin resonance
           spectroscopy
EXAFS   – extended X-ray absorption fine structure
FCC   – face-centered cubic lattice
FFT   – fast Fourier transformation
FITC   – fluorescein isothiocyanate (fluorescent label in biology)
FLY   – X-ray fluorescence yield mode – mode of measurement of X-ray
           fluorescence yield (one of the modes of measurements of XAFS
           spectra)
FPSN   – fluorescent porous silica nanoparticles
FTIR   – Fourier transform infrared spectroscopy
HCP   – hexagonal closed-packed cell
HOMO   – highest occupied molecular orbital
HRTEM   – high-resolution transmission electron microscopy
ICP MS   – inductively coupled plasma mass spectrometry
INDO   – intermediate neglect of differential overlap
IP   – imaging plate – photosensitive plate with optical memory (a type
           of X-ray detector)
IR   – infrared
ITO   – solid solution of indium and tin oxides.
LACVD   – laser-assisted chemical vapour deposition
LCEB   – the method of linear combination of energy bands
LDPE   – low-density polyethylene
LDS   – local density of states
LECBD   – low-energy cluster beam deposition technique
LPCVD   – vapor deposition of materials at a low pressure
LUMO   – lowest unoccupied molecular orbit
mc–PTFE – polytetrafluoroethylene microparticles
mc-Si   – silicon single crystal
MEG   – multi-exciton generation mechanism
MINDO   – modified intermediate neglect of differential overlap
MINDO3– Modified INDO, version 3
MOS   – metal–oxide–semiconductor (structure).
nc-Si   – silicon nanocrystals (particles with dimensions <0.1 μm) or
           nanocrystalline silicon
NEXAFS – near edge X-ray absorption fine structure – near-edge fine
           X-ray structure of the absorption spectrum (the spectrum in
           the range ±(30–50) eV relative to the excitation energy of the
           absorption threshold)
NMR   – nuclear magnetic resonance
n–-Si   – doped silicon with electronic conductivity
NW   – NanoWire – nanowires

OIHH   – organic–inorganic hybrid hydrogels.
OTCS   – octyltrichlorosilane
p+-Si   – doped Si with p-type conductivity
PC   – paramagnetic center
PCCS   – photon cross-correlation spectroscopy
PCS   – photon correlation spectroscopy
PDF   – pair distance (distribution) function
PE   – polyethylene
PECVD – vapor deposition of materials after they are processed by cold
            plasma
PEG   – polyethylene glycol
PET   – photoinduced electronic transition
PL   – photoluminescence
PLY   – photo-luminescence yield – yield of photoluminescence excited
            by X-rays (one of the modes of measurements of XAFS).
PMR   – proton magnetic resonance
P-Si   – porous silicon
PTFE   – polytetrafluoroethylene
PVP   – polyvinylpyrrolidone
QCE   – quantum constraint effect
QCSE   – Stark effect in quantum constraints
QD   – quantum dot
QELS   – quasi-elastic scattering – PCS with wavelength change
QMC   – the quantum Monte Carlo method
QWR   – quantum whiskers (nanowhiskers)
RBS   – Rutherford backscattering spectrometer
RDF   – radial distribution function
RGB   – red, green, blue – an additive colour model of colour synthesis
            based on red, green and blue
RPDF   – reduced pair distribution function – the reduced pair distance
            distribution function
RPECVD – remote PECVD – enhanced by indirect CVD plasma (CVD with
            a substrate outside the plasma discharge)
RS   – Raman scattering
RWL   – the natural width of the Raman line
SAED   – selected area electron diffraction
SAXS   – small angle X-ray scattering
sccm   – cm^3 per minute at standard temperature and pressure
SC-CO_2 – supercritical carbon dioxide
ΔSCF – self consistent field
SCF   – supercritical fluids
SDDF   – size distribution density function of size
SEM   – scanning electron microscopy
SEXAFS – surface EXAFS spectroscopy.
STEM   – scanning transmission electron microscopy

STEM–PEELS – method of spectral imaging in a scanning TEM with parallel electron energy-loss spectrometry
STM   – scanning tunneling microscopy
STS   – scanning tunneling spectroscopy
TEM   – transmission electron microscopy
TEY   – total electron yield – measurement mode of the total electron yield in irradiation with X-rays (one of the modes of measurement of XAFS spectra)
TFT   – thin film transistor
TGA   – thermogravimetric analysis
THF   – tetrahydrofuran
TMOS   – tetramethoxysilane
TOAB   – tetraoctylammonium bromide
TPSC   – thin film solar cell
TREOS   – triethoxysilane
UEC   – ultrafast electron crystallography
UEnC   – ultrafast electron nano-crystallography
UV   – ultraviolet
VLS   – vapor–liquid–solid – 'vapor–liquid–solid' growth mechanism'
VSS   – vapor–solid–solid – 'vapor–solid–solid' growth mechanism
XAFS   – X-ray absorption fine structure – the fine structure of X-ray absorption spectrum
XANES   – X-ray absorption near-edge structure – near-edge fine structure
XEOL   – X-ray excited optical luminescence – (one of the modes of measurements of XAFS spectra)
XPS   – X-ray photoelectron spectroscopy
XRD   – X-ray analysis

Part I

1

Some properties of semiconductors – terminology

As is known from solid state physics (Ashcroft and Mermin, 1979a and 1979b; Davydov, 1976; Kittel, 1978), the solids can be classified according to the nature of electrical conductivity as electrical conductors (usually metals), semiconductors and insulators (dielectrics). Appreciable conductivity is typical of conductors and semiconductors. In terms of the band theory of solids[1], semiconductors differ from metals in that their electrons are located in the valence band, separated from free conduction-band by the forbidden band (energy gap), the width of which can range from several tenths of eV to several eV. In conductors (metal) there is no forbidden gap between the valence and conduction bands. Another significant difference between semiconductors and metals is the strong direct dependence of conductivity on temperature, whereas in metals it is the opposite. In addition, the conductivity of semiconductors is strongly dependent on the purity of the material, i.e. the presence of impurities.

Semiconducting properties can be observed in crystalline materials, as well as in disordered systems – solid amorphous materials (glasses) and liquids (Bonch-Bruevich and Kalashnikov, 1977, Seeger, 1977; Yu and Cardona, 2002; Yacobi, 2004). The decisive factor is the nature of chemical bonding between the particles in the short-range order (in the first coordination sphere) and, consequently, the type of lattice in crystalline solids. This fact is confirmed experimentally. For example, semiconductors such as diamond, silicon, germanium under high pressure are converted into metals because the energy gap between the filled valence band and empty conduction band disappears. This is accompanied by the change of the interatomic bonding and crystal structure. The structure of diamond with

---

[1]Band theory of solids is a section of quantum mechanics studying the movement and state of electrons in a solid (see, for instance, Anselm, 1978; Ashcroft and Mermin, 1979; Bonch-Bruevich and Kalashnikov, 1977, Kaxiras, 2003; Moliton, 2009).

tetrahedral coordination[2] of the interatomic bonds changes to the structure of metallic tin with a higher coordination number. The covalent bond, which can be represented in the form of electronic bridges connecting the atoms in the diamond structure, becomes metallic, non-directional.

There are semiconductors with any type of chemical bond (covalent, covalent–metallic, covalent–ionic) with the exception of pure metallic and purely ionic, and the covalent component of the bond in semiconductors is usually dominant.

The conductivity of semiconductors is strongly dependent on temperature. An intrinsic semiconductor (or a semiconductor of type i, i.e. ultrapure single crystal semiconductor without impurities and free from defects of the crystal lattice) at the absolute zero temperature has no free carriers in the conduction band, in contrast to conductors, and behaves as an insulator. The permissible content of impurities at which a semiconductor can be regarded intrinsic (i.e. semiconductor of type i), depends on the forbidden band gap of the material. Thus, to achieve the level of intrinsic conductivity at room temperature, the permissible content of impurities in Ge should be $<5 \cdot 10^{-9}\%$, while in silicon $<5 \cdot 10^{-11}\%$.

Semiconductors differ from metals by a number of properties:

1) the electric current in semiconductors, in contrast to metals, is formed by two types of charge carriers – negatively charged electrons e^- and positively charged holes e^+;
2) the electrical conductivity of semiconductors is greatly lower than that of metals;
3) the electrical conductivity of semiconductors increases rapidly with increasing temperature, and, as a rule, in a wide temperature range increases exponentially, while in the metals it is reduced and is usually much less dependent on temperature;
4) the electrical conductivity of semiconductors very strongly depends on their purity (the impurity concentration) and usually grows with the introduction of impurities, whereas in the metal the dependence of electrical conductivity on the impurity content is weak and usually of the opposite sign (i.e. most often the introduction of impurities decreases the electrical conductivity of a metal);
5) The electrical conductivity of semiconductors affects exposure to light or ionizing radiation – for metals, this effect is virtually absent or very weak.

Interestingly, these properties of semiconductors are completely similar to the corresponding properties of insulators. The reason for this similarity is an equivalence of the electronic structures of atoms, semiconductors and

---

[2]The short-range order of the distribution of the atoms is quantitatively characterised by the coordination number – the number of the nearest neighbours of the atoms, and in the diamond lattice it is equal to 4, i.e. the atoms in the diamond lattice are characterised by tetrahedral coordination.

insulators, which are different from the electronic structures of metal atoms. For this similarity the semiconductors are sometimes referred to as semi-insulators. The features of the electrical conductivity of semiconductors, metals and insulators can be clearly explained and quantitatively described in terms of the band theory of solids (see, for example, Anselm, 1978, Ashcroft and Mermin, 1979, Bonch-Bruevich and Kalashnikov, 1977, Kaxiras, 2003; Moliton, 2009).

The main factors determining the fundamental properties of the semiconductors (e.g. optical and electrical) are associated with the chemical composition and crystallographic structure, the presence of various defects and impurities (both accidental and intentionally introduced), as well as with the size of a semiconductor or semiconductor structure. The chemical composition and crystal structure define the structure of electron bands (e.g. width and type of band gap and effective mass of charge carriers), which have a major impact on the semiconductor properties. The presence of various defects and impurities is expressed in the addition of various electronic states (both shallow and deep) in the band gap, which greatly alter the optical and electrical properties. Finally, as shown in sections 1.4 and 2.7, as well as in Chapters 4 and 11, the semiconductor properties may be strongly influenced by the effect of quantum-dimensional constraint when the size of the semiconductor becomes comparable with the de Broglie wavelength (in silicon this is usually observed with sizes less than ~10 nm).

1.1. The electrical conductivity of semiconductors from the viewpoint of band theory

Free electrons can have any energy – their energy spectrum is continuous. The electrons, belonging to isolated atoms, have certain discrete values of energy. In a solid, electrons are not only bonded with the atoms, but are also in a potential field of neighbouring atoms, so their energy spectrum is substantially different, it consists of some of the allowed bands separated by bands of forbidden energies.

Electrons in an atom are located at different orbital levels characterized by different distances from the nucleus and, accordingly, a different binding energy of the electron with the nucleus. In the formation of the crystal lattice of the solid the electron orbits are somewhat deformed, and, accordingly, the energy levels of electrons in them are shifted. This shift can be imagined in two ways. On the one hand, we can see that in a solid, an electron can not be subjected to electrical interference on the part of neighbouring atoms – it is attracted to their nuclei and is repelled by electrons. On the other hand, two electrons, by virtue of the Pauli exclusion principle[3], can not be on the same orbit in the same energy state. Therefore,

---

[3]The Pauli principle can be formulated as follows: the given quantum system in the given quantum state can contain only one particle, and the state of the other particle must differ by at least one quantum number.

in the formation of a solid body and in the crystallization of atoms into a rigid structure each energy electronic level in the atoms splits into a number of close sublevels combined into an energy band or zone. All the electrons in this energy band have very similar energies. At the orbits close to the nucleus the electrons are in a bound state: they are unable to break away from the nucleus, because although in theory the electron hopping from one atom to another – on the same orbit with respect to energy – is possible, all the lower orbits of the neighbouring atoms are occupied, and the actual migration of electrons between them is not possible.

The most important feature from the viewpoint of electrical conductivity is the valence band – the outer layer of the electron shells of atoms blurred and divided into sub-levels, which in the majority of substances is not filled (the exception – the inert gases, but they crystallize only at temperatures close to absolute zero). Since the outer layer is not saturated with electrons, it always contains free sub-levels, which can take electrons from the outer shell of neighbouring atoms. And the electrons, in fact, migrate within the valence band, and in the presence of an external electrical potential difference they collectively move in one direction, creating an electric current. That is why the region with lower binding energy in which there are freely moving electrons is called the conduction band.

The multiband theory of the structure of the solid can be used to explain the electrical conductivity of the substance. If the valence band of the solid is filled, and the next unfilled energy band is far away, the likelihood that electron transfers to it is close to zero. This means that the electrons are strongly bound to atoms, do not form a conductive layer, and can not leave their place, even under the influence of an external potential difference. This state is typical for insulators – substances that do not conduct electric current.

1.1.1. Direct- and indirect-gap semiconductors

A very important characteristic for the classification of semiconductors is their response to photon radiation. Absorption of light by semiconductors may lead to photon excitation of electrons in a semiconductor so that they pass from one level of the band to another, or rather from the valence band to the conduction band. However, the due to the Pauli exclusion principle, the electrons can move only from the filled energy level into an empty one. In an intrinsic semiconductor (i.e. in the semiconductor, free from impurities and extended defects), all states of the valence band are filled and all states of the conduction band are free, so transitions are possible only from the valence band to the conduction band. To implement such a transition, the electron must receive from the light energy not lower than the band gap width. A photon with lower energy does not cause transitions between electronic states of the semiconductor, so that the semiconductor is transparent to such photons. Transparency is observed for the photons from the frequency range $\omega < E_g/\hbar$, where E_g (energy gap) is the band gap width of the semiconductor, $\hbar$ is Planck's constant. This cut-off frequency $E_g/\hbar$ determines the fundamental absorption edge of a semiconductor. For semiconductors which are often used in electronics (silicon, germanium, gallium arsenide), the fundamental absorption edge lies in the infrared range.

Additional constraints on the absorption of light by a semiconductor are imposed by quantum selection rules, in particular the law of conservation of momentum. The law of conservation of momentum requires that the quasi-momentum[4] of the final state be different from the quasi-momentum of the initial state by the amount of the absorbed photon momentum. The wavenumber (absolute value of the momentum) of the photon is $2\pi/\lambda$, where λ is the wavelength, is very small in comparison with the modulus of the vector of the reciprocal lattice of the semiconductor, or, equivalently, the wavelength of a photon in the visible region is much larger than the characteristic atomic spacing in the semiconductor. As a consequence, the requirement of momentum conservation is to ensure that the quasi-momentum of the final state in the electronic transition should be almost equal to the quasi-momentum of the initial state. However, this condition at frequencies close to the fundamental absorption edge can be satisfied not for all semiconductors. The point is that there are direct-gap and indirect-gap semiconductors, and this condition is satisfied only for the former. The scheme of the dependence of the energy of current carriers on the quasi-momentum for the direct- and indirect-gap semiconductors is shown in Fig. 1.1.

Optical transitions in semiconductors, in which the electron momentum is almost constant, are called direct or vertical. However, the momentum of the final state may differ significantly from the momentum of the initial state, if in the process of absorption of a photon another, third particle, for example, a phonon, is involved. Such transitions are also possible, although less likely. They are called indirect transitions.

Direct-gap semiconductors such as gallium arsenide, are beginning to strongly absorb light when the light quantum energy exceeds the band gap. In the transition of an electron from the valence to the conduction band free charge carriers form in the semiconductor and hence photoconductivity. Upon return of the semiconductor excited by photons to a stable initial state reverse transitions of the electrons occur from the conduction band to the valence band, accompanied by the emission of a photon. These

---

[4]The quasi-momentum is a vector quantity characterising the state of the quasi-particle. The quasi-particle differs from a free particle by the fact that it is subjected to the effect (excitation) from a system in which it is located. The 'free' electron, situated in the solid with a variable potential, is a quasi-particle in this sense since it moves in the periodic field of the crystal lattice. In addition, the quasi-free electron (quasi-electron) has the effective mass m^* which can greatly differ from the mass of the actual free electron. This also applies to other quasi-particles.

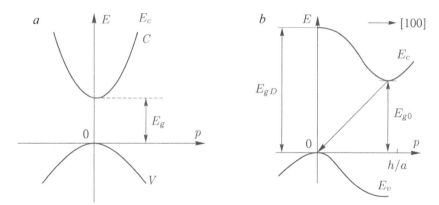

Fig. 1.1. The positions of the energy maximum in the valence band E_v and of the energy minimum E_c in the conduction band compared to the electron momentum p for photoexcited transition of the electron through the forbidden zone: a – for the direct-gap semiconductor, b – for the indirect-gap semiconductor (scheme for a single crystal of pure silicon), where E_{gD} is the width of the energy gap between the bands in direct and E_{g0} – in indirect transition.

semiconductors have effective photoluminescence and can be used in optoelectronics.

Indirect-gap semiconductors such as silicon, absorb in the light frequency range with the quantum energy slightly larger than the width of the band gap far less efficiently and only due to indirect transitions. The absorption intensity in such a case depends on the presence of photons and hence on temperature. If the indirect transition a photon is absorbed at the same time as a quantum of light, the energy of the absorbed light quantum may be less by the value of the phonon energy, which leads to absorption at frequencies slightly lower in energy than the energy of the fundamental absorption edge. The cutoff frequency of direct transitions of silicon (E_{gD} in Fig. 1.1b) is greater than 3 eV, i.e. it lies in the ultraviolet spectral region.

1.2. Quasi-particles in a solid

To describe the properties of solids we use such an important concept as 'collective disturbance of the system', in which a number of quasi-particles is introduced (see, for example, Brandt and Kul'bachinskii, 2010, Ashcroft and Mermin, 1979a,b; Kaxiras, 2003). In contrast to particles such as bosons, they are not similar to the particles that make up the real systems, and represent the collective (i.e. coherent) motion of many physical particles. Below is a list of quasi-particles and collective excitations, which are often used to describe the physics of solids, in the order of importance when considering the use of the basic material of this monograph.

1. **Quasi-electron**: a quasiparticle consisting of a real electron and an exchange-correlation hole, forming a cloud of the effective charge of opposite sign due to exchange and correlation effects arising from the interaction of all electrons in the system. If the electron is in a periodic potential, its motion is considered as the motion of the quasi-particles. The concept of the quasi-electron was introduced by L.D. Landau. The quasi-electron is a fermion with spin 1/2. Its Fermi energy (the highest occupied state) is about 5 eV and the Fermi velocity ($v_F = \hbar k_F/m_e$) about 10^8 cm/sec, i.e. belongs to the category of non-relativistic particles. The effective mass of the quasi-particles can significantly differ from the free electron mass.

2. **Hole**: a quasi-particle as a quasi-electron, but with a positive charge The hole corresponds to the absence of an electron in the single-particle state below the Fermi level. The concept of the hole is especially useful when the initial state is formed by the states of quasi-particles, which are fully occupied and separated by the energy gap from the unoccupied states. The concept of the hole is introduced in the band theory to describe the electronic effects in the valence band incompletely filled with electrons.

3. **Exciton**: a hydrogen-like quasi-particle, which is the electronic excitation in a dielectric or semiconductor, migrating through the crystal and not associated with the transfer of electric charge and mass. This is a collective excitation corresponding to the bound states of the conduction electron and the hole. There are two types of excitons: Frenkel excitons and Vanier–Mott excitons. If the electron and hole are located in the same lattice site, i.e. the exciton radius (the distance between the electron and the hole) is $a^* < a$, where a is the crystal lattice spacing, the quasi-particle is called a Frenkel exciton. But such states are usually observed only in molecular crystals, which are also described by Frenkel excitons. If the exciton radius is much greater than the atomic spacing in the crystal, i.e. $a^* \gg a$, then the quasi-particle is called the Vanier–Mott exciton. This type of exciton is typical for semiconductors, where the high dielectric constant leads to a weakening of the electrostatic attraction between the electron and the hole, with the result that it leads to a large electron radius. The binding energy of the electron–hole pair is proportional to $e^2/(\varepsilon \cdot a^*)$, where ε is the dielectric constant of the material (for semiconductors its value varies around 10), a^* is the distance between two particles forming the exciton (which is usually several lattice spacings and is measured in tens of angstroms), resulting in an energy of ~0.1 eV, or a few tenths of eV. The Vanier–Mott exciton binding energy is generally 0.02–0.005 eV in semiconductors and equals 0.014–0.015 eV in particular for a silicon crystal (Kaxiras, 2003; Yacobi, 2004). For example, for a silicon crystal at a temperature of 1.7 K, this energy is ~0.74 eV.

4. **Phonon**: a quasi-particle, which is a quantum of vibrational motion of the atoms in the crystal. The concept of the phonon was introduced by I.E. Tamm to describe the collective excitation corresponding to the coherent

oscillation of the thermal atoms of a solid. The phonon energy is of the order of $\hbar\omega \sim 0.1$ eV.

5. **Polaron**: a component quasi-particle (conductance electron + associated phonons), which can move through the crystal as an entity. The concept of the polaron is introduced to describe the polarization in the crystals, which occurs when a negatively charged electron, moving through a lattice of positively or negatively charged ions, causes perturbation of the ions around itself and distorts the crystal lattice. Since the electron motion in this case is associated with the displacement of the ions, the effective mass of the polaron is much greater than the free electron mass.

6. **Plasmon**: a quasi-particle corresponding to the quantization of the collective oscillations of the gas of free electrons relative to ions in the crystal. The presence of plasmons and of the collective states in the system described by them manifests itself in the long-range Coulomb interactions and is optically observed as the passage or reflection of light. The light with a frequency below the plasma frequency ω is reflected because the conduction electrons screen the electric field in the light electromagnetic wave, which is most pronounced in metals. The light with a frequency higher than the plasma frequency passes through the solid, because the electrons can not respond quickly enough to screen it; this is mostly characteristic of semiconductors and insulators. The role of the plasmons is particularly great in describing the optical properties of metals and doped semiconductors, and hence also of the transformations between metallic and semiconductor states of the solid, such as in the transition to the size of the particles, creating quantum size limitations. The plasmon energy is defined as $\hbar\omega \sim \hbar\sqrt{4\pi ne^2/m_e}$, where n is the number of charged particles per unit volume, and can be measured either by experiments or on the basis of energy losses of the electrons passing through the sample as the energy of the formation of plasmons, or when analyzing the spectrum of light radiation emitted by plasmons. For solids with the usual electron density the plasmon energy lies in the range from 5 eV to 20 eV. In most metals the plasma frequency is in the ultraviolet spectral region, making them shiny when viewed in visible light. In doped semiconductors, the plasma frequency is usually in the infrared region.

There are other quasi-particles used to characterize the properties of solids in condensed matter physics, but they are not as relevant for further understanding of the material of this book, but interested readers can find a detailed discussion of this subject, for example, in Brandt and Kul'bachinskii (2010).

1.3. The influence of defects on the electronic structure

Many of the major defects are electrically active (Yu and Cardona, 2002). Defects that can create free electrons in a crystal are called donors and defects which create a hole (i.e. destroying the free electrons) are called acceptors. Examples of defects-donors in Si are the substitutional atoms from the group V, such as P, As and Sb, or monovalent interstitial atoms such as Li and Na. The atoms of group V have one valence electron more than the Si atoms they replaced. In addition, the excess electron is weakly bound to an atom of group V in Si and can be easily excited into the conduction band of the Si crystal. Substitutional atoms from group VI, for example, S, Se and Te in Si, can give two electrons to the conduction band, so they are called double donors. Examples of acceptors in Si are the substitutional atoms of group III, such as B, AI, Ga and In. The substitutional atoms of Group II in Si (e.g. Be, and Zn) are double acceptors. If an impurity substitutional atom has the same valence as the host crystal atom, it is an isoelectronic or isovalent centre.

In the literature, one often finds the terms 'deep centre' and 'shallow centre' which can be interpreted as some defects with electronic levels located respectively near the middle of the band gap or near the allowed bands.

It is difficult to predict in advance whether a specific impurity atom will be a deep or shallow centre. This can be helped by the following symptoms, which, however are not absolutely immutable.

If the 'skeleton' of the impurity atom (the atom minus the outer valence electrons) is similar to the skeleton of its own atom (they can vary by a single nuclear charge), the impurity levels tend to be shallow. If an impurity atom creates a highly localized potential, the most likely result will be the emergence of a deep centre.

The defects are the centres of electron localization. As these electrons and holes are fixed, they are said to be linked. In contrast, electrons in the conduction band and the hole in the valence band of a semiconductor can carry electrical current. That is why they are called free carriers.

In some cases, the defect can produce both shallow and deep bound states. For example, electrons with the envelope wave functions with s-symmetry act as deep centres more frequently than the electrons from the envelope wave functions having p-symmetry.

1.3.1. Localized states

The most important point defects of the crystal structure of semiconductors are chemical impurities, vacant lattice sites (vacancies) and atoms embedded in the interstitials. A common feature for them is the ability to bind and release electrons (Madelung, 1985).

In terms of the influence of point defects on the band model, it turns out that delocalized band states and localized states, associated with defects, can be treated together in one scheme of the energy bands.

In describing the properties of crystalline solids we generally consider two systems: the electrons and the crystal lattice in which there are electrons. These two systems naturally interact with each other. The properties of electrons in a potential field of the crystal lattice, i.e. 'quasi-electrons', are described by the band model, while the properties of the lattice vibrations are described by the phonon spectrum. Originally the band theory assumes a strict periodicity of the structure of the lattice, which is actually violated by defects. Around some lattice defects some of the delocalized phonon states become localized and decoupled from the continuous spectrum in the form of branches (localized lattice vibrations), resulting in an additional potential around the defect and the reasons for the localization of electronic states. Taken together, this affects the optical properties of solids and the characteristics of the transfer in them.

To understand the influence of defects on the electronic structure of solids one can solve the Schrödinger equation (1.3) for electrons moving in a periodic lattice by adding to it the additional potential introduced by the defect. Examples of this type of analysis can be found in most textbooks on solid state and semiconductor physics (see, e.g. Madelung, 1985).

At large distances of the conduction electron (quasi-electron) from the positive charge, generated by atomic ions, the crystal lattice screens the Coulomb potential like a homogeneous medium with dielectric constant ε. In this case, the perturbing potential of the lattice defect is reduced to the potential of the hydrogen atom in a dielectric medium, i.e. in a continuum approximation, the problem of electron motion in the crystal lattice with a point defect is reduced to the well-known and long solved problem of solid state physics.

To limit the number of Bloch functions in the wave packet which participates in the solved equations, we can use the following approximation, in which we only consider the Bloch functions of the conduction band. This can be done if the energy with which the electron is bound in the defect is small compared with the binding energy of the valence electron in the lattice (the width of the energy gap E_g). This approximation corresponds to the defects which are called shallow impurities and allows us to work within the approximate theory of effective mass m^*. This condition is satisfied for the majority of donors and point defects in the crystal lattice. Otherwise (deep impurities) the approximation of effective mass m^* is inapplicable, and the solution should also take into account the Bloch functions of the valence band. Defects with the properties of the deep impurity whose binding energy is comparable to E_g, act as traps and centres of recombination. The binding energies of the most important donors in Si is less than 1% of the band gap, so they can be considered as shallow impurities (Madelung, 1985).

The results of solving the problem for an electron with effective mass m^* in the positive charge field in a medium with permittivity ε show that the bound states form a hydrogen-like spectrum, the lines of which (donor levels) lie under the lower edge of the minimum of the conduction band. The Bohr radius of the orbit of the ground state is increased as compared to its value for the free hydrogen atom (0.53 Å) ε (m/m^*) times. For silicon this leads to the value of the Bohr radius in the range from 20 to 50 Å.

Isolated defects modify not only modify the electronic states in the band model, but also the states in the vibrational spectrum of the crystal lattice. Here, as in the case of the electronic states, the influence of defects has only a small effect on the branches of the phonon spectrum, but in lattices with a basis (as in silicon, having a lattice with basis 2) leads to the appearance of localized states between acoustic and optical branches and above optical branches, and also creates conditions for the possibility of resonant states inside the branches (Madelung, 1985). Localized states are often active in the infrared range and can be found in the absorption spectrum of the crystal.

To describe the properties of solids within the framework of the band theory we use the concept of energy bands, the collective of electrons in the conduction band and collectives of holes in the valence band. To account for the real solid structures, we add to these concepts all the charged and uncharged defects of the crystal lattice, which are considered as separate groups. And all of these collective (each with its own chemical potential) are considered in the interaction, i.e. in reactions with each other. In this description, for example, the relaxation of the excited electron from the conduction band to the donor level means the reaction of the free electron with a positively charged donor atom with the formation of a neutral donor atom. In general, the interaction of the collectives is described similar to the kinetics of chemical reactions using the equality to zero of the total potential of the system in equilibrium and the law of acting masses, which describes the balance between the different collectives.

Of all the collectives of the considered defects two types of lattice defects – vacancies and interstitial defects – are fundamentally different from the impurity atoms. The difference is that the impurity atoms in the ideal case can be completely removed from the crystal, while the vacancies and interstitial defects are thermodynamically unavoidable.

In the conventional model of a semiconductor the photon absorption process implies the formation of electron–hole pairs, with the hole remaining in the valence band and the electron appearing in the conduction band. Accordingly, the reverse process is called the recombination of electron–hole pairs. For interband recombination, the electron returns to the valence band, again emitting a photon, and annihilate with the hole (radiative recombination, photoluminescence). However, radiationless recombination is also possible; in this case the electron returns to the valence band, giving the recombination energy to another electron (Auger recombination) or crystalline lattice by multiphonon processes.

In any recombination process the conservation laws of energy and momentum must be satisfied. Together with the probability of transition, these laws define the lifetime of the excited electron–hole pairs. The transition probability can be significantly higher if the recombination occurs via a two-step process through a defect in the crystal structure (recombination centre). In this case, the electron is first captured by the defect, and then released into the valence band. In order to implement this mechanism, the recombination centres should have sufficiently large effective cross sections for interaction with both the conduction and the valence band.

If we consider the kinetics of recombination processes generated by the external excitation of electrons, aloof from radiative or nonradiative transition, without the defects the only possible transitions are the transition of electrons from the conduction band to the valence band. Lifetime τ of the electron–hole pair in the steady state depends on the magnitude of excess concentration δn electrons, which in the case of small deviations from equilibrium concentration is in the ratio $\delta n = G\tau$, where G is the number of electron–hole pairs generated by the external force per unit time.

However, remember that along with the absorption-excited electron–hole pairs in the crystal there are always electron–hole pairs generated by thermal excitation, which constantly appear and recombine, but retained their equilibrium concentration.

1.4. Photoabsorption and luminescence of semiconductors

The peculiarity of the electronic structure of semiconductors is expressed in their specific response to photon radiation and excitation of the electron energy levels (Harrison, 1983; Yu and Cardona, 2002). This is due to the ability of many semiconductor for luminescence.

Luminescence is a glow of a substance that occurs after the absorption of excitation energy by the substance. Following the canonical definition luminescence will mean the excess over the thermal radiation from the body if this excessive radiation has a finite duration (order of 10^{-10} seconds or more), far exceeding the period of light fluctuations. The concept of duration in this definition was introduced by Russian scientist S. Vavilov[5]. The duration criterion makes it possible to separate the luminescence from other types of non-thermal radiation: scattering and reflection of light, Raman scattering, bremsstrahlung radiation and Cherenkov radiation. Their duration is less than the oscillation period the light wave (i.e. $<10^{-10}$ s).

An important feature of the luminescence is that it can occur at much lower temperatures than thermal radiation, since it does not use the thermal energy of the radiating system. Therefore, luminescence is often called 'cold light'. The physical nature of luminescence is the radiative transitions of electrons from the excited to ground state. In this case the cause of the initial excitation of the system can be represented by a variety of factors: external radiation, chemical reactions, etc.

For a substance to be able to luminesce, its spectra should be of discrete nature, i.e. its energy levels must be separated by zones of forbidden energies. This is just typical for semiconductors. At the same time, metals in solid and liquid states, which have a continuous energy range, do not give

---

[5]http://www.femto.com.ua/articles/part_1/2015.html

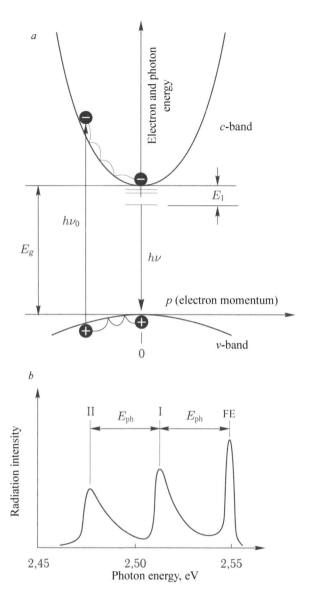

Fig. 1.2. The mechanism of photoluminescence in direct-gap semiconductors. a – scheme of the mechanism: $h\nu_0$ – energy of the exciting photon; $h\nu$ – the energy of the emitted photon, the wavy lines indicate the path of the cooling of electrons and holes generated by the exciting photon; E_1 – energy of formation of the free exciton. b – photoluminescene spectrum of free excitons in CdS crystal at $T = 60$ K: FE – the peak of radiative recombination of the free excitons without phonons; I and II – the free exciton emission peaks with the birth of respectively one and two phonons with energy E_{ph}. Figure from (Aghekyan, 2000).

luminescence. The excitation energy in metals is converted in a continuous way into heat, and metals can under X-ray fluorescence only in the short-range, i.e. under the influence of X-rays or electrons of sufficiently high energy to emit X-rays (characteristic X-rays).

For the excitation of luminescence the material must receive the energy exciting the electronic levels in its electronic structure. Depending on the form in which energy is supplied to the luminescent body (luminophor) we distinguish photo-, cathode-, X-ray-, electroluminescence, and so forth.

Photoexcitation of the semiconductor by the absorbed photon prior to emission of fluorescent radiation (the process of photogeneration) can be clearly explained by the diagram in Fig. 1.2, which shows the case of the direct-gap semiconductor.

In this scheme, the transition of an electron from the filled valence band (v-band) to the empty conduction band (c-band) in the absorption of a photon crystal is represented by a vertical arrow, because the transition takes place with virtually no change in momentum, because the momentum of the photon $p_{ph} = 2\pi/\lambda$ is very small compared to the length of the electron bands on the scale of p (i.e. as regards the modulus, the momentum of a light photon is much smaller than the length of the reciprocal lattice vector). If the energy extrema of both bands are located at one value of p (usually $p = 0$), the direct-gap semiconductor is called direct-gap, if at different p – it is an indirect-gap semiconductor. After absorption of a photon with energy $hv_0 > E_g$ the electron with momentum p_c appears in the c-band and a hole with momentum p_v forms in the v-band (i.e. photogeneration of free charge carriers takes place). The conservation laws of energy and momentum are satisifed in the process of light absorption, so that

$$p_v + p_c = p_{ph} \sim 0. \tag{1.1}$$

The free charge carriers, generated in absorption of a photon, naturally try to lower their energy to zero. But this opportunity in the separated bands separated is limited – for the hole by the maximum energy of the valence band, for the electron by the minimum of the conduction band. Therefore, the electron drops to the bottom of the c-band, and the holes floats to the ceiling of the v-band, and as a result in direct-gap semiconductors we have an electron and a hole with momenta close to zero. The excess energy released by this relaxation or cooling of the generated charge carriers is used to enhance the thermal vibrations of the crystal lattice, or in other words, leads to an increase in the number of thermal particles – phonons.

The luminous radiation itself is emitted in the process following excitation – recombination. As a result of recombination the electron returns to the valence band (and thus the hole also disappears) and the crystal will be in the original state. The transition of a conduction electron to the vacant site in the valence band (the annihilation of the electrons and the hole) is accompanied by emission of a photon with a frequency $\omega \leq E_g/h$. Such transitions in the emission and absorption spectra correspond to the broad (and relatively weak in intensity) bands. Thus, the photoluminescence spectrum of semiconductors is determined by the radiative recombination of non-equilibrium electrons and holes produced by the light.

1.4.1. Electroluminescence of semiconductors

In addition to photoluminescence, electroluminescence also occurs in semiconductors. In 1923, O.V. Losev (see Novikov, 2004) observed the glow of silicon carbide crystals when voltage was directly applied to them, and in 1936 G. Destriau reported the glow of fine crystalline zinc sulphide activated with copper (ZnS:Cu), stirred in a dielectric liquid and placed between the plates of a condenser to which alternating voltage was applied.

This means that in the case of semiconductors luminescence can be excited not only by electromagnetic radiation (photoluminescence), but also by the supply of electricity to the semiconductor (electroluminescence). The strong electric field, close to the breakdown value, can excite semiconductor crystals both due to tunnelling of electrons from the valence band and the luminescence centres the conduction band, and through the acceleration of electrons in the electric field to energies sufficient to ionise the crystal lattice and luminescence centres (impact ionization). In addition, there can be shock-excitation of the luminescence centres. The recombination of electrons and holes, both directly and through the centres of luminescence, as well as a return to the initial state of the excited luminescence centres, lead to the emission of luminescence light. This luminescence is called prebreakdown electroluminescence.

The electric field can also lead to electroluminescence (EL) by another mechanism. The point is that inorganic luminophors are wide- gap semiconductors with impurity conductivity. If they are in direct contact with the electrodes, then the current flow may inject into them additional minority charge carriers. The radiation produced by recombination of these carried with the main carriers, is the luminescence commonly called injection electroluminescence in the literature. This luminescence is observed, for example, the *p–n* transitions activated in the forward direction. This species of electroluminescence is important for light-emitting diodes.

1.5. Dimensional constraints and quantum effects

The features of the electronic properties that are manifested when the size of the particles approached nanometer range are explained in a simple and accessible way in, for example, review articles (Borisenko, 1997; Belyavsky, 1997; Demikhovskii, 1997, Schick, 1997). To learn more about the effects of quantum dimensional constraints in semiconductors, the reader should refer to, for example, textbooks (Yu and Cardona, 2002; Yu and Cardona, 2010; Hamaguchi, 2010). The consequences of the quantum-dimensional constraints will often be discussed in connection with nanosilicon in

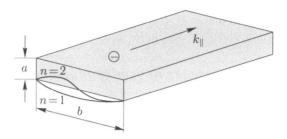

Fig. 1.3. Possibilities for movement of electrons in a nanostructure with a quantum-dimensional constraint.

subsequent chapters of this monograph. Here we consider only the physical essence of this phenomenon.

From the standpoint of quantum mechanics, an electron can be represented by a wave, described by the corresponding wave function (see, for example, Anselm, 1978, Ashcroft and Mermin, 1979, Bonch-Bruevich and Kalashnikov, 1977, Yu and Cardona, 2002, as well as many other textbooks on solid state physics, including physics of semiconductors). The distribution of this wave in nanoscale solid-state structures is controlled by effects associated with the quantum constraint, the interference and the possibility of tunnelling through potential barriers.

A wave corresponding to the free electron in a solid can be easily extended in any direction. The situation changes dramatically when an electron enters the solid state structure, whose size L, at least in one direction, is limited and its size is comparable to the wavelength of the electron. A classical analogue of such a structure is a string with fixed ends. Vibrations of the string can occur only in the mode of standing waves with a wavelength $\lambda_n = 2L/n$, $n = 1, 2, 3,...$. Similar patterns of the behaviour are also characteristic of a free electron in the solid state structure of a limited size or an area of a solid bounded by impenetrable potential barriers.

In Fig. 1.3 this situation is illustrated by the example of a quantum string, which has limited sizes of the cross section a and b. Only with the length divisible by the geometric dimensions can propagate in the directions a and b. The allowed values of the wave vector in one direction are given by $k = 2\pi/\lambda_n = n\pi/L$ ($n = 1, 2, 3, ...$), where L as shown in Fig. 1.3 can make values equal to a or b. For the corresponding electrons, this means that they can have only certain fixed values of energy, i.e. an additional quantization of energy levels takes place. This phenomenon is called the quantum constraint. In this case, the motion of electrons across the string is restricted and allowed for electrons with a certain energy state, and electrons with any energy can move along the string.

Locking the electron with effective mass m^*, at least in one of the directions, in accordance with the principle of uncertainty leads to an

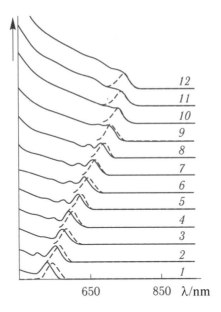

650   850   λ/nm

Fig. 1.4. Shift of the absorption (solid curves) and photoluminescene bands (dashed lines) with increase of the size of CdTe nanoparticles: $d = 3.2$ (*1*), 3.3 (*2*), 3.5 (*3*), 3.6 (*4*), 3.8 (*5*), 4.2 (*6*), 4.6 (*7*), 5.3 (*8*), 6.2 (*9*), 7.7 (*10*), 8.4 (*11*) and 9.1 nm (*12*). The figure from (Gubin et al., 2005).

increase ot its momentum by the value of h/L. Accordingly, the electron kinetic energy increases by the amount

$$\Delta E = \frac{\hbar^2 k^2}{2m^*} = \frac{\hbar^2}{2m^*} \cdot \frac{\pi^2}{L^2}. \tag{1.2}$$

Thus, the quantum constraint is accompanied by both an increase in the minimum energy of the locked electron, and by additional quantization of energy levels, corresponding to its excited state. Consequently, the electronic properties of nanoscale structures differ from the known bulk properties of the material from which they are made.

For sufficiently small semiconductor and metal nanoparticles, electrons in them can have only certain discrete values, i.e. the transition from the continuous energy spectrum of conduction electrons to discrete range takes place. With a decrease in the size of the semiconductor nanoparticles, starting from a certain size characteristic for each type of semiconductor, the energy band gap width increases (Fig. 1.4) and the optical absorption spectrum is shifted to shorter wavelengths; this phenomenon is often called 'blue shift'.

The main consequence of reducing the crystal size to the size of the quantum dot (cluster)[6] is an extension of the band gap of a semiconductor by an amount which is inversely correlated with the size of the cluster. In these

---

[6]For more details of clusters see Chapter 2.

clusters, the electrons need more energy to move during photoexcitation to free level in the conduction band. The photoemission process, following the absorption process is also characterized by a blue shift, because the excited electrons return to their ground state through a wider forbidden gap. The shift in the emission spectrum is due to the same changes in the available energy states which are created by the shift in the absorption spectrum.

These shifts of the absorption and emission spectra toward the high-energy side appear with decreasing size of the semiconductor particles to a value comparable to the diameter of the Vanier–Mott exciton in a bulk semiconductor. At such small particle sizes the excited electrons and holes associated with them are limited in all three directions, so this form of particles is called 'quantum dots'. The word 'quantum' is used to emphasize that the unusual electronic and optical properties are the result of limitations of the excited electrons to a finite number of available quantum states. To understand why this is happening, it is useful to turn to the example of the survey (Bley and Kauzlarich, 1998), reviewing changes in the energy spectrum of electrons in the formation of the crystal 'bottom up' – from individual atoms to the bulk material (Fig. 1.5).

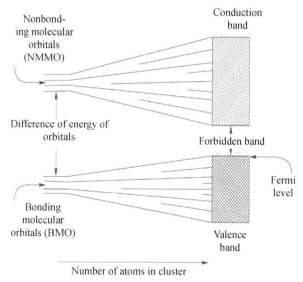

Fig. 1.5. The scheme of formation of zones of the binding (BMO/HOMO) and nonbonding (NMMO/LUMO) orbitals with increasing atomic number of the semiconductor material in the cluster. When there is a sufficient number of atoms in the cluster the discrete states overlap, forming a band structure that represents a continuum of possible states. At the beginning the Fermi energy coincides with the energy of the top of the valence band, but with an increase in the number of electrons in a quantum dot (i.e. the atoms in the cluster) it increases, and when a cluster of the macroscopic size forms the Fermi level is located in the middle of the forbidden gap (Bley & Kauzlarich, 1998).

A small cluster of several atoms of a semiconductor has only a few energy levels for any of the excited electrons (non-bonding molecular orbitals). This situation is similar to that which occurs in the molecules. Consequently, the optical and electronic properties of the cluster must be very similar to the properties of molecules. When the number of atoms in the cluster increases, new quantum states are added. This is illustrated in Fig. 1.5 by the extension of both the bands representing the occupied molecular orbitals binding in the valence band and the unoccupied antibonding bands of molecular orbitals in the conduction band.

The result of adding atoms to the cluster is to increase the total number of possible energy levels for excited electrons in the cluster in addition to a general decrease in the energy difference between the levels HOMO and LUMO (here we use the names of the energy states borrowed from the terminology commonly used in quantum-chemical literature)[7]. In terms of the quantum-mechanical description this means that the electrons (quantum-mechanical waves) are locked in this small area and will only have resolved energy spectra which correspond to standing waves that are available to them in clusters of different sizes. This process of increasing the number of available states with a decrease of the energy difference between the HOMO and the LUMO continues as more new atoms are added to the cluster as long as there is a necessary continuum of possible energy levels for excited electrons, and the energy difference between HOMO and LUMO is not equal to the width of the forbidden band of the bulk semiconductor. From this point the electronic and optical properties of the cluster are exactly the same as that of the bulk material. Adding an even greater number of atoms will only increase the size of the cluster, but no longer affects its optoelectronic properties.

It should be noted that the model of the quantization of the conduction band of the semiconductor and expansion of the band gap with transition from the bulk crystal to a 'quantum dot' was first substantiated by Al.L. Efros and A.L. Efros from the A.F. Ioffe Physico-Technical Institute in an article (Efros, Al. and A. Efros, 1982) and the citation index to date has reached tens of thousands. The authors explained the experimentally observed in (Ekimov and Onuschenko, 1981) the blue shift in CuCl nanocrystals of 30 Å, considering the quantum constraints in interband absorption of light in the semiconductor field, depending on the diameter of the sphere. This rather simple model shows exactly the picture of the band structure changes with decreasing particle size of the semiconductor, which is shown in Fig. 1.6 (size reduction occurs in the direction from right to left).

A unique property of quantum particles, including electrons, is their ability to penetrate the barrier, even when their energy is below the potential barrier corresponding to this barrier (tunnelling effect). Schematically this is shown in Fig. 1.7.

---

[7]The abbreviations HOMO (highest occupied molecular orbital) and LUMO (lowest unoccupied molecular orbital) are used widely in quantum chemistry.

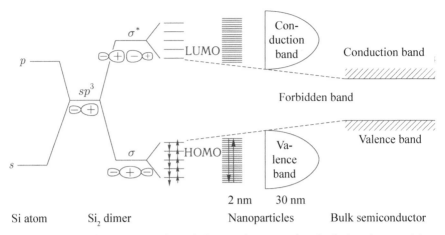

Fig. 1.6. Schematic representation of changes in energy levels during the transition from Si atom to clusters Si$_n$, nanoparticles with a gradual increase in size from 2 to 30 nm and a bulk semiconductor. Figure from (Gubin et al, 2005).

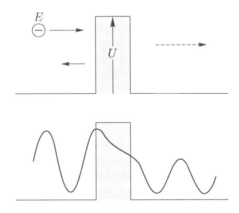

Fig. 1.7. Tunnelling of electrons with energy E through a potential barrier with height U, $U > E$.

If the electron were classical particle, having energy E, then the electron meeting an obstacle in its path and requiring higher energy U for overcoming the obstacle would reflect from this barrier. However, as a wave, the electron, though with a loss of energy, passes through this barrier. The corresponding wave function, Ψ and through it the tunnelling probability, is calculated from the Schrödinger equation:

$$-\frac{\hbar^2}{2m^*}\frac{d^2\Psi}{dx^2} = (E-U)\Psi(x). \tag{1.3}$$

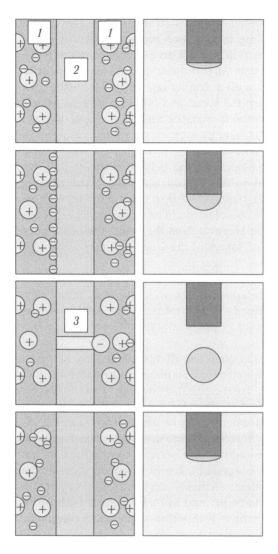

Fig. 1.8. Single-electron tunnelling in the Coulomb blockade: *1* – conductor layer *2* – layer of insulator, *3* – tunnelling electron. Figure from (Borisenko, 1997).

This probability is higher, the thinner the barrier and the smaller the difference between the energy of the incident electron and the barrier height. The quantum constraint, which manifests itself in nanoscale structures, imposes a specific imprint on the tunnelling. Thus, the quantization of energy states of electrons in very thin, periodically spaced potential wells leads to the fact that tunnelling through them becomes resonant, i.e. only electrons with a certain energy can tunnel through such a structure.

Another specific manifestation of the quantum constraint is single-electron tunnelling in Coulomb blockade conditions. To explain this term, consider illustrated in Fig. 1.8 an example of passing of the electron through the metal–insulator–metal structure. As an illustration, the figure show also an analogy with a droplet coming off the edge of a tube. Initially, the interface between the metal and the dielectric is electrically neutral. When applying a potential to metallic regions a charge will start to accumulate at this boundary. It lasts as long as its value can be no longer sufficient for the separation and tunnelling through of a single electron the dielectric. After the act of tunnelling, the system returns to its original state. When the external applied voltage is conserved the entire process is repeated again. Thus, the charge transport in this structure is carried out in portions, equal to the charge of one electron. The process of charge accumulation and the detachment of an electron from the metal–dielectric boundary is determined by the balance of forces of the Coulomb interaction of electrons with other mobile and fixed charges in the metal.

In the quantum constraint mode one can artificially create materials with the properties intermediate between those of molecules and bulk material. This gives tremendous flexibility in the design of materials with special properties and a lot of potentially useful new applications for semiconductor materials.

The considered quantum effects are already being used in the currently available nanoelectronic components for information systems (Koshida, 2009; Pavesi & Turan, 2010). However, it should be emphasized that they do not exhaust all the possibilities of utilizing the quantum behaviour of electrons in instruments. Active exploratory research in this area continues today. Recently developed nanoelectronic elements due its small size, speed and power consumption are serious competition to traditional solid-state transistors and integrated circuits based on them as the main elements of information systems. Already, technology is very close to the theoretical possibility to remember and to transmit data 1 bit of data (0 and 1) with a single electron whose location in space can be defined by a single atom.

1.5.1. Luminescence of indirect-gap semiconductors

The intensity of luminescence is high for direct-gap semiconductors which include, for example, compounds such as $A^{III}B^{V}$, widely used for the manufacture of LEDs. In contrast, in the indirect-gap semiconductors, which include silicon, the luminescence intensity is negligible, i.e. it is very difficult to make them glow. The reason for this can be understood by referring to the energy diagram of an excited indirect-gap semiconductor, i.e. the dependence of the energy of an electron in a crystal on its quasi-momentum (Kashkarov, 2001). For single-crystal Si, this dependence is well known and is shown in Fig. 1.9. In indirect-gap semiconductors, the

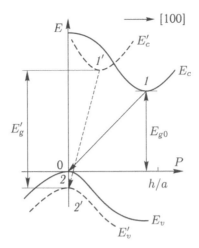

Fig. 1.9. The dependence of the electron energy E of the quasi-momentum P for single crystal (graphics $E_c(P)E_v(P)$) and a silicon quantum wire (graphics $E_c'(P)$ $E_v'(P)$) Figure from (Kashkarov, 2001).

energy extrema of the valence band and the conduction band are shifted as regards the momentum relative to each other (curves *2* and *1* in Fig. 1.9).

As with all semiconductors, the branches of allowed states – the conduction band $E_c(P)$ and the valence band $E_v(P)$ – are split by the band gap E_{g0} ($E_{g0} \approx 1.1$ eV). The emission of light quanta in the excited defect-free Si crystal at room temperature occurs at the transitions of electrons from the branch levels $E_c(P)$ to the branch levels $E_v(P)$. In the non-degenerate excited material the states at the bottom of the conduction band E_c are filled (point *1*), and those at the top of the valence band E_v are vacant (point *2*) – Fig. 1.9.

In the investigated transition the conservation laws of energy and quasi-momentum must be satisfied:

$$E_{c1} - E_{v2} = E_{ph},$$
$$P_{c1} - P_{v2} = P_{ph},$$

(1.4)

where E_{ph} and P_{ph} are the photon energy and quasi-momentum, as since $P_{ph} \ll P_{c1}$, we can consider $P_{ph} \approx 0$.

Silicon is a typical indirect-gap semiconductor, i.e, the absolute minimum of the *c*-band is shifted with respect to the absolute maximum of of he *v*-band along the momentum axis. Consequently, quasi-momentum conservation in the transition $1 \to 2$ is not possible and the transition will be banned. Such a process can be implemented in indirect-gap semiconductors can be only with the participation of the third particle – phonon, which reduces the probability of transition by two orders of magnitude (i.e. 100 times) compared with the case of direct-gap materials (e.g. GaAs). As a result, the quantum efficiency η of photoluminescence (PL) – emission of light under the influence of optical excitation – in c-Si at not too high excitation levels is only 10^{-4} %. Consequently, a single photon is emitted per million of exciting photons absorbed in Si, and the PL is usually observed at low temperatures ($T = 4$–80 K). The energy of the emitted photons is close to E_{g0} and falls on the near infrared range.

Fortunately, this is not the end for the applications of silicon as a luminescent material. There are ways in which silicon can be made to luminesce efficiently in the visible region. These methods are discussed, for example, in a review article (Kashkarov, 2001) or in a monograph (Khriachtchev, 2009).

The electronic properties of silicon can be changed by the formation of Si-based nanostructures – spatially separated Si regions with minimum dimensions of several nanometers. In this case, the charge carriers (electrons and holes) acquire additional energy due to the previously considered quantum size effect, leading to an increase in the band gap of the nanostructure compared to the single crystal of this material (E_{g0}), in particular for the quantum well:

$$E_g' = E_{g0} + \frac{\pi^2 \hbar^2}{2m_n d^2} + \frac{\pi^2 \hbar^2}{2m_p d^2}, \qquad (1.5)$$

where $h = 1.05 \cdot 10^{-34}$ J·s is Planck's constant, m_n and m_p are the effective masses of the electron and the hole, respectively, d is width of the well.

It is important to note that in silicon a decrease of the size of is accompanied by the shift of the absolute minimum of the conduction band to the left, thereby reducing the difference of the quasi-momenta in the initial $1'$ and the final $2'$ states (see Fig. 1.9). In addition, manifestation of the Heisenberg uncertainty becomes visible for nanostructures:

$$\Delta p_x \cdot \Delta x = \hbar, \qquad (1.6)$$

and the law of conservation of the quasi-momentum is no longer strict. These reasons substantially increase the probability of optical transitions and determine the shift of the PL spectrum to the visible region.

Unfortunately, the technology for creating nanostructures in the form of so-called quantum wells, wires and dots (or thin films, wires and tiny crystals embedded in another substance) is rather complicated and not yet developed to the level of mass production. In addition, the results obtained so far do not suggest any practical industrial application of these silicon structures in the light-emitting semiconductor devices. However, the available studies give grounds for optimism (Khriachtchev, 2009).

As early as 1956 A. Uhlir (Uhlir, 1956) produced the so-called porous silicon. This material was a single-crystal silicon (mc-Si) and its surface contained a huge number of tiny pores as a result of electrochemical etching. The density of pores in some samples was so large that they

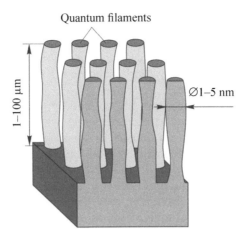

Fig. 1.10. Scheme of the formation of nanowires on the surface of a single crystal Si wafer by the merger of the surface pores created by electrochemical anodic etching of the wafer. Figure from (Kashkarov, 2001).

overlapped and the uneteched silicon sites were in the form of coral-like filaments of variable cross sections (Fig. 1.10).

The minimum dimensions of the cross section of silicon filaments and isolated areas (clusters) in the porous layer equal, according to electron microscopic data, the units of nanometers. Porous silicon is usually formed as a layer on the surface of the mc-Si wafer, which is very convenient for use in microelectronics. Thus, methods for producing porous silicon from single crystal wafers provide the technological prerequisites for the production of nanocrystalline silicon structures by a relatively simple method.

At the end of 1990 there was a real sensation in the science of semiconductors. Lee Canham, a researcher from the UK, reported (Canham, 1990) the observation of an effective red-orange photoluminescence from porous silicon at room temperature, the yield of which amounted to 1–10%. In this publication he presented graphs of the efficiency of the photoluminescence (PL) and its shift from the infrared to the red range with increasing porosity of the sample. This connection with porosity, and hence the size of particles allowed the author to suggest that the cause of the luminescence was associated with dimensional quantum constraints. The results were of considerable interest to experts and soon a report was published on the observation of electroluminescence (EL) in porous silicon (Canham et al, 1992; Halimaoui et al, 1991). In EL light emission occurs as a result of the passage of electric current in the semiconductor, leading to excitation of electrons and holes and their subsequent recombination.

1.6. The surface and surface states

Moving to the discussion of nanomaterials with particle sizes of a few nanometers, in which the surface area reaches up to hundreds of square meters per gram, we certainly can not ignore the consideration of problems associated with the surface.

The surface is different from the volume by the fact that, in principle, it is a defect in the crystal structure, along with such well-known two-dimensional crystal structure defects as stacking faults and twins. Planes of atomic positions in which the crystal ends or meets another plane of the crystal are also defective. In the first of these cases the defect is the surface, in the second – the interface or grain boundaries. Surface atoms do not have a four-sided environment, they have unsaturated dangling bonds, which leads to the appearance of the structure of the levels in the band gap of the semiconductor. These levels in the band gaps are called surface states (Seeger, 1977).

In a bounded crystal there are not only the quantum states of electrons moving in the crystal, but also additional states in which electrons are localized on the surface of the crystal (Bonch-Bruevich and Kalashnikov, 1977, Seeger, 1977). The presence of local surface levels of energy causes that the electrons and holes can 'stick' to the surface, forming the surface electric charge. The surface here is the induced charge in the bulk of equal magnitude and opposite in sign, i.e. there are enriched or poor near-surface layers.

Localized electronic states form on the crystal surface; they can be proper, i.e. caused by the breakage of the crystal lattice at the boundary, and improper, localized at impurities or defects on the surface or in a layer covering the surface (e.g. oxide layer). The proper surface states form permitted energy bands separated by forbidden bands, and the wave vectors of the wave functions of these states are located in the plane tangent to the surface. The permitted surface bands may be located at in energy ranges corresponding to both forbidden and permitted three-dimensional bands. The existence of surface states was explained by Tamm (1932) based on the difference between the heights of potential barriers for an electron in the wells on the surface and in the bulk of the crystal, so that their proper surface states are called Tamm states. The electron in the Tamm state resembles a float on the surface water: it can move freely along the surface, but it can not go into the interior of the crystal not escape from it. Electrons appear to adhere to the surface.

1.6.1. Reconstruction of the surface

In real crystals, the Tamm states correspond to dangling (unsaturated) valence bonds of the surface atoms. Usually breaking of the valence bonds leads to a restructuring referred to as surface reconstruction as a result of which the lattice symmetry of the surface layer is very different from the structure of the crystal lattice in the bulk of the single crystal (Kaxiras, 2003). It should be noted that the nature of the surface reconstruction depends on the characteristics of the system and is very different from metals and semiconductors. Restructuring can be expressed in a small displacement of the atoms in comparison with the positions characteristic of the crystal, and this restructuring is called 'surface relaxation'. However, the restructuring also called 'surface reconstruction' is expressed in a radical change in the atomic structure of the surface compared to the structure inside the crystal volume.

During the reconstruction the surface atoms are displaced in both the surface plane and perpendicular to the surface (Fig. 1.11); consequently, atomic structures with a period of several periods of the three-dimensional lattice or incommensurate with it form on the surface. The phenomenon of surface reconstruction is characterized primarily for crystals with covalent bonds, which include silicon.

On the clean surface of the crystal in a vacuum, immediately after its formation the atoms have broken bonds, which tend to saturate. If the surface contains no impurity atoms, which could saturate these dangling bonds, then the electrons in these bonds have only one opportunity for pairing: to form additional bonds between the surface atoms themselves. In the simplest case, the neighbouring atoms of the surface layer are combined into pairs, which are called dimers in solid state physics (by analogy with the terminology used in chemistry). The atoms of each dimer come closer to each other, while moving away from the adjacent surface atoms that are included in other dimers (Fig. 1.11). This changes the period of the crystal lattice on the surface.

In this simplest case, the surface aligns herself only after the shift to the distance between the centres of neighboring dimers, which is twice

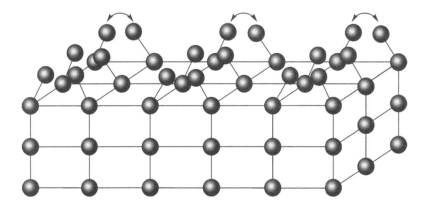

Fig. 1.11. An example of the formation of atomic dimers in the reconstruction of the surface of type (2 × 1). The arrows show the bonds in dimers.

the lattice period in the inner layers. This reconstruction is denoted by (2×1). The symbols $(n_1 \times n_2)$, where n_1 and n_2 are the ratios of the lengths of the periods of the identity of the axes 1 and 2 of the planar reconstructed surface lattice to the lengths of the periods of identity in the same direction in the ideal unreconstructed lattice in the bulk crystal, and are often used to describe the disparity of the epitaxial crystalline layers. In the above case, the symbol (2×1) indicates that the period along one of the directions on the surface has doubled, and along the other remained unchanged. Figure 1.11 shows that in reconstruction the atoms are displaced both in the surface plane and perpendicular to the surface.

The nature of the reconstruction depends on the crystallographic orientation of the surface and the method of its preparation, in particular, on the purity or impurity. For example, in the atomically pure silicon crystal (see Pikus, 1992; Kaxiras, 2003) the reconstruction leads to the formation on the surface (111) of a structure with periods (1×1), (2×1), (7×7) and on the surface (100) – structures (2×1), (4×2), (2×2). If the surface adsorbs hydrogen atoms, compensating dangling bonds, then the structures with periods (1×1), (3×1) form the (100) surface.

The nature of surface reconstruction has a significant influence on the spectrum of the surface states $E(\mathbf{k}_{\parallel})$ – dependence of energy on the wave vector parallel to the surface. The spectrum of surface states depends strongly on the orientation of the surface and its purity.

Surface energy levels can significantly alter the kinetics of electronic processes, since they create additional recombination centres and generation of charge carriers. Therefore, all the phenomena associated with non-equilibrium electrons and holes, such as photoconductivity and photovoltage, depend on the condition of the surface.

In addition, the surface usually accumulate surface excitons, the wave function of which is also localized near the boundary of the crystal. Due to the transition of electrons to surface states the electron concentration at the semiconductor surface is very different from the electron concentration in the crystal. The surface layer may have a conductivity much higher than the bulk conductivity. The change of electron concentration in the direction to the surface leads to the bending of energy bands, and sufficiently strong bending may change the very type of conductivity in the surface layer and the so-called inversion layer forms there.

Concluding remarks

In the following chapters of this monograph we will discuss in greater detail the properties of silicon nanoparticles characterized by the effects mentioned here determined by quantum constraints, methods for the preparation of nanoparticles of the required configuration and dimensions, as well as the areas of application of these materials that are already being implemented in practice, and those that may be developed in the near future. In the presentation of these materials we may introduce other necessary specific concepts from the theory of semiconductors, which were not covered in this introductory chapter.

References for Chapter 1

Aghekyan V.F. Photoluminescence of semiconductor crystals, Soros Educational Journal. 2000. V. 6. No. 10. pp. 101–107.

Anselm A.I. Introduction to Semiconductor Theory, Moscow, Nauka, 1978, 616.

Ashcroft N., Mermin N. Solid State Physics. V. 1, Academic Press, 1979a, 399 p.

Ashcroft N., Mermin N. Solid State Physics. T. 2, Academic Press, 1979b, 422.

Belyavski V. Excitons in low-dimensional systems, Soros Educational Journal. 1997, No. 5. pp. 93–99.

Bonch-Bruevich V.L., Kalashnikov S.G. Physics of Semiconductors, Moscow, Nauka, 1977, 678.

Borisenko V.E., Nanoelectronics – the basis of the information systems of the XXI century, Soros Educational Journal, 1997, No 5. pp. 100–104.

Brandt N.B., Kul'bachinskii V.A., Quasi-particles in condensed matter physics. 3rd ed, Moscow, Fizmatlit, 2010, 632 p.

Gubin S.P., Kataeva N.A., Khomutov G.B., Promising areas of nanoscience: chemistry nanoparticles of semiconductor materials, Izv, RAN, Ser. Khim., 2005, No. 4, 811–836.

Davydov A.S., Solid State Theory, Moscow, Nauka, 1976, 639.

Demikhovsky V.Ya., Quantum wells, wires, dots. What is it?, Soros Educational Journal, 1997, No. 5, 80–86.

Ekimov A.I., Onuschenko A.A., The quantum size effect in three-dimensional microscopic semiconductors, Pis'ma Zh. Eksp. Teor. Fiz., 1981, V. 34, No. 6, 363–366.

Seeger K., Semiconductor Physics, Academic Press, 1977, 615 p.

Kashkarov P.K., The unusual properties of porous silicon, Soros Educational Journal, 2001, V. 7, No. 11, 102–107.

Kittel S., Introduction to Solid State Physics, Moscow, Nauka, 1978, 791 p.

Madelung O., Solid State Physics. Localized states, Moscow, Nauka, 1985, 184 p.

Novikov M.A. Oleg Vladimirovich Losev – the pioneer of semiconductor electronics, Fiz. Tverd. Tela. 2004. V. 46, No. 1. pp. 5–9.

Pikus G.E. Surface states, Phys. Encyclopedia. T. 5, Moscow, Publishing House of the Large Russian Encyclopedia, 1992. 651–652.

Harrison W. Electronic Structure and Properties of Solids: Physics of the chemical bond. In 2 volumes, Springer Verlag, 1983.

Shik A.Ya. Quantum wires, Soros Educational Journal. 1997. No. 5. 87–92.

Efros Al.L., Efros A.L. Interband absorption of light in a semiconductor sphere. Fiz. Tekh. Poluprovod. 1982. V. 16. 1209–1214.

Yu P. Y., Cardona M. Fundamentals of Semiconductor Physics. 3rd ed, Moscow, Fizmatlit, 2002, 560 p.

Bley R.A., Kauzlarich S.M. Synthesis of Silicon Nanoclusters, In: Nanoparticles and Nanostructured Films: Preparation, Characterization and Applications, Ed. by J.H. Fendler, Wiley-VCH Verlag GmbH, 1998. Ch. 5. P. 101–118, Xx +468 p.

Canham L.T., Leong W.Y., Beale M.I.J., Cox T.I., Taylor L. Efficient visible electroluminescence from highly porous silicon under cathodic bias, Appl. Phys. Lett. 1992. V. 61. P. 2563–2565.

Canham L. T. Silicon quantum wire array fabrication by electrochemical and chemical dissolution of wafers, Appl. Phys. Lett. 1990. V. 57. P. 1046–1048.

Halimaoui A., Oules C., Bomchil G., Bsiesy A., Gaspard F., Herino R., Ligeon M., Muller F. Electroluminescence in the visible range during anodic oxidation of porous silicon films, Appl. Phys. Lett. 1991. V. 59. P. 304–306.

Hamaguchi C. Basic Semiconductor Physics. 2 ed, Springer, 2010, 570 p.

Kaxiras E. Atomic and electronic structure of solids. Cambridge University Press, 2003, 696 p.

Khriachtchev L. (Ed.). Silicon nanophotonics: Basic principles, present status and perspetives. World Scientific Publishing Co. Pte. Ltd., 2009, Xvii +452 p.

Koshida N. (Ed.). Device applications of silicon nanocrystals and nanostructures, Springer, 2009. Xii +344 p.

Moliton A., Solid-state physics for electronics, Wiley-ISTE, 2009, Xvi +389 p.

Pavesi L., Turan R. (Eds). Silicon Nanocrystals: Fundamentals, synthesis and applications. Wiley-VCH Verlag GmbH & Co, 2010, XXI +627 p.

Uhlir A., Electrolytic shaping of germanium and silicon, Bell System Technical Journal. 1956. V. 35. P. 333–347.

Yacobi B.G. Semiconductor materials: An introduction to basic principles, Kluwer Academic Publishers, 2004, Ix +228 p.

Yu P. Y., Cardona M. Fundamentals of Semiconductors, Graduate Texts in Physics. 4 ed., Berlin–Heidelberg, Springer–Verlag, 2010, 775 p.

2

Structure and properties of silicon

Silicon (Si) is a chemical element, located at number 14 in the periodic table of elements. The outer electron shell structure is $3s^23p^2$, so the silicon atoms are characterized by the state of sp^3-hybridization of the orbitals and the manifestation of the oxidation state +4 or −4 in all chemical compounds except silicon monoxide SiO. Si atoms usually crystallize in the cubic structure of the diamond – the symmetry of the unit cell – a face-centred cubic F, the space group $Fd\overline{3}m$ (No. 227), the edge of the cell $a = 0.54311$ nm (Int. Tables, 2006). In the crystalline form Si has a slightly shiny dark gray color, density 2.33 g/cm³ and a melting point of 1415°C. Since silicon has been used for a long time as the basic material of solid-state electronics, its properties in the crystalline and amorphous phases are very well studied, and the characteristics of these materials can be found in virtually all modern textbooks on solid state physics, where Si is considered as a reference sample, along with the classic diamond and semiconductors,

Group	IIB	IIIA	IVA	VA	VIA
Period					
2		5 B	6 C	7 N	8 O
3		13 Al	14 Si	15 P	16 S
4	30 Zn	31 Ga	32 Ge	33 As	34 Se
5	48 Cd	49 In	50 Sn	51 Sb	52 Te
6	80 Hg				

as well as in different directories of semiconductor materials. The properties of crystalline silicon are detailed, for example, in a large multi-author book, edited by R. Hull (Hull, 1999), dedicated only to this material, and the properties of amorphous silicon are described in detail in the book by Street (1991). Quite a lot of physical and chemical properties of silicon can be easily obtained, for example, on the Internet at the National Institute of Standards and Technology USA (http://srdata.nist.gov/gateway/, http://www.nist.gov/srd/), giving access to multiple databases. It should be said that the characteristics of silicon in the nanosized state are not so unambiguous and clear as the characteristics of the bulk crystals of Si, so they still continue to be studied and refined, which is reflected in numerous publications appearing in the last two decades in the periodic scientific literature, as well as in the continuously published monograph literature on this subject (see, e.g. Canham, 1997; Kumar, 2007; Khriachtchev, 2009; Pavessi & Turan, 2010).

As the prevalence of on the Earth is concerned, Si is the second chemical element after oxygen. It is interesting to note that despite the chemical inertness of pure silicon under normal conditions, the pure silicon is not common on the Earth, although about a quarter of the planet's mass (~28% by weight of the earth's crust). In nature, it is mostly distributed in the form of silica, or silicon oxide (IV) SiO_2 (river sand, quartz, etc.), which is about 12% by weight of the earth's crust, and is a component of several hundreds of different natural silicates and aluminosilicates.

As regards the electrophysical properties, silicon is located in the compact group of chemical elements in the periodic table with semiconductor properties. This group is indicated by the bold broken line shown on the section of the table of the elements. In addition to the chemical elements in it, the semiconductors also included the adjacent carbon in the modification of diamond and selenium, which has long served as the basic material for the manufacture of high-power semiconductor rectifiers of alternating current. The generality of these elements is that they are all p-elements, with the p-orbitals of the atoms are gradually filled with electrons. This is what leads to the peculiarities of their physical properties, crystal structures and the nearest coordination environment of atoms in the solid state.

According to its electron-optical characteristics the crystalline silicon in pure form[1] is a classic example of the indirect-gap semiconductor. The width of the indirect band gap of Si at room temperature is about 1.12 eV, and at $T = 0$ K ~1.21 eV. The concentration of majority charge carriers

---

[1]Pure or intrinsic are the semiconductors, in which the conductivity is not determined by residual impurities (which are impossible to remove completely). The conductivity of such materials is usually very low and due only by intrinsic electrons emitted into the conductivity into the conduction band by thermal fluctuations, and their resistance is close to the resistance of insulators. The demands for purity of the semiconductors with intrinsic conductivity are extremely stringent and for Si even more stringent. Pure Si with intrinsic conductivity may contain no more than 10^{-11} impurities.`

(electrons) in silicon with intrinsic conductivity at room temperature is ~$1.5 \cdot 10^{16}$ m^{-3}, the electron mobility is relatively high (~1400 cm$^2 \cdot$V$^{-1} \cdot$s^{-1}), so the band theory of solids can be applied for Si (Ashcroft and Mermin, 1979a; Bonch-Bruevich and Kalashnikov, 1977).

The electrophysical properties of crystalline silicon (c-Si) are strongly affected by microimpurities present in it. With the introduction of small amounts of impurities of other elements the intrinsic semiconductor can be converted to an impurity, and depending on the type of impurity a semiconductor with a hole (*p*-type) or electronic (*n*-type) conductivity can be produced. Single crystals of silicon with hole conductivity are produced by adding elements of group III – boron, aluminum, gallium and indium. The silicon with electronic conductivity is produced by adding the elements of group V – phosphorus, arsenic or antimony.

The semiconducting properties of silicon appear in all its forms – monocrystalline, polycrystalline and amorphous[2]. At present, single-crystal silicon (mc-Si) is the main material of microelectronic technology. mc-Si is used to produce a variety of semiconductor devices from discrete diodes and transistors to extremely complex integrated circuits and processors. In addition, the high photosensitivity of silicon (the change of electrical conductivity under illumination) is widely used, which allows one to convert light energy into electrical energy. This effect is used in silicon solar cells and photodetectors. However, the reverse process, i.e. sufficiently efficient conversion of electrical energy into visible light in silicon devices has so far failed. This is connected with the peculiarities of the electronic properties of silicon.

2.1. The structure of crystalline silicon

The structure of crystalline silicon is determined by the *sp*^{3}-hybridization of the outer electron shell in the formation of a bond (Si–Si)[3]. As a result, in the crystalline state a tetrahedron of four neighbouring atoms with the angle between the bonds of the central atom 109.47° forms around each silicon atom. The atomic structure of the silicon crystal (c-Si) can be represented by a unit cell in the form of a cube filled with four interconnected centred

[2]For brevity, we shall use the following symbolic notation of silicon phases: polycrystalline silicon with the crystallite size of 0.1 μm or more – c-Si; large single crystals of silicon – mc-Si; amorphous silicon – a-Si. Abbreviation will also be used: nc-Si to denote the fraction of crystalline silicon with a particle size less than 100 nm; P-Si – to refer to porous silicon. If there is no specific indication, then the symbol Si will continue to denote, depending on the context, either crystalline silicon with a large crystal size, or a silicon atom.

[3]The energy of the covalent Si–Si bonds is equal to 4.64 eV/atom (Kittel, 1978). For comparison, in the other elements of group IVA, having the structure of diamond (carbon, germanium and gray tin), the binding energy is equal to 7.36, 3.87 and 3.12 eV/atom. Under the binding energy we mean the energy required for separation of solid material into the individual atoms at a temperature of 0 K.

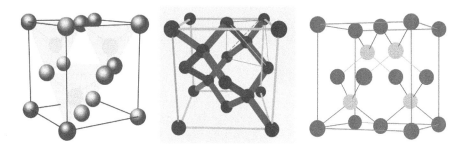

Fig. 2.1. Axonometric images of the elementary cell of crystalline silicon (spheres indicate Si atoms); a) representation of the structure of the elementary cell using coordination tetrahedrons; b) crystallochemical structure of diamond/silicon with indicated covalent bonds (the red lines show the coordination tehrahedron with the Si atom in the centre); c) explanation of the formation of the elementary cell of Si with mutual penetration of two FCC lattices where the grey colour indicates the Si atoms, forming the classic FCC cell, and the green colour the 'additional' Si atoms from the same cell inserted into it with the shift by 1/4 of the length of the spatial diagonal.

tetrahedra, as shown in Fig. 2.1 *a*, *b*.

This pattern reflects well the coordination environment and the chemical bonds in the crystal, but it does not make clear the assignment of the structure to the face-centred Bravais cell. In fact, such a structure, called the structure of diamond, is really face-centred cubic (FCC) and in the international crystallographic notations refers to the space group $Fd\bar{3}m$ (No. 227) – see International Tables for Crystallography, 2006. V. A1.

The primitive basis of the FCC unit cell of silicon consists of two identical quadratic atoms occupying the positions (0, 0, 0) and (1/4, 1/4, 1/4) *a*, where *a* is the period the unit cell[4]. This means that each node in the lattice unit cell *F* correspond to two Si atoms with such a position relative to each other. That is, the structure of the unit cell of silicon can be obtained by placing the above mentioned primitive diatomic basis in each

[4]It is important to mention the difference between the concepts of the crystal structure and the lattice, and that the face-centred cubic lattice cell *F* is not primitive, and is one of the 14 Bravais cells of the crystal lattice, giving the most balanced minimum volume of the atomic structure,which can describe the entire crystal with translational movements. Primitive (empty, i.e. the cell which does not contain lattice sites either within or on the edges of the parallelepiped) unit cell for the FCC is a rhombohedron, with two of its vertices being the vertices of the cube of FCC cells, located at opposite ends of a large diagonal, and the rest – at the points, centring the faces of the cube. More information about the principles of structural crystallography and crystal chemistry, including the structure of diamond and silicon, for example, see Ashcroft and Mermin (1979), Egorov-Tismenko (2005), Kittel (1978); Urusov and Eremin (2010); Glusker and Trueblood (2010); Grundmann (2010).

node of of the FCC cell by combining the initial (0,0,0) atom of the basis with the node, which confirms the validity of classifying the structure of the silicon as the FCC Bravais cell.

The structure of the silicon can be described in different words, like the FCC cell, in which out of the eight available tetrahedral voids whose centres are located at a distance of one quarter of the length of the diagonal of the cube from the vertex atoms, four are occupied by interstitial atoms of the Si, and the remaining four are unoccupied. In Fig. 2.1 in these 'implanted' atoms are depicted by green balls and blue balls depict the atoms of Si, forming the classic FCC cell. Because of the 'implantation' into the voids of the FCC cell of four additional atoms, the diamond unit cell of silicon has 8 atoms of Si, in contrast to the classical FCC cell with a monatomic basis, which accounts for only 4 atoms.

Sometimes the structure of the unit cell of diamond is described as the result of mutual penetration of two FCC lattice with a monatomic basis produced by shift of one of them to one quarter of the length of the body diagonal (Anselm, 1978, Ashcroft and Mermin, 1979). This description can also be seen in Fig. 2.1, where the blue atoms denote Si, forming the main classical FCC cell, and green – the 'extra' Si atoms of Si left from the shifted interpenetrating cell.

The packing density of atoms in the structure of diamond is quite low – only 0.34, which amounts to about 46% of the fill factor of the cube, which is characteristic for the close-packed structure of hard spheres of similar diameter. For this reason, silicon, along with the ice is a rare representative of solids whose volume decreases on melting (Kaxiras, 2003). In melting, the interatomic bonds that hold the Si atoms at a distance from each other, break down by the kinetic energy of thermal vibrations, and the packing of the atoms in the liquid phase is more dense, causing a decrease in volume.

Finally, the crystalline structure is often depicted with the imposition of the atomic nets of hard balls. With this method the classical structure with the FCC cell can be represented by repeating imposition of triples ABC ABC... of alternating flat layers (111), consisting of densely packed spheres, arranged so that the balls of each subsequent layer fall into the hole between the balls of the previous layer. In the structure of silicon, despite the fact that it is described by the FCC unit cell, because of the additional four atoms the planes of the {111} system are corrugated (Fig. 2.2a) and are actually two closely spaced planes (Fig. 2.2b). Therefore, the structure of diamond to which the crystalline silicon belongs, represent a sequence of periodic imposition of the three pairs of (111) layers of the packing, i.e.. aA bB cC, etc.

Interestingly, silicon crystallizes in the diamond structure much 'more happily' than the carbon, which is only diamond at ultrahigh pressures, and in other cases, 'prefers' the simple hexagonal structure of graphite. The reason lies in the fact that for Si the diamond structure is energetically advantageous to all others. Figure 2.3 shows the energy of a crystal

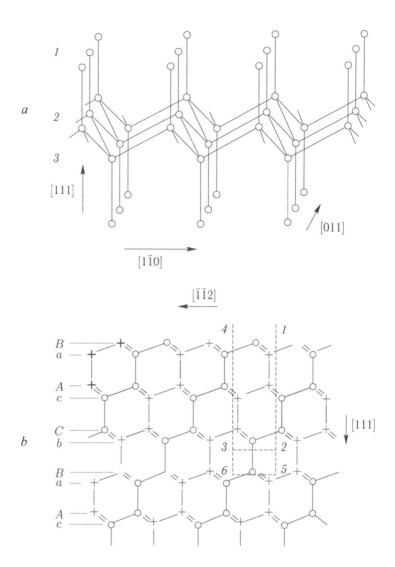

Fig. 2.2. A cubic lattice of diamond type: *a* – perspective view of the layer (111) – the atoms are located at the minima are shifted by half a period along the [011] direction relative to the atoms located at the maxima (from the textbook by Bonch-Bruevich and Kalashnikov, 1977), *b* – flat projection perpendicular to (1$\bar{1}$0), showing the cross sections of the layers (111); the atoms marked with ○ are located in the plane of the figure, the atoms denoted by + are below the plane of the figure. (111) plane is perpendicular to the plane of the figure and can be represented by a horizontal line. The picture by (Hirth and Lothe, 1972).

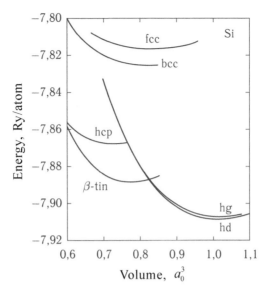

Fig. 2.3. The energy per Si atom in the structures of different types, calculated taking into account the dynamics of lattice vibrations. Energy is given in units of Ry – Rydberg (off-system unit of energy used in atomic physics and optics, 1 Ry = 13.60 eV – the energy of ionization of a hydrogen atom from the ground state); a_0 – the period of the unit cell; fcc – FCC cell; bcc – BCC cell; hcp – HCP cell; β-tin – tetragonal cell of the white tin type (β-Sn); hg – a cell of the type of the hexagonal structure of graphite; hd – the cell of the diamond type.From the article (Yin and Cohen, 1980).

consisting of atoms of Si, calculated for different types of crystal lattice (Yin and Cohen, 1980). In their paper (Bernstein et al, 2000) considered the energy of Si and of a number of other structures, which are well above the energy of the diamond lattice. According to thermodynamics, the phase with the lowest energy for given external conditions should be stable, and it follows from the results that for silicon it is the structure of diamond.

2.2. The electrical properties of silicon

The band structure of silicon depends on the phase state and packing density of the atoms. The dense packing of Si atoms in the solid state leads to overlapping of the allowed energy levels and causes splitting of each level into N allowed states, where N denotes the number of neighbouring atoms in the solid state (Fig. 2.4). External energy levels form two bands, called the valence and conduction bands. At a temperature of 0 K, the valence band is completely filled with electrons and the conduction band is completely free of electrons.

The band structure of monocrystalline silicon under normal conditions is shown in Fig. 2.5.

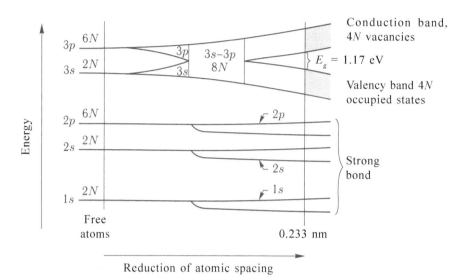

Fig. 2.4. The formation of the electronic structure of solid silicon as the atoms approach. On the left are the levels for the free atom, on the right the levels for the atom located at a short distance from the neighboring atom (0.233 nm in the figure denotes the distance between the centres of the nuclei of two covalently bonded Si atoms).

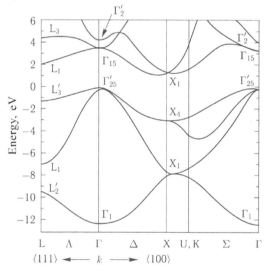

Fig. 2.5. Electronic band structure of single crystal Si, calculated by Chelikowsky and Cohen (1976) by the empirical pseudopotential method: k – the wave number of electron, characterizing its condition, and within the first Brillouin zone $-\pi/a < k \leq \pi/a$, where a is the unit cell parameter. Letters with subscripts denote the energy levels for the corresponding points of symmetry in the Brillouin zone. The minimum energy of the conduction band is located in the direction of $\langle 100 \rangle$ (Harrison, 1983; Harrison 1999).

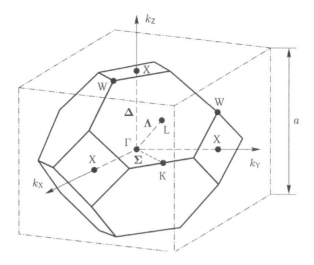

Fig. 2.6. Perspective view of the first Brillouin zone (bold line) single-crystal silicon (diamond structure) in comparison with the corresponding FCC unit cell (dash-dotted lines). The major axis of symmetry (marked by bold Greek letters) and the characteristic points of symmetry (marked by capital Latin letters) in the area are shown, Γ – is the centre of the Brillouin zone. The ratio between the area and the unit cell in the figure is only approximate, rather than dimensional, since the unit cell exists in real space (*a* – cell parameter), and the zone in the wave vector space (reciprocal space). Picture by (Sze, 1984).

The meaning of letter symbols in Fig. 2.5 is shown in Fig. 2.6.

This diagram show the points and axes of symmetry in the Brillouin zone generally used in the conventional band theory of semiconductors:

- Point Γ – centre of the Brillouin zone with coordinates $\dfrac{2\pi}{a}(0,0,0)$;
- point L with coordinates $\dfrac{2\pi}{a}\left(\dfrac{1}{2},\dfrac{1}{2},\dfrac{1}{2}\right)$ – the intersection of the axes $\langle 111 \rangle$ (line Λ) with the faces of the zone, the point X with coordinates $\dfrac{2\pi}{a}(0,0,1)$ – the intersection of the axes $\langle 100 \rangle$ (line Δ) with the faces of the zone;

- point K with coordinates $\dfrac{2\pi}{a}\left(\dfrac{3}{4},\dfrac{3}{4},0\right)$ – the intersection of the axes $\langle 110 \rangle$ (line Σ) with the edges of the zone.

The width of the energy gap E_g $X_1 - \Gamma'_{25}$ (indirect band gap) between the the conduction and valence bands in silicon depends on the temperature. This dependence for the temperature range 2–300 K is expressed by the empirical equation (Varshni, 1967):

$$E_g(T) = E_g(0) - \frac{\alpha T^2}{\beta + T}, \tag{2.1}$$

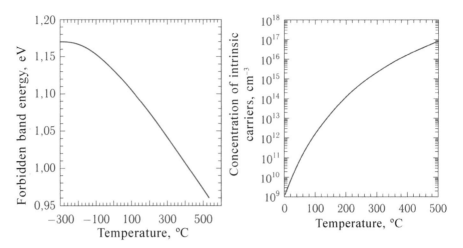

Fig. 2.7. Temperature dependence of the band gap of Si.

Fig. 2.8. Temperature dependence of the concentration (cm^{-3}) of intrinsic carriers in Si.

or the substitution of known constants for Si $E_g(0)$, α, β (Thurmond, 1975) relationship:

$$E_g(T) = 1.17 - \frac{4.73 \cdot 10^{-4} \cdot T^2}{\beta + T} \, (\text{eV}), \qquad (2.2)$$

with the graph shown in Fig. 2.7.

Intrinsic charge carriers in Si are electrons which 'do not hop' from the valence band to the conduction band under the influence of thermal excitation. The temperature dependence of the concentration of intrinsic charge carriers n_i (electrons) in pure Si is given by

$$n_i = 3.10 \cdot 10^{16} \cdot T^{3/2} \cdot \exp\left(-\frac{1.206}{kT}\right), \qquad (2.3)$$

where k is the Boltzmann constant. The graph of this function is shown in Fig. 2.8.

2.2.1. Summary of the basic physical properties of crystalline Si

Table 2.1 summarizes the basic physical properties (except mechanical) of crystalline silicon, as shown in Table 2.2. The basic properties of semiconductor Si are compared with those of other elemental semiconductors.

The above properties with sufficient accuracy remain unchanged for the semiconductor and do not depend on the size and shape of the crystal, if

Table 2.1. Summary of the main physical and electrical properties of crystalline silicon of semiconductor purity. The data are taken from the pages of the Internet http://www. el-cat.com/silicon-properties.htm (EL-CAT Inc. is a stocking distributor of silicon wafers, compound semiconductors and other crystal materials for use in electronics) and http://ru.wikipedia.org/wiki/silicon (silicon – the material from Wikipedia, the free encyclopedia), as well as from the Springer Handbook of Condensed Matter and Materials Data/Ed. by W. Martienssen & H. Warlimont. – Berlin: Springer, 2005, pp. 88–93, 578–603

Property	Value	Unit of measurement
Crystal properties		
Structure	Cubic (the structure diamond)	
The space group	$Fd\overline{3}m$	
Atomic mass (molar weight)	28.0855	a. amu. (g/mol)
Isotopes	28 (92.23%) 29 (4.67%) 30 (3.10%)	
Electronic shell	$1s^2 2s^2 2p^6 3s^2 3p^2$	
Conventional ion	Si^{4+}, Si^{4-}	
The radius of the atom	132	pm (=0.01 Å)
Covalent radius	111	pm
The radius of the ion	42 for +4 and 271 to −4	pm
The distance between neighbouring atoms at 300 K	0.235	nm
The lattice spacing (a_0) at 300 K	0.54311	nm
Density at 300 K	2.3290	g/cm^3
The number of atoms in 1 cm^3	$4.995 \cdot 10^{22}$	
The density of atoms on the surface with the crystallographic plane (hkl):	(100) 6.78 (110) 9.59 (111) 7.83	$10^{14}/cm^2$
Critical pressure	1450	atm
Critical temperature	5193	K
Thermal properties		
Melting point	1687	K
Boiling point	2628	K
Specific heat	0.7	$J/(g \cdot K)$

Thermal conductivity (300 K)	156	W/(m · K)
Thermal diffusivity	0,8	cm²/s
The linear coefficient of thermal expansion	$2.92 \cdot 10^{-6}$	K^{-1}
Debye temperature	640	K
Temperature dependence of the band gap width	$-2.3 \cdot 10^{-4}$	eV/K
Heat: melting/evaporation/ atomization	39.6/383.3/452	kJ/mol
Electrical properties		
Breakdown voltage	$\approx 3 \cdot 10^5$	V/cm
Refraction index	3.42	
Electron mobility	≈ 1400	cm²/(V·s)
Hole mobility (300 K)	≈ 370	cm²/(V·s)
The diffusion coefficient of electrons	≈ 36	cm²/s
The diffusion coefficient of holes	≈ 12	cm²/s
The thermal velocity of the electron	$2.3 \cdot 10^5$	m/s
Thermal velocity of holes	$1.65 \cdot 10^5$	m/s
The energy of optical phonon	0.063	eV
The work function	4,15	eV
The band structure		
The relative dielectric constant $(\varepsilon_s/\varepsilon_0)$ at 300 K	11.9	
The effective density of states (conductive, N_c at $T = 300$ K)	$2.8 \cdot 10^{19}$	cm⁻³
The effective density of states (valence, N_v at $T = 300$ K)	$1.04 \cdot 10^{19}$	cm⁻³
Affinity for electron	133.6	kJ/mol
The band gap E_g at 300 K (the difference between the energies of the minimum zone conduction and valence band maximum in an indirect band gap at 300 K)	1.12	eV
The width of the indirect gap E_g at 0 K	1.17	eV
The minimum width of the forbidden line zone at 300 K	3.4	eV
The concentration of intrinsic charge carriers	$1 \cdot 10^{10}$	cm⁻³

Intrinsic resistivity	$3.2 \cdot 10^5$	Ohm·cm
Auger recombination coefficient of electrons C_n	$1.1 \cdot 10^{-30}$	cm⁶/s
The coefficient of Auger recombination of holes C_p	$3 \cdot 10^{-31}$	cm⁶/s

Temperature dependence of the width of the band gap is described by equation (2.2)

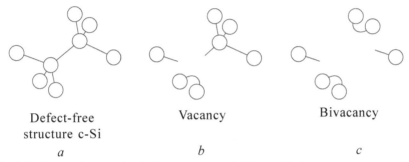

Defect-free structure c-Si	Vacancy	Bivacancy
a	*b*	*c*

Fig. 2.9. Types of vacancies in the structure of silicon. Circles are the atoms of Si, and the lines the chemical bonds between them. It is evident that the formation of vacancies leads to the formation of dangling bonds.

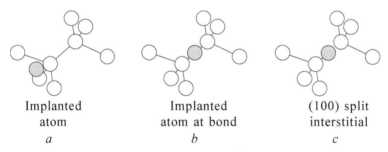

Implanted atom	Implanted atom at bond	(100) split interstitial
a	*b*	*c*

Fig. 2.10. Interstitial atoms in the crystal lattice of silicon. Open circles depict Si atoms of the crystal lattice, gray circles – interstitial atoms, lines are the chemical bonds between neighbouring atoms: *a* – interstitial atom in the tetrahedral void, does not form bonds with neighbouring atoms, *b* – interstitial atom in the bond between neighbouring tetrahedra Si; *c* – interstitial atom at the bond between the neighbouring Si tetrahedra, is associated with only one Si atom.

its size can be considered as macroscopic (ranging from few microns to larger). For silicon nanoparticles, as will be shown later, however, due to quantum-size restrictions, both optical and electronic properties depend on the particle size.

2.3. Crystal lattice defects

The properties of semiconductors are heavily influenced by lattice defects.

Table 2.2. Comparison of physical properties of Si at 300 K with the properties of other monoelement semiconductors (data from Physics of Semiconductor Devices, S.M. Sze, New York, Wiley Inc. 2nd ed., 1981; Madelung, 1967, Ugai 1975; Hislum and Rose-Innes, 1963)

Semi-conductor	A	ρ,g/cm³	a, Å	T_m, °C	ΔE_g eV	ρ_i, Ohm·cm	μ, cm²/(V·s)		m_d^*/m_0		$\dfrac{\varepsilon_s}{\varepsilon_0}$
							μ_n	μ_p	m_n	m_p	
β-B	10.81	2.34		2175	1.5	10^6	1	55			
C (diamond)	12.01	3.51	3.567	3700	5.47	$\sim 10^8$	1800	1200	0.2	0.25	5.7
one	28.09	2.33	5.431	1415	1.12	$2 \cdot 10^5$	1400	480	1.08	0.56	11.9
Ge	72.61	5.33	5.646	937	0.66	47	3900	1900	0.56	0.42	16
It	78.96	4.80		220	1.8	10^6-- 10^{12}		<1			6.3
α-Sn	118.7	5.75	6.489	232	0.06	$2 \cdot 10^{-4}$	1200	1200			
α-Te	127.6	6.25		452	0.35	0.5	1200				23

The following notations are used in the table: A and M – atomic or molecular weight, respectively; ρ – density; a – the crystal lattice spacing, T_m – melting point; ΔE_g – the band gap; ρ_i – resistivity of the intrinsic semiconductor; μ and m_d^*/m_0 – the mobility and the effective relative mass of the electrons n and holes p; $\varepsilon_s/\varepsilon_0$ – relative dielectric constant.

The crystal structure of pure silicon may show the formation of point and extended defects[5]. The point defects play an important role in the diffusion of intrinsic and extrinsic atoms and the formation of extended defects, as well as serve as pinning centres of the electronic states, change the band structure (Kaxiras, 2003, Bonch-Bruevich and Kalashnikov, 1977). Extended defects, firstly, affect the properties of a semiconductor like impurity centres, and secondly, can generate electricity and leakage paths leading to the degradation of the properties of semiconductor devices (Holt and Yacobi, 2007).

The main point defects in crystals are the vacancies and interstitial atoms. Point defects can interact to form complexes or even clusters in the crystal lattice. The main simplest defects are vacancies and interstitial atoms. Defects such as vacancies mean the absence of an atom in the corresponding position prescribed by the structural crystallography. Examples of vacancies in the structure of crystalline silicon is schematically shown in Fig. 2.9.

Interstitial defects in pure crystalline silicon formed by introduction of 'excess' Si atoms in the voids of the unit cell of silicon. Examples of such interstitial atoms are shown in Fig. 2.10.

---

[5]Presentation of the foundations of the theory of defects in crystals is given in almost all textbooks on solid state physics, and in more detail, with varying degrees of completeness and complexity of information about defects in the crystal lattice, their properties and effects on the properties of crystals can be obtained from special books on these subjects, for example, (Kelly and Groves, 1974; Hirth and Lothe, 1972; Kaxiras, 2003; Holt and Yacobi, 2007; Nabarro and Hirth, 2007).

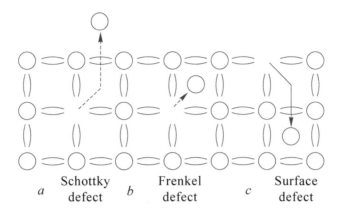

Fig. 2.11. Mechanisms of formation of point defects. Schottky mechanism creates a clean vacancy with removing of an atom from the crystal, the Frenkel mechanism creates a 'vacancy–interstitial atom' pair; the surface mechanism generates interstitial atoms, leaving 'dangling' bonds on the surface.

These atoms are also known as interstitial atoms. In the unit cell of silicon there are 5 free 'voids', into which interstitial atoms can be introduced. These cavities are located at positions (1/2, 1/2, 1/2), (1/4, 1/4, 1/4), (1/4, 3/4, 3/4), (3/4, 1/4, 3/4), (1/4, 3/4, 1/4).

There are many reasons for the formation of point defects in crystals, which can be internal, caused by the intrinsic dynamic characteristics of the atomic structure, or external factors, such as radiation exposure. But from thermodynamics it follows that even equilibrium perfect crystals at normal temperature contain inevitably a certain concentration of point defects (see, e.g. Ashcroft and Mermin, 1979b).

There are three mechanisms of the thermodynamic formation of point defects (Fig. 2.11).

- Schottky defect – a defect that occurs when an atom jumps into the interstitial position and subsequently diffuses to the surface, resulting in a vacancy in a crystal, which is not connected with the departed atom.
- Frenkel defect – a defect that occurs when the silicon atom jumps into the interstitial position and then a vacancy–interstitial atom pair is produced.
- The surface defect – a defect formed by diffusion of the atom from the surface into the interstitial position within the crystal.

The concentration of point defects in the crystal depends on the thermal fluctuations and vapour pressure. For solid silicon, vapour pressure is extremely low, so the thermal fluctuations in a wide temperature range are not important. The presence of defects changes the internal energy of the crystal and its entropy, so the equilibrium concentration of defects should depend on the defect formation energy and the equilibrium temperature. The concentration of Schottky (C_s) and Frenkel (C_f) defects can be calculated by the formulas:

$$C_s = \frac{N}{1 + \exp\left(E_s / kT\right)} \approx N \cdot \exp\left(-\frac{E_s}{kT}\right) - \text{for Schottky defects,} \quad (2.4)$$

$$C_f \approx \sqrt{N \cdot N'} \cdot \exp\left(-\frac{E_f}{kT}\right) = N \cdot \exp\left(-E_f / kT\right) - \text{for Frenkel defects,} \quad (2.5)$$

where N and N' are respectively the number of atoms and interstices in the crystal, E_s and E_f – the energy of formation of a Schottky and Frenkel defect, respectively. The point defects can capture free electrons and become charged.

2.3.1. Extended defects

The formation of extended defects in the crystal lattice – dislocations, twins and stacking faults – is not connected with thermodynamics, and in perfect equilibrium crystals they may be completely absent. However, real crystals due to the effects of stresses or because of the strong non-equilibrium process of formation almost always contain some or all of the defects of this type. The process of obtaining crystalline Si may be accompanied by the formation of three types of extended defects: twins, stacking faults and dislocations, and there is always the inevitable extended defect – the surface.

Dislocations in silicon. Dislocations are linear defects in the crystal lattice and form when stresses exceeding the yield stress form in the crystal.

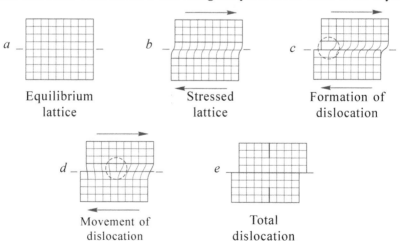

Fig. 2.12. Schematic representation of the formation of an edge dislocation in a crystal under the action of shear stress. Arrows indicate the direction of the forces in the crystal.

The Burgers vector is located in the slip plane. Figure 2.12 shows the mechanism of deformation formation of an edge dislocation in a crystal.

First, as shown in Fig. 2.12*b*, the influence of shear stress causes a distortion of the lattice; when the stress exceeds the yield strength (Fig. 2.12*c*) the interatomic bonds are broken and there is an inelastic displacement of atomic layers to a fraction of the lattice spacing with the formation of bonds between other pairs of atoms. The result is an 'extra' plane of atoms, which terminates within the crystal, and the planes adjacent to this plane pass through the entire crystal but become distorted. The ragged edge of the plane is called the dislocation line. If the dislocation under the stress continues to move (slide) and reaches the surface of the crystal, a step is formed on the surface (Fig. 2.12*d*), and a linear defect inside the crystal disappears.

The dislocations in silicon and the related planar defects, such as twins and stacking faults, are considered in detail, for example, in a review (George, 1999).

In crystalline silicon, sliding dislocations involved in plastic deformation, have the Burgers vector **b** of type $(a/2) \langle 110 \rangle$ (the length of the Burgers vector in silicon $|\mathbf{b}| = 0.384$ nm) and, in most cases, slip on planes $\{111\}$.

As confirmed by weak-beam TEM (Ray & Cockayne, 1971) and high-resolution TEM (Anstis et al, 1981; Olsen & Spence, 1981), the dislocations in the slip plane dissociate into two Shockley partial dislocations with Burgers vectors **b** = $(a/6) \langle 112 \rangle$ ($|\mathbf{b}| = 0.222$ nm), similar to a ribbon surrounding the stacking fault of the subtraction type. For a complete dislocation of this sign the sequence of partial dislocations is determined according to the topological rule given in (Hirt and Lothe, 1972).

Because of the double-layer packing of the planes of the $\{111\}$ system, as shown in Fig. 2.12, dislocations in silicon can be formed (Hirth and Lothe, 1972) through the shear, or between closely spaced planes, such as

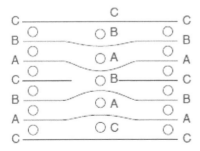
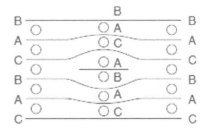

a) intrinsic stacking fault   b) extrinsic stacking fault

Fig. 2.13. Stacking faults in the structure of silicon: *a* – intrinsic stacking fault (missing part of the plane CC violated the alternation of planes and led to their deformation), *b* – extrinsic stacking fault (part of the plane AA is implanted, breaching the regular alternation of ABC and causing local deformation of the packing planes.

B|c (sliding set of dislocations), or between planes with a large interplanar spacing, for example, *b|B* (shuffled set of dislocations). In closely spaced planes the covalence of the bond is three times higher than that of the far away planes. In addition, since the low-energy stacking fault can be formed only between the closely spaced planes it is now believed that the dislocations in the crystalline silicon usually belong to the 'sliding set'.

Two-dimensional defects in c-Si *and* nc-Si. At the tetrahedral bond of the neighbouring atoms only low-energy stacking faults of two types can form in the Si structure (Fig. 2.13):

- subtracting the stacking fault type (intrinsic stacking fault – isf), corresponding to the removal of one of the alternating planes, for example ... cC aA bB | aA bB cC..., a stacking fault can also be obtained by appropriate shift of the upper half of the crystal (*a*/6) ⟨112⟩, resulting in a layer of cC falling into the position of the layer aA (naturally, all the layers above it will also be shifted);

- The introduction of the interstitial stacking fault (extrinsic stacking fault – esf), corresponding to the extraordinary addition of any of the double layers, for example ... cC aA bB | aA | cC aA bB cC ...; such a defect can be created by annealing in an oxidizing atmosphere, where the condensation of excess interstitial atoms, formed during the formation of the oxide, results in the formation of a stacking fault.

The density of stacking fault energy γ_{st} is determined by measuring the width of the appropriate dislocation configurations in which the stress of the stacking fault is the balanced well-known repulsive force acting between the surrounding dislocations. For Si such energies are considered (Carter, 1984)

$$\gamma_{st,isf} = 65 \pm 10 \, mJ \cdot m^{-2},$$

$$\gamma_{st,esf} = 60 \pm 10 \, mJ \cdot m^{-2}.$$

Electron doping has no noticeable effect on the value of γ_{st}, but the segregation of impurities in the plane of the stacking fault can lead to a marked decrease in stacking fault energy, increasing the dissociation of dislocations (George, 1999).

In addition to growth defects, real crystals often contain twins – stacking faults resulting from deformation of the crystal lattice by the external effects

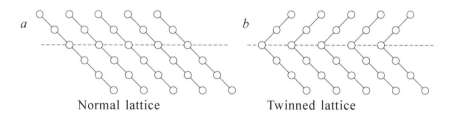

a   *b*

        Normal lattice   Twinned lattice

Fig. 2.14. The principle of the twin formation.

(see the diagram in Fig. 2.14). The presence of twins usually indicates a high density of dislocations in the crystal.

In the bulk deformed Si single crystals experiments usually reveal stacking faults in the planes {111}, formed as the result of complete dissociation of dislocations with Burgers vector $b = (1/2) \langle 110 \rangle$ into two Shockley partial dislocations with vectors $b = (1/6) \langle 112 \rangle$. Such defects have been observed, for example, in the study by Ray and Cockayne (1971) using TEM dark-field images in a low beam and in the study of Olsen and Spence (1981) by high resolution TEM. At the same time, stacking fault energy calculations performed by the authors (Ray and Cockayne, 1971), based on experimentally measured parameters of the dislocation showed that the energy of formation of extrinsic and intrinsic defects should be very close, i.e. in deformation of macroscopic crystals these defects form with almost equal probability.

The deformation and growth stacking faults in Si nanocrystals of different sizes, distributed in a matrix of SiO_2, were experimentally studied in Wang et al. (2005) by the method of dark-field images and high resolution TEM. The samples were prepared by irradiation of an amorphous SiO_2 film of the micron thickness with of 100 keV Si^+ ions until the excess concentration of ~$2.43 \cdot 10^{22}$ Si/cm^3 was produced, followed by annealing for 1 h at 1100°C in an atmosphere of pure nitrogen. After high-temperature annealing, the resultant Si nanoparticles were subjected to hydrogen passivation by exposure for 1 h at 500 °C in a nitrogen atmosphere with an addition of 5% H_2. As a result of this preparation procedure, the samples contained individual Si nanoparticles ranging in size from 5 nm to 22 nm. More than 200 particles of nc-Si were viewed, which allowed statistically valid conclusions to be made about the type and distribution of defects. It was found that in the particles size greater than 6 nm, the dominant defects were twins, observed in almost 90% of cases. Approximately 25% of cases were observed as stacking faults, which can coexist in a single crystal with twin nanostructures although some could be in nc-Si, and without twin structures. It is interesting that ~80% of the stacking faults found belonged to the intrinsic type, and only 20% were extrinsic stacking faults.

All stacking faults penetrated through nanocrystals and this distinguishes the nanocrystals from the bulk crystals, in which the stacking faults are usually localized within the crystal. Therefore, in the case of nc-Si we can not say which dislocation reactions could cause of the formation of the detected stacking faults, and we can only be interpreted the stacking faults as an excess or deficiency of {111} planes in the sequence ABCABC characteristic of defect-free silicon. However, in crystals with a size <5 nm, there were no twins or stacking faults (Wang et al, 2004).

2.3.2. Surface

The surface of the crystal is also an extended lattice defect, because the atomic structure of surface layers differs from the ideal crystal structure within the volume of the body. The nature of surface reconstruction (see 1.6.1) of the silicon crystals depends on the crystallographic orientation and surface preparation methods, in particular on purity or impurity. For example, in an atomically pure silicon crystal (see Pikus, 1992; Kaxiras, 2003), in the reconstruction of the (111) surface the structures with the periods (1×1), (2×1), (7×7) appear, and on the surface (100) – the structures (2×1), (4×2), (2×2). If the crystal adsorbs hydrogen atoms, which are often used in practice to saturate and compensate the dangling bonds, surface structures with periods(1×1) (3×1) form on the (100) surface. The nature of surface reconstruction has a significant impact on spectrum of surface states – the dependence of energy on the wave vector parallel to the surface $E(\mathbf{k}_{\parallel})$.

The structural model of surface reconstruction of the Si crystal, as shown in Fig. 2.15 (color inset), was developed and proposed by Kunio Takayanagi et al in 1985 (Takayanagi et al, 1985) to describe the reconstruction of the atomically clean surface of Si (111) 7×7, based on the analysis of experimental images in probe microscopy and theoretical calculations of energy. In this model, the process of reconstruction is accompanied by a shift of some of the atoms and by the appearance of a stacking fault between the surface layer and the volume of the crystal, so the authors called it the DAS model (dimer–adatom–stacking fault).

According to this model (see Fig. 2.15, colour inset), the surface unit cell of the reconstruction Si (111) 7×7 is composed of an angular pit and two triangular subcells, which are separated by dimer rows very apparent on the oxanometric model and the STM image in the filled electronic state; each subcell contains 6 atoms adsorbed on the crystal surface and not yet embedded in the crystal lattice (adatoms); the atomic layer below the layer of adatoms in one of the subcells is in state stacking fault state. The validity of this model was confirmed by the results of numerous studies, including images of scanning tunnelling microscopy in Fig. 2.15 *a*, *b*, and atomic force microscopy, which are discussed in detail further in Chapter 8 (see section 8.2.1).

If we consider the reconstruction process in the DAS model (Fig. 2.15, colour inset) purely geometrically, then the reconstructed two-dimensional surface unit 7×7 cell has 49 atoms of unreconstructed bulk silicon, and they all are on the surface in the (111) plane. As mentioned in section 1.1.3, the fresh surface contains a number of dangling bonds and tries to reconstruct. In this case, the fresh atomically clean (111) surface in an effort to reduce energy due to dangling bonds reduces the number of atoms in a plane cell. As a result, some of the atoms are pushed up and form a new upper layer containing 12 atoms on a flat cell (which Takayanagi et al called adatoms – 'extra atoms', but today they are called adatoms also in the Russian literature on the physics and chemistry of surfaces), and the layer beneath it consists of 42 of the remaining atoms, called rest-atoms, and

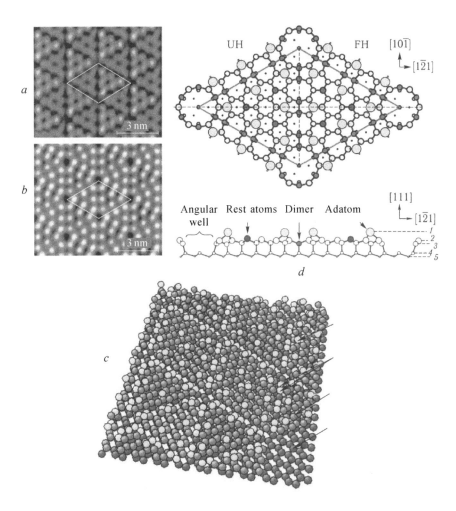

Fig. 2.15. The model of the surface Si (111) 7×7 (DAS model – the model for dimers–adatoms and stacking fault). The atomically clean surface Si (111) with the 7×7 reconstruction. STM images (a) of filled and unfilled (b) electronic states of the surface: c) a schematic representation of the surface (top view and side view) in accordance with the DAS model of Takayanagi; b) axonometric image of the model. The yellow circles in the flat projections indicate the 'additional' atoms (adatoms) of Si, red circles – dimerized Si atoms (in axonometry they are indicated by the white circles), the blue circles – Si atoms (rest atoms) of the second layer, remaining after departure of the adatoms in the upper layer (in axonometry they are indicated by the red circles). The rhombs on the flat projection and the STM images show the two-dimensional surface elementary cell 7×7. Half of the elementary cell, containing the stacking faults, is denoted as FH (faulted half), the half without the stacking forces indicated as UH (unfaulted half). It may be seen that on the STM image of the filled states (a) half of the cell with the stacking fault is brighter. The brightness maximum on the STM image corresponds to the adatoms. From http://thesaurus.rusnano.com/wiki/article14156.

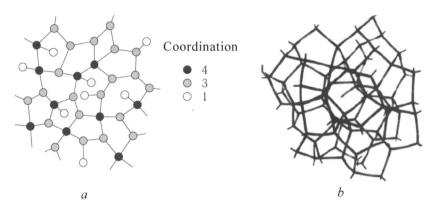

Fig. 2.16. An example of a continuous random network consisting of atoms with different coordination: a – flat projection of the irregular network, atoms with different coordination are identified different shading, b – a three-dimensional model of a continuous random network of a-Si, with the distortion of angles between the bonds less than 20% and containing unsaturated (dangling) bond (Brodsky, 1979).

a stacking fault forms between the layer of rest-atoms and the underlying layer of non-displaced atoms of the bulk crystal.

It should be borne in mind that for the diamond structure in which Si crystallizes the two-layer structure of crystallographic planes is a characteristic feature (see Fig. 2.2). Due to its specific location, the adatoms electronically differ from the other Si atoms in the crystal. Because of the stacking fault formed in this reconstruction in the surface unit cell (circled by the red line in Fig. 2.15) left and right triangles (half unit cells) are not equivalent, and they are called defective (left) and defect-free (right) halves of the cell.

In addition, the side view in Fig. 2.15c shows that the surface reconstruction is accompanied by very strong angular distortions of the structure. Dangling bonds are saturated by adatoms and together they form dimeric bonds. These surface distortions propagate from the surface into the depth of the third atomic layer.

Distortions of the structure of the surface layer, caused by the reconstruction of the surface of Si, can be eliminated if the dimers are removed or their formation is prevented (Sakurai & Hagstrom, 1976; Kaxiras, 2003). In silicon semiconductor technology the surface distortions of the lattice are usually to remove by the saturation of the surface dangling bonds at the expense of monovalent impurities, mostly by attaching hydrogen atoms to them. Since the Si–H bond (79.9 kcal/mol) is stronger than the Si–Si bond (51 kcal/mol), at high-temperature hydrogenation of the surface of the Si crystal formation of the Si–Si dimers on the surface is replaced by the formation of Si–H endings (Sakurai & Hagstrom, 1976). Since the saturation of each of the dangling bonds required only one H atom, the Si atoms do not have to shift from their normal positions and

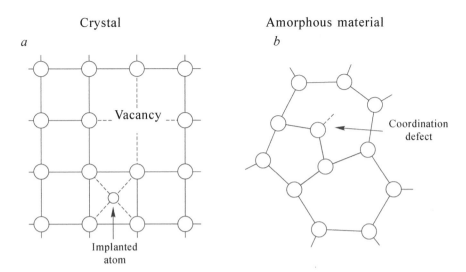

Fig. 2.17. The simplest types of defect (*a*) in the crystal and in the CRN of the amorphous material (*b*). In the crystal, point defects are vacancies and interstitial atoms, and in the amorphous material a point defect is a loss of coordination of the atom (the coordination defect).

the frequency of their distribution on the surface is the same as in the corresponding atomic planes within the crystal.

2.4. Amorphous silicon

In certain highly non-equilibrium solidification conditions we can obtain amorphous silicon (a-Si), which has no crystalline structure, but has semiconducting properties and is widely used in electronics, for example, in the production of solar photon–electronic converters (photoelements). The amorphous phase (as a structural component of the material) is often encountered in the process of obtaining nanocrystalline silicon (nc-Si) and coexists with it in different forms: amorphous nanoparticles, amorphous membranes of nc-Si particles or the matrix in which the nc-Si particles are distributed.

In comparison with the crystals, the amorphous materials have no long-range order, i.e. they are disordered on the macroscale, although they may have a short-range order on the atomic scale. The reason is that local quantum mechanics imposes stringent requirements on the length of chemical bonds for the nearest neighbour atoms, but places less stringent restrictions on the angles between the bonds, so the angles between the covalent bonds in the amorphous silicon may be different[6] and differ from the constant angle between the bonds in c-Si. As a result, the bond lengths

[6]In a-Si the angles between the bonds vary in the range of approximately ±7° around the valence angle of 109.47° in crystalline Si (Street, 1991).

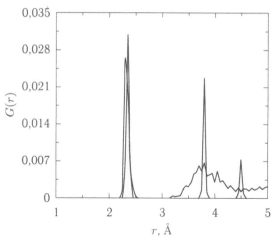

Fig. 2.18. The function of the radial distribution of atoms in c-Si (filled peaks) and a-Si (solid line with no fill), normalized to unit volume (Kaxiras, 2003).

create a local short-range order, and the freedom of choice of the angle between the bonds allows the destruction of the long-range order.

2.4.1. The model of the amorphous state

The short-range order and disorder of the amorphous materials at large distances lead to a model of a disordered network, which was proposed by Zachariasen (Zachariasen, 1932) to describe silica glass. A periodic crystal structure in this model is replaced by a random network, in which each atom has a specific number of bonds (coordination) with the nearest neighbours. Figure 2.16 shows a two-dimensional illustration of such a network containing atoms with different coordination (4, 3 and 1) and its three-dimensional version.

The continuous random network (CRN) has the property to accept easily atoms with different coordination. This distinguishes it from the crystal lattice which can be built only by atoms (guests) with the coordination equal to the coordination of host atoms, because only in this case can the long-range order be maintained. Usually, the network structure of amorphous silicon has a significant number of unsaturated bonds. Bonds try to make pairs, but if their total number is odd, some unsaturated bonds remain.

As the crystal lattice, CRN may also have defects, but the type of defects, compared with the crystal, is characterized in a different way. Any atom which is displaced from the correct position is a defect in the crystal as a defect, but in the CRN this condition is not considered defective.In the homoatomic CRN the only feature that distinguishes the atoms from each other is the coordination with its neighbors. Therefore, the elementary (simplest) structural defect in amorphous silicon is a coordination defect,

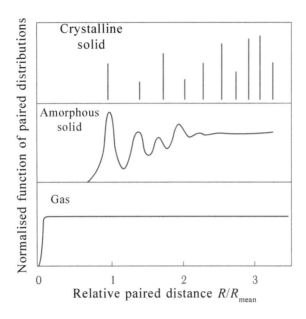

Fig. 2.19. Schematic representation of the function of the pair atomic distributions for the crystalline, amorphous solid and gas phases reduced to the scale of the average distance between nearest neighbour atoms R_{mean}, showing a difference in the degree of the structural order (Street, 1991).

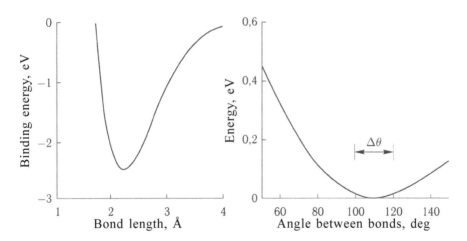

Fig. 2.20. Dependence of the energy of Si–Si bonds on the length and the angle between the bonds in amorphous silicon ($\Delta\theta$ indicates the average distortion of bond angles in a-Si:H). The results were obtained in quantum-chemical calculations in (Biswas & Hamann, 1987).

when an atom has too many or too few bonds with neighbours. The ability of the disordered network to adapt to any atomic coordination allows the presence in the network of isolated coordination defects that are not possible in the crystal. The difference between the simplest defects in the crystal and the amorphous material is shown in Fig. 2.17 (for details see, for example, Street, 1991).

The modern theory of the CRN for the covalent semiconductors (Kaxiras, 2003) including silicon, suggests the existence in the network of not only defects with insufficient threefold coordination, characterized by the presence of an unsaturated dangling bond, but also of fivefold coordination defects, characterized by the presence of a floating bond.

The disorder in the distribution of atoms in the CRN is difficult to classify from the position of point or linear defects. In the randomly disordered network, there are many configurations of the atomic arrangement, but if all the atoms have the same coordination, then all of the various structures in the CRN are equivalent and simply reflect natural variability (instability) of the material. Since there no single correct position for the atom, it is impossible to say whether this particular structure is defective or not. Moreover, the violation of the long-range order is inherent to amorphous materials and is described by the random variation of the disorder potential, which affects the electronic structure (Smith, 1982; Street, 1991).

Amorphous semiconductors, for example, a-Si, in fact are not entirely disordered. They covalent bonds between atoms in them are much the same as in crystalline silicon, and the number of the nearest neighbours is the same or even similar to the average values of bond lengths and bond angles.

The quantitative characterization of the ordering of the atomic structure of materials is carried out in many cases using the function of the radial distribution of atoms $G(r)$, which can be measured experimentally, for example, using X-ray diffraction (see sections 8.3.3.2–8.3.3.4) as a function of the radial distribution of electron density or by neutron diffraction as a function of the distribution of nuclear density. This function is also called the paired distribution density function (PDDF)[7], which shows the probability of finding pairs of atoms with the interatomic distance R.

Comparison of the experimentally measured radial distribution functions in c-Si and a-Si (Fig. 2.18) shows that in the first coordination sphere the amorphous and crystalline phases have the same coordination density in both phases. In the second coordination sphere, the peak of the radial distribution in a-Si is significantly blurred compared to c-Si, but the ordering of the

---

[7]In the Russian language literature this function is often referred to as 'the radiation distribution function' (Iveronova and Revkevich, 1978) and less frequently as the 'paired distribution correlation function'. Detailed and concise description of the current state of the PDDF method and areas of application can be found in, for example, books by Billinger and Thorpe (2002), Egami and Billinger (2003) and also in a review by Billinge (2008).

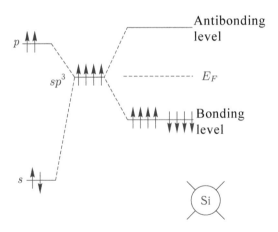

Fig. 2.21. Illustration of the bonded configuration of silicon atoms, consisting of hybridized molecular orbitals. The scheme shows the position of the Fermi energy E_F (Street, 1991).

atoms is still clearly seen, whereas in the third coordination sphere the peak is already absent, i.e. the ordering of Si atoms at this distance almost completely disappears.

An example of the typical density functions of pair distributions for different states of matter is shown schematically in Fig. 2.19. The figure shows that the relative positions of atoms in a dilute gas are completely random (except for very short distances), while in a perfect crystal, these provisions are strictly ordered to large distances. In the amorphous material the short-range order is almost the same as in crystals, but there is no long-range order. In the PDDF graph of the amorphous solid several neighbours can be distinguished fairly well, but the correlation between the pairs of atoms violates the structure and the order is lost over a few interatomic distances. Due to the similarity of the short-range order, the physical properties of amorphous Si and Ge are similar to the properties of these materials in the crystalline state.

The data in Fig. 2.20 shows that a strong covalent bond strongly limits the disordering of the lengths of Si–Si bonds in amorphous silicon. At the same time, the disordering of valence angles in a-Si can be large, reaching the amplitude of the order of ± 7° with respect to the exact values of the valence angle 109.47° in c-Si (Street, 1991).

2.4.2. The electronic structure of the amorphous state

Because of the tendency to form covalent chemical bonds, the amorphous phase Si and Ge, like the diamond-like crystal structure, is formed by groups of four atoms incorporated in the tetrahedra, which are destroyed only when melting. In the molten state, there is much more dense packing in which

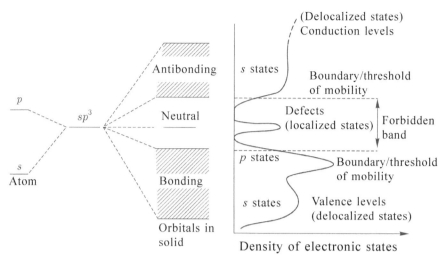

Fig. 2.22. Schematic representation of the electronic structure of amorphous silicon and the corresponding density of electronic states in the formation of a tetrahedron in which each atom is bonded to three neighbours. The calculation using the model of the molecular orbitals (Thorpe & Weaire, 1971). The remaining free bond sp^3, containing one valence electron, is electrically neutral and localized in the middle of the gap (which is called more appropriately an optical gap or mobility gap). The peak of the localized electronic states in the optical band gap (band gap of mobility) is due to their fixation on defects such as dangling or broken bonds. Compilation of (Street, 1991; Kaxiras, 2003; Moliton, 2009).

the number of nearest neighbours is equal to eight, and the molten Si and Ge become metals, as in the solid state at very high pressures. In a-Si the neighbouring tetrahedral groups of atoms are rather strongly linked by covalent bonds, which is confirmed by Raman scattering spectroscopy (RS). The main difference between the amorphous and crystal forms is that in the first these tetrahedra are randomly (nor regularly) oriented relative to each other. The fact that the relative position of the nearest neighbours, forming tetrahedra, in the amorphous state remains the same as in c-Si, plays a very important role, causing not too great a difference in the properties of the amorphous and crystalline forms of silica (Smith, 1982; Street, 1991; Kaxiras, 2003; Moliton, 2009).

The concept of the band structure, based on the crystal lattice, is not applicable to amorphous structures[8], but the presence of covalent chemical bonds, creating distinct local atomic configurations in semiconductors

---

[8]The classic band theory of solids based on the notion of Bloch waves in a periodic potential, which allows to consider the entire solid on an example of single elementary cell. And since there is no band structure in the amorphous material for lattice vibrations (phonons), the continuous phase of the Bloch wave loses its sense, and the band theory in the pure form is not applicable to amorphous materials (see, e.g. Chapter 9 in the book by Moliton, 2009, or Smith, 1982; Street, 1991).

of the Si type, allows to apply to them the theory of chemical bonding and molecular orbitals instead of the band theory. At the same time, the fundamental concepts of description of the electronic states of semiconductors, such as the sphere and the Fermi level, remain valid. There is also the concept of the band gap, which in the case of amorphous materials is not called the forbidden zone and instead is referred to as the energy gap or the gap of mobility of charge carriers between the valence electrons (binding orbitals) and the conduction electrons (bonding or antibonding orbitals). In amorphous semiconductors with covalent bonds we can also apply the concept of the phonon spectrum, although the distribution of phonons and their interactions are markedly different from the behaviour in the crystalline state (Street, 1991; Kaxiras, 2003; Moliton, 2009).

In an isolated Si atom the outer electrons, involved in the formation of chemical bonds, occupy two $3s$ and two $3p$ states. In the formation of a solid the interaction of these electrons in neighbouring atoms splits valence states into bonding and antibonding levels, as shown in Fig. 2.21. Chemical bonding occurs because the bonding state has lower energy than the isolated atomic levels, and the material acquires the lowest total energy when the maximum number of electron occupies the bonding condition. The number of such electrons is limited by the Pauli exclusion principle, which does not allow the presence of more than two electrons (with spins in different directions) in the same energy state. Because of this, the four valence electrons present in the Si atom combine to form four sp^3 orbitals. Each of these orbitals includes a $1/4$ s-state and $3/4$ of one of the three equivalent p-states. The resulting four orbitals form bonds between the four neighbouring atoms, and since the number of valence electrons is Si is equal to four, all bonds are filled, having received two electrons – one from each atom participating in bond formation. Since all bonds are saturated, and all bonding orbitals are completely filled, as a result of hybridization the system of the bonded Si atoms reaches a minimum total energy and is stable. Structural defects such as dangling bonds, also have their electronic states, which are located in the forbidden zone.

In amorphous semiconductors, as compared with the crystalline ones, there are new phenomena that are explained by the fact that their properties depend primarily on local chemical bonds, rather than the translational symmetry of the long-range order. The permissibility of various bonded configurations for each atom leads to the strong interaction between the electronic and structural states and to the phenomenon of metastability.

The main feature of semiconductors and insulators is the presence of an energy gap between the valence band and the band of conducting states. For crystalline semiconductors, according to the one-electron theory of solids, this energy gap between valence states and conduction states is a consequence of the periodicity of the crystal lattice. Studies have shown (Smith, 1982; Street, 1991; Kaxiras, 2003; Moliton, 2009) that a similar gap can also be explained by the theory of splitting of bonding and antibonding states of the covalent bond without involving the concept of the crystal lattice (Fig. 2.22). It turns out that the bonding and antibonding states, as well as the gap (optical gap) between them are most strongly influenced by the short-range order, formed by covalent bonds, and this short-range order is the same in crystalline and amorphous Si, except for the fact that in the crystalline state small perturbations of the periodic potential of the crystal lattice are superposed on this bond, and in a-Si this does not take place.

Quantum-chemical calculations of the electronic density of states conducted in the strong coupling approximation in the works (Weaire, 1971; Weaire & Thorpe, 1971; Thorpe & Weaire, 1971), confirmed that the amorphous silicon is also a semiconductor and has a band gap (Fig. 2.22) is similar to c-Si, and its structure differs only in minor details related to the lack of crystal periodicity and blurred boundaries of the zone.

In pioneering studies of the development of the concept of electronic states in disordered solids (Anderson, 1958, 1978; Cohen et al, 1969; Mott, 1967)[9] the theoretical calculations show that in amorphous semiconductors, there are localized electronic states (bound states), the wave function of which is localized in a region with dimensions of the order of the interatomic distance, and there may also be delocalized states, the wave function of which extends to the region of macroscopic size. In a crystalline solid, the dominant states are the delocalized states. It is the large number of localized states which is the main difference between the electronic structure of amorphous materials and their crystalline counterparts (Madelung, 1985; Brodsky, 1985; Kaxiras, 2003; Gantmakher, 2005; Drabold & Abtew, 2007; Moliton, 2009). The delocalized states may form energy bands very similar to the energy bands in the crystal. In this case, the energy levels corresponding to localized states are situated near the band edges (the thresholds of mobility of charge carriers)[10], extending these zones in the energy region in which the crystalline state in the material is not permitted energy levels. As shown in Fig. 2.22, resulting in a-Si transitions are formed of branches of the electron density of states p and s are the limits of mobility in the region of the band gap, allowing realized hopping mechanism of conductivity throughout the volume of a-Si, whereas in

---

[9]For fundamental theoretical studies of the electronic structure of magnetic and disordered systems, including the creation of the concept of localization, which explained the increase of the number of possible states at the expense of their localization on the defects in the system, PhilipWarren Anderson, Nevill Francis Mott and John van van Vleck were awarded in 1977 the Nobel Prize in physics.

[10]The additional levels localized on defects in the states that have a certain width in $\mathbf{k}$-space and are situated beneath the energy minimum of the conductivity band, were already mentioned in Sec. 1.3 in consideration of the band structure of c-Si with defects. In the case of the amorphous state, in comparison with the crystal, the number of these states is very large, so they merge into a nearly continuous band, which is shown in Fig. 2.22 by the penetration of the branch of the conduction level into the band gap.

crystalline semiconductors, the mechanism of conductivity is observed only at the surface (Gantmakher, 2005).

The presence of the same covalent interatomic bonds in the crystal and amorphous silicon leads to the similarity of electronic structures of these materials (Street, 1991). In particular, the amorphous and crystalline materials usually have a similar band gap (E_g = 1.12 eV at 300 K for the indirect band c-Si, and E_g = 1.6–1.8 eV for the mobility gap in a-Si)[11]. At the same time, the amorphous state results in a number of differences in the electrical and optical properties in comparison with the crystalline material. The disordering of the structure affects the length of the bonds and the angles between them and this widens the distribution of the electronic state and leads to localization of the electrons and holes and also marked scattering of the charge carriers.

Amorphous silicon has too many defects such as dangling bonds, significantly reducing its chemical stability and the stability of the electrical properties. Therefore, in practice it is used in the hydrated form (a-Si:H), when the dangling bonds are passivated by annexation of hydrogen atoms, which do not cause a radical change in the structure and electrical properties of a-Si but make it chemically stable (Street, 1991). Hydrogenation reduces the concentration of defects such as dangling bonds by about 4 orders of magnitude (from 10^{19} cm^{-3} in a-Si up to 10^{15} cm^{-3} in hydrogenated a-Si), while the states of the defects in the mobility gap are shifted to the valence levels (Fig. 2.22), and the optical gap widens to E_g = (1.6–1.8) eV. The mobility of the major charge carriers (electrons) in a-Si is much lower than in c-Si (μ_{De} ≈ 0.5 cm^2·V^{-1} s^{-1} in a-Si:H compared to μ_{De} ≈ 1400 cm^2·V^{-1} s^{-1} for c-Si), which is explained by the difference of the concentration of defects and by the corresponding difference between the conduction mechanisms. But due to their structural characteristics a-Si compared to c-Si has much higher photosensitivity (the strong dependence of photoconductivity/resistance to light exposure).

2.5. Silicon clusters

In the transition from individual atoms or Si melt to the massive solid material the solid phase a-Si is a metastable quenched state, whereas the equilibrium phase of Si is the crystalline state with the diamond structure, which is confirmed experimentally and by calculations of the minimum energy of the systems (see Fig. 2.3). Therefore, in high-temperature annealing a-Si is rapidly converted to c-Si.

---

[11]It is not possible to accurately determine the energy gap in a-Si, because, unlike c-Si, it has no sharp boundaries of the valence band with the conduction band (see Fig. 2.22). Therefore, E_g of amorphous silicon is measured approximately by the linear extrapolation of the dependence of the optical absorption coefficient on photon energy to the zero value of the coefficient (Street, 1991). Thus, the value E_g = 1.7 eV was obtained for pure a-Si:H – hydrogenated amorphous silicon.

The formation of nanocrystals/quantum dots of nc-Si in the synthesis process begins with the formation of nuclei, which are clusters of silicon atoms. Gradually, with increasing size the clusters acquire a crystalline structure. The difference between the structure and physico-chemical and electronic properties of clusters and nanocrystals is large, and the border between these two states of matter is quite arbitrary and is not exactly known. Thermodynamic calculations carried out in (Vepřek et al, 1982) show that the threshold of stability of the crystalline phase of silicon is at a particle size of about 3 nm. If the size is below this threshold the Si crystalline phase becomes disordered or amorphous, and vice versa, with increase of the size of the amorphous particles above this threshold size the crystalline phase is thermodynamically more favourable. Therefore, it is assumed that when the local ordered regions reach a diameter of 3 nm the formation of a stable region with translational crystal periodicity of can start around them, as around the centres of crystallization. But in any case, the formation of the diamond structure, characteristic of massive silicon and nc-Si, must bt preceded by the formation of nuclei with a pronounced short-range order of the atomic arrangement, i.e. clusters. It can be assumed that the clusters are a bridge from a single atom to a stable massive solid consisting of a large number of atoms.

Clusters are formation, consisting of different numbers of atoms – from a few to tens and hundreds of thousands, occupy an intermediate region between the individual atoms and the solid, possessing both the properties of molecules and of the solid body, and at the same time showing properties different from both.

At the moment, there is no single definition of the term 'cluster', and each branch of science understands it differently, often as objects very different in size and properties. The generality of different definitions[12] consists in the fact that clusters are nanoparticles that contain a small number of atoms (or molecules) linked together by chemical bonds.

In chemistry and materials science the clusters are often regarded as one of the intermediate states in the organization of matter between a single

---

[12]Cluster is a chemical compound containing a covalent bond between atoms or molecules. Clusters can be complex compounds, stabilized by ligands and neutral molecules. The term is also used in a wider sense, denoting any group of atoms intermediate in size between a molecule and a solid. (http://en.wikipedia.org/wiki/Cluster_chemistry; definition from Wikipedia). Cluster is a system consisting of a large number of weakly bound atoms or molecules. The cluster occupies an intermediate position between the van der Waals molecules, containing a few atoms or molecules, and fine particles (aerosols). If the cluster contains an ion, it is called a cluster ion or ion cluster. In this case, the binding energy per molecule is usually higher than in the van der Waals molecules. Clusters can be characterized by macroscopic parameters and as the number of particles in the cluster increases these parameters approach the corresponding characteristics of the particles of the dispersed condensed phase (Physical Encyclopedia. T. 2. – Moscow: Soviet Encyclopedia, 1990).

atom (molecule, ion) and a solid (nanoparticle). According to this view, the cluster is a group of a small, often variable, number of interacting atoms, ions or molecules. Depending on the type of united particles, clusters are divided into atomic, ionic and molecular. Clusters are also particles of the size for which the observed properties differ significantly from the properties of the macroscopic object, and vary considerably when adding another constituent element (atom or molecule). Their physical properties depend strongly on the type of their constituent atoms and their number N, where the increase of N is accompanied by the smoothing of this dependence, indicating a gradual transition of the material from the cluster to the bulk state.

When considering the clusters it must be remembered that we are talking about the particles with the size not greater than one or two thousand atoms, and often much less, which cannot be described by the concept of an infinite lattice, so that we can talk about the electronic spectrum of the discrete type, as in the molecules, rather than the band type as in crystals. Moreover, these formations can be characterized by the non-monotonic dependence of the properties of size, especially for small clusters, where for different sizes we may observe different competing structural types.

2.5.1. The difference between the clusters and amorphous and crystalline phases

Thus, clusters are a special form of matter, representing a group of interacting atoms or molecules, characterized by a specific short-range order, but this order may be different even in clusters of equal size (i.e. clusters may contain a large number of isomers).

This group of interacting atoms (clusters) can in most cases be stable only in the isolated state, which greatly complicates the experimental study of clusters. For example, a cluster of 10 Si atoms may not be stable if it has the opportunity to interact with another cluster of 10 Si atoms. These two clusters quickly merge into a new cluster of 20 silicon atoms, which may prove to be also unstable and will continue to merge with neighbouring clusters. This behaviour is very different from that of the clusters of interacting molecules of identical size. For example, if we consider the two benzene molecules, in interaction they will retain their integrity and in conjunction with each other will remain molecules[13]. Another difference between the clusters and the molecules is that the cluster may contain a number of atoms interacting with each other without any definite symmetry. This means that the clusters have a certain number of degrees of freedom of variation of both structural and electronic properties. The presence of

---

[13]It should be noted that under certain conditions, namely, at certain 'magic' numbers of atoms, clusters can be so stable that they can be called molecules, for example, cluster C_{60} is a fullerene molecule.

many variations extremely complicates the theoretical description of the cluster properties.

The properties of individual clusters in the materials research are absolutely unique, and they can not be obtained by a simple extrapolation of the large-scale properties of the solid of the same chemical composition. The best known example of the unique properties of the clusters compared to bulk materials of the same composition is the fullerene – a cluster of C_{60} in the form of a soccer ball, which can be regarded as a rigid body consisting of carbon atoms only. Thin films of C_{60} clusters are superconductors even at temperatures above 30 K, if they are doped with alkali metal atoms, which distinguishes this fullerene from other forms of carbon – graphite and diamond (Yeletsky and Smirnov, 1989; Masterov, 1997).

Another striking example are the clusters of gold atoms. It is known that gold is diamagnetic. However, direct measurements of the spin polarization using the method of magnetic circular dichroism at synchrotron radiation were conducted by Yamamoto et al. (2004) and show that clusters of gold with the size of about 1.9 nm, which contain about 210 atoms, exhibit ferromagnetic properties unusual for gold. There are many other examples of this kind, showing the change in the physical and chemical properties of systems with a decrease in their size to the cluster state.

In materials science the cluster materials are divided into three large categories which have different structural and physical characteristics: metallic clusters, carbon clusters and clusters of covalent elements or elements that are prone to the formation of covalent bonds.

Historically, the metallic clusters were the first to be produced and studied (Simpson, 1986). This can be explained by the relative simplicity of their preparation for the experimental studies, as well as by the applicability of the rather illustrative theoretical models and calculation methods (available already in 1970–1980) in the theory of many bodies to study their electronic structure and optical properties, processes of interaction among themselves and with other particles, and their formation and decay processes. An overview of the mechanisms of formation and properties of Si clusters can be found, for example, in publications (Bley & Kauzlarich, 1998; Chelikowsky, 2004).

The properties of clusters and their relation to the size is studied by both empirical research methods and theoretical approaches (modelling of Si clusters in the ground and excited states).

2.5.2. Theoretical studies of the structure and properties of clusters

The theoretical description of metallic atom clusters (Sugano, 1998; Brack, 1993) is often carried out (by analogy with nuclear physics) using 'jelly' model, in which the cluster is regarded as a single particle (the mega-atom) and is divided into two subsystems (Erkandt, 1984): a system of positive ions (nucleus) and a system of valence electrons (shell). The nucleus in this case is seen as a 'smeared' positive charge with some spherically symmetric density distribution $\rho(r)$, the field of which contains valence electrons forming electron shells of the mega-atom[14]. The validity of the shell model of atomic metal clusters has been confirmed experimentally by mass spectrometric measurements of clusters of alkali metals (Knight et al, 1984, 1985), where the distribution of the number of the formed clusters by the mass and their polarity correlated with this model. The experimentally measured (by mass spectrometry) distribution function of the number of clusters in dependence on the size showed distinctive peaks that corresponded to the clusters in which the outer electron shells of atoms were filled (Knight et al., 1984; de Heer et al, 1987). The relative number of clusters (Na)$_n$ resulted in prominent peaks at the number of atoms in them $n = 8, 20, 40, 58, 92$. The calculation of the single-particle electronic levels in the well of the limiting external spherical potential shows that these sizes correspond to the filling of the following electron shells: $n = 8$ $\rightarrow$ [1s 1p], $n = 20 \rightarrow$ [1s 1p 1d 2s], $n = 40 \rightarrow$ [1s 1p 1d 2s 1f 2p], $n = 58$ $\rightarrow$ [1s 1p 1d 2s 1f 2p 1g], $n = 92 \rightarrow$ [1s 1p 1d 2s 1f 2p 1g 2d 3s 1h], where the numbers 1, 2, ... denote the radial quantum number n_r, the letters s, p, d, f, g, h,... correspond to the orbital numbers $l = 0, 1, 2,$ . . .. This means that the stable magic clusters are those in which there is delocalization of $3s$ electrons and their collectivization in a potential well of the spherical potential with the formation of the conduction band, and the highest energy of interatomic bonds in the cluster is achieved. Thus, on the basis of the experimentally obtained dependence of the electronic structure of metal clusters on the number of atoms N and the theoretical calculations (Brack, 1993; de Heer, 1993) it was concluded that there is a certain similarity in the internal structure of the metallic clusters, atoms and atomic nuclei, which exhibit a periodic change of the properties with an increase in the number of particles. The number of atoms in stable clusters by analogy with nuclear physics is called 'magic numbers'.

The existence of multiple clusters of specific sizes can be explained by the exceptional stability of configurations with filled electron shells. In principle, the sequence of magic numbers can reach high values, but in experiments using mass spectrometry for large clusters this can not be traced, because in large clusters, the physical properties which are used to determine their size, become took weakly dependent on the number of atoms.

The same experimentally determined sequence of magic numbers for clusters with the filled outer electron shells of atoms is also confirmed by theoretical calculations discussed in the article (Bjornholm et al, 1990), and also by calculations carried out for clusters of different sizes up to the set

---

[14]A closer look at the 'jelly' model applied to describe the structure of bulk crystals, and its generalization to the nanoclusters can be found, for example, in the textbook (Kaxiras, 2003).

of atoms containing 4000 valence electrons in (Nishioka et al, 1990) in the mean-field approximation of the potentials or self-consistent calculations (using the 'jelly' model in the theory of the density functional and the approximation of local density).

The property of the valence electrons of the metal to leave their atoms (delocalize) and form the conduction band is the reason for the formation of the specific energy structure of the metallic atom clusters. It is the behaviour of the shell of the collectivized electrons that determines the majority of the unusual collective properties of the clusters. At present, the shell model of the electronic structure of clusters is common for the theoretical and experimental study of metallic clusters containing from a few to thousands of atoms (Sugano, 1998; Ivanov, 1999).

As shown by the energy calculations for Na clusters, which take into account both the geometric and electronic factors (Alonso et al, 1992), for the stability of clusters (with the minimum total energy) the predominant factor is the electronic structure, rather than the structure of the packing of atoms.

2.5.3. Classification of Si clusters

In the cluster the state of the atoms is similar to their states on the surface of a bulk sample, where they are deprived of part of the four-sided environment. That is, in the cluster we should observe the phenomenon of reconstruction of the surface, which is discussed in chapter 1.

The calculations show the kinetics of the structure formation by the simulated annealing algorithm (Binggeli & Chelikowsky, 1994; Bley & Kauzlarich, 1998), with formation and growth of the cluster, first, when the temperature is high, the interaction between the atoms is weak, but gradually the atoms begin to form bonds with each other and unite in a variety of configurations. Slow reduction of temperature should be accompanied by the formation of structures with minimum energy. This process can be simulated, for example, by molecular dynamics calculations, which take into account the quantum forces (Binggeli & Chelikowsky, 1994; Vepřek et al, 1982).

An example of such a simulation for Si is shown in Fig. 2.23. In the beginning at high temperature the seven Si atoms interact weakly with each other. In this state, they can form many different structural systems. As a result, these structures include tetramers and trimers, which are formed independently of each other. At this point, the atoms form polygonal clusters, whose formation leads to a sharp drop in the energy of the system.

The properties of Si clusters are classified by the size (Mélinon et al, 2008). This classification is generally recognized, at least for clusters with sizes below 10 nm. In earlier studies of crystalline, paracrystalline (pseudocrystalline), porous and amorphous silicon structures the experimentally justified assumption was made that in the size range of 2–3

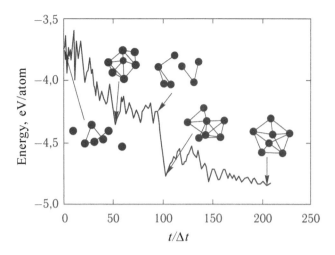

Fig. 2.23. Modelling of annealing by the molecular dynamics method for a cluster of seven atoms of Si. The structures obtained at different stages of lowering the temperature by using the search of configuration with minimal energy are shown. In the process of this modelling temperature decreased from $t = 5000$ K (left) to $t = 300$ K (right). The final structure of the cluster obtained by this method is in agreement with theoretical studies by other methods and with experimental data. The x-axis shows the steps in temperature (about 230 K for every 10 units of the scale). Figure from (Binggeli & Chelikowsky, 1994).

a phase transformation takes place(Iqbal & Vepřek, 1982). Silicon clusters or a mesh larger than this size are crystalline and at smaller size they are amorphous. The reason for this can be explained by the fact that at the sizes close to 5 nm the lattice shows, due to surface effects, deviations from the interatomic distances characteristic of the massive silicon (Hofmeister et al, 1999). If the particle size becomes smaller than 2–3 nm, the dominant role is played by the phenomenon of reconstruction of the particle surface, which is a violation of the lattice symmetry (the transition to the amorphous type). If the size is less than 1 nm, because of the Laplace law (melting point decreases as $1/D$, where D is the diameter of the particle) the cluster becomes a liquid at room temperature, which leads to the formation of a close-packed structure, as predicted by the standard phase diagram of Si.

Large Si clusters have similar characteristics to the characteristics of the crystalline phases of silicon: in them, as in crystalline silicon, fourfold hybridization and dense packing of atoms in the (111) and (100) planes take place, but, in contrast to c-Si, they, along with conventional six-membered rings that are characteristic of the macroscopic structure of silicon, there are five-membered cycles, as shown in (Erlandsson et al, 1996; Takayanagi et al, 1985) and in Fig. 2.24.

The presence in cluster structures of five-membered cycles, as well as similar types of its chemical bonds and valence with carbon, indicate the

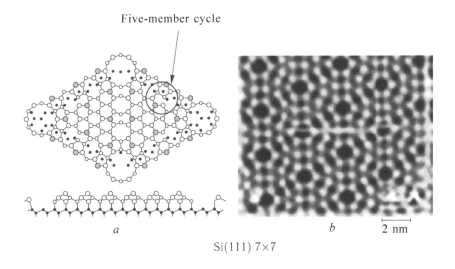

Five-member cycle

a

b   2 nm

Si(111) 7×7

Fig. 2.24. The result of the reconstruction of the surface of Si (111) 7 × 7. *a* – top view and side view on the the resulting stacking fault described by the dimmer-adatom model (Takayanagi et al, 1988), where the circled area indicates the five-membered ring. *b* – the image of the same surface, obtained in an atomic force microscope (Erlandsson et al, 1996). Adapted from (Mélinon et al, 2008).

possible existence of frame structures similar to fullerenes. Beginning in 1990's, the search for fullerene-like cluster Si structures was devoted to a number of experimental and theoretical studies.

2.5.4. Theoretical studies of small clusters of Si

The physical and chemical properties of Si clusters have been actively studied by the methods of theoretical calculations and computer modelling since the mid-1990s to the present time by many research groups. The most common energy calculations are carried out by the DFT, DMC and QMC methods. A historical overview of these studies can be found, for example, in section 1.9.2.5 of the publication (Nayfeh & Mitas, 2008). The focus was given to the study of changes in the properties of clusters of increasing size from a few atoms to several tens or hundreds of atoms, in particular, to the structural transition from an elongated structure to a flattened one, which was found experimentally and theoretically at the cluster size of ~20 atoms. In particular, thorough theoretical calculations (Nayfeh & Mitas, 2008) have shown that the energy of the Si_{25} cluster of the elongated shape is minimal among all possible isomers. It is possible to explain the cause of the structural transition from the original cell structure of the frame type with a small number of 'internal' atoms with an exaggerated coordination environment, to the structure of the close packing of atoms. The frame

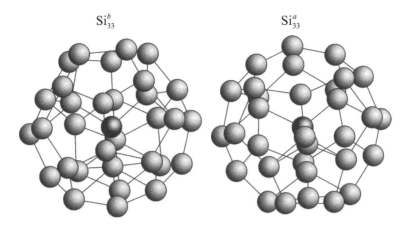

Fig. 2.25. Example of two isomers of the Si_{33} cluster. The featured isomers differ in the position of the Si_5 tetrahedron inside the fullerene shell. Figure from (Mélinon et al, 1997).

structure provides structural relaxation of by fast break up of bonds and forming new bonds with surface atoms of the cluster.

The Si_{33} cluster is interesting from the standpoint of stability and the size. The structure of this cluster, as well as of most chemically inert silicon clusters, belongs in the group of endoendric fullerene structures, i.e. with a fullerene frame, within which the atomic complex of a smaller size is enclosed. The frame Si_{33} structure has the form of the fullerene Si_{28}, formed from four hexagons and 12 pentagons, with the tetrahedral cell Si_5 inside the frame (Fig. 2.25). The structure of the Si_{28} frame is interesting as a building material for clathrates, which can be used to transfer, for example, biologically active molecules.

The position of the Si_5 cell in the Si_{33} cluster is not strictly defined, so several isomers of this cluster can exist (Kaxiras, 1990; Mélinon et al, 1997; Ramakrishna & Pun Pan, 1994; Rothlisberger et al, 1994). Figure 2.25 shows the structures of only two of many possible isomers of the Si_{33} cluster which differ from each other by the position and bonds of the Si_5 tetrahedron inside the fullerene shell Si_{28}. Compared with the diamond structure of the conventional crystalline silicon the closest environment of the Si atoms in the Si_{33} cluster is characterized by too high coordination numbers (greater than 4). For example, in structures in Fig. 2.25 the coordination numbers for the first spheres of short-range order around different Si atoms range from 3.6 to 4.2.

The conformation of this structure does not guarantee its excessive stability, however, a strong separation of the highest occupied and lowest unoccupied molecular orbitals (HOMO-LUMO) indicates the low reactivity of the cluster.

Meleshko (1999) carried out calculations of hollow silicon and Si_nH_n clusters with the structure of fullerenes with the number n of Si atoms up to 50. In the approximation[15] MINDO/3, taking into account interatomic interactions (Bingham et al, 1975) by the Monte Carlo method, the geometry was optimized by calculating the structure of fullerenes Si_n and Si_nH_n with dimensions $20 < n < 60$. The calculations showed the stability of spheroidal silicon clusters with the number of atoms of more than 36, and the rapid increase in binding energy due to an increase in their size. Such growth is not checked for clusters with compact structures, calculated as an alternative. The latter are inferior to fullerenes in the binding energy at the size $n > 40$–50 atoms. As a result of optimizing the geometry of the tetrahedral Si_{45} cluster a structure is obtained which is close to spheroidal, with a gain in binding energy.

In this paper (Yoo et al, 2008) by optimizing the functional density of electronic states at the given initial restrictions searched for geometric structures with minimum energy for the clusters Si_{39}, Si_{40}, Si_{50}, Si_{60}, Si_{70}, and Si_{80}. All the structures were found to be of the endoendric fullerene type. The calculations also showed that the energy of interatomic bonding almost linearly increases by > 10 meV for the addition to the cluster of next 10 Si atoms. *Ab initio* calculations to determine the electronic structure stable atomic conformation are carried mostly out for clusters of small and medium size with the number of atoms not more than 100, since the calculations of more cumbersome structures require powerful computational resources because the number of possible isomers increases dramatically when the number of atoms increases.

2.5.5. Structure and properties of large clusters of Si_n

Theoretical calculation of the structure and properties of large clusters is still a too complex and costly task, so the vast majority of the evidence for these objects is derived from experimental observations (Kumar, 2008). In clusters of Si with the number of silicon atoms $N > 100$ the most likely structure is the diamond structure, similar to the atomic structure of the surface of the planes (100) or (111) crystalline silicon. The temperature dependence of absorption in the range of Urbach tails in the fundamental absorption band of light that characterize the amplitude of the disorder in the structure and density of the unsaturated bonds (Anselm, 1978), indicates that the band gap narrows with increasing cluster size, although it is wider than that of crystalline silicon (Mélinon et al, 2008). This is similar to that observed in a-Si, in which the width of the the band gap is about 1.8–1.9 eV, whereas in c-Si with the diamond structure, it is 1.12–1.17 eV.

The structure and morphology of isolated clusters studied are experimentally difficult to study because of the difficulty of obtaining them, and therefore the number of such studies is not great, but there are a large

---

[15]MINDO = modified intermediate neglect of differential overlap.

number of studies carried out on ensembles of clusters of different size. Most often, studies are carried out on assemblies of clusters deposited on a substrate, which reduces the coalescence of particles to a minimum, and thus we can determine the characteristics of weakly interacting clusters and draw conclusions about the properties of a single cluster. Assemblies of a given cluster size can be prepared, for example, by low-energy molecular beam deposition of thin films[16]. The energy of the collision of clusters with the substrate in this method is insufficient for the destruction of their structure, and in a specific mode we can achieve minimal clumping of particles. Clusters for spraying are obtained from high-temperature plasma. Ionized particles are collected in a stream of helium and sorted by mass TOF spectrometry. Thus, it is possible to form thin films on substrates consisting of particles with a narrow mass distribution, which are suitable for further studies of the properties of individual clusters based on the probability of coalescence effects.

This method was used, for example, in experimental studies of the structure and properties of Si_n clusters with the size of 20 to 100 atoms, conducted by (Mélinon et al, 1997). Neutral nanoscale silicon clusters were produced by laser ablation and analyzed in a gas phase, and also after deposition on various substrates at room temperature in a very deep vacuum. The deposited nanostructured films, according to the TOF mass spectrometry, contain clusters ranging in size from 20 to 500 atoms, and to determine the electronic structure were analyzed by several complementary spectroscopic methods (XPS, AES and EELS). The properties these films were comparable to the properties of the disordered phase, but compared with amorphous or porous silicon have a number of features, such as bands of s-and sp-type electron spectra excited by x-rays (XPS), and the low density of valence electron states (DOS) near the Fermi level, which could not be interpreted by only one model of the quantum limit. According to the authors, as confirmed by the analysis of a number of variants of the structure of clusters using the strong-coupling model, the observed spectral features are largely attributable to the presence in external shells of the clusters of odd-membered cycles (five- or seven-member), with unsaturated bonds.

2.5.6. Experimental data on the growth kinetics of Si clusters

Direct experimental study of the formation of Si nanoparticles from its inception until the formation of a single crystal and the evolution of the physical and chemical properties in this process is very difficult because of the complexity of direct structural measurements on the particles, which consist of a small number of atoms, as well as correlating the results with the specific particle size. An additional difficulty is associated with

[16]This method is known as the low-energy cluster beam deposition technique (LECBD).

structural instability and high reactivity of the particles of nanometer and subnanometer sizes. For example, the pure Si clusters due to their high chemical activity can be produced and maintained only in conditions of deep vacuum or embedded in a protective inert matrix. Usually in practice we have to deal with clusters whose surface is passivated by some radicals, at best, attached to the surface by the hydrogen atoms. Hydrogen passivation is good because it does not violate the skeletal structure of the atomic clusters and allows us to investigate their electronic properties by conventional analytical tools.

The study of pure Si clusters of small size (7–65 atoms) by mass spectrometry, ion cyclotron resonance with Fourier transform of the signal[17] (Elkind et al, 1987) showed that, despite the ionization (presence of a positive charge), at some sizes the reactivity of the cluster is dramatically reduced. For example, the Si_n^+ clusters were inert consisting of $n = 20, 25, 33, 39$ and 45 atoms, whereas the clusters of $n = 18, 23, 30, 36, 43$ and 46 had very high chemical activity. A hypothesis was proposed on the basis of these facts that the observed difference is due to the properties of different structures of the clusters. Chemically inert are the clusters with the framed structure, which is energetically more stable than the dense packing of atoms in the clusters with high reactivity. The possibility of formation of clusters with the dense packing of atoms with 14–100 atoms has been confirmed by a number of experimental and theoretical studies of the structure, electronic and chemical properties (Mélinon et al, 2008), which also showed that the electronic structure and the band gap in pure isolated Si nanoclusters strongly depend on the cluster size, density and shape of packing of atoms in them. These studies also support the hypothesis of the existence of a critical size (30 atoms in the field), at which the restructuring of the clusters takes place.

In Honea et al. (1999) Raman spectroscopy scattering (SERS) was used to study Si clusters with a small number of atoms. Figure 2.26 shows the results for clusters formed at the start of formation of the Si nanoparticles.

The processing of the spectra in Fig. 2.26a shows that they correspond to the structures shown in Fig. 2.26b. The Si_4 cluster corresponds to the structure of a flat diamond, Si_6 – distorted octahedron, Si_9 – pentagonal bipyramid. In the last of these structures two atoms have the coordination environment of the fifth order and form dense packing. All the atoms in these clusters are located on the surface.

A further increase in the number of Si atoms in the cluster leads to the evolution of the skeleton structure and its transition to a dense packing, which is predicted by theoretical calculations (Nayfeh & Mitas, 2008).

The first work on the study of large Si clusters were associated with the hope to find stable skeleton-type formations of the fullerene molecule. However, the theoretical and experimental studies have shown that silicon clusters with magic numbers, characteristic of the carbon clusters, including

---

[17]the FTICR-MS method (Lehman and Bercy, 1980).

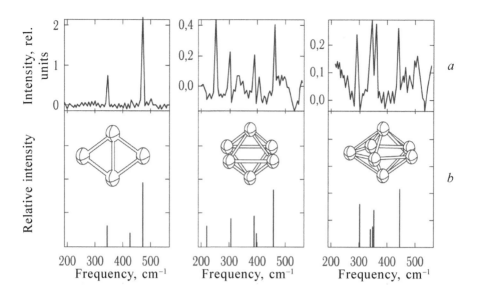

Fig. 2.26. The experimentally measured Raman spectra (*a*) for Si_4, Si_6, and Si_9 clusters, *b* – structures corresponding to these spectra. From (Honea et al, 1999).

the Si_{60} cluster, may not be stable. Nevertheless, the stable Si clusters, including the skeleton structure, are detected, but they do have other magic numbers than the carbon clusters.

It was established experimentally (Jarrold & Constant, 1991) that the shape of the Si clusters changes from elongated to spherical at the critical size of about 27 atoms. At the same time, theoretical calculations suggest that the reactivity of Si_{33}, Si_{39}, Si_{45} clusters should be about as low as that of the bulk silicon (Kaxiras, 1990; Ramakrishna & Pun Pan, 1994; Rothlisberger et al, 1994). Such features are characteristic for the fullerene-type clusters with internal shells completely filled with Si atoms. Using molecular dynamics simulations by the Parrinello–Kara method in (Rothlisberger et al, 1994) it was shown that in the case of silicon, such structures are formed by two shells of atoms, in which the outer shell ('cell') has a fullerene-like form and the inner shell (core) is composed of several atoms, saturating unsaturated bonds.

The authors of the publication (Bley & Kauzlarich, 1998) studied the morphology, structure and size distribution of Si nanoclusters, obtained by the method of chemical synthesis in solution at low temperature, and compared their characteristics with the characteristics of silicon nanoclusters formed by other methods. The final product was a colloidal suspension of Si nanoparticles in hexane, which were synthesized from Zintl salt KSi (Bley & Kauzlarich, 1996). To complete the reaction and ensure passivation of the surface of the nanoparticles, diglyme was added to the reaction mixture. This method also allows the synthesis of silicon nanoparticles to obtain methyl functional groups on the surface. FTIR, UV and photoluminescence spectroscopy methods showed that the nanoparticles contained functional groups H₃C–O– on the surface with which the Si atoms were linked through oxygen.

The study of nanoparticles obtained by solvent evaporation after HRTEM found that the residue contains fine particles with diameters of Si from 1.5 to 2.0 nm and larger particles and agglomerates of fine particles. Local electron diffraction showed the presence of diffraction rings from the systems of the {111} and {220} planes of crystalline silicon, but because of the small particles rings were badly eroded.

The optical absorption spectrum of this sol (Fig. 2.27) compared with the corresponding spectra of large Si crystals is strongly shifted into the ultraviolet region. The authors explained this by the quantum dimensional constraints as a result of which the energy gap of the small nanoparticles significantly increased. The measurement of photoluminescence spectra of sol particles with methoxy and methyl functional groups on the surface of Si nanoclusters revealed red and even blue photoluminescence peaks (Fig. 2.28).

A review of Dutta et al. (1998) discussed the kinetics of formation of Si powder nanoparticles during the synthesis of silane (SiH_4) by high-frequency plasma by the previously proposed method (Dutta et al, 1996) and subsequent crystallization at high temperature annealing. Compared with other technologies of Si nanopowder production, plasma-chemical processes provide the most opportunities for the study of particles, because the technology makes it relatively easy to control the product directly in the reactor, separating the particles by their masses. Much of the discussion in this review is based on research by the authors, but the experimental results and theoretical studies conducted by other researchers are also actively discussed. The main methods used to keep track of processes in

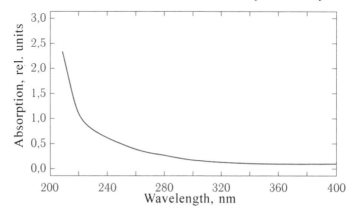

Fig. 2.27. The absorption spectrum of the sol of silicon nanoparticles with narrow size distribution of particles in the range of several nanometers. Figure from (Bley & Kauzlarich, 1998).

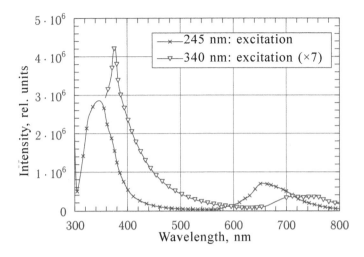

Fig. 2.28. The photoluminescence spectra of sols of Si clusters with methyl functional groups on the surface and the average diameter of 2.5 nm at different wavelengths of the exciting radiation. In both spectra there are red and blue peaks of luminescence. The blue shift of the luminescence is due to the contribution from smaller particles; the energy of exciting light with a wavelength of 340 nm was not enough for excation of these particles. Figure from (Bley & Kauzlarich, 1998).

the dynamics of particle formation (*in situ*) are vibrational spectroscopy, mass spectrometry and dynamic light scattering, giving information about the morphology and dispersion of powders, including the density function of the particle size distribution, as well as the perfection of their structure. At intermediate stages the morphology and structure of the particles by were studied by TEM.

Immediately after the ignition of plasma (SiH_4) with certain discharge characteristics, several tens of nanometer-sized particles form in the plasma, as seen from Fig. 2.29, which shows the dynamic curves of the evolution of the particle size and the number of the particles.

Already in the first seconds of the process the particle number density reaches value of 10^{15} m^{-3}, which is comparable to the density of electrons in the plasma. Measurements of the particle size by light scattering by the Rayleigh–Mie method show (Courteille et al, 1996; Courteille et al, 1995; Dorier et al, 1995; Dutta, Hofmann et al, 1997; Watanabe & Shiratani, 1993) that immediately after the formation of particles is immediately followed by their active agglomeration, resulting in the formation of particles with a diameter of 100–200 nm. The sintering process is accompanied by a noticeable decrease in the number of the particles (Fig. 2.29) – the volume density of the number of the particles decreases by several orders of magnitude. The study of intermediate samples by TEM shows that in the first stages of agglomeration the particles have a non-spherical shape, and the density function of their size distribution is described by the lognormal

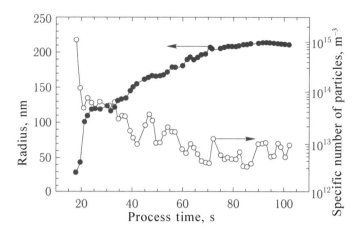

Fig. 2.29. The dependence of the radius of the powder particles of Si and their numerical density on the synthesis time in the high-frequency silane plasma. Figure from (Dutta et al, 1998).

distribution. Measurements of photoluminescence and infrared absorption spectra in the process of particle growth show that the growing amorphous particles at the stage of agglomeration have visible photoluminescence at room temperature (Courteille et al, 1995).

Measurements of light scattering in the synthesis process (*in situ*) show that the process of formation of the powder particles passes through three different stages: 1) initial clustering inherent in plasma-chemical processes; 2) the formation of larger primary spherical particles under the action of van der Waals forces; 3) aggregation of primary particles in agglomerates, which are actually powder particles. As can be seen from the results shown in Fig. 2.29, after a period of intensive growth of the agglomerates the particles reach the critical size, after which their growth is almost terminated. The number density of the particles at a very early stage of cluster formation, which occurs in the first 15–20 s, could not be recorded by the Rayleigh–Mie scattering method as this method by its nature does not allow to distinguish between small particles with sizes of a few nanometers. Since the plasma-chemical reactor in the decomposition of silane contains a large number of free hydrogen atoms, they can easily bond with the surface atoms of the particles of Si and hydrogenate their surface, which is confirmed by the study of IR absorption spectra.

Electron microscopic studies (Dutta et al, 1995) have shown (Figure 2.30) that the Si powders, obtained by plasma decomposition of silane, followed by annealing, are composed of individual nanoparticles and agglomerates with a very high porosity. The size of the particles in the powder is in the range from 8 nm to 200 nm, and due to the small particle size and high particle porosity the powder was characterized by a very large specific surface area (up to 162 m^2/g). The local diffraction electron

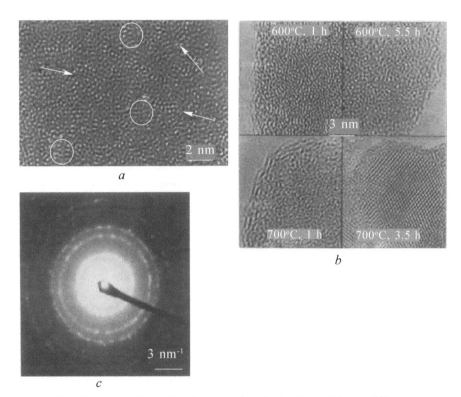

Fig. 2.30. High-resolution TEM images of a single Si particle at different stages of annealing, demonstrating the development of surface morphology: *a* – initial amorphous powder particle, the circles indicate the characteristic ring-shaped contrast areas and the arrows the banded region of contrast; *b* – a powder particle after annealing under different conditions (annealing time and temperature are shown in the images); *c* – local electron diffraction pattern of the powder after annealing for 1 h at 700°C. Images from (Dutta et al, 1998).

did not show diffraction rings typical of crystalline materials, and had the appearance of the diffuse halo in Fig. 2.30c, as the amorphous material. This indicates the amorphous structure of the powder particles.

In the high-resolution TEM photograph in Fig. 2.30a it can be seen that a single particle has a very rough surface, and the roughness consists of a series of approximately circular associations such as fingerprints with a diameter of about 1.5 nm, and banded like fringe areas, which can be clearly seen at high magnification in Fig. 2.31a. These features suggest that in the powder particles there are partially ordered zones smaller than the smallest crystallite observed after annealing. This ordering in the areas of the nanometre size is an intermediate structure between the amorphous state in which there is only a short-range order and the long-range translational order, as in crystals, and indicates that the particles could be formed by

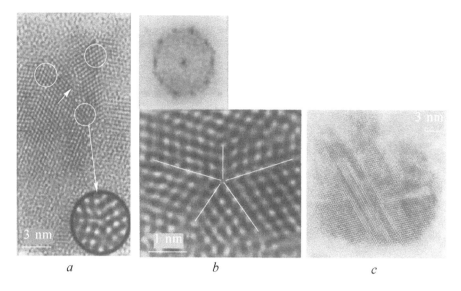

a   *b*   *c*

Fig. 2.31. Evolution of Si nanoparticles, depending on annealing temperature and time. High-resolution TEM images of nanoparticles of Si powder after annealing at 700 °C for 1 h. *a* – the image shows crystalline domains with the fivefold symmetry (circled) formed through the merger of the ordered crystalline subdomains annealed at 700 °C for 1 h. These areas in the particle are surrounded by parallel twin plates or twin plates azimuthally twisted relative to each other. The inset shows an enlarged image of one of the twinned regions. *b* – one of the crystalline regions annealed at 700 °C for 1 h, with clearly noticeable twinning. Next to it is the FFT diffraction pattern, showing the pseudo-symmetry of the formed crystal. *c* – high-resolution TEM image of the structure of the powder particles after annealing for 3.5 h at 800 °C, showing strongly twinned crystalline regions. Photographs from the review (Dutta et al, 1998).

the merger of clusters formed during synthesis. We can assume that in the process of crystallization at high temperatures twinned structures with fivefold symmetry show grow from these embryos.

The stability of clusters with the cell or cell–core configuration and the role of such clusters in the crystallization of materials with the cubic lattice of diamond (carbon, silicon and germanium) have been many times discussed in the literature (Matsumoto & Matsui, 1983; Okabe et al, 1991; Pun Pan & Ramakrishna, 1994; Rathlisberger et al, 1994; Ramakrishna & Pun Pan, 1994; Taylor et al, 1995). A common feature of the models proposed in these studies is the presence in them of non-crystallographic natural symmetries (pentagonal rings) which, due to increased probability of nucleation, can play a leading role in the formation of crystalline phases. For example, Okabe et al (1991) suggested that Ge_{15} cluster should be the nucleus of twinned structures with the symmetry of the fifth order, which are formed in thin films during deposition from the vapour phase. Such twins with the fivefold symmetry are among the first ordered structures formed at the beginning of the crystallization process of the materials with the diamond lattice, as shown by Hofmeister et al. (1996).

In the paper (Hofmeister et al, 1996) the effect of high-temperature annealing on the development of the atomic structure of Si nanopowders, grown by plasma-chemical decomposition of silane, was carefully studied by transmission electron microscopy. Electron microscopic study of the crystallization of the particles of amorphous Si found no significant formation of crystallites during annealing of the powders for 1 h at temperatures from 300 to 600°C. But as a result of annealing the inhomogeneous distribution of atoms with partially ordered domains, which manifested itself in the dark-field high-resolution TEM images of the original Si powders, became more orderly and the contrast of the ring morphology on the images became greater (Figure 2.30). With increasing annealing time the annular areas continued to slowly grow.

The studies by high-resolution TEM and local electron diffraction show that the apparent crystallization occurs in the process of annealing of the powder in the temperature range 700–800°C (see Fig. 2.30b,c). The diffraction pattern of the powder after 1 h annealing at 700°C shows clearly the interference points from the small particles of the crystalline phase on the background of diffuse rings from the amorphous Si phase. The high-resolution TEM images of this powder show quite large numbers of particles with distinctive areas of the crystalline order and the interference spots corresponding to the crystallographic planes {111} of silicon. In addition, some particles of this powder revealed twinned crystalline regions in which the twins are in compliance with the symmetry of the fifth order (Fig. 2.31), i.e. with an azimuthal rotation of 72° with respect to each other, as confirmed by electron diffraction patterns, on which 10 reflections of the (111) type appear from five mutually twinned sub-regions.

At annealing temperatures up to 800°C total crystallization did not occur, the areas of crystalline ordering occurred only in some particles, and most of the powder remained amorphous. Longer annealing at this temperature was accompanied by an increase in the size of crystalline regions (Fig. 2.32).

After annealing at 900°C the crystalline regions of a fairly large size were already observed in most of the powder particles. The powder particles without changing the size and signs of agglomeration almost completely crystallized in structures with a very high density of twins of different order. At the same time the areas of the amorphous phase still remained inside the particles, and the surface of the particles was clearly covered by a shell of the amorphous phase. Also, in annealing in these conditions there was active binding activity of the residual oxygen which according to the results of IR vibrational spectroscopy was also present in the original particles. Shells of silicon oxides with a thickness of 2 nm also formed on the surface. This oxide shell prevented the sintering of the powder particles during annealing.

 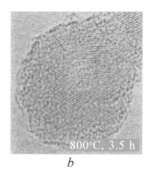 

| a | b | c |

Fig. 2.32. High-resolution TEM images showing crystalline grains in the initial particle powder (obtained by annealing for 1 h at 700°C), as well as the particles after annealing at 800°C for 3.5 h and 7.5 h (Dutta et al, 1998).

Experimental studies (Dutta et al, 1998) show that the size of crystallites, formed during the transformation of the amorphous to crystalline phase during annealing, is less than the size of the powder particles. Quantitative metallographic analysis and analysis of the width of the x-ray diffraction lines indicated that annealing for 1 h at 700 °C crystals with the average size of about 4.5 nm formed, and after annealing at 900°C particles with a size of 6 nm appeared. Electron microscopic study of the kinetics of growth of the crystallites with increasing annealing time (Fig. 2.32) shows that at 800°C primary crystallization centres with an average size of 4.5 nm after 7.5 h grow to 10 nm. At the boundaries of the crystal grains within the particles there was a surface layer of silicon oxide with a thickness 1.5–2.0 nm.

Based on these studies and thermodynamic calculations (Dutta, Houriet et al, 1997; Veprek et al, 1982; Veprek, Schopper et al, 1993) Dutta, Houriet et al (1997) proposed a three-stage growth mechanism of Si nanocrystals from the amorphous phase:

1) the formation of Si clusters and their growth to a critical size (the formation of an interim order);
2) formation of a crystal nucleus from the clusters (the formation of a long-range order);
3) the growth of the crystalline region to the size of the amorphous particle.

The first stage can occur both at high and at relatively low annealing temperatures ($<0.5T_m$), and the primary clusters are formed already in deposition from the plasma and are observed in the primary amorphous particles. The second stage is clearly observed at annealing temperatures above half the absolute melting point of Si. The growth of crystalline regions is observed at temperatures higher than $0,55T_m$.

Nuclei for the onset of crystallization of amorphous particles are Si clusters. Thermodynamic calculations carried out in the papers (Veprek et al, 1982) and (Veprek, Schopper et al, 1993) indicate that these nuclei are stable up to the size of the order of 3 nm (more than 1000 Si atoms in the cluster). The relatively large size of these nuclei explains why the Si nanocrystals with sizes smaller than 3 nm were not observed in the experiments. An amorphous silicon oxide layer, formed around each crystallized grain, inhibits the growth of grains due to their interaction (merger), which limits the maximum size of particles that can be obtained by crystallization of amorphous silicon powders. Moreover, after crystallization experiments did not detect fully single-crystalline powder particles and only crystallites with dimensions of 3–4 nm were observed. This suggests that the crystallized particles actually consist of 4–6 nanocrystalline Si grains with grain boundaries.

2.6. Silicon nanocrystals

Material properties, especially electro-optical, change dramatically with decreasing Si particle size to a value comparable with the deBroglie wavelength and the exciton Bohr radius. In silicon, these changes are strongest, so the properties of Si nanoparticles are radically different from the properties of bulk silicon. Thus, in the early 1990s, porous silicon nanoparticles (P-Si), produced by electrochemical etching of single crystal plates, showed bright photoluminescence in the visible region (Canham, 1990), which is not typical of macrocrystalline and coarse polycrystalline silicon (c-Si). Later, the method of crystallization from the gas phase during the pyrolysis of disilane (Littau et al, 1993) was used to synthesize by the sols of Si particles with dimensions of 3–8 nm. The suspension of these particles in ethylene glycol also showed a visible photoluminescence with a quantum yield of 5%.

According to TEM, X-ray diffraction and infrared spectroscopy, the particles had a core–shell-like structure (the shell of SiO_2 with a thickness of 1.2 nm was produced specifically by oxidation in hydrogen peroxide to protect the core), and the core had a crystalline structure similar to bulk silicon. Wilson et al (1993) obtained by a similar procedure Si particles by gel chromatography (exclusion gel filtration) which were divided into fractions, and it was shown that the energy of photoluminescent emission depends on the particle size. For particles of the finest fraction compared with c-Si the band gap increased by 0.9 eV, but they themselves behaved as indirect-gap semiconductors. This led to the assumption that the effect of the photoluminescence in nc-Si is associated with suppression of the probability of non-radiative recombination paths of the excited state by the dimensional constraints, i.e. there is the same mechanism that was previously justified by the results of spectroscopic research (Vial et al, 1992) for the luminescence of porous silicon. Later, numerous studies (Kumar, 2008; Pavesi & Turan, 2010) proved the connection of visible photoluminescence at room temperature for nc-Si with the quantum-size constraint. Gelloz et al. (2005) have shown that the extension of the band gap of nc-Si with decreasing crystallite size can result in the external output of photoluminescence at 23%, comparable to the quantum yield of luminescence in direct-gap semiconductors.

At the beginning of the XXI century the effect of optical amplification was discovered in nc-Si, which opened the prospect of a silicon laser (Pavesi, Dal Negro et al, 2000). The effect of spontaneous emission in silicon nanocrystals has been confirmed and studied by Dal Negro et al. (2003) using photoluminescence spectroscopy with time resolution (see measurement diagram in Fig. 2.33 and the results in Fig. 2.34).

Further work in this area, presented in the articles (Khriachtchev et al, 2001; Luterová et al, 2002; Nayfeh, Rao et al, 2002), along with the spontaneous emission, revealed the possibility of the effect of stimulated emission in the nc-Si, which occurs despite the key role of non-radiative recombination paths of recombination of the excited electronic states in conventional silicon.

The authors of studies (Akcakir et al, 2000; Belomoin et al, 2000) used electrochemical etching of crystalline silicon to produce P-Si particles with the size of < 1 nm which had a bright blue photoluminescence, and detailed studies showed that they had received a new phase of silicon which in the atomic structure differs from both the crystalline and amorphous silicon. These nanoclusters intensified the laser radiation passing through them (Nayfeh, Barry et al. 2001), and when excited by infrared laser radiation (765–835 nm) with a high power density the particles generated a bright light on the second harmonic (Nayfeh, Akcakir et al, 2000) – the effect, banned for c-Si by the centrosymmetry of its crystal structure.

The discovery of the above unusual optical effects in nanosilicon stimulated intensive investigations of the electron-optical properties of Si with the size and shape of the particles. In particular, properties such as Stokes shift[18] have been studied; the dependence of the photoluminescence energy on the size of nanocrystals, the doping properties, the lifetime of radiation processes, non-linear optical properties; the Stark effect at quantum constraints (QCSE), as well as many other physical and chemical properties that allow understand how to manage the optical response of composites based on nanocrystalline silicon.

Despite the many existing results, the answers to some questions could not yet be obtained (Bulutay & Ossicini, 2010). For example, the role of the surface has not been fully understood, in particular, the influence of surface oxidation and hydrogenation of the nc-Si particles, as well as their interactions at the interface with the matrix (if they are enclosed in a matrix) on their optoelectronic properties. So far, it is only clear that this effect is often a determining factor for the properties of nc-Si.

In addition to efficient photoluminescence of nc-Si in the visible region a a new exotic property has been detected. This property is the ability to

---

[18]Stokes shift is the difference between the energies of absorption and emission of the photon in luminescence of a semiconductor.

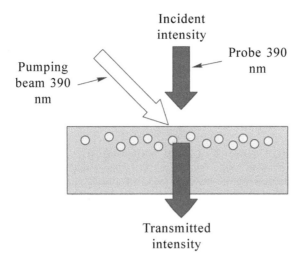

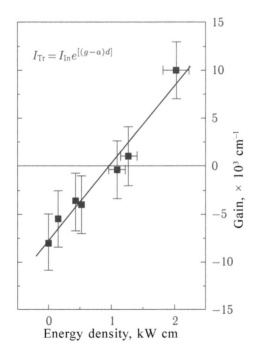

Fig. 2.33. The scheme of measuring the effect of optical amplification by the system of separately distributed nc-Si during pumping (Ray & Cockayne, 1971).

Fig. 2.34. The dependence of light amplification by the system of nc-Si particles on the pumping power density as shown in Fig. 2.33 (Kumar, 2008).

generate oxygen in the singlet state (Kovalev & Fujii, 2005; 2008) which is obtained under the influence of energy transfer from the excitons to triplet oxygen molecules on the surface of nanoparticles[19]. This feature appears due the molecule-like energy structure of Si nanocrystals and is a direct consequence of quantum constraint effects. Another exotic property of nanocrystalline silicon is the efficient photosensitization of ions of rare-earth elements by the Si nanocrystals.

2.7. Quantum size effects in nanosilicon

Due to the dimensional constraints, in the Si nanoclusters the valence electrons have a finite number of quantum states (Brus, 1986), so the nanoclusters possess electronic and optical properties unusual for bulk silicon. They do not have the usual continuous conduction band and their conduction band has only specific energy states at certain levels.

2.7.1. The luminescence of silicon nanoparticles

Judging only by the appearance of the photoluminescence effect in the visible region, we can assume that the Si nanocrystals behave as direct-gap semiconductors. But on closer study it becomes clear that this is not the case, and that even in the nanoscale scale Si is an indirect-gap semiconductor.

As shown by numerous theoretical and experimental studies, the first indication of the presence of the effect of the quantum-dimensional constraint in nc-Si is an increase in the optical band gap with decreasing particle size (see Fig. 2.35).

Reducing the particle size of nc-Si causes an exponential decrease of the radiative lifetime (Fig. 2.36), making the indirect-band semiconductor an efficient luminophor, despite the fact that the non-radiative recombination paths of the excited states, such as Auger recombination and carrier multiplication charge, still prevail in it (Sevik & Bulutay, 2008).

Numerous experimental and theoretical studies (see, e.g., (Delerue et al. In 1999; Soni et al, 1999; Altman et al, 2001; Glinka, 2001; Ranjan et al, 2002; Osicini et al, 2003; Belogorokhov et al 2006; Kumar, 2008; Khryschev, 2009; Pavesi & Turan, 2010)) have shown that when the size of the nc-Si crystal becomes smaller than the exciton Bohr radius (~4.9 nm in macroscopic Si crystals) then there are the quantum-size constraints, causing changes in the electronic band structure of silicon. The quantum-size effects are manifested in the form of the luminescence band shift toward higher energy and in enhancing the spontaneous emission (Fig. 2.37).

Figure 2.37 compares the luminescence spectra of nc-Si with the sizes up to 9 nm measured at room temperature and at the liquid helium

---

[19]This phenomenon and its practical applications are discussed in chapter 13.

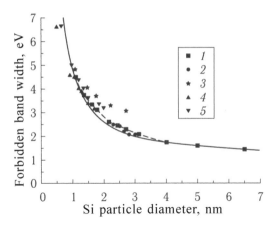

Fig. 2.35. The dependence of the width of the optical energy band gap o the size of Si particles – summary of results of different studies: markers *2–5* denote, respectively, the data taken from the studies (Furukawa, 1988; Garoufalis et al, 2001; Örüt et al, 1997; Vasiliev et al, 2001). Data *1* – the results of *ab initio* quantum chemical calculations (Bulutay & Ossicini, 2010) by the method of linear combination of the zones (LCBB method).

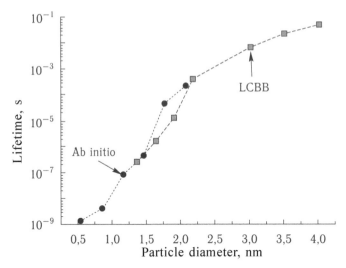

Fig. 2.36. Effect of particle size of nc-Si in the radiative lifetime of the excited states (Bulutay & Ossicini, 2010).

temperature (Kanzawa et al, 1997; Takeoka et al, 2000). At a temperature of 4 K together with the main peak of photoluminescence there is an additional low-energy peak. There is reason to believe (Hill & Whaley, 1995) that the nature of this peak is related to the recombination of electrons from the conduction band with holes in the deep surface traps at the

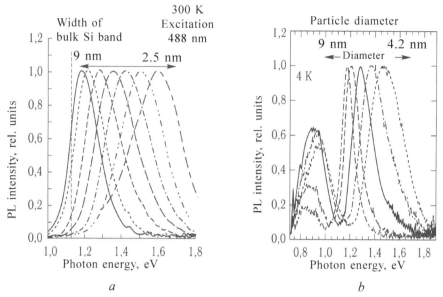

Fig. 2.37. The dependence of the position and width of the emission line on the size of nc-Si. The photoluminescence spectra of silicon nanocrystals embedded in SiO$_2$ are shown. Samples were obtained by combined ion sputtering of Si/SiO$_2$ and subsequent annealing at temperatures above 1100 °C. Figure *b* shows the spectra at 4 K. Photoluminescence was excited by the light of an argon laser with a wavelength of 488 nm. Figure from (Fujii, 2010).

Si–SiO$_2$ interface. The shift of this low-energy peak, and also of the main peak, is dependent on particle size, but its magnitude is approximately equal to half of the shift of the main peak (Takeoka et al, 2000; Heitmann et al, 2005; Hill & Whaley, 1995; Kanzawa et al, 1997; Meyer et al, 1993a; Meyer et al, 1993b; Mochizuki et al, 1992). It is believed (Fujii, 2010) that the size-dependent shift of the low-energy peak reflects the shift of the conduction band edge.

The studies (Cullis, 1997; Heitmann et al, 2005; Kovalev et al, 1999; Takeoka et al, 2000) show that the singlet-triplet splitting of the energy of the exciton states increases with a decrease in the crystal size. In addition, there are reports (Kanzawa et al, 1997; Takeoka et al, 2000) on the strong temperature dependence of the duration of the luminescence which at room temperature and at liquid helium temperature differs by two orders of magnitude (Fig. 2.38). Both of these phenomena can be qualitatively explained by the quantum constraint of excitons in the space smaller than the Bohr radius of the exciton in bulk Si.

The results, presented in Fig. 2.38, show that the temperature dependence of the lifetime of the photoluminescent states becomes stronger with the reduction of the nc-Si particles. This effect can be explained by the model proposed in (Calcott et al, 1993), in which the state of an exciton due to

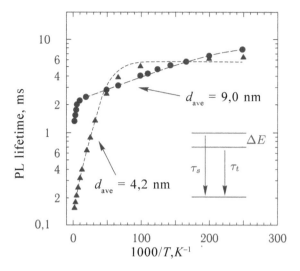

Fig. 2.38. Temperature dependence of photoluminescence lifetime for two nc-Si samples with different particle size distribution. The inset shows schematically singlet and triplet splitting of the exciton state. The lifetimes of the singlet and triplet states, respectively, are indicated by τ_s and τ_t, and ΔE is the energy of splitting. Figure from (Takeoka et al, 2000)

electron–hole exchange interactions is split into singlet and triplet states. By approximating the temperature dependence of the lifetime we can estimate both the triplet and singlet states, as well as the energy of splitting.

The lifetime of the triplet state is almost independent of the particle size, because the transition is forbidden, whereas the lifetime of the singlet state varies and becomes shorter with decreasing particle size. In indirect-gap semiconductors the geometric constraints of the charge carriers tend to increase the overlap of electron and hole wave functions in the momentum space and increases the strength of the exciton oscillator. Decrease in the lifetime of the singlet state is due to this effect.

The time and energy characteristics of nc-Si, obtained by measurements by photoluminescence spectroscopy with time resolution and by calculations (Fujii, 2010) are shown in Fig. 2.39. These results show that reducing the particle size of nc-Si the peak of the luminescence yield is gradually shifting from the energy close to the energy band gap of bulk silicon and reaches 1.6 eV. In particles with a surface passivated by oxygen, the high-energy shift is almost always limited to the value of 1.6 eV. The reason for this is usually assumed to be the formation of the band gap energy levels of Si nanocrystals associated with oxygen (Wolkin et al, 1999). At the same time in the nc-Si with the H atoms attached to the surface, the displacement field can be expanded up to ~2.3 eV (Reboredo et al, 1999).

If we take into account that the energy of the photoluminescence peak depends on the particle size of nc-Si (Fig. 2.37a), then Fig. 2.39b shows

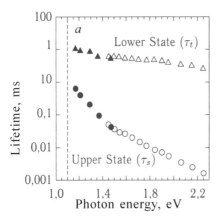
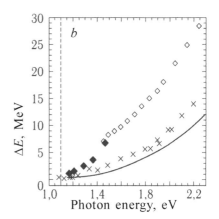

Fig. 2.39. Relationship of the energy of photoluminescent emission with lifetimes of the upper singlet and lower triplet exciton states (*a*) and with the exchange energy (ΔE) of splitting (*b*). Filled circles, triangles and squares correspond to data obtained by spectroscopic measurements of nc-Si in the matrix of SiO_2, prepared by ion sputtering (Takeoka et al, 2000). Open circles, triangles and squares – data from the oxidized porous Si surface (Calcott et al, 1993). Crosses – data from freshly prepared P-Si (Kovalev et al, 1999) (surface passivated by H atoms). The solid curve in Figure *b* – calculated dependence ΔE for nc-Si with the surface, passivated by hydrogen (Reboredo et al, 1999). The dashed line indicates the energy of the band gap in bulk Si. Figure from (Fujii, 2010).

that the energy of the singlet–triplet splitting increases monotonically with decreasing particle size. It is due to the stronger overlap of electron and hole wave functions due to space constraints (Fujii, 2010). The energy of splitting in Si nanocrystals, passivated by the H atoms, agrees well with the calculated results (Reboredo et al, 1999). In addition, it is clear that the energy of splitting in the case of nc-Si with the surface passivated by the O atoms is significantly greater than in the case of passivation with the H atoms. All the data in Fig. 2.39 vary continuously over a wide energy range, starting from the energy band gap of the crystalline bulk Si. The continuity of the change in these characteristics confirms that the photoluminescence forms here because of the quantum-constrained excitons.

There is reason to believe (Kovalev et al, 1999) that the law of conservation of momentum in nc-Si is only partially broken, and Si nanocrystals remain indirect-band, typical of bulk Si crystals. This is confirmed by the fact that even in very small Si nanocrystals, with visible photoluminescence, the radiative lifetime is much longer than in the case of direct-gap semiconductors (see Table 2.3).

The fact that the Si nanocrystals remain indirect-gap semiconductors is also confirmed by molecular laser spectroscopy (Kovalev et al, 1998) and resonant photoluminescence spectroscopy (Calcott et al, 1993; Kovalev et al, 1998). In resonant photoluminescence spectroscopy measurements when

Table 2.3. The radiative lifetime of minority carriers in some semiconductors (net and extrinsic) with tetrahedral bond and a different type of band structure at room temperature (Yu and Cardona, 2002, p. 312)

Semiconductor	Direct or indirect band	τ_{rad}	
		Intrinsic	10^{17} cm^{-3} main carriers
Si	indirect	4.6 h	2.5 ms
Ge	indirect	0,61 s	0.15 ms
GaAs	direct	2.8 µs	0.04 µs
InAs	direct	15 µs	0.24 µs
InSb	direct	0.62 µs	0.12 µs

photoluminescence is excited by photons with energies from the range of the wide photoluminescence band, only those nanocrystals in which the energy gap is smaller than the excitation energy are excited. This leads to the fact that the photoluminescence band significantly narrows and features associated with the phonons involved in the conservation of momentum appear.

It is well established that the cause of PL in Si nanoparticles are dimensional constraints which enhance the overlap of electron and hole wave functions and increase the rate of radiative relaxation processes. However, the high quantum efficiency of PL of nc-Si can not always be explained by the increase of the rate of radiative losses (Jurbergs et al, 2006; Ledoux et al, 2000; Mangolin & Kortshagen, 2007; Skryshevsky et al, 1996; Wilson et al, 1993; Vial et al, 1992). As the research results show, the main reason for increasing the photoluminescence quantum yield is the restriction of the transport of charge carriers in nanocrystals (Fujii, 2010).

If the movement of charge carriers is limited by the size of the nanocrystal, and there are no non-radiative recombination processes in it, the quantum efficiency of PL should be equal to unity. On the other hand, if the nanocrystal has at least one way of non-radiative recombination, the radiative processes with the duration in microseconds and milliseconds can not compete with the non-radiative processes and the quantum efficiency of PL in such nc-Si is equal to zero. This means that the nanocrystals, included in the nanocrystalline system, have the quantum efficiency equal to either zero or one, and the ratio of 'bright' and 'dark' nanocrystals determines the overall quantum efficiency of the entire assembly of nanocrystals. Such an explanation allows to estimate approximately the photoluminescence quantum efficiency of the nanocrystalline system, but not quite suitable for quantitative estimates, because in real systems of nc-Si the number of 'bright' Si nanocrystals does not always equal 100%.

Typically, optical transitions between quantum electronic states produce very narrow absorption and emission peaks, as in atoms. However, Fig. 2.37

shows that Si nanocrystals have a wide photoluminescence band (half width FWHM: 200–300 meV). The width of this band can not be substantially reduced, even at very low temperatures (Fujii, 2010; Takeoka et al, 2000). It follows from the assumption that the large width of the emission peak is not due to mechanisms of radiative processes but to the broadening of spectral lines due inhomogeneity of the material.

2.7.2. Measurements on a single nanocrystal

A team of researchers under the leadership of Linros were the first to measure the photoluminescence of a single Si nanocrystal at low temperatures (Sychugov et al, 2005). They have convincingly shown that each Si nanocrystal alone gives a very narrow photoluminescence peak (half-width of the peak at 35 K is equal to 2 meV) similar to atomic radiation, and the broadening observed in conventional measurements on ensembles of the nanoparticles is due to the distribution of particle size and shape. In these measurements a pair of peaks was observed in each nanocrystal. A high-energy peak corresponded to the zero phonon line, while the second peak was associated with the transition, accompanied by the emission with the conservation of the momentum due to the optical phonon in Δ minimum of the size-restricted acoustic phonon (Fujii et al, 1996). The observation of two peaks indicates that the conservation of momentum in Si nanocrystals is only partially broken, and even in the nanocrystalline state they behave as indirect-gap semiconductors.

It should be noted that measurements at room temperature (Valenta et al, 2002), the photoluminescence peak from a single Si nanocrystal was much wider (half-width 120–150 meV). This is explained by the fact that different phonons may take part in the optical transitions which produce photoluminescence.

The role of defects and the nature of luminescence of undoped Si nanoparticles, surrounded by a matrix of SiO_2, were studied in Godefroo et al. (2008) by joint research methods of photoluminescence spectroscopy and EPR. In this paper they showed that for measurements in strong magnetic fields, the dominant sources of PL of nc-Si are point defects. In addition, the possibility of controlling luminescence in the same sample if the nc-Si particles are passivated by hydrogen, was shown. The hydrogen atoms, attached to the surface of the particles, lead to the displacement of the defects from the nanocrystals, resulting in leaving only the photoluminescence from the levels determined by quantum constraints. If the system of such particles is affected by UV irradiation, defects reappear in the particles and the luminous contribution from them again forms.

2.8. Surface chemistry of Si nanoparticles and its effect on the properties

The luminescence in nc-Si has been studied for many years and a lot of experimental evidence has been found showing that the maximum yield of luminescence with decreasing particle size of nc-Si is shifted to higher energy and the lifetime of the luminescent state is shortened. However, there are experimental facts that are difficult to explain on the basis of only one model of quantum-size constraints. Obviously, these effects are strongly influenced by the surface condition of the particles. Such an effect was observed both in the nc-Si with a surface covered with H, and in particles with a surface passivated by O atoms, and significantly changed the value of the quantum-size effects in comparison with their values in the nc-Si with a clean surface. There are numerous quantitative estimates of the characteristics of the nanocrystals attached to the surface with ions of H and O, especially for very small particles. These estimates show that the optoelectronic properties of the nc-Si with the surface covered with H and O atoms are very similar. But these properties are very different from the nc-Si surface modified with N atoms (particles with SiN_x shells), and the mechanism of the photoluminescence of such nanoparticles is not yet entirely clear and is under discussion.

The main conclusion from a variety of experimental and theoretical studies of the photoluminescence of nanocrystalline Si is that although the PL emission of the nc-Si particles can be regulated by the size of the particles themselves, it is difficult to manage it with a single tool to obtain accurately the desired wavelength in the visible spectral region. This is due to the limited possibility of the dimensional adjustment of emission due to the surface states and defects in the clusters that make up the nanocrystalline system. For example (Fig. 2.40) in the silicon-enriched silica defects related to oxygen, such as bonds (Si=O), limit the emission of nc-Si to the near infrared region, whereas in the silicon-enriched silicon nitrides, the bonds of silicon with nitrogen also limit the emission, but only for some specific wavelengths in the infrared range, and allow the emission over the entire range of the visible spectrum from red to blue (Wolkin et al, 1999; Deshpande et al, 1995).

The current state of the problem of the influence of the chemical state of the surface on the structure and electro-optical properties of Si nanoparticles, based on the results of comprehensive experimental and theoretical studies, is presented in the review by Bulutay & Ossicini (2010).

For example, in studies of Littau et al. (1993), Wilson et al. (1993) it was shown that the position of the maximum of PL in P-Si with the natural oxidation of the samples can experience both a low-energy and a high-energy shift, while in accordance with the quantum-size model, the PL maximum can be shifted only to the high-energy side as a result of reducing the size of the crystallites. The 'red shift' of the PL band in nc-Si and in

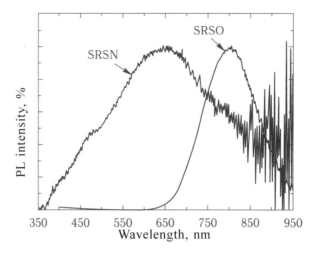

Fig. 2.40. Normalized PL spectra of nc-Si in the matrix of silicon nitride (SRSN) and silicon dioxide (SRSO), obtained upon excitation by ultraviolet radiation with a wavelength of 325 nm. PL emission in both materials is due to quantum size constraints, but the boundaries between the medium and nc-Si particles affect the spectra differently. In a system of nc-Si particles in SRSN several surface states allow PL in the entire visible light spectrum, whereas the surface states of nc-Si in SiO_2 quenched PL in the visible region (Roschuk et al, 2010).

P-Si after surface oxidation, indicating the important role of the surface of the Si nanoparticles in the mechanisms of light emission, has been observed by many researchers (Kumar, 2008; Pavessi & Turan, 2010).

In experimental studies of photoluminescence of Si nanoclusters in the silicon oxide matrix, annealed in different atmospheres, it is shown (Romaniuk et al, 2010) that the photoluminescence intensity significantly increases after low-temperature annealing (450°C), and most of all, after annealing in a mixture of oxygen and nitrogen. Experiments were performed on a material obtained by high-temperature (1150°C) decomposition of SiO_x films precipitated on the surface of a single-crystal wafer of Si(100) with the p-type conductivity. Spherical clusters of nc-Si with a diameter of 3.0 ± 0.4 nm, embedded in amorphous silica, were produced as a result. The PL spectrum of the material, measured immediately after producing the material, had the form of a widely asymmetric band approximated by two Gaussian peaks with maxima at wavelengths of $\lambda_{max} \approx 750$ nm (1.65 eV) and 880 nm (1.41 eV). After low-temperature annealing in a mixture of oxygen and nitrogen, the PL band changed its symmetry, extended to longer wavelengths ('red shift' was observed), and its maximum intensity formed at a wavelength of $\lambda_{max} \approx 880$ nm. In this paper, the analytical methods used included: photoluminescent spectral analysis, high-resolution electron microscopy and component analysis of films by mass spectroscopy of secondary post-ionized neutral particles (MCBH). The local chemical and structural analyses were not carried out. On the basis the behaviour of the PL spectrum and general considerations, the authors explained the luminescence enhancement effect and shift of its band observed after annealing in an atmosphere of O_2 + N_2 by the reconstruction of nc-Si/SiO_2 interface and by the formation at these interfaces of the energy levels involved in the recombination of non-equilibrium charge carriers.

Korsunskaya et al (2010) studied the photoluminescence and structural properties of porous silicon prepared by chemical etching in a mixture of hydrofluoric and nitric acids at room temperature. This P-Si by some characteristics differs from porous silicon obtained by anodic etching, in particular, the crystallites of nc-Si in it are more oxidized (Brandt et al, 1992; Korsunskaya et al, 2000) and this may influence their size and structure. The surface morphology and the characteristic dimensions of the particles and the structure of the layers in this study were determined using a tunnelling microscope. The PL spectra were studied at different temperatures in the T = 77–350 K range. At room temperature, the samples showed a bright orange photoluminescence. A significant broadening of the PL band and the complexity of the spectrum shape when cooled to low temperatures as well as the degradation with time of exposure to air were detected. At low temperatures, the maximum of the PL band shifted with increasing temperature to the low-energy side, respectively, in accordance with the the change of the band gap width. At temperatures close to room temperature the PL wavelength did not depend on temperature. On the basis of the temperature dependence of the luminescence spectra, the authors concluded that the PL band in this case consists of two components[20], one of which (high-energy) is due to the recombination of excitons in silicon nanoclusters smaller than 3 nm, and the other (low-energy) – the recombination of charge carriers through surface defects, presumably through defects in silicon oxide. In this case, the second recombination path was considered dominant at room temperature.

Wolkin et al. (1999) found that oxidation of the nc-Si particles created defects in their the band gap which determine the transition energy. The authors did concluded that the formation of the double bond Si=O leads to the formation of traps near the Fermi level and to its pinning in the band gap (pinning state). The same conclusion was drawn by the authors of other papers (Luppi & Ossicini, 2003a; Luppi & Ossicini, 2003b; Puzder et al, 2002; Vasiliev et al, 2002). However, these conclusions are based on indirect results and can be ambiguous. For example, in the article (Vasiliev et al, 2002) it is noted that a similar result can also be obtained at bonding of one atom of oxygen to two silicon atoms, i.e. the formation of single Si–O–Si bonds on the surface of nc-Si.

---

[20]Two PL peaks were observed in measurements on a separate Si nanocrystal in the already discussed study (Sychugov et al, 2005).

2.8.1. Effect of hydrogen passivation of Si nanoparticles

In silicon semiconductor technology, passivation of the surface of silicon single crystals with hydrogen is widely used to remove surface effects and helps to remove surface distortions of the atomic structure caused by the reconstruction of the surface (see Sec. 2.3.2). The expected effect of hydrogen passivation should be orders of magnitude higher in the case of nc-Si, where the surface layer has a significant fraction of the volume of the particles. Saturation of the free surface bonds by the addition thereto of H atoms should not only chemically stabilize the particles, but also restore the periodicity and long-range order in the atomic structure of the particles, destroying the surface dimers and the related deformation of the surface layer.

The effect of attachment of the hydrogen atoms to the surface of Si nanoparticles on their structure and electro-optical properties was studied by theoretical modelling by the authors of the review (Bulutay & Ossicini, 2010). Theoretical studies of the opto-electronic and structural properties of Si were carried out using the DFT method and the wave functions for silicon clusters in the ground and excited states. As a result of such modelling we can construct theoretically the absorption spectra, which can be used to restore the emission spectra and luminescence. Despite the proximity of the method, the model allowed the authors to track changes in the properties and structure that occur with increasing cluster size and the transitions between the ground and excited states. The analysis was carried out of the spectral, electronic and optical properties of oxidized and hydrogenated nc-Si particles, depending on the size and symmetry, and the main changes resulting from the excitation of the 'nanocrystal' were highlighted.

In the theoretical study of the hydrogenation of Si clusters growing from the monosilane (SiH_4) the position of all Si atoms in the model was fixed in accordance with their positions in bulk silicon crystals and passivating H atoms were located along the crystallographic directions in the bulk crystal at distances which are taken from the results of studies of the SiH_4 molecule. The initial symmetry of the H–nc-Si was defined as the symmetry T_d and attempts were made to maintain it the same at the relaxation of internuclear distances in the configuration of the ground state. Nevertheless, for the configurations of the excited state this symmetry is usually disrupted due to the population of energetically excited levels. The authors first studied the structural distortions caused by the relaxation of these structures in different electronic configurations. Analysis of structural properties revealed that the average length of the Si–Si bonds becomes close to the bond length in bulk silicon as the size of the cluster increases. In particular, if we move from the centre of the cluster to its surface, the outer shells of Si are compressed compared with the internal ones. The presence of electron–hole pairs in the clusters results in strong deformation of the structures relative to their configuration in the ground state and this becomes more noticeable the smaller the system, and also on the surface of H–nc-Si. In larger clusters, this effect is less pronounced because the density perturbations in them are distributed over a large volume. The importance of the distortion of the bond on the surface of nc-Si in the excited state for the creation of localized internal states, responsible for the photoluminescence emission, was emphasized earlier in the studies (Allan et al, 1996; Baierle et al, 1997).

The results of theoretical modelling of the energy states of Si_nH_m clusters depending on the cluster size are shown in Table 2.4.

Table 2.4 shows that the excitation of the electron–hole pair causes a change in the energy of the band gap and the magnitude of this change increases with the reduction of the cluster size. In the excited small clusters the HOMO and LUMO states are strongly localized due to distortions, which leads to a defect-like state reducing the width of the band gap. This distortion, induced by the excitation of the nanocluster, may provide a possible explanation for the observed Stokes shift in these systems.

The absorption of radiation by a cluster in the configuration of the ground state induces a transition between the levels of HOMO and LUMO, which is optically allowed for all such clusters. These transitions are followed by the relaxation of the excited state of the cluster, causing geometric distortions and leading to new LUMO and HOMO with the energy difference between them being smaller than that in the geometry of the ground state. The transition between these two latter states results in the emission which explains the Stokes shift. The magnitude of the shift depends on the size of nc-Si. Since the distortion is smaller for larger clusters, it is expected that the Stokes shift should decrease with increasing cluster size. This expectation is confirmed by the data in Table 2.4.

The widths of the absorption bands in the Si clusters, equal to 8.76 eV for Si_1H_4 and 6.09 eV for Si_5H_{12}, obtained by the authors (Bulutay & Ossicini, 2010) in modelling by DFT, are in good agreement with the values of the excitation energies experimentally measured in the study (Itoh et al, 1986) which for the clusters of SiH_4 and Si_3H_8 equal respectively 8.8

Table 2.4. Calculated values of the energy difference HOMO–LUMO between the ground and excited states and the energies of absorption and emission for silane and silicon clusters, covered with hydrogen atoms (Bulutay & Ossicini, 2010)

	Absorption energy	HOMO–LUMO gap, GS	Emission energy	HOMO–LUMO gap, EXC
Si_1H_4	8.76	7.93	0.38	1.84
Si_5H_{12}	6.09	5.75	0.42	0.46
$Si_{10}H_{16}$	4.81	4.71	0.41	0.55
$Si_{29}H_{36}$	3.65	3.58	2.29	2.44
$Si_{35}H_{36}$	3.56	3.50	2.64	2.74

GS – ground state; EXS – excited state. All values are given in units of eV.

and 6.5 eV. The values of the Stokes shift found in the study (Bulutay & Ossicini, 2010) are qualitatively consistent with the results of the previously performed theoretical calculations for small clusters in the studied (Franceschetti, 2003; Puzder et al, 2002).

2.8.2. Effect of surface oxygen

Recently, there have been many reliable experimental confirmations that the interaction of the surface of nc-Si with oxygen leads to a strong change of its optoelectronic properties (see, e.g. (Bulutay & Ossicini, 2010)). The nature of this influence can be analyzed also by means of theoretical calculations, as was done in the analysis of the influence of hydrogenation of the silicon surface. Bulutay & Ossicini (2010) conducted such studies by theoretical modelling using the Δ-SCF method (non-self-consistent approximation). Hydrogenated clusters (i.e. clusters with the passivated surface, which are stable under normal conditions) were studied. Initial models of the clusters in the study of the influence of oxygen were constructed exactly the same as in the studies of hydrogenation. Two systems of clusters were analyzed – nanoclusters with a core of Si_{10} and Si_{29}, and three types of bonds of silicon with oxygen at the surface of the cluster: Si–O reciprocal bond, the Si>O bridge bond, and the Si=O double bond. Calculations were performed for all combinations of these systems in stable and excited states. The resultant formation energies show that of the investigated bonds the reciprocal bond is the least suitable. From the results it also follows that the ion relaxation of the excited state is accompanied by structural changes compared with the initial geometry, which depend on the type of end bond of the cluster surface with oxygen. For example, in the case of the surface ending of the $Si_{10}H_{14}$= O type, the structural change is mainly localized near the O atoms, where, in particular, the angle between the O with the double bond, and the related atom with Si changes. In the case of a bridge bond, by contrast, the deformation is localized around the Si–O–Si bond, indicating a large deformation of the distance of the Si–Si dimer.

Similar results were obtained for larger clusters with a core of Si_{29}. In this case, the distortions induced by the electron transfer are less pronounced which is explained by averaging of the local distortions over a large volume. But in both cases the structural changes are reflected in the electronic and optical properties of the clusters, as shown in Table 2.5, where the values of the edges of the absorption bands and light emission for clusters of different sizes bonded with oxygen with different bonds are given.

It is seen that in the case of the double bond the red shift of the emission band with respect to the absorption band is almost independent of the cluster size, whereas at the bridge bond the size dependence is pronounced. This means that the double bond of oxygen with the cluster

Table 2.5. Energy gaps of absorption and emission and the Stokes shift, calculated as the difference of energies (Bulutay & Ossicini, 2010)

	Absorption energy	Emission energy	Stokes losses
$Si_{10}H_{14}=O$	2.79	1.09	1.70
$Si_{10}H_{14}>O$	4.03	0.13	3.90
$Si_{29}H_{34}=O$	2.82	1.17	1.65
$Si_{29}H_{34}>O$	3.29	3.01	0.28

All values are given in units of eV.

creates localized states within the energy gap which are not affected by the quantum constraint, as predicted previously (Luppi & Ossicini, 2003a).

Calculations performed in the studies (Bulutay & Ossicini, 2010; Luppi & Ossicini, 2005; Luppi & Ossicini, 2003b) show that the near-boundary oxygen affects the optoelectronic properties of Si nanoparticles also if they are implanted in a silicon oxide matrix. The effect is different for particles of nc-Si and a-Si, because the properties of these two phases of silicon also differ, as evidenced by the results of the calculations. In particular, the band gap in crystalline and amorphous states differs by a factor of 1.5. The calculated values of E_g for nc-Si and a-Si nanoparticles, distributed in the oxide matrix, are respectively 2.17 and 1.51 eV.

2.8.3. The effect of doping on the electro-optical properties of nc-Si

The properties of the nc-Si with intrinsic conductivity can be controlled by the size, shape and attachment to the surface of different functional groups. An additional degree of freedom in the design of materials based on nc-Si can be obtained at the expense of its doping by admixtures. The introduction of small amounts of admixtures to the semiconductor modifies the transport properties of charge carriers (Anselm, 1978). Of course, to achieve the desired effect in a system based on nc-Si, the key is the precise control of the profile of the admixture concentration.

Low doping of nc-Si is a difficult problem because of the difficulties of practical implementation of the certification of the received material. The main problem is associated with fluctuations in the number of admixture atoms in the crystal in the nanocrystalline assembly. For a crystal diameter of several nanometers, it is impossible to express the degree of doping by the value of the admixture concentration, so it must be expressed by the number of atoms in the crystal. For example, the introduction of a single atom in the crystal with a diameter of 3 nm ($1.4 \cdot 10^{-20}$ cm^3, 700 atoms) corresponds to the concentration of $7.0 \cdot 10^{19}$ atoms/cm^3 in bulk silicon. At this level of doping the bulk silicon becomes a degenerate semiconductor and behaves like a metal. Consequently, the adding a crystal of the nanometer size or removing from it only one admixture atom dramatically changing the electronic structure of the nanocrystal and hence its optical and electronic–transport properties.

Another difficult task is the growth of nanocrystals doped with Si. The difficulty is associated with the self-cleaning effect, in which the admixture atoms in nanocrystal growth are pushed out from it into the surrounding matrix. This effect is explained by the very high energy of formation of the doped Si single crystals (Ossicini, Degoli et al, 2005). Due to the self-cleaning effect the admixture concentration in different nanocrystals of the same sample can vary greatly. In the worst case, the admixture concentration in some of the particles can be zero, while the average admixture concentration in the sample is high.

Upon doping of the silicon nanocrystals with the admixture having 'shallow' levels the electronic states of the admixture are strongly modified compared to the bulk material if the size of the nanocrystal is close to the effective Bohr radius for the admixture (Ossicini, Degoli et al, 2005).

Conclusion

Silicon is one of the most common elements on earth – about a quarter of the planet's mass and about 12% of the mass of the crust. It is now the base material of semiconductor electronics and solar energy. However, the use of this most readily available semiconductor material in such important areas as optoelectronics and photonics is limited. Due to the nature of the band structure, in silicon there is virtually no photo- and electroluminescence in the visible light range. In the solid state (both crystalline and amorphous) solid silicon can not convert electricity to photon radiation. These effects were observed only in silicon nanoparticles, which gives hope for significant expansion of its application areas.

Bulk silicon is a typical indirect-band semiconductor with a rather wide gap between the minimum of the conduction band and the valence band maximum (~1.12 eV for pure crystalline Si). The band structure and conductivity type of crystalline silicon (in addition to the temperature dependence that is typical for all semiconductors) are very sensitive to admixtures and defects in the crystal lattice. Si in the solid state is semiconductor in all versions – monocrystalline, polycrystalline and amorphous; the band structure of these versions differs, though the indirect-band type conductivity is retained.

The electronic structure of silicon atoms and its valence properties are such that in the equilibrium conditions it always crystallizes in the diamond structure, and its crystals of the microscopic and macroscopic dimensions have a face-centred cubic unit cell F, the space group $Fd\overline{3}m$ (No. 227), the edge cell $a = 0.54311$ nm.

In addition to the inevitable point defects such as vacancies, interstitial atoms and their complexes, as well as surface defects such as surface reconstruction, the medium and high rates of crystallization result in the formation in Si crystals of linear and planar defects such as dislocations, stacking faults (growth defects) and twins (deformation defects), with parameters typical for the diamond structure. These extended defects affect the band structure of Si like dopant admixtures to form centres of pinning the electronic states.

The situation with lattice defects in Si crystals with the size of <10 nm (nc-Si) significantly differs from that in bulk single crystals (c-Si) and even crystals of microscopic size. First, the influence of planar defects on nc-Si is much stronger, since at the small size of the crystal even a single defect is characterized by a very high concentration. Second, for the same reason there is a marked difference between the electrical properties of different crystals of the same size, as they may have different numbers of planar defects strongly alter their concentration. Finally, nc-Si is characterized by the self-cleaning effect of removal of the dislocations, and stacking faults and twins permeate nanocrystals through, whereas in the bulk crystals and microcrystals they are usually located inside.

In the silicon crystals with the clean surface unsaturated bonds of the surface atoms are saturated by the formation of dimers from the neighbouring atoms, resulting in surface reconstruction, accompanied by the appearance on the surface of a new flat atomic lattice with the parameters very different from the parameters of the internal crystal lattice, and by the appearance of stacking faults. The difference between the parameters of the reconstructed surface lattice and the lattice of the volume of the crystal leads to distortions of the latter, spreading at least to a depth of 3–5 atomic layers. These distortions created in Si crystals a near-surface area for the capture of charge carriers and the associated surface electronic states, affecting the band structure. Perhaps this effect is not so evident in the bulk crystals, but is pronounced in the nanocrystals, consisting of only a dozen or a few tens of atomic layers, for which the surface layer of 3–5 atoms is a significant part of the volume. For example, the volume of the surface layer with a thickness of three bonded Si atoms for the spherical particles with diameters of 10, 5 and 3 nm is approximately 27, 49 and 70% of the particle volume, respectively, and such a particle should not be any longer considered a nanocrystal but as a two-phase material with the nc-Si core and the shell which has a different atomic and hence the band structure. Since the type of surface reconstruction depends on the type of crystallographic planes (*hkl*), the cells of the surface structures are different for different faces of the nc-Si, and the shell of the nanoparticle has a variable structure more similar to the structure of amorphous silicon, the electronic properties of which differ markedly from the properties of c-Si.

Usually, to eliminate the surface lattice distortions, silicon semiconductor technology uses saturation of surface dangling ties by the monovalent admixture, usually by attaching hydrogen atoms. Because the Si–H bond (79.9 kcal/mol) is stronger than the Si–Si bond (51 kcal/mol), high-temperature hydrogenation of the Si crystal surface is accompanied by the formation of terminal group Si–H on the surface, rather than of Si–Si dimers (Sakurai & Hagstrom, 1976). As the saturation of each of the dangling bonds required only one H atom, the Si atoms do not need to shift from their normal position, and the frequency of their location on the surface is the same as in the corresponding atomic planes within the crystal. In this case the regular structure of the nc-Si particles can be fully restored.

Silicon nanocrystals can be obtained in two ways: 1) from 'top to bottom' path when the nanoparticles are obtained from large single Si crystals by refinement (e.g. porous silicon); and 2) from 'bottom to top' path when nc-Si is synthesized by self-assembly of individual atoms or atomic clusters of Si. It was established experimentally that the nc-Si, obtained by both methods, has the properties unusual for bulk silicon: luminescence in the visible spectrum and laser properties (the effects of spontaneous and stimulated emission). In addition, the nc-Si also has other unusual properties such as the ability to generate singlet oxygen (Kovalev & Fujii, 2005; 2008), which is obtained under the influence of energy transfer from the excitons to the triplet oxygen molecules, adsorbed on the surface of nanoparticles, as well as an efficient photosensitization of ions of rare-earth elements.

In any synthesis of Si nanocrystals from the atoms the particle growth inevitably passes through a stage of cluster formation. Due to the nature of chemical bonding, silicon clusters differ significantly from the well-studied metal clusters by the fact that they do not have dense packing. However, they differ in structure from the carbon clusters, not only by the magic numbers of stable structures, but also by the fact that fullerenes of the well-known type C_{60} have not been found in them and all skeletal structures are known to be endoendric (centred) fullerenes. Packets with a diamond-type structure can form in the Si particle when the number of atoms is >100, but are usually only observed is clusters consisting of >1000 atoms (at a size of >3 nm), which is consistent with thermodynamic calculations (Veprek et al, 1982). Since the clusters have partly the properties of molecules, synthesis (e.g. plasma-chemical) is usually accompanied by their rapid agglomeration under the action of van der Waals forces, and agglomeration takes place in clusters of different sizes and with different structures, so that the resulting particle (agglomerate) can not be characterized by any specific certain structure and therefore has similarities with amorphous silicon. Full integration of the clusters in a nanocrystal with the diamond-like structure can occur only at high-temperature annealing. The majority of experimental and theoretical studies indicate that the crystallization of the particles with sizes <3 nm, is as a rule not complete and they have a structure similar to amorphous Si.

Numerous studies have confirmed that the unusual effects observed in nanosilicon are associated with the manifestation of quantum properties due to dimensional constraints. The magnitude of these effects depends on the particle size but is strongly influenced by the structure and morphology of the nanoparticles. In the last decade, many studies aimed at understanding mechanisms of this effect were conducted. Much has been explained but many outstanding issues still remain. In particular, the role of the surface, the influence of surface oxidation and hydrogenation of the nc-Si particles, and even the role of the atomic structure of the particles are still unclear. So far, only it is clear that the effect often turns out to be decisive for the properties of nc-Si.

The complexity of investigations of nc-Si is due to the lack of methods for obtaining monodisperse ensembles of particles with the same structure and the difficulty of studying individual particles with the size of 1–2 nm, consisting of several tens or hundreds of atoms. In addition, many classical methods of structural analysis do not have a suitable theory for the analysis of experimental data for such small particles. At the same time, it is clear that the study of such systems sensitive to the size, morphology and structure as the nc-Si compositions, is impossible without continuous monitoring of their particle size, structure, composition and properties. Therefore, work is being carried out around the world on adapting the existing instrumental methods for the study of nano-objects and creating new approaches to experimental studies of these materials[21].

In the following chapters of this book we discuss in detail the above-mentioned features of the properties of different forms of nanosilicon, methods of its preparation; traditional and modern methods for the study and control of its characteristics, as well as some important practical applications of nanosilicon.

References for Chapter 2

Anselm A. I., Introduction to Semiconductor Theory. Moscow: Nauka, 1978. 616.

Ashcroft N., Mermin N., Solid State Physics. T. 1. Academic Press, 1979a. 399 p.

Ashcroft N., Mermin N., Solid State Physics. T. 2. Academic Press, 1979b. 422.

Belogorokhov A. I., Tutorsky I. A., Storozenko P. A., Ischenko A. A., Bukanova E. F., Es'kova E. V., Mustafina M. R., Apectral and adsorption characteristics of plasma-chemistry silicon carbide, Dokl. AN RAN. 2006. V. 410, No. 3. pp. 354-356.

Bonch-Bruevich V. L., Kalashnikov S. G., Physics of Semiconductors. Moscow: Nauka, 1977. 678.

Gantmakher V. F., The electrons in disordered media. 2nd ed. Moscow: Fizmatlit, 2005. 232.

Gubin S. P., Cluster chemistry. Basis of classification and structure. Moscow: Nauka, 1987. 263.

Egorov-Tismenko Yu. K., Crystallography and crystal chemistry. Moscow: Publishing House of the KDU, 2005. 589.

Eletskii A. V., Smirnov B. M., Properties of cluster ions. Usp. Fiz. Nauk. 1989. V. 159. pp. 45-81.

Eletskii A. V., Smirnov B. M., Properties of cluster ions. Usp. Fiz. Nauk. 1995. V. 165. pp.

[21]See the chapters in part III of this book.

977-1009.

Zie S., Physics of Semiconductor Devices. In two volumes. Academic Press, 1984.

Ivanov V. K., The electronic properties of metal clusters, Soros Educational Journal. 1999. No. 8. pp. 67-102.

Iveronova V. I., Revkevich G. P., The theory of scattering of X-rays. Moscow: Moscow State University. 1978. 278 p.

Kittel C., Introduction to Solid State Physics. Moscow: Nauka, 1978. 791 p.

Korsunskaya N. E., Stara T. P., Khomenkova L. Yu., Svezhentsova E. V., Melnichenko N. N., Sizov F. F., Nature of the emission of porous silicon prepared by chemical etching. Fiz. Tekh. Poluprovod. 2010. V. 44, no. 1. Pp. 82-86.

Lehmann, T., Bercy M., Spectrometry and Ion Cyclotron Resonance. Wiley, 1980. 216.

Madelung O., Physics of semiconductor compounds of elements of groups III and V. Academic Press. 1967. 478.

Madelung O., Solid State Physics. Localized states. Moscow: Nauka, 1985. 184 p.

Masterov V. F., The physical properties of fullerenes. Sorosh Educational Journal. 1997. No. 1. pp. 92-99.

Meleshko V. P., Stability analysis of hollow spheroidal silicon clusters. Zh. Strukt. Khimii. 1999. V. 40, No. 1. pp. 21-28.

Mironov V. L., Fundamentals of the scanning probe microscopy. Textbook. Nizhny Novgorod: Institute of Physics of Microstructures, RAS, 2004. 110.

Petrov Yu. I., Clusters and small particles. Moscow: Nauka, 1986. 366 p.

Poole C., Owens, F. Nanotechnology. Moscow. Technosphere, 2004. 328 p.

Romaniuk B. N., Mel'nik V. P., Popov V. G., Khatsevich I. M., Oberemok A. Effect of low temperature annealing on the photoluminescence of silicon nanocluster structures. Fiz. Tekh. Poluprovod. 2010. V. 44, No. 4. pp. 533-537.

Smith, R. Semiconductors. Academic Press, 1982. 560 p.

Talanov V. M., Ereyskaya G. P., Yuzyuk Yu. I., Introduction to the chemistry and physics of nanostructures and nanostructured materials. Moscow: Publishing House of the Academy of Natural Sciences, 2008. 391 p.

Ugai Ya. A., Introduction to the chemistry of semiconductors. Text book. Moscow: Vysshaya shkola, 1975. 302.

Urusov V. S., Eremin N. N., Crystal chemistry. A short course. Moscow: Moscow State University Press, 2010. 258 p.

Fetisov G. V., XAFS spectroscopy for structural analysis, in.: Fetisov G.V., Synchrotron radiation. Methods for studying the structure of matter. Moscow: Fizmatlit, 2007. Chap. 5. pp. 488-579.

Harrison W., Electronic Structure and Properties of Solids: Physics of the chemical bond: In 2 volumes. M: Mir, 1983. V. 1. 381 p.

Hilsum K., Rose-Innes A., Semiconductors $A^{III}B^{V}$. Moscow: IL, 1963. 324.

Hirth J., Lothe I., Theory of dislocations. Oxford: Clarendon Press, 1972. 600.

Yu P., Cardona M., Fundamentals of Semiconductor Physics. 3rd ed. Moscow: Fizmatlit, 2002. 560 p.

Akcakir O., Terrien J., Belomoin G., Barry N., Muller J. D., Gratton E., Nayfeh M. H., Detection of luminescent single ultrasmall silicon nanoparticles using fluctuation correlation spectroscopy. Appl. Phys. Letters. 2000. V. 76. P. 1857-1859.

Allan G., Delerue C., Lannoo M. Nature of luminescent surface states of semiconductor nanocrystallites, Phys. Rev. Lett. 1996. V. 76. P. 2961-2964.

Alonso J. A., Glossman M. D., Iniguez M. P. Atomic structure of metallic clusters of medium size. Inter. J. of Modern Phys. B. 1992. V. 6, No. 23-24. P. 3613-3621.

Altman I. S., Lee D., Chung J. D., Song J., Choi M. Light absorption of silica nanoparticles.

Phys. Rev. B. 2001. V. 63. P. 161402 (4 p.).

Anderson P. W.. Absence of diffusion in certain random lattices. Phys. Rev. 1958. V. 109. P. 1492-1505.

Anderson P. W., Local moments and localized states. Rev. Mod. Phys. 1978. V. 50. P. 191-201.

Anstis G. R., Hirsch P. B., Yjmphreys C. J., Hutchinson J. L., Ourmazd A. Lattice images of the cores of 30 deg. partials in silicon. Inst. Phys. Conf. Ser. 1981. V. 60. P. 15.

Baierle R. J., Caldas M. J., Molinari E., Ossicini S. Optical emission from small Si particles, Solid State Commun. 1997. V. 102. P. 545-549.

Belomoin G., Terrien J., Nayfeh M. H., Oxide and hydrogen capped ultrasmall blue luminescent Si nanoparticles. Appl. Phys. Letters. 2000.V. 77. P. 779-781.

Bernstein N., Mehl M. J., Papaconstantopoulos D. A., Papanicolaou N. I., Bazant M. Z., Kaxiras E., Energetic, vibrational, and electronic properties of silicon using a non-orthogonal tight-binding model. Phys. Rev. B. 2000. V. 62. P. 4477-4487.

Bhushan B., Microscopy in Nanoscience and Nanotechnology. Springer, 2010. 987 p.

Billinge S., Local structure from total scattering and atomic pair distribution function (PDF) analysis. In: Powder diffraction theory and practice, Ed. by R. E. Dinnebier, S. J. L. Billinge. The Royal Society of Chemistry, 2008. P. 464-493.

Billinge S. J. L., Thorpe M. F. (Editors). Local structure from diffraction. NY-Boston-Dordrecht-London-Moscow: Kluwer Academic Publishers, 2002. 397 p.

Binggeli N., Chelikowsky J. R., Langevin molecular dynamics with quantum forces: Application to silicon clusters. Phys. Rev. B. 1994. V. 50. P. 11764-11770.

Bingham R. C., Dewar M. J. S., Lo D. H., Ground states of molecules. XXV. MINDO3. Improved version of the MINDO semiempirical SCF-MO method. J. Amer. Chem. Soc. 1975. V. 97. P. 1285-1293.

Biswas R., Hamann D. R. New classical models for silicon structural energies. Phys. Rev. B. 1987. V. 36. P. 6434-6445.

Bjornholm S., Borggreen J., Echt O., Hansen K., Pedersen J., Rasmussen H.D., Mean-field quantization of several hundred electrons in sodium metal clusters. Phys. Rev. Lett. 1990. V. 65. P. 1627-1630.

Bley R. A., Kauzlarich S. M., Synthesis of Silicon Nanoclusters. In: Nanoparticles and nanostructured films: Preparation, characterization and applications / Ed. by J.H. Fendler. Wiley-VCH Verlag GmbH, 1998. Ch. 5. P. 101-118. Xx +468 p.

Bley R.A., Kauzlarich S. M., A low-temperature solution phase route for the synthesis of silicon nanoclusters. J. Am. Chem. Soc. 1996. V. 118. P. 12461-12462.

Brack M., The physics of simple metal clusters: self-consistent jellium model and semiclassical approaches. Rev. Mod. Phys. 1993. V. 65. P. 677-732.

Brandt M. S., Fuchs H. D., Stutzmann M., Weber J., Cardona M., The origin of visible luminescence from 'porous silicon': A new interpretation. Sol. St. Commun. 1992. V. 81, No. 4. P. 307-312.

Brodsky M. H. (Ed.), Amorphous Semiconductors. Topics in Applied Physics. 2 ed. V. 36. - Springer, 1985. 347 p.

Brus L., Electronic wave functions in semiconductor clusters: Experiment and theory. J. Phys. Chem. 1986. V. 90. P. 2555-2560.

Brus L., Electronic wave functions in semiconductor clusters: experiment and theory. J. Phys. Chem. 1986. V. 90. P. 2555-2560.

Bulutay C., Ossicini S., Electronic and optical properties of silicon nanocrystals. In: Silicon nanocrystals: fundamentals, synthesis and applications, Ed. by L. Pavesi, R. Turan. Wiley-VCH Verlag GmbH & Co. KGaA, Weinheim, 2010. Ch. 2. P. 5-41.

Calcott P.D. J., Nash K. J., Canham L. T., Kane M. J., Brumhead D., Identification of radiative transitions in highly porous silicon. J. Phys.: Condens. Matter. 1993. V. 5. P. L91-L98.

Canham L., Properties of porous silicon. London: INSPEC, 1997. 414 p.

Canham L. T., Silicon quantum wire array fabrication by electrochemical and chemical dissolution of wafers. Appl. Phys. Lett. 1990. V. 57. P. 1046-1048.

Carter C.B., In: Dislocations / Ed. by P. Veyssi `ere, L. Kubin, J. Castaing. Editions du CNRS, Paris, 1984. P. 227.

Chang M., Chen Y. F.. Light emitting mechanism of porous silicon. J. Appl. Phys. 1997. V. 82 (7) 5 3514 p.).

Chelikowsky J. R., Cohen M. L., Nonlocal pseudopotential calculations for the electronic structure of eleven diamond and zinc-blende semiconductors. Phys. Rev. B. 1976. V. 14. P. 556-582.

Chelikowsky J. R., Silicon in all its forms. In: Silicon: evolution and future of a technology Ed. by P. Siffert, E. F. Krimmel. Springer, 2004. Ch. 1. P. 1-24.

Cohen M. H., Fritzsche H., Ovshinsky S. R., Simple band model for amorphous semiconducting alloys. Phys. Rev. Lett. 1969. V. 22. P. 1065-1068.

Courteille C., Dorier J.-L., Dutta J., Hollenstein C., Howling A. A., Stoto T., Visible photoluminescence from hydrogenated silicon particles suspended in a silane plasma. J. Appl. Phys. 1995. V. 78. P. 61-66.

Courteille C., Hollenstein C., Dorier J.-L., Gay P., Schwarzenbach W., Howling A. A., Bertran E., Viera G., Martins R., Macarico A., Particle agglomeration study in RF silane plasmas: In situ study by polarization sensitive laser light scattering. J. Appl. Phys. 1996. V. 80. P. 2069-2078.

Cullis A. G., Canham L. T., Calcott P. D. J., The structural and luminescence properties of porous silicon. J. Appl. Phys. 1997. V. 82. P. 909-965.

Dal Negro L., Cazzanelli M., Pavesi L., Ossicini S., Pacifici D., Franzo G., Priolo F., Iacona F., Dynamics of stimulated emission in silicon nanocrystals. Appl. Phys. Lett. 2003. V. 82. P. 4636 (3 p.).

de Heer W.A., The physics of simple metal clusters: experimental aspects and simple models. Rev. Mod. Phys. 1993. V. 65. P. 611-676.

de Heer W.A., Knight W.D., Chou M. Y., Cohen M. L., Electronic shell structure and metal clusters. In: Solid State Physics, V. 40, Ed. by H. Ehrenreich, F. Seitz, D. Turnbull. N.Y.: Academic Press, 1987. P. 93.

Deshpande S.V., Gulari E., Brown S.W., Rand S. C. Optical properties of silicon nitride films deposited by hot filament chemical vapor deposition. J. Appl. Phys. 1995. V. 77. P. 6534–6541.

Dorier J.-L., Hollenstein C., Howling A.A. Spatiotemporal powder formation and trapping in radiofrequency silane plasmas using two-dimensional polarization-sensitive laser scattering. J. Vac. Sci. Technol. 1995. V. A13. P. 918–926.

Drabold D.A., Abtew T.A. Defects in Amorphous Semiconductors: Amorphous Silicon. In: Theory of Defects in Semiconductors. Topics in Applied Physics. V. 104, Ed. by D.A. Drabold, S.K. Estreicher. — Springer, 2007. Ch. 10. P. 225–246.

Dutta J., Hofmann H., Hollenstein C., Hofmeister H. Plasma-produced silicon nanoparticle growth and crystallization process. In: Nanoparticles and Nanostructured Films: Preparation, Characterization and Applications, Ed. by J.H. Fendler. — Wiley-VCH Verlag GmbH, 1998. Ch. 8. P. 173–205. — xx+468 p.

Dutta J., Hofmann H., Houriet R., Hofmeister H., Hollenstein C. Growth, microstructure and sintering behavior of nanosized silicon powders. Colloids and Surfaces A: Physicochemical and Engineering Aspects. 1997. V. 127, No.1–3. P. 263–272.

Dutta J., Houriet R., Hofmann H., Dorier J. L., Howling A.A., Hollenstein C.. In: Proc. of

6th European Conference on Applications of Surface & Interface Analysis (ECA-SIA 95) / Ed. by H. J. Mathieu, B. Reihl, D. Briggs. — Wiley, Chichester, 1996. P. 483–486.

Dutta J., Houriet R., Hofmann H., Hofmeister H. Cluster-induced crystallization of nano-silicon particles. Nanostructured Materials. 1997. V. 9. P. 359–362.

Dutta J., Reaney I.M., Bossel C., Houriet R., Hofmann H. Crystallization of amorphous nanosized silicon powders. Nanostructured Materials. 1995. V. 6. P. 493–496.

Egami T., Billinge S. J. L. Underneath the Bragg peaks. — Elsevier Ltd., 2003. — 422 p.

Elkind J. L., Alford J.M., Weiss F.D., Laaksonen R. T., Smalley R. E. FT-ICR probes of silicon cluster chemistry: The special behavior of Si^+_{39} J. Chem. Phys. 1987. V. 87. P. 2397 (3 p.).

Erlandsson R., Olsson L. Force interaction in low-amplitude ac-mode atomic force micros-copy: cantilever simulations and comparison with data from Si (111) 7×7. Appl. Phys. A. 1998. V. 66. P. S879–S883.

Erlandsson R., Olsson L., Martensson P. Inequivalent atoms and imaging mechanisms in ac-mode atomic-force microscopy of Si (111) 7×7. Phys. Rev. B. 1996. V. 54. P. R8309–R8312.

Fendler J.H., Tian Y. Nanoparticles and nanostructured films: Current accomplishments and future prospects. In: Nanoparticles and Nanostructured Films: Preparation, Char-acterization and Applications, Ed. by J.H. Fendler. — Wiley-VCH Verlag GmbH, 1998. Ch. 18. Part 18.3. P. 429–461. — xx+468 p.

Franceschetti A., Pantelides S. T. Excited-state relaxations and Franck-Condon shift in Si quantum dots. Phys. Rev. B. 2003. V. 68. P. 033313 (4 p.).

Fujii M. Optical properties of intrinsic and shallow impurity-doped silicon nanocrystals. In: Silicon Nanocrystals: Fundamentals, Synthesis and Applications, Ed. by L. Pavesi, R. Turan. — Wiley-VCH Verlag GmbH & Co. KGaA, Weinheim, 2010. Ch. 3. P. 43–68.

Fujii M., Kanzawa Y., Hayashi S., Yamamoto K. Raman scattering from acoustic phonons confined in Si nanocrystals. Phys. Rev. B. 1996. V. 54. P. R8373–R8376.

Furukawa S., Miyasato T. Quantum size effects on the optical band gap of microcrystalline Si:H. Phys. Rev. B. 1988. V. 38. P. 5726–5729.

Garoufalis C. S., Zdetsis A. D ., Grimme S. High level ab initio calculations of the optical gap of small silicon quantum dots. Phys. Rev. Lett. 2001. V. 87. P. 276402 (4 p.).

Gelloz B., Kojima A., Koshida N. Highly efficient and stable luminescence of nanocrystal-line porous silicon treated by high-pressure water vapor annealing. Appl. Phys. Lett. 2005. V. 87. P. 031107 (3 p.).

George A. Core structures and energies of dislocations in Si. In: Properties of crystalline silicon, Ed. by R. Hull. — London: INSPEC publication, 1999. P. 108–112.

Glinka Y. D. Size effect in self-trapped exciton photoluminescence from SiO_2-based na-noscale materials. Phys. Rev. B. 2001. V. 64. P. 085421 (11 p.).

Glusker J. P., Trueblood K. N. Crystal structure analysis: A primer. 3 ed. — Oxford Univer-sity Press, 2010. — 288 p.

Godefroo S., Hayne M., Jivanescu M., Stesmans A., Zacharias M., Lebedev O., Van Tende-loo G., Moshchalkov V. V. Classification and control of the origin of photolumines-cence from Si nanocrystals. Nature Nanotechnol. 2008. V. 3 . P. 174–178.

Grundmann M. The Physics of Semiconductors: An introduction including devices and nanophysics. — Springer-Verlag, 2010. — xxxvii+864 p.

Harrison W.A. Elementary Electronic Structure. — World Scientific Publishing Company, 1999. — 817 p.

Heitmann J., Muller F., Zacharias M., Gosele U. Silicon nanocrystals: size matters. Adv.

Mater. 2005. V. 17. P. 795–803.

Hill N.A., Whaley K. B. A theoretical study of light emission from nanoscale silicon. J. Electron. Mater. 1995. V. 25. P. 269–285.

Hofmeister H., Dutta J., Hofmann H. Atomic structure of amorphous nanosized silicon powders upon thermal treatment. Phys. Rev. B. 1996. V. 54. P. 2856–2862.

Holt D. B., Yacobi B. G. Extended defects in semiconductors: Electronic properties, device effects and structures. — Cambridge University Press, 2007. — xi+631 p.

Honea E. G., Ogura A., Peale D. R., Felix C., Murray C. A., Raghavachari K., Sprenger W. O., Jarrold M. F., Brown W. L. Structures and coalescence behavior of size-selected silicon nanoclusters studied by surface-plasmon-polariton enhanced Raman spectroscopy. J. Chem. Phys. 1999. V. 110. P. 12161 (12 p.).

Hull R. (Ed.). Properties of crystalline silicon. — London: INSPEC publication, 1999. — 1042 p.

International Tables for Crystallography. 2006. V. A1. Maximal subgroups of space group, No.227. P. 387–389.

Iqbal Z., Vepřek S. Raman scattering from hydrogenated microcrystalline and amorphous silicon, J. Phys. C: Solid State Phys. 1982. V. 15. P. 377–392.

Itoh U., Toyoshima Y., Onuki H., Washida N., Ibuki T. Vacuum ultraviolet absorption cross sections of SiH_4, GeH_4, Si_2H_6, and Si_3H_8. J. Chem. Phys. 1986. V. 85. P. 4867 (6 p.).

Jarrold M. F., Constant V.A. Silicon cluster ions: Evidence for a structural transition. Phys. Rev. Lett. 1991. V. 67. P. 2994–2997.

Jarrold M. F. Drift tube studies of atomic clusters. J. Phys. Chem. 1995. V. 99. P. 11–12.

Jurbergs D., Rogojina E., Mangolini L., Kortshagen U. Silicon nanocrystals with ensemble quantum yields exceeding 60%. Appl. Phys. Lett. 2006. V. 88. P. 233116 (3 p.).

Kanzawa Y., Kageyama T., Takeoka S., Fujii M., Hayashi S., Yamamoto K. Size-dependent near-infrared photoluminescence spectra of Si nanocrystals embedded in SiO_2 matrices. Solid State Commun. 1997. V. 102. P. 533–537.

Kaxiras E. Atomic and electronic structure of solids. Cambridge University Press, 2003. — 696 p.

Kaxiras E. Effect of surface reconstruction on stability and reactivity of Si clusters. Phys. Rev. Lett. 1990. V. 64. P. 551–554.

Khriachtchev L. (Ed.). Silicon nanophotonics: Basic principles, present status and perspectives. — World Scientific Publishing Co. Pte. Ltd., 2009. — xvii+452 p.

Khriachtchev L., Rasanen M., Novikov S., Sinkkonen J. Optical gain in Si/SiO_2 lattice: Experimental evidence with nanosecond pulses. Appl. Phys. Lett. 2001. V. 79. P. 1249 (3 p.).

Knight W.D., Clemenger K., de Heer W.A., Saunders W. Polarizability of alkali clusters. Phys. Rev. B. 1985. V. 31. P. 2539–2540.

Knight W.D., Clemenger K., de Heer W.A., Saunders W., Chou M. Y., Cohen M. L. Electronic shell structure and abundances of sodium clusters. Phys. Rev. Lett. 1984. V. 52. P. 2141–2143.

Korsunskaya N. E., Kaganovich E. B., Khomenkova L. Yu., Bulach B. M., Dzhumaev B. R., Beketov G.V., Manoilov E.G. Effect of adsorption and desorption processes on photoluminescence excitation spectra of porous silicon. Appl. Surf. Sci., 2000. V. 166. P. 349–353.

Kovalev D., Fujii M. Silicon nanocrystal assemblies: Universal spin-flip activators?. In: Annual review of nanoresearch. V. 2 / Ed. by G. Cao, C. J. Brinker. — World Scientific Publishing Co. Pte. Ltd., 2008. P. 159–215.

Kovalev D., Fujii M. Silicon nanocrystals: photosensitizers for oxygen molecules. Adv. Mater. 2005. V. 17. P. 2531–2544.

Kovalev D., Heckler H., Averboukh B., Ben-Chorin M., Schwartzkopff M., Koch F. Hole burning spectroscopy of porous silicon. Phys. Rev. B. 1998. V. 57. P. 3741–3744.

Kovalev D., Heckler H., Ben-Chorin M., Polisski G., Schwartzkopff M., Koch F. Breakdown of the k-conservation rule in Si nanocrystals. Phys. Rev. Lett. 1998. V. 81. P. 2803–2806.

Kovalev D., Heckler H., Polisski G., Koch F. Optical properties of Si nanocrystals. Phys. Status Solidi B. 1999. V. 215. P. 871–932.

Kumar V. (Ed.). Nanosilicon. — Elsevier Ltd., 2007. — xiii+368 p.

Ledoux G., Guillois O., Porterat D., Reynaud C., Huisken F., Kohn B., Paillard V. Photoluminescence properties of silicon nanocrystals as a function of their size. Phys. Rev. B. 2000. V. 62. P. 15942–15951.

Littau K.A., Szajowski P. F., Muller A. J., Kortan A.R., Brus L. E. A luminescent silicon nanocrystal colloid via a high-temperature aerosol reaction. J. Phys. Chem. 1993. V. 97. P. 1224–1230.

Luppi M., Ossicini S. Ab initio study on oxidized silicon clusters and silicon nanocrystals embedded in SiO_2: Beyond the quantum confinement effect. Phys. Rev. B. 2005. V. 71. P. 035340 (15 p.).

Luppi M., Ossicini S. Multiple Si=O bonds at the silicon cluster surface. J. Appl. Phys. 2003. V. 94. P. 2130 (3 p.).

Luppi M., Ossicini S. Oxygen role on the structural and optoelectronic properties of silicon nanodots. Phys. Stat. Sol. (a). 2003. V. 197. P. 251–256.

Luterova K., Pelant I., Mikulskas I., Tomasiunas R., Muller D., Grob J.-J., Rehspringer J.-L., Honerlage B. Stimulated emission in blue-emitting Si+-implanted SiO_2 films?. J. Appl. Phys. 2002. V. 91. P. 2896 (5 p.).

Mangolini L., Kortshagen U. Plasma-assisted synthesis of silicon nanocrystal inks. Adv. Mater. 2007. V. 19. P. 2513–2519.

Matsumoto S., Matsui Y. Electron microscopic observation of diamond particles grown from the vapour phase. J. Mater. Sci. 1983, 18, 1785–1793.

Mélinon P., Blase X., San Miguel A., Perez A. Cluster assembled silicon networks. Chapter 2 in: Nanosilicon / Ed. by Vijay Kumar. — Elsevier Ltd., 2008. P. 79–112.

Mélinon P., Kéghélian P., Prével B., Perez A., Guiraud G., LeBrusq J., Lermé J., Pellarin M., Broyer M. Nanostructured silicon films obtained by neutral cluster depositions. J. Chem. Phys. 1997. V. 107. P. 10278 (10 p.).

Meyer B.K., Hofmann D. M., Stadler W., Petrova-Koch V., Koch F., Emanuelsson P., Omling P. Photoluminescence and optically detected magnetic resonance investigations on porous silicon. J. Lumin. 1993. V. 57. P. 137–140.

Meyer B.K., Hofmann D.M., Stadler W., Petrova-Koch V., Koch F., Omling P., Emanuelsson P. Defects in porous silicon investigated by optically detected and by electron paramagnetic resonance techniques. Appl. Phys. Lett. 1993. V. 63. P. 2120–2122.

Mochizuki Y., Mizuta M., Ochiai Y., Matsui S., Ohkubo N. Luminescent properties of visible and near-infrared emissions from porous silicon prepared by the anodization method. Phys. Rev. B. 1992. V. 46. P. 12353–12357.

Moliton A. Limits to classical band theory: Amorphous media. In: Solid-state physics for electronics / Ed. by A. Moliton. — Wiley-ISTE, 2009. Ch. 9. P. 301–333.

Moliton A. Solid-state physics for electronics. — ISTE Ltd. and John Wiley & Sons, Inc., 2009. 432 p.

Mott N. F. Electrons in disordered structures. Adv. Phys. 1967. V. 16. P. 49–144.

Nabarro F.R.N., Hirth J. P. (editors). Dislocations in Solids. V. 13. Elsevier B.V. 2007. — 668 p.

Nayfeh M.H., Akcakir O., Belomoin G., Barry N., Therrien J., Gratton E. Second harmonic generation in microcrystallite films of ultrasmall Si nanoparticles. Appl. Phys. Letters. 2000. V. 77. P. 4086–4088.

Nayfeh M., Rao S., Barry N., Smith A., Chaieb S. Observation of laser oscillation in aggregates of ultrasmall silicon nanoparticles. Appl. Phys. Lett. 2002. V. 80. P. 121 (3 p.).

Nayfeh M.H., Mitas L. Silicon Nanoparticles: New photonic and electronic material at the transition between solid and molecule. In: Nanosilicon / Ed. by V. Kumar. — Elsevier Ltd., 2008. Ch. 1. P. 1–78.

Nayfeh M.H., Barry N., Terrien J., Akcakir O., Gratton E., Belomoin G. Stimulated blue emission in reconstituted films of ultrasmall silicon nanoparticles. Appl. Phys. Letters. 2001. V. 78. P. 1131–1133.

Nishioka H., Hansen K., Mottelson B. R. Supershells in metal clusters. Phys. Rev. B. 1990. V. 42. P. 9377–9386.

Okabe T., Kagawa Y., Takai S. High resolution electron microscopic observation on a pentagonal nucleus formed in amorphous germanium films. Phil. Mag. Lett. 1991. V. 63. P. 233–239.

Olsen A., Spence J. C.H. Distinguishing dissociated glide and shuffle set dislocations by high resolution electron microscopy. Philos. Mag. A. 1981. V. 43. P. 945–965.

Ossicini S., Degoli E., Iori F., Luppi E., Magri R., Cantele G., Trani F., Ninno D. Simultaneously B- and P-doped silicon nanoclusters: formation energies and electronic properties. Appl. Phys. Lett. 2005. V. 87. P. 173120 (3 p.).

Pan J., Ramakrishna M.V. Magic numbers of silicon clusters. Phys. Rev. B. 1994. V. 50. P. 15431–15434.

Pavesi L., Turan R. (Eds.). Silicon nanocrystals: Fundamentals, synthesis and applications. — Wiley-VCH Verlag GmbH & Co. KGaA, Weinheim, 2010. — 652 p.

Pavesi L., Dal Negro L., Mazzoleni C., Franzo G., Priolo F. Optical gain in silicon nanocrystals, Nature. 2000. V. 408. P. 440–444.

Puzder A., Williamson A. J., Grossman J. C., Galli G. Surface chemistry of silicon nanoclusters, Phys. Rev. Lett. 2002. V. 88. P. 097401 (4 p.).

Qin G.G., Song H. Z., Zhang B. R., Lin J., Duan J.Q., Yao G.Q. Experimental evidence for luminescence from silicon oxide layers in oxidized porous silicon. Phys. Rev. B. 1996. V. 54. P. 2548–2555.

Ramakrishna M.V., Pun Pan J. Chemical reactions of silicon clusters. J. Chem. Phys. 1994. V. 101. P. 8108 (11 p.).

Ramakrishna M.V. Pun Pan J. Chemical reactions of silicon clusters. J. Chem. Phys. 1994. V. 101. P. 8108–8118.

Ranjan V., Kapoor M., Singh V.A. The band gap in silicon nanocrystallites. J. Phys.: Condens. Matt. 2002. V. 14. P. 6647–6655.

Ray I. L. F., Cockayne D. J.H. The Dissociation of Dislocations in Silicon. Proc. R. Soc. Lond. A. 1971. V. 325. P. 543–554.

Ray I. L. F., Cockayne D. J.H. The Dissociation of dislocations in silicon. Proc. R. Soc. London, Ser. A. 1971. V. 325. P. 543–554.

Reboredo F.A., Franceschetti A., Zunger A. Excitonic transitions and exchange splitting in Si quantum dots. Appl. Phys. Lett. 1999. V. 75. P. 2972–2974.

Roschuk T., Li J., Wojcik J., Mascher P., Calder I.D. Lighting applications of rare earth-doped silicon oxides. In: Silicon Nanocrystals: Fundamentals, Synthesis and Applications, Ed. by L. Pavesi, R. Turan. — Wiley-VCH Verlag GmbH & Co. KGaA, Weinheim, 2010. Ch. 17. P. 487–506.

Réthlisberger U., Andreoni W., Parinello M. Structure of nanoscale silicon clusters. Phys. Rev. Lett. 1994. V. 72. P. 665–668.

Röthlisberger U., Andreoni W., Parinello M. Structure of nanoscale silicon clusters. Phys.

Rev. Lett. 1994, 72, 665–668.

Sakurai T., Hagstrom H.D. Hydrogen chemisorption on the silicon (110) 5 × 1 surface. J. Vac. Sci. Technol. 1976. V. 13. P. 807–809.

Sevik C., Bulutay C. Auger recombination and carrier multiplication in embedded silicon and germanium nanocrystals. Phys. Rev. B. 2008. V. 77. P. 125414 (4 p.).

Shvartsburg A.A., Jarrold M. F., Liu B., Lu Z.-Y., Wang C.-Z., Ho K.-M. Dissociation energies of silicon clusters: A depth gauge for the global minimum on the potential energy surface. Phys. Rev. Lett. 1998. V. 81. P. 4616–4619.

Skryshevsky V.A., Laugier A., Strikha V. I., Vikulov V.A. Evaluation of quantum efficiency of porous silicon photoluminescence. Mater. Sci. Eng. B. 1996. V. 40. P. 54–57.

Soni R.K., Fonseca L. F., Resto O., Buzaianu M., Weisz S. Z. Size-dependent optical properties of silicon nanocrystals. J. Lumin. 1999. V. 83–84. P. 187–191.

Street R.A. Hydrogenated Amorphous Silicon. Cambridge University Press, 1991. xiv+ 417 p.

Sugano S. Microcluster Physics (Springer Series in Materials Science). Springer, 1991. — 236 p.

Takayanagi K., Tanishiro Y., Takahashi M., Takahashi S. Structural analysis of Si (111)-7 × 7 by UHV-transmission electron diffraction and microscopy. J. Vacuum Sci. Technol. A. 1985. V. 3. P. 1502 (5 p.); Takayangi K., Tanishiro Y., Takahashi S., Takahashi M. Structure analysis of Si (111)-7×7 reconstructed surface by transmission electron diffraction. Surf. Sci. 1985. V. 164. P. 367–392.

Takeoka S., Fujii M., Hayashi S. Size-dependent photoluminescence from surface-oxidized Si nanocrystals in a weak confinement regime. Phys. Rev. B. 2000. V. 62. P. 16820–16825.

Taylor P. R., Bylaska E., Weare J.H. Kawai R. J. Chem. Phys. Lett. 1995, 235, 558–563.

Thorpe M. F., Weaire D. Electronic Properties of an Amorphous Solid. II. Further Aspects of the Theory. Phys. Rev. B. 1971. V. 4. P. 3518–3527.

Thurmond C.D. The Standard thermodynamic functions for the formation of electrons and holes in Ge, Si, GaAs, and GaP. J. Electrochem. Soc. 1975. V. 122. P. 1133–1141.

Valenta J., Juhasz R., Linnros J. Photoluminescence spectroscopy of single silicon quantum dots. Appl. Phys. Lett. 2002. V. 80. P. 1070–1072.

Varshni Y. P. Temperature dependence of the energy gap in semiconductors. Physica. 1967. V. 37. P. 149–154.

Vasiliev I., Chelikowsky J. R., Martin R. M. Surface oxidation effects on the optical properties of silicon nanocrystals. Phys. Rev. B. 2002. V. 65. P. R121302 (4 p.).

Vasiliev I., Ögüt S., Chelikowsky J. R. Ab initio absorption spectra and optical gaps in nanocrystalline silicon. Phys. Rev. Lett. 2001. V. 86. P. 1813–1816.

Vepřek S., Iqbal Z., Sarott F.A. A thermodynamic criterion of the crystalline-to-amorphous transition in silicon. Phil. Mag. B. 1982. V. 45. P. 137–145.

Vepřek S., Iqbal Z., Sarott F. A. A thermodynamic criterion of the crystalline-to-amorphous transition in silicon. Phil. Mag. B. 1982. V. 45. P. 137–145.

Vepřek S., Schopper K., Ambacher O., Rieger W. Vepˇrek -Heijman M. G. J. Mechanism of cluster   formation in a clean silane discharge. J. Electrochem. Soc. 1993. V. 140. P. 1935–1942.

Vial J. C., Bsiesy A., Gaspard F., H´erino R., Ligeon M., Muller F., Romestain R., Macfarlane R.M. Mechanisms of visible-light emission from electro-oxidized porous silicon. Phys. Rev. B. 1992. V. 45. P. 14171–14176.

Wang Y.Q., Smirani R., Ross G.G. Nanotwinning in silicon nanocrystals produced by ion implantation. Nano Lett. 2004. V. 4. P. 2041–2045.

Wang Y.Q., Smirani R., Ross G.G. Stacking faults in Si nanocrystals. Appl. Phys. Lett. 2005.

V. 86. P. 221920 (3 p.).

Watanabe Y., Shiratani M. Growth kinetics and behavior of dust particles in silane plasmas Jpn. J. Appl. Phys. 1993. V. 32. P. 3074–3080.

Weaire D., Thorpe M. F. Electronic properties of an amorphous solid. I. A simple tight-binding theory. Phys. Rev. B. 1971. V. 4. P. 2508–2520.

Weaire D. Existence of a gap in the electronic density of states of a tetrahedrally bonded solid of arbitrary structure. Phys. Rev. Lett. 1971. V. 26. P. 1541–1543.

Wilson W. L., Szajowski P. F., Brus L. E. Quantum confinement in size-selected, surface-oxidized silicon nanocrystals. Science. 1993. V. 262. P. 1242–1244.

Wolkin M.V., Jorne J., Fauchet P.M., Allan G., Delerue C. Electronic states and luminescence in porous silicon quantum dots: the role of oxygen. Phys. Rev. Lett. 1999. V. 82. P. 197–200.

Yamamoto Y., Miura T., Teranishi T., Miyake M., Hori H., Suzuki M., Kawamura N., Miyagawa H., Nakamura T., Kobayashi K. Direct evidence for ferromagnetic spin polarization in gold nanoparticles. Phys.Rev.Lett. 2004. V. 93. P. 116801 (4 p.).

Yin M. T., Cohen M. L. Microscopic theory of the phase transformation and lattice dynamics of Si. Phys. Rev. Lett. 1980. V. 45. P. 1004–1007.

Yoo S., Shao N., Zeng X. C. Structures and relative stability of medium- and large-sized silicon clusters. VI. Fullerene cage motifs for low-lying clusters Si_{39}, Si_{40}, Si_{50}, Si_{60}, Si_{70}, and Si_{80}. J. Chem. Phys. 2008. V. 128. P. 104316 (9 p.).

Yu P. Y., Cardona M. Fundamentals of Semiconductors. Graduate Texts in Physics. 4 ed. — Berlin–Heidelberg: Springer-Verlag, 2010. — 775 p.

Zachariasen W.H. The atomic arrangement in glass. J. Am. Chem. Soc. 1932. V. 54, No.10. P. 3841–3851.

3

Porous silicon: luminescence properties

V.Yu. Timoshenko[1]

Porous silicon (P-Si) became known for the first time in 1956 in a publication by A. Uhlir, in which the material was obtained as a byproduct of photoelectrochemical etching of holes in the plates of crystalline silicon (Uhlir, 1956). Two years later, D. Turner (Turner, 1958) examined in more detail and described P-Si as a special kind of silicon formed by electrochemical etching of monocrystalline silicon (c-Si). After that, there was a lull in the studies of P-Si for almost thirty years, which was only occasionally broken by individual works. For example, in 1984 researchers at the British Royal Institute for Radar and Communication observed photoluminescence (PL) of P-Si in the visible spectrum at the liquid helium temperature (Pickering et al,1984). A detailed study of the properties of P-Si started after the discovery by L. Canham at the British Defense Research Agency of efficient photoluminescence of this material at room temperature (Canham et al, 1990). In his paper he proposed an interesting hypothesis that the PL in P-Si is related to the quantum size effect for charge carriers. Almost simultaneously a conclusion on the possibility of the quantum size effect in P-Si was proposed by Gössele and Lehmann who analyzed the mechanism of formation of this material (Lehmann & Gössele, 1991). By the time these pioneering works were carried out attention was already paid to the production, exploration and application of nanostructures, starting from quantum wells to quantum dots of various semiconductors. In this connection there was a tempting prospect of using a fairly simple way to obtain P-Si for the formation of silicon quantum wires and dots, which could then be used to create light-emitting silicon devices. But it has not yet been possible to achieve full success in the practical application of the light-emitting properties of P-Si because of the insufficiently high photoluminescence quantum yield.

One of the reasons for the increase in the quantum yield of luminescence of P-Si is the modification of the energy band structure of silicon with a

---

[1] Faculty of Physics, M.V. Lomonosov Moscow State University.

decrease in the size of crystals to a few nanometers. Another possible way to increase the quantum yield of PL is the introduction into a silicon matrix of luminescence activators, such as the molecules of organic dyes, quantum dots or ions of rare-earth elements (Kimerling et al, 1997; Lockwood, 1997; Schmidt et al, 2003).

At the present time, special interest is paid to the preparation and use of P-Si in various fields, including opto- and microelectronics, chemical synthesis, biomedical technology. The number of articles published in recent years on research into P-Si exceeds 800 per year and has not yet shown any significant reduction (see Fig. 3.1).

The aim of many studies of P-Si is the study of mechanisms of energy transfer between its constituent silicon nanocrystals (nc-Si) and their environment that is important to create a light-emitting device compatible with silicon technology Because P-Si has a large specific surface (up to 1000 m^2/cm^3), the environment has a significant influence on the optical and electrical properties of nanostructured semiconductors. Therefore, investigations of the effect of adsorption of various gases such as NO_2 (Timoshenko et al, 2001), CO_2 (Rocchia et al, 2003), NH_3 (Chiesa et al, 2003), O_2 (Kovalev et al, 2001, 2002), etc. on the properties of P-Si are of great scientific and practical interest to elucidate the mechanisms of energy exchange between semiconductor nanocrystals and their molecular environment.

It is the transfer of energy between the silicon nanocrystals and the oxygen molecules adsorbed on their surface that resulted in the detection around 10 years ago of the phenomenon of photosensitized generation of singlet oxygen in the P–Si/O_2 system (Kovalev et al, 2001), which opened the possibility of using this material in the photodynamic therapy of tumors (Timoshenko et al, 2006). At present, there are many other useful features of P-Si, which can be used in biomedical diagnostics and therapy

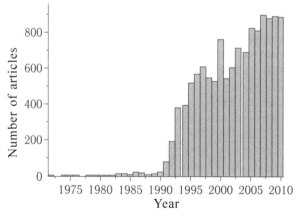

Fig. 3.1. The number of articles on the study of P-Si published in the year, according to the database of Thomson ISI Web of Knowledge, if 'porous silicon' is used in the search of titles and content of articles.

of various diseases (Salonen, Lehto, 2008). All this increases the interest in this material.

The following section describes the literature data on the basic patterns of production, physical properties and potential applications of P-Si. In this case, we focus on the P-Si produced by the electrochemical etching method. The properties of P-Si, produced by other methods, are also mentioned briefly.

3.1. Methods of production of porous silicon

The standard method of producing P-Si is the electrochemical anodic treatment of silicon wafers in an electrolyte consisting of hydrofluoric acid (HF), water and/or alcohol (Uhlir, 1956; Turner, 1958; Labunov et al, 1978). The simplest type of cell for the electrochemical production of P–Si is shown schematically in Fig. 3.2.

The silicon electrode (anode), in the form of a c–Si plate, receives the positive potential, and the cathode is typically a platinum electrode. One of the important conditions for obtaining laterally homogeneous films of P-Si is a good (preferably ohmic) contact between the back side of a silicon anode and a metallic contact electrode. If in the c-Si wafers with low resistivity (of the order of units and tens of milliohms $\cdot$ cm, i.e., p$^+$-Si and n$^+$-Si) this condition is easily implemented with conventional mechanical contact, for a high-resistance substrate (p-Si or n-Si) it is required to apply a special thin metal layer (e.g. aluminum), which provides low contact resistance. Another way to overcome the high contact resistance for low-doped silicon substrates is to replace the metal contact electrode by a liquid contact, for example, using the same solution of hydrofluoric acid. The latter can be implemented in special two-chamber cells, as schematically shown in Fig. 3.3.

Another interesting method for the electrochemical production of P-Si is the so-called capillary method (Timoshenko et al, 2010), in which the electrolyte is held by capillary forces between two closely spaced electrodes

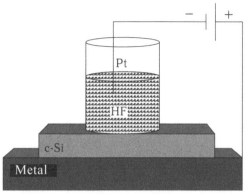

Fig. 3.2. Schematic representation of the elementary cell for the electrochemical production of porous silicon in a solution of hydrofluoric acid (HF).

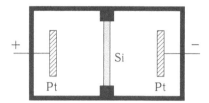

Fig. 3.3. The scheme of the dual-chamber cell for electrochemical production of porous silicon (Salonen, Lehto, 2008).

(see Fig. 3.4*a*), one of which is the anode made of c-Si and is used to produce P-Si. The other electrode (cathode) can also be of silicon or other material not liable to substantial dissolution in hydrofluoric acid at the cathode potential. Many successive electrodes can be used in the capillary cell (Fig. 3.4*b*). At the same time, the outer electrodes are directly attached to an electric circuit, and the electrodes located between them are actually virtual anodes and cathodes, the potential of which is formed due to the voltage drop in the capillary layer of the electrolyte/c-Si plate. Since the thickness of the capillary molecular layer is not more than 0.1–1.0 mm, the circuit is very economical as regards the consumption of the electrolyte. In addition, like the circuit shown in Fig. 3.3, it does not require application of special metal layers for improved contact with the silicon anode. The disadvantage of the capillary method is the high rate of depletion of the electrolyte, making it difficult to obtain thick layers of P-Si.

In all the above schemes for P-Si production the set of chemical reactions on the working silicon electrode (anode) is the same and determined by the concentration of fluoride ions in the electrolyte and by the applied voltage. The most common electrolyte is concentrated HF or its solutions in water–alcohol mixtures ($HF:H_2O:C_2H_5OH$) in certain proportions (Labunov et al, 1978; Smith, Collins, 1992). Another method is to produce P-Si in aqueous solutions of ammonium fluoride (NH_4F) with a specific degree of

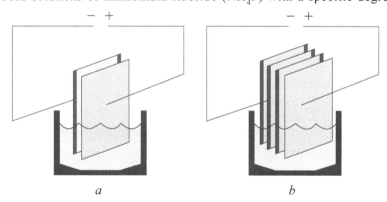

a   *b*

Fig. 3.4. Scheme of two-(*a*) and multi-electrode (*b*) capillary cells (Timoshenko et al, 2010).

acidity of pH = 2.5–4 (Dittrich et al, 1995). Since the dissociation of NH_4F molecules is accompanied by the formation of F^- and H^+, the character of the electrochemical processes occurring in the formation P-Si in general is similar to anodizing in HF solutions. Special features of the latter method include the relatively small thickness of the P-Si layers, which can be obtained in solutions of NH_4F, which is associated with low concentrations of fluoride ions and, consequently, a low rate of growth of the porous layer (Dittrich et al, 1995).

In forming P-Si in an electrolyte based on HF the following basic chemical reactions and physical processes occur at the silicon/electrolyte interface at the application of a positive potential to the silicon electrode (Labunov et al, 1978).

1. The formation of silicon bifluoride:

$$Si + 2HF + 2e^+ \rightarrow SiF_2 + H_2. \qquad (3.1)$$

2. Chemical reduction of silicon from silicon bifluoride:

$$2SiF_2 \rightarrow Si + SiF_4,$$
$$SiF_4 + 2HF \rightarrow H_2SiF_6. \qquad (3.2)$$

3. Chemical oxidation of silicon bifluoride to silica with subsequent dissolving of the latter in HF:

$$SiF_2 + 2H_2O \rightarrow SiO_2 + 2HF + H_2,$$
$$SiO_2 + 4HF \rightarrow SiF_4 + 2H_2O,$$
$$SiF_4 + 2HF \rightarrow H_2SiF_6. \qquad (3.3)$$

Depending on the processing conditions of silicon wafers, one of the reactions (3.2) or (3.3) is dominant. The result is either the formation of P-Si (reactions (3.1) and (3.2)), or electropolishing of the Si surface due to reaction (3.1) and (3.3). The reactions on the surface of the silicon electrode during the formation of P-Si are shown schematically in Fig. 3.5.

From this figure it is seen that the dissolution of silicon and the formation of molecular hydrogen is accompanied by the covering of the

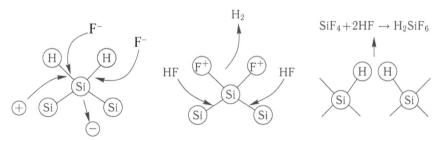

Fig. 3.5. Schematic representation of the process of etching the silicon electrode in obtaining porous silicon (Smith, Collins, 1991).

silicon surface with hydrogen (the so-called termination hydrogenation or H-termination). The latter process does not depend on applied voltage and is determined by chemical reactions on the silicon surface in a solution of HF (Yablonovitch et al, 1986).

Two main factors determine the formation of P-Si. First, the essential delivery of fluoride ions (F$^-$) in the reaction zone to form SiF$_2$ (reaction (3.1)). Second, the positive mobile charge carriers (holes) are required in the surface layer of the silicon anode. The first condition is associated with the properties of the electrolyte and the electrochemical treatment regime. The second factor causes that the processes of etching hole and electron-type conductivity (p- and n-type, respectively) are significantly different due to different concentrations of holes in them. For example, supply of ions F$^-$ is important for the p^+-Si substrate, while for n-Si it is not enough: an external factor stimulating the generation of holes is required. The latter condition can be achieved in several ways, such as heating, lighting or impact ionization. The process of growth of pores (pore formation) in n-Si is facilitated with increasing degree of doping with donor impurities in the presence of structural defects in it. With regard to the factor of the course of anode etching that is common for c-Si wafers of both types of conductivity – the presence of F$^-$ ions in the reaction zone – it is determined by the diffusion, convection, and migration processes. The exact quantitative description is complicated by the formation of an SiF$_2$ buffer anode layer (reaction (3.1)) which impedes the exchange of reagents. One should also take into account the mixing of the electrolyte due to gas release (reaction (3.1)) and thermal gradients in the Si–electrolyte system as a result of exo- and endothermal chemical reactions.

The current–voltage characteristics of the silicon–electrolyte system has a relatively complex form (Fig. 3.6). According to many studies, see e.g. reviews (Labunov et al, 1978; Smith, Collins, 1992), it is possible to allocate three sites. At low voltages U (section 1) the dependence is close to exponential at which pores form. The exponential dependence indicates the presence of a potential barrier at the silicon/electrolyte interface. Section 2 is the transition between the regions of pore formation and electrochemical polishing etching of the surface of the silicon electrode. At high values of voltage (section 3) the so-called electropolishing regime forms. It is this mode that has been studied and used in early work on P-Si (Uhlir, 1956; Turner, 1958).

Of practical interest are the dependences of the basic parameters of the P-Si films – thickness and growth rate – on the conditions of the process: the duration of anodic treatment, current density, HF content in the electrolyte, temperature, light intensity, the degree of doping of the initial Si.

It was found that the thickness of the film of P-Si generally increases with increasing etching time t (Labunov et al, 1978). Initially this ratio is governed by a linear law and the slope of the linear dependence increases

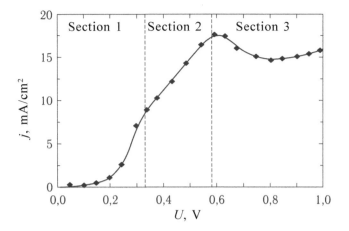

Fig. 3.6. Typical current–voltage characteristic of the P-Si/electrolyte system (HF solution in ethanol).

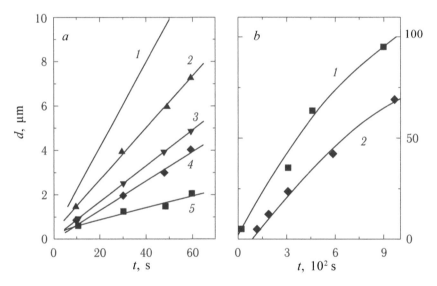

Fig. 3.7. The thickness of the film of P-Si *vs* the time of anodic treatment in concentrated HF at different illumination intensities: 52 mW/cm^2 (*a*) and 22 mW/cm^2 (*b*), as well as at different densities of the anode current of 100 mA/cm^2 (*1, 2*), 40 mA/cm^2 (*3*), 50 mA/cm^2 (*4*) 26 mA/cm^2 (*5*). c-Si substrates doped with donors at a concentration of $8 \cdot 10^{18}$ cm^{-3} (curves *1a* and *3a*) or acceptors at a concentration of $1.8 \cdot 10^{19}$ cm^{-3} (curves *2a, 4a, 5a*) and $1.2 \cdot 10^{19}$ cm^{-3} (curves *1b* and *2b*) (Labunov et al, 1978).

with current density j. In longer processing the film thickness in the first approximation is proportional to $t^{1/2}$ (Fig. 3.7). The rate of formation of the P-Si film increases linearly with increasing concentration of the electrolyte, and the temperature dependence of the latter is very weak.

The main influence on the rate is exerted by current density with almost a linear relationship for the P-Si substrates. At the same time for the n-Si substrates this dependence is more complicated. At the weak light intensity the rate depends weakly on the current intensity, and with increasing light intensity (i.e. with increasing concentration of holes) the rate of growth of the P-Si film increases. The rate of formation of P-Si films depends on the degree of doping of the substrate. Increasing the concentration of acceptors leads to an increase in the rate of growth and this dependence is linear. For n-Si, the P-Si layer growth rate first increases with increasing concentration of the donors, reaches a maximum at 10^{16}–10^{17} cm^{-3} and then decreases (Labunov et al, 1978).

Despite the fact that the first P-Si samples were produced for several decades ago, there is still no single point of view regarding the basic mechanism formation of pores. Thus, following the fundamental work of Smith and Collins (Smith, Collins, 1992), there are three most successful models.

The model of the depleted layer or the Beale model is based on the work of this author (Beale et al, 1985) who detected high specific electric resistance of P-Si ($\sim 10^6$ ohm·cm). Such a high resistance is due to the low concentration of free charge carriers in silicon nanocrystals which is due to the overlap of depleted layers formed in the etching process. As a consequence, this current is directed to the top of the pores. Description of the electrolyte–Si system is similar to the Schottky barrier, but, unfortunately, this does not justify this solution of the problem. In addition, the model is based on the assumption of 'pinning' the Fermi level in the middle of the band gap due to the high density of surface defects; this is not supported by experimental data.

The restricted diffusion model, proposed in (Smith, Collins, 1992) suggests that the basis of the mechanism of pore formation is the process of diffusion of holes/electrons to/from the surface of Si. Its main advantage over the previous model is that it initially contains no assumptions about the nature of the Si/electrolyte interface, the position of the Fermi level, etc., and operates only with the reagents required for the formation of pores (i.e. holes). Depletion of charge carriers is achieved in the ordinary course of the reactions of anodic dissolution. Computer simulation of pore growth is in good agreement with the data on the structure of porous silicon obtained by transmission electron microscopy. The disadvantages of the model include lack of attention to the specifics of surface chemistry, ignorance of the role of defects always present in silicon.

The quantum-confinement model (Lehman, Gössele, 1991) explains the formation of a porous layer, suggesting an increase in the width of the forbidden band (E_g) in the resultant material. Increase of E_g is due to the confinement of the charge carriers within small silicon remnants, formed in the anodization process, which leads to a decrease of the concentration of mobile charge carriers in the silicon 'skeleton' and creates an area of depletion, such as those described in the Beale model. Such a mechanism is reasonable for P-Si. However, it is unlikely to operate in p⁺-Si or n-Si because of the significant (~ hundreds of nanometres) cross sections of the produced silicon 'strings' and clusters, which prevents any significant manifestation of the quantum size effect.

It should be noted that in addition to the above-described electrochemical process of production of P-Si, there is also a method of chemical forming of P-Si by c-Si etching in solutions based on a mixture of hydrofluoric and nitric acids, without the application of the electric potential, which is sometimes called stain etching. This method has been known for a long time (Archer, 1960), at approximately the same time as the electrochemical production of P-Si. It has been established, for example, see reviews (Jung et al, 1993; Salonen, Lehto, 2008) that the formation of P-Si in this process also requires the participation of the holes and proceeds more efficiently for the heavily doped c-Si substrates. It is assumed that the mechanism of formation of the porous layer comprises local electrochemical reactions (3.1)–(3.3) on the silicon surface. The resulting P-Si compared to the electrochemically formed material has a limited thickness (up to several microns). Otherwise, its properties, including fluorescent, are close to the properties of the P-Si formed by the previously described standard method.

3.2. Main structural characteristics of porous silicon

The optical and electronic properties of P-Si strongly depend on the size and morphology of its constituent nanostructures, namely, nc-Si. In order to describe the structural properties of P-Si, we can use integral and local (microscopic) characteristics. One of the main integral characteristics of P-Si is the porosity of the sample p, defined by the relation:

$$p = 1 - \frac{\rho_{\text{P-Si}}}{\rho_{\text{Si}}}, \qquad (3.4)$$

where $\rho_{\text{P-Si}}$ is the density of P-Si, ρ_{Si} is the density of c-Si. The parameter p is the ratio of the removed agent to its original amount (up to pore formation) and is usually expressed as a percentage. One of the main ways to estimate the value p is the gravimetric method. It is based on weighing the silicon wafers prior to pore formation and also after delamination from of the P-Si film from the substrate (Herino et al, 1987; Svechnikov et al, 1994). In the experiment, the porosity can be calculated by the formula:

$$p = \frac{m_1 - m_2}{m_1 - m_3}, \qquad (3.5)$$

where m_1 – mass of the c-Si substrate prior to the etching of the specimen, m_2 – mass of the substrate with the P-Si film grown on it, m_3 – the mass of the substrate without the sample (after delamination of the P-Si film).

The porosity of P-Si samples depends on the parameters such as the type of conductivity and the doping level of the initial silicon wafer. It also depends on the conditions under which the process of electrochemical etching takes place: HF concentration in the electrolyte, current density (Lehmann et al, 2000). Typically, the value of p increases with increasing concentration of HF in the electrolyte (Svechnikov, et al, 1994). Figure 3.8 presents the scanning electron micrographs of the P-Si/c-Si interface for samples of n- and p-type, respectively, which are characterized by different levels of doping of the c-Si substrate, as well as the densities of the etching current (Lehmann et al, 2000). Figure 3.8 shows a tendency of the growth of porosity p with increasing etching current density and the doping level of silicon for both types of samples. In these experiments, the samples of P-Si were produced using the electrolyte HF (50%) : $C_2H_5OH = 1:1$.

Porous silicon is classified in accordance with the IUPAC (International Union of Pure and Applied Chemistry), determining the type of porous material depending on the size of pores (Rouquerol et al, 1994) (Table 3.1).

It was found that the microporous silicon (mp-Si) with a porosity of 80% has the smallest dimensions of nanostructures (~ few nm). In addition, mp-Si with even greater porosity (90%) can form, but these layers are very fragile and can crumble on drying due to surface tension forces. Therefore,

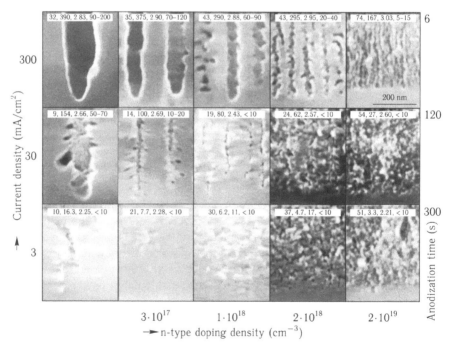

Fig. 3.8a. The scanning electron microscopy data for the c-Si/P-Si interface for n-type silicon with surface orientation (100) (Lehmann et al, 2000).

Table 3.1. Classification of P-Si on the basis of the size of pores (Rouquerol et al, 1994)

Type of porous material	Pore size
Microporous (nanoporous)	≤ 2 nm
Mesoporous	2–50 nm
Macroporous	> 50 nm

their formation requires a complicated procedure of supercritical drying (Canham et al, 1994; Cullis et al, 1997).

Studies of the microstructure of P-Si have shown that in the process of electrochemical etching of c-Si the formation of pores occurs mainly in the directions [100], see, e.g. (Cullis et al, 1997; Herino et al, 1987). This is due to the presence of the anisotropy of c-Si etching rates in HF for different crystallographic directions, as described in the following ratio: 15:10:1 = [100]:[110]:[111] (Cullis et al, 1997). This anisotropy of pore formation is well recorded in the layers on substrates of silicon with

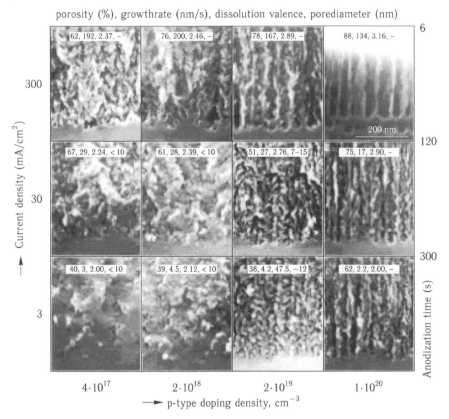

porosity (%), growthrate (nm/s), dissolution valence, porediameter (nm)

Fig. 3.8b. Scanning electron micrographs of the c-Si/P-Si interface for *p*-type silicon with surface orientation (100) (Lehmann et al, 2000).

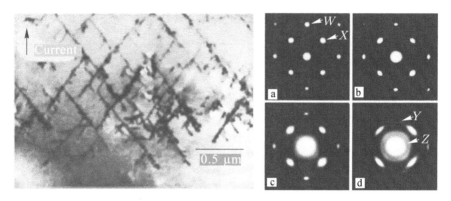

Fig. 3.9. TEM image of the cross-section of the macroporous Si layer formed on a (110)-oriented c-Si wafer (Cullis et al, 1997), showing growth of pores in the (110) direction (left) and the electron diffraction image (right) for samples with different porosity, increasing from *a* to *d* (Cullis, Canham, 1991).

heavily doped p-type conductivity (p^+-Si) or electron-type conductivity (n^--Si). In P-Si, produced on the lightly doped p-type silicon (p-Si), pore growth occurs with a greater degree of disorder. Figure 3.9 shows a snapshot of the cross-section of a macroporous layer of P-Si, produced by etching plates of c-Si (110), which indicates the preferential growth of pores along the [100] direction.

Images of electron diffraction patterns obtained in 'transmission' geometry indicate the retention of the crystal structure in P-Si (Canham et al, 1994). The electron diffraction patterns show the periodic location of specific spots. The smearing of spots in the picture indicates the increasing disorder of P-Si with an increase in its porosity (*p* increases from *a* to *d*). In all likelihood, this disordering is associated with some misorientation in Fig. 3.9 of various elements of the microstructure, namely silicon residues of nc-Si, which is obvious, given their nanometric dimensions.

However, within the limits of the individual nc-Si the ordering of atoms is apparently preserved, as in the crystal structure of c-Si. The conclusion on maintaining short-range of the arrangement of atoms in P-Si, which coincides with that in the crystal lattice of c-Si, also follows from the X-ray diffraction data (Cullis et al, 1997).

3.3. Luminescence of porous silicon

One of the interesting and most thoroughly investigated properties of P-Si is its efficient luminescence in the visible spectral region. As already noted, it was first discovered and described by Christopher Pickering et al (Pickering et al, 1984). Six years later, the British scientist Leigh Canham (Canham, 1990) reported effective PL of P-Si at room temperature, which was explained in terms of the quantum-confinement model.

Typically, the PL spectrum of P-Si is a broad (100–200 nm) band with a maximum range between 500 to 1000 nm, depending on the preparation conditions (Gardelis et al, 1991; Hamilton, 1995; Pickering et al, 1984). The position of the maximum of the luminescence spectrum depends on the porosity of the sample and, as a rule, is shifted to shorter wavelengths with increasing porosity (Canham, 1990; Gardelis et al, 1991; Hamilton, 1995; Pickering et al, 1984).

To explain the mechanisms of PL of P-Si, the analogy with the PL a-Si:H was proposed in (Pickering et al, 1984). Later, this view was held by a few more research groups (see review (Cullis et al, 1997)). Indeed, in the highly porous samples of P-Si a certain amount of the amorphous phase was fixed. But, apparently, the amorphous phase does not play a decisive role in the electronic properties of this material. The concept of the luminescence from the phase a-Si:H has not been generally accepted, because the largely crystalline structure of P-Si has been confirmed in a large number of studies.

Far more weighty arguments have been collected in favour of the explanation of the PL of P-Si by the change of the electronic energy spectrum of the material due to the quantum size effect (QSE) of nc-Si located in the porous layer nc-Si, which was proposed for the first time in (Canham, 1990). Indeed, this paper shows that with decreasing size of the nanostructure the shift of the PL band to shorter wavelengths takes place with an increase in its intensity (Fig. 3.10). The conclusion on the presence of QSE in highly porous P-Si also came as a result of the analysis of the process of formation of P-Si in another paper (Lehmann, Gössele, 1991). The hypothesis of the possibility of QSE in P-Si stimulated research studies, which both confirmed as well as challenged this hypothesis. Note that for all types of luminescent P-Si, obtained using electrochemical etching and the chemical etching method without passing electric current (*stain etching*), the presence of nc-Si with the size of the order of several nanometers (Jung et al, 1993) was recorded. This is an important argument for choosing an adequate model of the luminescence of this material. Consider the basic hypothesis known from the literature, explaining the photoluminescence of porous silicon obtained by electrochemical etching.

3.3.1. Models to explain the origin of the photoluminecence of porous silicon

Models of contributions of SiH_x *(x* = 1,2) *groups and amorphous hydrogenated silicon* (a-Si:H)

In the paper (Tsai et al, 1991) the authors compared the PL spectra and IR absorption in heating of P-Si, and found that at $T > 570$ K there is a sharp drop in PL intensity, which coincides with the desorption of hydrogen. It was concluded that a key role in the PL of P-Si is played by groups SiH_x. The authors of another study (Friedersdorf et al, 1992) also concluded that the possibility of PL is associated with the presence of SiH_x complexes,

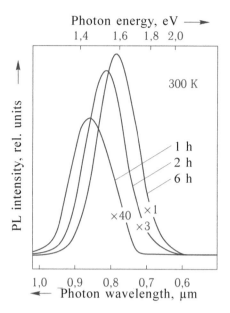

Fig. 3.10. The PL spectra of P-Si samples, obtained by electrochemical etching silicon wafers p-type after immersion in 40% aqueous solution of HF (Canham, 1990) for the times indicated.

as well as mechanical stresses in P-Si. Special experiments were carried out to clarify the question of whether hydrogen plays a key role in the possibility of effective photoluminescence. For example, in (Robinson et al, 1992) it was found that when heating P-Si its luminescence is quenched at lower temperatures than the desorption of hydrogen, namely, at $T \sim$ 550 K the PL disappears completely, but still remains around ~40% SiH_x groups. Moreover, the complete replacement of the hydrogen coating on the surface of P-Si nanostructures by the oxygen one increases the intensity of PL compared to the freshly prepared sample (Prokes et al, 1992; Xie et al, 1992).

In the study by Petrova-Koch et al. (1992), an analogy was drawn between a-Si:H and P-Si based on analysis of the characteristics of their photoluminescence. A number of studies by transmission electron microscopy (TEM) revealed the presence of an amorphous phase in the layers of P-Si and it was therefore concluded that the nature of PL is similar to that of a-Si:H (Jung et al, 1993). At the same time, in several studies of samples of P-Si the presence of an amorphous phase was not recorded but efficient photoluminescence was observed (Nishida et al, 1992).

Siloxenes (Si–O–H). Brandt and colleagues (Brandt et al, 1992) hypothesized about the siloxene nature of PL of P-Si. They found that the position of the maximum luminescence of P-Si can be changed as a result of purely chemical treatment (etching). Similarly, in the siloxenes the PL spectrum is varied by replacement of hydrogen in silicon rings by other ligands. Degradation processes of PL were observed in both P-Si and in Si–O–H. The relaxation time of PL in P-Si and siloxenes is several orders of magnitude larger than for c-Si. This result, according to the authors, is difficult to understand in the framework of the quantum size effect. Based on the similarity of the PL spectra, IR absorption and Raman scattering for siloxenes and P-Si, the authors concluded that the Si–O–H groups are responsible for luminescence. However, it is known that in complete oxidation P-Si, when the IR spectra indicate the absence of hydrogen (no Si–O–H groups), P-Si can luminesce efficiently (Yamada & Kondo, 1992). On the other hand, annealing of P-Si at 670 K completely quenches the PL, while the siloxenes continue to luminesce even after such treatment (Jung et al, 1993).

3.4. Quantum-size effect (QSE)

Currently, most researchers when explaining the origin of the PL of P-Si take into account the QSE. The QSE is considered most conveniently for model systems of reduced dimensionality, such as quantum wells, the filaments and dots for which the behaviour of the carriers will be radically different compared to the single crystal (Lutskii and Pinsker, 1983). Despite the fact that P-Si is a complex heterogeneous system which apparently involves nanostructures with different dimensionalities (Cullis et al, 1997), for analysis of the QSE in P-Si for simplicity we can consider a potential well with infinitely high walls (Bressler and Yassievich, 1993). Let d be the width of the quantum well for electrons or holes, then the allowed values of their quasi-momentum are determined from the following relationships well-known in quantum physics: $d = n\lambda/2$, $k = 2\pi/\lambda$, $p = \hbar k$, where λ is the de Broglie wavelength, k is the quasi-wave vector, p is the quasi-momentum, $\hbar$ is Planck's constant, $n = 1, 2,...$ is an integer. This implies the value of the quasi-momentum $p = \pi\hbar n/d$ and the additional kinetic energy $\Delta E_n = p^2/2m^*$, which can be written as:

$$\Delta E_n = \frac{\pi^2 \hbar^2 n^2}{2m^*} \frac{1}{d^2}, \tag{3.6}$$

where m^* is the effective mass of charge carriers in the crystal. The energy of the optical transitions of excited electrons and holes will be determined by the expression:

$$E = E_{g_0} + \Delta E_n, \tag{3.7}$$

where E_{g_0} is the band gap in the bulk material, ΔE_n is the quantum size correction, which in the simplest case is given by (3.6) and, generally speaking, consists of the sum of the QSE for electrons and holes.

Thus, the QSE is characterised by the shift in the optical absorption edge towards higher photon energies compared to the infinite crystal. It is this 'blue' shift that is registered in P-Si with respect to the crystalline silicon (Lockwood et al, 1994; Sagnes et al, 1993). The quantum-size effects should be manifested most markedly in the nc-Si, whose size is comparable to or smaller than the Bohr radius of the exciton in c-Si (~4 nm). These nanocrystals are present in the microporous P-Si (Cullis et al, 1997).

There are other, more accurate methods of accounting the QSE. In Buda et al. (1992) the calculations using the density functional theory in the approximation of local density show that the lowering of the symmetry of the object leads to the transformation of the energy spectrum of the original crystal. Calculations were made for quantum wires of square shape whose surface was passivated with hydrogen and the side length was d = 0.76, 1.14, 1.56 nm. This was followed by extrapolation for the filaments with the large size (Fig. 3.11). Despite the approximate nature of the calculation method, good agreement was obtained with the experimental data on the measurement of photoluminescence spectra and optical absorption in P-Si (Canham et al, 1990; Kamemitsu et al, 1993; Lee et al, 1993; Lehmann et al, 1993; Lockwood et al, 1993; Oswald et al, 1992; Sagnes et al, 1993; Lockwood et al, 1994).

Sanders and Chang (Sanders & Chang, 1992) conducted a more detailed numerical calculation of the band structure of P-Si in the approximation of the empirical tight-binding model with the two nearest neighbors taken into account. It was found that at small transverse filaments sizes $d < 3.1$ nm the oscillator strength of the radiative transition sharply increases and becomes comparable with the case of direct-gap semiconductors (e.g. GaAs).

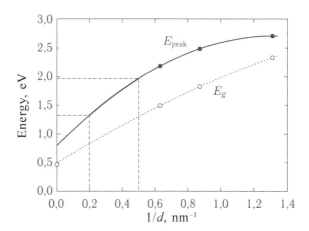

Fig. 3.11. Dependence on the forbidden band width (E_g) and the imaginary part of dielectric permeability (E_{peak}) on the reciprocal diameter of a silicon wire (Buda et al, 1992).

An important result of calculations carried out in Sanders & Chang (1992) was the confirmation of the possible existence of excitons in silicon nanostructures at room temperature. It was found that when the diameter of filaments changes from 3.0 to 1.5 nm the exciton binding energy increases from 60 to 140 meV. A similar conclusion about the possibility of existence in silicon quantum quantum wires with diameters of 2–4 nm of excitons with binding energies around 100 meV was obtained in the calculation in the effective mass approximation, taking into account the potential of image charges (Lisachenko, Timoshenko, 1999). Note that for c-Si the binding energy of the exciton is about 14 meV, which causes thermal dissociation of the excitons at 300 K. The high binding energy of the excitons in P-Si allows them to be relatively stable at room temperature and provide a significant contribution to PL. The exciton recombination mechanism in P-Si is confirmed by the experimental data obtained in several studies (Gardelis et al, 1991; Kovalev et al, 1999; Lisachenko et al, 2002), where the linear dependence of the intensity of luminescence on the excitation intensity was observed, which is different from the quadratic law, which is observed in the interband recombination of free electrons and holes in the lightly doped c-Si (Lisachenko et al, 2002). More detailed excitonic properties of PL of P-Si will be discussed in subsequent sections.

Experimental verification of QSE is the shift of the PL maximum (Canham, 1990) and of the absorption band of P-Si in the blue region with a decrease in the size of nanocrystals in etching of the sample in HF (Canham et al, 1990; Koyama et al, 1991; Voos et al, 1992). A similar effect of shifting the maximum of the luminescence was observed during the oxidation of (natural, thermal or chemical) P-Si, as described in the surveys (Prokes et al, 1992; Nakajima et al, 1992). In this case, the diameter of the filaments decreases due to formation of Si oxide on the surface of P-Si. At a suitable size of the nanostructures and low surface recombination rate such samples luminesce in the blue spectral region (Wang et al, 1993).

Nishida et al. (1992), using the high-resolution electron microscopy data, obtained the experimental dependence of the energy position of the PL peak of P-Si on the size of nc-Si, which coincides with that predicted in the QSE.

Thus, the models that explain the PL in P-Si on the basis of the QSE, give a good agreement between theory and experiment. However, not all laws of the PL of P-Si can be explained by the QSE. In particular, the rather long relaxation times of luminescence (from tens of microseconds to several milliseconds) may indicate the influence of capture processes of non-equilibrium charge carriers on surface states (Amato, 1994; Delerue et al, 1993). The full explanation of this property apparently requires the simultaneous consideration of radiative and non-radiative recombination channels, as well as the processes of capture and emission of carriers on the surface states. Kashkarov et al, (1998), attempted to take into account non-radiative recombination processes of free non-equilibrium charge carriers and the radiative recombination of the carriers bound in the excitons. The proposed phenomenological model was used to explain the intensity and the lifetime of the exciton PL of P-Si on temperature. An important factor that seems to affect the lifetime of the PL is the spin state of the excitons. Namely, in the state with total spin 0 the excitons are singlet and have relatively short lifetimes. In a state with spin 1 the excitons are triplet and their radiative lifetimes are in the millisecond range (Calcott et al, 1993; Kovalev et al, 1999). This question will be discussed in more detail in the next section.

It is important to mention another difficulty faced by the explanation of PL of P-Si by the QSE. In particular, it is well known that the position of the maximum of the PL bands of samples exposed to air for a sufficiently long time or specifically subjected to oxidation lies in the photon energy range 1.6–1.8 eV and practically does not depend on the size of nc-Si in the porous layer (Jung et al, 1993; Cullis et al, 1997). Such a stabilization of the spectral position of the PL line was explained in the work of Wolkin et al. (1999) by the capture of excitons on the surface Si = O bonds. The electron is localized on the p-orbitals of near the Si atoms, and the hole on the p-orbital in the vicinity of the O atom. As follows from *ab initio* calculations using the density functional theory, the energy of the trapped exciton is almost independent of the size of the nanocrystals when the size decreases to less than 2 nm (see Fig. 3.12a, zone III). At the same time, in the size range to 3 nm, the energy of trapped and free excitons did not differ and both were dependent on the size of nanocrystals (Fig. 3.12a, zone I). The calculated energies of the trapped exciton PL bands are in good agreement with the experimentally measured values (see Fig. 3.12b).

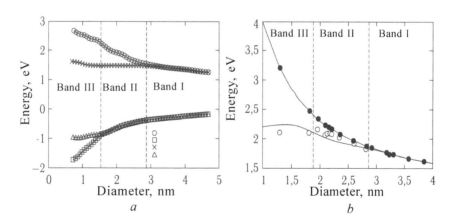

a   *b*

Fig. 3.12. *a* – energies of electronic states in silicon nanocrystals as a function of their size for the free excitons (circles and squares) and excitons bound on the Si = O (crosses and triangles), *b* – comparison between experimental (points) and theoretical (lines) dependences of the PL band on the size of the nanocrystals for models of their radiative recombination through the radiative annihilation of free (upper curve) and bound, on Si = O (lower curve), excitons (Wolkin et al, 1999).

With regard to luminescence of P-Si, arising under the influence of the applied voltage, it must be said that electroluminescence (EL) was observed for the first time during the anodic oxidation of porous layers, formed on the weakly doped P-Si, and was interpreted as the result of injection of charge carriers from the electrolyte into the P-Si layer (Halimaoui et al, 1991). Richter et al. (1991) reported the visible ($\lambda_{max} \sim 650$ nm) EL of the solid-state structure of Au/p$^+$-Si/n$^-$-Si/Au. The radiation was recorded by applying a large (~ 200 V) AC and DC voltage, and the EL properties of P-Si did not depend on the polarity of applied voltage. In the works mentioned the luminescence was short-lived and characterized by low quantum efficiency. Namavar et al. (1992) studied the structure of the type ITO/P-Si/P-Si/Al. The radiation in the visible spectrum ($\lambda_{max} \sim 580$ nm) was observed only at the forward bias, the voltage of several volts, and the continuous operating time was not less than 5 h. Note that the EL-structures based on P-Si may have a quantum efficiency of more than 1% and the efficiency of up to 3.7% (Gelloz et al, 2000). These values are higher than the literature values for other light-emitting silicon structures operating at room temperature (Daldosso, Pavesi, 2009).

3.5. Exciton effects in photoluminescence of porous silicon

With decreasing size of silicon nanocrystals, located in a medium with lower dielectric constant, for example, in vacuum or in air, the intensity of the Coulomb interaction between charge carriers increases. This can lead to changes in the parameters of the excitons. Initially, the existence of such an effect has been noted in theoretical studies (Babichenko et al, 1980; Rytova, 1967; Chaplik and Entin, 1971). Babichenko et al (1980) calculated the exciton binding energy E_{exc} and Bohr radii a_{exc} in semiconductor filaments surrounded by a medium with a lower dielectric constant. It has been shown that with a decrease in the diameter of the filament E_{exc} increases and a_{exc} decreases. Obviously, such effects should be observed in nanostructures with a network of silicon filaments, which are found in the samples of P-Si, formed in certain modes of electrochemical etching (see Fig. 3.8). We consider several aspects that relate to the exciton mechanism of PL in P-Si.

As follows from the theoretical work (Sanders & Chang, 1992) a decrease in the diameter of the silicon filaments d from 3.0 to 1.5 nm results is an increase of the binding energy of the excitons from 60 to 140 meV. The dependence of the exciton binding energy on the quantum wire diameter is well approximated by the expression: $E_{exc} \sim d^{-1.24}$. The calculated radiative recombination time of excitons varies from 60 ns to 170 μs for d is in the range from 1.5 to 3.0 nm.

Calculations by Sanders & Chang (1992) did not account for the presence of the influence of image charges, resulting in too low estimates of the magnitude of E_{exc}. Accounting for potential charges of images increases the energy values E_{exc}, which thus turns out to be in the range from 200 to 600 meV when changing d from 5 to 1.5 nm (Kashkarov et al, 1999). In this case, the dependence of the exciton binding energy on the quantum wire diameter is approximated another expression, namely: $E_{exc} \sim d^{-1.66}$. This indicates that in the presence of the potential charge images the interaction potential of charge carriers in the quantum wire is different from the Coulomb potential.

Large values of E_{exc} (about hundreds of meV) in silicon quantum filaments in vacuum, which greatly exceed the value $E_{exc} = 14$ meV for c-Si, ensure the stability of the excitons even at room temperature. Good passivation of non-radiative recombination centres at the surface of silicon filaments allows a significant part of the excitons to recombine radiatively (Fishman et al, 1993; Kashkarov et al, 1999, Bressler and Yassievich, 1993). The exciton mechanism of PL in P-Si nanostructures is also confirmed by the data on the adsorption of various molecules.

A huge number of works has been devoted to the analysis of the temperature dependence of photoluminescence of P-Si. The dependence was studied in a wide temperature range from 4 to 600 K (see, for example, Kashkarov et al, 1997; Narasimhan et al, 1993; Zheng et al, 1992). At lowering the temperature from room temperature to 150–100 K, there was a significant increase in PL intensity, which was replaced by a sharp fall with a further decrease of temperature to 4 K (Kashkarov et al, 1997; Zheng et al, 1992). Bayliss et al. (1994) noted the buildup of PL in the range from 80 to 300 K. These changes in the intensity of the PL signal were completely reversible: when the sample is heated to 300 K, the intensity returned to its original value. When increasing the temperature to 500 K the intensity of PL decreased, while at T of approximately 600 K it completely died out (Perry, 1992). Starting from 480 K, the change of intensity was irreversible. In explaining the temperature dependence of photoluminescence in the works (Bayliss et al, 1994; Perry, 1992; Zheng et al, 1992), it was only assumed that the excitons participate in the radiative recombination process, but the temperature dependence of the non-radiative recombination channel was not taken into account.

The temperature variations of the maximum of the PL spectrum were also mixed. For example, in Zheng et al. (1992) a decrease of temperature from 300 K to 4 K resulted in both 'blue' and 'red' shift of the PL peak for different samples of P-Si, whereas in the studies (Bayliss et al, 1994; Narasimhan et al, 1993; Perry, 1992) there was only a 'blue' shift of the PL spectrum. With increasing temperature to 570 K the PL maximum was shifted to longer wavelengths (Perry, 1992). To explain the results, in Zheng et al, (1992) the authors suggested the participation of phonons in the process of light emission, with the consideration that the phonon spectrum changes for different samples of P-Si. In other publications (Bayliss et al, 1994; Perry, 1992) it is indicated that the temperature dependence of the maximum of the PL spectrum of P-Si in the range from 4 to 300 K is well modelled by a temperature change of the band gap of silicon, and at $T > 300$ K there is a significant deviation from the case of c-Si, which in (Perry, 1992) is explained by the influence of the quantum size effect. Unfortunately, the authors did not explain why the above deviation is significant only at elevated temperatures.

Experiments showed the quenching of PL due to annealing in a vacuum of freshly prepared samples of P-Si at $T > 500$ K (Lavine et al, 1993; Tsai et al, 1992a). This treatment reduced the concentration of hydrogen groups on the surface, but no clear correspondence between the concentration of hydrogen and the intensity of photoluminescence was observed. Oxidation of P-Si thermal (Kumar et al, 1993; Petrova-Koch et al, 1992; Yamada & Kondo, 1992), chemical (Banerjee et al, 1994), electrochemical (Shin et al, 1993), air (Bao et al, 1994), in which almost all the hydrogen groups are replaced with oxygen, not only reduced the efficiency of the PL, but rather increased, and in addition, caused blue shift of the photoluminescence peak.

In addition to temperature, the PL intensity of P-Si can be significantly affected by the intensity of the exciting light I_e. In a vacuum, when I_e varies from 1 to 800 mW/cm^2 there was an almost linear increase in the PL, which slowed down when approaching a value of 1000 mW/cm^2 (Perry, 1992; Murayama et al, 1992) because of low thermal conductivity of P-Si. Deviations from linearity in the process of increasing the intensity of the exciting light in Murayama et al (1992) is explained by the increase in the contribution of non-radiative Auger recombination. The latter is unlikely, since the given process requires the third charge carried, and since P-Si is depleted by them (Polisski et al, 1999; Timoshenko et al, 2001), the probability of Auger recombination in P-Si must be less than in c-Si. In our opinion, the analysis (Murayama et al, 1992) does not take into account the thermal factor of the exciting radiation. The extremely low thermal conductivity of the microporous P-Si (about 0.25 10^{-3} W/(cm·deg) (Obraztsov et al, 1997)) may result in its heating and, consequently, a decrease in PL intensity.

3.6. The influence of molecular environment on the luminescent properties of P-Si

3.6.1. Effect of active molecules on the luminescence of P-Si

Porous silicon has a very developed surface with the value of specific surface area of up to 800 m^2/cm^3 (Bomchil et al, 1989), therefore, an important factor in its PL is the surface environment of silicon nanocrystals. The most complete and accurate analysis of the influence of the molecular environment of silicon nanocrystals on their PL was carried out in (Kashkarov et al, 1998; Kashkarov, etc. 1996; Kashkarov et al, 1997). From these studies the following main mechanisms of this effect can be defined:

- changes in the rate of radiative recombination of excitons in silicon nanocrystals, associated with defects on the surface of nanostructures;

- modification of the electronic properties of P-Si due to the dielectric screening of charges by polar molecules;
- establishment of the Coulomb centres associated with the adsorbed molecules.

Consider the basic properties of these manifestations and mechanisms of the effect of molecular–atomic coating on the emission characteristics of P-Si.

Effective PL of P-Si is possible if a significant suppression of the non-radiative recombination channel, which is associated with surface defects. Such defects can be silicon atoms with unsaturated or dangling bonds (DB) (Cullis et al, 1997). Freshly prepared P-Si is characterized by the nc-Si surface passivated with hydrogen (Cullis et al, 1997), which leads to a low rate of non-radiative recombination. Removal of hydrogen from the surface of nanostructures leads to an increase in the probability of radiationless processes. The quenching of PL due to the annealing in a vacuum of freshly prepared samples of porous silicon was observed in experiments (Lavine et al, 1993; Tsai et al, 1992a). This treatment reduced the concentrations of hydrogen groups on the surface, although no clear correspondence between the concentration of hydrogen and the intensity of photoluminescence was observed. Tsai et al (1992b) observed degradation of the PL of P-Si in air, which took place, starting from the intensity of excitation of a few mW/cm^2. A further increase in the intensity of the excitation increased the rate of degradation of the PL of P-Si. This effect can be explained by the photostimulated desorption of hydrogen and oxidation of the surface of nc-Si (registered by the IR spectra). The authors noted that the activation energy for thermal desorption of hydrogen is 1.86 eV and 2.82 eV for SiH$_2$ and SiH groups, respectively, which is close to the photon energy used in photoluminescence excitation radiation. In turn, Nishitani et al (1992) note that etching in HF of degraded P-Si samples does not lead to restoration of its PL, but at the same time, this can be achieved by exposure of P-Si in air or low-temperature (390–470 K) by annealing in a vacuum. The authors believe that the degradation of the PL is due to the photoinduced generation of centres of non-radiative recombination. Thus, in amorphous silicon the role of these centres is played by the DBs of silicon formed in the bulk of the sample. In the case of P-Si the DBs are likely to be localized near the surface of nc-Si, and actively interact with oxygen.

Zheng et al (1992) in studies of the degradation phenomena of PL in P-Si in a vacuum and in air (or oxygen) observed two possible mechanisms, namely, 1) Staebler–Wronski effect and 2) laser-stimulated oxidation. It should be noted that in addition to visible light, in the individual papers the authors also recorded infrared luminescence, which, according to the authors, originates from the silicon substrate. Note that in the presence of the DBs in P-Si there exists a non-zero probability of radiative recombination of non-equilibrium charge carriers in such defects, which may also contribute to the IR radiation. As regards the high intensity of the IR emission of light in P-Si compared to c-Si Perry (1992) linked it with a decrease in the concentration of defects at the P-Si/substrate interface which is a consequence of the anodizing process.

The effect of polar molecules on the PL of P-Si, which do not create stable chemical bonds on its surface, can be explained by the model of dielectric screening of charges in the silicon nanostructures (Konstantinova et al, 1996). The essence of this mechanism is to reduce the exciton binding energy or other related charge carriers in the area of the silicon nanocrystal in a medium with a dielectric constant much larger than 1. However, many initial studies, devoted to P-Si, contained only empirical evidence of a strong influence of polar molecules on the properties of silicon nanocrystals. For example, it was found that immersion of P-Si in a variety of alcohols (Friedersdorf et al, 1992; Lawerhaas & Sailor, 1993) or solutions organoamines (Coffer et al, 1993) led to quenching of the PL. A significant influence of water on chemically oxidized porous silicon samples, characterized by a hydrophilic surface, was also observed. Lawerhaas & Sailor (1993) found a significant (approximately 150 nm) blue shift of the PL spectrum of freshly prepared P-Si after placing the samples in deionized water for several hours. Experiments with Raman scattering show that the size of nanocrystals does not change. Therefore, the authors (Tamura et al, 1993) attributed the change of the PL spectra to the influence of surface defects. Ben-Chorin et al (1994) observed a reversible decrease in PL intensity, as well as an increase in the conductivity of P-Si after adsorption of methanol vapour. The adsorption of benzene has practically no effect on the PL and conductivity of P-Si (Kashkarov et al, 1996).

The influence of Coulomb centres on the PL of P-Si, apparently, is the destruction of the bound states of excitons by the electric field of such centres. The experiments revealed a strong quenching of the PL of P-Si and also the appearance of an EPR signal from radical anions upon adsorption tetracyanethylene which is a strong acceptor (Kashkarov et al, 1996). These data have only a qualitative explanation. A quantitative theory of the influence of the local Coulomb fields on the properties of silicon nanostructures is still lacking.

Consider the effect of adsorption of acceptor molecules of nitrogen dioxide on the PL of microporous P-Si samples, which have efficient photoluminescence at room temperature. Figure 3.13 shows the PL spectra of samples freshly prepared microporous silicon in a vacuum and in the NO_2 atmosphere at different pressures. The spectra are broad bands, which are superpositions of the contributions from the size distribution of nanocrystals (Kashkarov et al, 1998). With admission of NO_2 molecules into a cell with a sample PL quenching is observed in the entire pressure range studied. The intensity of photoluminescence (I_{PL}) in the atmosphere of nitrogen dioxide at $P_{NO_2} = 10$ Torr decreased by more than 20 times with respect to I_{PL} of freshly prepared samples.

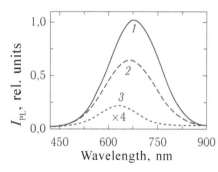

Fig. 3.13. PL spectra of microporous mc-Si in vacuum and at different pressures of NO_2: 1 – vacuum of 10^{-6} Torr, (freshly prepared samples), $2 - P_{NO_2} = 0.1$ Torr, $3 - P_{NO_2} = 10$ Torr.

In addition to the PL quenching, the adsorption of molecules of NO_2 is also accompanied by a shift of the maximum of the PL spectrum to shorter wavelengths. For example, for the nitrogen dioxide pressure of 10 Torr, a shift to shorter wavelengths with respect to its positions for the freshly prepared sample is $\Delta\lambda = 50$ nm (Fig. 3.13). In addition to PL quenching and the shift of the position of the maximum, the experiments also recorded the total narrowing of the photoluminescence lines (Fig. 3.13). In the admission of molecules NO_2 – subsequent evacuation cycle, the observed quenching of PL was irreversible.

As follows from the above mentioned results, the adsorption of NO_2 molecules causes significant changes in the PL of P-Si. Such changes may be related to variations in both radiative and non-radiative channels of recombination of charge carriers. According to literature data (Kashkarov et al, 1996), the latter is implemented through the defects on the surface of nanocrystals in P-Si layers.

According to the previously developed model of recombination processes in silicon nanocrystals (Kashkarov et al, 1996), the PL of P-Si is due to the radiative recombination of excitons in the size distribution of silicon nanocrystals. Calculations show (Lisachenko, Timoshenko, 1999) that such excitons for samples in vacuum or in air have large binding energies (of the order of hundreds of meV). The PL range contains contributions from nanocrystals of different sizes. The value of each contribution is determined by the probabilities of radiative and non-radiative recombination in the nanocrystal. Moreover, the adsorption of P-Si molecules can affect both recombination channels (Kashkarov et al, 1998).

Konstantinova et al (2004) have shown that the adsorption of NO_2 molecules is accompanied by the oxidation of the surface of the silicon nanocrystals, accompanied by the generation of defects – the silicon dangling bonds at the Si/SiO_2 interface. Moreover, the concentration of defects increases sharply with increasing NO_2 pressure, reaching a value of $3 \cdot 10^{19}$ cm^{-3} at a pressure of 10 Torr (Konstantinov et al, 2004). Thus, the decrease in I_{PL} of P-Si with increasing pressure of NO_2 (Fig. 3.13 and Fig. 3.14) can be explained by increasing the probability of non-radiative recombination (w_{nr}) due to defect formation.

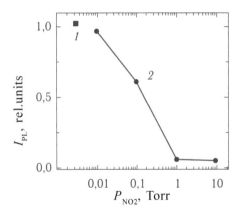

Fig. 3.14. The dependence of the measured spectrum at the maximum PL intensity of the mc-Si on NO_2 gas pressure: 1 – freshly prepared samples in vacuum of 10^{-6} Torr, 2 – NO_2 atmosphere (Konstantinov et al, 2004).

Another factor leading to the quenching of photoluminescence in an atmosphere of NO_2 is, we believe, the formation of Coulomb centres P_b^+– $(NO_2)^-$ on the surface of silicon nanocrystals. The excitons are destroyed in the electrical field of these centres. The mechanism of formation of these centres is as follows: the adsorption of a molecule near the defect results in capture of an electron with the silicon dangling bond in the molecule of NO_2 ($D^0 - NO_2 = D^+ - (NO_2)^-$, where D stands for the defect). The growth in the non-radiative recombination rate w_{nr} is most pronounced for nanocrystals with large cross-sections, luminescent in the long wavelength region of the spectrum, because the exciton binding energy decreases with increasing size of nanocrystals (Sanders & Chang, 1992). Thus, the shift λ_{max} to shorter wavelengths (Fig. 3.13) can be explained by the preferential decay of excitons in nanocrystals, luminescent at long wavelengths. Symmetrical quenching of PL at both short and long waves, which leads to a narrowing of the spectra, can be explained by the influence of electric fields of adsorption complexes (short-wave region) and low exciton binding energies (long wavelength region). Thus, the quenching and shift of the PL spectrum in the NO_2 atmosphere is caused by formation of charged adsorption complexes, as well as the concentration defects on the surface of silicon nanocrystals.

Consider the effect of adsorption of acceptor molecules of para-benzoquinon ($C_6H_4O_2$) on PL of P-Si. The PL spectrum of freshly prepared samples is a wide band with a maximum near 680 nm. As in the case of NO_2, the adsorption of $C_6H_4O_2$ molecules resulted in PL quenching of the sample. But unlike the adsorption of nitrogen dioxide, the shift of the maximum of the spectrum occurred in the region of large wavelengths (Fig. 3.15).

The decrease in PL intensity can be explained by the formation of donor–acceptor pairs ($C_6H_4O_2^- - P_b^+$) on the surface of silicon nanocrystals, which leads to the destruction of excitons by the localized electric fields of adsorption complexes. The shift of the maximum of the PL spectrum to higher wavelengths is due to the preferential destruction of excitons due

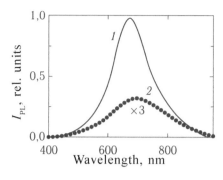

Fig. 3.15. PL spectra of samples of mc-Si: 1 – initial vacuum 10^{-6} Torr, 2 – in the atmosphere of $C_6H_4O_2$ at a pressure of $P_{C_6H_4O_2} = 0.1$ Torr (Konstantinova et al, 2004).

to a sharp increase in the electric field intensity in small nanocrystals, luminescing in the shortwave spectrum. The subsequent evacuation led to the partial reversibility of the PL signal. The maximum returned to its original position, which confirms the assumption of partial desorption of $C_6H_4O_2$ from the surface of silicon nanocrystals and is consistent with the data of EPR and IR spectroscopy (Konstantinova et al, 1996).

The effect of the adsorption of donor molecules of pyridine is shown in Fig. 3.16. The spectra of freshly prepared microporous P-Si samples at the adsorption of pyridine molecules at various pressures and subsequent evacuation are shown. The supply of C_5H_5N molecules on the P-Si sample resulted in a decrease of the photoluminescence intensity over the entire

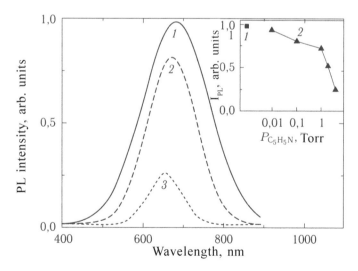

Fig. 3.16. The PL spectra of samples of P-Si in vacuum and at different pressures of C_5H_5N: 1 – initial sample in vacuum 10^{-6} Torr, $2 - P_{C_5H_5N} = 0.1$ Torr, $3 - P_{C_5H_5N} = 3.8$ Torr. The inset shows the dependence of the measured spectrum at the maximum PL intensity of P-Si on the gas pressure of C_5H_5N: 1 – initial value of I_{PL} in a vacuum of 10^{-6} Torr, 2 – in an atmosphere of C_5H_5N.

range of pressures used. For example, the PL intensity of P-Si in the C_5H_5N atmosphere at a pressure of its saturated vapor of the order of 3.8 Torr decreased approximately 4-fold compared with the freshly prepared sample. The moment of condensation of C_5H_5N molecules in the pores of the samples was monitored by increasing the reflectance of the beam of an He–Ne laser from the surface of P-Si.

For pyridine molecules in adsorption at low pressures ($P_{C_5H_5N} \leq 1$ Torr) the reduction of the PL intensity of P-Si is due to the destruction of excitons by localized electric fields of adsorption Coulomb centres $(C_5H_5N)^+$ on the surface of nc-Si (Kashkarov et al, 1996). This type of adsorption refers to the weak form of chemisorption, which is irreversible at room temperature. The formation of these complexes is possible due to the high affinity to the hole of the appropriate molecules (Wolkenstein, 1987). Reduction of I_{PL} at high pressures of the molecules of pyridine ($P_{C_5H_5N} > 1$ Torr) can be explained by the appearance of an additional mechanism for quenching of PL in addition to the formation of the above-mentioned complexes with charge transfer, namely, filling of the pores of samples with the dielectric liquid due to condensation of pyridine. As a result, the dielectric permeability of the medium increases and the value of E_{exc} decreases. This leads to an increase in the probability of non-radiative recombination of excitons and, consequently, to the quenching of photoluminescence (Lisachenko, Timoshenko, 1999).

In addition, the interaction of C_5H_5N molecules with the surface of P-Si, as in the case of NO_2, leads to a shift of the maximum of the PL spectrum to shorter wavelengths. In particular, the maximum of the PL spectrum in an atmosphere of pyridine at $P_{C_5H_5N} = 3.8$ Torr is shifted to shorter wavelengths by $\Delta\lambda = 20$ nm with respect to its position for the freshly prepared sample (Fig. 3.16). Most likely, this 'blue' shift of the PL spectrum in the C_5H_5N environment is due to the dielectric effect. The result is a decrease in E_{exc} for larger nanocrystals, which contribute to longer wavelengths.

The effect of the 'blue' shift of the maximum of the PL spectrum can also be related to the fact that the relaxation time of non-equilibrium charge carriers is reduced (and the effectiveness of non-radiative recombination, respectively, increases) mainly for the nanocrystals with large cross sections, fluorescent in the low-energy region of the PL spectrum (Kashkarov et al, 1998). Indeed, due to the low efficiency of the PL in the samples, the following relation for the time of radiative and radiativeless recombination is satisfied: $\tau_r > \tau_{nr}$. This ratio increases with the transition to large nanocrystals due to increase of τ_r (Kashkarov et al, 1997).

Note that the interaction of pyridine molecules with the surface of silicon nanocrystals is accompanied by the narrowing of the photoluminescence lines (Fig. 3.16, dependences *2* and *3*). This phenomenon can be explained as follows: the destructive effect of the fields of the generated Coulomb centres $(C_5H_5N)^+$ will be maximal for the nanocrystals with the minimal size, in which the electric field strength is the highest ($E \sim 1/r^2$). On the

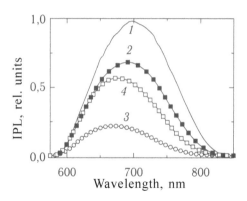

Fig. 3.17. The PL spectra of P-Si in vacuum 10^{-6} Torr – *1* and a curve at various pressures of ammonia: *2* – 1 Torr pressure of ammonia, *3* – 20 Torr pressure of ammonia, *4* – subsequent degassing.

other hand, the exciton binding energy decreases with increasing size of nanocrystals (Sanders, Chang, 1992). Thus, the symmetrical quenching of the entire spectral band of the PL is due, most likely, to the decrease in the concentration of excitons both in small nanocrystals (due to greater strength of the Coulomb field), and in large nanocrystals (due to the small binding energy of the excitons).

The results of studies of the effect of the adsorption of donor ammonia molecules on the PL of P-Si testify in favour of the above-mentioned mechanisms. Figure 3.17 shows PL spectra of freshly prepared samples of P-Si in vacuum and in an atmosphere of ammonia at different pressures. In the adsorption of ammonia quenching of the PL of the P-Si layers is observed over the entire range of pressures of the molecules. The intensity of the photoluminescence in the ammonia atmosphere at a pressure of 20 Torr decreased 4.5 times compared with the freshly prepared samples. Also, the maximum of the PL spectrum is shifted to shorter wavelengths with increasing pressure of the ammonia molecules. For the pressure of molecules of 20 Torr, the shift is equal to $\Delta\lambda = 20$ nm relative to the sample in a vacuum. In particular, it can be due to oxidation of the surface during adsorption. The subsequent evacuation of P-Si leads to incomplete recovery of the PL signal intensity, as well as a partial inverse shift of the spectrum. In addition to the above changes in the PL spectra of P-Si there was also narrowing of the photoluminescence lines with increasing pressure of the molecules. Thus, the adsorption of ammonia molecules affects the radiative and non-radiative channels of recombination of excitons in nc-Si, constituting the porous layer. In addition to quenching of PL of P-Si, the adsorption of ammonia is accompanied by a decrease of the lifetime of the PL, which also indicates an increase in the probability of non-radiative recombination of excitons in an ammonia atmosphere.

3.6.2. Luminescence of porous silicon with embedded dye molecules

Systems like P-Si may be of interest as a matrix for the molecules of dyes in order to obtain high efficiency of luminescence. There are have been studied in which the quantum efficiency of luminescence of the oxidized P-Si, impregnated with laser dye molecules, exceeded 1% (Canham, 1993). Thus, intensive PL was observed only in extensively oxidized samples so that it is possible to conclude that energy transfer from the dye to the crystallites takes place only in the case of fresh samples. Li et al. (1996) measured the quenching of photoluminescence at a wavelength of 575 nm for rhodamine B, with which porous silicon was saturated. An increase of the PL intensity of the dye–P-Si system with increasing annealing time of porous silicon was also recorded.

The luminescence of the dye is primarily dependent on its concentration: the relationship between the molecules varies with the distance D in accordance with the form of interaction, and can be written as D^{-n}, where n is associated with the interaction ($n \geqslant 6$). That is, the luminescence signal of the dye decreases with increasing concentration of the molecules.

Letant & Vial (1997) conducted experiments on photoluminescence of fresh and oxidized layers of P-Si, impregnated with the dye rhodamine 700 (LD 700). In irradiation of samples the absorption band of the dye showed luminescence of LD 700 with a maximum at 650 nm, and its intensity for the oxidized layer was higher than for the fresh samples. To determine the concentration of molecules of the embedded dye, the method of diffuse reflectance was used. Estimates of the number molecules, obtained by this method, were $5.4 \cdot 10^{24}$ m^{-3} for the oxidized layer, and $3 \cdot 10^{24}$ m^{-3} for the fresh layers. The found values agree quantitatively with the PL results. This means that the dye molecules penetrate easily into the porous matrix and dimerization of the dye was not observed. One of the possible explanations for the observed differences is that the oxidized layers of P-Si are hydrophilic, while the freshly prepared layers are hydrophobic. It is assumed that the hydrophilic surface of oxidized samples contains water molecules that contribute to dye penetration into the pores and, consequently, their luminescence, and vice versa, the hydrophobic surface of the fresh samples with a lower oxide prevents the penetration of the dye and contribute to the quenching of luminescence. In the excitation of samples the absorption band of porous silicon also showed the fluorescence of the dye at 625 nm, whose intensity was higher for fresh specimens than for oxidized ones, despite the fact that the dye molecules in the oxidized samples are larger than in the fresh ones. This indicates that the excitation of dye molecules in this case is the result of energy transfer from porous silicon crystallites to the dye molecules, and that the transfer may be more effective for the fresh samples than for the oxidized ones. This is explained by changing the distance D between the donors and acceptors of energy: as the interaction between the crystallites and the dye molecules can be expressed as D^{-n}, the energy transfer can be reduced with increasing D, which occurs during the oxidation of the sample. PL of the P-Si/dye system is more intense for the oxidized layers. This is good evidence for the energy transfer from the Si nanocrystallites to the dye molecules. Observation of the energy transfer to the dye (acceptor) from porous silicon (a donor) has a prerequisite for many applications, such as 'electroluminescence' of the dye, as a result of energy transfer from electroexcited nc-Si in the P-Si/ dye systems.

Note that the studies of P-Si with the introduced dye molecules are interesting in the light of the results obtained for systems such as porous titanium oxide with adsorbed dye molecules (Regan & Grätzel, 1991). These systems were used in the creation of a new type of solar cell – the so-called Grätzel cell whose efficiency reaches 10%, and the cost is significantly lower than that of standard solar cells based on crystalline silicon or A^3B^5 semiconductors. Note that although at present the use of P-Si for solar cells is limited to the use of its antireflection properties, a significant potential of this material, apparently, is the ability to fill the pores with different active light-absorbent materials such as dye molecules, quantum dots, etc.

3.7. Investigation of porous silicon by optical time-resolved spectroscopy

As follows from the above data, the optical absorption spectroscopy and photoluminescence yield important information about the structural and electronic properties of P-Si. More detailed studies of the electronic states and processes can be carried out using the methods of optical spectroscopy with temporal resolution. Consider first the data for absorption spectroscopy.

Klimov et al (1995) studied the differential transmission, i.e. the change in optical transmission $\Delta T/T_0$, where T_0 is the original transmission of P-Si layers, induced in them by the influence of intense laser pulses with a duration of 200 fs and an energy of light photons of 3.1 eV. Figure 3.18 shows the spectra of differential transmittance, measured for different delay times after the impact of the exciting laser pulse, which indicate the presence of resonance features in the spectrum of electronic states (denoted as the frequencies ω_1, ω_2, ω_3 in Fig. 3.18), while having different lifetimes. The kinetics of differential absorption, calculated from the curves $\Delta T/T_0$ (Fig. 3.19), indicate the presence of both fast (at times less than 10 ps), and slower processes. In this case, the ratio between the contributions of fast and slow processes depends on the energy of photons in the spectral region studied. The results were interpreted by the authors of the discussed study as a manifestation of the resonant processes of the photoinduced absorption of the electronic states of molecular-like complexes in the structure of P-Si. At the same time, the nature of these complexes remains unclear. As noted by the authors, their result is markedly different from their findings in a previous study, where there was a strong saturation of absorption in P-Si under the influence of pulses of light with the duration of 20 ps and photon energy of 2.33 eV (Klimov et al, 1994). In the latter case, the appearance of the absorption bands was interpreted as evidence of the filling of the levels

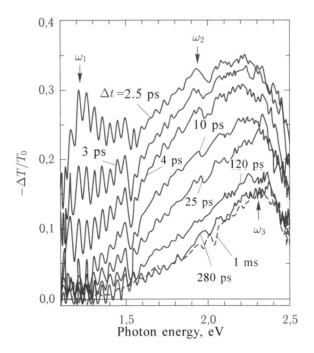

Fig. 3.18. Differential transmission spectra of the layers of P-Si, measured for different time delay of probing radiation (Klimov et al, 1995).

of the size quantization in nc-Si, present in P-Si. The difference between the results obtained in the two studies is explained by the authors of the work (Klimov et al, 1995) by the difference in the photon energies of the pumping radiation that could lead to the excitation of states of different nature in P-Si. All this indicates a rather complex composite character of P-Si in which the properties of the solid-state nanostructures and the molecular-type systems can be combined

Smith et al (2005), published 10 years after the articles by V. Klimov, et al, discussed above, studied the superfast kinetics of PL of silicon nanoparticles with the sizes of 1.0 and 2.85 nm, isolated from layers of P-Si. It was found that for smaller nanoparticles there has been a rapid decay in the kinetics of photoluminescence with a time of 1 ns (Fig. 3.20, approximation curve *1*). At the same time, for the larger nanoparticles, PL decay kinetics were recorded with substantially long times (Fig. 3.20, approximation curve *2*). The resulting time difference is explained by the authors by the difference in the nature of optical transitions in the investigated nanoparticles. In particular, it was assumed that the radiative transitions in silicon nanoparticles with dimensions of 1.0 nm is due to processes in molecular-like clusters of types similar to Si_{29} (see Fig. 4.1 in chapter 4), in which the force of the oscillator of the optical transition

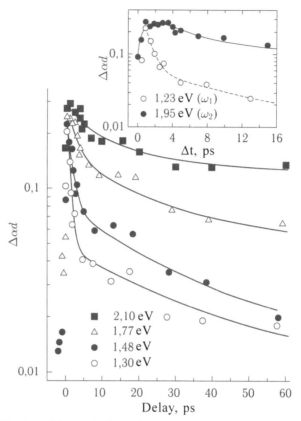

Fig. 3.19. Kinetics of induced absorption of layers of P-Si, measured for different energies of light quanta (Klimov et al, 1995).

is comparable to that of direct-gap semiconductors. It is assumed that the exciton is localized on the reconstructed Si–Si bonds on the surface of the cluster.

The high degree of localization causes a large oscillator strength of transition and high photoluminescence quantum yield, whose value was estimated by the authors as 50%. For larger nanoparticles, along with the fast component of the PL, the intensity slowly declined, which can be attributed to the radiative processes in the volume of nanoparticles (Smith A. et al, 2005), the probability of which was calculated by methods of strong coupling in Allan et al, (1996). Note that the exciton lifetimes calculated in the last work were 1–10 µs, which is close to the experimentally measured values of the lifetimes of PL in P-Si (Cullis et al, 1997; Kovalev et al, 1999; Bisi et al, 2000).

In discussing the data on measurements of the PL in P-Si with time resolution, we must mention the numerous experimental data on the measurement of the kinetics in the time range of more than 1 ns (Cullis et

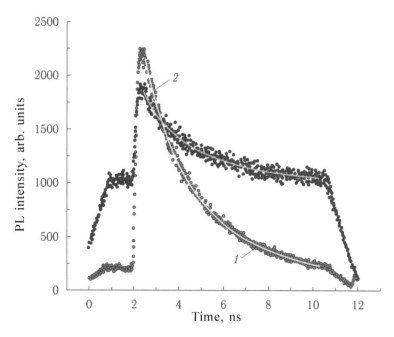

Fig. 3.20. Kinetics of the photoluminescence of silicon nanoparticles with the size of 1.0 nm (experimental points – dark symbols, approximation – curve *1*) and 2.850 nm (the experimental points – black symbols, approximation – curve *2*), isolated from layers of P-Si, after excitation at two-photon absorption of light with a wavelength of 780 nm and a pulse duration of 150 fs (Smith et al, 2005).

al, 1997; Kovalev et al, 1999; Bisi et al, 2000). It has been well established that the PL lifetime depends on the photon energy, and decreases with increase in the latter. In this case, the kinetics of decay of the intensity of PL is generally non-exponential. In some cases, such kinetics can be efficiently approximated by the so-called law of the stretched exponential decay:

$$I_{\mathrm{PL}}\left(t\right) = I_0 \exp\left(-\left(\frac{t}{\tau}\right)^{\beta}\right), \tag{3.8}$$

where τ is the average lifetime, and the parameter $0 < \beta < 1$ describes the extent of non-exponential (stretched) nature. The experimentally observed non-exponential kinetics of the PL in P-Si is due to energy relaxation processes, which include its transfer to a system of coupled nc-Si. In some cases, $\beta \approx 1$ and the non-exponential PL decay kinetics is neglected.

At room temperature the photoluminescence lifetimes of luminescent samples of P-Si are in the range from 1 to 100 μs for a range of energies of light quanta from 2 to 1.2 eV. These times can be attributed to radiative annihilation of excitons in indirect-band silicon nanocrystals (Kovalev et al, 1999). At the same time, the participation of phonons is required for satisfying the quasi-momentum conservation laws resulting in slower times with decreasing photon energy, i.e. an increase in the size of nc-Si. At the same time, for high-energy photons there is a significant contribution of photonless processes, due to a violation of selection rules for small nc-Si. Lowering the temperature leads to lengthening of the lifetimes of PL of P-Si, which reaches the values of 3–10 ms at liquid helium temperatures. Such a significant slowdown in the lifetime is well explained by a model of singlet–triplet splitting of the exciton states, which suggests a transition of the excitons at low temperatures in the long-lived triplet states, the energy of which is separated from the singlet states by the energy of exchange interaction of 1–20 meV, depending on the size of nc-Si. With decreasing size of the nanocrystals the exchange interaction energy increases, which can be understood by considering the increase in the degree of localization of the exciton, causing strengthening of the spin-spin interaction between the electron and the hole (Kovalev et al, 1999). Thus, in the small nanocrystals the exciton becomes similar to that in molecular crystals, and the process of PL can be considered as for molecular-like formations. Therefore, the description of the photoluminescence properties of P-Si, composed of small nanocrystals or silicon nanoclusters, can be carried out both from the viewpoint of solid-state physics, taking into account the quantum size effect and excitonic processes, and taking into account the bound excitons in molecular complexes with a strong localization of the excitation. Apparently, both approaches are complementary and necessary for a complete description of the properties of such a complex object as porous silicon.

References for Chapter 3

Babichenko V. S., Keldysh L. V., Silin A.P. Coulomb interaction in thin semiconductors and polymetallic filaments. Fiz. Tverd. Tela. 1980. V. 22. P. 1238-1240.

Bonch-Bruevich S. G. Kalashnikov S. G. Physics of Semiconductors. Moscow: Nauka, 1990. 688.

Bressler M. S. Yassievich I. I. Physical properties and photoluminescence of porous silicon. Fiz. Tekh. Poluprovod. 1993. V. 27, No. 5. P. 871-883.

Wolkenstein F. F. Electronic processes on semiconductor surfaces during chemisorption. Moscow: Nauka, 1987. 468 p.

Kashkarov P.K., Kamenev B. V., Konstantinova, E. A., Efimov E. I., Pavlikov A. V., Timoshenko V. Yu. Dynamics of non-equilibrium charge carriers in silicon quantum wires. Usp. Fiz. Nauk. 1998. V. 168, No. 5. P. 577-582.

Kashkarov P. K., Konstantinova E. A., Petrova S. A., Timoshenko V. Yu., Yunovich A. E. On the question of the temperature dependence of the photoluminescence of porous silicon. Fiz. Tekh. Poluprovod. 1997. V. 31, No. 6. P. 745-748.

Kashkarov P. K., Konstantinova E. A., Timoshenko V. Yu. Mechanisms of the effect of adsorption of molecules on the recombination processes in porous silicon. Fiz. Tekh. Poluprovod. 1996. V. 30, No. 8. P. 1479-1490.

Konstantinova, E. A., et al. The interaction of nitrogen dioxide molecules with the surface of silicon nanocrystals in porous silicon layer. Zh. Eksp. Teor. Fiz. 2004. V. 126, No. 4. P. 857-865.

Labunov V. A., Bondarenko V. P., Borisenko V. E. Properties and application of porous silicon. Zarubezh. Elektron. Tekh. 1978. No. 15. P. 3-27.

Lisachenko M. G., et al. Features of recombination of non-equilibrium charge carriers in porous silicon samples with different morphology of the nanostructures. Fiz. Tekh. Poluprovod. 2002. V. 36, No. 3. P. 344-348.

Lisachenko M. G., Timoshenko V. Yu. The influence of the dielectric environment on the exciton range of silicon quantum wires. Vestn. Mosk. Univ. Ser. Fizika. 1999. No. 5. P. 30-33.

Lutskii V. M., Pinsker T. N. Dimensional quantization. Moscow: Nauka, 1983. 56.

Obraztsov A. N. , Okushi H., Watanabe, H., Timoshenko V. Yu. Photoacoustic spectroscopy of porous silicon. Fiz. Tekh. Poluprovod. 1997. V. 31. P. 629-631.

Rytova N. S. Screened potential of a point charge in a thin film. Vestn. Mosk. Univ. Ser. Fizika. Astronomiya. 1967. No. 3. P. 30-37.

Svechnikov S. V., Savchenko A. V., Sukach G. A., Yevstigneev A. M., Kaganovich E. B. Light-emitting porous layers: Preparation, Properties and Applications. Optoel. i p/p Tekhnika. 1994. V. 27. P. 3-29.

Timoshenko V. Yu., et al. Silicon nanocrystals as photosensitizers of active oxygen for biomedical applications. Pis'ma Zh. Eksp. Fiz. 2006. V. 83. P. 492-95.

Chaplik A. V., Entin M.V. Charged impurities in very thin layers. Pis'ma Zh. Eksp. Fiz. 1971. V. 61. P. 2496-2503.

Allan G., Delarue C., Lanoo M. Nature of luminescent surface states of semiconductor nanocrystallites. Phys. Rev. Lett. 1996. V. 76. P. 2961–2964.

Amato G. A model for carrier de-excitation in light emitting porous silicon. Sol. St. Comm. 1994. V. 89, No.3. P. 213–217.

Archer R. J. J. Stain films on silicon. Phys. Chem. Solids. 1960. V. 14. P. 104–110.

Banerjee S., Narasimhan K. L., Sardesai A. Role of hydrogen and oxygen-terminated surfaces in the luminescence of porous silicon. Phys. Rev. B. 1994. V. 49, No.4. P. 2915–2918.

Bao X.-M., Wu X.-W., Zheng X.-Q., Yan F. Photoluminescence spectrum shifts of porous Si by spontaneous oxidation. Phys. Stat. Sol. (a). 1994. V. 141. P. K63–K66.

Bayliss S. C., Hutt D.A., Zhang Q., Danson N., Smith A. Local structure of porous silicon. Sol. St. Comm. 1994. V. 91, No.5. P. 371–375.

Beale M. I. J., Chew N.G., Uren M. J., Cullis A.G., Benjamin J.D. Microstructure and formation mechanism of porous silicon. Appl. Phys. Lett. 1985. V. 46, No.1. P. 86–88.

Ben-Chorin M., Kux A., Schechter I. Adsorbate effects on PL and electrical conductivity of porous silicon. Appl. Phys. Lett. 1994. V. 64, No.4. P. 481–483.

Bomchil G., Halimaoui A., Herino H. Porous Silicon: the Material and Its Applications in Silicon-on-Insulator Technologies. Appl. Surf. Science. 1989. V. 41/42. P. 604–611.

Brandt M. S., Fuchs H.D., Stutzmann M., Weber J., Cardona M. The origin of visible luminescence from porous silicon: a new interpretation. Sol. St. Comm. 1992. V. 81, No.4. P. 307.

Buda F., Kohanoff J., Parrinello M. Optical properties of porous silicon: a first-principles study. Phys. Rev. Lett. 1992. V. 69, No.8. P. 1272–1275.

Calcott P.D. J., Nash K. J., Canham L. T., Kane M. J., Brumhead D. Identification of radiative transitions in highly porous silicon. Journal of Physics: Condensed Matter. 1993. V. 5, No.7. P. L91.

Calcott P.D. J., Nash K. J., Canham L. T., Kane M. J., Brumhead D. Spectroscopic identifica-

tion of the luminescence mechanism of highly porous silicon. Journal of Luminescence. 1993. V. 57. P. 257–269.

Canham L. T., Cullis A.G., Pickering C., Dosser O.D., Cox D. I., Lynch T. P. Luminescent anodized silicon aerocrystal networks prepared by supercritical drying. Nature.
1994. V. 368. P. 133–135.

Canham L. T. Laser dye impregnated of oxidized porous silicon on silicon wafers. Appl.
Phys. Lett. 1993. V. 63, No.3. P. 337–339.

Canham L. T. Silicon Quantum Wire Array Fabrication by Electrochemical and Chemical
Dissolution of Wafers. Appl. Phys. Lett. 1990. V. 57, No.10. P. 1046–1048.

Chiesa M., Amato G., Boarino L., Garrone E., Geobaldo F., Giamello E. Effect of ammonia
adsorption on porous silicon surface. Angew. Chem. 2003. V. 42. P. 5031.

Coffer J. L., Lilley S. C., Martin R.A. Surface reactivity of luminescent porous silicon. J.
Appl. Phys. 1993. V. 74, No.3. P. 2094–2096.

Cullis A.G., Canham L. T., Calcott P.D. J. The structural and luminescence properties of
porous silicon. Appl. Phys. Lett. 1997. V. 82. P. 909–965.

Daldosso N., Pavesi L. Nanosilicon photonics. Laser & Photon. Rev. 2009. V. 3. P. 508–534.

Delerue C., Allan G., Lannoo M. Theoretical aspects of the luminescence of porous silicon //
Phys.Rev. B. 1993. V. 48. P. 11024–11036.

Dittrich Th., Rauscher S., Timoshenko V. Yu., Rappich J., Sieber I., Flietner H., Leverenz H.
J. Ultrathin Luminescent Nanoporous Silicon on n-Si: pH-dependent preparation in
aqueous NH_4F solutions. Appl. Phys. Lett. 1995. V. 67, No.8. P. 1134–1136.

Fishman G., Mihalcescu I., Romestein R. Effective-mass approximation and statistical description of luminescence line shape in porous silicon. Phys. Rev. B. 1993. V. 48,
No.3. P. 1464–1467.

Friedersdorf L. E., Searson P. C., Prokes S.M., Glembocki O. J., Macaulay J.M. Influence of
stress on the photoluminescence of porous silicon structures. Appl. Phys. Lett. 1992.
V. 60, No.18. P. 2285–2287.

Gardelis S., Rimmer J. S., Danson P., Hamilton B., Parker E.N. C. Evidence for quantum
confinement in the photoluminescence of porous Si and SiGe. Appl. Phys. Lett.
1991. V. 59, No.17. P. 2118–2120.

Gelloz B., Koshida N. Electroluminescence with high and stable quantum efficiency and
low threshold voltage from anodically oxidized thin porous silicon diode. J. Appl.
Phys. 2000. V. 88. P. 4319.

Halimaoui A., Oules C., Bomhill G., Bsiesy A., Gaspard F., Herino R., Ligeon M., Muller F.
Electroluminescence in the visible range during anodic oxidation of porous silicon films.
Appl. Phys. Lett. 1991. V. 59, No.3. P. 304–306.

Hamilton B. Topical review: Porous silicon. Semicond. Sci., Technol. 1995. V. 10. P. 1187–
1207.

Herino R., Bomchil G., Baria K., Bertrand C., Ginoux J. L. Porosity and pore size distribution of porous silicon layers. J. Electrochem. Soc. 1987. V. 134. — P. 1994–2000.

Jung K.H., Shih S., Kwong D. L. Developments in luminescent porous Si. J. Electrochem.
Soc. 1993. V. 140, No.10. P. 3016–3064.

Kamemitsu Y., Uto H., Masumoto Y., Matsumoto T., Futagi T., Mimura H. Microstructure
and optical properties of free-standing porous silicon films: size dependence of absorption spectra in Si nanometer-size crystallites. Phys. Rev. B. 1993. V. 48, No.4.
P. 2827–2830.

Kashkarov P.K., Konstantinova E.A., Efimova E.A., Kamenev B.V., Lisachenko M.G., Pavlikov A.V., Timoshenko V. Yu. Carrier Recombination in Si Quantum Wires Surrounded by Dielectric Medium. Physics of Low-Dimensional Structures. 1999. V.
3/4. P. 191–202.

Kashkarov P.K., Konstantinova E.A., Pavlikov A.V., Timoshenko V. Yu. Influence of ambient dielectric properties on the luminescence in quantum wires of porous silicon. Physics of Low-Dimensional Structures. 1997. V. 1/2. P. 123–130.

Kimerling L. C., Kolenbrander K.D., Michel J., Palm J. Light Emission from Silicon. Solid State Phys. 1997. V. 50. P. 333.

Klimov V., McBranch D., Karavanskii V. Strong optical nonlinearities in porous silicon: Femtosecond nonlinear transmission study. Phys. Rev. B. 1995. V. 52. P. R16989–R16992.

Klimov V. I., Dneprovskii V. S., Karavanskii V.A. Nonlinear transmission spectra of porous silicon: Manifestation of size quantization. Appl. Phys. Lett. 1994. V. 64. P. 2691–2693.

Konstantinova E.A., Dittrich Th., Timoshenko V. Yu., Kashkarov P.K. Adsorption induced modification of spin and recombination centers in porous silicon. Thin Solid Films. 1996. V. 276. P. 265–267.

Kovalev D., Gross E., Kunzner N., Koch F., Timoshenko V. Yu., Fujii M. Resonant electronic energy transfer from excitons confined in silicon nanocrystals to oxygen molecules. Phys. Rev. Lett. 2002. V. 89. P. 137401 (4 p.).

Kovalev D., Timoshenko V. Yu., Gross E., Kunzner N., Koch F. Strong explosive interaction of hydrogenated porous silicon with oxygen at cryogenic temperatures. Phys. Rev. Lett. 2001. V. 87. P. 068301 (4 p.).

Koyama H., Araki M., Yamamoto Y., Koshida N. Visible photoluminescence of porous silicon and its related optical properties. Jpn. J. of Appl. Phys. 1991. V. 30, No.12B. P. 3006–3609.

Kumar R., Kitoh Y. Hara K.. Effect of surface treatment on visible luminescence of porous silicon: correlation with hydrogen and oxygen terminators. Appl. Phys. Lett. 1993. V. 63, No.22. P. 3032–3034.

Lavine J.M., Sawan S. P., Shieh Y. T., Bellezza A. J. Role of Si–H and Si–Hx in the photoluminescence of porous Si. Appl. Phys.Lett. 1993. V. 62, No.10. P. 1099–1101.

Lawerhaas J.M., Sailor M. J. Chemical modification of the photoluminescence quenching of porous silicon. Science. 1993. V. 261. P. 1567–1568.

Lee H.-J., Seo Y.H., Oh D.-H., Nahm K. S., Suh E.-K., Lee Y.H., Lee H. J., Hwang Y.G., Park K.-H., Chang S.-H., Lee E.H. Correlation of optical and structural properties of light emitting porous silicon. Appl. Phys. Lett. 1993. V. 62, No.8. P. 855–857.

Lehmann V., Gössele U. Porous silicon formation: A quantum wire effect. Appl. Phys. Lett. 1991. V. 58. P. 856–858.

Lehmann V., Jobst B., Muschik T., Kux A., Petrova-Koch V. Correlation between optical properties and crystallite size in porous silicon. Jpn. J. Appl. Phys. 1993. V. 32. P. 2095–2099.

Lehmann V., Stengl R., Luigart A. On the morphology and the electrochemical formation mechanism of mesoporous silicon. Materials Science and Engineering. 2000. V. B69–70, No.11–12. P. 11–22.

Letant S., Vial J. C. Energy transfer in dye impregnated porous silicon. J. Appl. Phys. 1997. V. 82, No.1. P. 397–401.

Li P., Li Q., Ma Y., Fang R. Photoluminescence and its decay of the dye/porous silicon composite system. J. Appl. Phys. 1996. V. 80, No.1. P. 490–493.

Lockwood D. J. Light Emission from Silicon. Boston: Academic, 1997.

Lockwood D. J., Aers G. C., Allard L.B., Bryskiewicz B., Charbonneau S., Houghton D. C., Mc- Caffrey J. P., Wang A. Optical properties of porous silicon. Can. J. Phys. 1993. V. 70. P. 1184–1193.

Lockwood D. J., Wang A., Bryskiewicz B. Optical absorption evidence for quantum confinement effects in porous silicon. Sol. St. Comm. 1994. V. 89, No.7. P. 587.

Murayama K., Miyazaki S., Hirose M. Visible photoluminescence from porous silicon. Jpn. J. Appl. Phys. 1992. V. 31. Part 2, No.9B. P. L1358–L1361.

Nakajima A., Itakura T., Watanabe S., Nakayama N. Photoluminescence of porous silicon, oxidized then deoxidized chemically. Appl. Phys. Lett. 1992. V. 61, No.1. P. 46–48.

Namavar F., Maruska H., Kalkhoran N.M. Visible electroluminescence from porous silicon heterojunction diodes. Appl. Phys. Lett. 1992. V. 60, No.20. P. 2514–2516.

Narasimhan K. L., Banerjee S., Srivastava A.K., Sardesai A. Anomalous temperature dependence of photoluminescence in porous silicon. Appl. Phys. Lett. 1993. V. 62, No.4. P. 331–333.

Nishida A., Nakagawa K., Kakibayashi H., Shimada T. Microstructure of visible light emitting porous silicon. Jpn. J. Appl. Phys. 1992. V. 31. Part 2, No.9A. P. L1219.

Nishitani H., Nakata H., Fujiwara Y., Ohyama T. Light-induced degradation and recovery of visible photoluminescence in porous silicon. Jpn. J. Appl. Phys. 1992. V. 31. Part 2, No.11B. P. L1557–L1579.

Oswald J., Pastrnak J., Hospodkova A., Pangrac J. Temperature behavior of luminescence of free-standing porous silicon. Appl. Phys. Lett. 1992. V. 60, No.8. P. 986–988.

Perry C.H. Photoluminescence spectra from porous silicon (111) microstructures: temperature and magnetic-field effects. Appl. Phys. Lett. 1992. V. 60, No.25. P. 3117–3119.

Petrova-Koch V., Muschik T., Kux A., Meyer B.K., Koch F., Lehmann V. Rapid-thermal-oxidized porous silicon — the superior photoluminescent Si. Appl. Phys. Lett. 1992. V. 61, No.8. P. 943–945.

Pickering C., Beale M. I. J., Robbins D. J., Pearson P. J., Greet R. Optical studies of the structure of porous silicon films formed in p-type degenerate and non-degenerate silicon. J. Phys. C: Sol. St. Phys. 1984. V. 17, No.10. P. 6535–6552.

Polisski G., Kovalev D., Dollinger G.G., Sulima T., Koch F., Physica B. Where have all the carriers gone?. Physics of Condensed Matter. 1999. V. 273. P. 951–954.

Prokes M. S., Glembocki O. J., Bermudex V.M., Kaplan R., Friedersdorf L. E., Searson P. C. SiHx excitation: An alternate mechanism for porous Si photoluminescence. Phys. Rev. B. 1992. V. 45. P. 13788–13791.

Regan B., Grätzel M. A low-cost, high-efficiency solar cell based on dye-sensitized colloidal TiO_2 films. Nature. 1991. V. 353. P. 737–740.

Richter A., Steiner P., Kozlowski F., Lang W. Current-induced light emission from a porous silicon device. IEEE Electr. Dev. Lett. 1991. V. 12. P. 691–692.

Robinson M.B., Dillon A. C., Haynes D. R., George S.M. Effect of thermal annealing and surface coverage on porous silicon photoluminescence. Appl. Phys. Lett. 1992. V. 61, No.12. P. 1414–1416.

Rocchia M., Garrone E., Geobaldo F., Boarino L., Sailor M. J. Sensing CO_2 in a chemically modified porous silicon film. Phys. Stat. Sol. (a). 2003. V. 197, No.2. P. 365–369.

Rouquerol J., Avnir D., Fairbridge C.W., Everett D.H., Haynes J.H., Pernicone N., Ram-say J.D. F., Sing K. S.W., Unger K.K. Recommendations for the characterization of porous solids. Pure Appl.Chem. 1994. V. 66. P. 1739–1758.

Sagnes I., Halimaoui A., Vincent G., Badoz P.A. Optical absorption evidence of a quantum size effect in porous silicon. Appl. Phys. Lett. 1993. V. 62, No.10. P. 1155–1157.

Salonen J., Lehto V.-P. Fabrication and chemical surface modification of mesoporous silicon for biomedical applications. Chemical Engineering Journal. 2008. V. 137. P. 162–172.

Sanders G.D., Chang Y.-C. Theory of optical properties of quantum wires in porous silicon Phys. Rev. B. 1992. V. 45. P. 9202–9213.

Schmidt M., Heitmann J., Scholz R., Timoshenko V. Yu., Lisachenko M.G., Zacharias M. Er doping of ordered size controlled Si nanocrystals. Advanced Luminescence Materials and Quantum Confinement II, Electrochemical Society Proceedings. 2003. V. 2000–9. P. 83–92.

Shin S., Jung K.H., Yan J., Kwong D. L., Kovar M., White J.M., George T., Kirn S. Photoinduced luminescence enhanced from anodically oxidized porous Si. Appl. Phys. Lett. 1993. V. 63, No.24. P. 3306–3308.

Smith A., Yamani Z.H., Roberts N., Turner J., Habbal S. R., Granick S., Nayfeh M.H. Observation of strong direct-like oscillator strength in the photoluminescence of Si nanoparticles. Phys. Rev. B. 2005. V. 72. P. 205307.

Smith R. L., Collins S.D. Porous silicon formation mechanisms. J. Appl. Phys. 1992. V. 71, No.8. P. R1–R22.

Tamura T., Takazawa A., Yamada M. Blue shifts in the photoluminescence of porous Si by immersion in deionized water. Jpn. J. Appl. Phys. 1993. V. 32. Part 2, No.3A. P. L322–L325.

Timoshenko V. Yu., Dittrich Th., Lysenko V., Lisachenko M.G., Koch F. Free carriers in mesoporous silicon. Phys. Rev. B. 2001. V. 64. P. 085314 (2 p.).

Timoshenko V. Yu., Gonchar K.A., Maslova N. E., Taurbayev Ye. T., Taurbayev T. I. Electrochemical nanostructuring of semiconductor wafers by capillary-force-assisted method. International Journal of Nanoscience. 2010. V. 9, No.3. P. 139–143.

Tsai C., Li K.H., Campbell J. C., Hance B. V., White J.M. Laser-induced degradation of the photoluminescence intensity of porous silicon. J. Electr. Mater. 1992. V. 21, No.10. P. 589–591.

Tsai C., Li K.-H., Sarathy J., Shih S., Campbell J. C., Hanse B.K., White J.M. Thermal treatment studies of the photoluminescence intensity of porous silicon. Appl. Phys. Lett. 1991. V. 59, No.22. P. 2814–2816.

Tsai C., Li K.-H., Kinoski D. S., Qian R.-Z., Hsu T.-C., Irby J. T., Banerjee S.K., Tasch A. F., Campbell J. C., Hance B.K., White J.M. Correlation between silicon hydride species and the photoluminescence intensity of porous silicon. Appl. Phys. Lett. 1992. V. 60, No.14. P. 1700–1702.

Turner D. Electropolishing silicon in hydrofluoric acid solutions. J. Electochem. Soc. 1958. V. 5. P. 402–405.

Turner D. R. J. Electropolishing silicon in hydrofluoric acid solutions. J. Electrochem. Soc. 1958. V. 105. P. 402–408.

Uhlir A. Electrolytic shaping of germanium and silicon. BellSyst. Tech. 1956. V. 35, No.2. P. 333–347.

Voos M., Uzan P., Delande C., Bastard G., Halimaoui A. Visible photoluminescence from porous silicon: a quantum confinement effect mainly due to holes. Appl. Phys. Lett. 1992. V. 61, No.10. P. 1213–1215.

Wang X., Shi G., Zhang F. L., Chen H. J., Wang W., Hao P.H., Hou X. Y. Critical conditions for achieving blue light emission from porous silicon. Appl. Phys. Lett. 1993. V. 63, No.17. P. 2363–2365.

Wolkin M.V., Jorne J., Fauchet P.M., Allan G., Delerue C. Electronic States and Luminescence in Porous Silicon Quantum Dots: The Role of Oxygen. Phys. Rev. Lett. 1999. V. 82. P. 197–200.

Xie Y.H., Wilson W. L., Ross F.M., Mucha J.A., Fitzgerald E.A., Macaulay J.M., Harris T.D. Luminescence and structural study of porous silicon films. J. Appl. Phys. 1992. V. 71, No.5. P. 2403–2407.

Yablonovitch E., Allara D. L., Chang C. C., Gmitter T., Bright T. B. Unusually low surface-recombination velocity on silicon and germanium surfaces. Phys. Rev. Lett. 1986. V. 57, No.2. P. 249–252.

Yamada M., Kondo K. Comparing effects of vacuum annealing and dry oxidation on the photoluminescence of porous Si. Japan J. Appl. Phys. 1992. V. 31. L993-L996.

Zheng X. L., Chen H. C., Wang W. Laser induced oxygen adsorption and intensity degradation of porous silicon. Appl. Phys. Lett. 1992. V. 72, No.8. P. 3841–3842.

Zheng X. L., Wang W., Chen H. C. Anomalous temperature dependencies of photoluminescence for visible-light-emission porous silicon. Appl. Phys. Lett. 1992. V. 60, No.8. P. 986–988.

4

Quantum dots

Quantum dots (QDs) are often called 'artificial atoms', since the confinement of the electron motion in all spatial directions leads to a system of discrete energy levels, each of which corresponds to a localized in a quantum dot wave function. Quantum dots, which are the most important object of the physics of low-dimensional semiconductor heterostructures, are also called quasi-zero-dimensional systems (Fedorov, Baranov, 2005).

It is difficult to give a precise definition of quantum dots. This is due to the fact that in the literature quantum dots refer to a wide class of quasi-zero-dimensional systems which exhibit the quantum size effect of the energy spectra of electrons, holes and excitons (see sections 1.5, 2.7 and 3.4). This class primarily includes semiconductor crystals in which all three spatial size are of the order of the exciton Bohr radius, R_{ex} in the bulk material. This definition implies that the quantum dot is in a vacuum, gaseous or liquid medium, or limited by some solid material, different from the material out of which it is made. In this case, the three-dimensional spatial confinement of elementary excitations in quantum dots is determined by the presence of interfaces between different materials and media, i.e. existence of heterointerfaces. Such quantum dots are often referred to as nanocrystals. However, this simple definition is not complete, since there are quantum dots in which there are no heterointerfaces in one or two dimensions. Despite this, the movement of electrons, holes and excitons in such quantum dots is restricted in space due to the presence of the potential wells, arising, for example, due to mechanical stress or fluctuations of the thickness of the semiconductor layers.

Among the variety of different quantum dots there are several basic types which are most often used in experimental studies and numerous applications. First of all, nanocrystals in liquids, glass and in the matrices of wideband dielectrics. If they are grown in glass matrices, they usually have a spherical shape. It is in this system representing the quantum dots CuCl, embedded in a silicate glass, in which the study of single-photon absorption revealed for the first time the effect of three-dimensional size quantization of excitons (Ekimov & Onushchenko, 1981). This work has begun the rapid development of the physics of quasi-zero-dimensional systems. Other important types of quantum dots include the so-called self-organized quantum dots which are produced by the Stranski–Krastanov technique of molecular beam epitaxy. Their distinguishing feature is that they are linked through the hyperfine wetting layer the material of which is identical to the material of the quantum dot. Thus, in these quantum dots one of the heterointerfaces is missing. The same type, in principle, includes porous semiconductors, such as porous silicon (see chapters 3 and 13), as well as the potential wells in thin semiconductor layers arising from fluctuations in the thickness of the layers.

The quantum dots induced by mechanical stresses are treated as the third type. They form in thin semiconductor layers by mechanical stresses that arise due to the mismatch of the lattice constants of the materials of the heterointerfaces. These mechanical stresses lead to the appearance in a thin layer of a three-dimensional potential well for electrons, holes and excitons. These quantum dots do not have heterointerfaces in two directions.

In a joint paper (Karrai et al, 2004) researchers from Russia (Institute of Semiconductor Physics, Novosibirsk), Great Britain, Spain, Germany and the United States in the investigation of the luminescence spectra of quantum dots found that they have electronic states whose description requires going beyond the frame of the model of an artificial atom. These states are formed due to coherent hybridization of the localized levels with delocalized and have no analogues in atomic physics. They arise in photon emission from quantum dots. For a theoretical description of experimental data Karrai et al, (2004) used the modified Anderson model (Shklovskii, Efros, 1979).

One of the promising applications of semiconductor quantum dots is their use as elements of quantum logic. Quantum computing requires that the quantum states involved possess a sufficiently large coherence time. In Karrai et al. (2004) it is shown that at low temperatures large enough coherence times can be realized in quantum dots.

Manufacturing techniques of various quantum dots are described in a large number of scientific publications (see, e.g. chapters of monographs (Gaponenko, 1998; Bimberg et al, 1999; Lee, 2002; Fedorov, Baranov, 2005)).

4.1. Quantum dots based on nanosilicon

To date, the main direction of development of research – development of media with silicon nanocrystals – has been defined. In a sense, such a choice is the result of studies of porous silicon. Although research and development in this direction did not lead to the emergence of a competitive emitter, it is firmly believed in fact that silicon nanocrystals themselves are almost ideal for luminescence. The main obstacles to their application in optoelectronics are related to the difficulty of organizing an effective current pumping of individual nanocrystals in an insulating matrix.

Among the works of this trend the most noticeable is the work of Italian authors in the observation of optical gain in a silicon-based system. In the experiment performed by Pavesi et al (2000) the medium containing silicon nanocrystals was obtained by ion implantation of silicon ions in silicon dioxide (80 keV, 10^{17} cm^{-2}). Two types of samples with nanocrystals were obtained: in the surface layer of a bulk quartz substrate and in the layer of silicon dioxide grown by rapid thermal oxidation of the silicon substrate. The first type of samples was needed for measurement 'in transmission', the second – to demonstrate compatibility of the technology with the standards used in microelectronics. In both cases, the nanocrystals (3 nm, $2 \cdot 10^{19}$ cm^{-3}) formed in a layer with a thickness of ~100 nm. The distance between the nanocrystals – slightly greater than their own size. Thus, the authors (Pavesi et al, 2000) produced an active medium, which is close to optimum – with the maximum density of nanocrystals on the one hand, but these crystals were still quite isolated from each other. The latter, in turn, means a reduction of non-radiative recombination, increase in the homogeneity of the system, etc. The samples were characterized by fluorescent measurements, absorption in the visible and infrared regions; the time of ignition and quenching of luminescence was measured.

The main result which highlights the work (Pavesi et al, 2000) is the convincing demonstration of optical amplification. Excitation in the experiment was carried out by the second harmonic of a Ti:sapphire laser (λ = 390 nm, pulse duration 2 ps, frequency of 82 MHz). The illuminated area was a strip of variable length. This formulation of the experiment gave the authors the ability to separately change the intensity of exposure and length of the active region, demonstrating spectral narrowing of the luminescence band both with increasing intensity and with increasing length of the lighted spot. The spectral width of the luminescence depended also on the angle at which radiation is emitted from the active region – the narrowest spectrum was, as expected, for small angles of exit of light from the end of the film, when the radiation passed along the path of the maximum active layer. Pavesi et al. (2000) also demonstrated amplification of the passed probe beam – as they point out – a first for systems based on silicon. Independent experimental data on the characteristics of emissions allowed to obtain a reliable value of optical gain – about 100 cm^{-1} for the sample with nanocrystals in silica and about 15% smaller for the nanocrystals in the oxide layer on the surface of crystalline silicon.

The crucial point is to obtain electroluminescence from such environments. To do this, silicon dioxide is not the best environment – the band gap is too large for effective injection of charges in the nanocrystals. Similar results were previously obtained on partially oxidized porous silicon, where it was possible to obtain the external quantum efficiency of the order of a percent (Porjo et al, 2001). However, such high efficiency was obtained for very high resistance structures, and for low-resistance structures efficiency was not high.

Cheylan et al. (2001) used the CVD process of depositing layers of silicon and silicon dioxide. The latter type of layers was produced by adding oxygen to the monosilane, from which silicon layers were precipitated. By adjusting the ratio of gases incompletely oxidized layers were produced, i.e. layers of the dioxide containing quantum-size silicon dots. The authors also produced sets of pairs of layers of silicon/dioxide, i.e. a superlattice of alternating layers. The principal 'thin layer' form of the structures solved the problem that remained unsolved in the work of Nishimura, Nagao (2000) – current excitation of the luminescence of quantum dots with an average current density of 10–50 mA/cm². The work is interesting by the successful excitation of electroluminescence. In general, the idea of finding an effective medium in the Si–SiO$_2$ system is widespread enough (see, e.g. Park et al, 2001).

Nayfeh et al. (1999) also used the CVD deposition of silicon from silane but, in contrast to (Cheylan et al, 2001), nitrogen, not oxygen, was added to the silane. As a result, instead of silicon dioxide Cheylan et al.(2001) received silicon nitride with a significantly lower energy band gap, and, moreover, the deposition mode was such was that the silicon nanoparticles were amorphous. This is due to the fact that the amorphous silicon band gap is ~1.6 eV (~1.1 eV for crystalline silicon), which also facilitated the injection of charge into silicon nanoparticles. As a result, Nayfeh et al. (1999) were able to obtain particles with luminosity in any area of the visible spectrum (corresponding to the RGB signals of colour coding) and the layers with a white glow. Given the fact that in this case the external electroluminescence quantum efficiency was 0.2%, the resistance of the layers was of the order of tens of ohms, and the structures worked at a bias of less than 5 V. This result opens the way to getting a full-colour Si display. Moreover, it turns out that a transparent silicon dioxide as a medium for nanoparticles is not necessary. A number of papers have been devoted to the luminescence of implanted silicon – this material is also capable of efficient luminescence. Non-linear stimulated emission, directed blue light, the transformation of the second harmonic and a number of other effects (Nayfeh et al, 2000; Mitas et al, 2001) were observed in this material.

Mitas et al (2001) and Smith et al (2005) calculated the most probable structure of the cluster – Si$_{29}$H$_{24}$, which has a spherical shape and at the same time symmetry of the tetrahedron (Fig. 4.1). In the strictly tetrahedral environment such a cluster has only one central atom, which does not prevent the cluster to give the reflections corresponding to tetrahedral symmetry. The number of hydrogen terminations of the cluster is not accidental; the cluster completely covered with hydrogen has a band gap of about one electron volt less than Si$_{29}$H$_{24}$, while the cluster Si$_{29}$H$_{12}$ more than 2 eV, which agrees well with experimental data by Smith et al. (2005). Ng et al. (2001) proposed another version of the nanocluster of three interstitial Si atoms.

Belomoin et al. (2002) obtained nanosilcon clusters Si$_n$H$_x$, where $n >$ 20, which form a certain set (family) of the particles of discrete sizes: 1.0

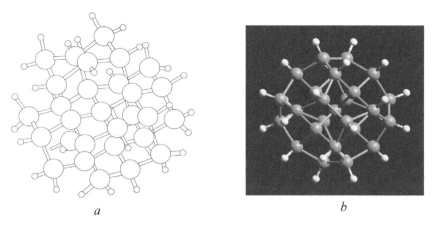

a b

Fig. 4.1. Configuration of Si atoms in the non-reconstructed $Si_{29}H_{36}$ nanoclusters with a diameter of 1.0 nm (a) and computer image of the reconstructed $Si_{29}H_{24}$ (b) nanoclusters (Smith et al., 2005)

Fig. 4.2. (Bottom, from right to left) Emission of the sols of the four members of the magic family of the silicon nanoclusters with the diameters of 1.0 (Si_{29}) (Si_{123}); 2.15 and 2.9 nm after their separation. Excitation was carried out with a commercial source with a mean wavelength of 365 nm (top view). The emission of blue, green and red colour by the sols with the magic dimensions − 1.0, 1.67 and 2.9 nm (Belomoin et al., 2002).

(Si$_{29}$); 1.67 (Si$_{123}$); 2.15; 2.9 and 3.7 nm. The particles were characterized using the methods of electron microscopy, optical absorption and emission spectroscopy, and crystallization of sols. The values of the band gap were measured. These particles have a very bright luminescence in blue, green, yellow and red spectral regions, in accordance with the increase of their diameter (Fig. 4.2). The ability to obtain particles of discrete sizes, luminescent in the red, green and blue (RGB) areas, can be used to generate biomarkers (see chapter 12), RBG displays, and flash memory.

However, the silicon nanoparticles is not the only path to a silicon light source. In Lin et al. (2001) electroluminescence was excited with high efficiency from the bulk silicon. The mechanism of luminescence is not clear. This however, offers a fundamentally new approach to the production of emitting silicon, compatible with standard silicon technology. By its parameters proposed in the work of Lin et al. (2001) this device is a transmitter for the near infrared range.

Hansson et al. (2001) showed that the electroluminescence with the photon energy corresponding to the energy of indirect E_g, can be produced, squeezing silicon with an ITO electrode. As a result, it may be possible to create a pressure sensor with the luminous output.

We single out the current direction – the development of emitters using the erbium impurity in silicon. Initially, the idea is great in its simplicity – using silicon as a transport medium for the current excitation of the erbium ion. The last by its luminescence wavelength of 1.5 μm falls in the band gap of silicon and at the same time is suitable for pumping fiber optic lines which have a transparency window in this region. It should be noted that the idea of the current pumping of ions of rare-earth elements in semiconductors, with slight changes, has been around for several decades. However, the practical implementation is associated with a number of difficulties. First, the Er^{3+} ion, well-studied without silicon, like all rare-earth ions has fluorescence in the environment of oxygen atoms or fluorine anions. Thus, in conjunction with erbium oxygen must be added to silicon, making the technology less than straightforward. In addition, the current excitation processes are also not as effective. As a result, in most works in this direction the authors note in one of the first paragraphs that the intensity of erbium luminescence is too small for practical applications. However, the specific parameters of LEDs (Du et al, 2001) and light transistors (Kazarinov et al, 1971) have been studied. In the latter case, a layer of silicon doped with erbium, was located between the base and the collector of the light transistor to allow one to create effective conditions for the excitation of Er^{3+} ions by current carriers. In Kazarinov et al. (1971), the effective radiation of erbium in the characteristic band of 1.54 μm was obtained at room temperature at a sufficiently low excitation current density (0.1 A/cm^2) and low applied voltage (3 V). Nevertheless, even in the best sample of the light transistor the efficiency remains at 10^{-5}.

Available wavelengths are not limited to visible or near infrared range. In silicon quantum-cascade lasers the radiation of the mean IR is used (Dehlinger et al, 2000). In general, such a device works as follows. Emitting layers (quantum wells), separated by barriers, form a superlattice. Application of external voltage shifts the potentials of the neighbouring wells so that the lower level of the radiative transition in a quantum well coincides with the upper level from which radiation takes place, but in this case for a neighbouring well. Carriers are transferred from the upper to the lower level in one of the wells, generating the radiation, then tunnel into the adjacent quantum well on the top state, once again make a radiative transition, and so on. This scheme allows circumvent the main difficulty – the indirect conduction band gap in silicon, since the entire process occurs within the same band.

One way to obtain superlattices with modulation of the potential in the valence area is to alternate layers of silicon and semiconductor solutions of SiGe. Such a system is implemented, for example, in Fridman et al. (2001). In this study, intraband electroluminescence was observed near the wavelength $\lambda \sim 10$ μm, which corresponds to transitions between subbands of heavy holes.The spectral width (FWHM) is 22 meV at 50 K and a volatge on the structure of 3.65 V. The time of non-radiative relaxation of carriers depends on the design of the layers. The intensity of the electroluminescence (at 50 K) was less than 10^{-11} W at a current through the structure of about 1 A.

A more efficient system is described in Bagraev et al (2000), in which transitions between the subbands of heavy and light holes in the area where the light-hole subband is substantially non-parabolic are used. This scheme is interesting because the energy gap between the emitting sub-bands is less than the optical phonon energy. As a result, the process of non-radiative relaxation is realized only by acoustic phonons, which leads to an increase in the time of non-radiative recombination by about three orders of magnitude compared to, for example, the data from work (Fridman et al, 2001). In the experiments by Bagraev et al (2000) the gain at liquid nitrogen temperature at a wavelength of $\lambda = 50$ μm was 170 cm^{-1}.

A number of studies carried out in the Ioffe Physics Institute, Russian Academy of Sciences (St. Petersburg) led to the development of new optoelectronic devices (Bagraev et al, 1998; Bagraev et al, 2000). Traditionally, the layers in semiconductor structures are grown as planar, parallel to the surface. However, the authors (Bagraev et al, 1998; Bagraev et al, 2000) produced layers in bulk crystalline silicon, directed perpendicular to the surface (i.e. into the sample). These layers may have different doping, can form a superlattice of the $p - n$ junctions and could once again break up into layers in the other direction, forming a self-organizing system of dots. Layers with thickness of ~20 nm can be 'cut' into the existing crystal lattice. The optical gain and the gain of the probe light in silicon systems was first obtained in the similar 'perpendicular' layers.

Effective Er^{3+} luminescence and narrowing of the electroluminescence line at room temperature were also obtained in these layers. Silicon emitters of the mid-IR range using this structure found unexpectedly use in medical practice and are produced in small numbers.

The use of quantum dots in such forward-looking applications such as quantum cryptography, the development of sources of single photons based on quantum dots and the creation of quantum computers has attracted the attention of researchers.

4.2. Investigation of structural-phase transformations and optical properties of composites based on silicon nanoclusters in the silicon oxide matrix

One major problem that arises when replacing metal interconnects in computers by optical waveguides to increase the speed of transmission of information by light pulses is the lack of light-emitting elements (LED, laser) the manufacturing technology of which would be compatible with silicon technology. One of the promising directions of construction of light-emitting elements is the development of MOS light-emitting diodes (metal–oxide–semiconductor), in which nc-Si/SiO_2 film is used as the oxide. This film is a matrix of amorphous SiO_2 with Si nanocrystals embedded in it. Seminogov et al (2009) presented the results of studies of structural transformations occurring in the films of SiO_x with $x \approx 1$ by thermal annealing at various temperatures, as well as research of their photoluminescent and electrical properties.

There are several methods for the deposition of initial films $SiO_x (x \leq 2)$. These include chemical vapour deposition at low pressure (LPCVD); plasma-induced chemical vapour deposition (PECVD); ion implantation of Si in the matrix of SiO_2, resistive evaporation and condensation in vacuum (or the thermal evaporation method, see chapter 5). Deposition on the substrates of Si, SiO_2 and Al_2O_3 films SiO_x with $x = 1.0–1.2$ is carried out from the original powder-Si monoxide (SiO – material for electronics) by resistive evaporation and condensation in vacuum in the vacuum system HAC-600 at a residual pressure of $P_{res} \leq 10^{-3}$ Pa. Film deposition modes were varied by the the substrate temperature $T_{sub} = 100–400°C$ and the deposition rate $V_{dep} = 2–20$ Å/s. Depending on the measurements, the thickness of SiO_x films in the case of the silicon substrate was $d = 50–500$ nm, and in the case of substrates made of KI grade fused quartz and sapphire $d = 250–500$ nm.

During thermal annealing some of the silicon atoms are released in the form of nanoclusters and, as a result, the original film of SiO_x ('as-deposited', for example, with $x \approx 1$) transforms into a nc-Si/SiO_y film with $y > x$, which is a matrix SiO_y with silicon nanoclusters implanted in it. The phase separation reaction can be described by the scheme

$$ySiO_x = (y-x)Si + xSiO_y. \tag{4.1}$$

Thermal annealing of the samples was carried out in an electrical furnace in a nitrogen atmosphere at a given temperature T_{ann} and then cooled with a natural rate. The annealing conditions were varied by varying the annealing temperature T_{ann} = 350–1200°C and annealing time t_{ann} = 0–4 h. The heating time of the furnace to the desired temperature T_{ann} and natural cooling was 1–1.5 h. Annealing time t_{ann} was the holding time of the sample at a given temperature T_{ann}. Thus, the zero annealing time at a given temperature T_{ann} corresponds to the annealing mode in which the furnace was switched off after heating to the required temperature.

IR spectroscopy studies were conducted on a batch (11 samples) of nc-Si/SiO$_y$ films with a thickness $d \approx 0.5$ mm, deposited on silicon substrates P-Si with a resistivity of ρ = 10 ohm · cm. All substrates of the batch were cut from the same plate, and all SiO$_x$ films with $x \approx 1$ were deposited on these substrate in a single process by resistive evaporation and condensation of granular SiO powder. The absorption spectra in the range 350–9000 cm^{-1} were measured using an FTIR-spectrometer Shimadzu-8400 S.

Figure 4.3 shows the absorption spectra (700–1400 cm^{-1}) of the nc-Si/SiO$_y$ films on P-Si substrates. Thus, the experimental dependence of the position of the absorption peak Ω_{peak} on T_{ann} was determined. Since silicon poorly absorbes in the range 990–1100 cm^{-1}, the infrared absorption spectrum in this region characterizes the state of the SiO$_y$ matrix: namely, increasing annealing temperature changes the composition (y) of the SiO$_y$ matrix which leads to a shift of the absorption peak frequency, Ω_{peak}. Absorption in the region 770–1400 cm^{-1} is due to (Nakamura et al, 1984; Bell et al, 1968) vibrational modes of (Si–O) and (Si–O–Si) bonds. Seminogov et al (2009) developed a method for determining the volume fraction of silicon contained in Si nanoclusters by measuring IR absorption spectra (see Fig. 4.4).

Figure 4.4 shows that in the case of the initial SiO$_x$ film the volume fraction of silicon V_{Si}, contained in the nanoclusters, at T_{ann} > 1000°C reaches V_{Si} = 0.31–0.32.

4.2.1. The hypothesis of the percolation nature of the structure of nc-Si/SiO$_y$ films

Figure 4.4 shows that in the case of initial SiO$_x$ films with x = 1 at annealing temperatures T_{ann} ≈ 580–600°C the volume fraction of silicon contained in the nanoclusters reaches V_{Si} ≈ 0.16. According to the data of Golubev et al (1997), the value V_{Si} ≈ 0.16 is the threshold for the effect of percolation. The essence of the percolation effect, which is covered in a large number of literature sources, with respect to the system considered by Seminogov et al (2009), is as follows. When the volume proportion of silicon, V_{Si}, contained in Si nanoclusters, exceeds the threshold value

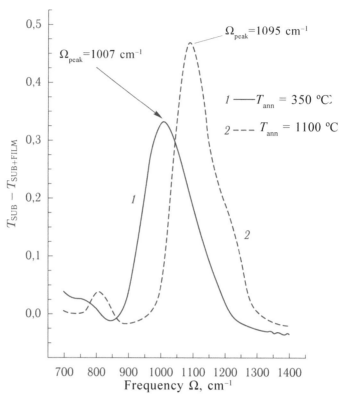

Fig. 4.3. The absorption spectra of the films nc-SiO/SiO$_y$, resulting from thermal annealing of SiO$_x$ films starting with $x \approx 1$ at temperatures T_{ann} = 350°C (curve *1*) and T_{ann} = 1100°C (curve *2*). Annealing time t_{ann} = 1 h, T_{sub}, $T_{sub+film}$ – transmittance of the substrate and of the film on the substrate, respectively, Ω_{peak} – the peak of absorption in the region 990–1100 cm^{-1}. The measured spectra are averaged over 25 neighbouring points (Seminogov et al, 2009).

V_{Si} = 0.16, part the isolated Si nanoclusters with a probability close to unity 'bonds' to Si nanowires, penetrating the entire thickness of the dielectric matrix of SiO$_y$. Consequently, at annealing temperatures T_{ann} > 600°C the initial SiO$_x$ film with x = 1 must be transformed into a nc-Si/SiO$_y$ film consisting of the SiO$_y$ matrix with the incorporated ensemble of isolated Si nanoclusters and an ensemble of Si nanowires. However, the films with V_{Si} < 0.16 should contain only one ensemble of isolated nc-Si. It was established experimentally that in the case of the original SiO$_x$ film with x = 1 at T_{ann} < 900–950°C the Si nanoclusters in these two ensembles are in the amorphous state, while at T_{ann} > 900–950°C there must exist a second percolation threshold of crystalline nc-Si.

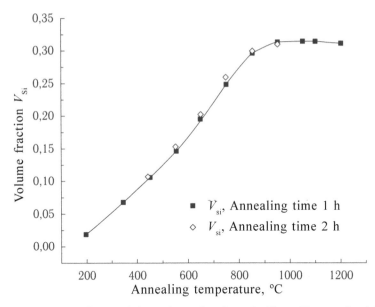

Fig. 4.4. The dependence of the volume fraction of silicon V_{Si} contained in the nanoclusters on annealing temperature T_{ann} when the original film is SiO_x with $x \approx 1$ (Seminogov et al, 2009).

Raman spectra were measured in MicroRaman LabRAM HR Visible equipment of Horiba Jobin Yvon. The excitation source was the focused radiation of an Ar-laser with a wavelength of 488 nm. The Raman signal was recorded in the reflection configuration. The detector was a digital camera with a silicon CCD matrix. The exposure during the measurements was 30 s, the number of averages – 5. The contribution of the substrate in the Raman spectra was taken into account by the additional measurement of the Raman signal from the substrate followed by its subtraction.

The Raman method was used study a batch (12 samples) of the films nc-Si/SiO$_y$ ($d = 0.5$ μm) on substrates of fused silica (grade KI) by thermal deposition and subsequent thermal annealing at one of the temperatures T_{ann} = 200, 350, 450, 550, 650, 750, 850, 950, 1000, 1050, 1100, 1200°C. A film of SiO_x with $x = 1$ ($d = 0.5$ μm) on a sapphire substrate (Al_2O_3) was also produced and was then annealed at a temperature T_{ann} = 1100°C. Annealing time was 1 h.

Figure 4.5 shows the shape of the Stokes Raman spectra (without subtracting the contribution from substrate) for nc-Si/SiO$_y$ films on quartz substrates. It also shows the shape of the Raman spectrum from a quartz substrate. The Raman spectra of the samples annealed at temperatures of 650°C and below in the presented range of the wave numbers are not fundamentally different from the Raman spectra of the sample annealed at a temperature 750°C. Figure 4.5 shows that at $T_{ann} \approx 900$–950°C the Raman spectra of the investigated

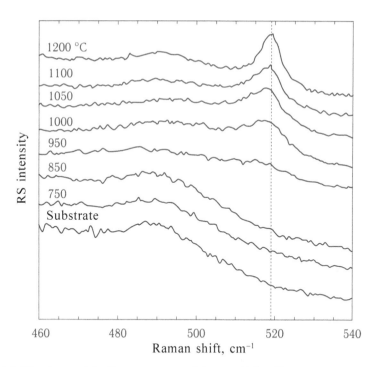

Fig. 4.5. The shape of the Raman spectra of the nc-Si/SiO$_x$ films on quartz substrates formed by thermal annealing of SiO$_x$ films with $x = 1$ at different temperatures. Spectra are presented without subtracting the contribution from the substrate (Seminogov et al, 2009).

structures start to show a band with a maximum of about 518 cm^{-1}, characteristic for crystalline silicon, the intensity of which increases with increasing annealing temperature. Note that the shift of the maximum of the Raman line (relative to the position of the maximum of the line or bulk crystalline silicon – 521 cm^{-1}) in the direction of smaller wave numbers is characteristic for nanocrystalline silicon structures (due to the effect of spatial confinement of the optical phonons) and serves as additional confirmation of the presence of nc-Si in the studied samples (Golubev et al, 1997; Richter et al, 1981; Campbell & Fauchet, 1986). Thus, Fig. 4.5 shows that when T_{ann} is lower than 900–950°C the nc-Si ensemble is in the amorphous state, and if the critical temperature is exceeded crystallization of Si nanoclusters starts and the volume fraction V_{c-Si} of the crystal nanoparticles increases with increasing annealing temperature. This is in agreement with experiment (Nesbit, 1985; Iacona et al, 2004) and the theory of crystallization of nanoclusters (Emelyanov, Seminogov, 2006).

From measurements of the Raman spectra it was found that the crystallization of the nanoclusters begins when $T_{ann} \approx 900$–950°C. With this in mind, on the basis of the percolation theory we can conclude that at temperatures $T_{ann} < 600$°C, when $V_{Si} < 0.16$, the nc-Si/SiO$_y$ film is an SiO$_y$

matrix with an embedded ensemble of isolated nanoclusters of amorphous Si, and at $T_{aaa} > 600°C$ ($V_{Si} > 0.16$) the SiO_y matrix contains an ensemble of isolated amorphous nanoclusters of amorphous Si and an ensemble of Si nanowires. With increasing annealing temperature the volume fraction V_{Si} of amorphous nc-Si and Si nanowires increases (see Fig. 4.4). At $T_{ann} \approx$ 900–950°C some of the nanoclusters of amorphous ensembles of isolated nc-Si and Si nanowires crystallize. Thus an additional interspersed ensemble of isolated crystalline Si nanoclusters forms.

The processing of the measured Raman spectra and IR absorption spectra shows that in the case of the initial SiO_x film with $x = 1$ at $T_{ann} \approx$ 1050–1100 °C the second percolation threshold ($V_{c-Si} = 0,16$) with respect to the concentration of silicon in the crystalline phase is reached. In terms of percolation theory this means that until the crystallization of all Si nanoclusters takes place, the nc-Si/SiO_y film will be a matrix with four ensembles embedded in it: two ensembles of isolated of amorphous and crystalline Si nanoclusters, and two ensembles of amorphous and crystalline Si nanowires. Thus the temperature $T_{ann} \approx$ 900–950°C, at which the crystallization process of nc-Si starts, is defined; the existence of the second percolation threshold in respect of the nanosilicon concentration in the crystalline phase is shown, and the percolation theory is used to described the process of structural and phase transformations with increasing temperature of thermal annealing of the initial SiO_x film with $x = 1$.

Percolation restructuring of the nc-Si/SiO_y films can lead to a percolation mechanism of coalescence, change of the time of non-radiative relaxation and the nature of the manifestation of the quantum size effect (nanodots, nanowires), and to a change in the mechanism of injection of electrons and holes in the thickness of the dielectric matrix in electroluminescence. If the hypothesis proposed by Seminogov et al (2009) is true, then the dependence of the photoluminescent optical, electrical and electroluminescent characteristics of the nc-Si/SiO_y films on the thermal annealing temperature should be have special features (extrema) at crystallization temperature and the temperatures corresponding to the first and second the percolation threshold.

4.2.2. Investigation of photoluminescence and electrical properties of nc-Si/SiO_y films at optical excitation

To study the evolution of the photoluminescent properties of the nc-Si/SiO_y films depending on the thermal annealing mode, Seminogov et al (2009) used the same batch (12 samples) of nc-Si/SiO_y films on a substrate of grade KI silica glass which was used in the studies of Raman scattering. Photoluminescence (PL) was recorded carried out with a spectrograph MS 3504 I (firm Solar TII) and a CCD-camera. The PL excitation source was an N_2-laser with a wavelength $\lambda = 337$ nm; with pulse time $\tau = 10$ ns, with

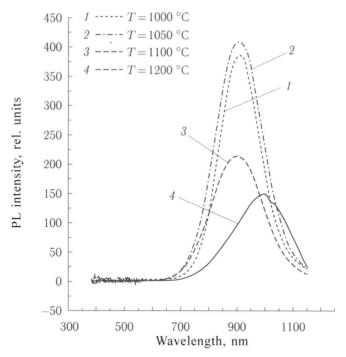

Fig. 4.6. The PL spectra of nc-Si/SiO$_y$ films formed during the annealing of SiO$_x$ films with $x = 1$ at different temperatures T_{ann} (curves 1–4). From the article by Seminogov et al (2009).

an energy $E = 1$ μJ and a repetition rate $v = 100$ Hz. The laser beam was focused on the sample to a spot with a diameter of 1 mm. The region of spectral sensitivity of the system was 200–1150 nm. The PL spectra were adjusted for the instrumental function of the measuring system.

Figure 4.6 shows the dependence of the PL intensity on the wavelength λ at various fixed temperatures. Figure 4.7 shows the dependence of the ratio of the peak intensities of PL on T_{ann} on a semi-logarithmic scale. THe non-monotonic dependence of photoluminescence intensity on thermal annealing temperature, as shown in Fig. 4.7, agrees well with the percolation picture of the structural transformations in the nc-Si/SiO$_y$ film. In the temperature range $T_{ann} = 200$–600 °C the increase in PL intensity is due to an increase in the volume fraction V_{Si} of silicon contained in the isolated amorphous silicon nanoparticles.

In the range 600°C < T_{ann} < 900–950°C the volume fraction of silicon V_{Si} contained in the isolated amorphous Si nanoparticles, decreases due to the coalescence of nanoclusters of amorphous Si nanowires. In the range 900–950°C < T_{ann} < 1050–1100°C an interspersed ensemble of the isolated crystalline Si nanoclusters forms, and with increasing temperature their volume fraction increases which leads to an increase in PL intensity. Increasing the annealing temperature in the range T_{ann} > 1050–1100°C is

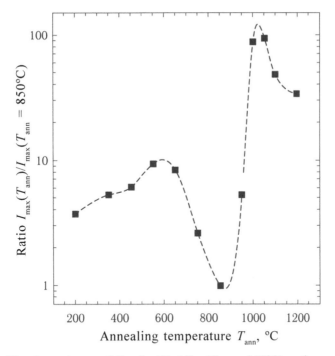

Fig. 4.7. The dependence of the $I_{max}(T_{ann})/I_{max}(T_{ann} = 850°C)$ ratio on T_{ann} on a semilogarithmic scale, obtained from the processing of the PL spectra. From the article by Seminogov et al (2009).

accompanied by reaching the second percolation threshold, an increase of the volume fraction of crystalline nanowires and a decrease in the volume fraction of isolated Si nanoparticles V_{Si}, and, consequently, a decrease in PL intensity. An additional factor leading to a decrease in PL intensity in this temperature range is the increase of the size of isolated crystalline Si nanoparticles by percolation coalescence and diffusion growth. This leads to a 'tightening' of the selection rules for indirect transitions in silicon due to weaker manifestation of the quantum size effect. A special feature when the second percolation threshold for $T_{ann} \approx 1050–1100°C$ is reached can also be seen in the electrical characteristics of the nc-Si/SiO$_y$ films. Thus, the experimental data on the photoluminescence and electrical characteristics of the nc-Si/SiO$_y$ films confirm, in general, the percolation picture of phase-structural transformations during thermal annealing.

4.3. The long coherence time in quantum dots

One of the promising applications of semiconductor quantum dots is to use them as elements of quantum logic. Quantum computing requires that the quantum states involved possess a sufficiently large coherence time. Experiments carried out in the work of Birkedal et al. (2001) show that

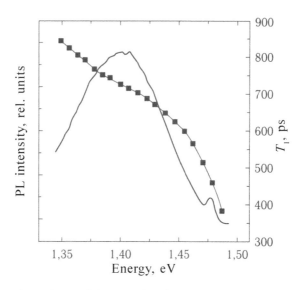

Fig. 4.8. Time dependence of the PL. Time-integrated photoluminescence spectrum of structures with quantum dots InAlGaAs/GaAlAs (solid line without dots) and spectral dependence of the photoluminescence quenching time of PL (graph with dots). Right hand axis – PL quenching time (Birkedal et al, 2001).

at low temperatures fairly large coherence times can be implemented in quantum dots. The photoluminescence spectrum of the structure with the InAlGaAs/GaAlAs quantum dots is shown in Fig. 4.8.

The use of quantum dots in such forward-looking applications such as quantum cryptography, quantum computers, is currently attracting a great attention of researchers. One of the main questions is: how fast is the loss of coherence in the system? In principle, the technique of manipulation of photoexcited carriers at ultrashort times (of the order of femtosecond) can be realized but, in any case, it is desirable that the coherence time should be longer if possible.

Electron–hole pairs form in the quantum dot under optical excitation (a bound state of an electron and a hole is called an exciton), which exist within a finite time, and the electron and the hole then recombine (annihilation). However, due to the fact that electronic excitations in solids interact with the crystal lattice as well as among themselves, the system quickly 'forgets' the initial phase (the higher the temperature, the faster). It would seem that there is a fairly simple way to find out how fast this occurs which is based on the fact that the finite coherence time is responsible for (homogeneous) linewidth of the quantum dot. However, the 'artificial atoms', as opposed to natural, characterized by a certain scatter of their parameters and, consequently, the ground state energy varies from dot to dot. As a result, the luminescence spectrum of the structure is a superposition of a set of emission lines of individual quantum dots (see

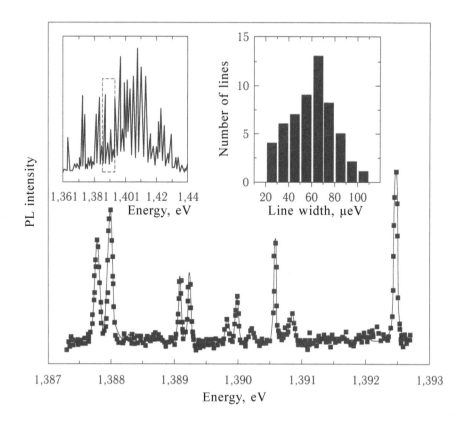

Fig. 4.9. The PL spectrum of the submicron area of the structure with quantum dots; a relatively small number of 'glowing' quantum dots leads to a 'collapse' of the structureless luminescence band into many separate lines (Birkedal et al, 2001).

Fig. 4.9), and the width of the total 'emission line' of the structure is equal to tens of meV. In principle, using various 'tricks', we can select a single (or several) quantum dot from a variety of quantum dots and record its radiation. Estimates of the time of dephasing obtained in this way give the values of the coherence time is not more than 50 ps (at low temperatures), which is about an order of magnitude smaller than the lifetime of an exciton in a quantum dot.

Birkedal et al. (2001) determined by four-wave mixing dephasing times in the structure with InAlGaAs/GaAlAs quantum dots and measured the linewidth of individual quantum dots. It was shown that the estimate of the coherence time on the basis the luminescence line width is low: the fact is that the signal from a single quantum dot accumulates enough long, and during this time the change of the electrostatic environment can lead to the variation of the energy level; a consequence of this is 'extra' line broadening. Experiments by Birkedal et al. (2001) showed that at temperatures close to zero the excitons in quantum dots can 'remember' their initial state virtually throughout the entire lifetime (about a nanosecond).

4.4. The dependence of the width of the optical gap of silicon quantum dots on their size

The optical properties of heterostructures with silicon quantum dots of a small size (several nanometers) have been actively studied in order to obtain the radiation in the near infrared (IR) or even in the visible spectrum. In connection it is interesting to calculate the energy of the main optical transition in these systems and analyze the dependence of this energy (or frequency) on the size of the quantum dots.

In the works by Takagahara, Takeda, 1992; Khurgin et al, 1996; Niquet et al, 2000 calculations were performed using the effective mass approximation (or the k–p-method) in a model with infinitely high potential barriers. However, many of authors have noticed (Niquet et al, 2000) that this approximation gives too high a result during the transition to smaller sizes of quantum dots (less than 6–8 nm). In this regard, quantum dots with sizes of a few nanometers and smaller were later studied using more sophisticated and powerful methods (which, however, are difficult to apply to larger quantum dots, including due to the volume of computer calculations sharply increasing with the size), such as the tight-binding model (Delerue et al, 2000), the pseudopotential method (Franceschetti, Zunger, 2000), the local density approximation (Delley & Steigmeier, 1995). In addition, some authors pointed out (see, e.g. Delerue et al, 1993) that the dependence of the width of the optical gap on the radius of the quantum dot R (for spherical dots) does not have the form R^{-2}, characteristic of the effective mass approximation, and is smoother. Delerue et al. (1993) pointed to the law $R^{-1.39}$.

Burdov (2002) calculated the optical gap of spherical Si quantum dots in a layer of fused silicon dioxide, using the k–p-method. It is shown that the k–p-method can also be used in areas of substantially smaller (than 6–8 nm) quantum dots, if we take into account more rigorously, compared to how it was done in (Takagahara, Takeda, 1992; Khurgin et al, 1996;. Niquet et al, 2000) the anisotropy of the law of dispersion of electrons and holes, as well as take into account the finite height of the potential barrier for both types of carriers (which have already been indicated in (Delley, Steigmeier, 1995) and the 'jump' of the value of the effective mass at the boundary of the quantum dot.

Burdov's calculations show (2002) that the spin-orbit interaction in silicon quantum dots, as well as in bulk silicon, has very little effect on the energy spectrum, giving rise to two energy levels in the valence band, separated by 0.04 eV. In the conduction band of silicon we commonly used the model of the isoenergetic surface in the form of an ellipsoid of rotation with the 'longitudinal' and 'transverse' effective masses resulting from the decomposition of the dispersion law in the vicinity of any one of six equivalent points of the energy minimum. However, this model has significant limitations, as the characteristic values of the energies of the quantum in a nanocluster with size of several nanometers are significantly higher than the difference of the energies of X-point at which two branches of energy intersect, and the point of the energy minimum. At these energies, the isoenergetic surface is quite different from the ellipsoid of revolution, and the non-parabolicity starts to play an essential role in the dispersion law.

The dependence of the band gap ε_g on the inverse radius of the quantum dot, $1/R$ at infinitely high potential barriers is shown in Fig. 4.10 by the solid line. The points *1* correspond to a model with a finite height of the barrier and the effective mass in the barrier region m^*, equal to the effective mass of the carrier in the semiconductor, i.e. m_e for the electrons and m_h for the holes. Points *2* shows the values of ε_g $(1/R)$ in the case of finite barrier height and m^*, equal to the mass of the free electron m_0 (which, apparently, is not far from the truth). The band gap in fused SiO_2 is 8.7 eV. The height of the potential barrier for electrons according to the data presented by Babic et al. (1992) is 3.2 eV; consequently, the barrier with a height of 4.3 eV remains for the holes.

It should be emphasized that even in the approximation of an infinitely high barrier the dependence of the gap width on R does not have the form const $+R^{-2}$ due to hybridization of the s- and p-states and provides an intermediate dependence between R^{-1} and R^{-2} for the width of the optical gap (Bourdieu, 2002). Generally speaking, the potential barriers for both electrons and holes are quite high and an order of magnitude higher than intrinsic energies ε_e and ε_h. However, as seen in Fig. 4.10, even such high barriers can not with reasonable accuracy be infinitely high: the difference

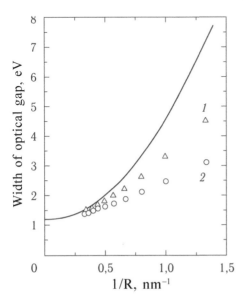

Fig. 4.10. The dependence of the width of the optical gap on the inverse radius of a quantum dot: solid curve – infinitely high potential barriers, *1* – the final potential barriers (3.2 eV for electrons and 4.3 eV for holes) and a constant effective mass; *2* – finite potential barriers (3.2 eV for electrons and 4.3 eV for holes) and the jump of the effective mass (Burdov, 2002).

of the energies of electrons and holes from their own values in the case of an infinitely high barrier (depending on R) ranges from 15 to 65%. A particularly strong influence of the barrier is, of course, observed in the case of smaller sizes – both points *1* and points *2* deviate more and more from the almost parabolic (for R^{-1}) dependence, increasing the magnitude of the correction (the terms R^{-3} and higher degrees are included) (Burdov, 2002). The position of the energy levels of the electrons and holes are also strongly influenced by the 'jump' of the effective mass at the boundary of the quantum dot. As can be seen from Fig. 4.10, the correction in the case of $m^* = m_0$ for all values of the radius of the quantum dot is approximately twice that in the case of constant effective mass.

Comparison of the results of Burdov (2002) with the calculations made by other, more sophisticated methods in the field of quantum dot sizes larger than 1.5 nm, shows very good agreement. Thus, the results of calculations of the width of the optical gap made by the method of strong coupling in the works (Ren, Dow, 1992; Delerue et al, 1993; Wang, Zunger, 1994), by the pseudopotential method (Franceschetti, Zunger, 2000) and in the approximation of the local density functional (Delley, Steigmeier, 1995) practically coincide with the values of $\varepsilon_g(R)$ of Burdov (2002) when $m^* = m_0$. The data of Hill, Whaley (1995) – the method of strong coupling, and Ogut, Chelikowsky (1997) – the pseudopotential method, differ by about 10–15%. The experimental data presented in the works (Mimura et al, 2000; van Buuren et al, 1998; Takeoka et al, 2000; Guha et al, 2000), also consistent with the calculations by Burdov (2002). The observed scatter in the experimental data can be due to different ways of obtaining quantum dots and this is difficult to take into account in the theoretical model.

Thus, even at fairly small sizes of quantum dots the envelope approximation is in good agreement with both experiment and with other calculation methods, comparing favourably to them that does not require such large amounts of computing and is, in fact, an analytic technique.

4.5. The characteristics of excitons and exciton photoluminescence of structures with silicon quantum dots

In recent years, special attention has been given to studies of silicon and germanium quantum dots (QDs) in the SiO_2-matrix, as well as germanium QDs in a silicon matrix (Kashkarov et al, 2003; Dvurechenskii et al, 2002; Brunner, 2002), which is associated with their unique photoluminescent and electroluminescent properties and the ability to efficiently emit light in the visible or near infrared bands at room temperature. However, the models, usually used for calculations of the energy spectrum and other properties of quasiparticles in the QD to date can not be considered sufficiently adequate to the real situation. They can be divided into two main groups.

The first group includes models that use cluster calculation methods. However, in this case there is a need for artificial closure of dangling bonds at the cluster boundary by atoms of hydrogen, oxygen or other neutralizing atoms or molecules to suppress the strong perturbations introduced into the energy spectrum by these bonds. In addition, because of the huge amount of computation, the use of this method in practice is often limited to clusters of up to 2 nm in diameter, even with complete disregard of the impact of the real environment on the energy spectrum. In fact, this effect can be quite essential, because a large (or even controlling) part of the power field lines of the Coulomb interaction between charges in the QD can be closed through the surrounding external environment. In particular, the nature of screening of the interaction between the charges by the barrier region should significantly influence also the electron–hole states in QDs.

The *second group* of models is based on the method of the envelope of the wave function and characteristics of the band spectrum, i.e. based on the solid-state approach in the description of nanocrystallites. These models are more suitable for describing QDs of relatively large sizes, when a prominent role is played by the factor of the crystal structure of the QD. In practice, this also occurs at crystal sizes of >2 nm, although the formal calculations in the framework of such models often extend to the area of smaller QDs. In this approach, it is also possible to a certain extent to take into account the barrier effect of the external environment, including the polarization of the heterointerface by charge carriers. In the case of a semiconductor quantum dot in a dielectric (e.g. silicon QD in silicon dioxide), the neglect of this effect can strongly affect the results of calculations of the energy characteristics of electron–hole pairs.

Kupchak et al (2006) calculated the main characteristics of excitons in spherical quantum dots, taking into account in the first approximation the polarization effect of the QD heterointerface with the final height of the barriers for the electrons and holes.

4.5.1. The model of a quantum dot and the spectrum of electron–hole excitations

Consider the case of a spherical semiconductor quantum dot (material 1) of radius R, located in a dielectric medium (material 2). The parameters of the QG material in the model by Kupchak et al (2006) are the dielectric constant ε_1, averaged over the directions of the effective electron mass m_{e1} and heavy hole m_{h1}. The same parameters of the dielectric matrix surrounding the QD are denoted as ε_2, m_{e2} and m_{h2}. In addition, the model uses power characteristics such as the gap width ε_g of material 1 and breaks of valence U_h conduction U_e bands at the semiconductor–insulator heterointerface. The model assumes that the states of the QD with the lowest energy come from the heavy-hole states of the Γ-valley of the Brillouin zone of silicon, whereas the electronic ones – from the states of the conduction band energy minima of silicon, i.e. the states of the X-valleys in the neighbourhood of the Brillouin zone of type $K_0 = 0.85 \cdot (2\pi/a)$ [1,0,0].

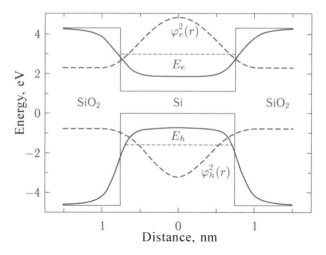

Fig. 4.11. Energy diagram of a silicon quantum dot in a silicon dioxide matrix (Kupchak et al, 2006).

The classical electrostatic potential of a point charge and, accordingly, the potential energy of the self-effect show unphysical divergence near the heterointerface:

$$U_s(z) = \frac{e^2}{4z} \cdot \frac{\varepsilon_1 - \varepsilon_2}{\varepsilon_1 \varepsilon_2}, \qquad (4.2)$$

where z – distance from the charge to the heterointerface. In real systems, in fact, there is always a transition layer from the material of the quantum well to the material of the barrier which is smooth, and a continuous change in environmental parameters (including the dielectric constant), so the actual electrostatic potential at the heterointerface will be continuous. Nevertheless, the error introduced by such a non-physical feature of the classical self-effect potential can be minimized, since in reality the contributions to the intrinsic energy shifts from the transition layers from opposite sides of the heterointerface largely cancel each other. In the simplest case, this can be done by linking the values of the total one-particle potential energies $U_{c(v)} + U_s$ on the borders of the transition layer on the opposite sides of the heterointerface of the linear approximation. The boundaries of the transition layer at the same time can be determined from the condition of the equality of the radial derivatives of U_s at these boundaries and coincidence at the heterointerface of the values of the total one-particle potential energy $U_{c(v)} + U_s$, obtained by linear extrapolation from the opposite boundaries of the transition layer to the heterointerface.

Figure 4.11 shows for the case of a spherical Si–SiO$_2$ QD with a diameter $D = 1.5$ nm potential wells $U_c(r)$ and $U_v(r)$, formed by rupture of conduction and valence bands at the boundary of QD (thin solid lines), as well as complete effective one-particle potential wells $U_c(r) + U_s(r)$ and $U_v(r) + U_s(r)$, taking into account the polarization interaction of charge carriers with this boundary (thick solid lines).

This figure also shows the lowest energy levels of the size quantization of the electron and the hole in the rectangular spherical potential wells modified by polarization interaction, the bottom of which is shifted with respect to the bottom of the original wells $U_c(r)$ and $U_v(r)$ by an amount

$$U_s(0) = \frac{e^2}{2R} \cdot \frac{\varepsilon_1 - \varepsilon_2}{\varepsilon_1 \varepsilon_2}. \qquad (4.3)$$

In the framework of the perturbation theory, the two-particle wave function $\Psi(r_e, r_h)$ is found in the form of an expansion of the products of one-particle wave functions of the size quantization of the electron and the hole in such rectangular spherical potential wells with the displaced bottom:

$$\Psi(r_e, r_h) = \sum_{ij} C_{ij} \varphi_{ei}^*(r_e) \varphi_{hi}(r_h), \qquad (4.4)$$

where i and j are the sets of quantum numbers that characterize the state of the size quantization of the electrons and holes in the QD (in each of these sets there are the radial quantum number n, orbital quantum number l and magnetic quantum number m). In this case, the main contribution to the one-particle intrinsic energy shifts gives the shift of the bottom of the potential well $U_s(0)$, which can easily be taken into account in the final result.

Figures 4.12 and 4.13 show the results of the calculation of the ground exciton state of the quantum dot for different barrier heights at the boundary with the matrix.

These figures clearly demonstrate the main results of this section:

1) the exciton binding energy in the quantum-sized $Si-SiO_2$-structure is strongly dependent on its dimension and the degree of spatial constraint;

2) differences in the dielectric permeabilities of the dot and the matrix provide the effect of dielectric enhancement;

3) the total energy of the ground exciton radiative transition is strongly dependent the height of the barriers for electrons and holes at the quantum dot–matrix interface.

In an idealized model of infinitely high barriers the transition energy is said to be grossly overestimated in comparison with the real situation (Kupchak et al, 2006).

4.5.2. Calculation of the characteristic time of radiative recombination of electron–hole excitations in the QD

The calculation of the probability of radiative electron–hole recombination in

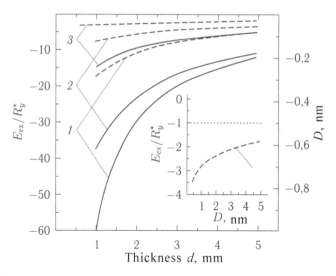

Fig. 4.12. The results of calculation of the exciton binding energy as a function of the size D of silicon quantum dots (curve 1), wires (2) and wells (3) in the SiO_2-matrix. The solid curves are obtained taking into account the effect of dielectric enhancement, and dashed ones – no such account. The inset – curve 3 on an enlarged energy scale the sector, important for the case of quantum wells in the absence of dielectric enhancement. R_y^* – the exciton binding energy in bulk silicon (Kupchak et al, 2006).

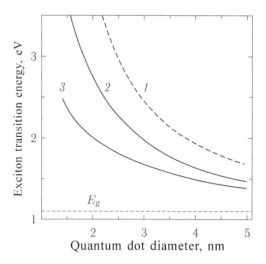

Fig. 4.13. Energy of the ground exciton transition as a function of diameter D of silicon quantum dots in the SiO_x-matrix. Curve 1 corresponds to the approximation of infinitely high barriers, curve 2 was obtained for the actual electron and hole barriers in the $Si–SiO_2$ structure, and curve 3 – in the $Si–SiO_x$ structure with $x = 1.5$ (Kupchak et al, 2006).

the QD by Kupchak et al (2006) was conducted by the standard procedure, taking into account two ground states of the electron system of the upper shells (valence electrons) and photons. We assume that the initial state of the system is characterized by unfilled states of the electromagnetic field and is populated by the exciton-like state with energy E_n and polarization σ, and the final state by the exciton-like unoccupied states and the populated photon state with energy $\hbar\omega$, the wave vector η and polarization λ.

It is evident that, as in the case of silicon quantum wells and quantum wires (Korbutyak et al, 2003), the probability of electron–hole radiative transition in silicon QDs is a non-monotonic function of their diameter D due to the oscillation of the overlap integral of the electron and hole wave functions due to the presence of a rapidly oscillating component of the electron wave function (4.4).

Figure 4.14 shows the dependence of the total complete characteristic radiative lifetime of excitons τ_{rx} in a Si–SiO$_2$ quantum dot on its diameter. The calculation was performed assuming the existence of two main channels of radiative recombination in silicon QDs: 1) the usual channel of radiative recombination of the excitons in silicon, with the participation of phonons with a characteristic time, and 2) the zero-phonon (pseudodirect) channel of radiative recombination with a characteristic time $\tau_{rx}^d = 1/W_0^r$, where W_0^r is the transition probability between the states of the system per unit time. The full characteristic time of the radiative exciton transition is defined in this case by the law of addition of inverse values:

$$\frac{1}{\tau_{rx}} = \frac{1}{\tau_{rx}^i} + \frac{1}{\tau_{rx}^d}.$$

Figure 4.14 shows that the time τ_{rx} in quantum dots, even very close in size, can differ by orders of magnitude. This may be one of the factors explaining well-known result of experimental microphotoluminescence studies of the nanostructures with silicon quantum dots, which consists in the fact that

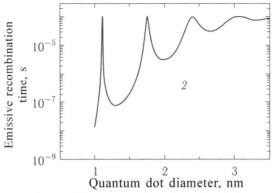

Fig. 4.14. The dependence of the characteristic time of radiative zero-phonon exciton transition in a Si–SiO$_2$ quantum dot on its diameter (Kupchak et al, 2006).

some quantum dots glow brightly enough, and others slightly or not at all.

4.5.3. Excitonic photoluminescence in silicon quantum dots

In future analysis, we restrict ourselves to the analysis of a range of QDs of a sufficiently small size, when the exciton binding energy is large (>0.3 eV) and the presence of quasi-free electron–hole pairs at room temperature can be neglected. Then the equation of generation–recombination balance for a single QD with a diameter D has the form

$$\frac{dn_x(D)}{dt} + \frac{n_x(D)}{\tau_x(D)} = c\alpha(D)D^3 I, \tag{4.5}$$

where $n_x(D)$ is the total number of excitons in QDs which are in the ground energy, I is the intensity of light, $\alpha(D)$ is the absorption coefficient, c is the form factor ($c \approx 1$ for the cubic QD; $c \approx 1/6$ for a spherical QD); $\tau_x(D)$ is the full lifetime of excitons in a QD:

$$\frac{1}{\tau_x(D)} = \frac{1}{\tau_{rx}^d(D)} + \frac{1}{\tau_{rx}^i(D)} + \frac{1}{\tau_{nx}(D)},$$

where $\tau_{nx}(D)$ is the non-radiative lifetime of the excitons.

In the stationary case we have:

$$n_x(D) = c\alpha(D)D^3 I \tau_x(D). \tag{4.6}$$

The integrated intensity of the exciton luminescence line corresponding to exciton transition energy $E(D)$ in this case is:

$$\mathcal{J}_{PL}(E) = \frac{c\alpha(D)D^3 I \tau_x(D)}{\tau_{rx}(D)}, \tag{4.7}$$

where $\tau_{rx}(D)$ is the full characteristic time of radiative exciton transition,

$$\frac{1}{\tau_{rx}(D)} = \frac{1}{\tau_{rx}^i(D)} + \frac{1}{\tau_{rx}^d(D)}.$$

In the case of an ensemble of nanocrystals, which is characterized by some (e.g. Gaussian) size distribution of QDs in around a few of the most probable sizes D_i, $i = 1, 2,..., n$, the spectral density of the exciton photoluminescence (PL), taking into account the additional (not related to the scatter of the size of QDs) broadening of the PL bands of quantum dots because of quantum effects of quantum–mesoscopic fluctuations we can be expressed in the form

$$\mathcal{J}_{PL}(E) = \int \sum_{i=1}^{n} a_i \mathcal{J}_{PL}(\xi) f_G\left(D(\xi), \bar{D}_i \sigma_i\right) \frac{\partial D}{\partial \xi} \frac{\Gamma(\xi)}{(E - \xi)^2 + \Gamma(\xi)^2/4} d\xi, \tag{4.8}$$

$$f_G\left(D,\bar{D}_i,\sigma_i\right)=\frac{1}{\sqrt{2\pi}\sigma_i}\exp\left[-\frac{\left(D-\bar{D}_i\right)^2}{2\sigma_i^2}\right],$$

where σ_i is the rms scatter of the thickness of QD in the vicinity of D_i, a_i is the weight coefficient of the corresponding Gaussian, $\Gamma(\xi)$ is the parameter of mesoscopic broadening which depends on the transition energy (i.e., on the diameter D). The D value in the integrand is expressed as a function of the energy of the exciton transition, i.e. as the function inverse to the the dependence $E(D)$.

The kinetics of the relaxation of the intensity of a single PL line with energy $E(D)$ after a short (compared to the characteristic time τ_x (D)) laser excitation pulse with t_i is given by

$$\mathcal{J}_{PL}\left(E,t\right)=c\alpha(D)D^3 I\left[\frac{t_i}{\tau_{rx}(D)}\right]\exp\left(-\frac{t}{\tau_{rx}(D)}\right). \tag{4.9}$$

As in the stationary case, in the case of some size distribution of the QD, the decay kinetics of the integrated intensity of PL is described by the formula

$$I_{int}^{PL}\left(t\right)=c\int_{D_{min}}^{D_{max}}\left[\alpha(\xi)\xi^3\exp\left(-\frac{t}{\tau_{rx}(\xi)}\right)\frac{t_i}{\tau_{rx}(\xi)}\right]f_G\left(\xi,\bar{D},\sigma\right)d\xi, \tag{4.10}$$

where f_G $(\xi,$ $D,$ $\sigma)$ is the corresponding distribution function. The spectral density of exciton PL of the ensemble of quantum dots, measured with a time delay t_d after the excitation pulse, can be written as:

$$\mathcal{J}_{PL}\left(E,td\right)=\int\sum_{i=1}^n a_i\mathcal{J}_{PL}\left(\xi,0\right)\tau_x\left(D\right)\left[\exp\left(-\left(\frac{T_d}{\tau_x(D)}\right)\right)-\exp\left(-\left(\frac{T_d+T_s}{\tau_x(D)}\right)\right)\right]\times$$

$$\times f_G\left(D(\xi),\bar{D}_i,\sigma_i\right)\frac{\partial D}{\partial\xi}\frac{\Gamma(\xi)}{\left(E-\xi\right)^2+\Gamma(\xi)^2/4}, \tag{4.11}$$

where the value of $\mathcal{J}_{PL}(\xi,$ 0) is given by (4.9) taken at $t = 0$, t_d is the time delay, T_s is the gate time.

4.5.4. Comparison with the experiment

First, we note that the results of analysis of peculiarities of the exciton photoluminescence of silicon quantum dots, surrounded by SiO_x (Kupchak et al, 2006), are in many respects similar to the results obtained for silicon filaments (Kaganovich et al, 2003).

In both cases, the indirect-gap semiconductor leads to an oscillatory nature of the dependence of the exciton zero-phonon radiative lifetime on the size of the quantum-dimensional object. In the experimentally obtained spectra of photoluminescence of this material we can observe not one but two or more major PL bands, and the relaxation kinetics of the PL is non-exponential. In addition, the photoluminescence of quantum dots should show the mesoscopic quantum effect due to the fact that in objects with a small number of particles the local fluctuations of physical quantities (including energy exciton states) can become quite large (e.g. due to various types of fluctuations of the atomic scale in the internal structure of the quantum-sized objects, their interfaces, or the immediate external environment).

Kupchak et al (2006) reviewed the experimental PL spectra of the quantum-size Si–SiO$_2$ structures obtained by different methods and with the structure similar to the 'silicon quantum dot in a matrix of SiO$_2$' model. These are: a) layers of nanoporous silicon obtained by electrochemical etching of (100)-oriented p-type silicon with resistivity ρ = 12 ohm · cm and subsequent post-anode processing for forming silicon oxide (Kaganovich et al, 2003); b) the structures with silicon quantum dots of different sizes obtained by laser deposition, c) silicon QDs obtained by sputtering of a silicon target with a beam of electrons in an electron accelerator (Efremov et al, 2004).

Fitting of the theoretical dependences to the experimental data in the majority of the studied structures identified two basic sizes of silicon nanocrystals. These dominant dimensions and other adjustable parameters are presented in Table 5.1. In the calculations (Burdov 2002) it was assumed that the mesoscopic broadening Γ, the characteristic time of radiative recombination of an exciton τ_{nx} and radiative recombination with the participation of phonons τ_{rx}^i are size-dependent:

$$\Gamma(D) = \Gamma_0 \left(\frac{D_x}{D}\right)^n, \quad \tau_{nx}(D) = T_n \left(\frac{D}{D_x}\right)^{n_1}, \quad \tau_{rx}^i(D) = T_f \left(\frac{D}{D_x}\right)^{n_2}, \quad (4.12)$$

where D_x = 3 nm. The analysis of parameters showed that in the case of highly oxidized, nanoporous silicon samples the formation of the low-energy PL band is caused by silicon nanocrystallites in the form of quantum wires with thicknesses of about D = 1.7 nm, and of the high-energy band – in the form of quantum dots with a diameter of about D = 1.8 nm.

In the case of nanostructures produced by laser deposition of silicon, the PL spectra, according to calculations by Burdov (2002), form by radiation of the quantum dots with a size of about 4 nm (low-energy band) and 2.2– 2.5 nm (high-energy band), and the differences in the spectra of different samples are mainly due to variations in the relative contributions of such QDs. This may indicate that laser deposition is accompanied by the formation of ensembles of quantum dots with the dimensions which are approximately multiples of the lattice constant of silicon; these ensembles emit the most intense radiation because of the minimization of the contribution of non-radiative recombination related mainly to the existence of dangling bonds at the interface (Koch et al, 1996). Finally, the dominant size for the quantum dots, obtained by sputtering of silicon by an electron beam, is $D = 2.2$ nm. It should be noted that in Patrone et al. (2000) the Raman spectra of Si nanostructures with quantum dots, obtained by sputtering a silicon target with the electron beam, the authors determined the average size of quantum points making the main contribution to the QDs. It turned out to be about 2 nm. Thus, the average sizes of silicon QDs, giving a decisive contribution to the photoluminescence and Raman spectra, are practically identical.

It is characteristic that the values of the exponent n in the dependence $\Gamma(D)$, determined by the matching of experimental and theoretical spectral dependence of PL for the structures with quantum dots (Fig. 4.15b and 4.15c), mainly appeared to lie between 1 and 2. The value of n for the low-energy PL band in Fig. 4.15a, which is interpreted as emission from quantum wires, was greater than 2. At the same time, the value of n is large ($n = 3.8$) also for the high-energy PL band (Fig. 4.15a), interpreted by Burdov (2002) as the emission of quantum dots.

We can assume that the high values of the exponent n are related with the non-spherical form of the quantum dots. It is known that the photoluminescence of porous silicon has a significant polarization memory – when excited by linearly polarized light it is polarized in the same way, but with a lower degree of polarization (Lavallard, Suris, 1995). This effect is explained by the non-spherical shape of the crystallites (Koyama & Fauchet, 2000). It is also well known that when in post-anode oxidation of porous silicon and the reduction of the size crystallites to the QD, depending on the method used the crystallites may become both more asymmetric and more close to spherical in shape (Koyama, 2003). In this regard, Burdov (2002) measured the polarization of the sample whose spectra is shown in Fig. 4.15a.

PL was excited by linearly polarized pulses from a nitrogen laser with a wavelength of $\lambda = 337$ nm, duration 10 ns and a repetition frequency of 100 Hz. The PL spectra were measured at normal incidence of the exciting light onto the sample using a computerized setup based on an MDR-2 monochromator, the V9-6 stroboscopic voltage converter with a gate width of 4 ns, the FEU-79 photoelectron multiplier, and polarizers. The degree of linear polarization was determined as $\rho = (I_{\parallel}^{PL} - I_{\perp}^{PL})/(I_{\parallel}^{PL} - I_{\perp}^{PL})$, where $I_{\parallel}^{PL}$ and $I_{\perp}^{PL}$ are the PL intensities, measured at the time of maximum laser pulse polarized in two mutually perpendicular directions – respectively, parallel and perpendicular to the polarization of the exciting light.

Figure 4.16 shows the spectral dependence of ρ for the sample (Fig. 4.15a, curve 1) and a control sample of porous silicon, made without special oxidation (curve 2). It is evident that the oxidized sample has a higher degree of polarization, which indicates that the silicon crystallites are highly elongated (or flattened) ellipsoids. The presence of significant

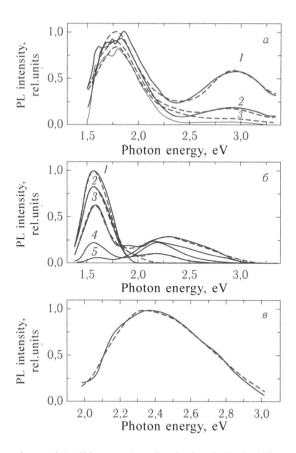

Fig. 4.15. Experimental (solid curves) and calculated (dashed line) PL spectra. *a* – PL spectra of porous silicon with a time delay of recording the moment of laser excitation by a pulse of a nitrogen laser with a wavelength $\lambda = 337$ nm, duration 10 ns and a repetition frequency of 100 Hz. Delay time, ms: *1* – 0.1, *2* – 0.2, *3* – 0.5. *b* – stationary PL spectra of five different parts of the sample with silicon quantum dots obtained by laser deposition (Kaganovich et al, 2003); with an increase in the number of the curve the proportion of QDs of smaller sizes increases (the studied section of the sample is at a greater distance from the area of deposition), and of large size – decreas. *c* – the spectrum of stationary photoluminescence of a group of silicon quantum dots from the work of Efremova et al (2004). Drawing from the work Kupchak et al (2006).

linear polarization of the high-energy PL band confirms the assumption that the strong oxidation of porous silicon the QDs, formed together with the quantum wires, are not spherical but ellipsoidal.

Thus, in the Si–SiO$_2$ structures with quantum dots with a diameter < 4 nm the concentration (amount) of the excitons is much higher than the concentration (number) of electrons and holes at room temperature. Accordingly, exciton PL dominates in these nanostructures. The main

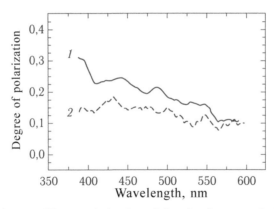

Fig. 4.16. The degree of linear polarization of photoluminescence in the region of the high-energy band of the strongly oxidized porous silicon (curve *1*) and unoxidized samples (curve *2*) (Burdov, 2002; Kupchak et al 2006).

mechanism of the broadening of the PL bands at such QD sizes if the quantum mesoscopic effect caused by a significant increase in the role of different kinds of fluctuations in systems with a small number of atoms. The developed theory can explain the experimentally observed spectra of both stationary Pl and PL with time resolution.

4.6. Silicon-based nanowires

Nanowires, nanoconductors or quantum whiskers (1–100 nm in effective diameter and micrometers in length) are objects that demonstrate the size quantum effect in the plane perpendicular to the major axis of the whisker, and the free behaviour of electrons along the main axis of the object. Typically, the nanowire (NW – NanoWire) are free-standing structures with a very large (practically – infinite) length to diameter ratio. Quantum whiskers (QWR), are usually (Avramov et al, 2008) embedded structures with a considerably smaller length to diameter ratio or aperiodic structures with one (with a substrate or a droplet of the metal catalyst) or more intermediate layers (when it comes to the axial heterostructures).

A typical way to produce the quantum nano-objects is to restrict the mobility of the electron gas in a semiconductor heterostructure with electrostatic barriers or using chemical etching technology. Such manipulations give rise to either the 0D-potential of the quantum dot or the 1D-potential of the quantum nanowhisker, which determines the behaviour of the electrons of the conductivity band (Fedorov et al, 2009). Applications of such structures intechnology revealed specific and, in general, have little-studied quantum effects.

The strong dimensional quantum effect allows to control the basic properties of objects and leads to new physics of objects in soft ambient conditions. Examples include the quantum Hall effect, ballistic conductance and Coulomb blockade (Brunner, 2002). In turn, the above-mentioned physical phenomena lead to a higher electron mobility in nanowhiskers, increase of the quantum yield, a lower lasing threshold of laser radiation with increase of the quantum size constraint and creation of new devices such as solar cells, based on nanowhiskers, infrared detectors or field effect transistors (Zunger, Wang, 1996). The recent development of new high-performance catalytic methods of synthesis with the use of metal nanoclusters has been stimulated by the study of nanowhiskers (Belyakov et al, 2008).

The main advantage of catalytic methods for the synthesis of nanostructures with the use of metal nanoclusters is the relative cheapness and the high quality of the structures. Typically, the semiconductor nanowhiskers are synthesized using the 'vapor–liquid–solid' (VLS) technology and metal clusters as catalysts. In this process, the metal nanoclusters are heated to a temperature above the melting point of the eutectic for the metal–semiconductor system in the presence of a vapour source of the semiconductor material, which leads to the formation of nanodroplets of the metal/semiconductor alloy. The continuous steam output of the semiconductor to the metal/semiconductor droplet leads to the crystallization of the semiconductor nanocluster. The 'liquid/solid' boundary leads to the formation of the boundary of nanocrystal growth with the drop of the alloy which determines the parameters of growth and structure of the whisker. In the CVD process metal nanoclusters are used as catalysts on the side on which the decomposition of precursors, generating the gaseous source of the semiconductor reagent, takes place. For example, silicon nanowhiskers are grown using monosilane and nanoclusters of gold, iron or aluminum deposited on crystalline silicon. The special, controlling role of the interfaces of mono- and nanocrystalline silicon with metal nanoclusters that serve as catalysts in the synthesis of various nanostructures has not as yet been studied by theoretical methods of quantum chemistry because of the complexity of these objects (Avramov et al, 2007; 2008, 2009).

Despite the fact that the basic electronic properties of ideal Si nanostructures have been investigated theoretically, the properties of real complex nanostructures consisting of two or more parts of different nature and composition are still theoretically unexplored. For example, the electronic and atomic structure of the following key nanoheterostructures have not as yet been calculated: nc-TM/QW, nc-TM/Si $\langle 111 \rangle$ (TM = Al, Fe, Au), the interfaces of crystalline silicon with Si–QW of different types, metal-doped silicon quantum dots and whiskers, and many other objects (Avramov et al, 2009). In particular, it was shown that the theoretical methods describe the size effect of photoluminescence spectra of complex silicon nanostructures with high accuracy. So, it was shown that the band structure $\langle 110 \rangle$ of oriented silicon nanowhiskers shows both the direct and indirect nature of the forbidden gaps of various widths, depending on the symmetry and size of objects (Sorokin et al, 2008).

The atomic structure of a new broad class of silicon quantum dots of the Goldberg type and conglomerate structures based on them were proposed (Avramov et al, 2009). It was shown that quantum dots of the Goldberg type have high-symmetry fullerene-like central cavities, surrounded by one or several layers of tetragonal silicon. The symmetry and electronic structure of the quantum dots themselves are determined by the symmetry of the central fragments. In fact, these nanoclusters are twin structures formed when connecting more than 20 tetrahedral fragments of the silicon crystal. Within the central cavities there can be guest atoms, ions and molecules forming a new class of endohedral complexes, arising, for example, when porous silicon is doped with rare earth element ions and in subsequent partial oxidation of the system (Avramov et al, 2009). These structures are of great interest for the development of new semiconductor materials and devices.

The structure of a new exotic class of silicon nanoclusters ('silicon nanocolours'), observed in the experiments, and also of silicon nanodots embedded in silicon whiskers, has also been proposed. It was shown that these structures can be formed when two or more identical nanostructures of the same of different symmetry are joined together. Electronic structure calculations have shown that the formation of QD/NW and NW/NW interfaces is energetically favourable, and the width of the band gap is determined by the size of the largest fragment of a complex system. The detected atypical effect of the quantum size constrain (with a plateau and/ or maxima) is determined by the interactions of the electronic states near the Fermi level, localized in different parts of the complex nanocluster (Sorokin et al, 2008).

4.7. Silicon quantum dots doped with boron and phosphorus

The radiation efficiency of silicon nanocrystals is low compared to the direct-gap III-V or II-VI materials. The indirect-band structure of bulk silicon is to some extent manifested also in the nanocrystals, which also hinders their use in optics. Therefore, the problem of introduction of silicon to the element base of modern optoelectronics as the main, or at least, a widespread component, is still far from being solved. This circumstance is caused by the search undertaken in the last decade for a controlled impact on the electronic structure of nanocrystals, effectively 'rectifying' the energy bands.

Doping with shallow impurities was proposed as one of the methods of modifying the optical properties of silicon nanocrystals. It was found that the photoluminescence intensity increased several times when nanocrystals doped with phosphorus (Tetelbaum et al, 1998). At the same time, no amplification took place when boron was embedded in the nanocrystals (Kachurin et al, 2006). The nature of this phenomenon is still debated. We can only assert that it is very likely it is caused by various processes and mechanisms that affect both the non-radiative and radiative recombination in nanocrystals. The role of the latter is practically unknown.

The electronic structure and optical properties of Si nanocrystals are studied using different calculation methods, such as the theory of density functional (Delley & Steigmeier, 1993), the pseudopotential method (Zunger & Wang, 1996), the strong coupling approximation (Niquet et al, 2000), as well as the envelope approximation (k–p-method), which are discussed in sections 4.4–4.6). Delerue et al, 2006; Belyakov, Bourdieu, 2007; Belyakov, Burdov, 2008 conducted a theoretical study of the effect of doping with shallow impurities on the emissivity of silicon nanocrystals.

Belyakov, Burdov, 2007 and Belyakov & Burdov, 2008 developed the method finding electronic states in a quantum dot with an impurity, which is based on the solution of the Schrödinger equation with the zero boundary condition for the envelope function. The Coulomb potential acting on an electron by an impurity ion is divided into the long-range and short-range parts:

$$V_c = V(r) + W(r), \tag{4.13}$$

where $V(r)$ is the long-range component, which describes the macroscopic Coulomb field in the medium, and $W(r)$ is the short-range potential existing only near the nucleus of the impurity on the scale of the order of the Bohr radius. According to the work of Baldereschi et al. (1974), the potential $W(r)$ for boron can be considered equal zero, which indicates that boron is a hydrogen-like acceptor and creates only a long-range field $V(r)$. The explicit form of the field $V(r)$ in the nanocrystal in the light of the polarization charges arising at the boundary of the quantum dot was determined. The method of calculation of the energy spectrum and wave functions within the envelope approximation was demonstrated for an undoped nanocrystal. The electronic structure in the valence band and in the conduction band, respectively, for the nanocrystal doped with a boron atom, was calculated It is shown that at the central position of the boron ion the ground state is degenerate twelve-fold in the conduction band and six times in the valence band and the wave functions of the ground states in both areas have predominantly envelopes of the s-type. For an arbitrary position of the impurity in the nanocrystal the degeneracy of the spectrum is almost completely removed (only two-fold spin degeneracy remains), and the wave functions of the ground states acquire big additions with the envelopes of the p-type.

Mikhaylov et al. (2008) examined the electronic states of nanocrystals doped with a donor impurity – phosphorus. The main difference from the problem with the hydrogen-like acceptor is the appearance of an additional short-range potential $W(r)$, which is also called the potential of the central cells created by the impurity. It is shown that for phosphorus in silicon when it is a substitutional donor

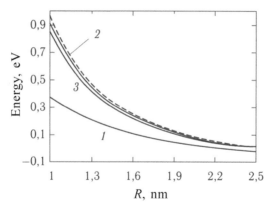

Fig. 4.17. The energies of singlet (*1*), doublet (*2*) and triplet (*3*) in the nanocrystal (phosphorus ion – in centre). Dashed line – the original six-fold degenerate level of a hydrogen-like donor (Belyakov et al, 2008)

$$W(r) = -\frac{e^2}{r}\left[A\exp(-\alpha r) + \left(1 - A - \frac{1}{\varepsilon_s}\right)\exp(-\beta r)\right], \qquad (4.14)$$

where ε_s is the dielectric constant of a spherical quantum dot, the constants α and β are equal to 0.82 and 5.0, respectively, of the inverse of the Bohr radius, and $A = 1.142$. The potential of the central cell differs from zero only in the unit cell containing the donor, so in the equation for the envelope the central cell potential can be replaced by the Dirac function, i.e. treat it as in the zero radius potential. It is shown that the short-range potential of the central cell causes the valley-orbital interaction in the conduction band, which leads to a strong splitting of the sixfold degenerate lower level into singlet, doublet and triplet with the symmetry of the irreducible representation A_1, E and T_2 of the point group T_d (see Fig. 4.17). Moreover, the degree of splitting increases dramatically with decreasing nanocrystal radius R.

The electronic states below the optical gap were studied. It is shown that in the valence band the potential of the central cell of phosphorus is not as effective and causes only small corrections in the energy spectrum and wave functions. Belyakov et al (2008) calculated the radiative recombination time τ_R for a quantum dot containing a hydrogen-like acceptor – boron. It is shown that regardless of the pumping method – continuous or pulsed, the intensity of photoluminescence is always dependent on the rate of radiative recombination, a key parameter that affects the efficiency of the generation of photons. The rate of zero-phonon radiative transfer in a silicon nanocrystal was calculated using the first-order of the perturbation theory with respect to time ('the Fermi golden rule'). The calculation results show that in the investigated range of sizes of nanocrystals (2–6 nm) doped with a hydrogen-like impurity (boron) the zero-phonon recombination rate is sufficiently small – less than 10^3 s^{-1} and more than two orders of magnitude lower than the rate of transitions involving a phonon. The radiative transitions, occurring with the participation of phonons, were studied. The rate of these processes were calculated in the second order of the perturbation theory. The interaction of the electron with the phonon is described by the model of rigid ions (Allen & Cardona, 1981). The dependence of reciprocal time (rate) of the radiative recombination in a quantum dot, doped with boron, at room temperature and different values of boron ion displacement from the centre of the quantum dot was also determined. The calculation results show that for any displacement of the acceptor from the centre the rate of radiative recombination is less than in the 'pure' nanocrystal. The recombination probabilities in the pure and doped nanocrystals are practically identical only in the case of the central position of the impurity.

The principal difference between phosphorus and boron (in addition to one of them being a donor and the other one an acceptor) is that the phosphorus ion produces a strong short-range potential in its small neighbourhood – the potential the central cell. This difference fundamentally affects the probabilities of the interband transitions. It is shown that the potential of the central cell is capable of, like phonons, effectively mixing the Bloch functions of Γ- and X-points of the Brillouin zone, and two of these mechanisms actually do not interfere. As a result, the total rate of radiative recombination in the nanocrystal doped with phosphorus is the sum of the contributions of these two processes:

$$\tau_R^{-1} = \tau_D^{-1} + \tau_{ph}^{-1}, \tag{4.15}$$

τ_D^{-1} – the rate of recombination induced by the potential of the central cell donor, and τ_{ph}^{-1} – the rate of recombination with the participation of a phonon.

It has been shown that the introduction of the phosphorus atom to a quantum dot almost always reduces the rate of radiative recombination taking place with the participation of phonons. Only in the case of the central position of the impurity in the nanocrystals with a radius greater than 2 nm we can observe a slight increase of τ_{ph}^{-1} (1–2%) compared with the undoped nanocrystal. In contrast, transitions, due to Γ–X mixing as a result of the short-range potential of the phosphorus ion, can be significantly accelerated. It is shown that for nanocrystals with the size of 2–4 nm with the central position of the impurity such transitions are faster than the transitions taking place with the participation of phonons (see Fig. 4.18). This leads to an increase in the total radiative recombination rate in comparison with the case of the undoped crystallites.

It was also found that τ_D^{-1} decreases with increasing displacement of the donor from the centre of the quantum dot. The fall sharply increases at $R \sim$ 2–3 nm. The reason is the impact of the weakening of the potential of the central cell on the state of the electron as it approaches the boundary of the

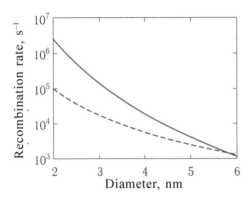

Fig. 4.18. The rate of radiative transition induced by the central cell potential τ_D^{-1} (solid line) as a function of the nanocrystal diameter at the central position of the donor. Dashed line – the rate of recombination in the undoped nanocrystal τ_{R0}^{-1} (Belyakov et al, 2008).

quantum dot due to a substantial reduction of electron density in this area. Analysis of the dependence of the rate of radiative recombination on the position of the donor enables us to introduce the concept of the effective region of doping – a region in the nanocrystal in which the phosphorus situated in this region is able to significantly increase the rate of radiative recombination and enhance the photoluminescence intensity in comparison with the case of the undoped quantum dots. The size of the effective doping region decreases with increasing size of the nanocrystal.

Concluding remarks

The restructuring of the energy spectrum of quantum dots in comparison with bulk materials and changes of the interactions of elementary excitations, induced by the spatial constraint, are manifested in the responses of such objects to external disturbances. In particular, large changes take place in the optical responses of quantum dots such as single- and multi-photon absorption, quasi-elastic, Raman and hyper-Raman scattering, as well as various types of luminescence (see detailed discussion of these issues in the monograph (Feder, 2005)). Changes also take place in the spectral lines, their widths and relative amplitudes, and significant changes are also observed in the rate of dephasing of the optical transitions and the relaxation rate of the excited states which determine the transient response of quantum dots to pulsed optical excitation (see chapter 11).

References for Chapter 4

Bagraev N. T., Buravlev A. D., Klyachkin L. E., Malyarenko A. M., Rykov S. A. Interference of current carriers in one-dimensional semiconductor rings. Fiz. Tekhnol.
Poluprovod. 2000. V. 34, No. 7. P. 846-855.

Belyakov V. A., Burdov V. A. The structure of the ground state of the electrons and holes in
silicon quantum dot with a small donor. Poverkhnost'. 2007. Np. 2. P. 40-43.

Belyakov V. A., Burdov V. A., Gaponova D. M., Mikhailov A.N., Tetelbaum D. I., Trushin
S. A. The radiative electron-hole recombination in silicon quantum dots with phonons. Fiz. Tverd. Tela. 2004. V. 46. P. 31-37.

Boychuk V. I., Kubai R. Yu., Effect of the intermediate layer with a variable dielectric
constant on the coordinates of the ground state energy of the electron in the spherical
complex nanoheterosystem. Fiz. Tverd. Tela. 2001. V. 43, No. 2. P. 226-232.

Burdov V. A. Dependence of the width of the silicon quantum dots on their size. Fiz. Tekhnol. Poluprovod. 2002. V. 36, No. 10. P. 1233-1236.

Geisler S. V., Semenova O., Sharafutdinov R. G., Kolesov B. A. The analysis of the Raman
spectra of amorphous-nanocrystalline silicon films. Fiz. Tverd. Tela. 2004. V. 46,
No. 8. P. 1484-1488.

Golubev V. G., Davydov V., Medvedev A. V., Pevtsov A.B., Feoktistov N. A., Raman
spectra and electrical conductivity of thin silicon films with mixed amorphous-crys-
talline composition: the definition of the volume fraction of nanocrystalline phase.
Fiz. Tverd. Tela. 1997. V. 39, No. 8. P. 1348-1353.

Emelyanov V. I., Seminogov V. N., Dependence of the fraction of the crystalline phase in
the system of Si nanoclusters in SiO_2 matrix on the annealing temperature. Pis'ma
Zh. Eksp. Teor. Fiz. 2006. V. 32, No. 24. P. 18-23.

Efremov M.D, et al.Visible photoluminescence of silicon nanopowder created evaporation
powerful electron beam. Pis'ma Zh. Eksp. Teor. Fiz. 2004. V. 80. P. 619-622.

Kaganovich E. B., Manoylov E. G., Bazylyuk I. R., Svechnikov S. V. The photoluminescence spectra of silicon nanocrystals. Fiz. Tekhnol. Poluprovod. 2003. V. 37. P.
353-357.

Kachurin G. A., Cherkova S. G., Volodin V. A., Marin J. M., Tetel'baum D. I., Becker H.
The influence of boron ion implantation and subsequent annealing on the properties
of nanocrystals of Si. Fiz. Tekhnol. Poluprovod. 2006. V. 40. P. 75-81.

Kashkarov P. K., Lisachenko M. G., Shalygina O. A., Timoshenko V. Yu., Kamenev B. V.,
Schmidt M., Khaitmann I., Zakharis M. Photoluminescence of Er^{3+} ions in the layers quasi-
ordered silicon nanocrystals in a silicon dioxide matrix. Pis'ma Zh. Eksp. Teor. Fiz.
V. 124. P. 1255-1261.

Kopylov A. A., Pikhtin A. N. . The electron–phonon interaction on 'small' donors in gallium
phosphide. Pis'ma Zh. Eksp. Teor. Fiz. V. 24, No. 4. P. 193-195.

Kupchak I. M., Korbutyak D. V., Kryuchenko Yu. V., Sachenko A. V., Sokolovskii I. O.
Sreseli O. M. Characteristics of excitons and excitonic photoluminescence of structures with silicon quantum dots. Fiz. Tekhnol. Poluprovod. 2006. V. 40, No. 1. P.
98-107.

Landau L. D., Lifshitz E. M. Quantum mechanics. V. 3. Moscow: Fizmatlit, 2001.

Lisovski I. P., Indutnyi I. Z., Gnennyi B. N., Litvin P. M., Mazunov D. O., Oberemok A. S.,
Sopinsky N. V., Shepelyavyi P. E., Phase-structural transformations in SiO_x films
in the process of vacuum thermal treatment. Fiz. Tekhnol. Poluprovod. 2003. V. 37,
No. 1. P. 98-103.

Pokutnyi S. I., Theory of excitons in quasi-zero-dimensional semiconductor systems. Odes-
sa: Astroprint, 2003. 154 p.

Pchelyakov O. P., Bolkhovityanov Yu. B., Dvurechenskii A. V., Sokolov L.V., Nikiforov A. I., Yakimov A. I., Foikhtelnder B. Silicon–germanium nanostructures with quantum dots: the mechanisms of formation and electrical properties. Fiz. Tekhnol. Poluprovod. 2000. V. 34, No. 11. P. 1281-1299.

Sachenko A. V., Kryuchenko Yu. V., Sokolovskii I. O., Sreseli O. M. The manifestation of quantum-dimensional oscillations of the radiative exciton recombination in the photoluminescence of silicon nanostructures. Fiz. Tekhnol. Poluprovod. 2004. V. 38, No. 7. P. 877-883.

Seminogov V. N., et al. The study of structural–phase transformations and optical properties of composites based on Si nanoclusters in a matrix of silicon oxide. Dynamics of Complex Systems. 2009. V. 3, No. 2. P. 3-16.

Tkach N. V., Golovatsky V. A., Voitsekhovskaya O. M., Mikhalkov M. Y., Fartushinskii R. B. The quasi-stationary states of electrons and holes in an open quantum wire. Fiz. Tverd. Tela. 2001. V. 43. P. 1315–1321.

Fedorov A. V., Baranov A. V. Optics quantum dots. In: Optics of Nanostructures. Ed. AV Fedorov. St. Petersburg: Nedra, 2005. Chap. 4. P. 181-274.

Shklovskii B. I., Efros A. L. Electronic properties of doped semiconductors. Moscow: Nauka. 1979. Chap. 5. P. 126-183.

Allen P.B., Cardona M. Theory of the temperature dependence of the direct gap of germanium. Phys. Rev. B 1981. V. 23. P. 1495–1505.

Avramov P.V., Fedorov D.G., Irle S., Kuzubov A.A., Morokuma K. Strong electron correlations determine energetic stability and electronic properties of Er-doped goldberg-type silicon quantum dots. J. Phys. Chem. C. 2009. V. 113. P. 15964–15971.

Avramov P.V., Fedorov D.G., Sorokin P. B., Chernozatonskii L.A., Ovchinnikov S.G. Quantum Dots Embedded into Silicon Nanowires Effectively Partition Electron Confinement. J. Appl. Phys. 2008. V. 104. P. 054305–1.

Avramov P.V., Fedorov D.G., Sorokin P. B., Chernozatonskii L.A., Gordon M. S. Atomic and electronic structure of new hollow-based symmetric families of silicon nanoclusters (Letter). J. Phys. Chem. C. 2007. V. 111. P. 18824.

Babic D., Tsu R., Greene R. F. Ground-state energies of one- and two-electron silicon dots in an amorphous silicon dioxide. Phys. Rev. B 1992. V. 45. P. 14150–14155.

Bagraev N. T., Chaikina E. I., Gehlhoff W., Klyachkin L. E., Markov I. I., Malarenko A.M. Infrared induced emission from silicon quantum wires. Solid State Electronics. 1998. V. 7–8. P. 1199–1204.

Baldereschi A., Lipari N.O. Cubic contribution to the spherical model of shallow impurity states. Phys. Rev. B 1974. V. 9. P. 1525–1539.

Bell R. J., Bird N. F., Dean P. The vibrational spectra of vireous silica, germania and beryllium fluoride. J. Phys. C: Solid State Phys. 1968. V. 1. P. 299–305.

Belomoin G., Therrien J., Smith A., Rao S., Chaieb S., Nayfeh M.H., Wagner L., Mitas L. Observation of a magic discrete family of ultrabright Si nanoparticles. Appl. Phys. Lett. 2002. V. 80. P. 841–843.

Belomoin G., Therrien J., Nayfeh M. Oxide and hydrogen capped ultrasmall blue luminescent Si nanoparticles. Appl. Phys. Lett. 2000. V. 77. P. 779–781.

Belyakov V.A., Burdov V.A. Anomalous splitting of the hole states in silicon quantum dot with shallow acceptor. J. Phys.: Condens. Matter. 2008. V. 20. P. 025213 (13 p.).

Belyakov V.A., Burdov V.A. Chemical-shift enhancement for strongly confined electrons in silicon nanocrystals. Phys. Lett. A. 2007. V. 367. P. 128–134.

Belyakov V.A., Burdov V.A. Fine splitting of electron states in silicon nanocrystal with a hydrogenlike shallow donor, Nanoscale Res. Lett. 2007. V. 2. P. 569–575.

Belyakov V.A., Burdov V.A. Valley-Orbit Splitting in Doped Nanocrystalline Silicon: k-p calculations. Phys. Rev. B. 2007. V. 76. P. 045335-1-045335–12.

Belyakov V.A., Burdov V.A., Lockwood R., Meldrum A. Silicon Nanocrystals: Fundamental Theory and Implications for Stimulated Emission. Adv. Opt. Tech. 2008. Article ID 279502. 32 p.

Bimberg D., Grundmann M., Ledentsov N.N. Quantum dot heterostructures. N.Y.: John Wiley, 1999. 301 p.

Birkedal D., Leosson K., Hvam J.M. Longe lived coherence in self-assembled quantum dot. Phys. Rev. Lett. 2001. V. 87. P. 227401 (4 p.).

Brunner K. Si/Ge nanostructures. Rep. Progr. Phys. 2002. V. 65. P. 27–68.

Campbell L.H., Fauchet P.M. The effects of microcrystal size and shape jn the one phonon Raman spectra of crystalline semiconductors. Solid State Comm. 1986. V. 58, No. 10. P. 739–741.

Canham L. T. Silicon quantum wire array fabrication by electrochemical and chemical dissolution of wafers. Appl. Phys. Lett. 1990. V. 57. P. 1046–1048.

Cheylan S., Elliman R.G., Gaff K., Durandet A. Luminescence from Si nanocrystals in silica deposited by helicon activated reaction evaporation. Appl. Phys. Lett. 2001. V. 78. P. 1670–1672.

Dehlinger G., Diehl L., Gennser U., Sigg H., Faist J., Ensslin K., Grutzmacher D., Muller E. Interband Electroluminescence from Silicon-Based Quantum Cascade Structures Science. 2000. V. 290. P. 2277–2280.

Delerue C., Allan G., Lannoo M. Theoretical aspects of the luminescence of porous silicon Phys. Rev. B. 1993. V. 48. P. 11024–11036.

Delerue C., Allan G., Reynaud C., Guillois O., Ledoux G., Huisken F. Multiexponential photoluminescence decay in indirect-gap semiconductor nanocrystals. Phys. Rev. B. 2006. V. 73. P. 235318-1-235318–4.

Delerue C., Lannoo M., Allan G. Excitonic and Quasiparticle Gaps in Si nanocrystals. Phys. Rev. Lett. 2000. V. 84. P. 2457–2460.

Delley B., Steigmeier E. F. Quantum confinement in Si nanocrystals. Phys. Rev B. 1993. V. 47. P. 1397–1400.

Delley B., Steigmeier E. F. Size dependence of band gap in silicon nanostructures. Appl. Phys. Lett. 1995. V. 67. P. 2370–2372.

Du C.-X., Duteil F., Hansson G.V., Ni W.-X. Si/SiGe/Si:Er:O light-emittting transistors prepared by differential molecular-beam epitaxy. Appl. Phys. Lett. 2001. V. 78. P. 1697–1699.

Dvurechenskii A.V., Nenashev A.V., Yakimov A. I. Electronic structure of Ge/Si quantum dots. Nanotechnology. 2002. V. 13. P. 75–82.

Ekimov A. I., Onushchenko A.A. Quantum size effect in three-dimensional microscopic semiconductor crystals. JETP Lett. 1981. V. 34. P. 345–348;

Fedorov D.G., Avramov P.V., Jensen J.H., Kitaura K. Analytic gradient for the adaptive frozen orbital bond detachment in the fragment molecular orbital method. Chem. Phys. Lett. 2009. V. 477. P. 169–173.

Franceschetti A. Nanostructured materials for improved photoconversion. MRS Bulletin. 2011. V. 36. P. 192–197.

Franceschetti A., Zander A. Electronic Phase Transitions with Diameter changes in Si and InP Nanowires. Bull. Amer. Phys. Soc. 2011. V. 56, No. 1. P. 73–79.

Friedman L., Sun G., Soref R.A. SiGe/Si THz laser band on transitions between inverted mass light-hole and heavy-hole subbands. Appl. Phys. Lett. 2001. V. 78. P. 401–403.

Gaponenko S.V. Optical properties of semiconductor nanocrystalls. Cambridge University Press. Cambridge, 1998. 260 p.

Giri P.K., Coffa S., Rimini E. Evidence for small interstitial clusters as the origin of photoluminescence W band in ion-implanted silicon. Appl. Phys. Lett. 2001. V. 78. P. 291–293.

Guha S., Quadri B., Musket R.G., Wall M.A., Shimizu-Iwayama T. Characterization of Si nanocrystals grown by annealing SiO films with uniform concentration of implanted Si, J. Appl. Phys. 2000. V. 88. P. 3954–3961.

Hansson G.V., Ni W.-X., Du C.-X., Elfing A., Duteil F. Origin of abnormal temperature dependence of electroluminescence from Er/O-doped Si diodes. Appl. Phys. Lett. 2001. V. 78. P. 2104–2106.

Hill N.A., Whaley K. B. Size dependence of excitons in silicon nanocrystals. Phys. Rev. Lett.1995. V. 75. P. 1130–1133.

Iacona F., Bongiorno C., Spinella C., Boninelli S., Priolo F. Formation and evolution of luminescent Si nanoclusters produced by thermal annealing of SiO_x films. J. Appl. Phys. 2004. V. 95 (7). P. 3723–3729.

Jackson J.D. Classical Electrodynamics. 3 ed. N.Y.–London: John Wiley & Sons Inc., 1999. 841 p.

Kanemitsu Y. Photoluminescence spectrum and dynamics in oxidized silicon nanocrystals: A nanoscopic disorder system. Phys. Rev. B. 1996. V. 53. P. 13515–13520.

Karrai K., Warburton R. J., Schulhauser C., H"ogele A., Urbaszek B., McGhee E. J., Govorov A.O., Garcia J.M., Gerardot B.D., Petroff P.M. Hybridization of electronic states in quantum dots through photon emission. Nature. 2004. V. 427. P. 135–138.

Kazarinov R. F., Suris R.A. Possibility of the amplification of electromagnetic waves in semiconductor with a superlattice. Sov. Phys. Semicond. 1971. V. 51. P. 77–85.

Khurgin J. B., Forsythe E.W., Tompa G. S., Khan B.A. Influence of the size dispersion on the emission spectra of the Si nanostructures. Appl. Phys. Lett. 1996. V. 69. P. 1241–1243.

Knox R. Theory of Excitons. N.Y.–London: Academic Press, 1963. 207 p.

Koch F., Kovalev D., Averboukh B., Polisski G., Ben-Chorin M. Polarization phenomena in the optimal properties of porous silicon. J. Luminesc. 1996. V. 70. P. 320–332.

Korbutyak D.V., Kryuchenko Yu.V., Kupchak I.M., Sachenko A.V. Characterization of confined exciton states in silicon quantum wires. Semicond. Phys., Quant. Electron. Optoelectron. 2003. V. 6, No. 2. P. 172–182.

Koyama H., Fauchet P.M. Anisotropic polarization memory in thermally oxidized porous silicon. Appl. Phys. Lett. 2000. V. 77 (15). P. 2516–2518.

Koyama H. Anisotropic photoluminescence from porous silicon layers made under polarized illumination: Origin of contradictory experimental observations. J. Appl. Phys. 2003. V. 93, No. 5. P. 2410–2413.

Lavallard P., Suris R.A. Polarized photoluminescence of an assembly of non cubic microcrystals in a dielectric matrix. Sol. St. Commun. 1995. V. 95. P. 267–269.

Lee J.-C. Grows of Self-organized quantum dots. In: Semiconductor quantum dots. Physics, Spectroscopy and Applicattions. Ed. By Y. Masumoto and T. Takagahara. Springer. 2002. P. 1–57.

Leung K., Whaley K. B. Electron-hole interactions in silicon nanocrystals. Phys. Rev. B. 1997. V. 56. P. 7455–7468.

Lin C.-F., Chen M.-J., Chang S.-W., Chung P.-F., Liang E.-Z., Su T.-W., Liu C.W. Electroluminescence at silicon band gap energy from mechanically pressed indium–tin–oxide/Si contact // Appl. Phys. Lett. 2001. V. 78. P. 1808–1810.

Linnros J., Lalic N., Galeckas A., Grivickas V. Analysis of stretched exponential photoluminescence decay from nanometer sized silicon crystals in SiO_2. J. Appl. Phys. 1999. V. 86. P. 6128–6134.

Mikhaylov A.N., Tetelbaum D. I., Burdov V.A., Gorshkov O.N., Belov A. I., Kambarov

D.A., Belyakov V.A., Vasiliev V.K., Kovalev A. I., Gaponova D.M. Effect of ion doping with donor and acceptor impurities on intensity and lifetime of photoluminescence from SiO_2 films with silicon quantum dots. J. Nanosci. Nanotechnol. 2008. V. 8. P. 780–789.

Mimura A., Fujii M., Hayashi S., Kovalev D., Koch F. Photoluminescence and free-electron absorption in heavy-doped Si nanocrystals. Phys. Rev. B. 2000. V. 62. P. 12625–12627.

Mitas L., Therrien J., Twesten R., Belomoin G., Nayfeh M. Effect of surface reconstruction on the structural prototypes of ultrasmall ultrabright Si 29 nanoparticles. Appl. Phys. Lett. 2001. V. 78. P. 1918–1920.

Nakamura M., Mochizuki Y., Usami K., Itoch Y., Nozaki T. Infrared absorption spectra and composition of evaporated silicon oxides (SiOx). Solid State Commun. 1984. V. 50, No. 12. P. 1079–1081.

Nayfeh M., Akcakir O., Therrien J., Belomoin G., Barry N., Gratton E. Second harmonic generation in microcristallite films of ultrasmall Si nanoparticles. Appl. Phys. Lett. 2000. V. 77. P. 4086–4088.

Nayfeh M., Akcakir O., Therrien J., Yamano Z., Barry N., Yu W., Gratton E. Highly nonlinear photoluminescence threshold in porous silicon. Appl. Phys. Lett. 1999. V. 75. P. 4112–4114.

Nesbit L.A. Annealing characterization of Si-rich SiO2 films. Appl. Phys. Lett. 1985. V. 46, No. 1. P. 38–40.

Ng W. L., Lorenco M.A., Gwillam R.M., Shao G., Homewood K. P. An efficient room-temperature silicon based light-emitting diode. Nature. 2001. V. 410. P. 192–194.

Niquet Y.M., Delerue C., Allan G., Lannoo M. Method for tight-binding parametrization: Application to silicon nanostructures. Phys. Rev. B. 2000. V. 62. P. 5109–5116.

Nishimura K., Nagao Y. Novel structures for porous silicon light-emitting diodes. J. Porous Materials. 2000. V. 7. P. 119–123.

Nozaki T., Iwamoto M., Usami K., Mukai K., Hiraiwa A. Helium-3 activation analysis of oxygen in silicon nitride films on silicon wafers. J. Radioanal. and Nucl. Chem. 1979. V. 52. P. 449–459.

Ogut S., Chelikowsky J. R., Louie S.G. Quantum confinement and optical gaps in Si nanocrystals // Phys. Rev. Lett. 1997. V. 79. P. 1770–1773.

Park N.M., Kim T. S., Park S. J. Band gap engineering of amorphous silicon quantum dots for light emitting diodes. Appl. Phys. Lett. 2001. V. 78. P. 2575–2577.

Patrone L., Nelson D., Safarov V. I., Sentis M., Marine W. Photoluminescence of silicon nanoclusters with reduced size dispersion produced by laser ablation. J. Appl. Phys. 2000. V. 87. P. 3829–3838.

Pavesi L., Dal Negro L., Mazzoleni C., Franzo G., Priolo F. Optical gain in silicon nanocrystals. Nature. 2000. V. 408. P. 440–444.

Porjo N., Kuustla T., Heikkila L. Characterization of photonic dots in Si/SiO thin film structures. J. Appl. Phys. 2001. V. 89. P. 4902–4906.

Reboredo F.A., Franceschetti A., Zunger A. Dark excitons due to Coulomb interactions in silicon quantum dots. Phys. Rev. B 2000. V. 61. P. 13073–13087.

Ren S. Y., Dow J.D. Hydrogenated Si clusters: Band formation with increasing size. Phys. Rev. B. 1992. V. 45. P. 6492–6496.

Richter H., Wang Z. P., Ley L. The one phonon Raman spectrum in microcrystalline silicon. Solid State Communications. 1981. V. 39, No. 5. P. 625–629.

Rinnet H., Vergant M., Burneau A. Evidence of light-emitting amorphous silicon clusters confined in a silicon oxide matrix. J. Appl. Phys. 2001. V. 89, No. 1. P. 237–243.

Smith A., Yamani Z.H., Roberts N., Turner J., Habbal S. R., Granick S., Nayfeh M.H.

Observation of strong direct-like oscillator strength in the photoluminescence of Si nanoparticles. Phys. Rev. B. 2005. V. 72. P. 205307 (1–5).

Sorokin P.B., Avramov P.V., Kvashnin A.G., Kvashnin D.G., Ovchinnikov S.G., Fedorov A. S. Density functional study of <110>-oriented thin silicon nanowires. Phys. Rev. B. 2008. V. 77. P. 235417–1.

Takagahara T., Takeda K. Excitonic exchange splitting and Stokes shift in Si nanocrystals and Si clusters. Phys. Rev. B. 1996. V. 53. P. R4205–R4208.

Takagahara T., Takeda K. Theory of quantum confinement effect on excitation in quantum dots in indirect gap materials. Phys. Rev. B 1992. V. 46. P. 15578–1581.

Takagi H., Ogawa H., Yamazaki Y., Ishizaki A., Nakagiri T. Quantum size effects on photo-luminescence in ultrafine Si particles. Appl. Phys. Lett. 1990. V. 56. P. 2379–2380.

Takeoka S., Fujii M., Hayashi S. Size-dependent photoluminescence from surface- oxi-dized Si nanocrystals in a weak confinement regime. Phys. Rev. B. 2000. V. 62. P. 16820–16825.

Tetelbaum D. I., Karpovich I.A., Stepikhova M.V., Shengurov V.G., Markov K.A., Gor-shkov O.N. Characteristics of photoluminescence in SiO_2 with Si nanoinclusions produced by ion implantation. J. Surf. Invest. 1998. V. 14. P. 601–608.

Valenta J., Juhasz R., Linnros J. Photoluminescence from single quantum dots at room tem-pera- ture. J. Luminesc. 2002. V. 98. P. 15–22.

van Buuren T., Dinh L.N., Chase L. L., Siekhaus W. J., Terminello I. J. Changes in the electronic properties of Si nanocrystals as a function of particle size. Phys. Rev. Lett. 1998. V. 80. P. 3803–3806.

Voos M., Uzan Ph., Delalande C., Bastard G., Halimaoui A. Visible photoluminescence from porous silicon: A quantum confinement effect mainly due to holes?. Appl. Phys. Lett. 1992. V. 61. P. 1213–1215.

Wang L.-W., Zunger A. Solving Schrödinger equation around a desired energy: Application to silicon quantum dots. J. Chem. Phys. 1994. V. 100. P. 2394–2397.

Zunger A., Wang L.-W. Theory of silicon nanostructures. Appl. Surf. Sci. 1996. V. 102. P. 350–359.

Part II

5

Synthesis, surface modification and characterization of nanosilicon

The essential difference between the properties of a material constructed, for example, from the individual atoms or molecules and the substance of the same composition, but consisting of nanoparticles, is associated not only with the manifestation of the quantum size effect, but also with the lack of the identity of structural elements in the nanomaterial. In contrast to the atoms or molecules, the nanoparticles are not completely identical. Strictly speaking, they are different not only in size but also in composition. It can be assumed that even nanoparticles of the same size are not identical due to possible differences in the amount, composition and distribution of impurities, which each contains. The presence and composition of the impurities are determined by the entire 'history' and depend on the nanoparticle synthesis method and then subsequent location of the substance in the environment.

One of the significant, inherent characteristics of the nanocrystal is the size distribution function. The main parameters of the distribution function – the position of the maximum, halfwidth, asymmetry and kurtosis – are significantly reflected, for example, in the optical properties of nanomaterials and composites on its basis. The directional change of these characteristics of the particles size distribution function to a correlated change in the spectral properties. The directional change in the particle distribution function is achieved primarily with the appropriate modification of the technological mode of their production, fractionation of the material, the chemical effects in etching and annealing in different media, by controlling the composition and amount of impurities in the nucleus and the shell of nanocrystalline silicon.

5.1. Classification of synthesis methods

The main methods for producing nanocrystalline silicon include the following (see also Heitmann et al, 2005):
  • reduction of silicon compounds in solutions;

- diffusion collection of silicon atoms in the gas phase;
- ion implantation in the SiO_2 matrix with subsequent high-temperature treatment;
- evaporation of silicon under the influence of the electron beam;
- laser ablation;
- laser chemical method;
- mechanochemical synthesis;
- spraying (including cathode, magnetron) followed by heat treatment for the crystallization of amorphous layers;
- molecular-beam epitaxy for producing crystalline Si films of nanometer thickness on the surface of the insulator (formation of Si/SiO_2 superstructures);
- annealing of artificial SiO/SiO_2 superlattices in the substoichiometric oxide film obtained by vapour deposition;
- deposition of films on a substrate from the gas phase with a controlled oxygen content or without oxygen to produce alternating semiconductor and insulating SiO/SiO_2 nanolayers (super-periodic structures);
- annealing of silicon monoxide;
- plasma–chemical method;
- reactive evaporation of silicon monoxide powder;
- transformation of the amorphous semiconductor to nanocrystalline during thermal exposure (annealing);
- electrochemical etching of single crystal Si in an electrolyte based on hydrofluoric acid (porous silicon);

These methods can be divided into physical, physico-chemical and chemical. Each method produces silicon nanoparticles, generally speaking, with different characteristics, for example, such as: the average particle size, size distribution function, the degree of crystallinity of the central nucleus, composition of the shell, the introduction of impurities, performance, price, and suitability for mass production. The method of obtaining, subsequent selection and protection of the surface of the nanoparticles largely determine the properties of the nanomaterial.

5.2. Physical methods of producing nanosilicon

Several different methods were developed in this class in the last decade. First, the Si nanoclusters were produced in a glass matrix and an SiO_2 matrix. One of the physical methods, which yields in a quartz matrix luminescent crystals of size ~3 nm, is the implantation of silicon ions, accelerated to high energies, in quartz and their subsequent diffusion self-assembly into nanocrystals with high-temperature annealing at 1100–1200°C. This is the so-called ion implantation method, which can be used to produce spherical inclusions of silicon nanocrystals in silicon oxide films.

There is also a method of obtaining some isolated nanocrystals (nanocrystalline powders) by erosion entrainment of particles of a single-crystal silicon wafer by a stream of inert gas at the destructive effect of special means on the wafer. These special means are spark erosion or laser heating. In the case of laser erosion the focused radiation from a powerful laser blasts the local area of the silicon wafer, which then can evaporate or scatter into small pieces. Torn pieces of the single crystal are carried away by a stream of inert gas which blows on the wafer, and are collected on a filter through which the gas leaves the chamber in which the process takes place. Unfortunately, this erosion technology of Si nanocrystal production has low productivity and is too expensive. Also, it is not possible to control the particle size to obtain the nanocrystals of the required size.

In addition to the methods for separating by destruction of large single crystals into individual nanoparticles, there are physical methods for the synthesis of silicon nanocrystals by diffusion gathering of silicon atoms into nanocrystals in the gas phase. For example, in (Efremov et al, 2007) this was carried out using an ELV-6 electron accelerator (INP, Novosibirsk), a powerful focused beam melted and evaporated silicon in an atmosphere in which silicon atoms are collected into spherical crystallites. The disadvantage of this method is exactly the same as in the above methods of destruction – the difficulty of managing the size of particles.

5.2.1. Ion implantation of silicon in SiO_2

The method of implantation (with subsequent crystallization of nanosilicon by annealing) is compatible with the existing microelectronic technologies, and can be used to implant the exact number of silicon ions at a controlled depth in the SiO_2 matrix which is achieved by regulating the energy and density of the ion beam. In one of the studies (Mayandi et al, 2007) an SiO_2 layer 250 nm thick was deposited on the surface of a monocrystalline silicon wafer. ^{28}Si ions with densities of $1 \cdot 10^{17}$ and $5 \cdot 10^{16}$ cm^{-2} at an energy of 100 keV were implanted in these layers. The distribution of Si atoms in SiO_2 at a depth of 140 nm reached a value of standard deviation of 43 nm and a Gaussian distribution with depth. The samples were annealed at 1050°C for 2 h under nitrogen. With a high density of implantation (10^{17} cm^{-2}) TEM and electron diffraction methods recorded nc-Si with the average size of 4 nm, and at a lower density ($5 \cdot 10^{16}$ cm^{-2}) nc-Si was not detected. It was found that in the adopted experimental conditions, the critical concentration of silicon in SiO_2, below which silicon nanocrystals are not observed is $4.5 \cdot 10^{21}$ cm^{-3}. The formation of nc-Si is also affected by the concentration of defects in SiO_2: implantation of each Si^+ ion is accompanied by disruption of the structure of the amorphous SiO_2, and the greater the density of the radiation, the greater the damage to the medium (SiO_2) in which nc-Si forms, and the easier formation of nc-Si.

These results are complemented by another study (Nicotra et al, 2007). A layer of SiO_2 with a thickness of 295 nm was thermally deposited on a substrate (silicon wafer in the orientation (100)) and a silicon was implanted in this layer at an energy of 83 keV and implantation density of $1 \cdot 10^{17}$ atom·cm^{-2}. Part of the sample was annealed at 1100°C for 1 h in nitrogen to form nc-Si. In this case, the implanted silicon atoms were arranged in a band whose thickness was approximately 150 nm and the maximum concentration of implanted Si atoms was achieved at a depth of 100 nm from the surface. The average size of nc-Si was found to be 2 nm. The concentration of silicon, forming the nc-Si, at any depth is two orders of magnitude lower than the concentration of excess silicon, i.e. the excess silicon is predominantly located in the amorphous state or does not form its own phase.

More information about dispersion of nc-Si in SiO$_2$ can be found in Wang et al. (2006). The dispersion was obtained by ion implantation of Si$^+$ into a layer of SiO$_2$ on the Si (100) surface. The implantation density was equal to $3 \cdot 10^{17}$ cm^{-2}. The sample was annealed in an inert atmosphere at 1100°C for 5, 15 and 25 min and 1 h, and after each annealing treatment transverse sections were prepared for TEM, HRTEM, EELS, and STEM.

It was found that after 1 h annealing, the size of nc-Si was in the range 2–22 nm, with the largest particles found in the middle layer, and small – closer to the SiO$_2$ surface layer or the silicon substrate. Most of the nc-Si were smaller than 5 nm and had no defects, while the large ones were often twins and contained errors of the position of the layers. Clusters of amorphous silicon with a diameter ~3 nm started to form at the beginning of annealing. Their coagulation started after 10–15 min of annealing and this was accompanied by the process of crystallization of amorphous silicon particles.

Amorphous silicon plays an important role (Barba et al, 2008) in the mechanism of the photoluminescence (PL) of nc-Si, dispersed in SiO$_2$, and also affects the dielectric properties of nc-Si. Its amount depends on the temperature and duration of annealing of the samples containing silicon in the SiO$_2$. Silicon ions with a density of $1 \cdot 10^{17}$ and $2 \cdot 10^{17}$ cm^{-2} were implanted in a wafer of fused SiO$_2$ with thickness of several millimeters and this was followed by annealing at 1100°C for 1 h in a nitrogen atmosphere. The methods of Raman scattering and TEM were used to investigate the distribution of silicon in the thickness of SiO$_2$. It is shown that the nucleation of large nc-Si begins in the area where the concentration of silicon atoms is greater than $1 \cdot 10^{22}$ cm^{-3}. The distribution of silicon in the thickness of SiO$_2$ is shown in Fig. 5.1. nc-Si particles 3 nm in diameter were located in the zone where the amount of excess silicon atoms was in the range $3 \cdot 10^{21}$ and $1 \cdot 10^{22}$ cm^{-3}, and the particles with a diameter greater than 6 and 12 nm were located at the depth in the volumes where the excess of silicon atoms is greater than $1 \cdot 10^{22}$ and $1.5 \cdot 10^{22}$ cm^{-3}, respectively. It was found that the proportion of amorphous silicon in SiO$_2$ can be as high as $23.5 \pm 0.2\%$ of the nc-Si, and this proportion increases with increase in the local concentration of crystalline silicon in SiO$_2$. Raman spectra do not

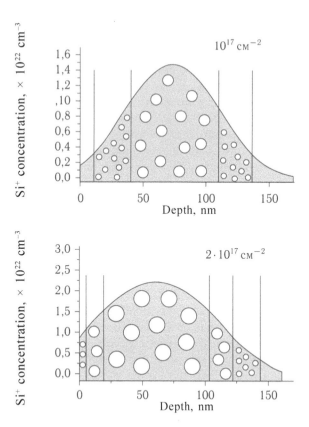

Fig. 5.1. Schematic representation of the sample obtained by implantation of silicon in a SiO$_2$ wafer with subsequent annealing: in the middle of the wafer nc-Si is larger and closer to the edges – smaller. The envelope curve reflects the change in the silicon content in the silica depending on the depth of penetration of silicon to the thickness of the SiO$_2$ wafer (Barba et al, 2008).

allow to determine whether amorphous silicon forms individual particles or forms with nc-Si a single whole. In the latter case, the amorphous silicon can affect the PL spectra of nc-Si, as the nc-Si/SiO$_2$ interface changes

The annealing mode of dispersions of Si in SiO$_2$ is important for the formation of nanoparticles. For example, in Mayandi et al. (2007) no nanoparticles were observed at a density of implantation of $5 \cdot 10^{16}$ ions cm^{-2}, despite two-hour annealing at 1050°C (see above). However, rapid preheating up to 1100°C at 50 deg/s followed by annealing in nitrogen for one hour at 1050°C increased the PL intensity by an order of magnitude compared with annealing without rapid preheating or after rapid heating without further annealing. Rapid heating creates many nc-Si nuclei, and the subsequent annealing allows the nuclei to grow. Rapid heating is less effective for samples implanted with higher density, since at high densities nuclei are created also without rapid heating.

PL intensity increased several times with increasing annealing time of the sample implanted with a density of $5 \cdot 10^{16}$ cm^{-2} from 1 h 15–20 h at 1050°C. In the opinion of Iwayama et al. (2006, 2007) 15–20-hour annealing at 1050°C is accompanied by Ostwald ripening of nc-Si: large nc-Si grow at the expense of small ones and this leads to an increase in distances between the nc-Si and, therefore, to a reduction of the intensity of interactions between them, which leads to an increase in the PL yield. PL intensity initially increases with increasing implantation density due to the increase of concentration of nc-Si, and then decreases, reaching a maximum at a density of implantation of $2 \cdot 10^{17}$ cm^{-2}. Increasing implantation density increases the size of nc-Si, which should lead to the experimentally observed red shift of the PL.

The resulting (from implantation of Si^{+}) dispersion of nc-Si in SiO$_2$ (after annealing at 1050°C for 2 h in nitrogen) had a PL spectrum with peaks at 775 and 825 nm at densities of the silicon ion fluxes of $5 \cdot 10^{16}$ cm^{-2} and $1 \cdot 10^{17}$ cm^{-2}, respectively (Serincan et al, 2007). As a result of implantation of additional Si^{+} ions carried out after annealing of the dispersion at 1050°C, the PL intensity dropped sharply because of the discontinuities of the (Si–Si) bonds in the nc-Si and formation of P_b-centres, and disappeared after a certain flux of ions Si^{+}, and PL, due to defects in SiO$_2$ (625 nm), formed and strengthened. After repeated annealing of the dispersion, the PL peaks at 775 and 825 nm recovered and reached the same intensity.

Implantation in the nc-Si of ions of boron and phosphorus (Kachurin et al, 2006), as well as and any other ions, results in PL quenching due to formation on the nc-Si surface of dangling bonds – P_b-centres, which serve as the non-radiative channel of the flow of the energy of excitons to nc-Si. In addition, immobile complexes of defects (vacancies and interstitials) can form in nc-Si, which, not being non-radiative recombination centres, contribute to the creation of centres of the P_b-type. The equilibrium structure of the nc-Si is restored as a result of annealing, and laser annealing is more efficient than thermal annealing.

It has been shown (Kovalev et al, 2006) that at the nc-Si/SiO$_2$ interface the surface of nc-Si is coated with a suboxide, and when doped with phosphorus by implantation the phosphorus atoms are collected in this layer, forming on the surface of nc-Si a group of atoms (P–O–Si).

Xie et al. (2007b) tested mixing with bombardment with Ar^{+} ions for used in the homogenization of substances used for alloying nc-Si alloy. Samples were produced by thermal evaporation (not by implantation) of SiO on the Si (100) substrate with annealing at 1100°C under nitrogen for 1 h. The layer thickness was 100 nm. At bombardment with Ar^{+} ions the upper layer of the surface evaporates, and the oxygen is easily detached and oxygen vacancies form in the residue. The PL intensity rapidly decreases, apparently due to formation of a set of dangling bonds. Annealing in oxygen restored the PL.

Levitcharsky et al. (2007) used P_b-centre passivation by hydrogen, i.e. to form Si–H bonds instead of P_b-centres. A Si(100) single crystal substrates was oxidized at 1100°C in a stream of oxygen to the formation on the surface of an SiO_2 layer with a thickness of 1 µm. This layer was implanted with Si^+ ions with an energy of 100 keV at a flux density of Si^+ ions in the range $2·10^{16}$ cm^{-2}–$3·10^{17}$ cm^{-2}. These flows correspond to a local excess of silicon of 3 at.% and 50 at.%, respectively. One sample was obtained at an implantation energy of Si^+ ions of 50 keV at a flux density of $1.9·10^{17}$ cm^{-2}. These samples were the annealed at 1100°C in a nitrogen atmosphere for 1 h and passivated at 500°C for 1 h with 5% H_2 + 95% N_2.

The highest photoluminescence was shown by the samples obtained at a flux density of Si^+ ions of $4·10^{16}$ to $6·10^{16}$ cm^{-2}, but the intensity of the photoluminescence of the samples prepared at a density of the Si^+ ion flux of $3·10^{17}$ cm^{-2} decreased 5 times at a strong red shift of the luminescence band. It was shown that when the flux density is less than $1·10^{17}$ cm^{-2} nc-Si with the average diameter of 3.0 ± 0.5 nm forms over the entire implanted layer, and the emergence of nc-Si less than 2 nm in diameter requires a local excess concentration of silicon about $3·10^{21}$ atom·cm^{-3}. It was found at the the concentration of silicon in excess of $2.3·10^{22}$ atom cm^{-3}, nc-Si agglomerate and contain many defects, which leads to a drop in the intensity of photoluminescence. Nanosilicon particles larger than 6 nm often contain twin stacking faults which quench the photoluminescence intensity. nc-Si is characterized by the formation of twins on {111} faces.

In addition to nc-Si, Levitcharsky et al. (2007) also investigated the SiO_2 matrix. It was found that the higher the density of the flux of Si^+ ions, the greater the loss of oxygen in SiO_{2-x} and the larger the number of unsaturated bonds at the silicon atoms in the composition of SiO_2, but the (Si–Si) bonds do not form. The local oxygen concentration was reduced by 50% where the maximum of the implanted Si^+ ions was found. After annealing, the oxygen deficiency persisted and no photoluminescence was observed in the annealed samples. Only passivation with hydrogen led to the emergence of a weak photoluminescence in samples with a high Si concentration.

A more complete study of passivation of P_b-centres by hydrogen was carried out by Wilkinson et al. (2006). The SiO_2 layers with a thickness of 1.25 mm were implanted with ions Si^+ with a flux density of $2·10^{17}$ cm^{-2} at room temperature up to the excess silicon content of 10 at.% at a depth of about 630 nm. Formation of nc-Si in SiO_2 was achieved by annealing at 1100°C for an hour in an argon stream. An aluminum layer with a thickness of 100 nm was deposited to remove the OH-groups from the SiO_2 surface of the samples obtained after implantation. The aluminum layer was then removed by etching at 90°C in phosphoric acid (85 wt.%) for 1 min. The desorption of the hydrogen, formed during the interaction of aluminum with surface hydroxyl groups, was carried out in a nitrogen atmosphere, but at the same time this hydrogen passivated silicon dangling bonds which penetrated into SiO_2 by implantation. Additional passivation of nc-Si was carried out by annealing in a forming gas (95% N_2 + 5% H_2).

The effect of hydrogen on nc-Si was manifested as: the PL intensity increased by more than an order of magnitude after annealing in forming gas for 1 hour at 500°C, but further annealing for up to 16 hours had almost no effect. However, the PL intensity was 1.5 times higher (if atomic hydrogen, which is formed under a layer of aluminum, was used) than the maximum intensity observed in the samples obtained by passivation with gaseous hydrogen. According to Wilkinson et al. (2006), this means that there are P_b-centres which react only with atomic hydrogen. Passivation with hydrogen is reversible: the Si–H bonds are destroyed when heated. In this regard, the bonds formed by atomic hydrogen are the first to break. It follows from the fact that annealing for 1 h at 525°C resulted in samples passivated by atomic and gaseous hydrogen to have the same PL intensity, despite the fact that prior to annealing the samples passivated by atomic hydrogen had greater PL intensity. The activation energy of desorption of hydrogen from the sites, passivated by atomic hydrogen only, is 2.4 eV, and by molecular hydrogen – 3.0 eV.

Implantation of Si^+ ions in SiO_2 is considered as the most promising for technological applications in electronics due to the presence of the native SiO_2 layer at the silicon surface. In Pavesi et al. (2000) nc-Si particles were obtained by Si^+ ion implantation in quartz or the SiO_2 layer on a silicon substrate followed by annealing at 1100°C for one hour. The particles of nc-Si, according to TEM, had an average diameter of 3 nm at a concentration of $2\cdot10^{19}$ cm^{-3}. The thickness of the nc-Si dispersion layer was 100 nm, and the middle of this layer was at a depth of 110 nm from the surface. The peak of the PL spectrum was at 800 nm. The authors attributed this radiation to the nc-Si/SiO$_2$ interface. nc-Si had a decay time of photoluminescence in the microsecond band. The surface state lies inside the forbidden zone of nc-Si and the electrons quickly relax from the conduction band to the surface state which has a long lifetime. This explanation is based on the work (Wolkin et al, 1999), which shows that the formation of new electronic states in the forbidden band of nanosilicon is associated with the formation of (Si = O) bonds.

nc-Si/SiO$_2$ dispersions, obtained by Si^+ implantation into SiO_2 (thickness 100–330 nm) showed (Barba et al, 2006) strong changes in the PL intensity due to optical interference effects of the incident and reflected, from the nc-Si/SiO$_2$ interface, radiation and also due to the spectral modulation of PL emission. The magnitude of these effects depends on the thickness of the nc-Si/SiO$_2$ dispersion, the wavelength and angle of incidence of the exciting radiation. These properties of the layers of the nc-Si/SiO$_2$ dispersion are observed regardless of the method of preparation of the layer.

The photoluminescence quantum efficiency of the dispersion of nc-Si in SiO_2, obtained by implanting Si^+ in SiO_2, reaches 59 ± 9% (Walters et al, 2006) in radiation at a wavelength of 750 nm at low exciting power. However, at high pumping energy the non-radiative PL quenching mechanisms lower the quantum efficiency.

Cathodoluminescence (CL) of dispersions of nc-Si in SiO_2 has significant differences from PL. In Dowd et al. (2006), the dispersion of nc-Si in SiO_2 was prepared by Si^+ ion implantation into SiO_2 and subsequent annealing at 1100°C for one hour. The samples were then passivated by heating at 500°C for an hour in a gas mixture of 95% N_2 + 5% H_2 to enhance the luminescence intensity. The spectrum of CL prior to annealing of the sample had two peaks at 470 nm and 670 nm, the first of which is due to the presence of silicon atoms with unpaired electrons. Each such atom is called the centre with oxygen deficiency. The second peak (670 nm) is usually related to to an oxygen atom with an unpaired electron. After annealing, the first peak (470 nm) rapidly lost its intensity due to the formation of (Si–H) bonds, and the second peak disappeared after the formation of (O–H) bonds. However, there is a peak at 800–860 nm, the position of which depended on the dose of implanted Si. This peak is due to emission of nc-Si.

In Liu et al. (2007) an SiO_2 layer with a thickness of 30 nm was grown on a p-type silicon substrate at 950°C, and Si^+ ions were implanted into this layer. The sample was then annealed at 1000°C for 100 min, and was coated with transparent electrically conductive layer of the solid solution of indium and tin oxides, which had a thickness of 130 nm. When the potential of 14–18 V was applied electroluminescence (EL) with a maximum at 500 nm (2.5 eV) formed and its intensity increased with increasing potential and EL was observed with the naked eye in the dark. The authors believe that EL is caused by defects arising as a result of implantation, rather than nc-Si. These defects are located at the nc-Si/SiO_2 interface and trap the hole. Electrons are trapped at the interface between the indium–tin–oxide electrode and the sample. Annealing the sample at 120°C for 40 s releases carriers from the traps, which means that the latter have little depth. Release of the charge carriers from traps is supported by the alternating electric field.

In Ding et al. (2008), a 30 nm thick SiO_2 film was grown on a p-type plate of Si (100) by oxidation in dry oxygen at 250°C. Implantation of Si^+ion produced samples at a flux density $1 \cdot 10^{16}$, $2 \cdot 10^{15}$ and $3 \cdot 10^{14}$ cm^{-2} at an implantation energy of 5 keV, as well as at three different implantation energies – 8 keV, 5 keV and 2 keV – but at the same flux densities of implantation – $1 \cdot 10^{16}$ cm^{-2}. Implantation was followed by annealing at 1000°C in a nitrogen atmosphere for 1 hour to coagulate the nc-Si in the SiO_2.

The TEM method showed that nc-Si have a diameter of about 4 nm. It turned out that at 8 keV implantation energy part of the Si^+ ions penetrate through the 30 nm thick SiO_2 layer, although the maximum concentration is reached at a depth of 15 nm. Si^+ ions with an implantation energy of 5 keV penetrate into the SiO_2 layer to a maximum depth of 23 nm and the maximum concentration of silicon atoms accumulates at a depth of 8 nm. The implantation energy of 2 keV allows maximum concentration of the Si atoms at a depth of ~5 nm and Si^+ ions do not penetrate into the layer of SiO_2 to a depth greater than 15 nm.

Electroluminescence is the result of the application of a negative potential to the transparent indium–tin–oxide electrode. Its intensity increases with increasing negative potential. The EL band consists of four components: most intense peaks at a wavelength of 600 nm, two peaks at 460 and 740 nm form its shoulders and a flat maximum at 1260 nm is separated apart from the first three. With the potential increasing to -15 V the intensity of the EL with a wavelength 600 nm is reduced by approximately 20%, and the intensity of the bands with 480 and 740 nm maxima increases by 50%. The intensity of the band with a peak at 1260 nm does not change. The EL intensity of the main peak at 600 nm in the specimen with an implantation density of $1 \cdot 10^{16}$ cm^{-2} is a third higher at -15 V than at the implantation dose of $2 \cdot 10^{15}$ cm^{-2} and 7 times higher than at the implantation dose of $3 \cdot 10^{14}$ cm^{-2}. In addition, the implantation energy of 8 keV produces at a voltage of -15 V the peak intensity EL at 600 nm a third higher than the implantation energy of 5 keV and 4 times higher than the implantation energy of 2 keV. The authors believe that the EL peak at 740 nm is generated by recombination of holes and electrons in the nc-Si, the peak at 460 nm assigned by them to the neutral oxygen vacancies ($O_3 \equiv Si–Si \equiv O_3$), and the peak at 600 nm is due to a hole in the non-bridging oxygen atom ($O_3 \equiv Si–Si–O^{\cdot}$). Infrared radiation came from the silicon substrate.

The EL of nc-Si, dispersed in SiO_2, is inhibited by good insulating properties of SiO_2 (Rafiq et al, 2006, 2008). A layer of nc-Si with a nanoparticle diameter of 8 nm and a layer of native SiO_2 with a thickness of 1.5 nm on the surface of nc-Si has a percolation mechanism of hopping conduction. Electroluminescent diodes were prepared in the form of MOS-structures with transparent layers of gold on top (SiO_2) and back (Si) planes of the plate, deposited by sputtering. The thickness of the gold layer was 20 nm, diameter 6 mm. EL was excited by constant voltage with a negative potential on the top transparent electrode. When an electric field of $10–12$ mV · cm^{-1} was applied, EL was visible by the naked eye in daylight, and lasted for several tens of seconds. EL was terminated due to degradation of the gold layer on the surface. Current through the diode increases exponentially with increasing voltage without breakdown up to 12 mV · cm^{-1}.

Gawlik et al. (2007) also investigated the PL spectrum. The spectrum showed bands whose position is determined by the conditions of sample preparation: the implantation energy and density of Si^+, etching with hydrofluoric acid or lack of it. The sample obtained at an Si^+ implantation energy of 25 keV and the sample obtained at 60 keV, but with the top layer of SiO_2 removed by etching with hydrofluoric acid, had the same PL spectra. Without etching, the sample obtained at 60 keV had an additional green PL band (500–600 nm). This green radiation appears in the layer where the concentration of nc-Si increases (cathode side). Blue light (400–500 nm) is emitted by a layer lying deeper than the maximum concentration of nc-Si, – where the concentration of nc-Si decreases, and, consequently, the nanosilicon particle size is smaller. The red PL was observed in samples with a high density of implantation and after prolonged annealing, which could not lead to growth of the particles, accompanied by a red shift of the PL band.

The enzyme ferritin was used by Nakama et al. (2008) for the manufacture of masks which made it possible to place nc-Si in SiO_2 more uniformly. The outer diameter of the ferritin globules was 12 nm, and the inner core, consisting of Fe_2O_3, has a diameter of 7 nm. The substrate was a Si (100) wafer with a 100 nm thick layer of SiO_2 on its surface, which was obtained by oxidation of the wafer at an elevated temperature. A self-organized layer of ferritin molecules was deposited on the clean substrate. To do this, a drop of solution of 0.5 mg ferritin in 1 ml (solvent not given) was placed on a paraffin film. The substrate was placed on a drop, which spread. After 30 min at 20°C the substrate was removed and the excess solution was removed by centrifugation. The substrate was then washed with water and heated for 3 min to 110°C. A layer of ferritin globules with a density of $6.2 \cdot 10^{11}$ cm^{-2} remained on the substrate, which is close to the densest packing in the layer. External shells of the ferritin globules were removed by annealing at 400°C for 10 min in an oxygen atmosphere. The substrate was then exposed amorphous silicon epitaxy with thickness of 2 nm and immersed for 30 min at room temperature in hydrochloric acid. When the thickness of the amorphous silicon layer was 2 nm the iron oxide was dissolved and a mask with holes 7 nm in diameter was produced. Si$^+$ ions at an accelerating energy of 0.6 keV were implanted through a mask in the SiO_2-substrate. Such a low implantation energy represents a 'record' and is due to the fact that the amorphous silicon layer with a thickness of 2 nm is impermeable to silicon ions with such low energy. The mask was etched from the surface of SiO_2 with the XeF_2 gas, and the silicon in the holes of the mask was protected by silicon dioxide, which had been implanted with Si$^+$ ions. The substrate was then annealed at 1050°C, and nc-Si with a diameter 3.0 ± 0.3 nm at distances of 5–6 nm apart was formed in SiO_2. The PL spectrum of nc-Si had a peak at 590 nm (2.11 eV).

5.2.2. Laser ablation of crystalline silicon

The method of pulsed laser ablation is characterized by extremely high temperature and pressure of the shock front, so in the process of laser ablation of silicon reacts with gases or liquids of the environment.

An evacuated reaction chamber with a single-crystal silicon wafer received hydrogen at 10 cm$^3 \cdot$min^{-1}, which created a certain amount of pressure inside the chamber (Umezu et al, 2008b). The nc-Si scattered by laser ablation were deposited on a horizontal silicon wafer or a quartz substrate placed at a distance of 23 mm from the area of ablation of the silicon wafer by the laser beam.

The primary nc-Si particles, produced by laser ablation, have an average diameter of 4–5 nm and their surface is covered with Si–H bonds, detected by FTIR, and the number of fragments ($Si=H_2$) and ($Si\equiv H_3$) is very small..

The expansion of the cloud of silicon particles formed by the impact of the laser pulse ($\sim 10^7$ W/cm^2, 40 ps, 266 nm), is characterized by three stages. Initially, the cloud is distributed as in a vacuum. Then the hydrogen compressed by the nc-Si cloud inhibits the rate of nc-Si and the result is a shock wave. Finally, the pressure inside the cloud of nc-Si decreases and approaches the hydrogen pressure inside the chamber. The second stage ends after 1 μs and a third stage begins, and the first stages changed into the second one after 200 ns. It is shown that within 50 ns after the laser pulse all primary particles in the cloud are nanodroplets – so high is their temperature, but the Si–H bonds do not yet form. They appear after 100 ns and are clearly observed in the shock wave stage.

Pulsed laser ablation of silicon in an inert gas atmosphere produces nc-Si of the same diameter (4–5 nm) as in the hydrogen atmosphere. This means that the diameter of the nanodroplets is determined by the first stage of expansion of the cloud of Si particles, formed by laser ablation. The gas pressure is important for the second stage (shock wave). It is known that the Si–H bonds are unstable above 700 K. Apparently, the Si–H bonds form on the surface of nc-Si at temperature below 700 K, i.e. in the third stage of ablation. If the hydrogen pressure in the reaction chamber is below 20 Pa, the structure of the film deposited on the substrate corresponds to the amorphous hydrogenated silicon. But if the hydrogen pressure exceeds 30 Pa, the nc-S form in the ablation cloud, since the rate of cooling in the cloud is lower than on the substrate.

Oxidation of nc-Si, obtained by laser ablation of silicon in hydrogen, has been studied by Umezu (2008a, 2006) in samples in which the size of nc-Si was the same and equal (according to TEM) to 4–5 nm, but the porosity of the samples increased with increasing hydrogen pressure in the reaction chamber – 270, 530 and 1100 Pa. A reference sample was a nc-Si layer without hydrogen ligands in a helium atmosphere under a pressure of 530 Pa. All samples were left in air at room temperature. This work was performed in Kobe, in the south of the Honshu island, Japan, with high humidity and temperature of the air. The hydrogen content in the samples was 20 at.%.

Oxidation of nc-Si/H in the air was recorded by FTIR. The peak of the infrared absorption of the Si–H bond was shifted from 2100 cm^{-1} in a freshly obtained still unoxidized sample toward higher wave numbers due to the increase of the electronegativity silicon atoms on the surface of nc-Si/H by the oxidation of the surface of nc-Si/H. Although the peak of

Si–H vibrations was shifted during oxidation from 2100 cm^{-1} to 2250 cm^{-1}, its integrated intensity remained almost constant. The authors suggest that oxidation of nc-Si/H does not affect the Si–H bonds.

The authors admitted that prior to oxidation the wave number 2100 cm^{-1} corresponds to the group $(Si)_3SiH$, and after oxidation absorption at 2250 cm^{-1} corresponds to the group $(O)_3SiH$ on the surface of nc-Si. By means of interpolation, the groups $(Si_2O)SiH$ and $(SiO_2)SiHN$ were classified as having wave numbers 2150 and 2200 cm^{-1}, respectively. On the contrary, the absorption bands, corresponding to the groups $SiH_2(SiO)$, SiH_2 and $(SiO_2)SiH_2$, were not found in the spectra. The rate of oxidation of nc-Si/H increases with increasing porosity of the nc-Si/H layer. Increasinng exposure time in air increases the share of $(O)_3SiH$ by reduction of the share of other groups – $(O_2Si)SiH$ and $(OSi_2)SiH$.

The PL for the samples, obtained in a helium atmosphere, was hardly noticeable. The EPR spectra showed that the amount of P_b-centres in such a sample is an order of magnitude greater than in the hydrogenated samples. The PL spectrum, measured in vacuum after deposition of the nc-Si/H layer, had a peak around 800 nm. After one minute of exposure to air the peak shifted to ~700 nm, and after 5 min to 660 nm. The same sample placed in a vacuum had the PL peak at 730 nm. Such shifts of the PL peaks were not observed in the argon atmosphere, but were observed in a mixture of Ar and water vapour. Thus, the reversible shift of the PL bands is caused by physical water vapour sorption on the surfaces of nc-Si/H. After exposure to air for 7 hours the PL peak was shifted to 600 nm and did not return back in the vacuum. The irreversible shift of the PL line coincides with the increase of intensity of the IR spectrum, characteristic for the (Si–O) bounds, regardless of the porosity of the nc-Si/H layers. With time, the PL spectra of nc-Si/H show three bands – 800, 700–600 and 500–400 nm.

It is assumed that the red light comes from the (Si=O) groups or non-bridging oxygen atoms with the hole charge on the surface. The blue light at about 400 nm occurs on the $(O_2=Si)$ groups and the defects associated with the (OH)-groups, but this has not been experimentally verified.

TEM and SEM (Umezu et al, 2007b) showed that the nc-Si have a diameter of 4–5 nm and their size is independent of hydrogen pressure in the reaction chamber, but the nc-Si agglomerate and the structure of the agglomerates depends on the pressure. At a pressure of 30 Pa or lower obtained amorphous silicon nanoparticles, and the nc-Si formed at the hydrogen pressure above 133 Pa. Laser ablation in vacuum leads to the deposition of a silicon film but not agglomerates, i.e. silicon vapours are cooled only by settling on the substrate. They form silicon nanoparticles prior to deposition on the substrate.

The particles of nc-Si, obtained by laser ablation, were annealed in the aerosol state in helium (Hirasawa et al, 2006), and it was found that the temperature required for crystallization of amorphous particles of nanosilicon decreases with decreasing particle size: 1000, 900, 800 and

700°C for particle diameters of 10, 8, 6 and 4 nm. There is no influence of the substrate or matrix as well as of the ligand environment.

Riabinina et al. (2006) investigated the effect of oxygen pressure in the gas phase on the formation of nc-Si. The layers of SiO_x ($0 \leq x \leq 2$) were formed by laser ablation on a silicon substrate coated with a native SiO_2, at an oxygen pressure of 0.01–1.5 mTorr. A silicon wafer was ablated by laser pulses with a duration of 17 ns and a repetition frequency of 20 Hz. The laser flux was $5 \text{ J} \cdot \text{cm}^{-2}$, the substrate was located at a distance of 40 mm from the ablation site. The thickness of the resulting SiO_x layer was 200 nm, according to SEM. The samples were then annealed in nitrogen for 1 hour at 1050°C. It was found that with increasing oxygen pressure the maximum in the PL spectrum is shifted from ~900 nm (at a pressure of 0.26 mTorr) to ~720 nm (1.1 mTorr), which corresponds to the reduction of the size of nc-Si from 6.5 ± 0.2 nm to 1.9 ± 0.1 nm, respectively.

nc-Si powders were obtained (Choi et al, 2007a) by means of plasma ablation as a result of the interaction of powerful proton pulses with silicon. Samples A and B were synthesized. The first – in a vacuum ($2 \cdot 10^{-4}$ Torr), and B – in helium at a pressure of 1 Torr. Both samples were investigated at the point of ablation ($r = 0$) and at a distance of 20 mm from it ($r = 20$). X-ray diffraction showed that nc-Si was produced. At the point of ablation the average particle size was equal 80 nm, and at $r = 20$ mm – 75 nm for sample A. For sample B at the point of ablation the size of nc-Si was 85 nm and at $r = 20$ mm it was 20 nm. The PL spectrum of sample B with the size of nc-Si of 20 nm has a maximum of ~420 nm and a plateau at 500–580 nm. This spectrum was stable for at least 4 months. All the nc-Si grains clearly melted. The holding time in the air between the end of the synthesis of nc-Si and the beginning of registration of the PL spectrum, was not reported.

We should also consider the results of laser ablation of silicon immersed in water. A plate of Si (100) was used (Du et al, 2007) as a target for producing nc-Si by laser ablation. The target was in the water and at the time of the effect of the laser pulse on the target the high temperature and pressure could lead to the interaction of water with the surface of nc-Si. The resulting nc-Si had a diamond-like crystalline structure. FTIR indicates the formation of (Si–O) bonds on the surface of nc-Si, such as those that exist in silicon-abundant oxides SiO_x. A week after the production of nc-Si, these lines increased significantly, indicating that the surface of nc-Si interacted with oxygen and moisture in the air. In addition to the Si–O bonds, the bonds Si–H (2100 and 890 cm^{-1}) were also found, but a week later they moved, which was attributed to the oxidation of (Si–Si) bonds with the formation of $HSi–O_3$ (2265–2240 and 927 cm^{-1}) and $H_2Si–O_2$ (2196–2180 and 973 cm^{-1}). These observations confirm the interaction nc-Si with oxygen and moisture in the air.

The particle size of nc-Si averaged 3.8 nm. The PL spectrum, excited by the radiation of 5.9 eV, contained a band with a maximum at 2.6 eV. A week later, the PL intensity slightly increased, indicating blocking of some of the P_b-centres. The analysis of the photoluminescence excitation spectra indicated the direct-gap behaviour of nc-Si. Comparison of PL spectra and PL excitation led the authors to conclude that recombination of electron–hole pairs takes place with the participation of surface states, namely the (Si–O) and (Si–H) bonds, where the electron–hole pairs recombine with the emission of visible light.

Attention was given not only to the laser ablation in water (Svrček et al., 2006a), but also aging of the nc-Si in water for six months at room temperature. During formation of nc-Si laser the radiation flux varied from 0.07 to 6 mJ/pulse. At the flow of 0.07 mJ/pulse spherical particles of 2 to 100 nm at the mean value of 60 nm were produced. Each particle with the diameter of 15–80 nm is an aggregate of a few crystalline particles 2–11 nm in size. Each nc-Si in the aggregate is surrounded by a film of amorphous SiO_2, which accounts for 25–30% of the volume of the clusters. At high laser radiation fluxes – up to 6 mJ/pulse – spherical aggregates were composed of small particles of irregular shape. The size of the aggregates was in the range 2–50 nm (average 20 nm). In the aggregate with a diameter of 20 nm and 50 nm the size of nc-Si ranged from 2 to 6 nm and occupyied 20% of the aggregate.

All the nc-Si particles have a diamond-like crystalline structure. Ageing in water led to the enlargement of the particles to an average size of 140 nm. The intensity of the photoluminescence spectra of the deposits aged in water was much stronger in the nc-Si obtained at the laser radiation flux of 6 mJ/pulse, in comparison with those that were obtained by ablation with the energy of 0.07 mJ/pulse. In addition, there was a shift of the maxima of the PL bands: 2.9 eV for 0.07 mJ/pulse, but the 3.1 eV to 6 mJ/pulse.

Since the band gap of 2.9 eV according to estimates (Svrček et al, 2006a) corresponds to the size of nc-Si of 1.5 nm, and the investigated nc-Si sizes were 2–5 nm, the authors, without rejecting the influence of the size effect on the recombination of electron–hole pairs, mentioned the influence of oxygen on the surface of the nc-Si, which broadens the band gap of nc-Si, and the stresses at the nc-Si/SiO_2 interface.

Increasing the laser pulse energy of 150 mJ did not cause any qualitative changes. A crystalline silicon (111) plate of the n-type (Yang et al, 2008) was immersed in 0.03 M aqueous solution of sodium dodecyl sulphate and was irradiated by laser radiation with a pulse energy of 150 mJ, 10 ns pulse width and a laser radiation wavelength of 1064 nm. The solution was continuously stirred during radiation. The resulting aqueous suspension of silicon nanoparticles was centrifuged and washed with ethanol. A powder was obtained which gave a clear X-ray powder pattern, since the size of the particles obtained amounted to 40 nm (average 20 nm). It was shown that each particle consists of misoriented blocks with an average diameter of 3 nm.

The nc-Si particles luminesce in the blue region with maxima of 415 and 435 nm. The PL intensity increased 130 times after 16 weeks of holding the samples in air at room temperature, but annealing of nc-Si at 400°C for an hour returned the PL intensity to the level of the PL of the freshly prepared sample. The authors believe that dodecyl sulphate prevents oxidation of nc-Si in the solution. The reaction with water was not mentioned.

ESR detected P_b-centres (broken bonds of silicon), but there was no signal that would correspond to the E'-centres (Si=O bonds). The blue photoluminescence (430 nm) was attributed to the size effect, and the authors emphasized that the blue PL was obtained by excitation with light having a wavelength of 360 nm, and the blue light of defects in SiO_2 is usually caused by the light with a wavelength of 275 nm. The authors believe that some of the electron–hole pairs recombine on the P_b-centres non-radiatively, and those couples which radiate recombine in the traps located at the Si/SiO_2 interface. During ageing in air nc-Si was oxidized by atmospheric oxygen, and this reduced the amount of P_b-centres that led to an increase in PL intensity. FTIR clearly showed an increase of the intensity of the (Si–O) and (O–H) peaks during the ageing of nc-Si, which indicated the interaction with water vapours in the air.

Laser ablation of crystalline silica in water was also been studied by Rioux et al. (2009). The result was a red–brown suspension, in which the nc-Si did not settle during several months after preparation. According to TEM, the particle size on average was equal to 2.4 nm. Some of the particles were amorphous, but there were many particles with the interplanar distance of 3.14 Å, which corresponds to the (111) silicon planes. The interaction of water with silicon during laser ablation was not studied, despite the fact that silicon at the point of ablation melts and partly evaporates and the water reacts with silicon even at room temperature. The generation of singlet oxygen on the produced particles of nc-Si was also reported.

Pulsed laser ablation was performed (Umezu et al, 2007a) in water, hexane and toluene. It turned out that the properties of the nc-Si depend on the fluid in which they are produced. For example, the width of the forbidden zones, evaluated on the basis of absorption spectra, for the nc-Si, synthesized in water and hexane, respectively, was equal to 2.9 and 3.5 eV, and the first of which shifted from 2.9 to 3.4 eV after 6 days being in the water after preparation of nc-Si. No changes were observed in hexane for 6 days. However, the PL peak of nc-Si, dried to remove hexane on a silicon substrate, varied from 3.2 in hexane to 2.4 eV in the dry state, which indicated the influence of the surface on the properties of the nc-Si or convergence of the particles. FTIR detected in all samples (Si–O) bonds on the basis of their vibrational modes at 1080 cm^{-1}. It was found that the nc-Si, obtained in hexane, have (Si–CH$_3$) bonds, indicating that the interaction of hexane with silicon at the time of the laser pulse. It was mentioned that laser ablation of silicon in a nitrogen atmosphere results in the formation of silicon nitride.

Further information can be found in (Yang et al, 2009). A Si (111) plate was placed in water, ethanol or a mixture thereof, and subjected to laser ablation. The liquid was stirred during ablation. The dispersions were centrifuged for 2 hours (4000 rpm), and the clear dispersion from the upper layer was removed for further centrifugation (14 000 rpm).

The dispersion in ethanol was more stable than in water. X-ray diffraction confirmed the presence of silicon with a diamond-like crystal structure in both water and alcohol dispersions. With increasing pulse energy, starting with 50 mJ/pulse, the average particle size in the water initially decreased (from 19 to 14 nm), and then increased (21 nm at 200 mJ/pulse), and in alcohol decreased monotonically from 6.8 nm at 50 mJ/pulse to 3.1 nm at 200 mJ/pulse (pulse duration 10 ns, laser wavelength 1064 nm). All silica particles, obtained in water, were polycrystalline, and those produced in ethanol – single crystal. The authors noted that answers to the questions of why the nc-Si particles in ethanol are smaller than in water, and why in ethanol there is no aggregation at the maximum pulse energy and in the water is, are not yet available. However, most of the material discussed above suggests that at high temperatures and pressures at the time of ablation in water we obtain silicon oxide, and in alcohol – the formation of (Si–OEt) and (Si–H) bonds, which leads to the agglomeration of nanoparticles in water and the formation of their dispersion in alcohol.

Laser ablation in a nitrogen atmosphere leads (Umezu et al, 2010) to the formation of nanoparticles of silicon in a nitride shell, which in contact with atmospheric air react with oxygen and moisture, acquiring the composition $SiO_{0.78}N_{0.33}$ with the shell thickness of 2 nm and a diameter of silicon nanoparticles of 12 nm.

5.2.3. Production of nanosilicon in evaporation of crystalline silicon by a beam of accelerated electrons

The G.I. Budker Institute of Nuclear Physics of the Siberian Division of the Russian Academy of Sciences designed and implemented a new method for producing nanosilicon by evaporation of crystalline silicon by beam of accelerated electrons in air (Lukashov et al, 1996; Bardakhanov, 1997; Bardakhanov, 2006) in the set up shown schematically in Fig. 5.2.

The high power density required for evaporation (up to 5000 kW/cm²) is provided using as a heat source a standard accelerator of the ELV series with a power up to 100 kW with the release into the atmosphere of a concentrated beam of electrons. The installation can vaporize any substance known to date and the process is carried out in a continuous mode with a capacity of tens of kilograms per hour with a high efficiency of transformation of electrical energy into thermal energy.

The process is described as follows: evaporation of a solid material by an electron beam; cooling and coagulation of the vapour–air mixture with the condensation of matter in the form of nanoparticles; capture and collection

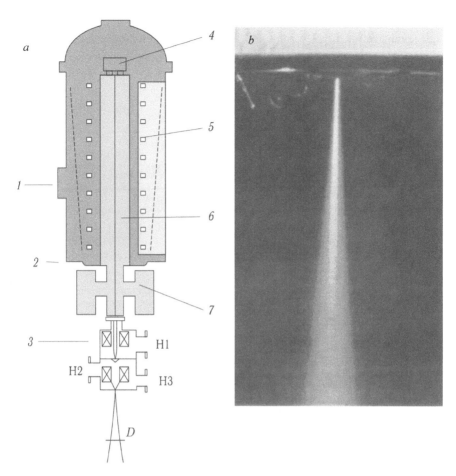

Fig. 5.2. *a* – general view of ELV electron accelerator: *1* – input voltage: *2* – water cooling, *3* – output of the electron beam, *4* – electron gun, *5* – high-voltage rectifier, *6* – accelerating tube, *7* – ion pump, H1–H3 – pumps for differential vacuum pumping; *b* – photo of the electron beam E = 1.4 MeV released into the air atmosphere (Korchagin et al, 2010).

of the resulting material to a powder. The raw material is silicon powder. The interaction of silicon with oxygen and nitrogen in air is not reported.

Efremova et al (2004) used the original radiation method of forming nanopowder of silicon with the evaporation of silicon ingots under the influence of a powerful electron beam in an inert gas atmosphere (argon), as well as under nitrogen or air flow. Silicon nanopowders, formed in this way, were studied by photoluminescence (PL) and Raman spectroscopy (RS). In the powders, consisting of silicon nanocrystals, the PL peak is detected at room temperature in the visible spectral range (Figure 5.3).

A strong blue shift of the PL peak can be explained as the effect of size quantization of electrons and holes in silicon nanocrystals with small sizes (up to 2 nm). From the Raman spectra (Figure 5.4) on the optical phonons the average size of the nanocrystals can be determined using the

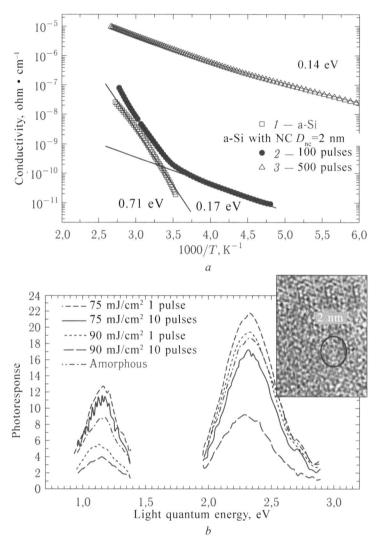

Fig. 5.3. *a* – the dependence of the conductivity of nc-Si on temperature, *b* – spectrum of the photoluminescence of silicon nanopowder. PL was excited by a pulsed N$_2$ laser, wavelength of 337 nm (energy ~ 3.68 eV), the temperature of 300 K. Insert – microphotograph of the research results by high-resolution TEM of nanosilicon obtained by evaporation of silicon by the electron beam (Efremov M.D., et al., Solid State Phenomena, 2005, Vol. 108–109, 65–70; http://www.scientific.net).

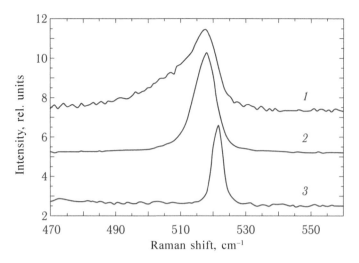

Fig. 5.4. Raman spectra of nanosilicon obtained by evaporation of Si ingots by the electron beam in argon (*1*) and nitrogen (*2*), *3* – RS peak of the substrate of single crystal silicon (Arzhannikova et al., 2005; Efremov et al., 2007).

method of convolution of the effective density of vibrational states. The thus determined average size coincided with the estimates obtained from the PL data. Depending on the evaporation conditions, the size of silicon nanocrystals ranged from ~1.5 to 5 nm. Evaporation in the air flow resulted in the formation of SiO_2 nanopowders.

The advantages of the chosen method of formation of nanopowders in comparison with the existing methods will be briefly discussed. First, there is the potential high performance of the method. Thus, the effectiveness of methods using laser ablation is determined by the average laser power, but even for modern, unique lasers the average output power is limited to units of kilowatts. The power of the electron beam in the ELV-6 accelerator at the Institute of Nuclear Physics of the Siberian Division of the Russian Academy of Sciences can reach 100 kilowatts. Second, the size of the silicon nanocrystals can be controlled by controlling their oxidation. Third, there is a fundamental possibility of separation of nanocrystals in size, which is very important for their potential applications.

5.2.4. Condensation of silicon vapours

Silicon nanocrystals with average sizes from 5 to 20 nm can be obtained by condensation of silicon vapours. For example, in Wang et.al. (2009) synthesis was carried out in vacuum evaporation of silicon from a tantalum ampoule. Silicon vapours were deposited on a silicon substrate, located on a stainless steel plate cooled with liquid nitrogen. The atmosphere in the nc-Si synthesis chamber was argon at a pressure of 67 Pa. The resulting

silicon nanocrystals showed photoluminescence in the range 600–900 nm
with a maximum of ~750 nm.

5.3. Physico-chemical methods

One of the widely used physical-chemical methods is picking up silicon
atoms in the rarefied atmosphere of precursors (monosilane) and the
carrier gas (argon, nitrogen) in which plasma is ignited (plasma-chemical
method). In this method, the formation of nanoclusters can occur both
in the atmosphere and on the surface during the growth of the film. The
characteristic temperature of the substrate may be about 50–300°C, which is
applicable for low-temperature plasma chemical technology of manufacturing
equipment. The structure of the embedded silicon clusters, obtained by this
method, can be modified using high-power excimer dimensional lasers.
The plasma-chemical method can be used to precipitate both amorphous
silicon and oxides, nitrides, silicon with inclusions of silicon at relatively
low temperatures (50–300°C). Then, using the laser effect the structure
of amorphous silicon can be transformed into a crystalline state. The
transformation process refers to the first-order phase transitions and takes
place with the original formation of new phase nuclei, appearing as a result
of thermal exposure. Pulsed laser processing allows one to locally heat
the film with nuclei, without overheating the substrate. This is particularly
important in the case of non-refractory substrates, including glass or plastic.
    Isolated nanoparticles may be obtained by vapour deposition from the
silanes. In this case, the method of thermal decomposition, the method of
microwave plasma, the gas deposition method or chemical vapour deposition
phase (CVD method) are used. One of the most effective methods is the
method based on the decomposition of silane in a CO_2 laser. All of these
technologies use the mechanism of formation of particles in rarefied gas
mixtures consisting of a very toxic silane, SiH_4, and their subsequent
collection on filters or substrates. As the physical methods discussed above,
the physico-chemical synthesis methods are characterized by difficult control
of the dimensions of produced particles.

5.3.1. Principles of plasma-chemical synthesis methods

Plasma-chemical processes are carried out in reactors in the environment
of the plasma with a temperature 10^3–10^4 K at pressures of 10^{-6}–10^{-4} atm
of the atmosphere which is obtained by special devices – plasma torches.
For non-equilibrium plasma, the plasma torches can used a variety of ways:
for example, methods of electric arc, glow, high-frequency and microwave
gas discharges, the methods of adiabatic compression, methods of reactive
discharge of a vapour jet from a nozzle at supersonic speed, etc.
    The peculiarity of the plasma is that it contains a high concentration
of reactive particles such as excited molecules, electrons, free atoms,

atomic and molecular ions, free radicals, which can be produced only in the plasma. The energy characteristics of the particles in plasma are very different from their physical characteristics under normal conditions. Thus, in the glow discharge plasma in gases at pressures of $10-10^3$ Pa the average electron energy is typically 3–10 eV, while the translational energy of heavy particles and the rotational energy of the molecules do not exceed 0.1 eV, and at the same the vibrational energy of molecules and radicals can be close to the average energy of the electrons (~1 eV). Thus, in the plasma their is a strong non-equilibrium of the component in which the subsystems of a single multi-component reactive system may have different translational temperature; the rotational, vibrational and electron temperatures can greatly differ; the Boltzmann law of the populations of energy levels is violated, etc. Therefore, plasma-chemical reactions take place usually in non-equilibrium conditions so that the mechanisms of chemical reactions in the plasma may be very different from the reactions in equilibrium. In particular, the excited and charged particles present in low-temperature plasma, such as vibrational and electronically excited molecules and electrons, initiate the dissociation reaction with the formation of free radicals. Electron impact accelerates the processes of vibrational relaxation and dissociation of molecules not only through the ground state, but in the electronically excited state. Dissociation and recombination through the electronically excited state increases the value of non-adiabatic transitions. The ion–molecule reactions, involving electronically excited ions, start to play a significant role in the dissociation process.

Plasma-chemical reactions are usually multi-channel processes and determine the variety of reactions carried out experimentally in low-temperature plasma. By changing the conditions of plasma generation and regulation of its composition the reaction can be directed to a particular channel. Among the reactions that occur in the non-equilibrium plasma, the most common are the dissociative ionization of molecules, the dissociation via the electronically excited state, dissociative electron attachment to molecules, stepped dissociation by electron impact, dissociative recombination in collisions of molecular ions with electrons and heavy particles with each other. The plasma is characterized by heterogeneous processes, such as formation (or etching) of films of different nature on the inner surface of the reactor or on substrates placed in the plasma.

Reactions in the conditions of non-equilibrium plasma are described by non-equilibrium chemical kinetics. Plasma-chemical reactions are quenched in the region of maximum desired product formation. Typically, the plasma-chemical processes are easy to manage, they are well modelled and optimized. In many cases, the plasma-chemical technology makes it possible to obtain materials (e.g. fine powders, films, coatings), and substances with very valuable properties (tungsten, for example, acquires resistance to recrystallization and creep and the anisotropy of the emission properties).

Silicon solar cells are widely produced by the thin-film technology, which uses thin layers of polycrystalline and amorphous silicon, and in the future nanocrystalline silicon layers will also be used. The choice of the method of thin-film silicon coating deposition is critical for obtaining high-quality photovoltaic cells. One of the main requirements which this procedure must satisfy is to minimize the influence of the boundary of the grains on the lifetime and mobility of charge carriers in a semiconductor. This condition can be satisfied by using nanocrystalline silicon which is a two-phase material consisting of silicon nanocrystals, distributed in the amorphous silicon matrix.

Thin films are also widely deposited using the method of plasma-vapour deposition in a high-frequency discharge plasma (known as PECVD method name – shortening of Plasma Enhanced Chemical Vapour Deposition). The main advantage of this method is the ability to create thin layers of high quality. However, the deposition rate in this method is very is low.

Acciarri et al. (2005) modified the PECVD method using plasma glow discharge generated by low-voltage direct current. With the modified method the authors were able to grow films of nanocrystalline Si with a thickness of 1–3 µm. This process involves the ions with low energy, so they called their method LEPECVD (Low Energy Plasma Enhanced Chemical Vapour Deposition). This method allows to increase the intensity of the plasma discharge in the area of film growth, which results in a significant increase in the rate of growth without ion damage to the growing layer.

High-temperature plasma-chemical method to obtain nanocrystalline silicon from the condensed phase

The main conditions for obtaining fine powders by plasma-chemical method is the reaction far from equilibrium and a high rate of nucleation of a new phase at the low rate of growth (A.I. Gusev, A.A. Rempel', 2000). In the real plasma chemical synthesis conditions the nanoparticles can efficiently produced by increasing the rate of cooling of the plasma flow in which condensation from the gas phase takes place. This reduces the size of the particles. Plasma chemical synthesis provides a high rate of formation and condensation of compounds and has a fairly high efficiency.

The main disadvantages of plasma-high-temperature synthesis nanosilicon a broad particle size distribution, and therefore the presence of fairly large particles, as well as high levels of impurities in the resulting nanocomposite. The temperature of the plasma, reaching up to 10 000 K, determines the availability in the plasma of ions, electrons, radicals and neutral particles in the excited condition. The presence of such particles leads to high rates of interaction and rapid (10^{-3}–10^{-6} s) reactions. The high temperature ensures move almost all the starting materials in the gaseous state, with their subsequent interaction and condensation products.

Plasma-chemical high-temperature synthesis involves several steps. The first step is the formation of active particles in the arc, high-frequency and microwave plasma torches. The highest power and efficiency have the arc plasma torches, but the materials produced by them are contaminated with products of erosion of the electrodes; electrodeless high-frequency and microwave plasma torches do not have this drawback. In the next stage quenching leads to the release of interaction products. The choice of the place and time of quenching yields nanocomposites with a given composition, shape and size of particles.

Synthesis of nanosilicon was carried out by the plasma chemical method in the reactor constructed at SSC RF GNIIChTEOS[1] (Dobrinsky et al, 1979). The process was performed in an argon plasma in a closed gas cycle. The data obtained for nanosilicon samples consisted of particles of the 'core–shell' type in which the core is the silicon coated with the shell formed in the process of passivation of the particles of oxygen and/or nitrogen. Deep cleaning of gas to remove the impurities of moisture and oxygen is carried out through the molten aluminum or special finish cleaners, irreversibly absorbing impurities to a level $(2–3) \cdot 10^{11}$ cm^{-3}. High-purity argon is supplied from the mains. The gas is circulated using a membrane compressor. Compressed gas from the receiver goes to a ramp of rotameters, through which the plasma is distributed over the plasma installation sites.

The reactor is a plasma evaporator-condenser, working in the low-frequency arc discharge mode. The raw material is silicon powder, which is fed to the reactor by the gas flow from the dispenser. In the reactor the powder evaporates at a temperature of 7000–10000 K. At the output of the high-temperature plasma zone the resulting vapour–gas mixture is subjected to rapid cooling by the gas streams, resulting in condensation of Si vapours to form aerosol.

The resulting aerosol with a temperature of 400–500 K travels to the refrigerator, where it is cooled to a temperature of 300–350 K. The large particles, including the unprocessed fraction, are separated from the soot in the classifier of the inertial type. The finished powder is captured in a fabric filter baghouse.

The ultrafine powders obtained in an inert atmosphere have pronounced pyrophoric properties. To prevent self-ignition, they should be stored and used in an inert atmosphere or particles should be coated with a special coating to prevent contact with the atmosphere. Several methods of encapsulation of the particle surface can be used. A simple method is partial oxidation – the formation of an oxide film. To reduce the thickness of the film and prevent self-ignition, oxidation is carried out in an inert atmosphere with small additions of oxygen or air at a temperature of 290–300 K.

From the filter the finished powder is unloaded in an inert atmosphere into a sealed container or moved to a microencapsulation system, where the surface of the powder particles is deposited with an inert protective layer that prevents the powder from inclement weather.

---

[1]The State Scientific Center of the Russian Federation 'State Research Institute for Chemistry and Technology of Organoelement Compounds'.

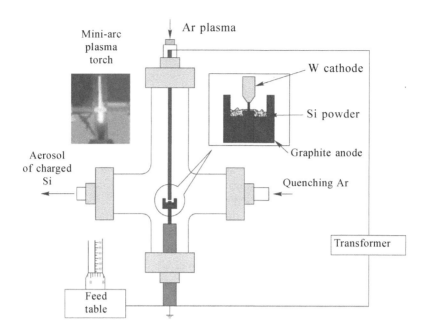

Fig. 5.5. The scheme of the arc plasma reactor for the synthesis of nc-Si (Liu et al, 2008).

Synthesis of nanosilicon in an electric arc

nc-Si particles are obtained also in an electric arc (Liu et al, 2008). The graphite anode was made in the form of a cup in which silicon is placed, and the cathode was a tungsten rod. Plasma was blown with a horizontal flow of argon which carried nc-Si away from the arc (see Fig. 5.5). The pressure inside the reactor was atmospheric. In addition to nc-Si, the nc-Si in the powder form was obtained by growing on the multiwall carbon nanotubes inserted in the reaction vessel. Both products had an impurity in the form of WO_3. The surface of the nc-Si was not described.

Kobayashi et al. (2006) used the electric arc discharge method in liquid nitrogen for producing Si nanoparticles. Equipment constructed for this purpose produces Si nanoparticles in large quantities, and the cost of production can be lower. Subsequent chemical etching of the produced particles in the solution reduced their size and led to passivation of the surface. As a result, under irradiation with ultraviolet light the particles emitted a vivid visible fluorescent light. Slow etching in an HF solution produces silicon nanoparticles with a narrow size range. Thus, the authors were able to obtain particles with a diameter of 3–6 nm. The diameter of the particles in the etching process was measured by Raman spectroscopy. The dependence of the photoluminescence spectrum on the particle size, which agrees well with the model of the quantum constraints, was determined.

Preparation of dispersions of nanosilicon in SiO$_2$ by the cold plasma method

Layers of SiO$_{1.16}$ 200 nm thick were deposited on a monocrystalline substrate Si (100) at 300°C using cold plasma (PECVD) (Jung et al, 2008). SiH$_4$ at a flow rate of 160 cm^3 · min^{-1} and N$_2$O at a flow rate of 50 cm^3·min^{-1} were used. The layer was then annealed at 1100°C for one hour in an atmosphere of nitrogen to form nc-Si. Passivation of the samples of nc-Si/SiO$_2$ by atomic hydrogen was carried out using a radio-frequency generator (290 mW/cm^2, 13.56 MHz) and pure hydrogen in a PECVD reactor where the pressure of the gas was 1 Torr, the flow rate of hydrogen 180 cm^3 · min^{-1}. The sample was heated to 250 °C. The average diameter of nc-Si was equal to 2.8 nm, and the density of nc-Si in the dispersion was 4.3 · 10^{12} cm^{-2}. Exposure to hydrogen plasma for 40 hours almost doubled the intensity of the photoluminescence of the sample due to passivation of defects at the nc-Si/SiO$_2$ interface quenching luminescence. If the passivated sample was annealed at 800°C for 1 h it was depassivated and the photoluminescence returned to the level characteristic of a given sample prior to hydrogen passivation. The defects quenching the luminescence formed in the process of separation of SiO$_{1.16}$ in two phases – nc-Si and SiO$_2$.

The lifetime of the PL-passivated samples, 52 ± 2 μs, is only slightly dependent on the passivation temperature, but the PL intensity increases with increasing passivation temperature of the sample (at a temperature of about 250°C). The main PL quenching traps are dangling bonds of silicon atoms, which are saturated with hydrogen in the passivation process.

To study cathode luminescence (CL), SiO$_x$ was deposited on a silicon substrate by LPCVD (Morana et al, 2008) using Si$_2$H$_6$ and O$_2$ as precursors and nitrogen as the carrier gas. The pressure was set in the range 185–300 mTorr, and the temperature was 250–400°C. The ratio of gas flows Si$_2$H$_6$/O$_2$ was in the range 2–5. The total flow rate was 102 cm^3 · min^{-1} (90 – nitrogen, 12 – silane + oxygen). The precipitated layer on the substrate was annealed at 1100°C for 1 hour to form nc-Si in SiO$_2$. Additional treatment consisted of heating the sample in a mixture of H$_2$–N$_2$ at 450°C for 1 hour. This increased the intensity of the CL by 30%. The excess silicon in the SiO$_x$ varied in the range 0–70% by volume. The FTIR method found Si–H bonds on the nc-Si surfaces and the number of bonds increased with the reduction of temperature at which the sample was produced. It was found that the hydrogen, passivating the surface of nc-Si, has little effect on CL of nc-Si, but the intensity of CL is maximal if an excess of silicon was 22%.

EL of samples obtained by means of cold plasma was studied. The SiO$_x$ layer was deposited by PECVD (Barreto et al, 2007) on a silicon p$^+$-type (10^{-3} ohm · m) substrate by the interaction of SiH$_4$ and N$_2$O at 300°C and a pressure of 26.66 Pa. The flow rate of the oxidizer at 100 cm^3·min^{-1} was ~9 times higher than the silane flow of 10.9 cm^3·min^{-1}. The sample was annealed at 1250°C for 1 h in nitrogen at atmospheric pressure. The selected mode corresponds to the strongest photoluminescence of the sample. Silicon excess reached 17.5%. Electric current was applied to the sample in pulses, and for these conditions electroluminescence was observed below 60 V (10 mV cm^{-1}) of the applied voltage and at constant voltage the maximum allowable voltage to the sample is 20–25 (~4 mV · cm^{-1}). The maxima of PL and EL spectra of about 800 nm were almost identical. The threshold value of bias voltage below which there is no EL is 15 V. The quantum efficiency was 0.03%. Comparing this result with studies of the EL mentioned above, it can be argued that the EL is highly dependent on synthesis conditions and sample preparation. Apparently, the threshold voltage depends not only on the material but also on the design of electroluminescent cells. For example, electroluminescence on the *p–i–n* diode with a wavelength of ~650 nm in the peak was observed (Fojtik et al, 2006b) in the layers of nc-Si, with successive layers of SiO$_2$ and amorphous silicon, at voltages of 6–16 V, and the intensity of EL increased with increasing voltage.

The effect of sample composition and synthesis conditions on the formation of nc-Si has been studied in (Hernandez et al, 2008). SiO$_x$ was deposited by LPCVD on the SiO$_2$ substrate at 600°C. The flow rates of N$_2$O and SiH$_4$ varied in the range 180–190 and 20–10 cm^3 · min^{-1}, respectively, in order to change the composition of SiO$_x$. The composition was determined by X-ray photoelectron spectroscopy, and showed an excess of silicon in the samples of 16, 19, 21 and 29 at.%. The samples were annealed at temperatures ranging from 800 to 1250°C for 1 h in nitrogen to form nc-Si.

Raman scattering showed that the samples before annealing contain a lot of amorphous silicon. Annealing at 800°C has little effect on the Raman spectrum, but after annealing at 1100°C the spectrum clearly indicates the reduction of the amount of amorphous silicon and the appearance of nc-Si. In general, crystallization was completed by annealing at 1250°C, but the samples with different contents silicon behaved differently. For example, in the samples annealed at 1100°C, the amount of nc-Si decreased rapidly with decrease in the amount of excess silicon. In other words, the SiO$_x$ samples lean in excess silicon must be at high temperatures, and the rich – at lower temperatures. However, the proportion of nc-Si even in the sample with an excess of 29 at.% reaches only 20% (the rest of the silicon remains amorphous). At an annealing temperature of 1250°C the opposite is true: in samples with excess silicon content of 16 and 19 at.% the share of nc-Si reaches 90%, and with 21–29 at.% – 40% (60% of silicon remain in the amorphous state). It was experimentally shown that with increasing content of excess silicon in SiO$_x$ from 16 to 29 at.%, the average diameter of nc-Si increases after annealing at 1250°C with 2.6 to 3.4 nm.

Increasing the annealing temperature of any sample increases the particle size of nc-Si. For example, when the content of excess Si reaches 21 at.% the average diameter of nc-Si after annealing at 1100°C is 2.7 nm and after annealing at 1250°C – 3.1 nm. Thus, with increasing annealing temperature increases and the proportion of nc-Si (amorphous by silicon) and the particle size of nc-Si.

The composition of the sample is affected by the composition of the gaseous medium. Using the PECVD (Forraioli et al, 2007), the substrate Si (100) was coated by SiO_x by the interaction of SiH_4 and N_2O. Operating pressure was 500 mTorr, the power 100 watts. The variation of the ratio N_2O/SiH_4 allowed to modify the contents of excess silicon in SiO_2 ranging from 52 at.% to zero.

Due to the use of N_2O the samples contained 1–2 at.% nitrogen. The dimensions of nc-Si increased with increasing annealing temperature from 1000 to 1100°C in an atmosphere of nitrogen or with increasing silicon content in the SiO_2.

PL is ineffective at the maximum and minimum silicon content in SiO_2: in the first case nc-Si was large that there was no size effect, and at a low concentration of silicon nc-Si was small and, at the same time, the fraction of Si remaining in the amorphous state, which contains the centres of non-radiative energy transfer, was large. It was found that the effectiveness of PL normalized in the thickness of the layers of nc-Si/SiO_2 dispersions was an order of magnitude higher in thick layers (200 nm) as compared with thin one (50 nm). This is associated with an increase in the fraction of amorphous silicon in thin layers.

The silicon content in the nc-Si/SiO_2 dispersions may decrease during annealing of SiO_x layers due to atmospheric oxygen. The protective layer of Si_3N_4 on the nc-Si/SiO_2 surface slightly inhibits the oxidation process. Small changes in the PL spectra as a function of excitation wavelength (360 and 488 nm) were noted.

The EL of the multilayer structure (MLS), which consists of layers of amorphous Si/SiO_2 (MLS Si/SiO_2), WAS studied by Chen et al. (2008). MLS was obtained on a silicon crystal substrate with hole conductivity. The silicon substrate was pre-etched with diluted HF for 20 s to remove the surface layer of oxides and metals. The layers of silicon and SiO_2 were deposited on the substrate by PECVD. SiH_4 was supplied at a rate of 5 $cm^3 \cdot min^{-1}$ to produce the Si amorphous layer and then the oxygen flow at a feed rate 20 cm^3 min^{-1} for the oxidation of silicon was activated. The substrate temperature was maintained at 250°C. 10 pairs of Si/SiO_2 layers were deposited at the amorphous silicon layer thickness of 3 nm and the SiO_2 – 2.5 nm.

The freshly deposited Si/SiO_2 MLS was dehydrogenated at 450°C for 1 h and annealed in a nitrogen atmosphere, first at 1000°C for 50 s in a special system for rapid annealing, and then at 1100°C for 1 h in a normal furnace for the formation of nc-Si (nc-Si/SiO_2 MLS) with a diameter of 3 nm.

In addition to the nc-Si/SiO_2 MLS *p–i–n* sandwich structures were also produced, namely, between the silicon substrate with hole conductivity, and nc-Si/SiO_2 MLS there was a layer of amorphous silicon with a thickness of 50 nm with hole conductivity (boron doped with a conductivity of 5 ·

10^{-5} ohm $\cdot$ cm^{-1}), and on the opposite side the nc-Si/SiO$_2$ MLS was coated with an amorphous silicon layer with a thickness of 50 nm with electron conductivity (doped with phosphorus, with the electronic conductivity of 3 $\cdot$ 10^{-3} ohm $\cdot$ cm^{-1}). The LED was obtained from this construction by spraying of aluminum electrodes.

The threshold voltage at which EL forms using the nc-Si/SiO$_2$ MLS (without the 50 nm layers of amorphous silicon) on a silicon substrate with hole conductivity was found to be 8 V, and at 12 V the intensity of the EL increased almost 6 times. In the EL spectrum there were two bands – at 520 and 650 nm. The first of these, according to the authors, is due to the relaxation of hot electrons, injected through the tunnel in the MLS. Hot electrons generate holes in SiO$_2$, and recombination of electrons and holes gives rise to the 520 nm band. The intensity of the 520 nm band increases with increasing bias voltage faster than the intensity of the 650 nm band, which arises from the recombination of the holes and electrons injected in the MLS, and recombination can occur both in the nc-Si and in the centres of luminescence at the nc-Si/SiO$_2$ interface.

The spectra of the MLS and of the p–i–n transition proved to be very different. In the spectrum of the p–i–n transition the 520 nm peak disappeared, which means that the EL is determined only by the electron-hole recombination responsible for the 650 nm peak. In addition, the threshold voltage decreased to 3 V by using a silicon substrate with a maximum concentration of holes, and to 5 V with the silicon substrate poorly doped with boron, and the intensity of EL increased by 50 times compared to the MLS using the substrate with maximum doping with boron. These improvements are explained by the authors by increasing the extent of injection of holes in the MLS. The intensity of the EL is proportional to the current passing through the p–i–n transition, which follows from the current–voltage characteristics of the samples.

Multilayer samples were synthesized (Boninelli et al, 2007) by PECVD. The substrate (silicon wafer) was heated to 300°C during the deposition of SiO$_x$ on it. The reagents were SiH$_4$ and N$_2$O in a ratio at which a SiO$_x$ film with 46 at.% silicon was produced. The composition of films was determined by RBS. The total layer thickness was 80 nm.

Samples were prepared, starting from with formation of the SiO$_2$ layer on the silicon substrate. SiH$_4$ was passed at a rate of 1 cm^3 $\cdot$ min^{-1}, and N$_2$O – 30 cm^3 $\cdot$ min^{-1} at a total pressure 3.5 $\cdot$ 10^{-2} Torr and 30 W RF power. A subsequent layer of silicon was deposited by passing SiH$_4$ at a speed of 20 cm^3 $\cdot$ min^{-1} under the pressure of 1.5 $\cdot$ 10^{-2} Torr and 25 W RF power generator. The thickness of the SiO$_2$ layer was 7 nm, the silicon layer 3 nm. This was followed by deposition of a layer of SiO$_2$ and the alternation of layers continued. After deposition, the samples were annealed at temperatures of 900–1250°C from 0.5 to 16 h under a nitrogen atmosphere in a horizontal furnace.

Comparison of the results obtained by DFTEM and EFTEM showed that already at 900°C Si inside the SiO_2 matrix forms amorphous clusters about 2 nm in diameter. The crystallization of silicon in the SiO_2 matrix requires a temperature of 1100°C and higher (recall that the SiO_x contained 46 at.% silicon). At a temperature of 1250°C Ostwald ripening takes place – growth of large silicon crystals at the expense of small ones. Crystallization of silicon at 1250°C ends within an hour.

The continuous layer of amorphous silicon breaks up at 1000°C to separate intrusions of silicon in SiO_2 isolated from each other. At 1100°C 10% of silicon crystallizes within an hour and 90% of the silicon remains in the amorphous state. At 1200°C 90% silicon crystallizes and at 1250°C 100%. After annealing SiO_x (46 atm.% Si) at 900°C photoluminescence was not observed, and a sample of the same composition annealed at 1100°C, showed a strong photoluminescence. It occurred at 1000°C, increased 15 times after annealing at 1150°C compared to 1000°C, but its intensity decreased three times after annealing at 1250°C as compared with 1150°C. The 10-layer sample (SiO_2/Si) did not show any photoluminescence after annealing at temperatures below 1100°C. The maximum intensity of photoluminescence was recorded after annealing at 1200°C, and the maximum of the luminescence band shifted from 850 nm for the sample annealed at 1100°C to 940 nm in the sample annealed at 1250°C. At low excitation power the photoluminescence intensity depends on the relaxation time and on the number of emitting centres. It turned out that the photoluminescence intensity varies exponentially with time:

$$I(t) = I_0 \exp[-(t/\tau)\beta],$$

where I_0 is the intensity at $t = 0$, β is the variance factor, τ is the relaxation time. It was experimentally shown that in the SiO_x samples annealed at a temperature of 1100–1250°C, the lifetime increases, respectively, from 20 to 30 μm, and β increases from 0.62 to 0.71. The values β show that the process of decay of photoluminescence is not described by a single exponent. Samples obtained by annealing at a temperature of 1000°C show two different mechanisms of quenching luminescence – one similar to that observed in the samples obtained at 1100–1250°C, and the other with $\tau < 1$ μs, due to scattering of the pumping energy by non-radiative recombination centres.

The lifetime of the photoluminescence of 10-layer samples annealed at temperatures of 1100–1250°C, increased from 15 to 70 μs, respectively, β – from 0.66 to 0.77. In both SiO_x and the 10-layer samples, the concentration of radiative centres grew up to annealing temperatures of 1200°C, and decreased for samples annealed at 1250°C, which coincides with the enlargement of nc-Si and a decrease in their concentration as a result of annealing at 1250°C, namely: at 1100°C the density of clusters was $6.7 \cdot 10^{11}$ cm^{-2}, at 1200°C it was $3.0 \cdot 10^{11}$ cm^{-2} (with the average radius of 3.7 nm), but at 1250°C it was $2.7 \cdot 10^{11}$ cm^{-2} (the average radius of 4.1

nm). Increasing annealing temperature increases the distance between the nc-Si particles: in samples annealed at 1100°C to 3.5 nm and at 1250°C to 11 nm, which leads to an increase in the value β, and hence to an increase in intensity, but the enlargement of nc-Si leads to an increase in the proportion of large nc-Si, exceeding the Bohr exciton radius, and it transforms large nc-Si to centres that do not emit in the visible spectrum.

Three-dimensional plasmon imaging is based on a combination of the possibilities of TEM and EELS (Yurtsever et al, 2006). The samples were prepared by cold plasma. The native oxide was removed from the substrate Si⟨100⟩ of n-type by pre-treatment with HF, and the silicon-rich oxide SiO_x was deposited on the substrate. The excess of silicon was ensured by a low $N_2O:SiH_4$ ratio in the reactor. The silicon content, 47 at.%, was experimentally defined. The thickness of the layer, defined by TEM of the cross section, was 186 nm. The sample was annealed for 3 h at 1100°C in nitrogen.

A diffuse signal is observed around the nc-Si which in some studies is interpreted as a shell of amorphous silicon around the nc-Si, but the same effect (boundary layer with a thickness of 1–3 nm) is the result of subtle physical effects at the time of measurement and not due to an amorphous layer on the surface of nc-Si, as suggested by the authors of the article.

Figure 5.6 shows that the form of nc-Si greatly varies which must inevitably affect the electronic and optical properties of nc-Si. A higher surface to volume ratio (0.55) than for a sphere of equivalent volume (0.35) leads to an increase of the density of states on the surface of the nc-Si, increasing their contribution to the lifetime of the excited states and recombination.

Synthesis of nanosilicon in cold plasma: modification of the surface of nc-Si with hydrogen

We have considered synthesis of nc-Si in a cold plasma with simultaneous partial oxidation of silicon. Now we consider the formation of nc-Si particles with hydrogen passivation of their surface.

The following method was used for such particles. A radiofrequency generator (13.56 MHz) creates cold plasma in a tubular reactor, through

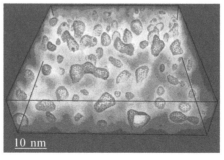

Fig. 5.6. Tomographic image reconstruction of nc-Si. Non-spherical particles dominate (Yurtsever et al, 2006).

which the reagents (usually SiH_4), mixed with a carrier gas (usually argon), are passed. The plasma electrons have a thermal energy of 3–5 eV (30 000–50 000 K) and dissociate under the influence of the reagents. Because of the low electron density the temperature of silane and argon remains close to room temperature, but the temperature of nc-Si is several hundred degrees above the temperature of the gases. Nanocrystals and the walls of the plasma reactor are negatively charged, preventing the aggregation of nc-Si (Mangolini et al, 2006b, 2007).

Formation of silicon nanoparticles (Cavarroc et al, 2006) by passing the Ar + SiH_4 mixture through the cold plasma zone passes through four stages: the growth of nanocrystals crystals (2–3 nm) from the products of plasmolysis of SiH_4, their accumulation, the coalescence and accumulation of the products of coalescence with simultaneous deposition on the substrate. At the same time the characteristics of the plasma (electron density, temperature, electric field, impedance, etc.) greatly change. This allows us to determine the time separating the accumulation phase of nc-Si from their coalescence. It turned out that after this moment the nc-Si grow to 5 nm, and over time reach 45 nm, and their specific surface area exceeds 100 $m^2 \cdot g^{-1}$. The size scatter of nc-Si size is large: 4–20 nm based on BET analysis and AFM (Meier et al, 2007). Sometimes the cold plasma is formed (Nguen-Tran et al, 2007) by the rectangular modulation of the discharge with a plasma time from 20 ms to 1 s, and then the particle size can be controlled. An important factor is the dilution of silane with hydrogen. Two gas mixtures were used: SiH_4 (1 $cm^3 \cdot min^{-1}$) + He (100 $cm^3 \cdot min^{-1}$) and SiH_4 (5 $cm^3 \cdot min^{-1}$) + Ar (50 $cm^3 \cdot min^{-1}$) + H_2 (50 $cm^3 \cdot min^{-1}$). The temperature of the electrodes was 150°C, gas pressure 1.4 Torr for the first mixture and 2.8 Torr for the second, with the power of the radio-frequency generator of 20 watts. The substrate for the collection of nc-Si in the first mixture was 1 cm from the plasma and 5 cm in the second mixture. The diameter of nc-Si was found to be 12 ± 1 nm. In the presence of hydrogen nc-Si grow 10 times faster than in the absence of H_2, and they are crystalline, and in the absence of hydrogen – amorphous. In the first 4 ms the process of nucleation takes place, and in the given synthesis conditions of synthesis the particles (such as crystalline and amorphous) were at least 5 nm in diameter. Crystallization can be affected not only hydrogen but also by argon, gas pressure and the distance from the plasma to the substrate. The crystallinity was checked by TEM. In the presence of hydrogen and argon the rate of growth of nc-Si was very high, hence the growth rate does not affect the crystallinity. The size of nc-Si decreased with the reduction of the plasma pulse time. Amorphous Si particles did not luminesce.

It has been shown (Anthony et al, 2009) that in the cold plasma method there is a threshold of the power introduced into the reaction volume, above which nc-Si is produced, and a-Si forms below this threshold. The photoluminescence quantum yield of the former reaches 40%, and of the latter only 2%, if not less. Sometimes amorphous silicon is produced intentionally by the cold plasma method. The Si (100) p-type substrate was cleaned and washed as usual and an oxide layer with a thickness of 3 nm was grown by heating to form a tunnel oxide layer for the production of memory cells (Chan et al, 2006). A layer of amorphous silicon 10 nm thick was grown on the top by PECVD at 400°C with a pressure of SiH_4 of 4.2 Torr at the deposition rate of 1.82 nm/s. One set of samples for transfer to the crystalline state was annealed in a furnace at 800 and 1000°C for 15 min in nitrogen. Another series of samples was annealed in nitrogen quickly – at 1000°C for 1 min. The third group of samples was annealed with a KrF-excimer laser. After annealing, all samples were etched in a solution of $NH_4OH:H_2O_2:H_2O = 1 : 2 : 5$ for isolating each nc-Si grain by dissolving the residue of amorphous silicon. nc-Si were found in the amorphous silicon prior to annealing. The highest density of nc-Si was achieved by laser annealing. The diameter of nc-Si was in the range 2–5 nm. The samples annealed in a furnace were crystallized more efficiently with increasing annealing temperature. PL was not observed prior to annealing the samples, but laser annealing produced the sample with the strongest photoluminescence. As a result of annealing in the furnace PL was weaker. The PL spectrum had two peaks: 670 nm (1.9 eV) and 710 nm (1.7 eV).

The dimensions of the particles in the aerosol are often measured by tandem differential analysis of the mobility of the particles (Holm et al, 2007a). A mixture of 0.5% SiH_4 and 99.5% argon was introduced into a tubular cold plasma reactor. The flow rate of the gas mixture was 5–10 $cm^3 \cdot min^{-1}$, and of additional argon 400 $cm^3 \cdot min^{-1}$. The 30 W powder of RF energy was fed to the two ring-shaped electrodes encircling a tubular reactor. The gas pressure inside the reactor was 20 Torr. At the reactor outlet the aerosol was diluted with nitrogen or argon, reducing the likelihood of coagulation. After that, the gas pressure was adjusted to atmospheric pressure. nc-Si in the aerosol was given a single positive charge, and the number of nc-Si particles was determined prior to passage of the aerosol through the tubular furnace and then through a second device for measuring the amount of nc-Si. The furnace temperature was varied from room temperature to 950°C. As a result, measurements were used to construct the dependence of nc-Si particles on their size. For comparison, similar dependences were measured in the initial aerosol before heat treatment. Figure 5.7 shows the distribution of the nc-Si diameters and Fig. 5.8 shows that after annealing at 500°C the concentration of nc-Si is much smaller than in the original aerosol.

Holm et al. (2007a) suggest that the reduction of the size of nc-Si in heating is due to the thermolysis of the $Si-H_x$ bonds. Reducing the diameter of the particles after heating to 500°C and the disappearance of the $Si-(H)_2$ and $Si-(H)_3$ bonds, detected by FTIR and mass spectrometry of secondary ions correlate well with each other. Heating up 700°C leads to the disappearance of the Si–H bonds and a slight decrease in the size of the particles. An attempt was made to add hydrogen to the surface of nc-Si with no Si–H bonds at 400°C.

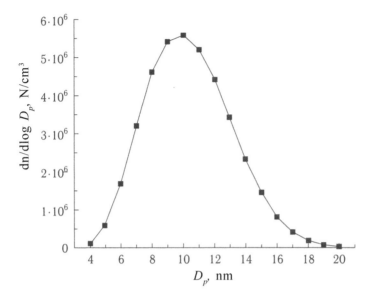

Fig. 5.7. The distribution of nc-Si particle size. D_p is the diameter of the nanoparticles (Holm et al, 2007a).

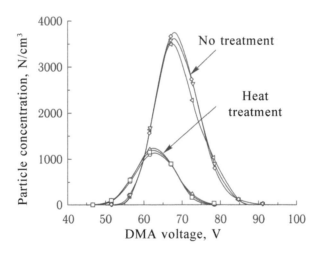

Fig. 5.8. The distribution of the concentration of nc-Si particles in the aerosol (measured in the differential mobility analyzer (DMA)) in nitrogen after passing through the furnace with a temperature of 500°C; for comparison, the measured distributions for the nanosilicon not subjected to annealing, i.e. for particles that have passed by the furnace, are shown. To confirm the reproducibility, the distributions were always measured three times (Holm et al, 2007a).

However, a negative result was obtained. Prior to heating the nc-Si particles showed red PL, after heating to 400–500°C the PL intensity decreased, and after heating at 600°C disappeared altogether.

The particles of nc-Si, obtained by cold plasma, cannot exist in the air without passivation of their surface. Silicon nanoparticles with a diameter of 40–60 nm were obtained (Roura et al, 2007) in the amorphous state from silane by the cold plasma method. Their surfaces were passivated with hydrogen but in the air they were rapidly oxidized and thus stabilized. The nanoparticles were crystallized by heating in argon. Heating at a rate of 30 K · min^{-1} at low temperature (770°C) resulted in the crystallization of the sample with composition $Si_1O_{0.28}H_{0.39}$, and at a temperature of 817°C – a sample containing less oxygen but more hydrogen – $Si_1O_{0.113}H_{0.53}$, i.e. the higher the silica content the lower the crystallization temperature or, equivalently, the rate of crystallization is higher. During crystallization the mass of the samples decreased by 1–2 wt.% due to loss of hydrogen. Crystallinity was confirmed by HRTEM and Raman spectroscopy. The crystallization rate is controlled by nucleation and does not depend on the growth rate. This is explained by the fact that the nucleus formed in the silicon nanoparticle rapidly expands to the borders of the particle, but any other silicon nanoparticles should 'wait' for its nucleus to appear. Most of the nanoparticles are single crystal, but aggregates also form.

The need for additional passivation was also noted in the work by Ligman et al. (2007). nc-Si were obtained from SiH_4 in cold plasma. They were dispersed in chloroform. The result was a dispersion with a concentration of 0.1 mg/ml. PL was not observed due to insufficient passivation, so the photooxidation under the influence of UV light (365 nm) produced a SiO_2 shell around nc-Si with a thickness of 1 nm due to oxygen and water, dissolved in chloroform. The PL spectrum of nc-Si in the solution after passivation with oxygen had a peak at 650 nm. A polymer with hole conductivity – poly-(9-vinylcarbazole) was added to the dispersion. Its concentration was 10 mg/ml. The choice of the polymer was determined by solubility in chloroform. Electroluminescence (EL) was investigated. The EL spectrum contained a peak at 640 nm from nc-Si and at 415 nm from the polymer. The current-voltage characteristics of the device, obtained at different temperatures, showed that the nc-Si particles are not directly involved in the transport of charge carriers and acted as their traps.

Oxidation of nc-Si was studied carefully Holm et al. (2007). The particles of nc-Si/H were synthesized in the low-temperature plasma from silane. The carrier gas was argon. The starting reagent was a mixture of gases – 0.5 vol.% SiH_4 in argon, which was further diluted with argon: 5–10 cm^3 · min^{-1} mixtures were mixed with 400 cm^3 · min^{-1} of argon. The decomposition of silane proceeded at a pressure of 18–20 Torr. The average particle diameter of nc-Si was equal to 6 nm. Next, the aerosol – nc-Si/H in argon– was diluted with nitrogen analyzer for passage in the analyzer of the number and size of nc-Si/H in the gas. A mixture of nitrogen and oxygen, in which the ratio of the gas could be varied within wide limits, was added to the aerosol and the aerosol was directed into a tubular furnace, at the output of which there was yet another analyzer of the number and size of solid particles. At any volume ratio of $O_2 : N_2$ from 0.004:0.996 to 0.86:0.14 the particle size after passing through the tubular furnace at a temperature of 400–500°C decreased in size from 6.0 nm to 5.8–5.9 nm, but with increasing temperature from 600 to 800°C their size increased rapidly, reaching a plateau in the temperature range 900–1100°C. The size of the particles eventually increased to 6.6–7.0 nm (the more oxygen in the gas mixture, the larger the size). FTIR of the nc-Si/H, obtained without oxygen, detected the absorption bands of the Si–H , $Si=(H)_2$ and $Si\equiv(H)_3$ bonds in all samples annealed at temperatures up to 400°C, but only Si–H bonds in the temperature range 500–700°C. Above 700°C all Si–H bonds were pyrolyzed. FTIR of nc-Si/H, obtained with 2.5 vol.% –10.0 vol.% oxygen in the gas mixture showed no oxidation of nc-Si/H in the temperature range 23–400°C. Only starting from 500°C nc-Si/H was significantly oxidized, but no absorption band corresponding to the Si–OH group was found in the spectra. FTIR showed that the oxide on the surface of nc-Si does not reach the stoichiometric SiO_2 composition. The thickness of the oxide layer on nc-Si was estimated to be 0.5 nm. Mass spectrometry of secondary positive ions with time resolution showed that with increasing temperature the number of SiHx bonds decreases, and that of the O–Si–H bonds grows. The spectra of negative ions are characterized by the dominance of H ($m/z = 1$) and O ($m/z = 16$), and there are also OH, Si, SiH_x and OSiH peaks. The intensity of the peaks of O and OH increases with increasing furnace temperature, and H is clearly present at any temperature. Only at 800°C the rate of oxidation of nc-Si/H is high to form O_2Si, O_2SiH, O_3Si and O_3SiH. X-ray high-resolution photoelectronic spectroscopy showed that the silicon nanoparticles are retained at any temperature. The results obtained by different methods indicate that at 400–500 °C pyrolysis of the Si–H bonds takes place and the particle size decreases slightly, while at higher temperatures, oxidation of silicon begins and the particle size increases, and the differences in the oxygen content in the gas mixture have a lesser effect on the oxidation process than temperature. The same pattern was observed for amorphous Si.

To reduce the size of nc-Si it is proposed to etch nc-Si by the dry method in the gas phase after synthesis in plasma (Pi et al, 2007a). A mixture of gases SiH_4/He (5%/95% by volume) and argon were added to the plasma. CF_4 was added to the gas flow in the immediate vicinity of the plasma zone. The flow rate of SiH_4/He was 9–12 $cm^3 \cdot min^{-1}$, Ar and CF_4 45 and 1.5–6 $cm^3 \cdot min^{-1}$, respectively. The gas mixture was fed into the second plasma region. The product was precipitated on a stainless steel filter. TEM showed that the nc-Si is crystalline. nc-Si had the PL bands in the red and infrared regions. It is assumed (Pi et al, 2007a) that their diameter was greater than 3 nm.

FTIR showed that the surface of nc-Si so coated with a fluorocarbon film which may interfere with etching of nc-Si. Oxidation in air of the etched and unetched samples over time leads to a blue shift of PL bands, which indicates a decrease in the size of nc-Si, but the PL intensity increases, i.e. the oxide layer blocks the P_b-centres on the surface of nc-Si. After 100 hours, oxidation of the nc-Si ends and changes in the spectra of nc-Si are terminated. This corresponds to the self-limiting oxidation in the nc-Si. The etched nc-Si are oxidized more rapidly than unetched n-Si, and changes in their PL spectra are the largest: increase in the intensities of the etched nc-Si is 7 times more than in the unetched. However, the oxidation does not completely block the P_b-centres on the surface of the etched nc-Si. Later, the same authors (Pi et al, 2008) specified that the etching of nc-Si reduces the size of nc-Si and, as a consequence of the size effect, the PL band maximum is shifted from 800 to 530 nm. In the air the wavelength of the nc-Si PL changes from 579 nm to 553 nm, but the intensity decreases by 15 times in 12 days because of the oxidation of the surface of the nc-Si by oxygen. At the same time there is blue PL, and on the 10th day its intensity is stabilized, exceeding the intensity of original PL 2–3 times. The source of blue light are surface defects of nc-Si, resulting from oxidation by atmospheric oxygen. The quantum yield of the red PL of nc-Si is 16%, yellow – 2%, and blue – 1%.

Attempts were made to etch the oxide layer on the surface of nc-Si with the vapour of hydrofluoric acid. nc-Si was produced by pyrolysis of a mixture of SiH_4 and argon in cold plasma (Pi et al, 2007). The powdered product was dispersed in methanol using an ultrasonic bath immediately after synthesis. The dispersion was placed on a silicon substrate and dried by heating. Vapours of 49% hydrofluoric acid were used to etch silicon oxide on the surface of nc-Si for 30 min. Measures were taken against condensation of hydrofluoric acid in the samples. FTIR and PL spectra were compared: a) for samples freshly deposited on a substrate, b) 12 days after exposure to air after sample preparation, c) 30 min after etching with the HF vapour; d) 3 days after etching with the HF vapours. It is shown that the oxidation of nc-Si by atmospheric oxygen ends after 2 days because of self restraint. The peaks in the photoluminescence spectra at 830, 746 and 656 nm were attributed to the nc-Si with the sizes of 5.0, 4.1 and 3.3 nm, respectively (obtained from TEM data). nc-Si with a diameter of 5.0 nm were produced immediately after deposition on a substrate, and 12 days later – 4.1 nm, and finally, after etching in the HF vapour and oxidation at air – 3.3 nm. The intensity of PL bands was the largest for the freshly precipitated samples, three times lower – after 12 days in the air, while in the samples etched in HF vapours, the PL intensity was almost disappearing and, finally, afted 3 days of oxidation in air after etching of in the HF vapours the PL intensity increased, but was an order of magnitude lower than that of freshly precipitated nc-Si. The latter, according to FTIR, did not contain oxygen, but their surfaces were passivated by hydrogen and contained SiH_2

and SiH_3 groups. After 12 days in the air no hydrogen was left, and the groups (Si–O–Si) formed on the surface. Etching in the HF vapour destroyed the oxide film, but restored hydrogen passivation. After 3 days the oxide film reappeared, and the presence of hydrogen atoms on the surface was strongly reduced. The authors suggest that, after etching in the HF vapours, P_b-centres were formed on the surface, and the subsequent oxidation of the surface in air reduced their number. The sample oxidized in air for 3 days after etching in HF vapours, contained O_3SiH groups on the surface, which indicated that the introduction of oxygen to the (Si–Si) bonds, adjacent to SiH groups. The oxide layer of the sample, oxidized in air three days after etching in HF, has the composition $SiO_{1.9}$. Oxidation and etching of nc-Si produced particles with a diameter of 1.5 nm with PL at 543 nm, but the PL intensity was halved. The decrease of intensity is due, according to the authors, not only to the decrease of the size of nc-Si, but also to the increase of the number of P_b-centres on the surface of the nc-Si 7 ± 1 times. This was confirmed by EPR spectroscopy.

For the synthesis of nc-Si by cold plasma it is not essential to use the monosilane as a starting material. In (Nozaki et al, 2007) nc-Si were obtained by cold plasma from a mixture of H_2 and $SiCl_4$ vapour in the carrier gas Ar. The Ar flow rate was 200 $cm^3 \cdot min^{-1}$, the content of $SiCl_4$ – 100 ppm. The hydrogen impurity content ranged from about 0–5 vol.% nc-Si/H was produced, and the natural oxidation at room temperature of the surface of nc-Si/H did not affect Si–H bonds, which showed surprising stability. The PL of the samples obtained at hydrogen content in the mixture of 0–0.5 vol.% was barely noticeable, but greatly increased at 0.7 vol.% hydrogen and the peak intensity was located at 670 nm. With the increase of the hydrogen content in the mixture to > 2 vol%. the PL maximum was shifted to 520 nm. Assessment the size of nc-Si/H, obtained by TEM at a concentration of about 0.7 vol.% H_2 was 2.8 nm in diameter. nc-Si/H, luminescent at 520 nm after oxidation in air for 24 hours, showed PL with a maximum located at 450 nm. The blue glow, the authors believe, is caused by an oxide film on the surface of the nc-Si/H.

In the works of Shen et al. (2007, 2010) argon was bubbled through the liquid $SiBr_4$ and $SiBr_4$ vapours travelled into the reaction zone. The admixture of hydrogen to argon was used as a reductant. Reactor temperature was 50°C. The wavelength of the PL of the produced nc-Si was regulated by the pressure of the gas mixture in the range 2–4 Torr. For example, nc-Si with blue PL were obtained under a pressure of 3 Torr. nc-Si were collected on a filter. The reaction yield was 18%. nc-Si from the filter was dispersed in ethanol at moderate sonification. The dispersion in ethanol was colourless, but exposure to UV at 340 nm resulted in blue photoluminescence (470 nm). SiO_2 could form in contact of nc-Si with air. The lifetime of PL was 2.3 ns. EDS spectra showed the presence of Si and O in samples. Raman spectroscopy showed that the nc-Si contains crystalline silicon. The FTIR spectrum contained absorption bands characteristic for bonds (Si–O–Si) and (Si–OH), but no Si–H bonds were detected. The authors believe that their nc-Si samples contained silicon oxides on the surface. The PL of the samples dispersed in alcohol and sealed in a glass vessel, were stable for several months. PL intensity decreased by 20% in the first 14 days and then remained constant for 200 days. The resulting dispersions of nc-Si in ethanol were sealed in ampoules after adding to them the 10-fold excess of HF on the basis that the entire mass of nc-Si consists of SiO_2. These sealed ampoules were used to measure the PL spectra. The quantum yield of PL of nc-Si was determined in comparison with rhodamine 6G. After etching in HF, th mean diameter of nc-Si was 3.4 nm. HRTEM registered an interplanar spacing of 3.1 Å, which corresponds to the interplanar spacing of (111) silicon. The authors believe that the freshly synthesized samples contained amorphous silicon. During 2 min – the loading time of the sample in the FTIR-spectrometer – silicon oxidized in air. The absorption bands of the (Si–O–Si) groups were clearly visible. After etching in HF, the content oxygen decreased strongly, judging from the EDS spectra. FTIR detected after etching Si–H and Si–F bonds, i.e. the surface of the nc-Si after etching changed from SiO_2 to Si–F and Si–H. The amount of nc-Si after etching was about 4 vol.% of freshly produced nc-Si. Before and after the etching of nc-Si the PL peak position was the same – 470 nm at excitation by light with a wavelength of 340 nm, but the PL intensity after etching increased by 30%. According to the authors, the blue PL comes from the nc-Si and not the surface oxides. The total pressure of the gas mixture of less than 2 Torr during synthesis (Shen et al, 2007, 2010) resulted in the formation of nc-Si with no PL, but at a pressure of 2.2 Torr nc-Si luminesce green–yellow light, and at a pressure of 3.5 Torr – blue. TEM showed that the nc-Si with the green–yellow photoluminescence have an average size of 3.9 nm and with blue PL – 3.4 nm, i.e. the size effect may be the cause of the change in the PL wavelength. Reliable surface passivation of nc-Si leads to an increase in the quantum yield of the PL. The increased quantum yield of the nc-Si PL is observed at surface passivation by organic ligands.

The cold plasma method in a mixture of Ar, H_2 and SiH_4 at a pressure of 1.4 Torr was used to produce nc-Si/H (Jurbergs et al, 2006). The surface was then passivated by hydrosilylation with octadecene. The photoluminescence quantum yield of the sample whose PL spectrum showed a peak at 789 nm was equal to 62 ± 11%, but at 765 nm, 46 ± 11%, and at 693 nm 1.8%. After 8 days the quantum yields of all samples fell. The peaks in the photoluminescence spectra shifted to the blue side, which was explained by partial oxidation. The authors concluded that the oxidation of nc-Si/H has an adverse effect on the quantum yield of PL.

Liao et al. (2006b) described the synthesis of nc-Si/H by cold plasma. A mixture of (Ar + SiH_4) under the pressure of 25 Torr was fed into the plasma zone. The particle diameter of nc-Si/H was equal to 12.2 nm – this fraction was selected from the polydisperse mixture. The diameter of the particles and their crystallinity were determined by TEM. The stream of Ar and nc-Si/H was injected a nitrogen stream saturated at room temperature by 1-hexene or 1-hexynyl. A mixture of gases and vapours was fed at atmospheric pressure into a tubular furnace where the temperature was set in the range 250–500°C. The length of stay of the mixture in the hot zone was not longer than 4 seconds. In the analyzers measurements were taken of the so-called mobile diameters before and after mixing with the vapour organic ligands (1-hexene and 1-hexynyl) reacting with the surfaces of particles of nc-Si/H in the hydrosilylation reaction, and it always turned out that the mobile diameter of the nc-Si/H particles in the second analyzer was 1.5 nm greater than the first, and the subtle differences in these increments of the mobile particle diameters coincided with the length of the alkyl chains of organic ligands. When the unsaturated hydrocarbons in the reaction with the nc-Si were replaced by alkylamines, (Si–N) bonds on the surface of nc-Si and hydrogen gas was released.

Synthesis in cold plasma was carried out in a mixture of Ar, SiH_4 and H_2 (Mangolini et al, 2006a, b). Particle sizes were controlled by the residence time of the mixture in the cold plasma zone (a few ms) by controlling the gas flow. All the operations were performed in the absence of oxygen and water vapour. nc-Si particles were deposited on a filter and then transferred to a dry organic solvent and dispersed with ultrasound. A turbid dispersion was obtained which was refluxed for several hours. The reaction of nc-Si/H lasted 5–10 min with octadecene and dodecene in mezithylene or toluene. Transparent sol was obtained. FTIR showed that prior to hydrosilylation the surface of nc-Si/H contained only SiH_3 and SiH_2. After hydrosilylation, they disappeared and alkyl groups appeared. A weak peak at 2086 cm^{-1} was assigned to the valency vibrations of (Si–H). In large nc-Si SiH_3 groups remain after hydrosilylation. It was found that the photoluminescence quantum yield of the samples with a diameter of nc-Si of 4.07 nm was up to ~70% at a peak wavelength of 800 nm, but vanished at ~700 nm. TEM revealed crystalline silicon nanoparticles. Their average diameter was ~5 nm. Atomic force microscopy showed that the particle diameter was 4.07 nm with a standard deviation of 1.55 nm.

Hot electrons in the plasma are very effective for the activation of organic molecules and reactions proceed rapidly. In the plasma reactor, the hydrosilylation reaction occurs within milliseconds. The hydrosilylation reaction in the gas stream was studied by Mangolini et al. (2007). A mixture of argon and silane was fed into the first plasma reactor where nc-Si, covered with hydrogen, formed. The gas flow transported nc-Si to the second plasma reactor into which argon, saturated with vapours of the reagent (1-dodecene, 1-hexene, 1-hexynyl, hexane, 1-pentene, hexanol) was also supplied. At the output of the second reactor there was a filter trapping nc-Si. The PL peak of the produced nanoparticles was found at about 860 nm in excitation by light with a wavelength of 390 nm. Synthesized particles of nc-Si easily produced clear dispersions in toluene. TEM confirmed by the crystal diamond-like structure of nc-Si and the determined average diameter of nc-Si was 4.1 nm with a standard deviation of 0.87 nm. The nc-Si produced by the plasma method had a quantum yield of PL of 10%, whereas hydrosilylation in a solution resulted in the quantum yield of 60–70%. FTIR discovered a group of SiH_3, SiH_2, and SiH, and EPR – a significant amount of P_b-centres. After heat processing in the liquid the quantum yield reached about ~50%, indicating the reorganization of the surface of the nc-Si reducing the amount of P_b-centres. The hydrosilylation reaction proceeded most successfully with 1-hexynyl somewhat less efficiently with 1-hexene, but hexanol and hexane reacted in the plasma equally acceptably and allow the formation   dispersion of nc-Si in organic solvents. This section does not discusses the effect of hydrogen, produced by decomposition of silane, on the hydrosilylation reaction (hydrogen can interact with unsaturated hydrocarbons) and on the photoluminescence quantum yield of the nc-Si.

In conclusion, this section refers to the formation of nc-Si in a plasma excited under the influence of direct electric current (Sankaran et al, 2005). A mixture of silane and argon (1–5 ppm SiH_4) under the pressure of just over 1 atm was passed through the plasma. As a result, nc-Si with an average diameter 1.6 nm formed , according to atomic force microscopy. The particle size did not exceed 6 nm.

When excited by light at 360 nm, the peak of the PL spectrum was at 420 nm, and the blue glow was clearly visible to the naked eye. In excitation with laser radiation with a wavelength of 405 nm the PL peak was at 511 nm for the film on a substrate, but the suspension of the same substance in octanol with excitation with the light at 405 nm has a PL spectrum with a peak at 465 nm. The authors explain the blue shift of PL by the reaction of nc-Si with the air during the preparation of the sample to determine the PL spectrum. When exciting by a light with a wavelength of 457.9 nm measurements were taken of the PL decay, which turned out to be $\tau = 30$ ns. This is longer than for the PL generated by electron–hole radiative recombination on the surface states ($\tau < 1$ ns) but shorter than the interband recombination of the indirect-gap semiconductor (10 –100 ms). The quantum yield was measured in comparison with 9,10-diphenylanthracene in cyclohexane and was found to be 30%.

In (Wiggers et al, 2001), silane (10–40 vol.%) was mixed with argon and passed through a tubular furnace heated up to 1000°C. Thermolysis took place not only in volume but also on the walls of the furnace. The flow rate was 185 cm^3 s^{-1}. The average diameter of the nc-Si was equal to ~20 nm. The surface of the nc-Si/H was then modified with 1-octene (Nelles et al, 2007). The hydrosilylation reaction had the greatest efficiency – 61% – when H_2PtCl_6 was used as a catalyst. The efficiency of other methods, used for the hydrosilylation reactions (e.g. heating or exposure to UV-radiation), was 10% or less. H_2PtCl_6 catalyzes the oxidation reaction of nc-Si/H. The lowest efficiency of the hydrosilylation reaction – 5% – was noted when exposed to UV radiation. These results were obtained in two-stage synthesis: in the first stage nc-Si/H interacted with HF in ethanol to remove the silica, the product was extracted with an organic solvent and separated from the alcohol fraction, and in the course of the second stage nc-Si/H was subjected to hydrosilylation. In the one-step synthesis after etching the nc-Si/H by the hydrofluoric acid HF evaporated and alkene and the catalyst were added. The efficiency of the reaction remained the highest when using H_2PtCl_6, but the oxidation rate was also very high.

Synthesis in cold plasma and a microwave reactor: doping nc-Si films

Production of thin-film transistors on a flexible (usually polymer) base is possible only at temperatures below 200°C due to thermal instability of the substrate at higher temperatures. Cold plasma produces nc-Si:H films at temperatures of 250–300°C, where the starting material is SiH_4, and the carrier gas – hydrogen. At temperatures around 150°C films are porous, with stresses, and the mobility of charge carriers is reduced. Addition of argon to the carrier results in a considerable increase of the mobility of charge carriers in the films prepared at temperatures of 165–200°C, but the stability of the electrical properties of such films remains low.

In Kandoussi et al. (2008) a nc-Si film was deposited on glass substrate by cold plasma at a substrate temperature of 165°C, a gas pressure of 0.9 mbar, RF power of 15 W and the inter-electrode distance of 4.5 cm. The concentration of SiH_4 in the gas flow was equal to 1%, the feed rate of the mixture of the gas carriers (Ar + H_2) was constant, 100 $cm^3 \cdot min^{-1}$, but the ratio of Ar:H_2 was varied. The thickness of the deposited films was ~200 nm. The films were annealed at a temperature of 200°C. For n-doping of nc-Si 7 $\cdot$ 10^{-3} $cm^3 \cdot min^{-1}$ AsH_3 without argon was added to the mixture of gases. At 50% of argon in the mixture (Ar + H_2) the rate of formation of the nc-Si/H film was maximum: it was 2.6 times higher than the rate of formation of the nc-Si/H film in a stream of pure hydrogen. Optimal value of the gas mixture (50% Ar) were used in further experiments, in which the layer thickness varied from 25 to 36 nm. It was found that at the silicon layer thickness of 25 nm the material was predominantly amorphous. If in pure hydrogen the increase of the layer thickness to 150 nm is accompanied by a monotonic increase of the crystalline silicon fraction to 70%, then in the (Ar + H_2) mixture the fraction of nc-Si/H reaches 73% at a layer thickness of 50 nm, and at a layer thickness of 150 nm – 77% and at 360 nm – 79%, but the surface roughness of the film increases rapidly with the layer thickness (much faster than when using pure hydrogen). Optical emission spectroscopy showed that an increase in the proportion of argon in the gas mixture up to 50% causes the dissociation of hydrogen molecules to reach a maximum, and the dissociation of SiH_4 molecules is maximal in pure argon. In the next series of experiments by Kandoussi et al. (2008), the thickness of the films of pure nc-Si/H was fixed at 50 nm and of the films doped with arsenic at 70 nm. The mobility of charge carriers μ reached its maximum value (> 8 cm$^2 \cdot$ V^{-1} s^{-1}) at 50% of argon in the (Ar + H$_2$) mixture, which is 10 times higher than that of the films produced in pure hydrogen.

An important role in the dissociation of H$_2$ and SiH$_4$ is played by the metastable excited state of Ar* with a lifetime of 1 s and the energy level of 11.55 eV. The argon atom transforms to an energy accumulator and leads to the formation of chemically active molecules:

$$SiH_4 + Ar^* \rightarrow SiH_3 + H + Ar \quad (k = 1.4 \cdot 10^{-10}\, cm^3\, s^{-1});$$

$$SiH_4 + Ar^* \rightarrow SiH_2 + H + Ar \quad (k = 2.6 \cdot 10^{-10}\, cm^3\, s^{-1});$$

$$SiH_3 + Ar^* \rightarrow SiH_2 + H + Ar \quad (k = 1 \cdot 10^{-10}\, cm^3\, s^{-1});$$

$$SiH_2 + Ar^* \rightarrow SiH + H + Ar \quad (k = 1 \cdot 10^{-10}\, cm^3\, s^{-1});$$

$$SiH + Ar^* \rightarrow Si + H + Ar \quad (k = 1 \cdot 10^{-10}\, cm^3\, s^{-1}).$$

Atomic hydrogen covers the substrate, contributes to the migration of silicon atoms (in the form of lower hydrides, such as, for example, SiH$_3$) on the surface of the growing film prior to insertion of the silicon atom in the lattice of nc-Si/H, as well as the dissociation of Si–Si bonds with the weakest bond to the surface of the silicon atoms, which leads to the perfection of nc-Si/H crystal. For hydrogen, the following reaction is important

$$H_2 + Ar^* \rightarrow 2H + Ar \quad (k = 7 \cdot 10^{-11}\, cm^3\, s^{-1}).$$

and contributes to the quality of TFT (improving the stability of the electrical properties). There is competition between the processes of dissociation of H$_2$ molecules with Ar* and reduction of the proportion of atomic hydrogen due to lower proportion of H$_2$ in the mixture. Optimum in this competition is achieved at 50% Ar in the (Ar + H$_2$) mixture. Excess argon leads to the degradation of the nc-Si/H film and the deterioration of the stability of its electrical properties.

In (Stegner et al, 2008) nc-Si was obtained by the decomposition of SiH$_4$ in the plasma. The particles were round and covered with a layer of oxide in the air after formation of nc-Si. The diameter of the samples was 4–50 nm, as measured by BET. The authors could adjust the size of the particles so that the standard deviation did not exceed 1.5 nm, but the method of regulation was not mentioned. Doping was achieved by admixture of PH$_3$ to SiH$_4$. The nominal doping level was determined by the amount of PH$_3$ in the gas mixture, but the actual concentration of phosphorus in the nc-Si did not necessarily correspond to the nominal value. EPR showed that phosphorus atoms were embedded in the crystal lattice of nc-Si with a diameter of 46 ± 1 nm, 11 nm and 4.3 nm, but with decreasing particle diameter the concentration of phosphorus in the nc-Si decreased at the same nominal level of doping. nc-Si layers were fabricated as undoped and doped. To do this, nc-Si was milled with ethanol in a ball mill and deposited by spreading on the rotating surface on a polymeric substrate with gold contacts. After drying, the thickness of the nc-Si layer was 500 nm. Such a layer does not conduct electricity (electrical conductivity $< 10^{-14}$ ohm·cm^{-1}), but after etching in 5% HF for 15 s and washing off the layer with water it acquired electrical conductivity ($8 \cdot 10^{-14}$ ohm-cm^{-1} for the undoped and $2 \cdot 10^{-8}$ ohm·cm^{-1} for doped layers at a concentration of $[P] = 1.5 \cdot 10^{20}$ cm^{-3}), since all (Si–O) bonds were removed and only the Si–H bonds passivating the surface remained, as shown by FTIR. The activation energy of thermally activated transport of charge carriers for the undoped samples reached values of ~0.5 eV, but decreased with increasing doping with an increase in the electrical conductivity, induced by the increase in the concentration of charge carriers due to the degree of doping. The method of electrical detection of magnetic resonance (EDMR), which detects only the paramagnetic states which are involved in the mechanism of electron transfer, was used to study a sample with a particle diameter of ~30 nm, and $[P] = 1.5 \cdot 10^{20}$ cm^{-3} at 7 K. In addition to a strong increase of conductivity with increasing level of doping nc-Si particles with phosphorus, the role of P donors in the conduction mechanism was also determined. It is shown that the P donors and the dangling Si bonds contribute to the dark conductivity by spin-dependent hopping conductivity, whereas in the photoconductivity these states act as spin-dependent recombination centres, generated by the light of electrons and holes.

Boron- and phosphorus-doped nc-Si were produced (Lechner et al, 2008) separately by the same process in a 2.45 GHz microwave reactor under a gas pressure of 50–500 mbar and the power output of $1000-2000$ W. The carrier gas was a mixture of argon and hydrogen, and a precursor – SiH$_4$ and B$_2$H$_6$ or PH$_3$. The gas flow rate was 10 000–20 000 cm$^3 \cdot$ min^{-1}. The average size of nc-Si could be changed by specifying the conditions for obtaining nc-Si, and the standard deviation of the particle diameter was 1.3–1.5 nm with the lognormal distribution of the particle size. Pure nc-Si and P-nc-Si (phosphorus doped) were obtained with average particle sizes in the range of 4.3–45 nm, and B-nc-Si (doped with boron) – 20 nm. The powders, obtained in the microwave reactor, were milled for four hours in a ball mill into which they were loaded in the form a suspension of 6 wt.% of nc-Si in ethanol. The resulting dispersion was applied by spreading on the rotating surface on a flexible polyimide substrate, giving a porous layer of nc-Si with a thickness of 700 nm. Oxides from the surface of nc-Si were removed by immersion in 10% HF solution for 20 s and then rinsing with deionized water. The layers were then recrystallized by laser annealing in a conventional environment. The porosity of a layer of P-nc-Si prior to laser annealing and after treatment with minimal laser power density (40 mJ $\cdot$ cm^{-2}) was 60–70%, but after treatment with a specific power of 60 mJ $\cdot$ cm^{-2}, some nc-Si melted, others were sintered with each other. The predominant particle size reached 100 nm. After processing with a more powerful laser (80 and 100 mJ $\cdot$ cm^{-2}) the size of individual grains reached

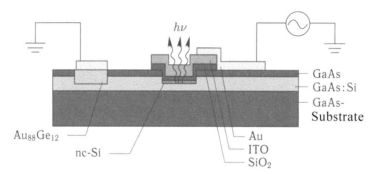

Fig. 5.9. Section of an electronic device that contains an optically active layer of nc-Si, placed between the substrate of highly doped GaAs with a transparent top electrode made of indium and tin oxides (Theis et al, 2010).

up to 400 nm, the grains were spherical and their surface was molten, regardless of the size of nc-Si prior to laser treatment. Similar results were obtained for B-nc-Si. It was shown that the large grains of 0.2–2.5 mm, produced after laser annealing, are agglomerates of crystals with the size of 20–50 nm. It was found that the concentration of holes in the B-nc-Si, $5 \cdot 10^{20}$ cm^{-3}, is consistent with the concentration of boron in the B-nc-Si (10^{21} cm^{-3}), but the electron concentration ($3 \cdot 10^{19}$ cm^{-3}) in the P-nc-Si is an order of magnitude lower than the concentration of phosphorus ($5 \cdot 10^{20}$ cm^{-3}). The electrical conductivity of the nc-Si doped with the maximum amount of B and P compared to the undoped nc-Si prior laser annealing increases by up to three orders of magnitude. Annealing the undoped sample at the irradiation power 100 mJ $\cdot$ cm^{-2} increases the electrical conductivity of 10.10 ohm $\cdot$ cm^{-1} to $5 \cdot 10^{-8}$ ohm $\cdot$ cm^{-1}, doped – to 10 ohm $\cdot$ cm^{-1}. It was found that 90% of the phosphorus, introduced into the doping process, concentrated on the surface of P-nc-Si, as well as on the interface with SiO$_2$. Boron is distributed more evenly on B-nc-Si.

Nanoparticles of silicon, coated with alkyl ligands, were used to prepare dispersions, potentially applicable for the manufacture of electronic devices using inkjet printers. These nanosilicon particles are stable at temperatures up to 300°C. In Nelles et al (2010), a mixture of silane and argon was fed into a microwave reactor and silane pyrolysis products were subjected to hydrosilylation in the presence of 24% hydrofluoric acid. The role of the solvent has executed by one of the compounds – n-octene, n-dodecyl, allyl mercaptan, allyl amine – used at the same time as reagents, formed the ligand coating on the nanosilicon surface. The average nanosilicon particle size was 42 nm with a standard deviation of 16%. After exposure to air the nanoparticles were coated with an amorphous layer of native oxide with a thickness of 0.8 nm, which could be registered with EELS. DRIFT confirmed that the silicon nanoparticles immediately after synthesis are predominantly covered with hydrogen. After hydrosilylation the layer of alkyl ligands inhibited the oxidation of silicon in the air, but did not prevent it entirely. To study the electrical properties of the samples, dense layers of the alkylated nanosilicon about 600 nm thick were prepared. Electrical conductivity was measured by impedance spectroscopy. A layer of nanosilicon was a disordered system in which the electrical conductivity is usually based on the percolation mechanism of hopping conductivity. The activation energy of the hopping conductivity of freshly prepared nanosilicon was 0.49 eV – it is the energy that must be expended in order to enable an electron to jump from one silicon nanoparticle to another through the barrier of hydrogen ligands. After 3 months of stay in the air the activation energy increased to 0.59 eV, i.e. the electron overcame the native oxide layer. It was found that the conductivity of silicon nanoparticles coated with native oxide does not have long-term stability. However, after etching with hydrofluoric acid the activation barrier decreased to 0.50 eV. For nanoparticles coated with n-octyl and n-dodecyl, the activation energy was 0.66 eV and 0.68 eV, respectively, and for allyl amine and allyl mercaptan 0.52 eV and 0.56 eV, respectively.

Silicon nanoparticles were obtained (Theis et al, 2010) in the gas phase from SiH_4 in a low-pressure microwave reactor. After storage in the open air the nanoparticles had an average diameter of 3.9 nm and were coated with native SiO_x with a thickness of 1–2 nm. The design concept of the cell, which was used to study the EL of nanosilicon, is shown in Fig. 5.9.

Orange PL of nanosilicon was visible to the naked eye. The EL peak was at 740 nm. The threshold voltage, above which EL appeared and intensified, was 4 W. The authors suggest that the EL arises in nanosilicon under the influence of ionization caused by the flux of electrons travelling from the upper electrode to the substrate, and the resulting electron–hole pairs radiatively recombine.

Nanosilicon in matrices other than SiO_2

Around each nc-Si, dispersed in SiO_2, there is a transition layer with a thickness of about 1 nm (the 'stress' oxide), which strongly influences the optoelectronic properties of the dispersion. But the matrix, which contains nanosilicon, may consist of silicon nitride and the transition layer can have different properties.

Synthesis of the dispersion by PECVD from SiH_4 and NH_3 was carried out in (Parm et al, 2006) at 300°C and a pressure of 6 Pa. The SiH_4 flow rate was 6 cm^3 · min^{-1}, NH_3 – 2 cm^3 · min^{-1}. The substrate for the reaction products was a plate of Si (100). The thickness of the silicon nitride on the substrate was 60 nm. The layer was annealed at temperatures between 600–900°C for 10 min by rapid annealing. The highest PL intensity was in the sample annealed at 600°C. The PL spectrum at ~750–800 nm had 5 peaks and, in addition, there were peaks around 500 and 400 nm. The authors concluded that crystalline silicon nanoparticles form in a matrix of silicon nitride during synthesis, as the annealing temperature – 600°C – was insufficient to convert the amorphous silicon to crystalline.

The peaks in the 750–800 nm range belonged to nc-Si in the matrix of SiN_x. Interpretation of the peaks at ~500 nm and ~400 nm was described in (Wang et al, 2007), in which a mixture of NH_3, N_2, and SiH_4 was transformed by PECVD into a non-stoichiometric silicon nitride deposited on a substrate of silicon or quartz, heated to 300°C. The samples were then annealed at 1100°C in a nitrogen atmosphere for 60 min. The NH_3:SiH_4 ration was varied from 1:1 to 3:1 and further to 6:1 and 10:1. At a ratio of 3:1 and 6:1 silicon nanoclusters 5–6 nm in size were found, but they were not crystalline. All PL spectra, excited by 325 nm radiation, had flat maxima at 410 and 520 nm. According to the authors, the first of them is due to dangling bonds on the nitrogen atom, and the second – dangling bonds on the silicon atom. The PL spectra, excited by 514.5 nm radiation, were recorded in samples with a NH_3:SiH_4 ratio of 6:1, 3:1 and 1:1 after annealing at 1100°C, but not at 10:1 due to the small content of silicon in silicon nitride. The PL spectra of samples in a series of 6:1, 3:1, 1:1 had a red shift of the peaks: 630, 670 and 720 nm, respectively. Apparently, it was due to increase in the size of nc-Si. The photoluminescence, excited by 514.5 nm radiation, was due to the recombination of electrons and holes in nc-Si (Wang et al, 2007).

The dispersion of nc-Si was in SiN_x was obtained (Tsai et al, 2007) by PECVD, combined with LACVD for more efficient decomposition of SiH_4, mixed with argon and NH_3. The dispersion was deposited at temperatures above 85°C and the sample was not annealed for the coalescence of excess silicon. In such circumstances, samples with the composition $SiN_{0.64}$, $SiN_{0.54}$, $SiN_{0.48}$, $SiN_{0.42}$ $SiN_{0.36}$ contained nc-Si inclusions with a clearly expressed crystalline diamond-like structure. The samples, obtained using only PECVD, but without LACVD, did not contain such inclusions and had compositions with about twice the nitrogen content. The PL intensity of the samples producyed by the combination PECVD-LACVD, was significantly higher than the PL intensity of the sample that have been obtained using only PECVD. Radiation maxima not only grew in intensity with increasing nitrogen content in the samples, but also shifted to the blue region in the range 1.6–2.6 eV, whereas the interval for the PECVD was 2.0–2.4 eV. THe EL 2.12 eV peak was shifted relative to the maximum PL 2.55 eV to the red side. The EL quantum yield was not reported.

Comparing the photoluminescence properties of dispersions of nc-Si in SiN_x, synthesized in an atmosphere of nitrogen and ammonia, the authors of Ma et al. (2006) concluded that the SiN_x matrix does not completely block the P_b-centres, and the presence of Si–H bonds reduces the amount of the latter and leads to an increase in the intensity of PL. The presence of hydrogen in the reaction mixture enables even chlorosilanes to be used as a precursor.

The dispersion of nc-Si in the SiN_x was synthesized by PECVD (Lopez-Suarez et al, 2008). They used a mixture of NH_3/H_2/$SiCl_2H_2$/Ar, varied the amount of NH_3 in the mixture and recorded changes in the photoluminescence. Hydrogen atoms passivated non-radiative defects, increasing the photoluminescence. The dispersion layer was deposited on the (100) plate of n-type silicon at a gas pressure of 300 mTorr, a substrate temperature of 300°C and RF power of 150 W. The flow rates of H_2, Ar, and SiH_2Cl_2 were 20, 150 and 5 $cm^3 \cdot min^{-1}$, respectively. The NH_3 flow rate was varied in the range 0–50 $cm^3 \cdot min^{-1}$. Chlorine reacted with hydrogen and was bonded with ammonia in NH_4Cl, however, judging by the results of elemental analysis, chlorine at 7–10 at.% remained in the dispersion at any feed rate of NH_3. At an NH_3 flow rate of 50 $cm^3 \cdot min^{-1}$ nc-Si with a diameter 6.9 nm was found, and at a flow rate of 10 $cm^3 \cdot min^{-1}$ – 26.1 nm, and therefore photoluminescence was not found in the dispersions formed at NH_3 flow rates less than 25 $cm^3 \cdot min^{-1}$, but was observed at 2.06 eV at a flow rate of NH_3 equal to 25 $cm^3 \cdot min^{-1}$ and was shifted in the range 2.10–3.06 eV at NH_3 flow rates of 30–50 $cm^3 \cdot min^{-1}$. Thus, the smaller the nc-Si particles, the shorter the wavelength of photoluminescence wavelength; the authors explained this by the size effect.

The PECVD method was used to deposit the SiN_x layer (Benami et al, 2007) on the high-ohmic substrate Si (100) of n-type, purified to remove the native oxide of silicon. The precursor was a mixture of $SiCl_4$ vapour and gases NH_3 and H_2. The $[SiCl_4]/[NH_3]$ ratio remained constant, but the flow rate of hydrogen was varied in the range 0–60 $cm^3 \cdot min^{-1}$. For any hydrogen content in the mixture the pressure in the reactor was kept constant – 40 Pa – by adding argon. Temperature of the substrate was 200°C. The thickness of the resulting layer was adjusted to 1000 nm. After deposition the samples were annealed in nitrogen for 1 h at 500–1000°C. In the course of the reaction ammonolysis of $SiCl_4$ took place. PL intensity increased with increasing flow rate of H_2 from 0 to 20 $cm^3 \cdot min^{-1}$ and then decreased with a further increase of the flow rate, but the band of the PL spectrum was the same. PL became more intensive after annealing at 1000°C. The PL intensity of the sample annealed at 750°C was an order of magnitude lower and almost equal to the PL intensity of the freshly prepared sample. After annealing at 1000°C PL was white, and after 750°C – green. FTIR showed no N–H bonds and Si–Cl, but confirmed the presence of Si–N and Si–H bonds. The samples, obtained without any admixture of hydrogen, retain the Si–Cl bonds, which hydrolyzed in air. With increasing hydrogen content in the reaction mixture the concentration of chlorine in the sample decreased. The samples also contained oxygen, and the nitrogen and oxygen content after annealing at 1000°C in all samples increased, and the chlorine content decreased.

Several authors have studied the dispersion of nc-Si in the oxide–nitride matrices. When using N_2O as a reagent for the synthesis of the dispersion of nc-Si in SiO_2 the amorphous SiO_x precursor contained nitrogen atoms, which created defects actively causing the capture of holes in traps and their transport, as well as the balance of radiative and non-radiative processes.

Upon annealing SiO_x even at temperatures of 1250°C part of silicon forms nc-Si, and a certain fraction of Si remains amorphous and bonded oxygen and nitrogen atoms, possibly in the boundary layer around the nc-Si.

Samples of SiO_xN_y were synthesized by PECVD (Daldosso et al, 2007) from SiH_4 and N_2O. The reaction product was deposited on Si(100) in a layer with a thickness of 200–1000 nm. The product contained 10% nitrogen. Samples with 39, 42, 44 and 46 at.% of Si annealed for 1 hour at temperatures of 500–1250°C were studied. The absorption spectra of X-rays were obtained by synchrotron radiation. They showed the presence of Si–N and Si–Si bonds in the SiO_2 matrix. The result of annealing was to reduce the content of the former and ensure growth of the latter. The (Si–N) groups disappeared after annealing at 1000°C. FTIR showed that the layer of freshly deposited SiO_xN_y contains Si–H and N–H bonds in addition to Si–O–Si and Si–N, but when heated to 1000°C and above the first two and last disappear, and the intensity of the absorption of Si–O–Si increases. Raman spectra showed the formation of amorphous silicon at annealing temperatures of 650–1000°C, but at higher annealing temperatures amorphous silicon was not found.

Annealing of the silicon-rich mixture of silicon oxide and silicon using a CO_2 laser to form nc-Si in the oxide–nitride matrix is very useful due to the fact that its infrared radiation (10.6 μm) can be effectively absorbed by the silicon oxide and provides local heating, which leads to high temperatures, even when using lasers of moderate power. Annealing is carried out locally within the controlled time interval that can not be achieved in the furnace. The mixture of gases (2% SiH_4 + 98% N_2) reacted (Tewary et al, 2006) with N_2O in the cold plasma reactor. The temperature and reaction pressure were 350°C and 650 mTorr, respectively. The ratio N_2O:SiH_4 = 6:7, and the rate of deposition on a quartz substrate was chosen such that the resulting layer had a thickness of 350 nm. The resulting samples were irradiated with a CO_2-laser. For comparison, one sample was annealed in a furnace at 1100°C for an hour in Ar. Some samples were additionally annealed in forming gas (95% N_2 + 5% H_2) at 400°C for 30 min. The composition of the samples was determined by spectroscopy RBS: $SiO_{1.08}N_{0.32}$. Samples with PL contained nc-Si, as detected by TEM.

The PL spectra were recorded with a time resolution of 5 ns. The freshly deposited samples prior to annealing showed a peak in the PL spectrum at 570 nm. Annealing with CO_2-laser radiation (330 W · cm^{-2}) caused the appearance of a peak at 750 nm, which shifted to 800 nm and became stronger with increasing power density up to 480 W · cm^{-2}. These conditions corresponded to annealing at 900°C and 1300°C, respectively, and these temperatures were reached within a few seconds. Further increase in power density to 580 W · cm^{-2} resulted in a drop in the PL intensity and damage of the sample (at a temperature of 1600°C). At a specific laser power of 480 W · cm^{-2} the maximum intensity at 570 nm was achieved after 10 s of annealing by the CO_2-laser, but the intensity of the peak at 750–

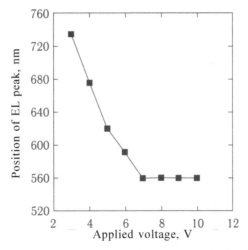

Fig. 5.10. The dependence of the EL peak position on the applied voltage for devices with a-Si in a partially oxidized matrix $SiN_x:O_y$ (Wang et al, 2010).

800 nm increased during 60 s of annealing, and then disappeared at the same time with damage to the sample. Annealing in a hydrogen atmosphere extinguished photoluminescence at 570 nm, but PL remained at 800 nm. The lifetime of the PL of nc-Si at 800 nm was 45 ms, at 570 nm it was less than 5 microseconds. PL at 570 nm is caused by defects in either the matrix or defects at the interface $Si–SiO_xN_y$. Hydrogen is totally quenches a PL, but promotes the growth of the PL intensity nc-Si.

It was found (Wang et al, 2010) that amorphous Si nanoparticles in the matrix of partially oxidized SiN_x change not only the intensity but also the wavelength of the EL as a function of the applied voltage (EL light changes from red to yellow–green, see Fig. 5.10).

Noted are also controversial results. For example, the samples were prepared by PECVD (Caristia et al, 2007) using a mixture of N_2O and SiH_4 as reagents. The layers were deposited at 400°C at a power of RF emitter of 125 W under the pressure of 2.8 Torr. The thickness of the deposited layers was about 280 nm. The substrate for the deposited layers was the (100) plane of the silicon crystal of p-type. The samples were annealed at 1100°C under nitrogen for 30 min. The freshly deposited layers contained bands in the IR absorption spectra corresponding to the fluctuations of the Si–H bond; they completely disappeared as a result of annealing at 1100°C. It was shown that the excess silicon crystallized to less than 50%. The conclusion of the authors on the inhibition of crystallization of nc-Si with nitrogen is questionable. The authors calculated the excess silicon, considering only the silicon bound in SiO_2, and nitrogen, which was part of the samples, was considered free and did not bond with silicon. In fact, Si_3N_4 inevitably formed after annealing at 1100°C, and if this was taken into account the findings authors may be revised.

The main disadvantage of gas-phase synthesis methods is the agglomeration of Si nanoparticles nanosilicon. PECVD provides a low degree of agglomeration and narrow particle size distribution. In Chai et al (2010) plasma was generated in the gas mixture 5% SiH_4 and 95% Ar. The size and shape of nanoparticles (agglomerates of crystallites) were studied by TEM, and the proportion of crystallinity or amorphous state and the crystallite size in the agglomerate – by Raman light scattering. With increasing power input into the plasma reactor the radius of the resulting polycrystalline silicon nanoparticles did not change and amounted to ~3.5 nm, but the size of the crystallites in this agglomerate increased from 0.9 nm to 1.4 nm, and the proportion of crystalline phase – from 58% to 84%. Increasing the argon pressure from 34 mTorr to 51 mTorr increased the radius of the nanoparticles from 3.5 nm to 4.5 nm, with the size of the crystallites in the agglomerates reduced from 1.4 nm to 0.9 nm, and the proportion of crystalline fractions – from 84% to 69%. In plasma, pyrolysis of silane was followed by nucleation, coalescence and growth of new particles during the pyrolysis of silane on the surface of the particles. Due to the increase of partial pressure of hydrogen in the gas mixture from zero to 9 mTorr (at a fixed pressure of the SiH_4/Ar mixture) the size of the particles did not change and remained equal to the radius of 3.5 nm, but the size of crystallites in the agglomerates decreased from 1.4 nm to almost zero, while the proportion of crystalline fractions – from 80% to ~8%, i.e. in the presence of hydrogen with the partial pressure of 9 mTorr almost amorphous silicon was deposited. Thus, the result contradicts the existing literature views of increasing the crystallinity of nanosilicon with increasing hydrogen partial pressure.

The particles of nc-Si, distributed in a matrix of SiO_2, have luminescence spectra which are difficult to interpret, since there are very small differences in the photoluminescence from defects in the SiO_2 matrix and the nc-Si, located in the matrix. Therefore, in (Kovalev et al, 2008) Si^+ ions were implanted in a sapphire plate and a film pf amorphous Al_2O_3 ~300 nm thick deposited on a silicon substrate by evaporation of Al_2O_3 powder by an electron gun. The energy of implanted Si^+ ions was 100 keV with densities of $5 \cdot 10^{16}$–$3 \cdot 10^{17}$ cm^{-2}. The samples were then annealed at 500–1100°C in an atmosphere of dry nitrogen for 2 h. nc-Si formed during annealing. In the sapphire at a depth of 20–40 nm the size of nc-Si was 1–2 nm, but their number was small. nc-Si were mostly in sapphire at a depth of 84–100 nm and were larger these. However, the nc-Si in the sapphire did not luminesce. The authors believe that the structural mismatch between nc-Si and sapphire, and the stresses arising in the process of implantation of Si^+ and annealing and leading to the emergence of defects at the nc-Si/ sapphire, result in luminescence quenching.

In amorphous Al_2O_3 the PL of nc-Si is observed due to the relaxation of mechanical stresses arising from the implantation of Si^+, due to 'looseness' of the structure of amorphous Al_2O_3, weak adhesion of Al_2O_3 to the Si

Table 5.1. The mean values of diameters of nc-Si (nm) implanted into the α–Al$_2$O$_3$ matrix (Yerei et al, 2006)

Implantation density	$1 \cdot 10^{17}$ см$^{-2}$	$2 \cdot 10^{17}$ см$^{-2}$
1100 °C	$D = 7.0 \pm 0.2$	$D = 7.2 \pm 0.2$
1000 °C	$D = 3.9 \pm 0.2$	$D = 5.1 \pm 0.2$

substrate, the formation of the SiO$_x$ transition layer at the nc-Si/Al$_2$O$_3$ interface. Al$_2$O$_3$ is a promising material for microelectronics as its dielectric constant is twice as high as that of SiO$_2$, and the band gap slightly higher (9.2 eV instead of 8.7 eV for SiO$_2$). The nc-Si particles were formed in α-Al$_2$O$_3$ by ion implantation and subsequent annealing at 600–1100°C in nitrogen for 2 h (Yerei et al, 2006). The diameters of nc-Si, calculated from X-ray powder using the Scherrer formula, were found to be (depending on the density of implantation and annealing temperature) equal to the values given in Table 5.1.

The density of implantation of $2 \cdot 10^{16}$ cm^{-2} is not sufficient for the crystallization of nc-Si even at 1100°C. At a density of implantation of 10^{17} cm^{-2}, the annealing temperature which nc-Si starts to form, is equal to 900°C, and at a density of implantation of $2 \cdot 10^{17}$ cm^{-2} – 800°C. In the PL spectrum there is a very broad band with a maximum ~570 nm. Its intensity decreases with increasing implantation dose, but increases with increasing annealing temperature at all implantation doses. This PL band does not arise as a result of recombination of electron–hole pairs in nc-Si. It is assumed that the dispersion of nc-Si in Al$_2$O$_3$ does not have the transition layer of oxides found in dispersion of the nc-Si in SiO$_2$, and this leads to the formation of P_b-centres on the surface of nc-Si in Al$_2$O$_3$ which quench the photoluminescence. At the same time, the Si = O bonds, responsible for the PL in the blue-green region of the spectrum cannot form in the Al$_2$O$_3$ matrix. The sources of the ~ 570 nm band are defects in the oxygen sublattice of Al$_2$O$_3$, F_2^{2+} (two oxygen vacancies with two trapped electrons).

5.3.2. The method of laser-induced dissociation of silane

High-performance gas-phase synthesis methods also include the laser-chemical method (Kononov et al, 2005; Vladimirov et al, 2010, 2011). Laser heating provides a controlled, homogeneous nucleation and eliminates the possibility of contamination. The size of the nanocrystalline particles decreases with increasing density of the laser power as a result of increasing the temperature and heating rate of gas–reagents.

Samples of nanosized silicon formed in the gas-dynamic flow reactor (Fig. 5.11) at SiH$_4$ flow rates of 100 cm^3/min and argon flow rate (as a buffer gas) of 1000 cm^3/min. The average particle size of nc-Si, obtained at the buffer gas temperature of 20°C, was ~10 nm.

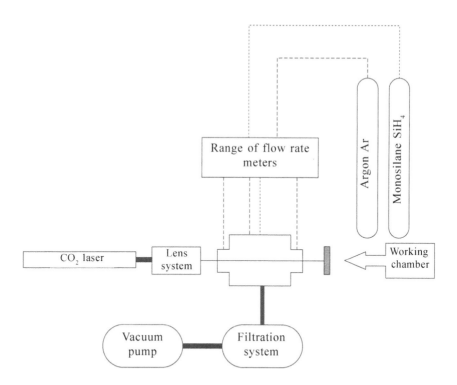

Fig. 5.11. Schematic representation of the setup for the synthesis of powders of silicon nanocomposites (Vladimirov et al, 2010, 2011).

The heating of the silane jet by continuous CO_2-laser (laser-induced chemical vapour deposition – LICVD) was possible due to the fact that the strongest line of generation of the CO_2-laser, such as $P(18)$ and $P(20)$ with wave numbers $v_{18} = 945.98$ cm^{-1} and $v_{20} = 944.19$ cm^{-1}, enter the absorption path of the SiH_4 molecule corresponding deformation vibrations and centred near wavenumber $v_3 = 970$ cm^{-1} (Nakamoto, 1966). However, due to significant detuning of the centre of this band from strong lasing lines, the absorption in the silane is not strong enough, which reduces the efficiency of laser heating. This difficulty can be overcome by using a diffraction grating installed in the CO_2-laser cavity to ensure lasing on other rotational levels of the vibrational transition 0001–1000 of the CO_2 molecule, but in this case there inevitable losses of laser power and the corresponding increase of the cost of the process.

   In order to provide a noticeable absorption of CO_2-laser radiation, it is necessary to maintain a sufficiently high pressure in the silane jet (Korovin et al, 1999, Kononov et al, 2005; Vladimirov et al., 2011). Since the cross expansion of the jet in LICVD is limited by the coaxial flow of the buffer gas (argon, helium), propagating along the jet, and participating in the formation of silicon particles, the pressure of the buffer gas in the reactor chamber must be maintained sufficiently. For this reason, the technology for producing nanosized particles of Si, associated with CO_2-laser decomposition of silane, can be realized only when the buffer gas pressure in the reactor chamber ≥ 100 Torr. However, as shown by numerous studies (Roca et. Al, 1998) the average particle size, *ceteris paribus*, increases with increasing buffer gas pressure in the reactor chamber. Therefore, the use of the described technology makes it possible to obtain Si powders, in which the maximum of the particle size distribution is in the range ≥ 10 nm.

In order to reduce the buffer gas pressure and thus prevent reducing the absorption of laser radiation in the silane jet, the buffer gas in the synthesis of the nanoparticles was heated to temperatures of 500–600 K. The result of this was an increase in the initial heating temperature of the silane jet and a significant increase of absorption CO_2-laser radiation in the jet. As a result, at the intensity of the focused laser beam of about $\sim 10^3$ W/cm^2 it was possible to light the optical discharge in the buffer gas at pressures ≥ 50 Torr.

Raising the temperature of buffer gas in synthesis makes it also possible to control the structure of the Si particles. This is primarily due to increased diffusion mobility of Si atoms, which leads to the formation of nanocrystals with a lower concentration of defects (Shirai et al, 1997). In addition, at higher temperatures of the buffer gas the hydrogen atoms which are presented in silane dissociation products are more rapidly incorporated into the structural network of the Si nanoparticles, passivating lattice defects such as mono- and divacancies, resulting in the synthesis of particles.

The buffer gas near the zone of synthesis of nanoparticles could be heated to a temperature 573 K with a nichrome coil enclosed in a high-temperature ceramic casing, mounted on a vertical cylindrical quartz vessel. The gas temperature was measured with a thermocouple located on the inner wall of the vessel (see Fig. 5.11). On the opposite walls of the vessel there were holes through which CO_2-laser radiation, heating the silane jet, was focused.

The synthesis methods: the emission of a 100 W CO_2 laser was focused by a lens made of NaCl in a spot of diameter ~ 2 mm. The axis of the laser radiation caustic was held at a distance of ~ 1 mm below the gas nozzle. The gas jet of monosilane SiH_4 (99.9%) was formed by the gas-dynamic nozzle and intersected a focused laser beam in the perpendicular direction. The flow of argon, which increased the rate of synthesis of silicon particles and served as a carrier gas, propagated coaxially to the SiH4 gas jet. nc-Si powder was collected by a special collector at the bottom of the reaction chamber at the normal air atmosphere. The temperature of the flame in the reaction zone was measured by a pyrometer, the rate of SiH_4 and Ar flows was regulated by rotameters. The argon pressure in the reaction chamber was maintained at 350 Torr. According to IR spectroscopy data, the

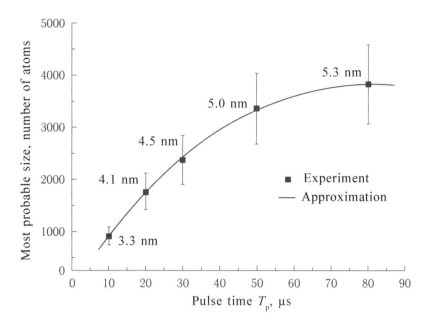

Fig. 5.12. The most probable size of the nanoparticle Si (the number of atoms per particle), depending on of the laser pulse duration (Sublemontier et al, 2009).

produced Si nanocomposites contained a significant amount of chemically bound hydrogen.

The synthesis of nc-Si by pyrolysis of silane in the CO_2-laser (Sublemontier et al, 2009) allows one to fine-tune the equipment (flow rate of the gas mixture and the proportion of silane in a mixture with the carrier gas, the power of the laser pulse, its duration, length of time between two successive pulses, etc.) for a certain size of nc-Si (Fig. 5.12). Apparently, the surface of nc-Si is almost free from hydrogen and nanosilicon thus obtained is not luminescent and easily oxidized.

These properties are illustrated below. nc-Si particles were obtained (Lacour et al, 2007) by pyrolysis of SiH_4 using a CO_2-laser with a wavelength of 10.6 μm. SiH_4 was mixed with helium. The argon flow carried from the reaction zone the nc-Si deposited on the filter. Helium reduced the residence time of silane in the reaction zone (up to 0.1–1 ms) and reduced the likelihood of collisions of atoms of silicon, which limited the growth of nc-Si. nc-Si particles were washed off from the filter with ethanol and the suspension was further dispersed with ultrasound. Particle sizes were determined by HRTEM and BET, and the results of measurements were identical.

The most important parameters that control the size of nc-Si were the degree of dilution of silane with helium, gas pressure and laser power. By varying these parameters it was possible to obtain nc-Si with the sizes from 4 to 9 nm. The yield fell with decreasing particle size: for particles with a diameter of 4 nm the yield was 0.08 g/h, for 5 nm – 0.12 g/h, for 7.5 nm – 0.2 g/h and for 9 nm – 0.9 g/h.

In air, nc-Si was covered with a layer of SiO_2, the thickness of which reached 10% of the diameter of the particles. This is the easiest way to passivate the nc-Si without which PL does not appear. Storage in air of the nc-Si produced by this method leads to an increase in PL intensity.

The dispersion in ethanol was produced using nc-Si without the silica. The dispersion of nc-Si in ethanol luminesced red light only after 2 weeks of its preparation; the authors interpreted this by passivation of nc-Si with alcohol. In acetone, this process was not observed. The alcohol dispersion of nc-Si was dried by spontaneous evaporation in air. The size of the particles of nc-Si, according to the HRTEM, decreased from 4.2 nm to 3.5 nm and a maximum in the PL spectrum shifted from 760 nm to 700 nm. This shift does not correspond to the decrease of the size of the nanosilicon particles which also proved to be amorphous rather than crystalline. The PL curve is very similar to the one that was received from the hollow particles of amorphous SiO_2. Apparently, in drying of ethanol nc-Si oxidized to SiO_2.

Quenching of photoluminescence defects seems to be the most important phenomenon, leading to uncontrolled and non-reproducible properties of nc-Si. Therefore, the control of the surface is the key question of synthesis and application of nc-Si. One possible way of control is to cover nc-Si with a layer of SiO_x, which occurs spontaneously in the air over time and its duration depends on humidity, temperature, etc. The properties of the SiO_x layer depend on the rate of growth and the reaction conditions. nc-Si particles with an average diameter of 3 nm were obtained by thermolysis of silane in a CO_2-laser beam (Vicent et al, 2010). The powder from the reactor was collected in the air. The nc-Si particles did not have PL, but after 2 years of ageing in the air photoluminescence appeared in the red and infrared bands with nc-Si dispersed in absolute ethanol with sonification for 10 min. The concentration of the dispersion was equal to $10 \text{ mg} \cdot \text{l}^{-1}$. No sedimentation or flocculation was observed in it for several years. 5% deionized water was added to the dispersion. It was heated in a closed vessel at a temperature of 120°C from 4 min up to 20 h. The authors suggest that the heating only affects the reaction rate. In some experiments, few tens of ml of 0.03 M HCl per liter of dispersion or a few drops of 37% HNO_3 was added. Excess amounts of these acids led to sedimentation half an hour after adding the acid. After heating the nc-Si remained in the same solvent (ethanol/water).

The size of freshly produced nc-Si determined by line broadening of the powder was 4.9 nm, and of the nc-Si aged in the air – by TEM (~6 nm). PL was observed only in air-aged nc-Si; it was not observed in the freshly prepared samples. However, the SiO_2 layer was observed on the surface of both. FTIR clearly registered the (Si–O) bonds as well as links (Si–H) bonds only in the freshly prepared nc-Si. The dispersion of the air-aged nc-Si in ethanol showed a maximum at 763 nm in the

PL spectrum. This spectrum was used for comparison with all other substances. In FTIR the spectrum had a distinct band at ~1650 cm^{-1}, usually interpreted as the (OH)-group. It presence, along with the (Si–H) bonds indicates that the surface of nc-Si undergoes a disproportionation reaction involving moisture in the air.

Judging from the absorption spectra, heating the dispersion at 120°C for 1 hour led to a complete transformation of nc-Si to SiO$_2$ of both the freshly prepared and aged in air samples.. The PL spectra were excited by UV radiation with a wavelength 254 nm. In the PL spectra of freshly prepared nc-Si heating to 120°C led to appearance and intensification of the peak at 430 nm, reaching a maximum intensity after 15–20 hours of heating the dispersion. This peak is attributed by the authors to defects in SiO$_2$.

Ageing in air of nc-Si due to the heating in a water–ethanol medium changed the PL spectrum over time: the peak initially observed at 763 nm after 8 min of heating is shifted to 744 nm and dramatically loses its intensity, 14 min from the beginning of heating is shifted to 640 nm, and after 30 min – to 610 nm. At the same time the peak at 430 nm, characteristic of defects in SiO$_2$, also increased. After 1 h of heating all the peaks disappeared, except the one at 430 nm. The authors emphasize that the absence of photoluminescence in the freshly prepared samples is the evidence of defects affecting the non-radiative recombination of charge carriers in the nc-Si.

In the next series of experiments, the dispersion of aged nc-Si in the ethanol mixture was irradiated with ultraviolet light from 8 min to 20 h. Immediately after dispersion the PL maximum shifted from 763 nm to 782 nm, but during exposure it steadily moved in the direction of the blue side: 8 min – 750 nm, after 14 min – 742 nm, and after 30 min – 710 nm, and the intensity was almost doubled in comparison with the initial intensity. Apparently, the size of nc-Si decreased due to the interaction with water, which led to the blue shift, but the amount of P_b-centres also decreaes, and this caused increase of the PL intensity. It is noteworthy that there were no changes in PL in anhydrous ethanol. After 1 h of irradiation the intensity of the peak began to fall without changing its position, as the peak detected after 8 min of irradiation at 430 nm continued to systematically increase its intensity. Thus, both heating in the presence of water and exposure to UV radiation led to the same the result – the transformation of nc-Si to SiO$_2$.

Films with a thickness of 5–30 nm were prepared (Lioutas et al, 2008) by precipitation (LPCVD) of silicon vapor at reduced pressure – 300 mTorr and 610°C on a quartz substrate. The substrate was dissolved in hydrofluoric acid. All samples had diamond-like crystal structures. The curves of size distribution of nc-Si, regardless of the dependence on the thickness of the deposited film, had a maximum at about 8–10 nm. At large thicknesses of the films the maximum size could reach or be greater than 25 nm. The minimum size of nc-Si in all the samples was limited at 4–5 nm. The study of transverse sections of the layers showed that the layers consist of a mixture of nc-Si and amorphous silicon in thin films (less than 30 nm) and almost polycrystalline layers in the film with a thickness of about 30 nm.

The films had no photoluminescence. However, after oxidation in a dry atmosphere at 250°C for 17 min 2.8 nm remained from the original 5 nm silicon film, the surface of the layer opposite to the substrate transformed into SiO_2. The maximum of the particle size distribution was shifted from 10 to 4 nm. These nc-Si photoluminesced, emitting light with a wavelength of ~725 nm.

The optical properties of nc-Si strongly depend on surface effects. The presence of impurity atoms or radicals on the surface strongly affects the electronic structure of nc-Si and hence the optical properties.

Agglomerates of nc-Si – the main obstacle to most applications nc-Si. Salivati et al. (2009) – showed that the agglomerated nc-Si have no PL, so we need studies of the effects of surface effects should be carried out on non-agglomerated nc-Si free from the influence of the substrate. In Trave et al. (2006), the authors removed the SiO_2 layer from the surface of nc-Si by using hydrofluoric acid or alkali. The authors suggest that after etching in HF the surface was passivated by hydrogen and was not oxidized by air, acquiring the hydrophobic properties, and alkaline etching by KOH leads to a hydrophilic surface due to its coverage with OH groups. nc-Si were obtained by pyrolysis in a CO_2-laser beam. Ar was passed along the walls of the reactor coaxially with the flow of SiH_4, diluted with helium. The walls of the reactor remained cold. The size of nc-Si changed with the time of stay of SiH_4 in the area of pyrolysis and depending on the laser power. nc-Si was collected at the outlet of the reactor. nc-Si was dispersed in ethanol using ultrasound.

The resulting nc-Si particles were covered with a layer of native SiO_2. The average particle size was 12.2–12.7 nm with a standard deviation σ = 5.2–6 nm. PL was not observed. nc-Si was annealed in a stream of air at 700–1000°C, was covered with a thick layer of SiO_2, and began to fluoresce due to the reduction of the size of nc-Si. However, the agglomerates formed, which limited the opportunity to explore optical properties.

The nc-Si particles oxidation of their surface for 5 min were dispersed by ultrasound in a solution of 10% HF in ethanol. As shown by TEM, after 10 min after the SiO_2 shell was completely dissolved, and the size of nc-Si decreased from 10.8 to 5.3 nm. In addition, nc-Si was dispersed in a 0.1% solution HF in ethanol. Within 2 hours, the PL intensity remained constant, then began to decline, and the PL disappeared after 5 hours of etching with hydrofluoric acid. The authors explained the disappearance of photoluminescence by the complete dissolution of the SiO_2 shell of nc-Si. After aging for several hours, the nc-Si particles from which SiO_2 had been removed, began to aggregate and therefore, a surfactant (trioctylphosphine oxide) was added which stabilized the dispersion for 2 weeks.

Etching was carried out with tetramethylammonium hydroxide (TMAH). nc-Si was dispersed in methanol using an ultrasonic bath for 30 min. In the dispersion there were aggregates up to 100 nm in diameter. TMAH was added to the dispersion to dissolve the oxide film on the surfaces of the nc-Si. If the concentration of TMAH was greater than $5 \cdot 10^{-3}$ M, the almost complete precipitation of silica occurred in few minutes, and if less, the dispersion remained transparent for several days. Surfactants increased the stability of the dispersion. FTIR showed the presence of OH-groups on the surfaces of the nc-Si, but only a few Si–H bonds remained.

Modification of the surface of nc-Si, obtained by pyrolysis using CO_2-laser radiation, is often carried out after oxidation and etching of nc-Si to reduce their size to those at which the nc-Si begins to luminesce. Etching was combined with coating the surface of nc-Si with hydrogen atoms, so the hydrosilylation reaction could be used in subsequent stages. Thus, for example, the nc-Si surface was hydrophylized, using the reaction of hydrosilylation with acrylic acid (Li et al, 2004) or undecylenic acid (Li et al, 2003).

nc-Si was etched either with a mixture of HF and 20% HNO_3, or a mixture of H_2SO_4 + 30% H_2O_2 (7:3 by volume). nc-Si with an average diameter of 5 nm was produced. FTIR showed that the etched nc-Si contained, in addition to Si–H bonds, quite a lot (Si–O) fragments, and after hydrosilylation with octadecene the Si–H bonds did not entirely disappear, and the number of (Si–O) bonds increased. Hydrosilylation with undecylenic acid significantly increased the number of the (Si–O) bonds, because of its boiling in ethanol with a reflux cooler in ethanol (absolutized ethanol was not reported). The authors suggest that nc-Si was oxidized in the course of this operation and this happens quite often. It was found that 3% and 5% HF form Si–H bonds, and 15% HF – (> SiH_2) and (–SiH_3) groups. In addition, 15% HF catalyzes the formation of (Si–OH) bonds due to the interaction of water with the nc-Si. The instability of the PL on the nc-Si not covered by alkyls the stability of the PL of the nc-Si coated with the alkyls was compared. The rapid degradation of nc-Si in water and slow degradation in alcohols were described. The highest stability was achieved in toluene. The authors suggest that the cause of degradation is oxygen, but water was not excluded.

Hydrophilization of the surface of the nc-Si already covered with alkyl ligands is possible (Rosso et al, 2010) by oxidation with oxygen in the plasma: the end methyl groups of alkyl ligands are oxidized to alcohol, aldehyde and carboxyl groups.

The particles of nc-Si with an average diameter of ~5 nm, were obtained (Hua et al, 2005) by thermolysis of silane mixed with hydrogen and helium, by means of CO_2-laser radiation, followed by etching in a mixture of HNO_3/water to reduce size and hydrogen passivation. The nc-Si particles with green and yellow luminescence, were unstable in air: a decrease in intensity and red shift of the PL were detected. All reagents deaerated. nc-Si was added in a mixture of 48 wt.% HF:HNO_3 (69 wt.%) = 10:1 by volume. The dimensions of nc-Si decreased, and methanol was added when the desired PL colour was reached. The mixture was then filtered on a membrane of polyvinylidene fluoride (pores d = 100 nm) and washed with a 'methanol-water' mixture. Next, nc-Si was hydrosilylated with octadecene or undecylenic acid and divided into individual size groups by chromatography. For the nc-Si of different sizes separated from nc-Si with yellow PL three PL peaks were obtained by excitation at 355 nm: 650 nm, 550 nm and 425 nm. Prior to etching of nc-Si bonds FTIR did not reveal even Si–H and (Si–O) bonds. But after etching and rinsing in water there was a small number of Si–H bonds and many (Si–O) bonds. Rinsing with a mixture of methanol and water gives a lot of Si–H bonds but a few (Si–O) bonds. Thus, not only etching but also subsequent rinsing determined the composition of the surface of nc-Si. NMR and thermogravimetry showed that for different samples pf nc-Si coated with n-pentylene, n-hexyl, n-octyl, n-dodecyl, and n-octadecyl, the ratio of the amount of the ligands and the silicon atoms on the surface of nc-Si is in the range 0.30–0.45, i.e. alkyls do not cover the entire surface of the nc-Si, and part of it is occupied by Si–H and (Si–O) bonds.

The same authors (Hua et al, 2006) continued to study the properties of nc-Si with the alkylated surface. The oxidation and etching conditions were the same, but 2–3 ml of one of the reactants – styrene, 1-octene, 1-dodecene, 1-octadecene – was added for hydrosilylation of nc-Si The dispersion was turbid even after sonication. The mixture was degassed in vacuum. The reaction proceeded under the influence of UV radiation (5 min for styrene, 45 min for 1-octene or 1-dodecene, 60 min for 1-octadecene). After the dispersion became transparent, the excess of alkene was evaporated in vacuum. FTIR spectroscopy showed that after hydrosilylation part of the hydrogen atoms always remains on the surface of nc-Si.

Hua et al, 2006 studied in detail the oxidation of functionalized nc-Si/alkyl. It should be noted that after the below mentioned procedures of oxidation and formation of blue photoluminescence the authors could not detect the nanocrystals of silicon by any of three methods: HRTEM, TEM, XRD, i.e. the oxidation of silicon probably led to amorphous SiO_2. In particular, the dried particles of nc-Si/alkyl were dispersed in 3 ml of chloroform or toluene. The solution was closed to prevent evaporation of the solvent and subjected to UV irradiation in the presence of the air. The solution was not degassed and, apparently, was saturated with air. For PL to change from yellow or green to the blue region ethylbenzol ligands required 2 hours and long alkyl ligands 12 hours. Alkylated nc-Si in the dry state were placed in a furnace at 140°C in air. A day later, they were cooled and dispersed in chloroform or toluene. When excited by ultraviolet radiation at 355 nm the dispersion showed blue PL. PL intensity in any part of the spectrum except blue decreased with time. Blue photoluminescence from any dispersion was stable for at least 8 months. After 6 hours of oxidation in toluene with UV light at 254 nm the intensity FTIR of the Si–H band decreased markedly, and (Si–O) increased significantly, but even after 12 h small amounts of Si–H remained. All C–H and C–C bands retained their intensity in the FTIR spectra. The stability of alkyl radicals on the surface of nc-Si was confirmed by the ^{1}H and ^{13}C NMR methods. Oxidation can lead to blue PL when heating nc-Si in air at 140°C for 24 h, but longer (4 days) heating at 70°C did not give the same effect.

It has been shown (Sublemontier et al, 2009) that the pyrolysis of silane by CO_2 laser radiation produced nc-Si particles of the desired size, if the duration of the laser pulse is carefully controlled, while maintaining all other process parameters constant. Electroexcited chemiluminescence (ECL) is much more sensitive to surface states of nanoparticles than the PL. It turns out that the band gap, as measured by the electrochemical method, is greater than the energy of the PL peak, which is explained by the participation of surface states in the excitation of photoluminescence.

The nc-Si particles were synthesized by laser pyrolysis of silane (Bae et al, 2006). The average particle size was 15 nm and there was no PL. The nc-Si was then etched with a mixture of HF and HNO_3 acids to the size of ~3.5 nm, resulting in bright visible photoluminescence. Hydrogen-passivated nc-Si in the process of the hydrosilylation reaction was refluxed in 1-octadecene to passivate the surface by octadecyl groups, which, according tot the authors, protects the nc-Si against oxidation and chemical etching. The quantum yield of PL was equal to 20–30%. The yield did not change after exposure to air for many months. TEM showed the diamond-like structure of the nc-Si and the average diameter particles of 3.4 nm. nc-Si was deposited on substrate in the form of a ITO layer of thickness ~500 nm. Electrochemical and ECL-experiments were carried out in an aqueous solution of 0.1 M KOH, with the addition of 0.1 M solution of $K_2S_2O_8$ or without it. The counter electrode was made of Pt, and the reference electrode of Ag|AgCl. The solutions were not degassed. Only the presence of potasium peroxodisulphate with the negative potential applied to the sample resulted in the following reactions:

$$S_2O_8^{2-} + e \rightarrow SO_4^{2-} + SO_4^- \text{ (anion radical)},$$

$$SO_4^- \rightarrow SO_4^{2-} + h^+,$$

$$Si + e \rightarrow Si^- \text{ (on the cathode)}$$

$$Si^- + h^+ \rightarrow Si + \text{light}.$$

But even the positive potential applied to the nc-Si resulted in ECL, even in the absence of peroxodisulphate ions, however, its intensity was an order of magnitude weaker than that of the cathodic ECL. The wavelength of the PL at the peak intensity was equal to 670 nm (1.85 eV). ECL was attributed by the authors to surface states. When scanning the potential from 0 to –1.8 V peak cathodic current occurred at –1.53 V and continued to strongly grow toward the value of -1.8 V. The authors attributed this to the restoration of water on the nc-Si or ITO. When scanning in the opposite direction, the anode current peak was found at -1.07 V, which was interpreted as the injection of an electron into silicon by an electron donor. These phenomena were recorded in a KOH solution.

The authors assumed that oxidation of nc-Si takes place on surface areas that are not protected by octadecyl groups and on which the Si–H bonds were preserved and did not react during the hydrosilylation reaction. ECL with the positive potential on the nc-Si was regarded by the authors as unexpected since there is no reduction agent in the KOH solution. They believe that the emergence the ECL at the positive potential corresponds to the oxidation of an unknown substance, whereas oxidation of the nc-Si can be occur at potentials less positive than those which are characteristic for the unknown substance. This reaction is used to explain a significant difference in the potentials at which ECL appears. We can assume that the authors observed the disproportionation reaction of silicon activated by rapid hydrolysis of of the resultant silane formed in an alkaline medium (see the beginning of section 6.4).

5.3.3. Electrochemical etching

Electrochemical anodizing is based on the galvanostatic processing of mono- or polycrystalline silicon in hydrofluoric acid at a moderate current density, resulting in the surface of the substrate is converted into a layer of porous silicon (P-Si). Surfaces of P-Si particles are passivated by hydrogen atoms formed during anodization. The anodization conditions (the concentration of HF, current density, luminosity, temperature of the solution) should be selected in accordance with the electrical and structural characteristics of the silicon substrate (type of conductivity, resistivity, crystal orientation). Anodization of the silicon substrate of n-type should take place under illumination for the formation of holes. As anodization, in all 'dry' methods of obtaining nc-Si, the particle size distribution is Gaussian and has the same half-width of the peak (Koshida et al, 2007).

It was found that the etching of silicon with hydrofluoric acid is anisotropic: the etching rate depends on the crystallographic direction. It is minimal in the direction $\langle 111 \rangle$, as the acid has to overcome the (111) layers with the densest packaging. The (100) plane has a much lower density of packing of atoms and, therefore, the reaction rate in the $\langle 100 \rangle$ direction is 10–15 times higher.

A possible mechanism for the formation of porous silicon in the initial stage on all facets of the silicon is the same: the influence of the electric potential leads to uniform etching of silicon and chaotic deposition of silicon atoms in separate locations on the crystal. The sediment, because of its looseness, has a high electrical resistance and does not dissolve at moderate current densities, in contrast to the crystal where an etching mask forms gradually on the surface. In the next phase the mask determined the formation of pores, and the etching rate depends on the crystallographic direction. The crystallographic face of the crystal also determined the morphology of the pores. Etching of the (111) face leads (Timokhov and others, 2009) to the formation of a spongy pore system, and the pores on the (100) face have a shape close to cylindrical and are elongated along the fourfold axis. The particles of nc-Si, obtained by etching the (111) face under otherwise identical conditions, are obtained smaller than those after etching the (100) face.

It has been shown (Belomoin et al, 2002) that the electrochemical etching of single-crystal silicon wafers in an electrolyte solution of a mixture of HF and H_2O_2 and subsequent ultrasonic treatment of samples produces almost spherical stable over time Si nanoparticles with a discrete size distribution, giving the red, green and blue photoluminescence necessary for the production of colour displays by RGB technology.

Electrochemical etching was the first method of obtaining nc-Si. Its fundamentals were described in many publications (see chapter 3). A review of methods for producing porous silicon nanostructures using electrochemical etching can be found, for example, in the article (Golovan et al, 2007). Although detailed studies of the method are continuing (Wijesinghe et al, 2008), in the last five years (2006–2010) studies have concentrated on the surface modification of the nc-Si particles, obtained by electrochemical anodization, nc-Si etching in order to reduce their size and the practical applications of nc-Si (Stupca et al, 2007; Nayfeh et al, 2009).

Among the methods of surface modification of particles nc-Si, obtained by anodization, the dominant method is hydrosilylation, which is natural given the passivation of the nc-Si surface with hydrogen in the process of electrochemical etching. For example, nc-Si was obtained (Chao et al, 2006) by anodic etching plates of Si (100) of n-type in a mixture of equal volumes of 48% HF and ethanol with the subsequent modification of the surface. The resulting porous silicon subjected to a hydrosilylation reaction with 1-undecene by refluxing in dry toluene. The solid residue was filtered and the solvent and excess reagent removed under vacuum. A waxy substance soluble in non-polar organic solvents remained. TEM microscopy, atomic force microscopy and small-angle scattering of X-rays showed the size of the nc-Si to be 2.20 ± 0.04 nm. PL has been studied with the synchrotron beam in the energy range 5.1–23 eV at temperatures of 8–300 K. Below 150 K, two bands were observed in the photoluminescence spectrum: blue (430 nm) and orange (600 nm). According to the authors, the first was due to the (Si = O) and (Si–O–Si) bonds on the surface of nc-Si. This follows from the fact that the blue PL band occurs when the excitation energy exceeds 8.7 eV, i.e. above the threshold energy of photoemission of the electrons captured by traps in SiO_2. The orange PL band comes from the silicon core of nc-Si and occurs because of the size effect which agrees well with the diameter of nc-Si, 2.2 nm.

The reactions of surface modification of nanosilicon described above, were performed by a group of authors (Lie et al, 2002). It was established by FTIR that only half of the hydrogen atoms are replaced by alkyls, and hydrosilylation leads to oxidation of nc-Si. It should be borne in mind that the nc-Si is already oxidized during electrochemical etching (Wijesinghe et al, 2008).

Alkylated nc-Si are considered as fluorescent labels for biological applications (see chapter 14). They are more stable than molecular dyes and their red radiation is weakly absorbed by biological objects. They are not as toxic as CdSe and InAs. In contrast to molecular dyes the luminescence of nc-Si does not depend on the pH of the medium. However, biomarkers should have a hydrophilic surface. Alkylated nc-Si can be used for intracellular studies. It is known that the hydrophobic particles can be converted to hydrophilic by shaking them with tetrahydrofurane, DMSO or ether, which perform the role of surfactants. The resulting sol is stable for at least six months. DMSO does not have acute cytotoxicity at the dilutions used.

Porous silicon was obtained (Dickinson et al, 2008) by galvanostatic anodization of boron-doped (100) silicon wafers in a mixture of 48% $HF:C_2H_5OH = 1 : 1$ (by volume). The sample was dried under vacuum for 15 minutes, refluxed for 2 hours in 5 ml of toluene (dried with sodium) with a mixture of 0.1 ml of 1-undecene. A yellow dispersion of nc-Si in toluene formed. Then the toluene and an excess of 1-undecene were removed vacuum. Their remains were removed by evaporation combined with 5 mL of CH_2Cl_2, and then with 5 ml of pentanol to obtain a waxy residue, which was mixed with THF, DMSO, toluene etc. The ether proved to be harmless to the cells and a good dispersion medium for the alkylated particles of nc-Si. The diameter of nc-Si was found to be 2.5 nm, and together with alkyl ligands – 5 nm. According to the authors, alkylated nc-Si oxidized to form a layer on the surface responsible for the blue PL. They do not react with 1 M HCl, although there were destroyed in a solution of 1 M NaOH. The dispersion in water has a bright orange luminescence. The dispersion in water was obtained by premixing alkylated nc-Si with a surfactant and subsequent dilution with water. The authors noted that the produced nc-Si, appears to slowly react with water to form on the surface non-radiative recombination centres. Toxicity of nc-Si was not detected. A similar method of synthesis of biomarkers based on nc-Si is described in (Choi et al, 2008).

In order to clarify the question of the oxidation of the alkylated (by hydrosilylation) surface of the nc-Si, obtained by electrochemical etching, half of the samples immediately after synthesis were placed in a vacuum, while the other half remained in the air for a few hours (Chao et al, 2005). The authors suggest that their samples were covered by alkyls to only 30–50%, while the rest of the surface was coated with hydrogen. They believe that the specimens, held in air for several hours, during exposure to vacuum ultraviolet radiation were oxidized from the surface to suboxide SiO_x ($x \leq 2$) by water and/or oxygen adsorbed during the stay in the air. Authors believe that in a few hours of stay in the air the remains of the solvent change to water, which upon irradiation leads to the appearance of oxides of silicon. However, the PL spectrum of the sample, immediately placed in a vacuum, contained two peaks – 390 ± 20 nm and 480 ± 20 nm. The source of the first of the two peaks, was considered by the authors to be the (Si–O) bonds, resulting in the oxidation of the surface of the nc-Si in hydrosilylation of nc-Si and/or electrochemical etching.

The article by Scheres et al. (2010) answered the important question of the density of the coating of the hydrocarbon ligands on the surface of the nc-Si. It was found that alkyl ligands cover 50–55% of the Si (111) surface, and the use of alkenyl ligands increases the proportion of the coating increases with the length of the hydrocarbon chain from 55% for C_{12} to 65% for C_{18}, i.e. almost to the theoretical maximum achievable in coating with hydrogen ligands the same (111) plane of the crystal silicon.

Rogozhina et al. (2006) described the reaction of hydrosilylation of the surface of nc-Si/H, and it is reported that the hydrosilylation has the advantage over other methods of surface alkylation of nc-Si, which is not affected by the (Si–Si) bond: Grignard and lithium–alkyl reagents break the (Si–Si) bonds on the surface of nc-Si. Even the reaction with alcohols can lead to the rupture of (Si–Si) bonds with the formation of Si–H and Si–OR bonds.

However, alkylation of nc-Si/H with alkyl lithium is found in the literature (eg. Guiliani et al, 2007). Nanoporous silicon, P-Si/H, was obtained by electrochemical etching of single crystal silicon plates doped with boron. Removal P-Si/H from the surface of the plate was carried out by sonification in hexane, resulting in a brown sol from which hexane was removed and dry degassed hexane was added to the sediment whoich was redispersed with ultrasound. Bromine, replacing hydrogen on the surface of the obtained nc-Si/H, was then added to the dispersion. The reaction was carried out in a stream of argon at reflux to the bleaching mixture. The vacuum line sucked away the excess volatile substance, and the mixture again heated with the reflux condenser for 4 hours. The brown dispersion was then cooled to 25°C. Nanoparticles nc-Si/DMAP were obtained by the reaction of nc-Si/Br with 3-(dimethylamino) propyl lithium (DMAP) in dry hexane at 0°C for 6 h. Particles of nc-Si/DMAP were extracted from hexane with water and lithium bromide was removed by dialysis.

The dimensions of the resulting nc-Si/DMAP particles were within 10–50 nm, and the maximum bandwidth of the photoluminescence spectrum – in the region of 300 nm. The IR spectra did not show absorption in areas of (Si–O), Si–OH and SiH_x bonds, i.e. hydrogen atoms have been replaced without a trace of contamination with oxygen. This result was confirmed by ^{1}H, ^{13}C and ^{29}Si NMR spectroscopy. nc-Si are stable over a year and can be applied in biology and medicine.

Noteworthy are two observations in the literature about the properties of alkylated nc-Si, obtained from the number obtained by electrochemical etching. First, nc-Si with alkyl ligands can sublime without decomposition in high vacuum at 200°C (Caxon et al, 2008). Second, by chromatography we divide the nc-Si on the basis of the size, and by capillary electrophoresis to separate electroneutral nc-Si from those whose ligands which carry the charge.

In a number of publications the nc-Si produced by anodizing were subjected to further processing in order to reduce the size of the particles and the differences in size. For example, in (Choi et al, 2007b), the dispersion of the porous silicon in methanol was etched a mixture of 10 volumes of 48% HF and 1 volume of HNO_3 (69–70%) in a nitrogen atmosphere, and the cell was irradiated with ultraviolet light (340 nm) at 20°C. As a result of UV irradiation the reaction rate increased by 20 times. The following reaction occurs in the process of nc-Si etching:

$$3Si + 4HNO_3 + 18HF = 3H_2SiF_6 + 4NO + 8H_2O,$$

and the size of nc-Si decreased. The initial dispersion of porous silicon in methanol (before the addition of acids) had a weak photoluminescence with a peak around 660 nm, but after 80 min etching the PL intensity reaches a maximum, and the peak position shifted to ~570 nm. Further etching reduced the PL intensity due to dissolution of porous silicon, while the peak shifted to 500 nm. The photoluminescence quantum yield, measured by the standard (quinine sulphate), and in the dispersion with yellow PL it was 60%. PL quenching time in the blue region of the spectrum was 6 ns, but in the samples with the PL red to green, it was equal to $5 \cdot 10^{-8}$–$5 \cdot 10^{-5}$. Samples with blue PL were 1 nm in size for the hydrogen passivation of the surface.

A method for obtaining nc-Si with blue photoluminescence (Svrček et al, 2006b) by pulsed laser ablation of porous silicon grains was described. Porous Si was obtained by electrochemical etching, dispersed in a commercial SOG-solution (solution of polymer C_2H_5O [$SiO(C_2H_5O)_2$]$_n$$C_2H_5$ in ethanol and ethyl acetate), and there subjected to laser ablation. Increasing laser intensity resulted in melting and evaporation of nc-Si.

In (Svrček et al, 2006b) a plate of Si (100) of p-type was subjected to electrochemical etching. Porous silicon was mechanically scraped off. These micrograins were dispersed in ethanol by ultrasound for 10 min and then micrograins were deposited over a period of 4 h. Ethanol with the smallest particles was decanted and the residue was dried at 50°C in air. This procedure was repeated 4 times to remove small particles of porous silicon. The liquid was used as the SOG medium for dispersing 0.01 wt.% micrograins of by silicon ultrasound for 10 min. The dispersion was irradiated by a laser at room temperature. The laser beam was focused on the surface of the dispersion in a spot with a diameter of 1 mm. During irradiation, the glass container was closed and moved. The exposure time was 2 hours. The laser energy ranged from 0.06 to 5.9 mJ/pulse. After irradiation, the dispersion was treated with ultrasound for 10 min and dried in air at 323 K for 24 h to get the SOG film with nc-Si dispersed in it. For comparison, similar samples were prepared in deionized water. The size of the particles of porous silicon prior to laser ablation reached a few microns. They contained a few small particles, in spite of attempts to clean up, so the deposit had a weak PL, but the maximum of its spectrum was shifted toward the red by 0.2 eV compared to those particles which were decanted with alcohol and had a maximum in the photoluminescence spectrum at 1.7 eV.

The dispersion of porous silicon (0.01 wt.%) in SOG (5 mL) was light yellow in colour and was turbid. During laser irradiation, the colour changed to dark brown. When the laser energy was < 1.13 mJ/pulse, the dispersion remained unchanged throughout the whole period of exposure. However, at an energy of > 3 mJ/pulse, the solution during irradiation became transparent. When the energy was 5.9 mJ/pulse, the dispersion became transparent after 2 h. In a control experiment, the hydrophobic porous silicon remained on the surface after the second ultrasound treatment and laser radiation with an energy of 5.9 mJ/pulse for 2 h. The yellow colour of porous silicon did not change.

Below the energy of 1.13 mJ/pulse, the dispersion showed no changes, but above this threshold there was visually observed radiation from the plasma generated by laser irradiation. Stresses appeared in porous silicon particles caused by heating and cracking of the porous silicon particles. Part of silicon evaporated to the atomic state. In addition, the solid particles were affected by high pressure. During the explosion, caused by the laser pulse, electrons and ions appeared which, according to the authors, helped to disperse porous silicon to the individual nanocrystals. TEM showed that the structure of the nc-Si was diamond-like. The nc-Si particles were larger than 1.5 nm, which corresponds to the band gap width of 2.9 eV. However, the PL spectrum of the sample obtained at a laser energy of 5.9 mJ/pulse, had a peak at 2.9 eV and the initial porous silicon (before laser irradiation) had a PL peak at ~1.7 eV, and an order of magnitude weaker than the peak at 2.9 eV after pulsed irradiation with an energy of 5.9 mJ/pulse. nc-Si was covered an SiO_2 film during irradiation, and the stresses formed at the SiO_2/nc-Si interface resulted in blue photoluminescence. Defects and non-stoichiometry in SiO_2 give their contribution to the blue glow. As the SOG at its core is SiO_2, it is possible to suspect the influence of SOG on the PL. However, the experience with the SOG rejected this hypothesis: without nc-Si the PL was very weak. Blue photoluminescence, according to the authors,is due to the combination of the size effect, surface states at the Si/SiO_2 interface and defects associated with oxygen.

Following the discovery of ultrafast charge transfer in the 'fullerene–electrically conducting polymer' tests were carried out on the 'nc-Si-polymer' composite to find material for solar panels, which would ensure effective separation of electrons and holes without recombination, improved charge transfer and increased the stability of the polymer. It was found that boron-doped nc-Si improves the transfer of charges and the stability of the photoconductivity of thin films of methoxy-ethylhexyloxi–polyphenylenevinylene (MEH-PPV). nc-Si was obtained by electrochemical etching (Svrček et al, 2008) a boron-doped silicon wafer in a mixture of HF: $C_2H_5OH = 1:4$. nc-Si was mechanically scraped off. The authors suggest that at concentrations above 50 wt.% nc-Si nanocrystals in the polymer matrix form a continuous chain, and the electron finds its way to the electrical contacts. 10 mg of polymer was dissolved in 10 g of chlorobenzene and 400 mg of this solution was mixed with 2 mg of nc-Si. This mixture was applied with a thin film (~300 nm) on a glass substrate 15×15 mm^2 in size coated with platinum contacts. At the same time, samples without nc-Si for were prepared for comparison. The samples were dried at 140°C for 30 min in a vacuum. The content of nc-Si in the composite was evaluated as 83 wt.%. It was shown (Svrček et al, 2008) that the nc-Si/polymer composite has a greater stability of dark conductivity and photoconductivity as compared to the pure polymer. It is known that said the polymer is oxidized by atmospheric oxygen with the release of aromatic aldehydes, which leads to a decrease in the mobility of charge carriers. nc-Si decreases the intensity of diffusion of oxygen into the polymer. The film of the nc-Si/polymer composite had dark conductivity and photoconductivity twice that of the pure polymer film.

The electron–hole pairs, formed under the effect of light, are divided at the nc-Si/polymer interface; the electron remains on the nc-Si, and the hole – on the polymer. The authors (Svrček et al, 2008) believe that in the PL spectra the band of 500–700 nm with a maximum of 590 nm forms due to the quantum size effect nc-Si, and the band 650–850 nm with a maximum at 740 nm – due to surface defects.

nc-Si passivation by hydrogen, which arises in the process of anodization of silicon, makes the surface of nanosilicon hydrophobic but does not prevent the interaction with oxygen and water. For example, the nc-Si, obtained by electrochemical etching (Froner et al, 2006), were held in air for different times (from 3 hours to 30 days) and then sonicated in water for 1 hour. The aqueous suspension was investigated. The control sample was always in the air.

Figure 5.13 shows that the control sample was gradually oxidized in air and, therefore, decreased in size, and the peak of its PL spectrum showed blue shift reaching a constant value of 650 nm, which corresponds to the size of nc-Si/H of 3.4 nm. Sonification in water for 1 hour resulted in a disproportionation reaction, and a sharp decrease in the size of nc-Si/H in the beginning of the experiment, but over time the growing SiO$_2$ layer

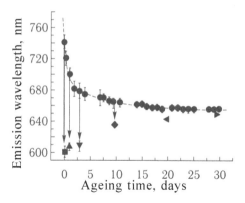

Fig. 5.13. The dependence of the luminescence peak wavelength of the reference sample of P-Si on time spent in the air (●) and the wavelength of the luminescence peaks of water solutions of different crystals obtained from P-Si, after ageing in air for different times (■ – 3 hours, ▲ – 1 day, ▼– 3 days, ♦ – 10 days, ◄ – 20 days, ► – 30 days). The dotted line shows the approximating curve drawn through the experimental points for the standard model P-Si (Svrček et al, 2008).

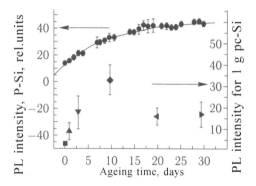

Fig. 5.14. The dependence of the PL intensity of standard sample of P-Si on the aging time in air (●), and the dependence of the PL intensity of aqueous solutions of different crystals, derived from P-Si, after ageing in air for different times (■ – 3 hours, ▲ – 1 day, ▼– 3 days, ♦ – 10 days, ◄ – 20 days, ► – 30 days) (Froner et al, 2006).

inhibited this reaction and differences in the wavelengths of the PL of the reference sample, stored in the air and treated with ultrasound in the water, decreased, until they merged at the PL wavelength of 650 nm. The PL intensity of the reference sample, stored at the air, increased nine times (Fig. 5.14), while in the samples which reacted with water in an ultrasonic bath the increase was 10-fold after 10 days of aging, but then the intensity of PL fell. The formation of new chemical bonds at the surface nc-Si/H as a result of oxidation by oxygen and the interaction with water was studied by FTIR. The formation of (Si–O–Si) bonds reached saturation within 20 days. The density of Si–H$_x$ bonds gradually decreased and that of (O$_y$–Si–H$_x$) bonds increased. (O$_y$–Si–H$_x$) fragments formed most rapidly from fresh samples of nc-Si/H treated with water. The surface of the samples which were for a long time in the air showed the formation of SiO$_2$ layers and no interaction with water was detected.

In (Bychto et al, 2008) porous silicon was obtained by electrochemical etching of silicon wafers doped with boron in a mixture of 48% HF:ethanol = 1:1. The porous layer was washed with ethanol, hexane and scraped from the substrate. The authors paid special attention to the pulsed regime of formation of P-Si due to the change of the morphology of the sample. One-third of the sample was used later in the unaltered form, and another was ground in a mortar. The third portion was covered with a surfactant – undecylenic acid. The behaviour of the samples in air, in hexane and water (with the surfactant) was studied. According to the authors (Bychto et al, 2008), nc-Si was embedded in the spongy mass of porous silicon. Adding hexane shifted the PL band in the blue region from ~700 to ~640 nm in comparison with air. In air, the PL intensity did not change during 30 days but decreased in hexane. In water, the PL peak position shifted in 5 days from 700–730 nm to 650–620 nm. PL intensity in the water increased over 5 days at 20–40 times and then fell by 3–4 times in the samples obtained in the pulsed mode and by 20% in the sample obtained by DC anodization.

The behavioural differences in the PL of -Si in water due to the authors (Bychto et al, 2008) as follows: in water the P-Si aggregates, hydrophilized with the surfactants are divided into individual nc-Si, which improves the efficiency of absorption of the exciting radiation. This is confirmed by SEM imaging studies: first aggregates of a few microns are visible, and 3 days later – some nc-Si particles with a diameter tens to hundreds of nanometers. The number of (Si=O) groups on the surface increased. It is noted that PL quenching by the molecular oxygen, included in the surface centres, is more efficient than the PL of free excitons in nc-Si/H.

Since the early 1990s, studied have been conducted of the rapid blue glow of the F-band and slow red glow of the S-band of porous silicon with the particle size of 1.5 nm, prepared by electrochemical etching of crystalline silicon. But up to now there is no clarity as to whether there is a link between them, and if so, what. Usually, one of them was studied. For example, Wolkin et al. (1999) studied only the S-band. The bands F-and S- were simultaneously observed by Hua et al. (2006) and several other researchers.

Dohnalova et al. (2010) carried out the oxidation of nc-Si with hydrogen peroxide to reduce the particle size and the shift of PL in the blue direction. nc-Si were obtained by electrochemical etching in hydrogen peroxide with the addition of HF (50%). After synthesis, nc-Si were treated with hydrogen peroxide in different modes, which led to the luminescence of samples from red to blue (see Table 5.2).

Table 5.2 Effect of the composition of the solution for etching and subsequent treatment on the wavelength of S-band of nanosilicon (Dohnalova et al, 2010)

Sample	Composition of the solution for etching	Post-processing	Wavelength of S-band
Standard (red)	13 ml of 50% HF+37 ml ethanol	not done	680
Yellow	13 ml of 50% HF+7 ml 3% H_2O_2+32 ml ethanol	5 min 3% H_2O_2	645
White	13 ml of 50% HF+2 ml 30% H_2O_2+37 ml ethanol	10–20 min 30% H_2O_2	590–600
Blue	13 ml of 50% HF+2 ml 30% H_2O_2 + 37 ml ethanol	10–20 min 30% H_2O_2+maturation for several days at a humidity 50–70% and a temperature of 25–30°C	560 (weak)

The sample, with red luminescence, was washed with alcohol and slowly dried in air (the porous surface by capillary action could get water from the air). The sample immediately after preparation had green PL and then red PL after oxidation (the sample was named 'standard' by the authors). If the porous silicon was scraped off from the substrate and processed in H_2O_2 (3%) for 5 min, changes took place in the PL spectrum: the photoluminescence of the S-band was shifted to the blue side and the blue glow of the F-band appeared and its intensity increased with increasing treatment time with hydrogen peroxide and with increasing concentration of H_2O_2. FTIR showed the existence of (Si–O–Si), (Si–OH) bonds, and the (Si–OH) bonds became stronger after additional processing with hydrogen peroxide. The authors (Dohnalova et al, 2010) suggest restructuring of nc-Si with the size of 1.5–2 nm and 1–1.5 nm. That is, reducing the size of nc-Si and changes of the composition of the surface are accompanied by structural changes of silicon nanoparticles. 'Yellow' particles, according to TEM, had a diameter of 3.5 nm, and in accordance with Raman measurements 2.5–2.7 nm. The dimensions of 'blue' particles could not be measured because they were beyond the sensitivity of the available measurement methods.

The PL spectra with time resolution showed the blue spectral band at around 430 nm (F-band), decaying in nanoseconds, and the red band at approximately 590–850 nm (S-band), decaying in microseconds. In the PL spectrum of 'blue' sample S-band low-intensity and was shifted to 560 nm. The authors suggest that blue shift of the PL is associated with switching from the S-band to the F-band and with the gradual decrease in the intensity of the S-band and increasing intensity of the F-band. Such a change in the

PL spectrum was observed Belomoin et al. (2002) and Sato et al. (2006) for a particle diameter of 1.5 nm. However, the blue PL with a wavelength shorter than 500 nm can be in nc-Si with organic ligands (Heintz et al, 2007; Warker et al, 2005). The authors believe that the smallest particles are amorphous. The F-band is often detected only in pulsed excitation. If the nature of the S-band is known, then the origin of the F-band so far has no explanation. Inmost cases, it is believed that the blue PL originates from the defects of the SiO_x/SiO_2 layer, covering nc-Si. However, it is not clear why the S-band stops at about 590 nm.

A special area of current research is the production of porous from crystalline nanosilicon by catalytic etching and not electrochemical etching. Nanoparticles of Pt, Au, Ag serve as catalysts. It is shown (Büttner et al, 2009) that silicon etching with hydrofluoric acid in the presence of gold is possible only in the presence of an oxidizer – atmospheric oxygen that is in the solution.

A silicon powder with a grain diameter of 80 mm was loaded for 15 m in a mixture of HF and H_2PtCl_6 (Nielsen et al, 2007). Hydrofluoric acid dissolved the surface layer of SiO_2, and platinum in the form of colloidal metal particles was deposited on the surface of the silicon grains. The grains were transferred to a mixture of $HF:H_2O_2:CH_3OH= 1:3:2$ (concentration not specified) and after etching were washed with isopropyl alcohol. The residue has the PL spectrum with a maximum of ~600 nm. After treatment of the residue dispersed in isopropyl alcohol with ultrasound for 2–5 min a fine suspension was separated it which luminesced under irradiation with ultraviolet light and the remainder lost luminescence.

The particles of colloidal silver were used (Asoh et al, 2007) as a catalyst for chemical etching of the silicon surface with the orientation (100) with a mixture of HF and H_2O_2. The surface, studied by atomic force microscopy, was etched to a depth of ~200 nm per minute. The particle sizes of silver ranged from 10–250 nm, but in general in the range 50–100 nm. Similar results were obtained using colloidal gold (Wan et al, 2008).

Anodic etching of silicon in the presence of catalysts of the oxidation of silicon nanoparticles – Ag, Pt, Pd – was considered in Chourou et al.(2010).

In Xie et al. (2007) amorphous silicon nanoparticles were obtained in two stages. Initially, magnetron sputtering was used to obtain a layer of amorphous silicon (A-Si), and this was followed by electrochemical etching in a mixture of $HF:H_2O_2:C_2H_5OH = 1:1:2$ to produce porous amorphous silicon with the surface passivated by hydrogen, which was transferred into an aqueous dispersion in an ultrasonic bath. For comparison a dispersion of nanoparticles of amorphous silicon was prepared in chloroform. TEM showed that the maximum of the particle size distribution is 1.7 nm. The maximum of the wide band of photoluminescence showed a blue shift in the spectra of dispersions of a-Si nanoparticles in chloroform with a decrease in the wavelength of exciting radiation, ranging from 380 to 300 nm. However, the dispersion of the same particles in water showed only a blue shift in the range 380–320 nm, and starting from 320 nm the blue shift was not observed in the photoluminescence spectra. The authors explained these differences in the PL spectra of the two dispersions by the fact that in water Si–OH groups form on the surface of a-Si/H instead of hydrogen passivation, and in the chloroform hydrogen passivation is preserved. The authors suggest that the energy of localization of Si–OH surface states stabilizes the blue emission of 2.9 eV (430 nm), even if the particle size is reduced, because the oxygen atoms on the surface of a-Si nanoparticles induce localization of excitons near the surface, and the magnitude of the localization energy of surface states of a-Si, passivated by hydrogen, is much less than in the particles passivated by oxygen.

5.4. Chemical methods of synthesis
5.4.1. Nanosilicon produced from silicon monoxide

Silicon monoxide, SiO, is stable in the gas phase above 1000°C. With rapid cooling it condenses into an amorphous product of light brown colour. In ageing and annealing SiO decomposes into SiO_2 and silicon clusters containing up to 10^{20} cm^{-3} paramagnetic centres. Thin fresh layers of SiO are an independent phase. It is stable up to 300 °C. In the temperature range 400–800°C amorphous silicon clusters form, and the crystallization of silicon begins at 800–900°C. Increasing annealing temperature to 1000–1100°C increases the size of nc-Si. The surface of silicon nanoparticles is in contact with both SiO, and SiO_2 (Wang et al, 2007).

The layers of different compositions SiO_x (x =1–1.9) were deposited (Kang et al, 2006) by thermal evaporation of SiO. The substrate was Si (100) or quartz at 50°C. Deposition was carried out at a pressure of 3 · 10^{-6} mbar in the presence of an oxygen flow which varied within 0–35 cm^3 min^{-1}. A quartz microbalance was used to control the deposition process. The layer thickness was 300 nm. The samples were then annealed at 300–1100°C from 5 min to 72 h in nitrogen or a mixture of 96% N_2 and 4% H_2.

It was found that after annealing the sample, whose composition corresponds to the formula SiO_x (x = 1.0), for 1 hour at 1100°C nanoparticles with a diameter of about 20 nm formed on the sample surface, but at x = 1.3 they did not exceed 5 nm, and nc-Si with an average diameter of ~4 nm were uniformly distributed in the matrix. With x increasing to 1.5 and 1.8 the size of nc-Si decreased, respectively, to ~3 and ~2 nm, as showed by TEM.

SiO_x (x < 1.75), calcined at 1200°C, showed PL in the red and infrared regions in excitation with the light at 275–380 nm. With the increase of x the PL peak shifted from 840 to 745 nm by reducing the size of nc-Si. The maximum PL intensity was obtained for the sample produced in an oxygen stream of 10 cm^3 · min^{-1}: the sample had the best the size of nc-Si and their density in the dispersion.

PL intensity increases with increasing annealing time in nitrogen up to 72 hours (3 times in comparison with annealing for 1 h), and the maximum shifts from 780 to 745 nm. The authors explained this by the reduction in the size of nc-Si due to reaction with nitrogen. Annealing in a mixture of 96% N_2 and 4% H_2 for an hour at 1100°C led to an increase in PL intensity by 2.5 times as compared with annealing in nitrogen. The quantum yield in excitation with ultraviolet light with a wavelength of 275 nm amounted to about 4%, and 370 nm ~7%.

The influence of the content of excess silicon in a thin layer of SiO_2 on the properties of dispersions of silicon in silica was studied by Thogersen et al. (2008). An SiO_2 layer with a thickness of 3 nm was deposited on a crystalline silicon substrate and SiO_x layers with the excess silicon of 4–46 at.% were then deposited on this substrate. The samples were heated in a nitrogen atmosphere at 1000–1100°C 30 to 60 minutes. HRTEM and EFTEM techniques showed no silicon clusters in the sample with 18 at.% of excess silicon. In the sample with 28 at.% there were only amorphous silicon nanoclusters. In the samples 50–70 at.% there were nc-Si in the SiO_2, located 5 nm from the silicon substrate and 4 nm from the outer surface of SiO_2. The thickness of the layer containing nc-Si was equal to 4 nm. The structure of the nc-Si was diamond-like.

Prior to annealing the sample with 46 at.% silicon did not contain the nc-Si. After annealing at 800°C for 1 h small clusters of amorphous silicon appeared, and annealing at 1100°C resulted in the formation on nanocrystals. It was found that oxygen from the air diffuses into the surface of SiO_2 and silicon atoms – from the substrate into the SiO_2 layer. There were errors in applying layers of silicon atoms: ABCBAC. In the nc-Si smaller than 3 nm HRTEM revealed noticeable distortion, for example, bending of the atomic rows. These stresses lead to errors in deposition of layers and twinning in the growth of nc-Si. High silicon concentrations, long annealing and increasing annealing temperature of nc-Si resulted in the formation of elongated crystals, which grow along $\langle 100 \rangle$. Dislocations were not observed in them.

The simultaneous deposition of Si and SiO_2 in an argon atmosphere (Huy et al, 2006) provided samples that luminesced brightly after preparation. The maximum of the PL band was about 440 nm (2.82 eV) with 1.3% Si in SiO_2. The authors suggest that excess Si should create defects with respect to oxygen in SiO_2 and these defects give PL. Heating at 600°C for an hour led to relaxation of SiO_2 which reduces the number of defects in SiO_2 and the sample is no longer capable of photoluminescence. After annealing at 800°C the Si in SiO_2 forms amorphous clusters which again luminescence with maxima at 370 nm (3.35 eV), 445 nm (2.79 eV) and 537 nm (2.31 eV). This PL is attributed to the traps on the surface of nanoclusters of amorphous silicon. Annealing at 1100°C led to crystallization of nc-Si and consequently to a change in the surface states of charge carriers in the nc-Si, which can lead to extinction of the photoluminescence at a small excess of silicon in SiO_2, as in this paper (0.5–1.8%).

Annealing of SiO_x at 500–900°C (Szekeres et al, 2008) produced amorphous silicon clusters in SiO_2, and at 1000°C – nc-Si with a certain amount of amorphous silicon. The PL intensity originating from clusters of amorphous silicon in SiO_2, in samples annealed at 700°C, was 5–10 times higher than in samples annealed at 1000°C and containing nc-Si. The initial stage of formation of silicon clusters in SiO_x ($x < 2$) was investigated in Fitting et al. (2010).

Nanosilicon in SiO_2 changes the refractive index, and a layer of nc-Si in the SiO_2, located between two layers of SiO_2, is a flat strip-like waveguide which has the property of a PL filter: narrow, polarized spectral peaks are filtered out of the broad PL bands, if the spectrum is recorded at the edge of the waveguide which is located along the strip waveguide. Oxide SiO_x ($x < 2$) was deposited (Khriachtchev et al, 2008; Butulay et al, 2008) on the SiO_2 substrate. Samples were annealed at 1100, 1150 and 1200°C under nitrogen for 1 h. For samples annealed at 1100°C, the intensity of the polarized lines of the PL spectrum was much higher than in the samples that were annealed at 1200°C. In addition, the samples annealed at 1100°C showed birefringence, explained by the non-spherical form of the nc-Si particles. Annealing of the samples at 1200°C eliminated the birefringence.

Examination by X-ray photoelectron spectroscopy (XPS) showed that at the greatest thickness of the deposited layer and the maximum content of silicon in it, the layer contains ~10 at.% silicon, about 25 at.% SiO, and ~65 at.% SiO_2, and the outer surface of the deposited silicon layer is depleted in Si but enriched with SiO_2 with the uniform distribution of the SiO concentration in the depth. The sample, annealed at 1100°C, contained half of the silicon nanoparticles in the crystalline form, and the other half – in the amorphous. TEM showed that in the samples annealed at 1100°C, the silicon nanoparticles were elongated and they were oriented on the substrate so that the direction of elongation was perpendicular to the strips of samples prepared by depositing SiO_x on the SiO_2 substrate. At an annealing temperature of 1200°C silicon nanoparticles were either isometric or arbitrarily misoriented. In addition, the proportion of amorphous silicon was reduced and that of the crystalline silicon increased.

The PL intensity of the dispersion of nc-Si in SiO_2 increases not only after heat treatment with hydrogen, oxygen or high-intensity UV irradiation. It is shown in (Lisowski et al, 2008) that the same effect is exerted by small doses of γ-radiation. The dispersion of nc-Si in SiO_2 was obtained by evaporation of SiO in vacuum and by vapour deposition on a silicon substrate which had a temperature of 150°C. The thickness of the SiO was 450 nm. Annealing of the samples at 1100°C lasted 15 min. The average particle diameter of nc-Si was ~3 nm. Subsequent γ-irradiation of the produced dispersions did not change the composition and the structure of the oxide matrix, but the PL intensity grew by 40% without changing the position and the half-width of the size distribution, i.e. the effect of γ-radiation did not change the size of nc-Si. A dose of $2 \cdot 10^4$ rad was optimal; a smaller dose $(10^3$ rad) did not increase the intensity, but at a higher dose $(10^7$ rad) the PL intensity decreases. The authors suggest that the optimal dose of γ-radiation leads to a structural ordering at the nc-Si/SiO$_2$ interface and a decrease in the number of P_b centres of non-radiative recombination of electron–hole pairs.

Unusual synthesis conditions led to unexpected results (Lu et al, 2007). A dispersion of nc-Si in SiO$_2$ was obtained alternately by sputtering of Si and SiO$_2$ at an argon pressure of 2.5 Pa. The p-type Si(100) substrate was used in three experiments at a voltage of 0, 0.9 or 1.1 kV. The substrate temperature was 400°C. The samples were not annealed. It turned out that in the absence of the electric potential silicon was amorphous, while at 0.9 and 1.1 kV electron diffraction (SAED) and HRTEM showed that silicon has an FCC structure and the maximum of the PL spectrum is in the blue field. The authors suggest that the given experimental conditions resulted in the formation of ions Ar$^+$, which bombarded the sample, which led to the formation of the FCC and not diamond-like structure of nc-Si.

In (Das et al, 2008) it is argued that the proposed method of obtaining dispersions of nc-Si in SiO$_2$ is much cheaper compared with others and, therefore, attractive to technology, even though the photoluminescence of these samples is not better than in the case of the nc-Si obtained by traditional methods. Triethoxysilane (TREOS) was dissolved in ethanol (TREOS: EtOH = 1:2) and hydrolyzed with water acidified with hydrochloric acid (pH = 5; H$_2$O:TREOS = 2:1). To produce the sol, the acid hydrolysis of triethoxysilane was inhibited by an ice bath, adding water in droplets with fast stirring for 1 h. The film of the partially hydrolyzed triethoxysilane on the silicon substrate (001) with a thickness of 800 nm was deposited by by spreading on a rotating substrate at 3000 rpm for 15 s. The sample was then kept at 80°C and heated in a nitrogen atmosphere for one hour at each annealing temperature. At temperatures below 1100°C pyrolysis of TREOS took place, and in the range 1100–1300°C nc-Si formed in the SiO$_2$ matrix. At 1100°C FTIR spectra showed no Si–H and C–H bonds.

The samples calcined at 1000 and 1100 °C showed photoluminescence spectra with maxima at 550 and 600 nm, respectively. This radiation, the authors believe, is generated by recombination of charge carriers captured by oxygen vacancies. The photoluminescence band at 705 nm occurs at the annealing temperature of 1150°C was shifted to 790 nm and 860 nm with the annealing temperature increasing up to 1200 and 1300°C, respectively. This band reflects the fact of the emergence and growth of nc-Si in the SiO$_2$ matrix and its nature is connected with the size effect.

The special properties, acquired by nanosilicon as a result of rapid heating, were mentioned above. Ultrafast heating in an electric arc in an argon atmosphere (Okada et al, 2008) was experimentally tested. The arc was

formed between a tungsten cathode and a copper anode spaced at a distance
of 1 mm. A layer of SiO_x, containing 49 at.% of silicon, with the thickness of
280–780 nm was prepared on a quartz substrate by vacuum evaporation
of powdered SiO. The composition of SiO_x was determined by X-ray
photoelectron spectroscopy (XPS). The sample was passed through the
plasma at a rate of 700–1200 mm/s. At a pass speed of the sample through
the plasma of 700 mm/s the substrate temperature reached 1670 °C, and the
Raman spectrum clearly showed the formation of crystalline silicon. The
size of nc-Si reached 300 nm. The closer the temperature of the substrate
to the melting point of silicon (1687 K), the greater the size of nc-Si in
SiO_2 due to premelting of silicon in SiO_2. TEM found that the layers of
SiO_x, passed above the plasma at a speed of 1200 mm/s, contain nc-Si with
a diameter of 3–5 nm with a distinct crystal structure. The PL spectra had
the maximum intensity at a rate of passage of the sample of 1200 mm/s
and a substrate temperature of 1210 K, but with decreasing rate of passage
and, therefore, with increasing annealing temperature, the PL intensity
decreased dramatically.

Interesting results were obtained in the study of the composition,
structure and properties of multilayer dispersions of nc-Si in SiO_2. The
dispersion of nc-Si in SiO_2 is discussed in the literature as the material for
the flash memory of the next generation. At smaller sizes it can contain
more information. Information is recorded by injection of charge carriers
in nc-Si through a layer of SiO_2. Thick SiO_2 layers require high-energy
electrons which leads to an increase in recording time. Thin SiO_2 layers are
inefficient because they reduce the information storage time. The properties
of the dispersion of nc-Si in SiO_2 also depend on the size distribution of
nanosilicon particles and their density in SiO_2. The distances between the
nc-Si should be sufficiently large enough to prevent tunnelling of charges
between the nc-Si.

The n-type Si(100) substrate, treated with an HF solution to remove
silicon oxides was deposited (Lu et al, 2006) with an SiO_2 layer with
a thickness of 4 nm, and three pairs of SiO_x (4 nm thick)/SiO_2 (3 nm
thickness). The composition of SiO_x was fixed for each sample, but
differed from sample to sample. The SiO_x layers were deposited by
thermal evaporation at a controlled pressure of oxygen. In the sample
with the highest possible density of nc-Si, annealing of the dispersion
at 900°C in a nitrogen atmosphere for 0.5 h resulted in the formation
of clusters of amorphous silicon. The clusters were then transformed
into nc-Si by additional annealing at 1100°C for 0.5 h. TEM method
was used to determine the diameter of spherical nc-Si 3.8 ± 0.5 nm. All
particles were almost of the same size, as determined by the thickness
of deposited SiO_x layers. The intensity of the photoluminescence band at
770 nm decreased monotonically with the decrease of the silicon content
in the nc-Si/SiO_2 dispersion.

'Capacitance – voltage' (C–V) hysteresis curves, obtained on the layers of the nc-Si/SiO$_2$ dispersion, correspond to the 'injection of electrons in the nc-Si–storage of electrons in the nc-Si – removal of electrons from the nc-Si' cycle. The higher the potential the greater the number of layers into which the electrons are injected. The lower the density of nc-Si in the SiO$_2$, the smaller the shift of the C–V curve for the injection of electrons relative to the same curve obtained by removing electrons from the nc-Si. Since a decrease in the density of nc-Si in SiO$_2$ decreases the PL intensity, the shift of the C–V curves and the PL intensity are linked by a linear relationship with any of the selected voltage. It was determined that the average distance between nc-Si is equal to 2.2 nm in the sample with the highest silicon content and in the sample with the minimum Si content it is 6.7 nm.

The superlattice of 10 pairs of SiO and SiO$_2$ layers on the Si(100) substrate was obtained (Berenstorff et al, 2007; Kovacevic et al, 2007) by annealing in vacuum at a temperature of 1050°C for 2 hours. The samples were studied using two-dimensional sliding the small-angle X-ray scattering. It was found that nc-Si in the cross sections of the layers of the sample were situated located under each other through the layer, i.e. arranged in a checkerboard pattern, but the distribution of the nc-Si in the plane of the layer is closer to disordered. Annealing the samples in air instead of vacuum destroys any order in the distribution of nc-Si in SiO$_2$.

The superlattice was synthesized (Zimina et al, 2006) by gradual deposition on a quartz 45 substrate of SiO(d)/SiO$_2$ (5 nm) layers, where d = 5, 4, 3, 2 nm. The package was then annealed to form nc-Si. The density of states of electrons in the nc-Si was measured by emission spectroscopy using soft x-rays (SXE) for the occupied states and by the method of absorption spectroscopy of soft X-rays (SXA) for the vacant states with synchrotron radiation. SXE resolution was 300 meV, and SXA – 50 meV, so that it was possible to analyze the electronic structure of the nc-Si, silicon suboxides and SiO$_2$.

One can define an intermediate layer between the surface of nc-Si and the matrix SiO$_2$ matrix as Si$_x$O$_{4-x}$. It has been shown that they are suboxides of composition Si$_x$O$_{4-x}$, where $x \leq 2$. Table 5.3 presents the diameters of the nc-Si for different thicknesses of the SiO layers and the thickness of suboxide layers on the surface of nc-Si, namely: at a diameter of nc-Si the thickness of the transition layer is 0.5 nm, and at a diameter of 1.6 nm it is 0.2 nm. Thus, not only the thickness of the SiO$_2$ layer decreases with decreasing diameter of the nc-Si because of self-restriction of oxidation of silicon, but also the transition layer becomes thinner at the nc-Si/SiO$_2$ interface.

The molecular dynamics method was used to construct a model of nc-Si with a diameter of ~2.5 nm, surrounded by the α-SiO$_2$ oxide (~4.3 nm), without affecting the periodicity of the boundary layer whose thickness does not exceed 7 Å and inside of which less than 20% atoms

Table 5.3. The diameters of silicon nanocrystals in the shell of suboxides L_{SiO} and without it D_{Si}, as well as the thickness of the suboxide shell (Zimina et al, 2006)

L_{SiO}, nm	D_{Si}, nm	Thickness of suboxide shell, nm
5	4.0	0.5
4	3.2	0.4
3	2.4	0.3
2	1.6	0.3

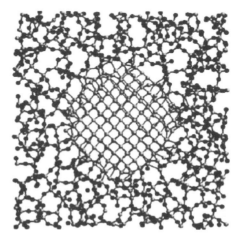

Fig. 5.15. The model of nc-Si (~2.5 nm diameter) embedded in amorphous SiO_2. Gray and black circles denote the Si and O atoms (Djurabekova et al, 2008).

have a defective coordination environment (Djurabekova et al, 2008). Insufficiently coordinated atoms clearly dominate in comparison with the atoms with excess coordination (Figure 5.15).

Samples of nc-Si in SiO_2 were prepared (Godefroo et al, 2008) by the superlattice method: 37 SiO/SiO_2 double layers were deposited on a silicon substrate and then annealed to form nc-Si. The thickness of the SiO layer was 1.5 nm and of the SiO_2 layer 3 nm. TEM was used to determine the average diameter of nc-Si as ~3 nm in the layer plane and ~1.5 nm in the perpendicular direction. The nuclei of nc-Si were clearly crystalline.

The EPR method detected various defects on the Si/SiO_2 interface in the surface layer of nc-Si, including non-radiative recombination centres P_b. After irradiation with ultraviolet light in a hydrogen atmosphere ESR signals from the defects disappeared due to the formation of Si–H bonds. Hydrogenated samples had photoluminescence.

Possible paramagnetic defects in the nc-Si, studied by the EPR method, were discussed in (Jivanescu et al, 2007). A Si (100) substrate was coated with layers of SiO and SiO_2 with a thickness of 2 and 4 nm, respectively, a total of 45 pairs. The total thickness of the sample reached 270 nm. The samples were then annealed at 1100°C in a nitrogen atmosphere, resulting in nc-Si formed with a diameter of about 2 nm in the form of a dispersion in SiO_2 (based on TEM). P_b-paramagnetic centres are groups $(Si_3 \equiv Si^{\cdot})$ at the $(111)Si/SiO_2$ interface and perhaps also at $(110)Si/SiO_2$. P_{b0}- and P_{b1} are paramagnetic centres – they are the same groups as P_b, but at the (100) Si/SiO_2 interface. The D-paramagnetic centre is the same group $(Si_3 \equiv Si^{\cdot})$, but it is not located on the the crystal surface and is located on the surface of amorphous silica particles. The following defects were found in SiO_2: E'_{γ} – $(O_3 \equiv Si^{\cdot})$ groups typical for amorphous SiO_2, and EX – a cluster consisting of four oxygen atoms, each of them has contact with three Si atoms, one electron is delocalized over all oxygen atoms. A large number of P_b and D defects were found.

As mentioned above, the Si = O and Si–O bonds exist at the nc-Si/SiO_2 interface and significantly reduce the band gap, creating local states and impurity levels inside the forbidden zone, which leads to shifts in the PL peaks towards lower energies than expected from the quantum size effect. Kim et al, 2007, obtained samples of two types: 1) 50 pairs of SiO_x/SiO_2 layers with a SiO_x layer thickness of 4 nm, 2) single-layer film of SiO_x, thickness 4 nm. The layers were deposited on the Si(100) substrate at room temperature with an Ar^+ ion beam. The composition of SiO_x was controlled using X-ray photoelectron spectroscopy. The samples were annealed at 1100°C in nitrogen. Some samples were annealed for a further 1 hour in a hydrogen flow at 650°C to block P_b centres. According to FEM, nc-Si had a mean diameter of 3.79 ± 0.95 nm. The density of nc-Si was $3.5 \cdot 10^{12}$ cm^{-2}. The PL peak of the hydrogenated multilayer samples was shifted to the blue region with increasing x in SiO_x, i.e. with decreasing size of nc-Si. At 77 K the spectrum of cathodoluminescence (CL) contained two peaks ~1.66 eV (750 nm) and 2.18 eV (570 nm) at x = 1.0. The first is called low-energy (LE), and the second – high-energy (HE). Annealing in hydrogen leads to a strong increase in the signal/noise ratio that can be explained by blocking of P_b-centres. No CL was observed in multilayer unannealed samples.

Changing the composition of SiO_x from x = 1.0 up to x = 1.8 leads to a shift in the spectra of CL: LE from 2.18 to 2.64 eV, and HE from 1.66 to 2.16 eV, and the shift in the CL is twice that in the PL. These changes were observed in both the hydrogenated samples and in the samples which had not been annealed in hydrogen. The integrated PL intensity has a maximum at the SiO_x composition (x = 1.2) in both the hydrogen-annealed and unannealed samples. However, in the CL spectra the intensity is rising steadily with increasing x from 1.0 to 1.8 for the hydrogenated samples (in the non-hydrogenated samples the intensity of the NE and HE bands varies slightly). The authors suggest that the quantum efficiency of the CL is higher than that of the PL for small nc-Si because their share is less than for large x, and high efficiency results in greater CL intensity in comparison with PL for large x, regardless of the reduction of the number of nc-Si in the samples. The type of CL spectra at 300 K is quite different. A strong band appears at 2.70 eV (460 nm), and its position is almost completely independent of the value of x. This band is attributed by the authors to localized defects at the nc-Si/SiO$_2$ interface.

EL dispersions of nc-Si in the SiO$_2$, obtained by disproportionation of SiO, also have their own peculiarities. The nc-Si/SiO$_2$ dispersion was obtained by evaporation of SiO in a vacuum on the Si(100) substrate, followed by annealing in nitrogen at 1100°C for an hour (Chen et al, 2007). The layer thickness was equal to 80 nm. The maximum voltage applied to the sample was 7 V. The intensity of EL increased with increasing voltage 3 to 4, and then fell to the entry level with a further increase in voltage to 7 V. The EL spectrum contained two components – from the nc-Si and from the oxygen vacancies in the SiO$_2$-matrix.

In (Sato et al, 2006) nc-Si were produced by the dissolution of the SiO$_2$ matrix for the study of nc-Si. For example, the suboxide SiO$_x$, containing nc-Si, was dissolved in a mixture of HF (49 wt.%) and HNO$_3$ (69 wt.%) taken in the bulk ratio 10:1. The reaction proceeded in an ultrasonic bath. The resulting nc-Si/H, whose surface was modified by hydrogen, was filtered and washed three times in methanol. nc-Si/H interacted with acrylic acid which under UV radiation underwent the hydrosilylation reaction. This reaction was reproduced consistently if four volumes of acrylic acid (99.5 wt.%) were combined with one volume of HF (49 wt.%). The hydrofluoric acid was added to remove the silicon oxide which formed due to the presence of water in the reaction mixture. Irradiation with UV light (254 nm) lasted 2–12 hours. The PL spectra did not reach the blue area, even for nc-Si with a diameter of 1.5 nm. The blue PL appeared after the ultrasonic treatment of the dispersion in water for 5 hours. FTIR found at the same time increase in the (Si–O) absorption band, which indicated the oxidation of the surface of the nc-Si.

In another paper (Hessel et al, 2008) a commercial product called Si$_2$O$_3$ was restored by the flow of the gas mixture 5% H$_2$ and 95% Ar for 1 hour. The result was an X-ray amorphous material, which was heated up to 1100°C for disproportionation. TEM and X-ray analysis revealed the nc-Si with a diamond structure with a diameter of 4 nm. This sample the authors called nc-Si/SiO$_2$. SiO$_2$ was removed with a mixture of 10 ml 49% HF:water: ethanol = 1:1:1 and 300 mg of mechanically milled nc-Si/SiO$_2$. The mixture was stirred at a speed of 300 rpm to remove the SiO$_2$. nc-Si/H was extracted with 10 ml of pentane. The dispersion of nc-Si/H in pentane was poured in droplets on an aluminum disc in a nitrogen atmosphere and after drying the pentane it was evacuated to high vacuum (10^{-8} Torr).

The resulting nc-Si was studied by synchrotron radiation. The XEOL spectrum had a peak at 780 nm, which is due to the radiative recombination of holes and electrons in the nc-Si with the particle diameter of 4 nm in accordance with the quantum size effect, and the peak at 540 nm is the radiation of the interface of nc-Si and silicon oxide of indefinite composition. XEOL allows to divide the radiation of different substances. After etching nc-Si/SiO$_2$ with hydrofluoric acid the resulting nc-Si/H in the XEOL spectra had no 540 nm band, which once again confirms that the radiation with a wavelength of 540 nm is generated by the nc-Si/SiO$_x$ interface.

Separate studies investigated the nc-Si, formed in an SiO$_2$ matrix but isolated from it. The H$_8$Si$_8$O$_{12}$ compound was annealed at 1100°C for hours in a stream of 5% H$_2$ + 95% Ar (Kelly et al, 2010). As a result, nc-Si with a diameter of 3.5 nm, as determined by TEM, formed in the SiO$_2$ matrix. Silicon dioxide dissolved in a mixture of HF:ethanol:water = 1:1:1 by volume with stirring. nc-Si/H was extracted with toluene. The dispersion of nc-Si/H in toluene had a peak in the PL spectrum at 650 nm.

The hydrosilylation reaction with styrene CH$_2$=CH–Ph under illumination by UV radiation produced nc-Si with organic ligands on the surface and with a maximum PL at 600 nm. The K-edge NEXAFS nc-Si with 2-phenyl-ethyl ligands had 3 signals at 1841, 1844 and 1848 eV. The 1841 eV signal has been attributed to Si(0) atoms in the nc-Si, and the 1884 eV and 1848 eV signals were assigned to the atoms of silicon with (Si–C) and (Si–O) bonds. FTIR confirmed the presence of (Si–C) and (Si–O) bonds. The XEOL spectra differed for nc-Si in different environments: the nc-Si in the SiO$_2$ matrix had the main peak at 780 nm and a secondary one at 535 nm; the nc-Si precipitated from SiO$_2$ and coated with hydrogen had a peak at 680 nm; the nc-Si with 2-phenyl-ethyl ligands at 590 nm. Thus, the replacement of ligands without decreasing the particle size of nanosilicon led to the blue shift of the S-band, but it was not the F-band.

The same procedure was used in (Hessel et al, 2006). Thermal dissociation of RSiO$_{1.5}$ (R – hydrogen, alkyl, aryl, silyl, etc., Fig. 5.16 – scheme with the formula RSiO$_{1.5}$) in a mixture of nitrogen and 96% N$_2$ and 4% H$_2$ at 500–1000°C for an hour gives the dispersion of nc-Si in SiO$_2$ at a heating rate of 20°C/min.

The dispersion was etched in a mixture of 49% HF:H$_2$O:ethanol = 1: 1: 1 for 1.5 hours. The remainder contained powders whose colour became darker with increasing annealing temperature. nc-Si, rather than amorphous silicon, was observed only after heating at 1100°C. The photoluminescence quantum yield was ~4%. The wavelength of the PL peak depended on the etching time, with other conditions being equal: 50 min – 705 nm (red), 85 min – 612 nm (orange), 115 min – 585 nm (yellow), 135 min – 512 nm (green). FTIR detected Si–H$_x$ bonds and weak absorption lines originating from the Si–O–Si bonds. TEM was used to determine the size distribution. For example, after 115 min etching, the average size of nc-Si was equal to 3.41 nm at 2σ = 1.40 nm.

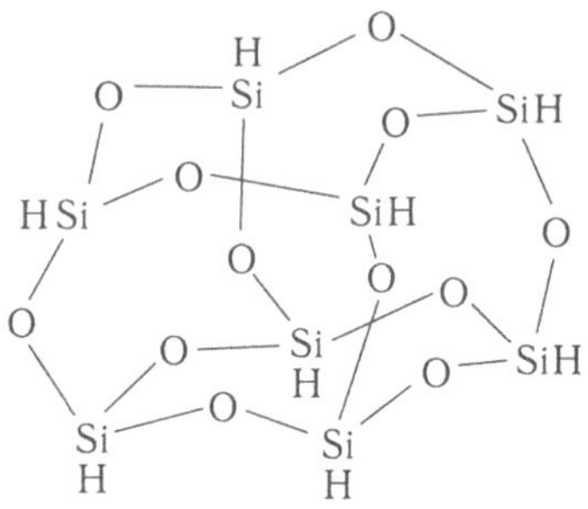

Fig. 5.16. The structural formula of $H_2Si_2O_3$ (Hessel et al, 2006).

Disproportionation of SiO (Nozaki et al, 2008) into Si and SiO_2, and the formation of particles of amorphous silicon or nc-Si is possible not only at high but also low temperatures in a laser beam with a wavelength of 325, 532 and 1064 nm. The optimal exposure time was 5–10 s and the average particle size increased with the reduction the wavelength of laser radiation and with increase of its power.

5.4.2. Effect of temperature in synthesis of nc-Si from monoxide on the optical properties of nanosilicon

Dorofeev et al (2009) developed a technique to obtain nc-Si from SiO powders at temperatures of 25°C to 950°C and subsequent hydrosilylation of the nanoparticle surface for a time not exceeding 10 min. The yield of nanoparticles that are capable of bright photoluminescence in the red spectral region obtained using this technique is ~16% of the theoretical value. The photoluminescence of nanoparticles is stable for a long time after the hydrosilylation of the surface. The method of thermogravimetry and differential scanning calorimetry established the stability of the particles in the air up to 220°C. The data obtained in transmission electron microscopy showed that the nuclei of the produced nanoparticles consist of crystalline silicon with the maximum size of less than 5 nm. The method for analyzing small-angle scattering of X-rays determined the density distribution function from the size of the particles synthesized by heating the silicon monoxide to 950°C. The proposed method can be used for producing large quantities of nc-Si with bright luminescence.

Silicon nanocrystals (nc-Si) were synthesized (Dorofeev et al, 2009) from the powder of silicon monoxide SiO, pre-refined to a particle size of about ~100 μm. The crushed SiO powder was placed in a corundum crucible with a lid and heated from room temperature to 950°C in air for 2–4 hours. The process of disproportionation of silicon monoxide took place:

$$2SiO \rightarrow SiO_2 + Si.$$

Usually, the portion of the heated powder was 2 grams. The products of the reaction were then etched for 20 min in concentrated hydrofluoric acid (HF, 49 wt.%) in order to remove the SiO_2 formed in heating the starting powder. Etching occurred in the conditions of heating up to 50°C and stirring in an ultrasonic bath. At the end of the etching process the resultant silicon nanoparticles were washed with distilled water to stop etching and remove them from the HF solution. They were then washed with methanol to remove water and with hexane to remove methanol. The resulting nc-Si particles were hydrophobic because their surface was passivated by hydrogen atoms. Therefore, in water and methanol, these particles were collected in agglomerates which were deposited by centrifugation at 2000 g for 5 min.

The procedure of the thermally-initiated hydrosilylation of silicon nanoparticles, saturated with the (Si–H) bonds is based on the fact that at sufficiently high temperature (ultraviolet irradiation or under the influence of the catalyst), some Si–H bonds are broken: $Si–H \rightarrow Si^{\cdot}+H^{\cdot}$. Subsequently, the ruptured bonds can interact with alkenes having terminal double bonds (see section 5.5). 1-octadecene was used as organic molecules with a terminal double bond in Dorofeev et al (2009). Silicon nanoparticles, collected after centrifugation, were placed in a quartz tube to which 1-octadecene (4 ml). The tube was heated to the boiling point of 1-octadecene (315°C) and kept at this temperature for 5 min. Boiling of 1-octadecene ensure intensive mixing of initially insoluble particles. After the first minute of hydrosilylation the initially turbid solution becomes transparent and acquires a reddish-brown colour. At the completion of the process the reaction mixture was diluted twice with chloroform to prevent separation of the liquid phase. The diluted reaction mixture was centrifuged to remove unreacted SiO. The solution formed after centrifugation was diluted with methanol. The agglomerates formed during this process were deposited by centrifugation. The deposited nc-Si was washed with methanol, dried and then dispersed in a solvent such as toluene, chloroform or hexane. The obtained nanoparticles formed a stable sol with each of these solvents. The silicon nanoparticles, which were synthesized at a temperature of 25°C to 950 °C, with be denoted as Si25,. . . , Si950, respectively.

Hydrosilylation of the nc-Si surface led to a substantial increase of the photoluminescence intensity. The procedure that was used in the work of Dorofeev et al (2009) differs from a similar procedure described in the papers (Liu et al, 2006 and Li et al, 2004) by higher temperature at which hydrosilylation took place and by a significantly shorter time: 5 min instead of 3 and 48 hours respectively.

The samples of nc-Si pre-dried at 120°C the surface which was covered with 1-octadecene were studied by differential scanning calorimetry (DSC) and thermogravimetric analysis (TGA) at a rate of heating 5°C/min in the air atmosphere from 20°C to 400°C. The mass of the sample, placed in the corundum crucible, was 6 mg. The nc-Si samples were investigated transmission electron microscopy. The density distribution function (DDF) of the particle size of nc-Si was determined by Fourier analysis of the received signal in small-angle X-ray scattering using GIFT computer program (Bergmann et al, 2000).

A small weight loss in the temperature range 170–220°C is connected, apparently, with the loss of the residual solvent (hexane, from sol in which were the nanoparticles were isolated) situated in the shell 1-octadecene. Rapid weight loss started at temperatures above 220°C accompanied by a significant endothermic effect due evidently to the dissociation of the shell. Thus, it can be concluded that the investigated nanoparticles are fully stable in air at temperatures below 220°C.

Analysis of TEM microphotographs (Fig. 5.17) showed that the maximum particle size of synthesized nc-Si in all the samples did not exceed 5 nm. From this analysis it follows that the diffraction rings nearest to the centre correspond to the 111), (220), (311) and (331) atomic planes of Si.

The density distribution function of of the particle size was determined by analyzing the data obtained in small-angle X-ray scattering by colloidal solutions of Si in hexane. The results are shown in Fig. 5.18.

DDF maxima of the particle sizes correspond to the values (nm): 2.15 (Si200), 2.50 (Si300), 2.60 (Si450) and 2.35 (Si950). Figure 5.18 shows peaks most of which are located close to the values that are multiples of the values of the first maximum. It can be assumed that these peaks can be attributed to dimers $[(Si)_n]_2$ and trimers $[(Si)_n]_3$ in sols of nanocrystalline silicon, whose concentration increases with increasing concentration of the starting material.

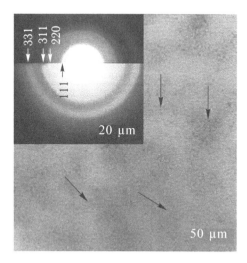

Fig. 5.17. TEM micrograph of nc-Si (for the sample obtained at 300°C). Insert shows a portion of the diffraction pattern obtained with an increased exposure time. Arrows indicate the largest particles with the size in the range of 5 nm (Dorofeev et al, 2009).

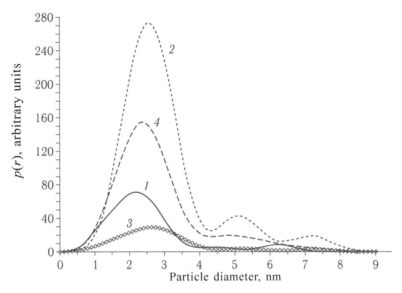

Fig. 5.18. Density distribution function of the particle sizes determined from analysis of small-angle X-ray scattering. Numbers *1–4* denote the samples obtained at 200, 300, 450 and 950°C, respectively (Dorofeev et al, 2009).

Dorofeev et al (2009 and 2010) presented the results of spectral studies of nc-Si, synthesized from silicon monoxide at temperatures ranging from 25°C to 950°C. The effectiveness of the proposed procedure for hydrosilylation is shown. The photoluminescence quantum yield of nc-Si increased with increasing synthesis temperature and reached a maximum of 11.7% at 950°C. The photoluminescence of the obtained sols of nanocrystalline silicon remained stable during a long time – more than seven months after their synthesis.

The absorption spectra of nc-Si sols in hexane were recorded in the range of 200 nm to 1100 nm for the samples prepared at different temperatures of 25°C up to 950°C. The recorded transmission spectra, taking into account the transmission of the cell with pure hexane, were used to calculate the corresponding absorption coefficients of nc-Si. The mass concentration of nc-Si in all the studied solutions was similar and amounted to $C = 75$ mg/ml. To calculate the effective absorption length of nc-Si in the solutions, the authors used the fact that the maximum size nc-Si does not exceed 5 nm. Consequently, at the mass concentration of 75 µg/ml the concentration of the nanoparticles in the solution was not lower than $5.5 \cdot 10^{14}$ cm^{-3}. Accordingly, at this concentration the average distance between neighbouring nanoparticles was not greater than 120 nm. Therefore, in the calculation of the absorption coefficient in the wavelength range from 200 nm to 1100 nm the nc-Si sol with discretely distributed nanoparticles located in it can be considered as a continuous medium with an effective thickness, determined by the formula:

$$d = \frac{C \cdot L}{\rho},$$

where ρ is the density of single-crystal silicon and L is thickness of the cell in the direction parallel to the direction of probing radiation which in the experiments in (Dorofeev et al, 2009; 2010) was 1 cm.

The absorption coefficients, calculated for the samples Si25, Si300, Si600 and Si950, are shown in Fig. 5.19. The same figure shows for comparison the absorption spectrum of monocrystalline silicon (c-Si). It is seen that near the fundamental edge of silicon absorption for all four samples obtained from the SiO, there is blue shift of absorption with respect to c-Si, which is most pronounced for the Si25 nanoparticles. However, if in the case of the nanoparticles Si25 and Si300 in the incident photon energy range 1 eV the absorption behaviour is similar to the absorption spectrum of monocrystalline silicon, then the samples Si600 and Si950 showed increase of absorption.

The increase in absorption in microcrystalline and amorphous silicon is usually associated with the presence of structural defects in it (Paruba et al, 2000; Klein et al, 2007). Therefore, it can be argued that the concentration of defects in silicon nanoparticles, synthesized at temperatures greater than

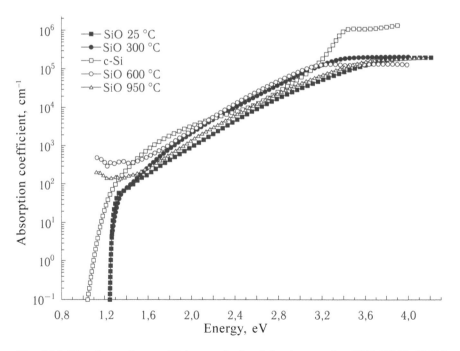

Fig. 5.19. The absorption coefficients calculated for the samples Si25, Si300, Si600 and Si950. For comparison, the absorption spectrum of monocrystalline silicon (c-Si) is shown. From an article by Dorofeev et al (2010).

600°C, is significantly higher than in the nanoparticles formed at lower temperatures.

In Fig. 5.20 the absorption spectra are shown in the coordinate axes $\sqrt{\alpha(hv) \cdot hv}$ and hv, where $\alpha(hv)$ is the absorption coefficient of the sample for photons with energy hv. It is seen that at the energy of incident photons in the range $2.6 \leq hv$ 3.4 eV for the Si300 and Si600 samples and $3.1 \leq hv \leq 3.6$ eV for the samples Si25 and Si950 the graphs are well approximated by a linear dependence:

$$\sqrt{\alpha \cdot hv} = A\left(hv - E_g\right). \qquad (5.1)$$

Graphically, the value of E_g is defined as the point of intersection of the linear approximation of the form (5.1) with the axis hv. The value of E_g for the samples Si25, Si300, Si600 and Si950 E_g is 2.58, 2.48, 2.20 and 2.50 eV, respectively.

The dependence of the type (5.1) is characteristic of indirect transitions near the fundamental absorption edge of crystalline and amorphous silicon (Sze, 1981; Tauc et al, 1966). Accordingly, the value of E_g determines the width of the fundamental gap $\Gamma'_{25}-X_1$. In this case, such a dependence is justified in the energy range $2.6 \leq hv \leq 3.6$, i.e. away from the fundamental absorption edge of monocrystalline

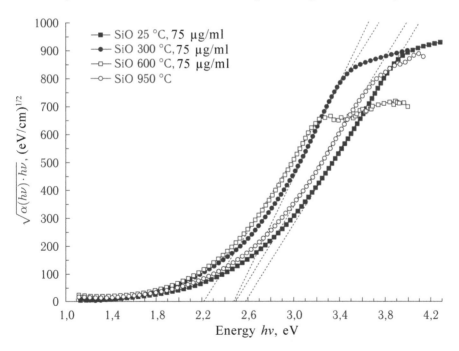

Fig. 5.20. The absorption spectra of nc-Si, synthesized from SiO at temperatures of 25°C up to 950°C (samples Si25, Si300, Si600 and Si950). Dotted lines correspond to linear approximation of the type (5.1). From an article by Dorofeev et al (2010).

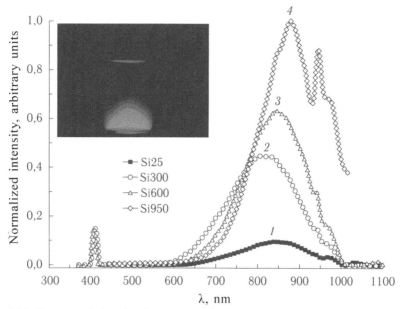

Fig. 5.21. Spectra of photoluminescene excited in Si nanoparticles in the samples Si25, Si300, Si600 and Si950 by radiation with a wavelength of 405 nm. The insert shows the photograph of the glow of nanoparticles in the Si300 sample (Dorofeev et al., 2010).

silicon. However, in the works by Dorofeev et al (2009) and Ben-Chorin et al. (1996) it is shown that the relation (5.1) can be observed near the indirect transition of silicon $\Gamma'_{25}-L_1$. In accordance with this concluded that the measured values of E_g 2.58; 2.48; 2.20 and 2.50 eV correspond to the band gap of nc-Si for indirect transition $\Gamma'_{25}-L_1$ For c-Si at room temperature, $E_g(\Gamma'_{25}-L_1) \equiv E_g(\text{c-Si}) = 1.8$ eV (Forman et al, 1974; Delerue et al, 1993). Consequently, for the nanoparticles in the samples Si25, Si300, Si600 and Si950, $E_g(\Gamma'_{25}-L_1)$ is greater than $E_g(\text{c-Si})$ at 0.78, 0.68, 0.40 and 0.70 eV, respectively (Dorofeev et al, 2010).

It should be noted here that the measured absorption spectra for each sample are associated with an ensemble of silicon particles. Therefore, the calculated values of the energies of band gaps are averages over the ensemble of the corresponding sample.

Photoluminescence spectra excited in the silicon nanoparticles in the samples Si25, Si300, Si600 and Si950 radiation with a wavelength of 405 nm are shown in Fig. 5.21. The inset in the same figure shows a photograph of the glow of nanoparticles in the sample Si300.

All registered photoluminescence spectra were corrected taking into account the spectral apparatus function of the device. To determine the centre of gravity of the photoluminescence peak of each signal is

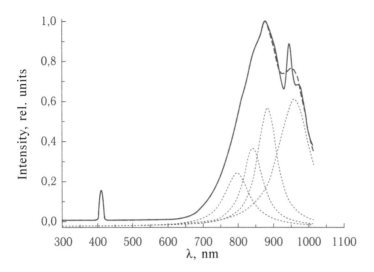

Fig. 5.22. Example of approximation of the spectral data of the sample nc-Si (Si950) by Lorentz functions. The thin dashed line – Lorentzians used in the approximation; bold dashed line – the resulting approximating curve (Dorofeev et al, 2010).

approximated by four Lorentzian contours. A typical example of such an approximation for the sample Si950 is shown in Fig. 5.22.

The centre of gravity of each experimental peak of luminescence on the wavelength axis was calculated as a weighted average of the position of the Lorentz contours with a weight corresponding to the integrated intensity in each circuit.

The values of the position of the centre of gravity of the photoluminescence peak E_{PL}, depending on the temperature at which the nanoparticles are synthesized in the samples of Si25, Si300, Si600 and Si950, are shown in Table. 5.4. This table also shows the band gap $E_g(\Gamma'_{25} - L_1)$ of the silicon nanoparticles, calculated from the absorption spectra, as well as the Stokes shift, which is defined as the energy difference the band gap E_g and the photon energy at the the maximum photoluminescence signal E_{PL}. The same table shows the average diameters of ensembles of silicon nanoparticles, calculated in the framework of the the quantum constrain method (Ledoux et al, 2000; Wolkin et al, 1999), in accordance with the experimental values of E_g (Dorofeev et al, 2010).

Table 5.4 shows that the values E_g decrease monotonically for the samples Si25, Si300 and Si600, while for the sample Si950 there is a noticeable increase in the band gap. In accordance with the values of E_g the average diameters in ensembles of silicon nanoparticles initially increase to 2.5 nm and then in the sample Si950 drop to 2.1 nm. Simultaneously, the value of E_{FL} decreases monotonically with increasing synthesis temperature, which corresponds to displacement of the centre of gravity

Table 5.4. Optical characteristics and average particle size of nc-Si as a function of synthesis temperature (Dorofeev et al, 2010)

Synthesis temperature, °C	25	300	600	950
The band gap E_g, eV	2.58	2.48	2.22	2.50
The average particle diameter of nc-Si, nm	2.0	2.1	2,5.	2.1
The position of maximum intensity of PL, eV	1.57	1.52	1.48	1.38
Stokes shift, eV	1.01	0.96	0.72	1.12

of the photoluminescence signal to the red, regardless of the diameters of the nanoparticles. Accordingly, the Stokes shift first decreases, reaching a minimum of 0.72 eV for the sample Si600, and then increases to 1.12 eV in the sample Si950. Such a large Stokes shift indicates that the photorecombination in the samples does not take place through the states of free excitons and it takes place through the impurity energy levels in the forbidden zone.

The problem of the large red shift, which occurs at PL in the visible spectrum in silicon nanoparticles has been studied by Puzder et al. (2002 and 2003), where it was shown that the most likely source of localized levels in the forbidden zone are the nanoparticles with a diameter smaller than 3 nm. Oxygen atoms are located at the particle surface and connected to the silicon atom by a double bond (Si = O). In papers Puzder et al. (2002 and 2003) also shows that the red shift of the luminescence can reach values of ~ 1 eV. As noted above, we study the surface of nanoparticles passivated silicon molecules of 1-octadecene. As a result of broken bonds surface of the silicon atoms to be saturated with carbon atoms, forming a connection (Si–C). If we assume that the observed Stokes shift is determined by the presence of oxygen atoms on the surface of nc-Si, then passivation is not complete, and a number of oxygen atoms are located on the surface of nc-Si.

The E_{PL} values for the samples Si600 and Si950 equal 1.48 and 1.38 eV, respectively (Table 5.4). However, the absorption spectra shown in Fig. 5.19 indicate that in the energy range of incident photons $1.0 \leq hv \leq 1.8$ eV, the absorption behaviour in these samples is significantly different from the absorption in the Si25 and Si300 samples. This difference is associated with a higher concentration of defects in the samples Si600 and Si950, which also determine the fundamental absorption in this range. The authors suggest that the main source of these defective states in the forbidden zone of nc-Si are the dangling bonds on the surface of nanoparticles, i.e. covalent bonds of silicon atoms not saturated with hydrogen atoms or oxygen or (Si–O) bonds.

Thus, in the Si600 and Si950 samples the recombination of the photoexcited 'electron – hole' exciton pair takes place not just through radiation dispersal of the levels determined by the (Si=O) double bonds,

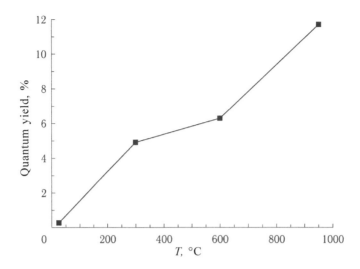

Fig. 5.23. The dependence of quantum yield on synthesis temperature of silicon nanoparticles (Dorofeev et al, 2010).

but also by collisional recombination via defect levels associated with the dangling bonds of the surface silicon atoms. This situation, in principle, should reduce the efficiency of photoluminescence in the resultant nc-Si samples.

The quantum yield of photoluminescence of the sols of silicon nanoparticles in hexane was obtained from the comparison of the photoluminescence signal of nc-Si and rhodamine 6G, dissolved in ethanol. Figure 5.23 shows the dependence of the quantum yield on the synthesis temperature of the silicon nanoparticles. The figure shows that with increasing synthesis temperature from 25°C to 950°C the quantum yield increases from 0.3% to 11.7%.

The quantum yield for the Si300 sample was measured for two wavelengths of photoexcitation: violet, λ_1 = 405 nm (photon energy 3.06 eV) and green, λ_2 = 532 nm (photon energy 2.33 eV). The power of the radiation sources in this case amounted to P_1 = 30 mW and P_2 = 100 mW, respectively. The photoluminescence spectra, excited by these sources, are shown in Fig. 5.24.

The figure shows that the shape of the photoluminescence signal is virtually independent of the wavelength and power of the exciting radiation. The ratio of the photoluminescence quantum yield, excited at wavelengths λ_1 and λ_2, is defined by the following equation:

$$\frac{\eta_2}{\eta_1} = \frac{I_{Ph2}I_1\lambda_1\left[1-\exp\left(-D_1\right)\right]}{I_{Ph1}I_2\lambda_2\left[1-\exp\left(-D_2\right)\right]}. \tag{5.2}$$

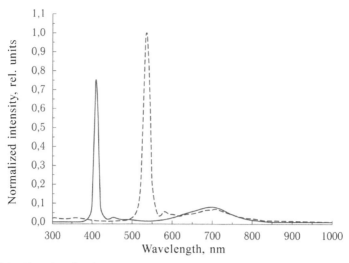

Fig. 5.24. The photoluminescence spectra of the sample Si300, excited by violet (λ_1 = 405 nm, photon energy 3.06 eV) and green (λ_2 = 532 nm, photon energy 2.33 eV) light (Dorofeev et al, 2010).

Here I_{Ph1} and I_{Ph2} are the integral with respect to spectrum energies of photoluminescence (areas under the corresponding peaks) that were converted to the number of photons in excitation by the radiation with wavelengths λ_1 and λ_2; I_1 and I_2 are the intensities of the incident pumping radiation; D_1 and D_2 are the optical densities of the sample at wavelengths λ_1 and λ_2. Using the expression (5.2) and the corresponding values of D_1 and D_2, determined from the absorption spectrum of the Si300 sample, gives the ratio of quantum yields equal to 0.92. Thus, with the luminescence excited by photons with an energy of 3.06 eV and 2.33 eV, the quantum yields are approximately equal. Since the photon energy of 2.33 eV practically coincides with the band gap of the ensemble of the Si300 particles (see Table 5.4.), determined from optical absorption, the proximity of the values of the quantum yields at wavelengths λ_1 and λ_2 leads to (Dorofeev et al, 2010): 1) the loss of the energy of the photoexcited nanoparticle in the stage of relaxation from the highest state excited by the photon energy of 3.06 eV to the state defined by the (Si=O) double bond is very small, and 2) the major loss of energy, resulting in a decrease in the photoluminescence quantum yield is related with the collisional recombination via defect levels lying below the level of (Si=O).

5.4.3. Recovery of silica with magnesium

In the study (Kravitz et al, 2010) a fine dispersion of SiO_2 was restored with a magnesium powder at recovered at elevated temperatures:

$$Mg + SiO_2 = MgO + Si.$$

For this purpose, gelatin was dissolved in an aqueous dispersion of amorphous silica (average particle diameter of 11.5 nm, concentration 40 wt.%) to 0.5 wt.% of the solution to which was the magnesium powder was added. The mixture was immediately frozen in liquid nitrogen and subjected to freeze-drying overnight. The molar ratio of SiO_2:Mg = 1:20, the gelatin served as a homogenizing mixture formed after freeze-drying. The reaction mixture was heated in a ceramic crucible for 5 h at various temperatures in the range 520–750°C in a nitrogen atmosphere. All by-products (MgO, $Mg(OH)_2$, Mg_2N_3) were removed with concentrated hydrochloric acid. The residue was washed with water three times and left to dry at room temperature overnight. The reaction yield was 18%. The product obtained at 750°C contained 11 ± 6% crystalline silica (the rest – the amorphous phase). HRTEM clearly registered the nanoparticles of crystalline silicon with a diameter of 4–10 nm, but according to the broadening of the peaks of the X-ray diffraction powder pattern the average particle size of nanosilicon was ~40 nm, reflecting a wide range of the size distribution of silicon particles. No crystals formed at a reaction temperature of 520°C. Their presence was observed only from 600°C and above. The reaction time from 1 to 5 hours had no effect on the size of the nanocrystals and the amount of the crystalline phase. HRSEM found a significant amount of agglomerates. The composition of the sample was determined by EDS analysis: 75% Si and 20% O, i.e. the sample contained silicon oxides. The resulting nanosilicon had a bright photoluminescence with a maximum at 710 nm and the branch at 820 nm (excitation wavelength 320 nm).

5.4.4. Recovery of silicon dioxide magnesium silicide

Ischenko et al (2010) proposed a new method for the synthesis of nanocrystalline silicon, which has a bright stable photoluminescence in the blue–green region of the spectrum.

Silicon nanocrystals were synthesized from a powder of magnesium silicide, Mg_2Si, pre-refined to a particle size of ~100 µm, and Aerosil AC-200. The mixture was placed in a corundum crucible with a lid and heated to 950°C in air for 2–4 hours. The following reaction took place:

$$Mg_2Si + SiO_2 \rightarrow 2Si + 2MgO.$$

Usually, the portion of the heated powder is 2 grams. The products of the reaction were then etched for 20 min in concentrated hydrofluoric acid (HF, 49 wt.%) in order to remove the remainder of SiO_2 and the resulting magnesium oxide. Etching took place in the conditions of heating up to 50°C and stirring in an ultrasonic bath (PULSE 270 M-002 935) at an ultrasonic power of 150 W at a frequency of 37 kHz. At the end of the etching process the resultant silicon nanoparticles were washed with distilled water to stop the etching and removal from the HF solution. Then they were washed with methanol to remove water and hexane to remove methanol. The resulting nc-Si particles were hydrophobic, because their surface was passivated by hydrogen atoms. Therefore, in water and methanol, these particles were collected in agglomerates that precipitated under centrifugation at 2000g for 5 min. 1-octadecene was used as an organic molecule with a terminal double bond. The silicon nanoparticles, collected after centrifugation, were placed in a quartz tube to which 1-octadecene was added (4 ml). The tube was heated to the boiling point of 1-octadecene (315°C) and kept at this temperature for 5 min. Boiling of 1-octadecene provided intensive mixing of the initially insoluble particles. After the first minutes of hydrosilylation the initially cloudy solution became clear and acquired a reddish–brown colour. At the completion of the reaction the mass was diluted twice with chloroform to prevent the separation of the liquid phase. The diluted reaction mixture was centrifuged to remove unreacted SiO_2. After centrifugation the resulting solution was diluted with methanol. The agglomerates were deposited by centrifugation. The deposite nc-Si was washed with methanol, dried and then dispersed in one of solvents such as toluene, chloroform or hexane. The produced nanoparticles form a stable sol with each of these solvents. Hydrosilylation of the nc-Si surface leads to a significant increase in the intensity of photoluminescence.

Figure 5.25 shows the results of studies of photoluminescence spectra of nc-Si, synthesized from magnesium silicide and silicon dioxide at temperatures from 400°C to 950°C.

The effectiveness of the proposed hydrosilylation procedure was confirmed. The photoluminescence of the sols of nanocrystalline silicon

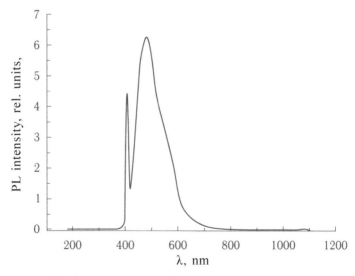

Fig. 5.25. Photoluminescence spectra of nanocrystalline silicon produced in (Ischenko et al, 2010). The exciting line is ~405 nm.

remained stable for a long time – more than seven months after their synthesis.

5.4.5. Synthesis of nanosilicon in solution

nc-Si is usually synthesized from silicon halides or Zintl salts containing silicon. Synthesis of nc-Si from the Zintl salts is very effective for producing dopes samples of nc-Si (Zhang et al, 2007a). For example, the nc-Si, doped with manganese, can be produced. This sample has magnetic properties inherent in manganese, and the PL characteristic of nc-Si:

$$NaSi_{1-x}Mn_x + NH_4Br \rightarrow Si_{1-x}Mn - H + NH_3 + NaBr.$$

It is proved that the 5% Mn is included in the composition of nc-Si.

Nanosilicon with particles with a diameter of 30–500 nm was produced (Bux et al, 2010) by the self-sustaining exothermic reaction of magnesium or calcium silicide with SiI_4, initiated by a drop of alcohol. The mixture ignites spontaneously in air due to the release and self-ignition of silane.

Silicon halides are recovered in most cases by lithium aluminum hydride due to the fact that the obtained nc-Si, the surface of which is covered with hydrogen atoms, is easily processed by hydrosilylation to modify the surface of nc-Si with alkyl ligands – most strongly bonded with silicon in comparison with other ligands. However, the attempt to reproduce the synthesis of nc-Si, as described in (Wilcoxon, 1999a,b), led to a surge of silane (Veinot et al, 2005), which is synthesized by interaction of $SiCl_4$ and $LiAlH_4$, and the authors (Huber et al, 2008) stress the risk of fire when using this method of synthesis. Nevertheless, the resulting nc-Si, passivated by hydrogen, had dimensions 1.8–10 nm and, according to the authors, their PL spectra suggest that the PL originates from the surface layer of nc-Si, and not from the silicon oxide layer that covered the nc-Si in contact with air.

Restoration of silicon halides with the lithium aluminum hydride is usually carried out in micelles, which inhibit the growth of nc-Si, reducing the size of nc-Si and the variations in size. Tetraoctylammonium bromide (TOAB) is often used as the surfactant. For example, in (Tilley et al, 2006, 2005) a solution of 1 M $LiAlH_4$ in THF in a two-fold excess was added in droplets under a nitrogen atmosphere to a mixture of $SiCl_4$, toluene and TOAB.. The mixture was stirred for 3 hours, after which dry methanol was added to quench excess reductant. The solution became transparent. A catalyst – H_2PtCl_6 – in isopropanol and either 1-heptene or allylamine were added to this mixture. The mixture is stirred for 3 hours. The samples were then removed from the dry chamber and the nc-Si sol, modified with 1-heptane, was mixed with hexane, which was then washed with DMF three times. nc-Si was transferred to the hexane fraction. From a mixture of nc-Si with allylamine, the solvent was removed on a rotary evaporator, leaving a white powder, which contained mostly TOAB. This was followed by adding water which dispersed nc-Si, but not TOAB, so the latter was filtered off. The average size of the resulting nc-Si, passivated 1-heptane, was found to be 1.8 ± 0.2 nm. The peak in the PL spectrum at 335 nm was explained by the authors only by the action of the size effect.

Similar conditions of synthesis of nc-Si with passivation with heptyl ligands (Warner et al, 2006a) gave the same average size of nc-Si 1.8 ± 0.2 nm, i.e. the reproducibility was very good. It is stressed (Warner et al, 2006b, 2005a) that the addition of drops of the LiAlH$_4$ solution over 5 min leads to monodispersed nc-Si/H (1–2 nm), but if the entire solution is added at once, this gives polydisperse nc-Si/H (1–7 nm with a maximum particle size distribution of 3 nm), as determined by TEM. HRTEM clearly showed the crystal structure of nc-Si/H. The difference between the two options is reduced to the concentration of Si(0): in the monodisperse state it is always low, and in the polydisperse state – maximum at the beginning of recovery.

The PL spectra of nc-Si with the surface modified with allylamine showed peaks of 460 and 470 nm when excited with ultraviolet light with wavelengths of 300 and 400 nm, respectively. The authors believe that the passivation of nc-Si with alkyl ligands leads to the dipole-allowed direct transitions and, consequently, to the blue PL at high recombination speeds (10^{-8}–10^{-9} s) for the nc-Si with the size of 1–2 nm. The measured PL decay time was found to be 4 ns.

The same conclusion was reached by another group of authors (Rosso-Vasic et al, 2008, 2009, 2010). TOAB was dispersed in dry deareated toluene by sonification. SiCl$_4$ was added to the mixture and the mixture was again sonified for homogenizing. Only after was the LiAlH$_4$ solution added to THF, followed by sonification of the dispersion. The excess hydride was removed with methanol. 1-dodecene and the solution of H$_2$PtCl$_4$ in methanol were added to the reaction environment without dividing it into liquid and solid fractions. The mixture was sonified and the solvent was distilled off. nc-Si/C$_{12}$H$_{25}$ was extracted with hexane. An excess of 1-dodecene was removed by dialysis. The same procedure was used for the synthesis of 1-decene, 1-tetradecene and 1-hexadecene.

According to TEM, the average particle size was equal to 1.57 ± 0.21 nm, but the evidence of crystallinity of nc-Si/C$_n$H$_{2n+1}$ was not presented in the study. FTIR discovered nc-Si/C$_n$H$_{2n+1}$, where n = 10, a number of (Si–O) bonds, which are absent in nanoparticles with n = 12, 14, 16. X-ray electron spectroscopy confirmed these data. It was also shown that the extent of coating the surface of the nanoparticles with alkyls is ~50%.

The PL spectra with ligands of different lengths of the hydrocarbon chains have peaks in the range 308–322 nm when excited with ultraviolet light with a wavelength of 280 nm, and the wavelength of the PL does not depend on the length of the chain of the alkyl ligand. The quantum yield of PL decreases from 23% to 13% in the sequence of ligands hexadecyl → decyl which indicates of the interaction of silicon nanoparticles surrounded by short ligands with the solvent, which reduces the photoluminescence quantum yield. The lifetimes of two excited states of silicon nanoparticles were measured: 4.0 ± 0.1 ns and 0.75 ± 0.15 ns. These values, according to the authors, suggest direct-band transitions in the resulting particles of nc-Si.

The same procedure was used for the synthesis of nanosilicon with allylamines, hexyl-5en-1-amine, and undecyl-10-en-1-amine instead of 1-decene and related compounds discussed above. TEM was used to determine the particle diameter of nanosilicon $-$ 1.57 ± 0.24 nm with any of the three amine ligands. However, no confirmation of crystallinity was presented. The issue of the possible amorphous form of nanosilicon remained open.

The maxima of the curves of the optical absorption of silicon nanoparticles with alkylamine ligands are shifted toward the red size compared with the spectra of nc-Si, protected by the alkyl ligands. Just as in the particles with the alkyl ligands, the oxygen content on the surface of nanosilicon decreases with increase of the length of the ligand and, therefore, the red shift of the curves of the optical absorption decreases with increasing length of the ligand. The band gap was estimated at the individual silicon nanoparticles by scanning tunnelling spectroscopy. For 60% of the measured particles its value was in the range 4–5 eV, and for 40% – in the range 3–3.75 eV. These values are consistent with those obtained from the spectra of optical absorption. The PL spectra contained peaks in the range 350–600 nm in excitation by light with wavelengths of 280–460 nm. The photoluminescence quantum yield was independent the length of the ligand and equal to ~12% at pH = 7.5. The lifetimes of the photoexcited state of silicon nanoparticles were shorter than less than 5 ns, i.e. the produced nanoparticles have the properties of the direct-gap conductor. PL intensity at pH = 7.5 was almost constant for 150 hours, but at pH = 0.6, it increases at the same time 6%, and at pH = 13.3 decreased by 38%. Protonation of amine groups in the acidic medium led to the appearance of charges on the ligands, and this contributed to greater dispersion of silicon nanoparticles, accompanied by the growth of the PL intensity, and in an alkaline aqueous medium increased the intensity of hydrolysis of Si–H bonds, which accelerates the disproportionation reaction of silicon in water, whereby silicon transformed to silicon dioxide and the PL intensity decreased.

If polyethylene glycol (PEG) is placed on the surface of nc-Si as ligands, this nc-Si yields stable dispersions in water and in organic solvents. In (Sudeep et al, 2008), 1.5 g of tetra-n-octylammonium bromide was added in a dry chamber to 100 ml of toluene and the mixture was stirred for 5 min. 100 ml $SiCl_4$ was added to the mixture which was then stirred for 1 hour. 2 ml of monomolar solution of $LiAlH_4$ in THF was slowly added to the mixture and the mixture was stirred for 3 hours. This was followed by adding 20 ml of methanol, 5 mg of H_2PtCl_6 as a catalyst and 500 mg of PEG-alkene and the mixture was stirred for 3 hours. The resulting nanoparticles were extracted with water during sonification and the unreacted PEG-alkene was removed by dialysis.

Another synthetic process proceeded at the same time. nc-Si was placed in a quartz tube, containing 50 mg of PEG-alkene. The mixture was irradiated with ultraviolet light for 2 hours. The solvent was removed by evaporation and the residue was sonified with 10 ml of water. Tetra-n-octylammonium bromide was filtered off and nc-Si formed. TEM showed that the nc-Si had a diameter of ~2 nm and a narrow size distribution. The peak of the PL spectrum was in the range 350 nm. The relative sharpness of the peak indicates, according to the authors, the absence of oxygen on the surface of nc-Si.

The process of interaction with oxygen on the surface of nc-Si, passivated with alkyls, was studied in (Shiohara et al, 2010). All operations were carried out in a dry chamber filled with nitrogen. $SiBr_4$ either dissolved in anhydrous toluene with the addition of of tetraoctyammonium bromide or in anhydrous hexane with an admixture of $C_{12}E_5$ (monododecyl ether of pentaethyleneglycol). Restoration was carried out with lithium aluminum hydride resulting in the nc-Si surface covered with hydrogen atoms. Hydrosilylation of nc-Si was carried out in isopropanol in the presence of a catalyst – H_2PtCl_6 – and either allylamine or 1,5-hexadiene. nc-Si with the alkene ligands were separated into fractions. First, the solvent was evaporated and then hexane was added, and the dispersion was subjected to sonification for homogenizing. The mixture was washed three times in a separatory funnel with N-methylformamide. The organic layer was washed with water three times. The organic layer was concentrated, giving the nc-Si with the alkene ligands on the surface. Amine and diol ligands were used in addition to alkene ligands. According to TEM, the average sizes of nc-Si was 3.7 ± 0.9 nm (with amine ligands), 3.6 nm (with alkene ligands) and 3.4 nm (with diol ligands).

FTIR contained absorption bands in the range 1080–1125 cm^{-1}, which indicated the presence of (Si–O) bonds which could form during the washing of the organic fraction with water. However, the absorption bands of the (Si–O) bonds were registered most clearly in the nc-Si with amine ligands (processing with water was not mentioned). This example is a reminder of the complexities that accompany the synthesis of nc-Si.

Zintl salts were used as precursors for the synthesis of nc-Si in the work of Zhang et al. (2007c). For example, 100 ml of N, N-dimethylformamide (DMF) were added to 0.1 g (2 mmol) Na_4Si a three-neck flask and heated with a reverse refrigerator until a black suspension formed. 0.2 g (2 mmol) of was then added NH_4Br and the mixture was stirred for 48 hours, forming a brown dispersion. The mixture was cooled, the black precipitate was separated, and 40 ml 0.072 M of the H_2PtCl_6 solution and 2 ml of octene were added to the dispersion. The solution was stirred for 12 hours and the solvent was removed by vacuum evaporation. The product was dissolved in chloroform, and the solution was purified by extraction with a mixture of water and hexane to remove impurities – NaBr, H_2PtCl_6 and excess NH_4Br. The hexane fraction was centrifuged and turned into a transparent, yellow dispersion. Hexane can be removed from the dispersion, receiving a light yellow solid, dispersible in hexane, ethanol, or chloroform.

According to TEM, the diameter of nc-Si is equal to an average of 3.9 ± 1.3 nm. The PL spectra in hexane and water (in the presence of the surfactant) have a band on the same wavelength (~435 nm) with the same half-widths of the peaks, but the quantum yield in water was half that in hexane. In the same series of experiments (Zhang et al, 2007b), the hydrosilylation reaction was carried out not only with octene, but also with propargylamine producing hydrophilic nc-Si. The photoluminescence spectra, obtained in the aqueous dispersion, had maxima in the region 460–470 nm.

The particles of nc-Si with photoluminescence in the green spectral region were prepared so that the size of nc-Si was larger than that of the particles whose synthesis was described above. For this, the solvent, in which Na_4Si and NH_4Br interacted, changed to a high-boiling dioctyl ether, and the mixture was heated at 180°C for 2 days. The precipitate was separated, etched with a 10% aqueous solution of hydrofluoric acid for 10 min and the resulting dispersion of nc-Si was filtered off. This was followed by hydrosilylation.

Zintl salts can react with $SiCl_4$, giving nc-Si. For example, in (Liu et al, 2002) 200 mg of Mg_2Si and 1.2 ml of $SiCl_4$ were added to 100 ml of glyme. the mixture was boiled in a reflux cooler for 2 days under argon, cooled and the solvent with an excess of $SiCl_4$ was removed under vacuum. Dry glyme, 2.6 ml of $LiAlH_4$ solution in THF at 0°C were added to the residue. The mixture was stirred in an ice bath at 0°C and allowed to heat to room temperature. 7.4 ml of 1.4 M solution CH_3Li in diethyl ether was then added. The mixture was stirred for 1 day and the solvent was then removed. To remove the salts, the residue was washed with a 1 M HCl solution in water. nc-Si was extracted with hexane. The hexane solution was washed 10 times with water. The hexane layer was transparent and colourless. Hexane was removed by evaporation to give an oily nc-Si residue. FTIR showed that part of the registered Si–H bonds were hydrolyzed to (Si–O–Si). The luminescence intensity decreased when excited by light with wavelengths of 280, 300, 320, 340 nm, with the maximum of the PL band at the same timne shifted toward the red: 330, 350, 365, 375 nm, respectively. This suggests that the radiation comes from the defects of SiO_2.

Very important results for understanding the relationship of between the composition, structure and properties of the nc-Si were obtained in a series of studies by Tanaka and co-workers (Tanaka et al, 2006, 2007, 2008b), as well as (Saito et al, 2006; Kamikake et al, 2007). Mg_2Si and Br_2 were mixed with n-octane, stirred for 2 hours, the mixture was heated with a backflow condenser for 60 hours, then the solvent and excess Br_2 were removed in vacuum. Butyllithium was added to the residue and the mixture was stirred at room temperature for 2 hours. The solvent was then distilled off on a rotary evaporator. An aqueous solution of HCl was added to dissolve the salts. Hexane was then added; after extraction of nc-Si with hexane the mixture was divided in a separating funnel and the organic fraction washed three times with water. Ultrafiltration produced nc-Si of approximately the same size – 2 nm or less, as determined by TEM. The average particle size was 1.4 nm, and HRTEM indicated the diamond-like structure of such small nc-Si. To study by synchrotron photoelectron spectroscopy, nc-Si was deposited on a graphite substrate in a dry chamber filled with nitrogen and connected with the device. Therefore, the object was placed in the device without contact with atmospheric air. FTIR showed the presence of butyl, but there was no trace of (Si–O) bonds. The maximum range of PL lies in 3.3 eV with no bands below 2 eV. Due to the fact that the maximum of the band in the PL spectra is shifted to the blue region of the spectrum with the shift of the exciting photon energy in the same direction, the authors conclude that the source of PL is the recombination of electron–hole pairs, which is affected by the quantum size effect rather than by the defects or surface oxide groups and butyl groups. To confirm this conclusion, the electronic structure of the nc-Si near the Fermi level was studied by the photoemission excited by synchrotron radiation. The band gap of nc-Si with a diameter of 2 nm, passivated by butyl groups, was found to be 4 eV, which agrees well with the resonance energy of the PL spectra and, therefore, confirms the conclusion about the recombination of electron-hole pairs as a source of radiation. After 180 days the same sample was studied by FTIR. The result shows that on the surface of nc-Si there are (Si–O) bonds, i.e. the samples in air interacted with oxygen and/or moisture in the air. Theoretical calculations showed that the difference of the HOMO–LUMO energies for the nc-Si with only butyl groups on the surface is 4 eV, with one additional oxygen atom (Si=O) – 2.7 eV but with many (Si=O) bonds on the surface – 2.3 eV, i.e. the red shift should appear in the photoluminescence spectrum. Experiments show (Tanaka et al, 2008a) that the partial oxidation of alkylated nc-Si reduces the band gap from 4.4 eV (for samples without oxygen) to 2.8 eV (for samples containing silicon–oxygen bonds on the surface due to oxygen along with Si–C bonds).

Sodium naphtalide in glime (Zou et al, 2006) was added during stirring to a solution of $SiCl_4$ in 1.2-dimethoxyethane (glime) at room temperature in an inert atmosphere. A brown suspension was formed to which methanol was then added. The suspension became yellow. A yellow oily product remained after removing the solvent and the naphthalide. Water was added to this dispersion and the nc-Si, covered with hydroxyl groups, formed Finally, octyltrichlorosilane was added, which gave the final coating of nc-Si. A wax-like substance was produced as a result. For the monodisperse particles the number of Si atoms in the volume (V) and on the surface (S) was estimated and the values of $m = V/S$ were calculated (Tables 5.5 and 5.6).

It is assumed that on the surface each atom has only one (Si–Cl) bond. Consequently, the following relation is satisfied

Table 5.5. Estimates of the relations between the diameters of nanoparticles: [a]total number of Si atoms; [b]the number of surface Si atoms; [c]m, the ratio of the total number of atoms to the number of atoms on the surface of nc-Si; [d]X, the calculated concentrations of the reactants (Zou et al, 2006)

Diameter nc-Si (nm)	The total number of Si atoms (V)[a]	The number of Si atoms on the surface (S)[b]	$m = \dfrac{V}{S}$ [c]	$X = \dfrac{m}{4m-1}$ [d]
1	25	21	1.18	0.3175
2	204	112	1.82	0.2899
3	688	275	2.50	0.2778
4	1630	505	3.23	0.2710
5	3184	828	3.85	0.2674
6	5501	1210	4.55	0.2646
7	8736	1660	5.26	0.2625
8	13040	2217	5.88	0.2611
9	18567	2785	6.67	0.2597
10	25469	3566	7.14	0.2591

Table 5.6. Effect of initial ratio of the concentrations $X = m/(4m-1)$ on the average size of silicon nanoparticles (Zou et al, 2006)

Sample	Ratio of reactants, X	The average size of nanoparticles, nm	Standard deviation, %
A	0.2668	3.28	26.2
B	0.2747	4.54	24.2
C	0.2774	5.00	18.9
D	0.2992	8.85	18.1

$$\frac{\text{the number of Si atoms}}{\text{the number of atoms Cl}} = m.$$

Then, the reaction proceeds by the scheme:

$$m\mathrm{SiCl}_4 + (4m-1)\mathrm{Na} \rightarrow \mathrm{Si}_m - \mathrm{Cl} + (4m-1)\mathrm{NaCl}.$$

The influence of the initial concentration ratio $X = m/(4m-1)$ on the average size of silicon nanoparticles was studied. The size of the silanized particles was determined on 800–900 of spots by TEM.

SAED and HRTEM showed the crystallinity of nc-Si. In the given synthesis method, nucleation ends at an early stage of particle formation and is controlled by the growth of produced nuclei. The number of nuclei is associated with the diffusing monomer, which is regulated by the concentrations of reactants with other parameters being constant. The nuclei have a short nucleation stage, defined by the kinetics of their formation and then reach the stage of growth (determined by thermodynamics), consisting first in Ostwald ripening and then saturation. Since the Si–Cl bonds are thermodynamically stable but kinetically labile, dissolution of small particles and their growth due to the large particles are possible. Thus, the size of of the nc-Si is controlled by the ratios of the agents. The quantum yield of the nc-Si of the minimum size is 12%.

The properties of nc-Si in the SiO_2 shell are affected by both self-trapping of excitons on the surface of nc-Si and by the size effect. In (Pan et al, 2009) the authors consider the surface states, where the holes and electrons are captured separately. The scheme of synthesis of nc-Si with two different methods of surface passivation is as follows:

$$SiCl_4 = [Na\text{-}napthalide] \Rightarrow (Si)_m - (Cl)_n = [H_2O_2] \Rightarrow (Si)_{m-n}(SiO_2)_n, \qquad (a)$$

$$SiCl_4 = [Na\text{-}napthalide] \Rightarrow (Si)_m - (Cl)_n = [EtOH] \Rightarrow (Si)_m - (OEt)_n. \qquad (b)$$

Upon heating a mixture of H_2O_2 (30%) with (Si) – (Cl)$_n$ for an hour with the reflux refrigerator (method (a)), the orange colour of the dispersion in glime changed to white. The solvent and naphthalide were removed under vacuum, the residue was washed several times with water and the nc-Si/SiO_2 was dissolved in chloroform. After treatment of nc-Si/Cl with absolute alcohol (method (b)) and removing the solvent and napthalide, the residue was also dissolved in chloroform. According to TEM, the mean particle size of nc-Si/SiO_2 was 3.4 mm, but their powder X-ray patterns were very clear, whereas nc-Si/OEt gave a pattern with much lower intensity. When excited by light with a wavelength of 488 nm the nc-Si/SiO_2 in chloroform showed two PL peaks of 535 nm (2.32 eV) and 578 nm (2.14 eV), the first more intense than the second. PL was not observed in nc-Si/OEt. According to the size of nc-Si/SiO_2, the band gap should be equal to 4.48 eV. This means that the PL comes from the surface states. The double peak in the PL spectrum is explained by the separate localization of the electron and the hole (Fig. 5.26).

The electron is localized because of the greater length of the wave function and lower mass compared to the hole. The excitons with the localized hole emit at 535 nm (2.32 eV), and with the localized electrons – at 578 nm (2.14 eV). The PL spectra with time resolution showed that the lifetime of the PL is 4.1–4.2 ns. Such short a time suggests that the PL comes from surface states.

In Zou et al, (2004), the produced nc-Si with chlorine atoms on the surface were treated with methanol and the alcoholate groups were then hydrolyzed and the surface nc-Si covered with hydroxyls. This was followed by adding alkyltrichlorosilane that formed (Si–O–Si-Alk) bonds with the release of hydrogen chloride. The chlorinated nc-Si can not be processed with water (instead of alcohol) because a sticky gel would form. TEM

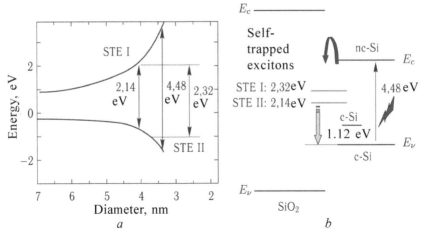

Fig. 5.26. The dependence of the electronic states of nc-Si on the size of the nanocrystal. a – scheme of the forbidden gap of nc-Si, and b – the localized states at the surface. There are two different bands of localized states STE I with a width of 2.32 eV and STE II with a width of 2.14 eV, which due to the localization of holes and electrons, respectively (Pan et al, 2009).

showed the presence of particles with an average size of 4.51 nm and a standard deviation of 1.10 nm. nc-Si with a diameter of 4 nm contains 1630 silicon atoms and 505 points on the surface that are associated with ligands (see Table 5.5). The reaction yield is 30%. FTIR showed the presence of C–H and (Si–O–Si) bonds, which was expected, but also the presence of Si–H bonds.

The PL spectra of nc-Si had a maximum at 400 nm for any coating (chlorine, methoxy, silanized surface) and in different solvents (chloroform, hexane and THF) under excitation by light with a wavelength of 360 nm. The PL peak of nc-Si depended on the excitation wavelength: at the changes of the excitation wavelength from 320 to 390 nm the PL maximum shifted from 390 to 450 nm. The stability of the PL was highest in silanized nc-Si: PL intensity did not change for 60 days, whereas the nc-Si coated with alcoholate lost PL intensity almost completely by the end of the 10th day, and with chloride – 4th. These results are confirmed and supplemented later (Zou et al, 2009).

Octyltrichlorosilane (OTCS) was used to stabilize the nc-Si (Pradhan et al, 2008). First, nc-Si was synthesized and its surface was covered with chlorine atoms by reduction of $SiCl_4$ with sodium naphthalide. An excess of methanol was added to this dispersion and the mixture was thoroughly stirred for 12 hours, during which methanol is reacted with the surface atoms of chlorine, replacing them by ($-OCH_3$) groups. Water in glime was then added and after that OTCS. nc-Si contained a silicon core and a SiO_xH_y shell which reacted with $Cl_3Si–C_8H_{17}$. The mixture was heated at 60°C for 30 min and was left at room temperature for 12 hours. In the end, the product in which the nc-Si core is bonded with the octyl through $(Si)_n \equiv O_3 \equiv Si-C_8H_{17}$ bridges was collected in chloroform and repeatedly extracted with a water–hexane mixture in the form of a waxy yellow solid mass able to disperse in non-polar organic solvents.

A monolayer of nc-Si was prepared by the Langmuir–Blodgett method, producing a sample with certain distances between the nc-Si at room temperature. The dispersion of nc-Si in chloroform was mixed with water in the Langmuir–Blodgett bath. The resulting monolayer was removed, transferred to a hydrophobic electrode, washed with ethanol and dried under a nitrogen atmosphere. The particle sizes of nc-Si were in the range 2–5 nm (average of 3.86 ± 0.85 nm). The electronic conductivity of the monolayer of nc-Si in vacuum and in the dark at temperatures above 200 K obeys the Arrhenius law, which indicates the existence of the thermal activation mechanism for charge transfer between the nc-Si. In the temperature range 100–200 K the electrical conductivity does not depend on temperature and, apparently, is due to tunnelling effect. When illuminated, the conductivity increased rapidly with decreasing wavelength of light, but this effect diminished with rise in temperature.

Sodium naphthalide was used in the above studies for maximum dispersion of sodium. Potassium intercalates in graphite are used for tdhe same purpose. Tetrahydrofuran (THF) was used as the solvent (Arquier et al, 2007); in the reaction the THF cycle was opened, i.e. THF is not suitable for the synthesis of nc-Si, as the products of opening of the cycle are uncontrollably attached to the surface of the nc-Si. $SiCl_4$ in THF was added at 0°C to freshly prepared C_8K. The mixture was stirred for 24 hours and boiled at reflux for 1 hour. Graphite and KCl filtered off and the solvent was distilled off under vacuum. Colourless oil was distilled off from the brown oily residue at 100°C leaving an insoluble brown solid denoted by SO. The SO suspension in THF reacted with a nucleophilic reagent for 30 min and was then at reflux for 1 hour. The solid colour changed from dark brown to green when using propane-2-ol and to yellow in the case of the Grignard reagent CH_3MgCl or $LiN(i-Pr)_2$, and $H_2N(CH_2)_3Si(OMe)_3$. The best results were obtained at a ratio of $SiCl_4/C_8K = 0.28$. The SO sample contained on the nc-Si surface chlorine atoms which in holding in air were replaced by the OH-group.

The average particle diameter of nc-Si was equal to 13.0 ± 3.4 nm. The photoluminescence spectra of nc-Si have maxima in the range 410–430 nm in excitation with light with a wavelength of 334 nm. For such a large nc-Si the wavelength of photoluminescence was unexpectedly short. According to the authors, this feature of the substance obtained by them is determined by the composition of the ligands on the surface of nc-Si, including products of ring opening of the THF cycle, but it can not be explained by the quantum size effect.

A method was developed (Aslan et al, 2010, Kamyshny et al, 2010) for synthesis of nc-Si of the required size by stabilization with nitrogen heterocyclic carbenes, namely imidazole-2-ilidene. With increasing size of the substituents in positions 1 and 3 of the imidazole-2-ilidene the average particle size decreases from 3 nm in diameter up to two-atom clusters of silicon, i.e. to molecular organosilicon compounds.

nc-Si sintered is at a temperature of 162°C, and large areas of polycrystalline silicon are formed at 350°C. nc-Si with the size of 2 nm melts at 500°C. Before producing a film of polycrystalline silicon, nc-Si can be doped. Baldwin et al, 2006 described a method of phosphorus doping of nc-Si, obtained in the solution:

$$8SiCl_4 + nPCl_3 + (15+3/2n)Mg \rightarrow (15+3/2n)MgCl_2 + (2Si_4P_{n/2}) - Cl,$$

$$Si_4P_{n/2} - Cl + RMgCl \rightarrow (Si_4P_{n/2}) - R + MgCl_2, \quad R - octyl.$$

The salt was filtered off and the solvent was evaporated under vacuum at 100°C. That left an orange oil. NMR spectroscopy found phosphorus atoms in nc-Si and on their surfaces. EPR also recorded phosphorus atoms. The PL spectra had a peak in the range 400–430 nm with excitation at 270–300 nm. RTEM found nc-Si with an interplanar spacing of 3.1 Å, which corresponds to the (111) plane of the diamond crystal lattice. Energy dispersive X-ray spectroscopy allowed to determine the phosphorus content in the nc-Si as ~6%. Phosphorus was evenly distributed in the polycrystalline silicon film obtained by heating the nc-Si layer on a germanium substrate at 600°C, phosphorus was evenly distributed.

There are also quite rare methods of synthesis of nc-Si, starting from $SiCl_4$. For example, the nc-Si in $LaCl_3$ was obtained (Hahn et al, 2008) by direct reduction of $SiCl_4$ with hydrogen at high temperature. The product was rubbed in a mortar. $LaCl_3$ was dissolved in boiling ethanol, thus releasing nc-Si/(OEt), – a brown finely dispersed material. It was treated with 1% solution of HF in a mixture of water:ethanol = 1 : 1 – in an ultrasonic bath. The result, according to the authors, was nc-Si:H_x. All operations were performed before etching in argon, and etching – in the air. Si–H bonds were detected by FTIR and NMR ^{1}H–MAS–NMR. It was found that in nc-Si:H_x there are no silicon dangling bonds, in nc-Si/$LaCl_3$ and nc-Si (OEt) such dangling bonds were detected by EPR, and photoconductivity measurements at $T = 300$ K. Photoluminescence was found in the nc-Si: H_x, but not in nc-Si/$LaCl_3$, as the unsaturated bonds are known to be strong non-radiative recombination centres. From the photoluminescence spectra it follows that the average size of nc-Si is 2.8 nm, and from the Raman spectra 4.1 nm.

5.4.6. Synthesis in supercritical fluids

nc-Si particles were obtained (Holmes et al, 2001) in supercritical anhydrous and deaerated octanol (T_c = 385°C, p_c = 34.5 bar) and hexane (T_c = 235°C, p_c = 30 bar) at 500°C and a pressure of 345 bar. The nc-Si precursor was diphenylsilane. Synthesis produced a yellow dispersion. Thermolysis of diphenylsilane in supercritical ethanol resulted in the rapid transformation of the red dispersion to a brown mixture in which contained polydisperse silicon with particles from the nano to micron size; the solvent at the same time became clear. The reaction proceeded for 2 h. Chloroform was used to remove the nc-Si from the reaction capsule. Chloroform evaporated and the residue was dispersed in ethanol. Small particles (up to 1.5 nm diameter) were dispersed in ethanol, but larger ones remained in the sediment.

In another series of experiments the supercritical solvent was hexane, and octanol in the supercritical state is used as an additive modifying the surface of nc-Si. TEM unambiguously confirmed the presence in the reaction product of nc-Si with diameter of 2–3 nm and a wide particle size distribution. The reaction yield was from 0.5% to 5%. X-ray photoelectron spectroscopy showed that for a single particle of nc-Si with a diameter of 1.5 nm there are 16 octanol ligands, i.e. a ligand per 44 Å^2, or 50% of the expected density of the coating. However, the particles with a diameter of 2.0 nm represent 70% of the densest cover by the octanol groups. FTIR showed that the surface is partly passivated by hydrogen, but the oxide and the ligand are also possible (Si–(R) C = O). Hydrogen is a pressure of 345 bar could remain from the silane, but the emergence of the radical (–(R) C=O) indicates that the alcohol in the supercritical state in the presence of nanosilicon is oxidized to an aldehyde with elimination of hydrogen. The PL spectrum of nc-Si with a diameter of 1.5 nm in hexane at 320 nm light excitation had a maximum of ~419 nm and a branch at 467 nm, and the nc-Si with 2.5–4.0 nm in diameter had a peak at 510 nm. The lifetime of the radiation was equal to 2 ns. The PL maximum was reached during excitation with the light with a wavelength of 363 nm. The authors attributed the 419 nm peak to the size effect in the nc-Si, and the 457 nm peak to the (Si = O) bonds. The photoluminescence quantum yield was 23% at room temperature.

In (English et al, 2002) nc-Si was also obtained in the supercritical hexane. The Precursor was diphenylsilane and octanethiol was used as a stabilizer. Diphenylsilane, octanethiol and hexane were deaerated and loaded in a nitrogen atmosphere in the reactor. The mixture was heated to 500°C and was kept 30 min under a pressure of 83 bar. The product was extracted with chloroform and precipitated in an excess of ethanol to remove byproducts. nc-Si redispersed in various organic solvents. The photoluminescence quantum yield was measured in comparison with rhodamine 6G. According to the AFM, the size distribution of nc-Si was 1–9 nm with an average diameter of 4.35 ± 2.02 nm, and according to the

TEM it was 4.65 ± 1.36 nm. Studies of the optical properties of single nc-Si showed a stochastic 'flicker' of the emitted light. When excited with the light of 488 nm single nc-Si had PL peaks from 525 to 700 nm. Shorter exciting radiation was not available to the authors. The nc-Si stabilized with octanethiol had one component of the PL with a lifetime of ~300 ns, and two components with 2 and 6 ns, i.e. 3 orders of magnitude faster than that of porous silicon and the nc-Si/SiO$_2$ dispersion. It turned out that the first, the fastest component of the PL spectrum corresponds to the radiative 'electron–hole' recombination and its quantum yield is 5.5%.

The authors suggest that the radiation is strongly influenced by the surface, and the state the inner core can 'borrow' the oscillator strength from the dipole-allowed surface states, which leads to an increase in the probability of interband transitions in the nc-Si. And even the local vibrational mode of surface centres manifests itself in PL emission. An alternative explanation is that radiation comes from the oxide surface centres of nc-Si, which may capture an exciton. The authors argue that the octanethiol-stabilized nc-Si have dipole-allowed optical transitions even for particles with a diameter of ~6.5 nm. These transitions are the result of strong coupling between the states of the crystalline core and the surface.

5.4.7. Mechanochemical synthesis

Mechanochemical synthesis (Heintz et al, 2007) led to the nc-Si, and the surface of some particles was covered with hydrocarbon groups. Crushed silicon particles, not exceeding 1 mm in size, were loaded in a stainless steel ball mill to which 1-octene or 1-octine was added. The volume was filled with nitrogen in a dry chamber and hermetically closed. Silicon was not only crushed, but on the fresh surfaces silicon reacted with 1-octene or 1-octine whose molecules formed a ligand environment of the silicon particles. 24 h of mechanochemical treatment produced a dispersion with a very broad particle size distribution, ranging from the particles with the diameter substantially less than 4 nm. This is evidenced by the TEM results and a distinct opalescence of the dispersion. FTIR spectra showed noth only the (Si–C) but also (Si–O) bonds. ^{13}C NMR method confirmed the presence of the (Si–C) bonds. The strong blue luminescence was attributed by the authors to the size effect, but in the presence of oxygen on the surface of nc-Si PL one must also consider the PL formed at the the Si/SiO$_2$ interface.

It was found (Pereira et al, 2007) that nc-Si grinding in a ball mill generates a set of P_b-centres (Si dangling bonds), and the rupture of the Si–H bonds, remaining on the surface of nc-Si from the moment of synthesis, occurs at 400–500°C in a vacuum of 10^{-8} mbar, while P_b-centres are formed on the surface (Stegner et al, 2007).

Typically, the PL comes from the core of silicon nanoparticles, however, as has been mentioned many times, the significant role is played by the surface states, which depend on the nature of the ligands. For example, the native silicon oxide can significantly reduce the effectiveness of PL, whereas the (Si–C) bonds protects from the reduction of PL efficiency due to the passivation of surface states.

In (Heintz et al, 2010) a stainless steel ampoule was filled with silicon, two stainless steel balls and one of the following reagents: 1-octine, 1-octene, 1-octyl acid, 1-octyl aldehyde and 1-octanol. All operations were carried out in a dry chamber in a nitrogen atmosphere. The ampolues were sealed and mechanochemical synthesis of nanosilicon was carried out for 24 hours. Products were dissolved in organic solvents. FTIR showed the presence on the surface of the nanosilicon particles of ligands as well as groups of atoms (Si–O–Si) and (Si–OH) when using 1-octanol, 1-octyl octyl aldehyde and octyl acid as ligands. The PL spectra of all samples had peaks in the blue region, and the maximum brightness was registered for arbitrary from 1-octine (210 arb. units), and 1-octene – 72 arb. units, and the rest – 48 arb. units is or less. Four samples (with 1-octine, 1-octene, 1-octanol, 1-octyl acid) have absorption spectra typical for indirect-gap semiconductors, but with 1-octyl aldehyde the absorption spectrum has a maximum around 290 nm, which is typical for direct-gap semiconductors.

5.4.8. Interaction of precursors in the gas phase

The particles of nc-Si in the SiO_2-shell were obtained (Fojtik et al, 2006a) by burning SiH_4 with a lack of oxygen. They did not have PL. The resulting powder was added to a mixture of cyclohexane and propanol-2 (1:1), and then 40% hydrofluoric acid was added for etching of SiO_2. Cyclohexane and propanol-2 stratified, with water and HF contained mainly in the alcohol layer where SiO_2 was mainly removed from nc-Si and the surface of nc-Si was passivated by hydrogen and became hydrophobic and, therefore, nc-Si passed to the cyclohexane fraction within a few hours. The opalescent cyclohexane fraction was ultrasonicated and stirred. All the operations were carried out in air as under argon no PL was detected in nc-Si. The cyclohexane fraction was separated from the alcohol. Its PL spectrum showed two peaks – at 650 nm and 480 nm. Energy dispersive analysis showed that the nc-Si with red PL contains 2% oxygen (OH-groups). The quantum yield for the nc-Si with red PL was 35%, while with blue PL – 18%. The diameter of nc-Si particles with red PL was ~4 nm, while blue PL – 1.5 nm. These particles were separated by centrifugation (4000 rpm, 30 min).

At atmospheric pressure SiH_2Cl_2 and N_2O reacted (Yang et al, 2008) in a ratio of 1.56 (the carrier gas was nitrogen). SiO_x was deposited on the Si (111) substrate with hole conductivity at 950°C. The samples were then annealed in nitrogen at 1150°C for 30 min for the nucleation of nc-Si. The samples were also annealed in the temperature range 875–950°C for 1 hour in an atmosphere of CO_2. The diameters of nc-Si were in the range 4–5 nm. PL intensity increased three times after annealing at 950°C in an atmosphere of CO_2, while the peak in the spectrum shifted from 1.7 to 1.75 eV. As shown by TEM, the size of nc-Si decreased. The authors suggest annealing in CO_2 was accompanied by the dissociation of $CO_2 = CO + (1/2)$ O_2 and oxidation of the surface of nc-Si with oxygen. Estimate calculations have shown that the diffusion of oxygen through SiO_2 was possible. Oxygen blocked P_b-centres and PL increased. Groups (Si=O) and (Si–O–Si) formed at the nc-Si/SiO_2 interface which, according to the authors (Yang et al, 2008) could recombine photoinduced electron–hole pairs. In confirmation This hypothesis can be confirmed by the following experimental fact. The PL spectra of the dispersion of nc-Si in SiO_2 (Vandyshev et al, 2007) were studied in an electric field with a strength of up to 450 kV cm^{-1}. When the electric potential was applied to the sample PL intensity fell by 14% and a red shift was observed. These features prompted the authors to conclude that the PL arises in the recombination of electrons and holes captured by traps located at the interface of nc-Si and SiO_2.

5.4.9. Silicon nanowires

A particular direction in the experiments with nanosilicon has been the study of nanowires. A silicon nanowire (nw-Si) can be used for the manufacture of electronic components based on field transistors placed on flexible substrates. There are two notions of growth mechanisms of nw–Si: vapor–liquid–solid (VLS) and vapor–solid–solid (VSS).

In (Qi et al, 2008) SiH_4 was decomposed at a temperature of 450°C and a pressure of 100 Torr, and the decomposition products were transferred by gas (hydrogen was used as a carrier) into the growth zone of nw-Si. At the diameter of the grains of an Au catalyst of 60 nm the length of nw-Si decreased from 20 μm to 5 μm from the reduction of growth temperature from 400 to 350°C. At 320°C nw-Si did not form. However, with the Au catalyst Au, with the particle diameters of 10 nm, nw-Si grew to a length of 3 μm during the same period. The authors believe that at temperatures above 340°C with a diameter of the Au-catalyst particles of 10 nm nw-Si grows by the VLS mechanism, and down to 340°C – by the VSS mechanism, since at temperatures below 340°C there is no the liquid phase in the system. nw-Si does not grow on large gold particles (60 nm) at 320°C because diffusion of silicon atoms from the gas phase at the Au/nw-Si interface should take place over long distances (from the surface to the axis of the nanowire along the Au/Si interface) – up to 30 nm, and the gold particles with small diameter (10 nm) require diffusion of silicon atoms from the gas phase along the same interface only to a distance of 5 nm or less. In addition, at temperatures >340°C the rate of growth of nw-Si is more sensitive to temperature (Fig. 5.27). Boles et al, 2009 developed a mathematical model of growth of silicon nanowires with a gold catalyst.

The nw-Si structure was grown (Heitsch et al, 2008) by the solution–liquid–solid method at atmospheric pressure using trisilane (Si_3H_8)

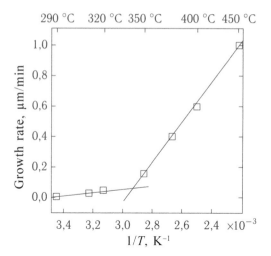

Fig. 5.27. Temperature dependence of the rate of growth of Si nanowires. From the work of Qi et al. (2008).

as a starting reagent, octacosane ($C_{28}H_{58}$) or squalane ($C_{30}H_{62}$) as solvents, and nanocrystals of gold or bismuth as the seeds for growth nw-Si. Gold and silicon form a eutectic at 363°C, and bismuth and silicon at 264°C. The boiling points of the solvents $C_{28}H_{58}$ and $C_{30}H_{62}$ – 430 and 423°C respectively, – are higher than the indicated eutectic temperatures. The mechanisms of growth of the nanowires at the contact of the seed and nw-Si were confirmed by TEM, SEM, EDS. nw-Si diameters were within the range of 20–30 nm and lengths reaching more than 1 μm (some samples reached 3 nm in 10 min of growth at a temperature of 430°C). nw-Si had a diamond-like crystal structure that is registered by X-ray diffraction. Basically, nw-Si grew in the direction $\langle 111 \rangle$, but 20% nw-Si grew in the $\langle 110 \rangle$ direction. The limiting stage of growth of nw-Si was the dissociation of trisilane.

The VLS growth method catalyzed with a nanocluster is very promising for producing single-crystal silicon nanowires with controlled diameters, length and the electronic properties (Park et al, 2008). The process can be broken into several stages: 1) catalytic adsorption of the gas precursor on the surface of a liquid nanoparticle, 2) the diffusion of silicon through the molten alloy to the sink, and 3) crystallization at the liquid–crystal interface. The catalyst are the gold nanoparticles. It was shown that the kinetics of decomposition of SiH_4 is more important than transport of gas, and therefore Si_2H_6 should be used as a precursor instead of SiH_4, since in Si_2H_6 the (Si–Si) bonds breaks more easily than Si–H in SiH_4, i.e. the beginning of silane decomposition is easier.

Compared with the growth of nw-Si under conditions in which SiH_4 was used as a precursor, the application of Si_2H_6 per unit of time produces nw-Si 130 times longer, namely, the average length of 1.8 mm in one hour at a thickness of 30 nm, and the diameter of nw-Si is absolutely identical throughout. The maximum length of nw-Si, produced in an hour, is 3.5 mm. The single crystal wires grew along the $\langle 110 \rangle$ direction at any diameter.

The electrical and optical properties of the nanowires strongly depend on the diameter, crystallographic orientation and defects of nw-Si. Lungstein et al (2008) prefer the VLS (vapor–liquid–solid) growth mechanism and gold as a catalyst for nw-Si. Au and Si form a liquid alloy which has a eutectic at 363°C. The alloy, after supersaturation, produces a nucleus that starts the growth of nw-Si. For the VLS mechanism the diameter of nw-Si is crucial when selecting the direction of growth. If the diameter nw-Si > 40 nm, the preferred direction is $\langle 111 \rangle$, and if less than 20 nm, in most cases $\langle 110 \rangle$. In the range 20–40 nm the direction $\langle 112 \rangle$ can also form. Silicon oxide was removed from the Si(111) substrate with hydrofluoric acid, leaving the hydrogenated surface. Gold as a catalyst was deposited on the substrate at room temperature in the form of a thin film (thickness 2 nm) by thermal evaporation. nw-Si was grown at a pressure of 3 and 15 mbar at a flow rate of SiH_4 of 100 cm^3 min^{-1} (2% helium admixture) and 10 cm$^3 \cdot$ min^{-1} H_2; growth temperature was equal to 500°C. Growth was interrupted by switching off the gas supply and the product was cooled in vacuum. At a gas pressure of 3 mbar nw-Si grew perpendicular to the (111) plane of the silicon substrate, but there was also nw-Si, growing at an angle of 19.5° with respect to the substrate, which corresponds to the $\langle 112 \rangle$ direction. The length of nw-Si reached about 3.3 μm for 100 minutes of growth. TEM showed that if the growth direction is $\langle 111 \rangle$, then nw-Si is usually free of dislocations and covered with a thin amorphous oxide layer. At a gas pressure of 15 mbar nw-Si was abundantly supplied with silicon and the rate of growth nw-Si increased 7 times, with the nw-Si section constant along the whole length, whereas at a gas pressure of 3 mbar nw-Si was tapered. At a gas pressure of 15 mbar the growth direction changed to $\langle 112 \rangle$ and almost all the nw-Si grew at an angle of 70° to the substrate plane. The pressure in the range 3–15 mbar resulted in the growth of a mixture of orientations $\langle 111 \rangle$ and $\langle 112 \rangle$ of nw-Si. Thus, the growth direction of nw-Si depends on the partial pressure of SiH_4.

Under certain conditions (Cao et al, 2006) nw-Si was replaced by the growth of conical silicon formations when using gold nanoparticles as growth catalysts.

It was found that silicon dioxide does not have to be necessarily removed from the substrate before growth of nw-Si on it. The Au-catalyst for the growth of silicon nanowires (Albuschies et al, 2006) was deposited on the SiO_2 layer by vacuum evaporation of gold in a vacuum of $7 \cdot 10^{-7}$ mbar. Gold atoms on the dielectric surface in the monatomic layer thickness of the metal gathered in clusters, forming a patchy covering of SiO_2. The size of the spots was ~5 nm spots, but the nw-Si diameter reached 70 to 250 nm (the higher the temperature and pressure of SiH_4, the thicker the nw-

Si), i.e. the individual gold spots coalesce under the influence of silicon pyrolytically obtained from SiH_4.

The growth conditions of silicon nanowires in the above examples are such that the use of nanowires in the electronics industry is problematic (Zou et al, 2007). First, the nanowire grows not in the plane of the substrate, but aside from it, while thousands of items on the board must be connected in the plane of the substrate. Secondly, the gold catalyst easily diffuses into silicon and greatly reduces the lifetime of charge carriers, and without the catalyst the growth of the nanowires requires high temperatures and long periods of time.

Aluminium is desirable for the growth of nanowires, as it is the alloying material for silicon and is used to make contacts on silicon. nw-Si up to 3 μm long and about ~0.1 μm thick was obtained at the interface layer of amorphous silicon deposited with an aluminium layer.

Silicon nanowires were grown with an iron catalyst (Chang et al, 2006). A mixture of $SiO_2 : C = 1:1$ was processed with the cathodic arc plasma in an argon atmosphere, which led to the formation of nc-Si with a diameter about 25 nm, which were dispersed in a solution of $Fe(NO_3)_3$ in ethyl alcohol. The alcohol was evaporated and the mixture of nc-Si and $Fe(NO_3)_3$ calcined at 980°C for 1 hour in a stream of hydrogen. Iron nitrate decomposed to the oxide and was reduced with hydrogen to iron. Iron particles catalyzed the growth of nw-Si.

Experiments were carried out (Tuan et al, 2006) with the growth of nw-Si in supercritical toluene. Growth seeds were nanoparticles of Co, Ni, CuS, Ir, Mn, $MnPt_3$, Fe_2O_3, FePt. Growth temperature was 300–1000°C below the temperature of the eutectics formed by silicon and metals included in the composition of the seeds, so the growth mechanism of nanowires in supercritical toluene was of the solid phase type. Gold was not used as a seed because of the above reasons. In supercritical toluene silicon nanowires grew at 500°C and a pressure of 10.3 MPa. Monophenylsilane was used as a silicon-containing precursor, and catalysts for the growth of nw-Si were CuS nanocrystals, which in the course of the reaction transformed to copper. nw-Si had a diameter of 5–20 nm and the average length 7.4 μm. 75% nw-Si were elongated along the ⟨111⟩ crystalline direction; 20% along ⟨110⟩, and 5% along ⟨112⟩ (Tuan et al, 2008).

Nanowires can be obtained by etching silicon wafers in the presence of metal catalysts (Sivakov et al, 2010). The native SiO_2 layer was removed from the Si(111) and Si(100) wafers of p-type by 40% HF and the wafers were washed with 1% solution of HF, rinsed in water and dried in a stream of nitrogen. They were deposited with silver nanoparticles whose morphology strongly depended on the duration of their preparation. After that, the wafers were immersed in a mixture of 5 M HF and 30% H_2O_2 with a volume ratio 10:1 for one hour at room temperature for etching silicon. The wafers were then washed with deionized water and dried at room temperature. The resulting porous surfaces were washed with 65% HNO_3

to produce silver nanoparticles. The morphology of the pores depended on the morphology of silver nanoparticles and the orientation of silicon wafers. For example, the nanowires on the Si(100) plate surface were straight and in the case of the plate Si(111) – zigzag-shaped. nw-Si can be straight also on the surface of the Si(111) wafer, but is required to cover the Si(111) surface almost completely with silver nanoparticles almost completely. Bending of the silicon nanowires is explained by the authors by the change of the orientation of the etching of the wafers due to changes in temperature.

5.5. Passivation of surface and functionalization of nc-Si

The nanoparticles are characterized by an enormous ratio of surface area to volume. For example, a silicon particle of the icosahedral shape with a diameter 2.0 nm has about 280 Si atoms of which 120 atoms (43%) are located on the surface (Hua et al, 2005). Considering this point, it is natural to assume that the properties of the surface, including chemical, will significantly affect the characteristics of the material.

Surface modification of nanosilicon is based on the formation of (Si–C) and (Si–N) polar covalent bonds, as well as (Si–O) bonds. Different approaches to the formation of the (Si–C) bond are described in the literature, for example, hydrosilylation (see, for example, Andrianov et al, 1979; Pavesi, Turan, 2010). However, there is still a set of outstanding issues related to the efficiency of surface passivation and stability of the functionalized particles of nc-Si.

Effective functionalization of the silicon nanoparticles is based on reaction centres on the surface of nc-Si. Such a surface can be obtained by numerous methods, including the formation of a reactive surface of the nanoparticles in the preparation process (for example, etching of silicon plates, the reactions of Zintl salts), during the release of particles from the oxide matrix when etching in hydrofluoric acid and, finally, etching of silicon nanoparticles with various reagents, for example, HF/HNO_3, H_2O_2/H_2SO_4, $LiAlH_4$, halogens. The subsequent step is the surface modification using various procedures developed in the chemistry of silicon (see, e.g. Lukevics, Voronkov, 1964, Andrianov et al, 1979).

One of the first attempts to obtain surface-functionalized silicon nanoparticles was proposed by Heath (1992), based on a heterogeneous reaction mixture containing dispersions of sodium, tetrachlorosilane, and n-octyltrichlorosilane at high temperature and pressure:

$$SiCl_4 + RSiCl_3 + Na \rightarrow (Si)_n + NaCl.$$

The reaction resulted in the formation of silicon nanoparticles, but studies by IR spectroscopy revealed almost no absorption bands corresponding to alkylation of the surface, which can be explained by destruction of the Si–C bonds, and the prevalence of Si–O, Si–Cl and Si–H bonds in the reaction conditions.

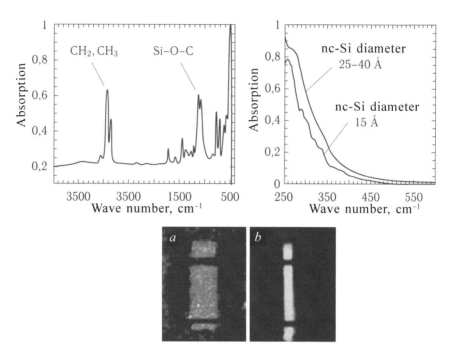

Fig. 5.28. Top left – Fourier IR spectrum of alkoxylated nc-Si. Top right – absorption spectrum of nc-Si. Bottom – photograph of luminescent nc-Si in hexane at excitation at ~320 nm (a); photoluminescence of particles with a diameter of 1.5 nm (b) (Holmes et al, 2001).

In 2001, Holmes et al. reported on the synthesis of stabilized nc-Si by thermal decomposition of diphenylsilane in supercritical octanol. As a result, particles with diameters ranging from 1.5 to 4.0 nm were produced. Analysis of the IR spectra confirmed that the surface organic layer was indeed connected via alkoxide (Si–O–C) bonds (see Fig. 5.28).

At present there is no doubt that IR spectroscopy is an invaluable tool in the study of the surface chemistry of nc-Si. However, this method is not comprehensive and requires a number of additional methods to characterize the surface of the nanoparticles. Thus, the use of the XPS method in this case allowed to determine the Si:C ratio of the shell and determine that the particle with a diameter of 1.5 nm has about 16 ligands on the total surface area and each ligand occupies an area of ~44 $Å^2$.

The resulting nanoparticles easily formed a sol in ethyl alcohol and had a bright photoluminescence. The molar absorption coefficient in the range 250–600 nm should depend on energy by the quadratic dependence, indicating the indirect-gap transition in the synthesized nanosilicon particles (Holmes et al, 2001).

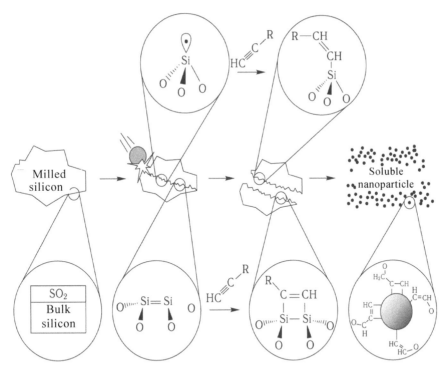

Fig. 5.29. Preparation of the silicon surface during mechanochemical synthesis of alkylated particles of nc-Si (Heintz et al, 2007).

Heintz et al. (2007) obtained nc-Si by refined high-purity silicon in a ball mill in a mixture of 1-octine. The mechanochemical process was carried out in an inert atmosphere (Fig. 5.29).

It is assumed that using this method produces a clean surface of silicon nanoparticles and in subsequent stages there is a fairly easy interaction with unsaturated hydrocarbons with the formation of (Si–C) covalent bonds. This assumption is confirmed by the study of IR and NMR spectra (^{1}H) (see Fig. 5.30). The IR spectra showed absorption bands in the region < 1257, < 806 and < 796 cm^{-1}, characteristic of the (Si–C) bonds, which are usually masked by the absorption of (Si–O) groups usually present in large numbers in these materials.

5.5.1. *Functionalization of nc-Si in a spray*

The nanocrystalline silicon, obtained by the plasma chemical method, may be modify *in situ*, by introducing functionalized organic reagents directly into the reactor chamber. The first study in this direction was conducted by Liao and Roberts (2006), who studied gas-phase reactions for surface modification of nc-Si. Nanoparticles were obtained from monosilane in

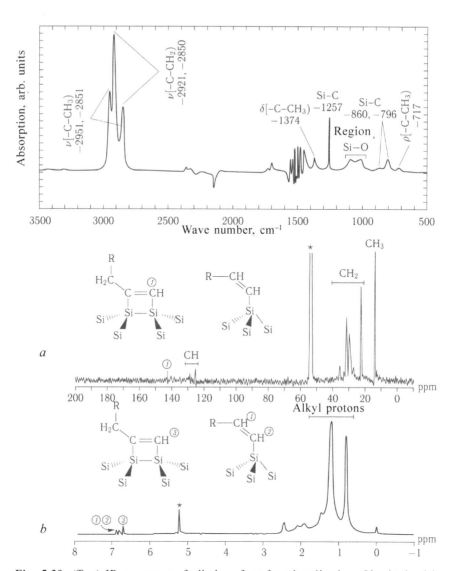

Fig. 5.30. (Top) IR spectrum of alkyl surface-functionalized nc-Si, obtained by mechanochemical synthesis. $^{13}C\{^1H\}$ (*a*) and ^1H (*b*) NMR spectra of the nc-Si. From the work of Heintz et al. (2007).

a non-thermal reactor and transferred into the chamber of the reactor containing organic reagents (e.g. amines, alkenes, alkynes). Heating the mixture in the gas phase led to the modification of the surface of nc-Si, which was confirmed by Fouried IR spectroscopy.

Mangolini and Kortshagen (2007) proposed a modification of this approach, using inoculation of different groups in the plasma. The reactor is shown schematically in Fig. 5.31.

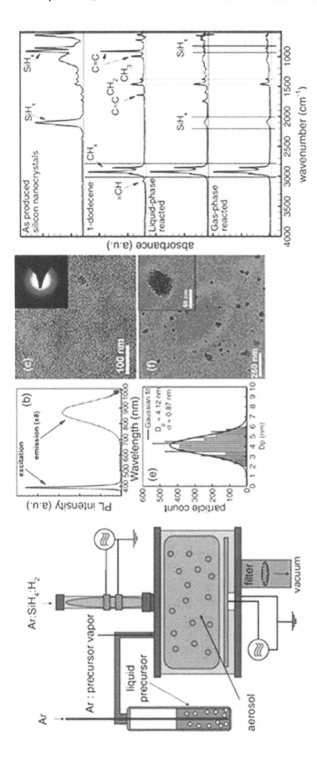

Fig. 5.31. On the left the diagram of the 'Plazma/Aerozol' appartus used to prepare the particles of surface-functionalized nc-Si. In the middle is the fluorescence spectrum (b) and size distribution (e) and TEM micrographs (c, f) of nc-Si, functionalized in an aerosol. Right, the FTIR spectra of Si–H, octadecene, Si–R (R – octadecyl) of nc-Si, functionalized in solution, and Si–R (R – octadecyl) of nc-Si, functionalized in an aerosol (Mangolini, Kortshagen, 2007).

Using the reactor shown in Fig. 5.31, allows various functional groups to be attached to the surface of nc-Si. Nanosilicon particles obtained in this way, are crystalline, have excellent solubility, without the need for the dissolution of ultrasound exposure, and have the photoluminescence quantum yield up to 60% (Jurbergs et al, 2006). The Fourier IR spectra of isolated particles are remarkably similar to the spectra of modified particles in solution, and clearly show the presence of characteristic absorption bands, which can be attributed to the alkyl groups and the residual Si–H bonds. In the region of vibrations of the (C = C) fragment there were no absorption bands which indicates the occurrence of a fairly complete hydrosilylation process. However, in Jurbergs et al. (2006) no absorption bands were found in the vibrational frequency range of the (Si–C) fragments.

There is extensive literature both in the field of chemistry of organometallic compounds of silicon and the chemistry of bulk silicon surface which describes many analogues of the reactions of attachment of various radicals to the surface of nc-Si. There are preparative methods of functionalization of silicon surfaces, (e.g. oxidation, attachment $-H$, $-OH$, halogen). This section describes methods that allow for direct attachment to the surface of nanosilicon of different groups, and briefly discusses modifications in the formation of the oxide film on the surface of the nc-Si. A more detailed description of this issue is contained, for example, in the monograph by Pavesi and Turan (2010).

5.5.2. Hydrosilylation

Without a doubt, the most widely studied method is the method of changes in the surface chemistry of nc-Si by hydrosilylation of terminal alkenes. It is the result of the huge amount of research relating to functionalization of porous silicon and bulk silicon, which uses similar methodology to generate a sufficiently strong covalent Si–C bond on the surface of nc-Si. So far, using this method, nanoparticles with the highest fluorescence quantum yield have been obtained. This is due to the fact that hydrosilylation can be carried out without the access of atmospheric oxygen and thus oxidation of the surface of the nc-Si (Jurbergs et al, 2006). This observation also emphasizes the importance of surface chemistry and the efficacy of binding of alkyl groups to protect the core of nanosilicon.

The hydrosilylation reaction is the attachment of the monomers or polymers, containing the Si–H bond, to unsaturated compounds. The reaction uses mainly hydrosilanes (hydride silanes) H_nSiX_{4-n}, where $X = $ Cl, CH_3, OC_2H_5, and others, $n = 1–3$, and hydro-organosiloxanes (hydride organosiloxanes) $[-RHSiO–(-R'_2SiO-)n-]x$, where R and $R' - CH_3$, C_2H_5, etc., $n > 0$. The most widely investigated has been the hydrosilylation of compounds with multiple carbon–carbon bonds (Lukevics, Voronkov, 1964).

The thermal hydrosilylation temperature is usually greater than 250°C. The temperature range of the reaction, initiated by transition metals or their compounds (e.g. Pt on carbon or Al_2O_3, H_2PtCl_6 solution in isopropanol – so-called Speier catalyst), peroxides, amines, UV or gamma-radiation and others, lies in the range from -15 to $300°C$. The catalysts of the reaction are specific. For example, the Speier catalyst is highly efficient in the hydrosilylation of lower olefins or acetylene with alkyldichlorosilanes (the reaction proceeds at room temperature with a virtually quantitative yield), and in the hydrosilylation of acrylonitrile – a system containing amines and salts of copper (II) (Andrianov et al, 1979).

The procedure for the thermally-initiated hydrosilylation of silicon nanoparticles, (Si–H) saturated bonds, lies in the fact that at sufficiently high temperatures, some (Si–H) bonds are broken: $Si–H \rightarrow Si^\bullet + H^\bullet$. Subsequently, the silicon dangling bonds can interact with alkenes having terminal double bonds. This chain reaction is described by the following scheme:

$$
\begin{array}{c}
\overset{H}{\underset{|}{\overset{|}{\text{Si}}}} - \overset{H}{\underset{|}{\overset{|}{\text{Si}}}} - \overset{}{\underset{|}{\overset{|}{\text{Si}}}} - \quad \xrightarrow{\quad\wedge R\quad} \quad -\overset{R}{\underset{|}{\overset{|}{\text{Si}}}} - \overset{H}{\underset{|}{\overset{|}{\text{Si}}}} - \overset{H}{\underset{|}{\overset{|}{\text{Si}}}} - \quad \longrightarrow \quad -\overset{R}{\underset{|}{\overset{|}{\text{Si}}}} - \overset{}{\underset{|}{\overset{}{\text{Si}}}} - \overset{H}{\underset{|}{\overset{|}{\text{Si}}}} - \Rightarrow
\end{array}
$$

Platinum and similar catalysts seem to form intermediate compounds of the ionic type with the reacting compounds. Hydrosilylation of carbon-carbon bonds may lead to the formation of isomers, for example:

$$
\overset{\diagdown}{\diagup}SiH + HC{\equiv}CR \left\{
\begin{array}{l}
\longrightarrow \ \overset{\diagdown}{\diagup}SiCH{=}CHR \\[2ex]
\longrightarrow \ CH_2{=}C(R)Si\overset{\diagup}{\diagdown}
\end{array}
\right.
$$

In hydrosilylation by the bonds $=C=O$, $=C=N$, $-C{\equiv}N$, $-N=O$, etc. the silyl group is attached to the more electronegative atom, such as:

$$
\overset{\diagdown}{\diagup}SiH + O{=}CHR \longrightarrow \overset{\diagdown}{\diagup}SiOCH_2R.
$$

Secondary reactions during hydrosilylation are: isomerization, hydrogenation, telomerization of olefin, disproportionation of the hydrosilylation agent, splitting of the C–O bond, silylation of the group N–H (Andrianov et al, 1979).

Before discussing details of the hydrosilylation reaction and its role in the chemistry of silicon nanoparticles, it is useful to note that the chemistry of the surface of nc-Si, from the starting point – the formation of a surface covered by Si–H, has some fine moments by which nanocrystalline silicon differs from porous and bulk silicon. For example, the' etching rate of the nanoparticles in HF is much smaller than the etching rate of silicon wafers: < 1 nm/min for the nanoparticles in comparison with mm/min for silicon wafers. In this regard, it was suggested that the fluoride on the surface may affect the etching process. In addition to this, it was reported that an aqueous solution of HF does not remove all the oxide from the surface (Li et al, 2004). However, the addition of ethyl alcohol to the mixture used for etching increases the wettability of the surface and provides a system with no oxide (Hessel et al, 2008).

There are only a few reports of detailed and systematic studies of the etching process of nanosilicon particles in aqueous solutions of HF and its impact on the surface of the particles. Li et al. (2004) changed the concentration of an aqueous solution of HF, used for etching the surface of the nanoparticles separated from the primary product by the HF/HNO$_3$ reaction mixture. The results of Fourier IR spectroscopy of nanoparticles after etching in 3 wt.% HF acid showed intense peaks in the region 2040–2231 cm^{-1} corresponding to the stretching vibrations of Si–H and a weak peak in the fluctuations range of Si–O (1070 cm^{-1}), which may be the result of subsequent oxidation after etching. Increasing the HF concentration to 5–15 wt.% complicates the structure of the peaks of stretching vibrations of Si–H (inhomogeneous broadening of peaks) that can be attributed to the emergence of =SiH$_2$–SiH$_3$ groups on the particle surface. One also can not exclude the formation of (Si–O–H) groups, which leads to a broadening of the absorption band of Si–H. In addition, higher concentrations of HF lead to an increase in the relative intensity of (Si–O) and the appearance of a broad absorption band in the area of 3430 cm^{-1}, which can be attributed to the silanol group. In this regard, the authors suggest that higher HF concentrations catalyze the reaction of the surface of the silicon atoms with water, which leads to the formation of unwanted (Si–OH) fragments.

As one might expect on the basis of extensive studies of porous silicon and bulk silicon surface chemistry, as well as the relatively easy formation of hydrogen atoms covering the surface of the nc-Si, thermally activated (see Fig. 5.32) and photoinduced (see Fig. 5.33) attachment of Si–H to the unsaturated carbon–carbon bonds is an effective method in the chemistry of silicon nanoparticles. This process is also effective when using catalysts based on transition element compounds (Tilley et al, 2005; Gutsulyak et al, 2010).

The hydrosilylation process yields significant opportunities to control hydrophilicity nc-Si, eventually making them soluble in organic solvents (Liu et al, 2005) and in some cases, in aqueous solution (Tilley et al, 2005; Hua et al, 2005). FTIR spectra of functionalized nanoparticles clearly show the characteristic absorption bands associated to the surface of organic radicals, the residual Si–H groups, and oxide groups, which substantially mask the absorption band (Si–C) in 1083 cm^{-1} (see Fig. 5.34) (Li et al, 2004; Hua et al, 2005).

In accordance with the generally accepted notions of the radical mechanism of the photoinduced hydrosilylation process, spectral analysis of the functionalized particles of nc-Si, obtained by this reaction, indicates

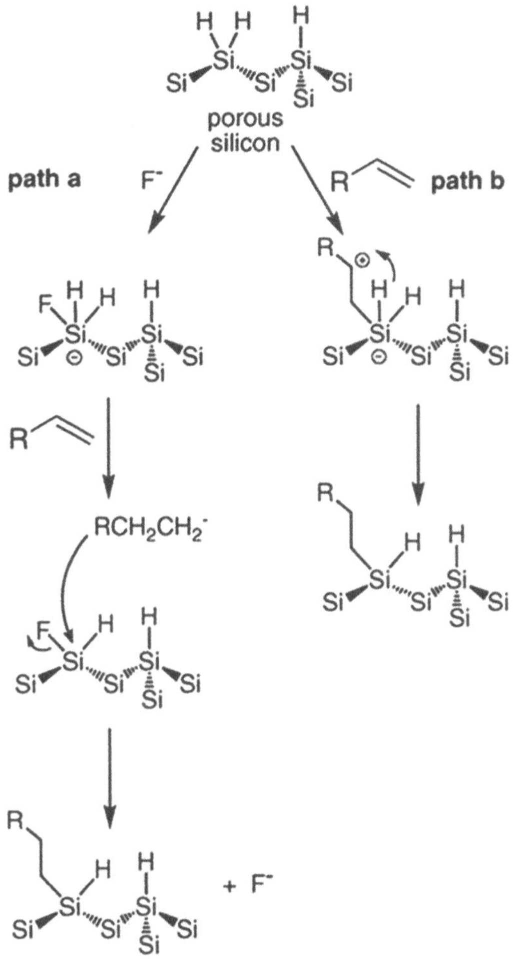

Fig. 5.32. Two possible mechanisms for the thermally activated process of hydrosilylation of Si–H groups on the surface of nc-Si groups with terminal alkene groups (Buriak, 2002).

different channels of attachment of the alkyl groups to their surface (see Fig. 5.35). It was found that depending on the steric characteristics of the attached ligand, binding of both α- and β-carbon atoms can take place (Hua et al, 2005). Moreover, the α-attachment is characteristic of large molecules, as β-attachment – for small olefin molecules. Similar observations were not recorded in the thermal activation of the hydrosilylation reaction.

The water solubility of functionalized particles of nc-Si is a prerequisite, for all applications of these particles in biological objects, and this property attracts the attention of a large number of research groups. For this reason many studies have focused on the attachment of hydrophilic surface groups, such as propionic acid (Sato et al, 2006), hydrolyzable esters (Rogozhina

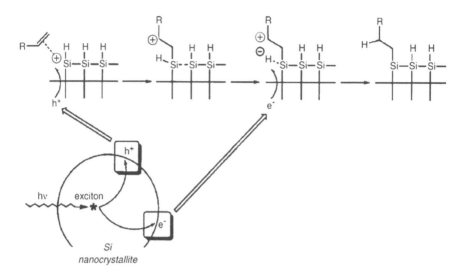

Fig. 5.33. The proposed mechanism of photochemically initiated hydrosilylation reaction of Si–H groups on the surface of nc-Si groups with terminal alkene groups (Buriak, 2002).

et al, 2006) or allylamines (Tilley, Yamamoto, 2006; Warner et al, 2005), polyacrylic acid (Li, Ruckenstein, 2004) in hydrosilylation. The products of attachment of propionic acid were obtained by etching in a HF/HNO$_3$ mixture of commercial SiO$_x$ oxide with a high content of silicon and stabilized by acrylic acid in the presence of HF. The authors report that the hydrosilylation reaction does not occur efficiently in the absence of HF. Its observation suggests a useful alternative to the tedious stages of removal of the solvent and degassing. The samples of solutions of particles were stable, not opalescent and brightly photoluminescent suspensions over a long period of time after their transport into the water by dialysis. This contrasts sharply with the initial mixture of nc-Si with acrylic acid, in which particles are easily deposited (see Fig. 5.36). It should be noted that the isolated powder particles of nc-Si require a significant effect of ultrasound to resuspend them in water, due to the oxidation of the surface of nanosilicon.

Analysis of these materials by TEM is complicated by the agglomeration of particles and their small size. However, HRTEM confirmed the crystal structure of nc-Si with an interplanar distance of 3.1 Å, corresponding to the distance (111) in crystalline silicon. As expected, the Fourier IR spectra nc-Si/propionic acid show the characteristic frequency of propionic acid and the absence of lines corresponding to the unsaturated acrylic acid bonds. Unfortunately, we cannot exclude the presence of oligomeric or polymeric impurities resulting from the interaction of several molecules of acrylic acid and having almost identical absorption bands. Such water-soluble nc-Si/

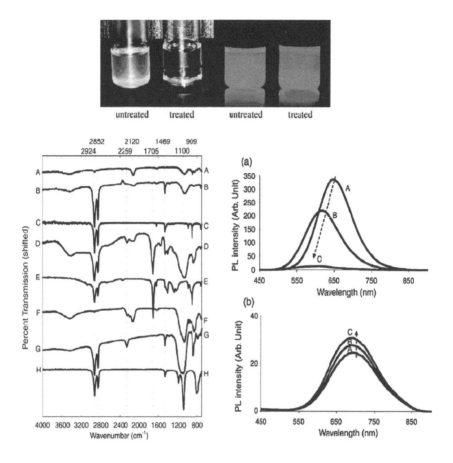

Fig. 5.34. Top – alkylated nc-Si, produced by the photostimulated hydrosylilation reaction of the solution and their photoluminescent properties (two sample on the left – in daylight, two on the right – under UV radiation). Treated – particles were treated with octadecene. Bottom left – Fourier IR spectra of nc-Si/Si–H (A), octadecyl (B), nc-Si/1-octadecene (C), nc-Si/undecylenic acid (D), undecylenic acid (E), nc-Si/Si–OH (F), nc-Si/octadecyl trimethoxysilane (G), octadecyl trimethoxysilane (H). Bottom right – stability of photoluminescence of particles of nc-Si/Si–H (top) and nc-Si/alkyl (Li X., et al., 2004).

polyacrylic acid samples were observed in the work of Li and Ruckenstein (2004).

Warner et al. (2005) studied the problem of obtaining water-soluble particles in the hydrosilylation reaction of nc Si and acrylamine using a platinum catalyst H_2PtCl_6 (Fig. 5.37). The Fourier IR spectra give characteristic absorption bands of alkyl groups and amino groups, scissor vibrations ($Si–CH_2$), symmetric bending vibrations in the range 1420 and 1260 cm^{-1}. The particles obtained in this way have the photoluminescence with the

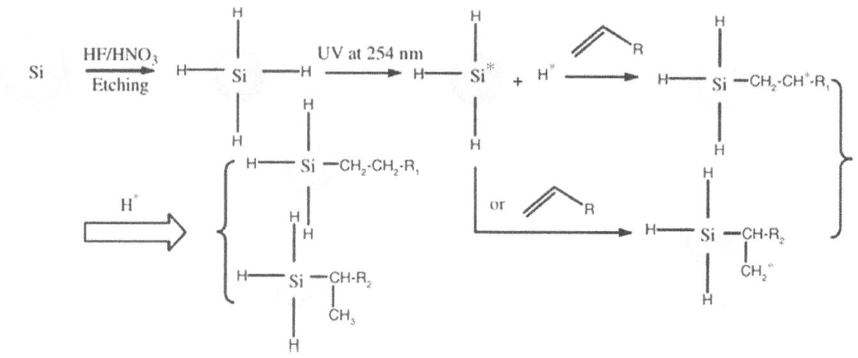

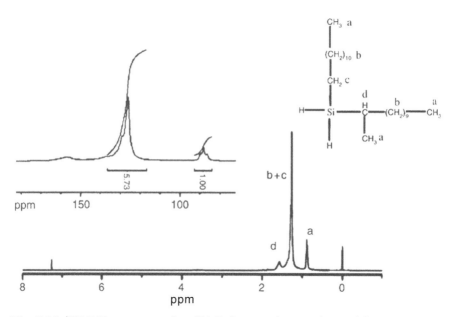

Fig. 5.35. Possible ways of joining the alkyl groups to the surface of nc-Si in the photoinduced hydrosilylation process (Hua et al, 2005).

Fig. 5.36. ^{1}H NMR spectrum of nc-Si/alkyl at attachment of α- and β-carbon atoms of alkenes (Hua et al, 2005).

maximum of 480 nm with excitation at the 300 nm line and were used for imaging HeLa cells (see Chapter 12).

The third approach to obtain water-soluble nc-Si with a protected surface was presented in Rogozhina et al. (2006). The hydrosilylation reaction was carried out with methyl-4-pentenoate in a THF solution with heating. At the end of the process the hydrosilylation product was hydrolyzed in a solution containing methanol, water and NaOH. Fourier IR spectra confirm the presence of ester and acid groups on the surface. However, the presence of Si–C bonds was not confirmed.

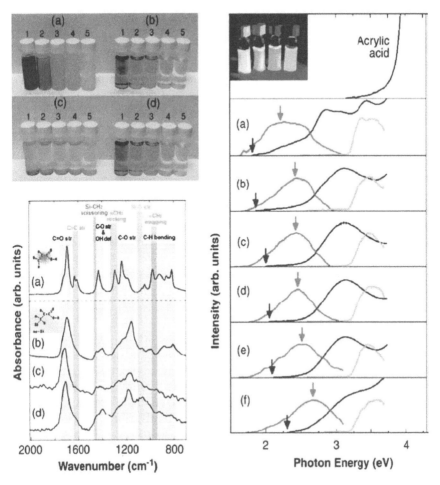

Fig. 5.37. Top left – comparison of nc-Si in acrylic acid (a) without effect of UV radiation, (b) after UV irradiation; (c) and (d) show the samples after storage for 1 week. Bottom left – Fourier IR spectra of propionic acid (a), nc-Si functionalized by propionic acid (b), nc-Si functionalized by propionic acid and isolated by dialysis (c) and resuspended nc-Si, functionalized by propionic acid (d). On the right – luminescence of nc-Si functionalized by propionic acid (Sato, Swihart, 2006).

5.5.3. Functionalization of surface by substitutional reactions

One of the convenient methods for functionalization of the surface of the nc-Si is to use halogen (especially, Cl_2 and Br_2). The surfaces of such a composition can be directly derived from synthesis reactions, such as the use of Zintl salts as a precursor or formed as a result of modification following the synthesis of nanoparticles in the interaction of Si–H and Cl_2.

The reactive surface of the nc-Si, covered with groups Si–Cl groups, opens great possibilities for the modification of nanoparticles by using different reagents for reactions in the solution. First Bley, Kauzlarich (1996) demonstrated the possible reaction of chlorinated silica nanoparticles in the study of interaction between $SiCl_4$ and Zintl salts. FTIR spectra of the isolated product of the reaction have characteristic absorption bands of Si–O and C–H bending and stretching vibrations of saturated hydrocarbons. The authors suggested that the initial reaction product is the nanosilicon having chlorinated surface.

The subsequent interaction between these particles and methanol was used to remove impurities of salt, which eventually led to functionalization of the surface of the particles by the groups – OMe. Hydrophobic nanoparticles soluble in organic solvents were obtained as a result. Based on the observed electrophilic behaviour of the surface layer, (Mayeri et al, 2001; Yang et al, 1999) used standard nucleophilic reagents, such as alkyl lithium and Grignard reagents, which allowed to alkylate the surface of the nanoparticles. A similar approach was used in Rogozhina et al. (2001), where the reaction was carried out of the surface Si–Cl groups with butylamine that led to the formation of Si–N bonds on the surface of nanosilicon.

Clearly, the use of the electrophilic properties of the Si–Cl bonds on the surface of nc-Si and the demonstration of their reactivity, as well as the fact that similar reactions are possible for the Si–H and Si–Br under the influence of lithium aluminum hydride, $LiAlH_4$, allows us to offer a fairly comprehensive scheme of possible reactions (see Fig. 5.38). Apparently, so far, this method of surface modification has significant untapped potential in the chemistry nanosilicon (Pavesi, Turan, 2010).

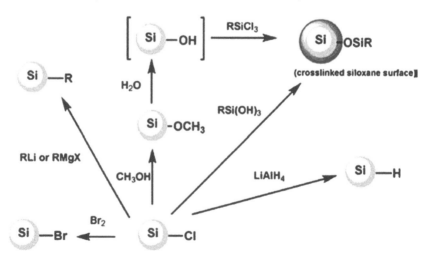

Fig. 5.38. Possible reactions at the surface of nc-Si, comprising Si–Cl bonds (Pavesi, Turan, 2010).

5.5.4. Using the oxidized surface layer

Some researchers suggest a protective layer of nc-Si based on the oxidized layer. Zou et al. (2004); Zou, Kauzlarich (2008) modified the surface of the chlorinated nanosilicon and produced very stable particles of nc-Si, coated with siloxane groups. Two independent methods were proposed for this. In the first case, Si–Cl bonds were converted to Si–OMe under the action of methanol. Further, the resulting product was subjected to hydrolysis. After this, the interaction of alkoxy groups with trihydroxy silanes produced silicon nanoparticles the surface which was bonded with cross-linked siloxane groups (see Fig. 5.39). The same product can be obtained by other methods. In this case, the direct reaction of alkoxy groups with trihydrosilanes took place. The modified particles of the nc-Si, obtained by this method in the isolation form formed a waxy product light yellow in colour, which was resuspended in organic solvents (such as chloroform or hexane) to give a yellow solution. Spectral analysis using Fourier IR spectroscopy and NMR fully confirmed the structural features of the surface – the presence of oxide and organic groups (see Fig. 5.38) (Zou et al, 2004; Pavesi, Turan, 2010).

The blue photoluminescence of the nc-Si particles, modified by the siloxane groups, depends on the wavelength of the exciting radiation. The particles are stable to photobleaching and environmental conditions for at least 60 days. This is in sharp contrasts with the behaviour of particles whose surface is covered with Si–Cl or Si–OR (OR=alkoxy), the luminescence of which stopped after 15 days.

How many groups are attached on the surface of the particles ? Despite all efforts made to study the surface of nc-Si, the effectiveness of its coverage is still open. Numerous estimates, based on the number of surface Si atoms of the nanoparticles, have been published to date. The number of surface atoms is the starting point in evaluating the approximate value of the coverage of alkylated nc-Si (Holmes et al, 2001; Zhao et al, 2004).

Holmes et al. (2001) used the XPS method to determine the number of groups attached to the silicon nanoparticles having an average diameter of 1.5 nm. The particles were produced by one-step synthesis in interaction

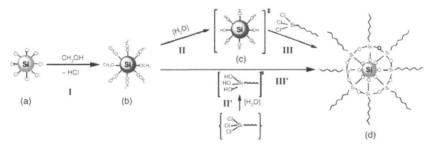

Fig. 5.39. Two independent methods of synthesis of alkylated nc-Si/SiO$_2$ particles with stable optical properties (Zou et al, 2004).

of nc-Si with diphenyl silane in supercritical octanol. According to the estimated, the average number of attached groups for one nanoparticle is 16. This is approximately half the amount that is required in order to create a close-packed monolayer.

Hua et al (2005, 2006) used the thermogravimetric method and standardized ^{1}H NMR solutions for nanoparticles of diameter ~2 nm and concluded that the particles obtained in photoinduced hydrosilylation contain an almost close-packed surface layer which contains 124 groups. This result suggests that the single-stage process *in situ*, used by Holmes et al. (2001), is a less effective method of functionalization of the surface compared to the surface hydrosilylation of nc-Si, coated with the Si–H groups after etching of the particles. However, it is known that if hydrosilylation is conducted after etching of the particles the surface retains $=$ SiH$_2$ groups and partial oxidation of the surface Si atoms with the formation of (Si–O) takes place.

Concluding remarks

One of the main tasks of nanophotonics and optoelectronics based on the nc-Si is to create quantum dots with the optically or electrically controlled luminescence properties. The luminescence of the ns-Si quantum dots smaller than the radius of the free Bohr exciton (4.3 nm), due to radiative recombination of carriers situated in the volume of the nanocrystallite. Thus, the change in the particle size distribution, in principle, allows the directional change of the spectrum of photoluminescence. However, the amplitude of the wave function of the charge carriers at the surface of the nc-Si quantum dot depends on the state of its surface, which contains silicon oxide (SiO$_x$, $0 < x < 2$), formed in the interaction of nanoparticles with the oxygen of the environment. The natural passivation of the shell by the amorphous layer of SiO$_x$ leads to the formation of defective structures which in general do not possess optical activity. Structural defects can effectively capture the charge carriers and thereby reduce the quantum yield of luminescence, since the recombination of electron–hole pairs in this case is a non-radiative process. One of the main problems to create stable and efficient luminescent composites based on nc-Si is the saturation of dangling bonds, the elimination of structural defects, and the protection of the surface from natural oxidation. One of the most effective methods of surface protection is the hydrosilylation by the molecules of organic compounds containing a terminal double bond.

Studies of nanosilicon in the last decade have been very intensive. Modified nanoparticles soluble in organic solvents and in water have been produced. It is shown that the properties of the produced material depend on the size of the central silicon core and the protective shell. However, only a few studies clearly show the influence of the shell on the optical properties of nanosilicon. Based on the analysis of several works presented in this chapter, it can be concluded that many obtained results are ambiguous and, in some cases, contradicting.

The chemical composition of the surface layer significantly affects the LUMO energy and, consequently, the value of the band gap of nc-Si with a diameter less than 5 nm (Zou et al, 2004; Reboredo, Galli, 2005). This allows us to understand the mismatch of the data from various studies of the photoluminescent properties of nanoparticles, which would seem to have identical dimensions.

In this regard, despite significant progress in the synthesis and chemistry of the surface of nc-Si, it is essential to develop methods to ensure investigation of the structure and the dynamics of the core and the surface of nanoparticles. These effective methods that can clarify the relationship of the central core of nc-Si and the surface layer, are femtosecond spectroscopy and 4D diffraction methods, the use of which to study nanosilicon and nanoobjects as a whole is described in Chapter 11.

References for Chapter 5

Abrosimova G. E., Aronin A. S. Influence of size on the perfection of the structure of nano-crystals based on Al and Ni. Fiz. Tverd. Tela. 2008. V. 50, No. 1. P. 154-158.

Aghekyan V. F. Photoluminescence of semiconductor crystals. Soros Educational Journal. 2000. V. 6, No. 10. P. 101-107.

Andrianov K. A., J. Soucek I., Khananashvili L. M. Hydride joining organohydroxylanes to compounds with multiple carbon–carbon bonds. Usp. Khimii.1979. V. 48, No. 7. P. 1233-1255.

Aslanov L. A., Zakharov V.N., Zakharov, M.A., Kamyshnyi A. L., Magdassi S., Yatsenko A.V. The stabilization of nanoparticles by carbenes. Koord. Khimiya. 2010. V. 36, No. 5. P. 330-332.

Astrova E. V., Nechitaylov A. A., Zabrodskii A. G. Silicon technology for microfuel elements. ISJAEE. 2007. No. 2 (46). P. 60-65.

Astrova E. V. ,et al. Silicon technology for hydrogen energy. Proc. Reports. Intern. Forum Hydrogen technologies for production of energy. M., 2006. P. 188-190.

Bagratashvili V. N., Belogorokhov A. I., Ischenko A. A., Storozenko P. A., Tutorsky I. A. Management of the spectral characteristics of multiphase ultrafine systems based on nanocrystalline silicon in the UV wavelength range. Dokl. RAN. Fiz. Khimiya. 2005. V. 405. P. 360-363.

Bardakhanov S. P. Radiation and nanostructures. Nauka v Siberi. 2006. No. 5.

Bardakhanov S. P., Gindulina V. Z., Lienko V. A. The possible use of nanosized powder in the manufacture of ceramics. Proc. Reports. V All-Russian conference. PPhysical chemistry of ultrafine systems. Novoural'sk, 9-13 October 2000. pp. 371-372.

Bielavski V. I. Excitons in low-dimensional systems. Soros Educational Journal. 1997. No. 5. P. 93-99.

Borisenko V. E. Nanoelectronics – the basis of the information systems of the XXI century. Soros Educational Journal. 1997. No. 5. S. 100-104.

Bresler M. S., Gusev O. B., Terukov E. I., Froitzheim A. A boundary-value silicon electroluminescence: heterostructure amorphous silicon-crystalline silicon. Fiz Tverd. Tela. 2004. V. 46, No. 1. P. 18-20.

Galaev I. Yu. Smart polymers in biotechnology and medicine. Usp. Khimii. 1995. V. 64, No. 5. P. 505-524.

Georgobiani A. N., Electroluminescence of semiconductors and semiconductor structures. Soros Educational Journal. 2000. V. 6, No. 3. P. 105-111.

Golovan' L. A., Timoshenko V. Yu., Kashkarov P. K. Optical properties of nanocomposites based on porous systems. Usp. Fiz Nauk. 2007. V. 177. P. 619-638.

Gubin S. P., Kataeva N. A., Khomutov G. B. Promising areas of nanoscience: chemistry nanoparticles of semiconductor materials. Izv. RAN. Ser. Khim. 2005, No. 4. P. 811-836.

Gusev A. I. Nanomaterials, nanostructures, nanotechnology. Moscow: Fizmatlit, 2005. 416 p.

Demikhovsky V. Y. Quantum wells, thread point. What is it?. Soros Educational Journal. 1997. No. 5. P. 80-86.

Dobrinskiy E. K., Uryukov B. A., Friedberg A. E. Research to stabilize the plasma jet by a swirl. Izv. SO AN SSSR. Ser. Tekh. Nauk. 1979. No. 3, P. 42-48.

Dorofeyev S. G. Optical and structural properties of thin films deposited from sol of nanoparticles of silicon. Fiz. Tekh. Poluprovod. 2009. V. 43, No. 11. P. 1420-1427.

Dorofeyev S. G., et al. Nanocrystalline silicon obtained from SiO. Nanotekhnika. 2010. No. 3 (23). P. 3-12.

Ekimov A. I., Onuschenko A. A. The quantum size effect in three-dimensional microscopic semiconductors. Pis'ma Zh. Eksp. Teor. Fiz. 1981. V. 34, No. 6. P. 363-366.

Efremov M. D., et al. Nanometer clusters and silicon nanocrystals. Bulletin of the Novosibirsk State. Univ. 2007. Volume 2. No. 2. P. 51-60.

Efremov M. F., et al. Visible photoluminescence of silicon nanopowder created by evaporation of silicon a powerful electron beam. Pis'ma Zh. Eksp. Teor. Fiz. 2004. V. 80. P. 619-622.

Zabrodskii A. G., et al. Micro-and nanotechnologies for portable fuel cells. International Scientific Journal for Alternative Energy and Ecology ISJAEE. 2007. No. 2 (46). P. 54-59.

Zimin S. P. Porous silicon - a material with new properties. Soros educational Journal. 2004. V. 8, No. 1. P. 101-107.

Ischenko A. A., Dorofeyev S. G., Kononov N. N., Eskova E. V., Ol'khov A. A. Application for issuance of patent, 16.12.2009. Registration No. 2009146715.

Ischenko A. A., Sviridov A.A. Sunscreen. II. Inorganic UV filters and their compositions with organic protectors. Izv. VUZ. Ser. Khim. Khim. Tekhnol. 2006. V. 49, No. 12. P. 3-16.

Kachurin G. A., et al. The influence of boron ion implantation and subsequent annealing on the properties of nanocrystals of Si. Fiz. Tekh. Poluprovod. 2006. V. 40, No. 1. P. 75-81.

Kashkarov P. K. The unusual properties of porous silicon. Soros Educational Journal. 2001. V. 7, No. 11. P. 102-107.

Korsunskii V. I., et al. A study by the method of radial distribution function of the local structure of nanocrystalline particles of CdS. Zh. Strukt. Khim. 2004. 45, No. 3. P. 452-461.

Korchagin A. A.. New method for nanoscale powders. Dynamics of Continuous environment. Novosibirsk, 2001. 8.

Krutikova A. A. Spectral analysis of composite materials based on nanocrystalline silicon. Author. dissertation. Moscow: MITHT. 2007.

Lisovski I. P., et al. Luminescence enhancement of structures with nanocrystalline silicon stimulated by low-dose irradiation. Fiz. Tekh. Poluprovod. 2008. V. 42, No. 5. P.

591-594.

Lukashov V. P., et al.A method for producing ultrafine silica, a device for implementation and ultradispersive silica. Patent of the Russian Federation No. 2067077. Bull. No. 27. 27.09.96.

Lukevits E. Ya, Voronkov M. G., Hydrosilylation. Riga, 1964. P. 5-73, 87-99, 101-319, 320-345.

Meitin M. Photovoltaics: Materials, Technology, Perspectives. Electronics: Science. Technology Business. 2000. No.. 6. Pp. 40-46.

Nakamoto K. Infrared Spectra of Inorganic and Coordination Compounds. Moscow: Mir, 1966. P. 147.

Neizvestnyi I. G., et al. Phase transformations in amorphous silicon films at low temperature. Poverkhnost'. 2007. No. 9. P. 95-102.

Ol'khov A. A., Goldshtrah M. A., Ischenko A. A. Application for issuance of a patent, 04.12.2009, Registration No. 2009145013.

Pool C., Owens, F. Nanotechnology. M. Tekhnosfera, 2004. 328 p.

Radtsig V. A., Rybaltovsky A. O., Ischenko A. A., Sviridova A. A. Thermo-oxidative processes in nanoscale powders of silicon. II. Paramagnetic centers. Nanotekhnika. 2007b. No. 3 (11). P. 116-121.

Radtsig V.A., et al. Thermo-oxidative processes in the nanoscale powders of silicon. I. Spectral manifestations. Nanotekhnika. 2007a. No. 3 (11). P. 110-116.

Rybaltovsky A.O., et al. Methods of experimental study of the spectral characteristics of water-emulsion composite media containing silica nanoparticles. GSSSD certificate from DOE 131-2007 July 12, 2007.

Rybaltovsky A.O., et al. The spectral features of water-emulsion composite media containing nanoparticles of silicon. Optika Spektroskopiya. 2006. V. 101, No. 4. P. 626-633.

Rybaltovsky A.O., et al. The spectral features of water-emulsion composite media containing nanoparticles of silicon. Optika Spektroskopiya. 2006. V. 100. P. 626-633.

Sviridova A. A., et al. Correlation technology for producing new UV-shielding composite materials based on nanocrystalline silicon. II. Scientific and Technical Conference of Young Scientists 'Science-intensive chemical technology'. M., 2007. P. 88.

Sviridova A. A., et al. Cosmetic protection from UV based on nanocrystalline silicon. Proc. Reports. XIII International Conference "New information technologies in medicine, bioogy, pharmacology and ecology, May 31-June 9, Ukraine, Crimea, Gursuf. Ed. A. N. Kurzanov. -Zaporozhye: Zaporizhye National University, 2005. P. 146-148.

Sviridova A. A., Ischenko A. A. Sunscreen. I. Classification and mechanism of action of organic UV filters. Izv. VUZ. Ser. Khim. Khim. Tekhnol. 2006. V. 49, No. 11. P. 3-14.

Smirnov B. M. Clusters and phase transitions. Usp. Fiz. Nauk. 2007. V. 177 (4). P. 233-274.

Smirnov B. M., et al. Nanocomposite proton-conducting membranes for microfuel elements. Alternative Energy (ISJAEE). 2007. No. 6 (50). P. 24-30.

Suzdalev I. P. Nanotechnology: Physics and chemistry of nanoclusters, nanostructures and nanomaterials. Moscow: KomKniga 2006. 592 p.

Timokhov D. F., Timokhov F. P. The influence of the crystallographic orientation of the silicon to form silicon nanoclusters in the anode for electrochemical etching.Fiz. Tekh. Poluprovod., 2009. V. 43, No. 1. P. 95-99.

Timoshenko V.A. et al. Silicon nanocrystals as a photosensitizer-active oxygen analyzers for biomedical applications. Fiz. 2006. V. 83, No. 9. P. 492-495.

Fetisov G. Synchrotron radiation. Methods for studying the structure of matter. Moscow: Fizmatlit, 2007.

Harris P. Carbon nanotubes and related structures. New materials for the XXI century. M.

Tekhnosfera, 2003. 336.

Shik A.Ya. Quantum wires. Soros Educational Journal. 1997. No. 5. P. 87-92.

Efros Al.L., Efros A. L. Interband absorption of light in a semiconductor sphere. Fiz. Tekh. Poluprovod. 1982. V. 16. P. 1209-1214.

Acciarri M., Binetti S., Bollani M., Fumagalli L., Pizzini S., von Känel H. Properties of nanocrystalline silicon films grown by LEPECVD for photovoltaic applications. Solar Energy Mat. and Solar Cells. 2005. V. 87. P. 11–24.

Albuschies J., Baus M., Winkler O., Hadam B., Spangenberg B., Kurz H. High-density silicon nanowire growth from self-assembled Au nanoparticles. Microelectronic Engineering. 2006. V. 83. P. 1530–1533.

Allan G., Delerue C., Lannoo M. Nature of luminescent surface states of semiconductor nanocrystallites. Phys. Rev. Lett. 1996. V. 76. P. 2961–2964.

Anthony R., Kortshagen U. Photoluminescence quantum yields of amorphous and crystalline silicon nanoparticles. Physical Review B. 2009.V. 80. P. 115407 (6 p.).

Arquier D., Calleja G., Cerveau G., Corriu R. J. P. A new solution route for the synthesis of silicon nanoparticles presenting different surface substituents. C.R. Chimie. 2007. No. 10. P. 795–802.

Arzhannikova S.A., Efremov M.D., Kamaev G.N. et al. Laser assited formation on nano-crystals in plasma-chemical deposited SiN$_x$ films. Solid State Phenomena. 2005. V. 108–109. P. 53–58.

Asoh H., Arai F., Ono S. Site-selective chemical etching of silicon using patterned silver catalyst. Electrochemistry Communications. 2007. V. 9. P. 535–539.

Bae Y., Lee D. C., Rhogojina E.V., Jurbergs D. C., Korgel B.A., Bard A. J. Electrochemistry and electrogenerated chemiluminescence of films of silicon nanoparticles in aqueous solution. Nanotechnology. 2006. V. 17. P. 3791–3797.

Bahruji H., Bowker M., Davies P. R. Photoactivated reaction of water with silicon nanoparticles. International Journal of Hydrogen Energy. 2009. V. 34. P. 8504–8510.

Baldwin R.K., Zou J., Pettigrew K.A., Yeagle G. J., Britt R.D., Kauzlarich S.M. The preparation of a phosphorus doped silicon film from phosphorus containing silicon nanoparticles. Chem. Commun. 2006. P. 658–660.

Barba D., Martin F., Ross G.G. Evidence of localized amorphous silicon clustering from Raman depth-probing of silicon nanocrystals in fused silica. Nanotechnology. 2008. V. 19. P. 115707 (5 p.).

Barba D., Martin F., Dahmoune C., Rossa G.G. Effects of oxide layer thickness on Si-nanocrystal photoluminescence intensity in Si$_+$-implanted SiO$_2$/Si systems. Appl. Phys. Lett. 2006. V. 89. P. 034107 (3 p.).

Bardakhanov S. P. Large scale nanopowders production by electron beam at atmospheric pressure. Proc. of Int. Conf. on Advanced Materials (ICAM 2008), Feb. 18–21, Kottayam, India. P. 44.

Bardakhanov S. P., Korchagin A. I., Kuksanov N.K., Lavrukhin A.V., Salimov R.A., Fadeev S.N., Cherepkov V.V. Nanopowder production based on technology of solid raw substances evaporation by electron beam accelerator. Materials Science and Engineering: B. 2006. V. 132. P. 204–208.

Bardakhanov S. P., Volodin V.A., Efremov M.D., Cherepkov V.V., Fadeev S.N., Korchagin A. I., Marin D. V., Golkovskiy M.G., Tanashev Yu. Yu., Lysenko V. I., Nomoev A.V., Buyantuev M.D., Sangaa D. High Volume Synthesis of Silicon Nanopowder by Electron Beam Ablation of Si Ingot at Atmospheric Pressure. Jpn. J. Appl. Phys. 2008. V. 47. P. 7019–7022.

Bardakhanov S. P. The formation of fine silica powder after vaporization of quartz. In: Abstracts of V International Conference on Computer-Aided Design of Advanced Materials and Technologies (CADAMT-97). Baikal Lake, Russia. 1997. P. 88–89.

Barreto J., Per´alvarez M., Rodr´ıguez J.A., Morales A., Riera M., L´opez M., Garrido B., Lechuga L., Dominguez C. Pulsed electroluminescence in silicon nanocrystals-based devices fabricated by PECVD. Physica E. 2007. V. 38. P. 193–196.

Beard M. C., Ellingson R. J. Multiple exciton generation in semiconductor nanocrystals: Toward efficient solar energy conversion. Laser & Photonics Reviews. 2008. V. 2, No. 5. P. 377–399.

Beard M. C., Knutsen K.K., Yu P., Luther J., Song Q., Ellingson R. J., Nozik A. J. Multiple exciton generation in colloidal silicon nanocrystals. Nano Lett. 2007. V. 7. P. 2506–2512.

Belomoin G., Therrien J., Nayfeh M. Oxide and hydrogen capped ultrasmall blue luminescent Si nanoparticles. Appl. Phys. Lett. 2000. V. 77. P. 779–781.

Belomoin G., Therrien J., Smith A., Rao S., Chaieb S., Nayfeh M.H. Spatially selective electrochemical deposition of composite films of metal and luminescent Si nanoparticles. Chem. Phys. Lett. 2003. V. 372, No. 3–4. P. 415–418.

Belomoin G., Therrien J., Smith A., Rao S., Twesten R., Chaieb S., Nayfeh M.H., Wagner L., Mitas L. Observation of a magic discrete family of ultrabright Si nanoparticles. Appl. Phys. Lett. 2002. V. 80. P. 841–843.

Benami A., Santana G., Monroy B.M., Ortiz A., Alonso J. C., Fandino J., Aguilar-Hernández J., Contreras-Puente G. Visible photoluminescence from silicon nanoclusters embedded in silicon nitride films prepared by remote plasma-enhanced chemical vapor deposition. Physica E. 2007. V. 38. P. 148–151.

Ben-Chorin M., Averboukh B., Kovalev D., Polisski G., Koch F. Influence of quantum confinement on the critical points of the band structure of Si. Phys. Rev. Lett. 1996. V. 77. P. 763–766.

Bergmann A., Fritz G., Glatter O. Solving the generalized indirect Fourier transform (GIFT) byBoltzmann simplex simulated annealing (BSSA). J. Appl. Cryst. 2000. V. 33. P. 1212–1216.

Bernstorff S., Dubcek P., Kovacevic I., Radic N., Pivac B. Si nanocrystals in SiO$_2$ films analyzed by small angle X-ray scattering. Thin Solid Films. 2007. V. 515. P. 5637–5640.

Bi L., He Y., Feng J. Y. Effect of post-annealing in oxygen atmosphere on the photoluminescence properties of nc-Si rich SiO$_2$ films. J. Cryst. Growth. 2006. V. 289. P. 564–567.

Bitten J. S., Lewis N. S., Atwater H.A., Polman A. Size-dependent oxygen-related electronic states in silicon nanocrystalls. Appl. Phys. Lett. 2004. V. 84. P. 5389–5391.

Bley R.A., Kazlarich S.M. F low-temperature solution phase route for the synthesis of silicon nanoclusters. J. Am. Chem. Soc. 1996. V. 118, No. 49. P. 12461–12462.

Boles S. T., Fitzgerald E.A., Thompson C.V., Ho C.K. F., Pey K. L. Catalyst proximity effects on the growth rate of Si nanowires. J. Appl. Phys. 2009. V. 106. P. 044311 (4 p.).

Boninelli S., Iacona F., Franzo G., Bongiorno C., Spinella C., Priolo F. Formation, evolution and photoluminescence properties of Si nanoclusters. J. Phys.: Condens. Matter. 2007. V. 19. P. 225003 (24 p.)

Bulutay C. Electronic Structure and Optical Properties of Silicon Nanocrystals along their Aggregation Stages. Condensed Matter. 2008. arXiv:0804.2016v1. P. 1–5.

Buriak J.M. Organometallic chemistry on silicon and germanium surfaces. Chem. Rev. 2002. V. 102. P. 1271–1308.

Büttner C. C., Langner A., Geuss M., Müller F., Werner P., Gösele U. Formation of Straight 10 nm Diameter Silicon Nanopores in Gold Decorated Silicon. ACS Nano. 2009. V. 3. P. 3122–3126.

Bux S.K., Rodriguez M., Yeung M. T., Yang C., Makhluf A., Blair R.G., Fleurial J. P., Kaner

R. B. Rapid Solid-State Synthesis of Nanostructured Silicon. Chem. Mater. 2010. V. 22. P. 2534–2540.

Bychto L., Balaguer M., Pastor E., Chirvony V., Matveeva E. Influence of preparation and storage conditions on photoluminescence of porous silicon powder with embedded Si nanocrystals. J. Nanopart. Res. 2008. V. 10. P. 1241–1249.

Canham L. T. Silicon quantum wire array fabrication by electrochemical and chemical dissolution of wafers. Appl. Phys. Lett. 1990. V. 57. P. 1046–1048.

Cao L., Garipcan B., Atchison J. S., Ni C., Nabet B., Spanier J. E. Instability and Transport of Metal Catalyst in the Growth of Tapered Silicon Nanowires. Nano Lett. 2006. V. 6. P. 1852–1857.

Caristia L., Nicotra G., Bongiorno C., Costa N., Ravesi S., Coffa S., De Bastiani R., Grimaldi M.G., Spinella C. The influence of hydrogen and nitrogen on the formation of Si nanoclusters embedded in sub-stoichiometric silicon oxide layers. Microelectronics Reliability. 2007. V. 47. P. 777–780.

Cavarroc M., Mikikian M., Perrier G., Boufendi L. Single-crystal silicon nanoparticles: An instability to check their synthesis. Appl. Phys. Lett. 2006. V. 89. P. 013107.

Chai K.-B., Seon C.R., Moon S. Y., Choe W. Parametric study on synthesis of crystalline silicon nanoparticles in capacitively-coupled silane plasmas. Thin Solid Films. 2010. V. 518. P. 6614–6618.

Chan M. Y., Lee P. S. Fabrication of silicon nanocrystals and its room temperature luminescence effects. International Journal of Nanoscience. 2006. V. 5. P. 565–570.

Chang J. B., Liu J. Z., Yan P.X., Bai L. F., Yan Z. J., Yuan X.M., Yang Q. Ultrafast growth of single-crystalline Si nanowires. Materials Letters. 2006. V. 60. P. 2125–2128.

Chao Y., Houlton A., Horrocks B. R., Hunt M. R. C., Poolton N. R. J., Yang J., ˜Siller L. Optical luminescence from alkyl-passivated Si nanocrystals under vacuum ultraviolet excitation: Origin and temperature dependence of the blue and orange emissions. Appl. Phys. Lett. 2006. V. 88. P. 263119.

Chao Y., Krishnamurthy S., Montalti M., Lie L.H., Houlton A., Horrocks B. R., Kjeldgaard L., Dhanak V. R., Hunt M. R. C., ˜Siller L. Reactions and luminescence in passivated Si nanocrystallites induced by vacuum ultraviolet and soft-x-ray photons. J. Appl. Phys. 2005. V. 98. P. 044316.

Chen D. Y., Wei D. Y., Xu J., Han P.G., Wang X., Ma Z. Y., Chen K. J., Shi W.H., Wang Q.M. Enhancement of electroluminescence in p–i–n structures with nano-crystalline Si/ SiO₂ multi- layers. Semicond. Sci. Technol. 2008. V. 23. P. 015013 (4 p.).

Chen D., Xie Z.-Q., Wu Q., Zhao Y.-Y., Lu M. Electroluminescence of Si nanocrystaldopped SiO₂. Chin. Phys. Lett. 2007. V. 24. P. 2390–2393.

Choi B.-J., Lee J.-H., Yatsui K., Yang S.-C. Preparation of silicon nanoparticles for device of photoluminescence. Surface & Coatings Technology. 2007a. V. 201. P. 5003–5006.

Choi J., Wang N. S., Reipa V. Conjugation of the Photoluminescent Silicon Nanoparticles to Streptavidin. Bioconjugate Chem. 2008. V. 19. P. 680–685.

Choi J., Wang N. S., Reipa V. Photoassisted Tuning of Silicon Nanocrystal Photoluminescence. Langmuir. 2007б. V. 23. P. 3388–3394.

Chourou M. L., Fukami K., Sakka T., Virtanen S., Ogata Y.H. Metal-assisted etching of ptype silicon under anodic polarization in HF solution with and without H₂O₂. Electrochimica Acta. 2010. V. 55. P. 903–912.

Cimpean C., Groenevegen V., Kuntermann V., Sommer A., Kryschi C. Ultrafast exciton relaxation dynamics in silicon quantum dots. Laser and Photonics Rev. 2009. V. 3, No. 1–2. P. 138–145.

Coffin H., Bonafos C., Schamm S., Cherkashin N., Ben Assayag G., Claverie A., Respaud M., Dimitrakis P., Normand P. Oxidation of Si nanocrystals fabricated by ultralow-energy ion implantation in thin SiO2 layers. J. Appl. Phys. 2006. V. 99. P. 044302.

Coxon P.R., Chao Y., Horrocks B. R., Gass M., Bangert U., Šiller L. Electron energy loss spectroscopy on alkylated silicon nanocrystals. J. Appl. Phys. 2008. V. 104. P. 084318 (8 p.).

Cuilis A.G., Canham L. T., Calcoct P. The structural and luminescence properties of porous silicon. J. Appl. Phys. 1997. V. 82. P. 909 (57 p.).

Dai Y., Huang B., Yu L., Han S., Dai D. Effects of surface oxygen on the electronic propeties of silicon nanoclusters. International Journal of Nanoscience. 2006. V. 5, No. 1. P. 13–21.

Daldosso N., Das G., Larcheri S., Mariotto G., Dalba G., Pavesi L., Irrera A., Priolo F., Iacona F.,

Rocca F. Silicon nanocrystal formation in annealed silicon-rich silicon oxide films prepared by plasma enhanced chemical vapor deposition. J. Appl. Phys. 2007. V. 101. P. 113510.

Das G., Ferraioli L., Bettotti P., De Angelis F., Mariotto G., Pavesi L., Di Fabrizio E., Soraru G.D. Si-nanocrystals/SiO2 thin films obtained by pyrolysis of sol–gel precursors. Thin Solid Films. 2008. V. 516. P. 6804–6807.

Delerue C., Allan G., Lannoo M. Theoretical aspects of the luminescence of porous silicon. Phys. Rev. B. 1993. V. 48. P. 11024–11036.

Dickinson F.M., Alsop T.A., Al-Sharif N., Berger C. E.M., Datta H.K., ˙Siller L., Chao Y., Tuite E.M., Houlton A., Horrocks B. R. Dispersions of alkyl-capped silicon nanocrystals in aqueous media: photoluminescence and ageing. Analyst. 2008. V. 133. P. 1573–1580.

Ding L., Chen T. P., Yang M., Zhu F. R. Influence of nanocrystal distribution on electroluminescence from Si⁻-implanted SiO2 thin films. Proc. of SPIE. 2008. V. 6898. P. 68980H.

Djurabekova F., Björkas C., Nordlund K. Atomistic modelling of the interface structure of Si nanocrystals in silica. Journal of Physics: Conference Series. 2008. V. 100. P. 052023.

Dohnalová K., Kusová K., Pelant I. Time-resolved photoluminescence spectroscopy of the initial oxidation stage of small silicon nanocrystals. Appl. Phys. Lett. 2009. V. 94. P. 211903.

Dowd A., Johansson B., Armstrong N., Ton-That C., Phillips M. Cathodoluminescence as a method of extracting detailed information from nanophotonics systems: a study of silicon nanocrystals. Proc. of SPIE. 2006. V. 6038. P. 60380J.

Du X.-W., Qin W.-J., Lu Y.-W., Han X., Fu Y.-S., Hu S.-L. Face-centered-cubic Si nanocrystals prepared by microsecond pulsed laser ablation. J. Appl. Phys. 2007. V. 102. P. 013518.

Ellingson R., Beard M., Johnson J., Murphy J., Knutsen K., Gerth K., Luther J., Hanna M., Micic O., Shabaev A., Efros A. L., Nozik A. J. Nanocrystals generating >1 electron per photon may lead to increased solar cell efficiency. 2006. SPIE Newsroom, Article No. 10.1117/2.1200606.0229 (4 p.).

Ellingson R. J., Beard M. C., Johnson J. C., Yu P., Micic O. I., Nozik A. J., Shabaev A., Efros A. L. Highly efficient multiple exciton generation in colloidal PbSe and PbS quantum dots. Nano Lett. 2005. Vol: 5. P. 865–871.

English D. S., Pell L. E., Yu Z., Barbara P. F., Korgel B.A. Size Tunable Visible Luminescence from Individual Organic Monolayer Stabilized Silicon Nanocrystal Quantum Dots. Nano Lett. 2002. V. 2, No. 7. P. 681–685.

Faraci G., Gibilisco S., Pennisi A.R., Franzo G., La Rosa S., Lozzi L. Catalytic role of adsorbates in the photoluminescence emission of Si nanocrystals. Phys. Rev. B. 2008.

V. 78. P. 245425 (6 p.).

Ferraioli L., Wang M., Pucker G., Navarro-Urrios D., Daldosso N., Kompocholis C., Pavesi L. Photoluminescence of Silicon Nanocrystals in Silicon Oxide. Journal of Nanomaterials. 2007. Article ID 43491. 5 p.

Fitting H.-J., Kourkoutis L. F., Salh R., Zamoryanskaya M.V., Schmidt B. Silicon nanocluster aggregation in SiO$_2$:Si layers. Phys. Status Solidi A. 2010. V. 207, No. 1. P. 117–123.

Fojtik A., Henglein A. Surface Chemistry of Luminescent Colloidal Silicon Nanoparticles. J. Phys. Chem. B. 2006a. V. 110. P. 1994–1998.

Fojtik A., Valenta J., Stuchl´ıkov´a T.H., Stuchl´ık J., Pelant I., Kocka J. Electroluminescence of silicon nanocrystals in p–i–n diode structures. Thin Solid Films. 2006б. V. 515. P. 775–777.

Forman R.A., Thurber W. R., Aspens D. E. Second indirect band gap in silicon. Solid State Comm. 1974. V. 14. P. 1007–1009.

Froner E., Adamo R., Gaburro Z., Margesin B., Pavesi L., Rigo A., Scarpa M. Luminescence of porous silicon derived nanocrystals dispersed in water: dependence on initial porous silicon oxidation. J. Nanoparticle Research. 2006. V. 8, No. 6. P. 1071–1074.

Fujii M., Hayashi S., Yamamoto K. Growth of Ge microcrystals in SiO$_2$ thin film matrices: A Raman and electron microscopic study. Jpn. J. Appl. Phys. 1991. V. 30. P. 687–694.

Furukawa S., Miyasato T. Quantum size effects on the optical band gap of microcrystalline Si:H. Phys. Rev. B. 1988. V. 38. P. 5726–5729.

Furukawa S., Miyasato T. Three-dimensional quantum well effects in ultrafine silicon particles. Jpn. J. Appl. Phys. 1988. V. 27. P. L2207–L2209.

Gawlik G., Jagielski J. Visible light emission from silicon dioxide with silicon nanocrystals. Vacuum. 2007. V. 81. P. 1371–1373.

Gelloz B., Koshida N. Highly enhanced photoluminescence of as-anodized and electrochemically oxidized nanocrystalline p-type porous silicon treated by high-pressure water vapor annealing. Thin Solid Films. 2006. V. 508. P. 406–409.

Giuliani J. R., Harley S. J., Carter R. S., Power P. P., Augustine M. P. Using liquid and solid state NMR and photoluminescence to study the synthesis and solubility properties of amine capped silicon nanoparticles. Solid State Nuclear Magnetic Resonance. 2007. V. 32. P. 1–10.

Godefroo S., Hayne M., Jivanescu M., Stesmans A., Zacharias M., Lebedev O. I., Van Tendeloos G., Moshchalkov V.V. Classification and control of the origin of photoluminescence from Si nanocrystals. Nature Nanotechnology. 2008. V. 3. P. 174–178.

Gutsulyk D. V., van der Est A., Nikonov G. I. Facile Hydrosilylation of Pyridines. Angew. Chem. Int. Ed. 2010. V. 50. P. 1384–1387.

Hahn T., Heimfarth J. P., Roewer G., Kroke E. Silicon nano particles: surface characterization, defects and electronic properties. Phys. stat. sol. (b). 2008. V. 245, No. 5. P. 959–962.

Heath J. R. A liquid-solution-phase synthesis of crystalline silicon. Science. 1992. V. 258. P. 1131–1133.

Heintz A. S., Fink M. J., Mitchell B. S. Mechanochemical Synthesis of Blue Luminescent Alkyl/Alkenyl-Passivated Silicon Nanoparticles. Adv. Mater. 2007. V. 19. P. 3984–3988.

Heintz A. S., Fink M. J., Mitchell B. S. Silicon nanoparticles with chemically tailored surfaces. Appl. Organometal. Chem. 2010. V. 24. P. 236–240.

Heitmann J., Muller F., Zacharias M., Gosele U. Silicon nanocrystals size matters. Adv. Mater. 2005. V. 17 (7). P. 795–803.

Heitsch A. T., Fanfair D.D., Tuan H.-Y., Korgel B.A. Solution-Liquid-Solid (SLS) Growth of Silicon Nanowires. J. Am. Chem. Soc. 2008. V. 130. P. 5436–5437.

Hernandez S., Martinez A., Pellegrin P., Lebour Y., Garrido B., Jordana E., Fedeli J.M. Silicon nanocluster crystallization in SiO_x films studied by Raman scattering. J. Appl. Phys. 2008. V. 104. 044304 (5 p.).

Hersam M. C., Guisinger N. P., Lyding J.W. Silicon-based molecular nanotechnology. Nanotechnology. 2000. V. 11. P. 70–76.

Hessel C.M., Henderson E. J., Kelly J.A., Cavell R.G., Sham T.-K., Veinot J.G. C. Origin of luminescence from silicon nanocrystals: a near edge x-ray absorption fine structure (NEXAFS) and X-ray excited optical luminescence (XEOL) study of oxide-embedded and free-standing systems. J. Phys. Chem.: C. 2008. V. 112. P. 14247–14254.

Hessel C.M., Henderson E. J., Veinot J. G. C. Hydrogen Silsesquioxane: A Molecular precursor for nanocrystalline Si–SiO_2 composites and freestanding hydride-surface-terminated silicon nanoparticles. Chem. Mater. 2006. V. 18. P. 6139–6146.

Hirasawa M., Orii T., Seto T. Size-dependent crystallization of Si nanoparticles. Appl. Phys. Lett. 2006. V. 88. P. 093119.

Hodes G. When Small Is Different: Some Recent Advances in Concepts and Applications of Nanoscale Phenomena. Adv. Mater. 2007. V. 19. P. 639–655.

Holm J., Roberts J. T. Surface Chemistry of Aerosolized Silicon Nanoparticles: Evolution and Desorption of Hydrogen from 6-n Diameter Particles. J. Am. Chem. Soc. 2007a. V. 129. P. 2496–2503.

Holm J., Roberts J. T. Thermal Oxidation of 6 nm Aerosolized Silicon Nanoparticles: Size and Surface Chemistry Changes. Langmuir. 20076. V. 23. P. 11217–11224.

Holmes J.D., Ziegler K. J., Doty R. C., Pell L. E., Johnston K. P., Korgel B.A. Highly Luminescent Silicon Nanocrystals with Discrete Optical Transitions. J. Am. Chem. Soc. 2001. V. 123. P. 3743–3748.

Hua F., Erogbogbo F., Swihart M. T., Ruckenstein E. Organically capped silicon nanoparticles with blue photoluminescence prepared by hydrosilylation followed by oxidation. Langmuir. 2006. V. 22. P. 4363–4370.

Hua F., Swihart M. T., Ruckenstein E. Efficient Surface Grafting of Luminescent Silicon Quantum Dots by Photoinitiated Hydrosilylation. Langmuir. 2005. V. 21. P. 6054–6062.

Huy P. T., Thu V.V., Chien N.D., Ammerlaan C.A. J., Weber J. Structural and optical properties of Si-nanoclusters embedded in silicon dioxide. Physica B. 2006. V. 376–377. P. 868–871.

Iwayama T. S., Hama T., Hole D. E. RTA effects on the formation process of embedded luminescent Si nanocrystals in SiO_2. Microelectronics Reliability. 2007. V. 47. P. 781–785.

Iwayama T. S., Hama T., Hole D. E., Boyd I.W. Enhanced luminescence from encapsulated silicon nanocrystals in SiO_2 with rapid thermal anneal. Vacuum. 2006. V. 81. P. 179–185.

Jivanescu M., Stesmans A., Godefroo S., Zacharias M. Electron spin resonans of Si nanocrystals embedded in a SiO_2 matrix. J. Optoelectr. Adv. Mater. 2007. V. 9, No. 3. P. 721–724.

Jung Y.-J., Yoon J.-H., Elliman R.G., Wilkinson A.R. Photoluminescence from Si nanocrystals exposed to a hydrogen plasma. J. Appl. Phys. 2008. V. 104. 083518.

Jurbergs D., Rogojina E., Mangolini L., Kortshagen U. Silicon nanocrystals with ensemble quantum yields exceeding 60%. Appl. Phys. Lett. 2006. V. 88. P. 233116.

Kamikake T., Imamura M., Murase Y., Tanaka A., Yasuda H. Electronic structure of alkyl-passivated Si nanoparticles: synchrotron-radiation photoemission study. Transactions of the Materials Research Society of Japan. 2007. V. 32, No. 2. P. 433–436.

Kamyshny A., Zakharov V.N., Zakharov M.A., Yatsenko A.V., Savilov S.V., Aslanov L.A., Magdassi S. Photoluminescent silicon nanocrystals stabilizes by ionic liquids. J. Nanopart. Res. 2011. V. 13. P. 1971–1978.

Kandoussi K., Simon C., Coulon N., Belarbi K., Mohammed-Brahim T. Nanocrystalline silicon TFT process using silane diluted in argon–hydrogen mixtures. Journal of Non-Crystalline Solids. 2008. V. 354. P. 2513–2518.

Kanemitsu Y. Light emission from porous silicon and related materials (Review Article). Phys. Rep. 1995. V. 263. P. 1–91.

Kanemitsu Y. Light-emitting silicon materials. J. Luminescence. 1996. V. 70. P. 333–342.

Kang Z. T., Arnold B., Summers C. J., Wagner B.K. Synthesis of silicon quantum dot buried SiO_x films with controlled luminescent properties for solid-state lighting. Nanotechnology. 2006. V. 17. P. 4477–4482.

Kelly J.A., Henderson E. J., Hessel C.M., Cavell R.G., Veinot J.G. C. Soft X-ray spectroscopy of oxide-embedded and functionalized silicon nanocrystals. Nuclear Instruments and Methods in Physics Research. B. 2010. V. 268. P. 246–250.

Khriachtchev L., Räsäanen M., Novikov S., Lahtinen J. Tunable wavelength-selective waveguiding of photoluminescence in Si-rich silica optical wedges. J. Appl. Phys. 2004. V. 95, No. 12. 104316.

Kim S., Park Y.M., Choi S.-H., Kim K. J. Origin of cathodoluminescence from Si nanocrystal/SiO_2 multilayers. J. Appl. Phys. 2007. V. 101. P. 034306.

Klein S., Finger F., Carius R., Dylla T., Klamfass J. J. Relationship between the optical absorption and the density of deep gap states in microcrystalline silicon. J. Appl. Phys. 2007. V. 102. P. 103501–6.

Koch C. C. (Ed.). Nanostructured Materials. Processing, properties, and applications. N.Y.: William Andrew Publishing. 2009. 752 p.

Korchagin A. I., Kuksanov N.K., Lavrukhin A.V., Fadeev S.N., Salimov R.A., Bardakhanov S. P., Goncharov V.B., Suknev A. P., Paukshtis E.A., Larina T.V., Zaikovskii V. I., Bogdanov S.V., Bal'zhinimaev B. S. Production of silver nano-powders by electron beam evaporation. Vacuum. 2005. V. 77, No. 4. P. 485–491.

Korovin S. B., Pustovoi V. I., Krinetskii B. B., Fadeeva S. Optical properties of metal-coated silicon nanocrystals. Proc. SPIE. 1999. V. 4070. P. 465–471.

Korovin S. B., Pustovoi V. I., Orlov A.N. High nonlinear susceptibility of the silicon based nanostructures. Proc. SPIE. 1999. V. 4070. P. 472–478.

Koshida N., Gelloz B. Photonic and relateddevice applications of nano-crystalline silicon. Proc. of SPIE. 2007. V. 6775. P. 67750N (9 p.).

Kovacevic I., Dubcek P., Duguay S., Zorc H., Radic N., Pivac B., Slaoui A., Bernstorff S. Silicon nanoparticles formation in annealed SiO/SiO_2 multilayers. Physica E. 2007. V. 38. P. 50–53.

Kovalev A., Wainstein D., Tetelbaum D., Mikhaylov A., Pavesi L., Ferrarioli L., Ershov A., Belov A. The electron and crystalline structure features of ionsynthesized nanocomposite of Si nanocrystals in Al_2O_3 matrix revealed by electron spectroscopy. Journal of Physics: Conference Series. 2008. V. 100. P. 072012.

Kovalev A., Wainstein D., Tetelbaum D., Mikhailov A. The peculiarities of electronic structure of Si nanocrystals formed in SiO_2 and Al_2O_3 matrix with and without P doping. Surf. Interface Anal. 2006. V. 38. P. 433–436.

Kravitz K., Kamyshny A., Gedanken A., Magdassi S. Solid state synthesis of water-dispersible silicon nanoparticles from silica nanoparticles. J. Solid State Chemistry. 2010. V. 183. P. 1442–1447.

Kumar V. (Ed.). Nanosilicon. Elsevier Ltd. 2008. xiii+368 p.

Kuntermann V., Cimpean C., Brehm G., Sauer G., Kryschi C., Wiggers H. Femtosecond transient absorption spectroscopy of silanized silicon quantum dots. Phys. Rev. 2008. V. B77. P. 115343–1.

Lacour F., Guillois O., Portier X., Perez H., Herlin N., Reynaud C. Laser pyrolysis synthesis and characterization of luminescent silicon nanocrystals. Physica E. 2007. V. 38. P. 11–15.

Lechner R., Stegner A.R., Pereira R.N., Dietmueller R., Brandt M. S., Ebbers A., Trocha M., Wiggers H., Stutzmann M. Electronic properties of doped silicon nanocrystal films. J. Appl. Phys. 2008. V. 104. P. 053701.

Ledoux G., Guillois O., Porterat D., Reynaud C., Huisken F., Kohn B., Paillard V. Photoluminescence properties of silicon nanocrystals as a function of their size. Phys. Rev. B. 2000. V. 62 (23). P. 15942–15951.

Ledoux J., Gong J., Huisken F., Guillois O., Reynaud C. Photoluminescence of size-separated silicon nanocrystals: Confirmation of quantum confinement. Appl. Phys. Lett. 2002. V. 80. P. 4834–4836.

Levitcharsky V., Saint-Jacques R.G., Wang Y.Q., Nikolova L., Smirani R., Ross G.G. Si implantation in SiO_2: Stucture of Si nanocrystals and composition of SiO_2 layer. Surface & Coatings Technology. 2007. V. 201. P. 8547–8551.

Li X., He Y., Talukdar S. S., Swihart M. T. Process for Preparing Macroscopic Quantities of Brightly Photoluminescent Silicon Nanoparticles with Emission Spanning the Visible Spectrum. Langmuir. 2003. V. 19. P. 8490–8496.

Li X., He Y., Swihart M. T. Surface functionalization of silicon nanoparticles produced by laser-driven pyrolysis of silane followed by HF-HNO_3 etching. Langmuir. 2004. V. 20. P. 4720–4727.

Li Z. F., Ruckenstein E. Water-Soluble Poly(acrylic acid) Grafted Luminescent Silicon Nanoparticles and Their Use as Fluorescent Biological Staining Labels. Nano Lett. 2004. V. 4. P. 1463–1467.

Liao Y.-C., Nienow A.M., Roberts J. T. Surface Chemistry of Aerosolized Nanoparticles:Thermal Oxidation of Silicon. J. Phys. Chem. B. 2006a. V. 110. P. 6190–6197.

Liao Y.-C., Roberts J. T. Self-Assembly of Organic Monolayers on Aerosolized Silicon Nanoparti- cles. J. Am. Chem. Soc. 2006б. V. 128. P. 9061–9065.

Lie L.H., Duerdin M., Tuite E.M., Houlton A., Horrocks B. R. Preparation and characterisation of luminescent alkylated-silicon quantum dots. Journal of Electroanalytical Chemistry. 2002. V. 538–539. P. 183–190.

Ligman R.K., Mangolini L., Kortshagen U. R., Campbell S.A. Electroluminescence from surface oxidized silicon nanoparticles dispersed within a polymer matrix. Appl. Phys. Lett. 2007. V. 90. P. 061116.

Lioutas Ch. B., Vouroutzis N., Tsiaoussis I., Frangis N., Gardelis S., Nassiopoulou A.G. Columnar growth of ultra-thin nanocrystalline Si films on quartz by Low Pressure Chemical Vapor Deposition: accurate control of vertical size. Phys. stat. sol. (a). 2008. V. 205, No. 11. P. 2615–2620.

Liu H. I., Biegelsen D.K., Ponce F.A., Johnson N.M., Pease R. F.W. Self-limiting oxidation for fabricating sub-5 nm silicon nanowires. Appl. Phys. Lett. 1994. V. 64. P. 1383–1385.

Liu M., Lu G., Chen J. Synthesis, assembly, and characterization of Si nanocrystals and Si nanocrystal–carbon nanotube hybrid structures. Nanotechnology. 2008. V. 19. P. 265705 (5 p.).

Liu Q., Kauzlarich S.M. A new synthetic route for the synthesis of hydrogen terminated silicon nanoparticles. Materials Science and Engineering. 2002. V. B96. P. 72–75.

Liu S.M., Sato S., Kimura K. Synthesis of luminescent silicon nanopowders redispersible to various solvents. Langmuir. 2005. V. 21. P. 6324–6329.

Liu S.-M., Yang Y., Sato S., Kimura K. Enhanced Photoluminescence form Si Nano-organosols by Functionalization with Alkenes and Their Size Evolution. Chem. Mater. 2006. V. 18. P. 637–642.

Liu Y., Chen T. P., Ding L., Yang M., Wong J. I., Ng C. Y., Yu S. F., Li Z.X., Yuen C., Zhu F. R., Tan M. C., Fung S. Influence of charge trapping on electroluminescence from Si-nanocrystal light emitting structure. J. Appl. Phys. 2007. V. 101. 104306.

Lopez-Suarez A., Fandino J., Monroy B.M., Santana G., Alonso J. C. Study of the influence of NH3 flow rates on the structure and photoluminescence of silicon-nitride films with silicon nanoparticles. Physica E. 2008. V. 40. P. 3141–3146.

Lu T. Z., Alexe M., Scholz R., Talalaev V., Zhang R. J., Zacharias M. Si nanocrystal based memories: Effect of the nanocrystal density. J. Appl. Phys. 2006. V. 100. P. 014310.

Lu Y.W., Du X.W., Sun J., Hu S. L., Han X., Li H. Formation and luminescent properties of face-centered-cubic Si nanocrystals in silica matrix by magnetron sputtering with substrate bias. Appl. Phys. Lett. 2007. V. 90. P. 241910.

Lu Z.H., Lockwood D. J., Baribean J.M. Quantum confinement and light emission in SiO_2/Si superlattices. Nature. 1995. V. 378. P. 258–260.

Lugstein A., Steinmair M., Hyun Y. J., Hauer G., Pongratz P., Bertagnolli E., Pressure-Induced P. Orientation Control of the Growth of Epitaxial Silicon Nanowires. Nano Lett. 2008. V. 8, No. 8. P. 2310–2314.

Ma K., Feng J. Y., Zhang Z. J. Improved photoluminescence of silicon nanocrystals in silicon nitride prepared by ammonia sputtering. Nanotechnology. 2006. V. 17. P. 4650–4653.

Mangolini L., Jurbergs D., Rogojina E., Kortshagen U. High efficiency photoluminescence from silicon nanocrystals prepared by plasma synthesis and organic surface passivation. Phys. stat. sol. (c). 2006a. V. 3, No. 11. P. 3975–3978.

Mangolini L., Jurbergs D., Rogojina E., Kortshagen U. Plasma synthesis and liquid-phase surface passivation of brightly luminescent Si nanocrystals. Journal of Luminescence. 2006б. V. 121. P. 327–334.

Mangolini L., Kortshagen U. Plasma-Assisted Synthesis of Silicon Nanocrystal Inks. Adv. Mater. 2007. V. 19. P. 2513–2519.

Martin J., Cichos F., Huisken F., von Borczyskowski C. Electron-phonon coupling and localization of excitons in single silicon nanocrystals. Nano Lett. 2008. V. 8. P. 656–660.

Mayandi J., Finstad T.G., Foss S., Thøgersen A., Serincan U., Turan R. Ion beam synthesized luminescent Si nanocrystals embedded in SiO_2 films and the role of damage on nucleation during annealing. Surface & Coatings Technology. 2007. V. 201. P. 8482–8485.

Mayeri D., Phillips B. L., Augustine M. P., Kauzlarich S.M. NMR stydy of the synthesis of alkyl-terminated silicon nanoparticles from the reaction of $SiCl_4$ with the Zintl salt. Chem. Mater. 2001. V. 13. P. 765–770.

Meier C., Gondorf A., Lüttjohann S., Lorke A., Wiggers H. Silicon nanoparticles: Absorption, emission, and the nature of the electronic band gap. J. Appl. Phys. 2007. V. 101. P. 103112 (8 p.).

Morana B., de Sande J. C.G., Rodriguez A., Sangrador J., Rodriguez T., Avella M., Jimenez J. Optimization of the luminescence emission of Si nanocrystals synthesized from non-stoichiometric Si oxides using a Central Composite Design of the deposition process. Materials Science and Engineering. B. 2008. V. 147. P. 195–199.

Nakama Y., Nagamachi S., Ohta J., Nunoshita M. Position-Controlled Si Nanocrystals in a SiO_2 Thin Film Using a Novel Amorphous Si Ultra-Thin-Film 'Nanomask' due to a

Bio-Nanoprocess for Low-Energy Ion Implantation. Applied Physics Express. 2008. V. 1. P. 034001.

Nayfeh O.M., Antoniadis D.A., Mantey K., Nayfeh M.H. Uniform delivery of silicon nanoparticles on device quality substrates using spin coating from isopropyl alcohol colloids. Appl. Phys. Lett. 2009. V. 94. P. 043112.

Nelles J., Sendor D., Ebbers A., Petrat F.M., Wiggers H., Schulz C., Simon U. Functionalization of silicon nanoparticles via hydrosilylation with 1-alkenes. Colloid Polym. Sci. 2007. V. 285. P. 729–736.

Nelles J., Sendor D., Petrat F.-M., Simon U. Electrical properties of surface functionalized silicon nanoparticles. J. Nanopart. Res. 2010. V. 12. P. 1367–1375.

Nguyen-Tran Th., Roca i Cabarrocas P., Patriarche G. Study of radial growth rate and size control of silicon nanocrystals in square-wave-modulated silane plasmas. Appl. Phys. Lett. 2007. V. 91. P. 111501.

Nicotra G., Franzo G., Spinella C. Evaluation of the excess and clustered silicon profiles in a silicon implanted SiO_2 layer. Nuclear Instruments and Methods in Physics Research. B. 2007. V. 257. P. 104–107.

Nielsen D., Abuhassan L., Alchihabi M., Al-Muhanna A., Host J., Nayfeh M.H. Current-less anodization of intrinsic silicon powder grains: Formation of fluorescent Si nanoparticles. J. Appl. Phys. 2007. V. 101. P. 114302.

Nozaki S., Chen C. Y., Kimura S., Ono H., Uchida K. Photoluminescence of Si nanocrystals formed by the photosynthesis. Thin Solid Films. 2008. V. 517. P. 50–54.

Nozaki T., Sasaki K., Ogino T., Asahi D., Okazaki K. Microplasma synthesis of tunable photoluminescent silicon nanocrystals. Nanotechnology. 2007. V. 18. P. 235603 (6pp).

O'Farrell N., Houlton A., Horrocks B. R. Silicon nanoparticles: applications in cell biology and medicine. International Journal of Nanomedicine. 2006. V. 1, No. 4. P. 451–472.

Okada T., Higashi S., Kaku H., Yorimoto T., Murakami H., Miyazaki S. Growth of Si crystalline in SiOx films induced by millisecond rapid thermal annealing using thermal plasma jet. Solid-State Electronics. 2008. V. 52. P. 377–380.

Okano T. (Ed.). Biorelated Polymers and Gels. San Diego: Acad. Press, 1998. 568 p.

Pailard V., Puech P., Laguna M.A., Carles R., Kohn B., Huisken F. Improved one-phonon confinement model for an accurate size determination of silicon nanocrystals. J. Appl. Phys. 1999. V. 86. P. 1921 (4 p.).

Pan X.-W., Shi M.-M., Zheng D.-X., Liu N., Wu G., Wang M., Chen H.-Z. Room-temperature solution route to free-standing SiO_2-capped Si nanocrystals with green luminescence. Materials Chemistry and Physics. 2009. V. 117. P. 517–521.

Park W. I., Zheng G., Jiang X., Tian B., Lieber C.M. Controlled Synthesis of Millimeter-Long Silicon Nanowires with Uniform Electronic Properties. Nano Lett. 2008. V. 8, No. 9. P. 3004–3009.

Parm I.O., Yi J. Exciton luminescence from silicon nanocrystals embedded in silicon nitride film. Materials Science and Engineering. B. 2006. V. 134. P. 130–132.

Pavesi L., Dal Negro L., Mazzoleni C., Franzo G., Priolo F. Optical gain in silicon nanocrystals. Nature. 2000. V. 408. P. 440–444.

Pereira R.N., Stegner A.R., Klein K., Lechner R., Dietmueller R., Wiggers H., Brandt M. S., Stutzmann M. Electronic transport through Si nanocrystal films: spin-dependent conductivity studies. Physica B. 2007. V. 401–402. P. 527–530.

Pi X.D., Liptak R.W., Campbell S.A., Kortshagen U. In-flight dry etching of plasma-synthesized silicon nanocrystals. Appl. Phys. Lett. 2007a. V. 91. P. 083112.

Pi X.D., Liptak R.W., Deneen Nowak J., Wells N. P., Carter C.B., Campbell S.A., Kortshagen U. Air-stable full-visible-spectrum emission from silicon nanocrystals synthesized by an all-gas-phase plasma approach. Nanotechnology. 2008. V. 19. P. 245603

(5 p).

Pi X.D., Mangolini L., Campbell S.A., Kortshagen U. Room-temperature atmospheric oxidation of Si nanocrystals after HF etching. Physical Review. B. 2007̄. V. 75. P. 085423.

Poruba A., Fejfar A., Remes Z., Springer J., Vanecek M., Kocka J., Meier J., Torres P., Shah A. Optical absorption and light scattering on microcrystalline silicon thin films and solar cells. J. Appl. Phys. 2000. V. 88. P. 148–160.

Pradhan S., Chen S., Zou J., Kauzlarich S.M. Photoconductivity of Langmuir-Blodgett Monolayers of Silicon Nanoparticles. J. Phys. Chem. C. 2008. V. 112. P. 13292–13298.

Puzder A., Williamson A. J., Grossman J. C., Gally G. Computational studies of optical emission of silicon nanocrystals. J. Am. Chem. Soc. 2003. V. 125. P. 2786–2791.

Puzder A., Williamson A. J., Grossman J. C., Gally G. Surface chemistry of silicon nanoclusters. Phys. Rev. Lett. 2002. V. 88. P. 097401–4.

Qi P., Wong W. S., Zhao H., Wang D. Low-temperature synthesis of Si nanowires using multizone chemical vapor deposition methods. Appl. Phys. Lett. 2008. V. 93. P. 163101.

Qiu J., Shen W., Yu R., Yao B. Solvothermal Preparation of Silicon Nanocrystals. Chemistry Letters. 2008. V. 37, No. 6. P. 644–645.

Rafiq M.A., Durrani Z.A.K., Mizuta H., Hassan M.M., Oda S. Field-dependant hopping conduction in silicon nanocrystal films. J. Appl. Phys. 2008. V. 104. P. 123710.

Rafiq M.A., Tsuchiya Y., Mizuta H., Oda S., Uno S., Durrani Z.A.K., Milne W. I. Hopping conduction in size-controlled Si nanocrystals. J. Appl. Phys. 2006. V. 100. P. 014303.

Reboredo F.A., Galli G. Theory of alkyl-terminated silicon quantum dots. J. Phys. Chem. B. 2005. V. 109. P. 1072–1078.

Riabinina D., Durand C., Chaker M., Rosei F. Photoluminescent silicon nanocrystals synthesized by reactive laser ablation. Appl. Phys. Lett. 2006. V. 88. 073105.

Rioux D., Laferriere M., Douplik A., Shah D., Lilge L., Kabashin A.V., Meunier M.M. Silicon nanoparticles produced by femtosecond laser ablation in water as novel contamination-free photosensitizers. Journal of Biomedical Optics. 2009. V. 14, No. 2. 021010.

Rogozhina E., Belomoin G., Smith A., Barry N., Akcakir O., Braun P.V., Nayfeh M.H. Si-N linkage in ultrabright, ultrasmall Si nanoparticles. Appl. Phys. Lett. 2001. V. 78. P. 3711–3713.

Rogozhina E.V., Eckhoff D.A., Gratton E., Braun P.V. Carboxyl functionalization of ultrasmall luminescent silicon nanoparticles through thermal hydrosilylation. J. Mater. Chem. 2006. V. 16. P. 1421–1430.

Rosso M., Giesbers M., Schroen K., Zuilhof H. Controlled Oxidation, Biofunctionalization, and Patterning of Alkyl Monolayers on Silicon and Silicon Nitride Surfaces using Plasma Treatment. Langmuir. 2010. V. 26, No. 2. P. 866–872.

Rosso-Vasic M., Spruijt E., Popovic Z., Overgaag K., van Lagen B., Grandidier B., Vanmaekelbergh D., Dom´ınguez-Guti´errez D., De Cola L., Zuilhof H. Amine-terminated silicon nanoparticles: synthesis, optical properties and their use in bioimaging. J. Mater. Chem. 2009. V. 19. P. 5926–5933.

Rosso-Vasic M., Spruijt E., van Lagen B., De Cola L. Han Zuilhof. Alkyl-Functionalized Oxide-Free Silicon Nanoparticles: Synthesis and Optical Properties. Small. 2008. V. 4. P. 1835–1841.

Rosso-Vasic M., Spruijt E., van Lagen B., De Cola L. Han Zuilhof. Alkyl-Functionalized Oxide-Free Silicon Nanoparticles: Synthesis and Optical Properties. Corrigendum. Small. 2009. V. 5. P. 2637.

Roura P., Farjas J., Pinyol A., Bertran E. The crystallization temperature of silicon nanoparticles. Nanotechnology. 2007. V. 18. P. 175705 (4 p).

Saito R., Kamikake T., Tanaka A., Yasuda H. Synthesis and spectroscopic study of alkyl-terminated silicon nanoparticles. Transactions of the Materials Research Society of Japan. 2006. V. 31, No. 2. P. 545–548.

Salivati N., Shuall N., Baskin E., Garber V., McCrate J.M., Ekerdt J. G. Influence of surface chemistry on photoluminescence from deuterium-passivated silicon nanocrystals. J. Appl. Phys. 2009. V. 106. P. 063121.

Sankaran R.M., Holunga D., Flagan R. C., Giapis K. P. Synthesis of Blue Luminescent Si Nanoparticles Using Atmospheric-Pressure Microdischarges. Nano Lett. 2005. V. 5, No. 3. P. 537–541.

Santana G., Monroy B.M., Ortiz A., Huerta L., Alonso J. C., Fandino J., Aguilar-Hern´andez J., Hoyos E., Cruz-Gandarilla F., Contreras-Puentes G. Influence of the surrounding host in obtaining tunable and strong visible photoluminescence from silicon nanoparticles. Appl. Phys. Lett. 2006. V. 88. P. 041916.

Sato K., Fukata N., Hirakuri K., Murakami M., Shimizu T., Yamauchi Y. Flexible and Transparent Silicon Nanoparticle/Polymer Composites with Stable Luminescence. Chem. Asian J. 2010. V. 5. P. 50–55.

Sato K., Kishimoto N., Oku T., Hirakuri K. Improvement of luminescence degradation in pure water of nanocrystalline silicon particles covered by a hydrogenated amorphous carbon layer. J. Appl. Phys. 2007. V. 102. P. 014302.

Sato S., Swihart M. T. Propionic-Acid-Terminated Silicon Nanoparticles: Synthesis and Optical Characterization. Chem. Mater. 2006. V. 18. P. 4083–4088.

Scheer K. C., Rao R.A., Muralidhar R., Bagchi S., Conner J., Lozano L., Perez C., Sadd M., White B. E. Jr. Thermal oxidation of silicon nanocrystals in O_2 and NO ambient. J. Appl. Phys. 2003. V. 93, No. 9. P. 5637–5642.

Serincan U., Kulakci M., Turan R., Foss S., Finstad T.G. Variation of photoluminescence from Si nanostructures in SiO_2 matrix with Si+post implantation. Nuclear Instruments and Methods in Physics Research. B 2007. V. 254. P. 87–92.

Shen P., Uesawa N., Inasawa S., Yamaguchi Y. Stable and color-tunable fluorescence from silicon nanoparticles formed by single-step plasma assisted decomposition of $SiBr_4$.. J. Mater. Chem. 2010. V. 20. P. 1669–1675.

Shiohara A., Hanada S., Prabakar S., Fujioka K., Lim T.H., Yamamoto K., Northcote P. T., Tilley R.D. Chemical Reactions on Surface Molecules Attached to Silicon Quantum Dots. J. Am. Chem. Soc. 2010. V. 132. P. 248–253.

Silalahi S. T.H., Yang H. Y., Pita K., Mingbin Y. Rapid Thermal Annealing of Sputtered Silicon-Rich Oxide/SiO_2 Superlattice Strucrture. Electrochem. And Solid-Lett. 2009. V. 12. P. K29–K32.

Sivakov V.A., Brönstrup G., Pecz B., Berger A., Radnoczi G. Z., Krause M., Christiansen S.H.

Realization of Vertical and Zigzag Single Crystalline Silicon Nanowire Architectures. J. Phys. Chem. C. 2010. V. 114. P. 3798–3803.

Stegner A. R., Pereira R.N., Klein K., Lechner R., Dietmueller R., Brandt M. S., Stutzmann M., Wiggers H. Electronic Transport in Phosphorus-Doped Silicon Nanocrystal Networks. PRL. 2008. V. 100. P. 026803.

Stegner A.R., Pereira R.N., Klein K., Wiggers H., Brandt M. S., Stutzmann M. Phosphorus doping of Si nanocrystals: Interface defects and charge compensation. Physica B. 2007. V. 401–402. P. 541–545.

Stupca M., Alsalhi M., Al Saud T., Almuhanna A., Nayfeh M.H. Enhancement of polycrystalline silicon solar cells using ultrathin films of silicon nanoparticle. Appl. Phys. Lett. 2007. V. 91. 063107.

Sublemontier O., Lacour F., Leconte Y., Herlin-Boime N., Reynaud C. CO_2 laser-driven pyrolysis synthesis of silicon nanocrystals and applications. Journal of Alloys and Compounds. 2009. V. 483. P. 499–502.

Sudeep P.K., Page Z., Emrick T. PEGylated silicon nanoparticles: synthesis and characterization. Chem. Commun. 2008. P. 6126–6127.

Svrček V., Fujiwara H., Kondo M. Improved transport and photostability of poly(methoxyethylexyloxyphenylenevinilene) polymer thin films by boron doped freestanding silicon nanocrystals. Appl. Phys. Lett. 2008. V. 92. P. 143301.

Svrček V., Sasaki T., Shimizu Y., Koshizaki N. Aggregation of Silicon Nanocrystals Prepared by Laser Ablation in Deionized Water. Journal of Laser Micro/Nanoengineering. 2007. V. 2, No. 1. P. 15–20.

Svrček V., Sasaki T., Shimizu Y., Koshizaki N. Blue luminescent silicon nanocrystals prepared by ns pulsed laser ablation in water. Appl. Phys. Lett. 2006a. V. 89. P. 213113.

Svrček V., Sasaki T., Shimizu Y., Koshizaki N. Silicon nanocrystals formed by pulsed laser-induced fragmentation of electrochemically etched Si micrograins. Chemical Physics Letters. 2006b. V. 429. P. 483–487.

Sychugov I., Juhasz R., Valenta J., Zhang M., Pirouz P., Linnros J. Light emission from silicon nanocrystals: Probing a single quantum dot. Applied Surface Science. 2006. V. 252. P. 5249–5253.

Sze S.M. Physics and properties of semiconductors – a review. In: Physics of Semiconductor Devices, Ed. by S.M. Sze, Kwok K. Ng. – John Wiley & Sones Inc., 2007. P. 7–78.

Szekeres A., Nikolova T., Paneva A., Lisovskyy I., Shepeliavyi P. E., Rudko G. Yu. Effect of Si nanoparticles embedded in SiO_x on optical properties of the films studied by spectroscopic ellipsometry and photoluminescence spectroscopy. Optical Materials. 2008. V. 30. P. 1115–1120.

Tanaka A., Saito R., Kamikake T., Imamura M., Yasuda H. Optical and photoelectron spectroscopic studies of alkyl-passivated silicon nanoparticles. Eur. Phys. J. 2007. V. D43. P. 229–232.

Tanaka A., Saito R., Kamikake T., Imamura M., Yasuda H. Electronic structures and optical properties of butyl-passivated Si nanoparticles. Solid State Communications. 2006. V. 140. P. 400–403.

Tanaka A., Takashima N., Imamura M., Kitagava T., Murase Y., Yasuda H. Surface chemistry of all-passivated silicon nanoparticles studied by synchrotron-radiation photoelectron spectroscopy. J. Phys. Soc. Japan. 2008a. V. 77, No. 9. 094701.

Tanaka A., Takashima N., Imamura M., Murase Y., Yasuda H. Electronic structure and surface chemistry of alkyl-passivated Si nanoparticles. Journal of Physics: Conference Series. 2008b. V. 100. P. 052086.

Tauc J., Grigorovici R., Vancu A. Optical properties and electronic structure of amorphous Germanium. Phys. Status Solidi. 1966. V. 15. P. 627–635.

Tewary A., Kekatpure R.D., Brongersma M. L. Controlling defect and Si nanoparticle luminescence from silicon oxynitride films with CO_2 laser annealing. Appl. Phys. Lett. 2006. V. 88. P. 093114.

Theis J., Geller M., Lorke A., Wiggers H., Wieck A., Meier C. Electroluminescence from silicon nanoparticles fabricated from the gas phase. Nanotechnology. 2010. V. 21. P. 455201.

Thogersen A., Mayandi J., Finstad T.G., Olsen A., Christensen J. S., Mitome M., Bando Y. Characterization of amorphous and crystalline silicon nanoclusters in ultra thin silica layers. J. Appl. Phys. 2008. V. 104. P. 094315.

Tilley R.D., Warner J.H., Yamamoto K., Matsui I., Fujimori H. Micro-emulsion synthesis of monodisperse surface stabilized silicon nanocrystals. Chem. Commun. 2005. P.

1833–1835.

Tilley R.D., Yamamoto K. The Microemulsion Synthesis of Hydrophobic and Hydrophilic Silicon Nanocrystals. Adv. Mater. 2006. V. 18. P. 2053–2056.

Trave E., Bello V., Mattei G., Mattiazzi M., Borsella E., Carpanese M., Fabbri F., Falconieri M., D'Amato R., Herlin-Boime N. Surface control of optical properties in silicon nanocrystals produced by laser pyrolysis. Applied Surface Science. 2006. V. 252. P. 4467–4471.

Trojanek F., Neudert K., Maly P., Dohnalova K., Pelant I. Ultrafast photoluminescence in silicon nanocrystals studied by femtosecond up-conversion technique. J. Appl. Phys. 2006. V. 99. P. 116108 (3 p.).

Tsai T.-C., Yu L.-Z., Lee C.-T. Electroluminescence emission of crystalline silicon nanoclusters grown at a low temperature. Nanotechnology. 2007. V. 18. P. 275707 (5 pp).

Tuan H.-Y., Ghezelbash A., Korgel B.A. Silicon Nanowires and Silica Nanotubes Seeded by Copper Nanoparticles in an Organic Solvent. Chem. Mater. 2008. V. 20. P. 2306–2313.

Tuan H.-Y., Lee D. C., Korgel B.A. Nanocrystal-mediated crystallization of silicon and germanium nanowires in organic solvents: The role of catalysis and solid-phase seeding. Angew. Chem. Int. Ed. 2006. V. 45. P. 5184–5187.

Umezu I., Kimura T., Sugimura A. Effects of surface adsorption on the photoluminescence wavelength of silicon nanocrystal. Physica B. 2006. V. 376–377. P. 853–856.

Umezu I., Minami H., Senoo H., Sugimura A. Synthesis of photoluminescent colloidal silicon nanoparticles by pulsed laser ablation in liquids. Journal of Physics: Conference Series. 2007a. V. 59. P. 392–395.

Umezu I., Nakayama Y., Sugimura A. Formation of core-shell structured silicon nanoparticles during pulsed laser ablation. J. Appl. Phys. 2010. V. 107. P. 094318.

Umezu I., Sugimura A., Inada M., Makino T., Matsumoto K., Takata M. Formation of nanoscale fine-structured silicon by pulsed laser ablation in hydrogen background gas. Physical Review B. 2007b. V. 76. P. 045328.

Umezu I., Sugimura A., Makino T., Inada M., Matsumoto K. Oxidation processes of surface hydrogenated silicon nanocrystallites prepared by pulsed laser ablation and their effects on the photoluminescence wavelength. J. Appl. Phys. 2008a. V. 103. P. 024305.

Umezu I., Takata M., Sugimura A. Surface hydrogenation of silicon nanocrystallites during pulsed laser ablation of silicon target in hydrogen background gas. J. Appl. Phys. 2008b. V. 103. P. 114309.

Vandyshev E.N., Zhuravlev K. S. Effect of electric field on recombination of self-trapped excitons in silicon nanocrystals. Phys. stat. sol. (c). 2007. V. 4. P. 382–384.

Veinot J., Fok E., Boates K., MacDonald J. Chemical safety: LiAlH4 reduction of SiCl4 // Chemical & Engineering News. 2005. V. 83. P. 4–5.

Veinot J.G. C. Synthesis, surface functionalization, and properties of freestanding silicon nanocrystals. Chem. Commun. 2006. P. 4160–4168.

Vincent J., Maurice V., Paquez X., Sublemontier O., Leconte Y., Guillois O., Reynaud C., Herlin-Boime N., Raccurt O., Tardif F. Effect of water and UV passivation on the luminescence of suspensions of silicon quantum dots. J. Nanopart. Res. 2010. V. 12. P. 39–46.

Vladimirov A., Korovin S., Surkov A., Kelm E., Pustovoy V. Synthesis of Luminescent Si Nanoparticles Using the Laser-Induced Pyrolysis. Laser Physics. 2011. V. 21, No. 4. P. 830–835.

Vladimirov A., Korovin S., Surkov A., Kelm E., Pustovoy V. Tunable Luminescence of Silicon Nanoparticles. In: Breakthrough in nanoparticles for Bio-Imiging / Ed. by E. Borsella. BONSAI Project Symposium, AIP, 2010. P. 58–62.

Walters R. J., Kalkman J., Polman A., Atwater H.A., de Dood M. J.A. Photoluminescence quantum efficiency of dense silicon nanocrystal ensembles in SiO_2.. Physical Review B. 2006. V. 73. P. 132302.

Wan D.H., Chen H. L., Chuang S. Y., Yu C. C., Lee Y. C. Using self-assembled nanoparticles to fabricate and optimize subwavelength textured structures in solar cells. J. Phys. Chem. C. 2008. V. 112. P. 20567–20573.

Wang D., Dong H., Chen K., Huang R., Xu J., Li W., Ma Z. Low turn-on and high efficient oxidized amorphous silicon nitride light-emitting devices induced by high density amorphous silicon nanoparticles. Thin Solid Films. 2010. V. 518. P. 3938–3941.

Wang J., Wang X. F., Li Q., Hryciw A., Meldrum A. The microstructure of SiO thin films: from nanoclusters to nanocrystals. Philosophical Magazine. 2007. V. 87, No. 1. P. 11–27.

Wang M., Li D., Yuan Z., Yang D., Que D. Photoluminescence of Si-rich silicon nitride: Defect-related states and silicon nanoclusters. Appl. Phys. Lett. 2007. V. 90. P. 131903.

Wang Y.Q., Smirani R., Ross G.G. The formation mechanism of Si nanocrystals in SiO_2. Journal of Crystal Growth. 2006. V. 294. P. 486–489.

Warner J.H., Hoshino A., Shiohara A., Yamamoto K., Tilley R.D. The Synthesis of Silicon and Germanium Quantum dots for Biomedical Applications. Proc. of SPIE. 2006a. V. 6096. P. 609607.

Warner J.H., Hoshino A., Yamamoto K., Tilley R.D. Water-Soluble Photoluminescent Silicon Quantum Dots. Angew. Chem. Int. Ed. 2005a. V. 44. P. 4550–4554.

Warner J.H., Rubinsztein-Dunlop H., Tilley R.D. Surface Morphology Dependent Photoluminescence from Colloidal Silicon Nanocrystals. J. Phys. Chem. B. 2005б. V. 109, No. 41. P. 19064–19069.

Warner J.H., Tilley R.D. Photonics of Silicon Nanocrystals. Proc. of SPIE. 2006b. V. 6038. P. 603815.

Wiggers H., Starke R., Roth P. Silicon Particle Formation by Pyrolysis of Silane in a Hot Wall Gasphase Reactor. Chem. Eng. Technol. 2001. V. 24, No. 3. P. 261–264.

Wijesinghe T. L. S. L., Teo E. J., Blackwood D. J. Potentiostatic formation of porous silicon in dilute HF: Evidence that nanocrystal size is not restricted by quantum confinement. Electrochimica Acta. 2008. V. 53. P. 4381–4386.

Wilcoxon J. P., Samara G.A. Tailorable, visible light emission from silicon nanocrystals. Appl. Phys. Lett. 1999a. V. 74, No. 21. P. 3164–3166.

Wilcoxon J. P., Samara G.A., Provencio P.N. Optical and electronic properties of Si nanoclusters synthesized in inverse micelles. Physical Review B. 1999б. V. 60, No. 4. P. 2704–2714.

Wilkinson A.R., Elliman R.G. Maximizing light emission from silicon nanocrystals – The role of hydrogen. Nuclear Instruments and Methods in Physics Research B. 2006. V. 242. P. 303–306.

Wolkin M.V., Jorne J., Fauchet P.M., Allan G., Delerue C. Electronic states and luminescence in porous silicon quantum dots: The role of oxygen. Physical Review Letters. 1999. V. 82. P. 197–200.

Xie Y., Wu X. L., Qiu T., Chu P.K., Siu G.G. Luminescence properties of ultrasmall amorphous Si nanoparticles with sizes smaller than 2 nm. Journal of Crystal Growth. 2007a. V. 304. P. 476–480.

Xie Z.-Q., Li Z.-H., Fan W.-B., Chen D., Zhao Y.-Y., Lu M. Ar+ irradiation of Si nanocrystal-doped SiO_2: Evolution of photoluminescence. Applied Surface Science. 2007б. V. 253. P. 5501–5505.

Yang C.-S., Bley R.A., Kauzlarich S.M., Lee H.W.H., Delgado G. R. Synthesis of alkyl-terminated silicon nanoclusters by a solution route. J. Am. Chem. Soc. 1999. V. 121.

P. 5191–5195.

Yang M.D., Chu A.H.M., Shen J. L., Huang Y.H., Yang T.N., Chen M. C., Chiang C. C., Lan S.M., Chou W. C., Lee Y. C. Improvement of luminescence from Si nanocrystals with thermal annealing in CO_2. Journal of Crystal Growth. 2008. V. 310. P. 313–317.

Yang S., Cai W., Zeng H., Li Z. Polycrystalline Si nanoparticles and their strong aging enhancement of blue photoluminescence. J. Appl. Phys. 2008. V. 104. 023516.

Yang S., Cai W., Zhang H., Xu X., Zeng H. Size and Structure Control of Si Nanoparticles by Laser Ablation in Different Liquid Media and Further Centrifugation Classification. J. Phys. Chem. C. 2009. V. 113. P. 19091–19095.

Yerci S., Serincan U., Dogan I., Tokay S., Genisel M., Aydinli A., Turan R. Formation of silicon nanocrystals in sapphire by ion implantation and the origin of visible photoluminescence. J. Appl. Phys. 2006. V. 100. P. 074301.

Yurtsever A., Weyland M., Muller D.A. Three-dimensional imaging of nonspherical silicon nanoparticles embedded in silicon oxide by plasmon tomography. Appl. Phys. Lett. 2006. V. 89. P. 151920.

Zacharias M., Heitmann J., Scholz R., Kahler U., Schmidt M., Bläsing J. Size-controlled highly luminescent silicon nanocrystals: A SiO/SiO_2 superlattice approach. Applied Physics Letters. 2002. V. 80. (4). P. 661–663.

Zhang X., Brynda M., Britt R.D., Carroll E. C., Larsen D. S., Louie A. Y., Kauzlarich S.M. Synthesis and characterization of manganese-doped silicon nanoparticles: bifunctional paramagnetic-optical nanomaterial. J. Am. Chem. Soc. 2007a. V. 129. P. 10668–10669.

Zhang X., Neiner D., Wang S., Louie A. Y., Kauzlarich S.M. Fabrication of Silicon-Based Nanoparticles for Biological Imaging. Proc. of SPIE. 2007б. V. 6448. 644804.

Zhang X., Neiner D., Wang S., Louie A. Y., Kauzlarich S.M. A new solution route to hydrogen-terminated silicon nanoparticles: synthesis, functionalization and water stability Nanotechnology. 2007в. V. 18. P. 095601 (6 p.).

Zhao Y., Kim Y.-H., Du M.-H., Zhang S. B. First principles prediction of icosahedral quantum dots for tetravalent semiconductors. Phys. Rev. Lett. 2004. V. 93. P. 015502.

Zhou Z., Brus L., Friesner R. Electronic structure and luminescence of 1.1- and 1.4-nm silicon nanocrystals: Oxide shell versus hydrogen passivation. Nano Lett. 2003. V. 3. P. 163–167.

Zimina A., Eisebitt S., Eberhardt W., Heitmann J., Zacharias M. Electronic structure and chemical environment of silicon nanoclusters embedded in a silicon dioxide matrix. Appl. Phys. Lett. 2006. V. 88. P. 163103.

Zou J., Baldwin R.K., Pettigrew K.A., Kauzlarich S.M. Solution Synthesis of Ultrastable Luminescent Siloxane-Coated Silicon Nanoparticles. Nano Lett. 2004. V. 4, No. 7. P. 1181–1186.

Zou J., Kauzlarich S.M. Functionalization of Silicon Nanoparticles via Silanization: Alkyl, Halide and Ester. J. Clust. Sci. 2008. V. 19. P. 341–355.

Zou J., Sanelle P., Pettigrew K.A., Kauzlarich S.M. Size and spectroscopy of silicon nanoparticles prepared via reduction of $SiCl_4$. Journal of Cluster Science. 2006. V. 17, No. 4. P. 565–578.

Zou M., Dorey S., Cai D., Song Y., Premachandran-Nair R., Cai L., Brown W. Self-assembly of Si nanoparticles produced by aluminum-induced crystallization of amorphous silicon film Electrochemical and Solid-State Letters. 2007. V. 10, No. 2. P. K7–K9.

6

Thermal oxidation processes in nanosilicon powders

The optical and electrical characteristics of nanosized silicon are determined not only by the size of the particles, but also the composition and structure of their surfaces (Hadj Zoubir et al, 1995; Jia et al, 1997; Kononov et al, 2005; Rybaltovsky, etc. In 2006). As shown in (Delerue et al, 1999; Prusry et al, 2005; Meier et al, 2006), changing the particle size of nc-Si or chemically modifying the structure of their surface layer can be used to control the characteristics of media containing these particles.

In most cases, the synthesized nc-Si powders are some time in contact with the atmosphere. Due to the high reactivity of the freshly formed surface of fine-dispersion silicon relative to oxygen molecules and water, the particles are covered by an oxide film – this is uncontrolled oxidation. But it is possible to organize the process of oxidation intentionally and with its help manage the properties of nanoparticles. Controlled oxidation of the silicon particles, accompanied by the formation of SiO_2 films, is one of the stages of the real production process of electronic devices. On the other hand, the oxidation is the most easiest way to change in a controlled manner the particle size of nanosilicon (Jia et al, 1997).

The influence of components of the atmosphere on the chemical composition of the surface was studied in some detail in the case of bulk samples of crystalline silicon (see, e.g. Jolly et al, 1999). Several models were proposed of the processes leading to the formation of an oxide layer on the surface of crystalline silicon, depending on the composition of the gas phase and the processing conditions (Konstantinova et al, 2005; Cerofolini & Meda, 1997, and references therein). For the nc-Si powder such information is scarce. We know only a few works which addressed issues related to the study of chemical processes in the surface layers and their influence on the spectral characteristics of nc-Si powders. These include the study by Jia et al. (1997), where the authors examined changes of the spectra of diffuse reflection of light from a powder obtained by laser decomposition of silane after heat treatment in air at temperatures 600–800°C. Later, Kononov et al (2005) used IR spectroscopy for the powders obtained by the same method to study the transformation of hydrogen-containing groups located both in the amount of particles and on their surface during heat treatment of samples in an inert gas atmosphere. The effect of the chemical composition of the surface layer formed in the process of plasma chemical synthesis of nanoparticles or in their heat treatment in air, on the characteristics of the transmission spectra of emulsion materials, containing these particles, was studied by Rybaltovsky et al (2006). Initial information was obtained on the laws of thermal oxidative processes involving silicon nanoparticles and the possibility was analyzed of obtaining from the spectral and kinetic experimental data information on the size distribution function particles in the sample (Radtsig et al (2007).

6.1. Spectral manifestations

Radtsig et al (2007) used nanosilicon powders synthesized by different technologies. First, a nanocrystalline silicon powder was prepared in the discharge of high-frequency induction plasma in interaction of the plasma with samples of crystalline silicon on a special plasma torch[1]. Synthesis was carried out under an argon atmosphere with the addition in the final stage of oxygen (powders of type I) or a mixture of oxygen and nitrogen (powders of type II). Powders obtained in an argon atmosphere, without adding any other special gases, were regarded as type IV. This method allows to obtain silicon powders in which silicon particles have a different shell – oxide shell for samples of type I and oxynitride for samples of type II. The spectral characteristics of these materials are described in detail in the article by Rybaltovsky et al (2006).

Another method for producing silicon powders is the method of laser-induced decomposition of silane with a powerful CO_2 laser. Silicon nanoparticles were formed in a flow gas-dynamic reactor in interaction of laser radiation with a coaxial flow of silane cross and a buffer gas (argon) at the pressure which in principle can be varied to vary the average size of the particles. A detailed description of this process can be found in Kononov et al (2005) and in Chapter 5 this monograph. The resulting silicon powders (samples of type III) according to IR spectroscopy contained a significant amount of chemically bonded hydrogen.

The synthesized nc-Si powders were stored for a long time (about a year) at room temperature in contact with the air atmosphere and were then investigated by electron paramagnetic resonance (EPR) and optical spectroscopy, including the measurement of absorption and Raman spectra. The particle size of the powders was monitored by two methods – measuring the specific surface area of the material and by the method of Raman scattering (RS), analyzing the position and shape of the band near 520 cm^{-1}. The specific surface area of powders S (m^2/g) was determined from adsorption isotherms of argon by the silicon

---

[1]The process of plasma chemical synthesis of the nanosized Si powder was described in detailed in Chapter 5 of this book.

samples at 77 K (the site area for the argon molecule was assumed to be 0.16 nm^2). The resulting isotherms were analyzed in the framework of the BET theory, which allows to determine the magnitude of the specific surface of the sample and the parameter C, characterizing the heat of adsorption of gas molecules on a solid surface. For the powders synthesized by different methods, yielded the following values of specific surface – 70 m^2/g for sample of type III and 96 m^2/g for samples of type IV. The value of the constant C was in the range of 30–40, which indicates the absence of capillaries and micropores in the synthesized samples (Branauer, 1948). On the assumption that the particles have a spherical shape and differ little in size, their diameters $d = 6/S \cdot \rho$ ($\rho = 2.3$ g/cm^3 – the density of silicon) can be estimated at 35 and 25 nm, respectively. Raman spectra were measured, as in Rybaltovsky et al (2006), with a T-64 000 monochromator (Jobin Yvon) with the exciting radiation from an argon laser ($\lambda = 514.5$ nm).

The optical absorption spectra of the heat treated powders were recorded in a SPECORD M40 spectrophotometer (Carl Zeiss, Jena). In contrast to the method of measurement of the absorption spectra as described in Kuz'min et al. (2000), where the silicon powder of type IV was deposited on a plate of MgF$_2$ or transparent sticky (Scotch) tape with the thickness less than 5 µm, a uniform layer of powder with a thickness of 10 µm was deposited in this case and the prepared sample was placed into the working channel of the instrument for measurement (Rybaltovsky and others, 2007a). The sticky tape was highly transparent in the whole measured wavelength range from 200 to 900 nm, nevertheless, in the processing of the spectra the contribution from the absorption of pure Scotch tape was also taken into. We can assume that with this method of measurement the results qualitatively correctly reflect the changes in spectral characteristics of the investigated powders.

Volumetric measurements were performed as follows (Radtsig et al, 2007). A nanosilicon powder portion was placed in a quartz vessel connected with a high-vacuum unit. The heating element was used to control the sample temperature in the range 300–1250°C. The amount of adsorbed gas was determined by monitoring the change in gas pressure in the system (volume was 255 cm^3). High-temperature oxidation of silicon was carried out in an atmosphere of pure oxygen at a pressure of 90 Torr. The total change in pressure due to heating of the sample up to 1100°C was 28 Torr. The composition of gaseous products formed during pyrolysis of silicon powders was determined by the chromatographic method from the time of retention of molecules of different types by the adsorbent. Figure 6.1 (curve 1) shows the experimental data on the temperature dependence of the number of the oxygen molecules chemisorbed by nc-Si (type III) (total number of chemisorbed molecules was $2.0 \cdot 10^{22}$ g^{-1} and taken as the unity).

As can be seen from Fig. 6.1, the rise of the sample temperature to 600°C leads to a slight uptake of oxygen, not more than 20% of the total absorption. The most intense absorption of oxygen and, consequently,

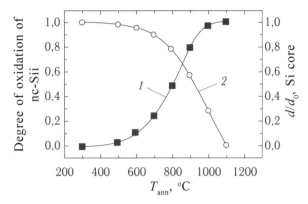

Fig. 6.1. The absorption of oxygen in the oxidation of silicon powder (P (O_2) = 90 Torr, the exposure of 15 minutes at each temperature): 1 (■) – The temperature dependence of the relative share of chemisorbed oxygen molecules, and 2 (○) relative change in diameter (d) of the core of the Si particles in oxidation (for spherical particles of the same size) (Radtsig et al, 2007) calculated from curve 1.

the formation of silicon oxide occurs at temperatures above 700°C. At a temperature of 1000°C oxygen consumption virtually ceases, indicating completion of the process. The resulting powder has a snow-white colour characteristic of highly dispersed aerosol. Formation of SiO_2 is also indicated by the infrared spectrum of the reaction product. The value of its specific surface determined by the BET method. It was 80 ± 20 m²/g at a value of constant $C = 30–40$. This means that oxidation is not accompanied by sintering of the resultant silica, i.e. silicon particles are oxidized independently of each other. Note that in the above-mentioned paper (Jia et al, 1997) the authors also observed complete transfer of silicon particles to SiO_2 particles after two hours of annealing in air at 800°C.

The formation of silica leads to a weight gain of the powder. If the initial silicon sample was free from the oxide phase, with full oxidation the weight increased 2.143 times. However, the measured increase in mass of the sample after oxidation was lower – 1.98 times. This difference in behaviour (Radtsig et al, 2007a) is associated with the partial oxidation of the initial sample. From the presented experimental data, we can estimate the degree of oxidation of the initial sample. It is $7.5 \pm 2.5\%$ of silicon atoms from their total number. With an average diameter (d) of silicon particles in the sample III, equal to 35 nm (determined from the specific surface of the powder, see above), the thickness of the oxidized layer (δ) can be estimated from the relationship: $\pi d^2 \delta \approx 0.075$ ($\pi d^3/6$), i.e. $\delta \approx 0.075$ ($d/6$) = 0.44 nm. Consequently, Oxidation took place in approximately the top two monolayers of nanosilicon particles. This result seems reasonable. According to available literature data, the top two layers of atoms of silicon particles are rapidly oxidized, after which the oxidation of the material slows down sharply. Figure 6.1 (curve 2) shows the calculated (from curve *1*) dependence of the dimensionless parameter d/d_0 (d – diameter of the silicon core) on the depth of oxidation of the particles assuming their spherical shape and uniform size

In the oxidation process, there are two limiting flow regimes: kinetic, where the reaction rate is determined by the oxidation reaction at the silicon/silicon oxide interface, and diffusion, in which the limiting stage is the supply of oxygen to the interface. In the first case, in oxidation of the spherical particles the reaction rate is:

$$\frac{dN(O_2)}{dt} = k \cdot s \cdot [S \cdot P(O_2)], \qquad (6.1)$$

where k is the reaction rate constant, s is the surface of the silicon core, $P(O_2)$ is the partial pressure of oxygen in the gas phase above the sample, and S is the solubility of oxygen in SiO_2. The rate of the diffusion-controlled oxidation reaction:

$$\frac{dN(O_2)}{dt} = \frac{D \cdot [S \cdot P(O_2)]}{(d_0 - d)}, \qquad (6.2)$$

is determined by the diffusive flux of oxygen molecules ($D \cdot S$ is the permeability silica for oxygen molecules, D is the diffusion coefficient, and S the solubility of oxygen in silica) through the formed oxide layer with the thickness $d_0 - d$. The amount of absorbed oxygen $dN(O_2) \sim s \cdot \delta d$ is determined by the change in the diameter of the silicon core δd. If the reaction proceeds in the first mode the $\delta d/dt \sim k \cdot S$, and from the temperature dependence $d(T)$ we can estimate the effective activation energy of the oxidation reaction. It was 18 ± 2 kcal/mol (Fig. 6.2, curve

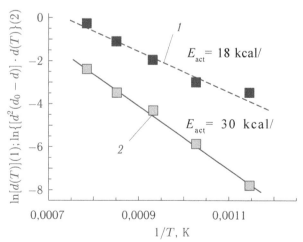

Fig. 6.2. Analysis of the temperature dependence of the oxidation process of silicon by oxygen for the two limiting regimes: (*1*) kinetic and (*2*) diffusion (Radtsig, et al., 2007) (details in text).

1). According to Norton (1961), for the solubility of oxygen in silica $\Delta H = 5$ kcal/mol, which leads to the activation energy of the elementary the act of oxidation (the rate constant k) of 23 ± 2 kcal/mol.

For the diffusive regime $[d^2(d_0 - d)]$ $(\delta d/dt) \sim D \cdot S$, and from the temperature dependence of the left-hand side of (Fig. 6.2, curve *2*) for the permeability the activation energy is 30 ± 4 kcal/mol, and for diffusion of 35 ± 4 kcal/mol (Radtsig et al, 2007a). Analysis of a large number of, usually, contradictory experimental data concerning the oxidation of silicon with oxygen also allows us to suggest that in this case the reaction of oxidation is diffusion-controlled. The activation barrier of ~ 23 kcal/mol is too low for the process of high-temperature oxidation of silicon (Jolly et al. 1999). The heating of silicon powder which is in contact with the atmosphere (containing water vapour) at room temperature up to 700°C is accompanied by molecular hydrogen. This procedure – pyrolysis, the sample cooled to room temperature, its contact with the atmosphere, exposure under these conditions for about a day, subsequent pyrolysis, etc., can be consistently carried out several times. Below are the results of such measurements. The amounts of evolved hydrogen per unit surface of the silicon powders are as follows: for sample IV – $2.3 \cdot 10^{14}$ (molecular H_2)/cm², $2.0 \cdot 10^{14}$ (molecules H_2)/cm² for sample III – $2.5 \cdot 10^{14}$ (molecular H_2)/cm², $2.4 \cdot 10^{14}$ (molecules of H_2)/cm² (the results of two consecutive cycles). The obtained values are close to the number of hydrogen atoms corresponding to monomolecular filling of the surface. It is obvious that the source of hydrogen in the final analysis is water and the Si–H and (or) Si–OH groups formed as a result of interaction with the silicon atoms (Dai et al, 1992). The collapse of these groups is accompanied by the release of molecular hydrogen, while the oxygen of the water molecule is used for the oxidation of silicon, i.e. the water acts as a donor of atomic oxygen. The stage of high-temperature annealing is required to activate the surface – this is accompanied by the rearrangement of the surface layer of particles, removal and modification of the hydrolysis products formed at contact of the surface with the atmosphere at room temperature. Thus, it is possible to carry out layer-by-layer oxidation of the surface of the silicon particles.

The increase in the thickness of the oxide shell and, consequently, reduction of the size of the silicon core during heat treatment of the powder in the air were also evident in the results of optical experiments (Rybaltovsky et al, 2007)). Figure 6.3 shows the results of the changes of the absorption spectra of the silicon powder of type III, deposited in a thin layer on the transparent polypropylene Scotch tape for the initial sample and subjected to heat treatment in air at temperatures 600, 700 and 800°C for 15–20 min at each temperature. In addition, this figure shows a similar absorption curve of the aerosil powder, consisting of pure SiO_2 nanoparticles with specific surface area of about ~ 300 m²/g.

The presented absorption curves are qualitative in nature, since the thickness of the applied layer of the powder on the Scotch tape and

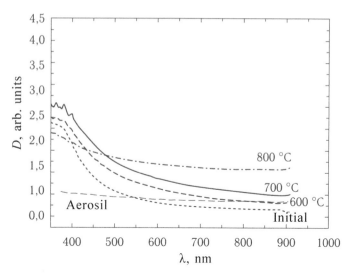

Fig. 6.3. The absorption spectra of samples obtained by laser decomposition of silane (samples of type III), the initial and treated in air at 600, 700, 800°C. The spectrum of aerosil powder is also shown (Radtsig et al, 2007).

the distribution density of the particles on it have not been determined. Comparing these spectra it may be noted that after annealing the starting powder at 800°C the absorption in the short-wavelength region decreases and thus the absorption curve is similar to that of the aerosil curve. The results obtained in Radtsig et al, 2007a as well as the results of the study by Rybaltovsky et al (2006) can be explained by the growth of reflection of particles associated with an increase in the thickness of the oxide layer of the shell after heat treatment.

The use of Raman scattering in studies of silicon nanoparticles is based on the dependence of the position of the absorption band near 520 cm^{-1} on their physico-chemical characteristics. Owing to the small half-width of bands in the Raman spectrum its rather small (less than 1 cm^{-1}) shifts can be experimentally recorded that allows one to monitor even small changes in the characteristics of the nanoparticles. The position of this band depends on the crystallite size. There are other factors that have an impact on the position and shape of this band – the internal stresses, the shape of the particles, the degree of their defectiveness, availability of free charge carriers, etc. Unfortunately, their identification and separation are still a problem that has not been completely solved.

If the individual particles in the sample differ in size or in some other physico-chemical characteristics, this leads to an inhomogeneous broadening of the absorption band contour and the corresponding effects are also apparent in the Raman spectra. Radtsig et al, 2007 proposed a simple (semi-empirical) method of analysis of the Raman spectra of nc-Si powders

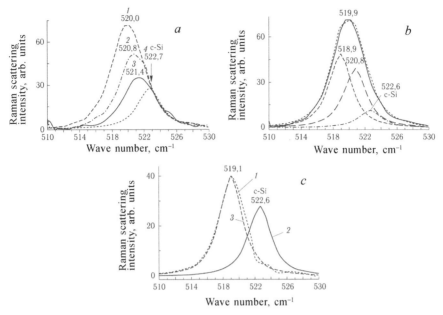

Fig. 6.4. Raman spectra of silicon samples (Radtsig et al, 2007). *a* – initial sample (*1*); the powder after annealing at 500°C (*2*); after annealing at 1000°C (*3*); single-crystal silicon plate (*4*). *b* – expansion of the original spectrum into three fractions. c – comparison of the shape of the components of the Raman spectrum upon heating the nc-Si sample at 500°C (*1*) with the spectrum of Raman scattering of crystalline silicon (*3*).

which allows to identify the contribution of inhomogeneous broadening to the recorded signal.

Figure 6.4*a* shows the spectra of Raman scattering (RS) of the silicon powder (a sample of type IV not previously been subjected to any treatments (curve *1*) and after heating in air at temperatures 500°C (curve *2*) and 1000°C (curve *3*), heating time 5 min). For comparison, the figure shows the RS spectrum of a single crystal silicon wafer (line *4*). The intensities of the individual spectra are selected in such a way as to achieve the greatest possible proximity to the high-frequency wing of the Raman spectra.

The inhomogeneously broadened line $F(\lambda)$ is a superposition of signals from the individual groups (fractions) of the particles or centres differing in their spectral characteristics:

$$F(\lambda) = \int \psi(\lambda_0) f(\lambda, \lambda_0) d\lambda_0. \tag{6.3}$$

In this equation $f(\lambda, \lambda_0)$ is a function describing the shape of the signal from the fraction of the particles $\lambda = \lambda_0$ (form of the individual components), and $\psi(\lambda_0)\, d\lambda_0$ is the number of particles in this fraction. The distortion (broadening) of the signal, introduced by the recording system (instrument broadening) is non-uniform by its nature. However, because it is the same for all fractions, it is convenient to include it in the function $f(\lambda, \lambda_0)$. If the form of the function $f(\lambda, \lambda_0)$ is known, it is possible to determined also the form of the distribution function $\psi(\lambda_0)$, which determines the line shape $F(\lambda)$. In the analysis of Raman spectra of the nc-Si powders it is proposed to use the Raman spectrum of single crystal silicon as $f(\lambda, \lambda_0)$. Indeed, as can be seen from Fig. 6.4a, it is possible to combine the high-frequency wing of the Raman spectra of crystalline silicon and nc-Si. Since the distribution function $\psi(\lambda_1) = 0$ is in this field, it means that the line shape of the single-crystal sample can indeed be used as a 'basic' function that performs the role of the individual component to represent the inhomogeneously broadened spectrum (as evidenced by their differences in the low-frequency part) of the nanosilicon powder.

Figure 6.4b shows the result of decomposition of the original signal into three components which are derived from the c-Si line by its shift (along the horizontal axis) and changes of the amplitude (ordinate). The satisfactory proximity of the form of the experimentally recorded and calculated spectra is clearly visible. The following result should also be noted (Radtsig et al, 2007a). Figure 6.4c (spectrum 1) shows the difference Raman spectrum of the sample after heating at 500°C. Spectrum 2 in the same figure is the 'basic function' shifted along the horizontal axis (the spectrum of crystalline silicon (spectrum 3 in the figure). Thus, for the low-frequency fraction of the inhomogeneously broadened spectrum the RS line of single-crystal silicon satisfactorily reproduces the form of individual components.

Thus, the presented experimental results (Radtsig, etc. 2007) suggest that the Raman spectrum of the investigated sample of the silicon powder is inhomogeneously broadened and is a superposition of signals with different position of the maximum: 518.9 ± 1.0, 520.8 ± 0.9 and 522.6 ± 0.4 cm^{-1} respectively.

If it is assumed that the magnitude of the frequency shift dY depends only on the particle size

$$D = \frac{35}{dY},$$

where D is the diameter of the particles in nm, dY is the frequency shift with respect to single-crystalline silicon in cm^{-1}. For the three fractions of silicon particles with the shift of 1.3, 1.9 and 2.7 cm^{-1}, according to the above formula their size as 30, 20.5 and 15 nm, respectively. Assuming that the number of particles included in a single fraction, is proportional to the area under the corresponding component of the expansion, we find that their contributions are 0.1:0.4:0.5. These values are used to estimate the specific surface of the sample:

$$S = 0.1 \, S_1 + 0.4 \, S_2 + 0.5 \, S_3,$$

where $S_i = 6/D_i^{2.3}$ is the specific surface of a separate fraction of the distribution function.

Substituting the above values of D_i, gives $S = 150$ m²/g. The measured (by the BET method) specific surface area of the powder was 100 m²/g. Considering the number of assumptions made in this estimate, the results can be regarded satisfactory. However, one should also note the following. As it can be seen from Fig. 6.3a, the almost complete destruction of the particles belonging to the low-frequency fraction (annealing in air at 500°C), did not lead to a change in the shape of the high-frequency component of the expansion. On the other hand, annealing the sample at $T = 500$°C is not accompanied by any noticeable oxidation of the sample (see Fig. 6.1, curve 2). These results suggest that changes in the particle size is not the only reason for the shift of the RS lines. It can be assumed that in this case the position of the RS line is determined not only by the size of the particles, but an important role is also played by their defectiveness, including the state of the oxidized surface layer. As a result, the size of the particles is underestimates and this leads to an overestimation of the specific surface area of the powder determined by this procedure.

A similar situation is encountered in X-ray diffraction when contributions to the observed width of the diffraction lines are made both by the size of regions of coherent scattering and by microdistortions caused by violations of the crystal structure of the material (defects, impurity atoms, dislocations, etc.). In this case, the observed changes in the Raman spectra of the samples during annealing can be caused not only by their oxidation but also by relaxation process leading to the ordering of the material structure and, consequently, to a narrowing of the RS band and the shift of its maximum to the high-frequency range. Thus, the question about the nature of the effects, responsible for the experimentally recorded phenomena, remains open.

6.2. Paramagnetic centres

The phenomena discussed in the previous section are associated with the processes of education and growth of the SiO_x oxide shell on the particles of nc-Si in the conditions of their high-temperature processing (500–1100°C) in an oxygen atmosphere or air, when the direct reactions of oxidation of silicon with oxygen are initiated (Sato et al. 2003). There have been studies (see, for example, Hadj Zoubir et al, 1995; Cerofolini & Meda, 1997; Szymanski et al, 2001; Kostantinova, 2007) devoted to the study of oxidative processes involving nanoporous and single crystal silicon, and occurring in the interaction of the surface of these materials with an air atmosphere at room and higher temperatures. In this case, the formation of oxygen-containing groups ($\equiv$Si–O–Si$\equiv$) in the surface layer of silicon is a more complex multistep process in which the molecules of oxygen and water from the surrounding atmosphere take part. At the same time one of the stages of this process is the formation of hole centres in the structure of silicon and electrophilic centres in oxygen (Szymanski et al, 2001; Kostantinova, 2007). However, in this field of research questions remain about the mechanisms of reactions leading to chemical modification of the structure of the surface layer of silicon nanoparticles. This also should include participation in these processes of compounds present in the ambient atmosphere. Useful information can be obtained from the studies by electron paramagnetic resonance (EPR), which allows to fix some changes in the structure of the surface layer of nc-Si on microlevel, i.e. the formation of certain defects, such as dangling bonds on silicon (Sato et al, 2003; Kostantinova, 2007). Studies of oxidative processes of nc-Si powders at relatively low-temperature heat treatment (from room temperature to 500°C), are not described in the literature. The aim of the work (Rybaltovsky et al, 2007b) was to conduct studies of these processes in nc-Si powders, obtained by different synthesis methods, using the EPR method.

For registration of EPR spectra samples of a silicon powder were placed in a special quartz ampoule connected to a vacuum system. This allowed measurements to be taken in vacuum and in a controlled atmosphere of the test gas. These ampoules can be used in annealing of powders at temperatures up to 1100°C in vacuum or in the atmosphere of the required gas. EPR signals were recorded radio spectrometers of the 3 cm band at 77 or 300 K. All EPR signals were recorded at microwave power levels far from saturation. A reference sample, allowing to register changes in the concentration of paramagnetic centres (PC), was a sample of Mn^{2+} in MgO with a known content of the PC and the values of the g-factors for the individual lines of Mn^{2+}.

The EPR spectra of paramagnetic centres in the surface layers of silicon samples, both in single crystals and in nc-Si, were analyzed in the works (Konstantinova et al, 2006; Sato et al, 2003). According to Sato et al. (2003) the value of the g-factor for the PC in the nc-Si particles corresponds to a signal from the dangling bonds on the atoms of Si which have different structures of their first coordination sphere. The signal with the highest g-factor, equal to 2.0055, refers to the PC, located inside the nanocrystal, where the low-coordination silicon atom is surrounded by three silicon atoms. Signals with $g = 2.0040$ are attributed to the silicon atoms in the surface layers of nanoparticles, in an environment where there may be oxygen atoms, i.e. the amorphous phase of SiO_x. For comparison, recall that the g-factor of the EPR signal of radicals with an unpaired electron is localized on the silicon atom in the amorphous SiO_2, the first coordination sphere of which consists of three oxygen atoms, is anisotropic with $g_\perp = 2.0005$, and $g_\parallel = 2.0018$ (Radzig, 2007). In papers devoted to the study of the PC in silicon single crystals (e.g. Cerofolini & Meda, 1997) the authors reported, as a rule, on the anisotropic signals such as P_b-centres with g-factors strongly differing in size: $g_\parallel = 2.0019$ and $g_\perp = 2.0089$. These centres include dangling bonds at the silicon dimers and are located at the boundary of the crystal phases of silicon and the amorphous phase SiO_2.

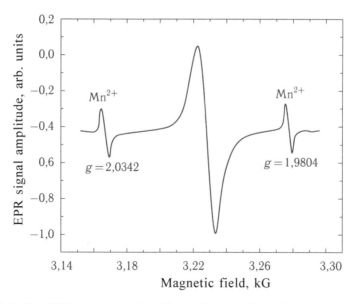

Fig. 6.5. The EPR spectrum of nc-Si, obtained by laser decomposition of S:H₄ (sample III type) (Rybaltovsky et al, 2007). The lines on the left and right of the EPR signal with g-factor of 2.0047 ± 0.0003 belong to a reference sample MgO:Mn²⁺.

All original powder samples of type III and IV studied in the work by Rybaltovsky et al (2007) showed sufficiently intense ESR signals with the g-factor equal to 2.0047 ± 0.0003 (Fig. 6.5). The value of the g-factor was determined from the value of the magnetic field at the point of intersection of the derivative of the EPR signal with the zero line. In the above paper by Sato et al. (2003), where the transformation of the EPR signals in the Si/SiO₂ films in annealing in an argon atmosphere was investigated, the authors also reported the appearance of a signal with a similar g-factor after annealing the films at 700°C. The authors of this work did present any conclusions on the affiliation of this PC to any environment. But it is possible to suggest that this signal also relates to dangling bonds on the silicon atoms, which are surrounded by other silicon atoms. Note that a signal with a similar g-factor but with the intensity an order of magnitude higher was observed in the silicon powders obtained by mechanical milling of crystalline silicon in a ball mill, and where the minimum size of particles was not less than 100 nm, and the specific surface area was 1–2 orders of magnitude smaller than in the powder of type I–IV (Butyagin, 2001). This fact may serve as an indirect argument in favour of the EPR signal observed by us belonging to the PC situated in a silicon environment. The concentration of paramagnetic centres in the samples of the type III and IV was ~10¹⁸ particles/g.

According to models of the process proposed in several papers (see, for example, Kostantinova, 2007; Konstantinova et al, 2006; Szymanski et al, 2001; Cerofolini & Meda, 1997), similar centres in single crystals are formed as a result of recharging of silicon atoms and oxygen molecules falling from the atmosphere into the outer layer. Cerofolini & Meda (1997) proposed a model of a multistage process of formation and decay of the centres with dangling bonds on the silicon which takes into account the role of water molecules, also falling from the atmosphere, in stabilizing the hole centres in the silicon lattice. Participation of H_2O molecules in this process thus stimulates the formation of the PCs. Note that the mechanism of recharging with the formation of the PCs on silicon may involve not only the molecules of O_2, but also the molecule of NO_2, if they are present in the ambient atmosphere (Konstantinova et al., 2005). In general, the development of this mechanism is far from complete and, in particular, some details of the process associated with the collapse of the PCs and the formation of eventually oxide groups ($\equiv$Si–O–Si$\equiv$) in the surface layer remain unclear.

Annealing the samples in air, even for a few minutes at relatively low temperatures (100–200°C), leads to a change in the concentration of the observed PCs. Figure 6.6 shows the curves of annealing the samples of the types III and IV, corresponding to the change in intensity of the ESR signal with the temperature of heating. The exposure of samples at each temperature did not exceed 5 min. It is seen that the curve for the annealing of the sample of type III is very different from the curve for the type IV sample. The first curve, except for the signal intensity decay plot after 200°C, is observed to rise, and then after annealing at 600°C – again declines. The second curve shows a monotonic decrease of the EPR signal in the whole annealing temperature range.

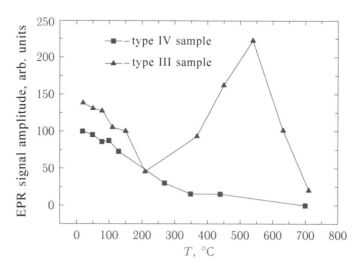

Fig. 6.6. Curves of changes in the intensity of the EPR signal in dependence on the temperature of heating of type III and IV samples, located in an air atmosphere (Rybaltovsky et al, 2007).

The data indicate that the concentration of paramagnetic centres in both samples decreased by 3–5 times after treatment in air at temperatures 200–300°C. The observed effect of increasing the concentration of paramagnetic centres in a sample of type III in the temperature range 200–600°C can be interpreted as inclusions of another source of PC formation in the mechanism of thermally stimulated reactions.

To determine the channels through which the reactions with participation of the PCs in heat treatment of the powders take place, a series of experiments was carried out with annealing in vacuum at certain temperatures and then exposing them to air at room temperature. The gas generation products were recorded along with the vacuum annealing of the samples. It was found that the gas impurity, released during the annealing of both types of samples at 300°C and 700°C, is hydrogen, and the amount of hydrogen released from the type III samples was several times greater than from the type IV samples.

Annealing of the samples under vacuum for 15 min at 300°C, accompanied by release of hydrogen molecules, led to a decrease in the intensity of the ESR signal by almost 7 times for a sample of type IV and 3 times for a sample of type III (as shown in Fig. 6.6). However, when air was suplied into the ampoule with the powder at room temperature there was a gradual increase in the EPR signal, which lasted for 24 hours. The intensity of the reconstructed signal in the type III samples was even higher than in the original sample, and for the type IV sample type it was somewhat lower.

It is interesting to note that the samples annealed in vacuum but at 700°C, showed a more intensive process of hydrogen evolution. In this case, the amount of released hydrogen was several times larger with respect to that value which corresponds to annealing at 300°C, and the concentration of paramagnetic centres was found to be more than in the initial samples. So, a type IV sample after annealing showed almost doubling of the ESR signal in comparison with the initial sample. After letting the air of the atmosphere into the ampoule with the powder there was an increase in the PC concentration for 24 hours as in the previous case (when the sample was annealed at 300°C), but the effect was considerably smaller and consisted, for example, in a type IV sample of only 20–25% of the concentration in the initial sample. If such a sample is heated in vacuum up to even higher temperatures (800°C) then the effect disappears completely. Exposure of the sample in air for 24 hours did not cause any change in the EPR signal intensity. Moreover, if this sample is now subjected to heat treatment in air in accordance with the procedure applied to the initial sample (see the experimental results in Fig. 6.5), we can find a strong 'temperature gap' in the process of breakdown of the PCs. Thus, for the temperature point of 260°C the concentration of the PCs in the initial sample was reduced almost four times, and in the heat-treated sample by only 20%. Even after heating to 700°C, when the concentration of the paramagnetic centres in the initial sample dropped to zero, the heat-treated sample retained the PCs in the amount of 10% of the initial value.

These results, as well as data from other studies (Kononov et al, 2005; Konstantinova et al, 2006; Hadj Zoubir et al, 1995; Chang & Lue, 1995) indicate the presence of complex multichannel reactions taking place during annealing in the surface layers of silicon nanoparticles. Indeed, counting the number of hydrogen molecules H_2, separated from the samples of types III and IV as a result of annealing in vacuum at a temperature of 700°C, shows that this value is one or two orders of magnitude greater than the number of the PCs observed in these samples. This fact indicates that chemical reactions involving the PCs reflect only certain part of the processes that occur in the surface layers of nc-Si during annealing. In the hydrogen-containing samples of type III the data obtained in infrared spectroscopy (Kononov et al, 2005) indicated the disintegration of the structural groups incorporating a silicon atom and two hydrogen atoms – ($=SiH_2$), at an annealing temperature lower than 300°C. At higher temperatures, up to 700°C, decay also takes place in a group with one hydrogen atom – ($\equiv Si–H$). It can be assumed that, firstly, the destruction of these groups is accompanied by the liberation of hydrogen, as observed in the experiment, and secondly, leads to the appearance of new structural defects. Note that the two stages of separation of H_2 molecules in the temperature range 400 and 500°C in heated porous silicon samples were reported in (Hadj Zoubir et al, 1995). The authors observed in the IR spectra a correlation between the release of hydrogen and disappearance of ($=SiH_2$) and ($\equiv SiH$) groups.

Thus, the reaction:

$$= SiH_2 \rightarrow = Si: + H_2,$$

taking place in the sample in heating to a temperature of 300°C is accompanied by the formation of a defect of the 'two-coordinated silicon atom bound to two neighboring silicon atoms' type. Also, heating to higher temperatures may lead to defects of the 'three-coordinated silicon atom with a dangling bond', according to the reaction:

$$\equiv SiH \rightarrow \equiv Si + H.$$

In the latter case, the resulting centre is paramagnetic, and its signal is apparently determined by the same g-factor as the signal in Fig. 6.5. It can be concluded that the above mechanism of formation of the PCs with the recharging of silicon and oxygen, is connected with the mechanism of formation of these centres by homolytic rupture of bonds in the structural groups ($\equiv SiH$). In general, based on the above considerations, the process of formation and decay of the PCs, involved in oxidative reactions, is described as follows. The effective formation of the PCs, as already noted above, requires not only the presence of oxygen in the ambient atmosphere, but also the presence of water molecules. This conclusion is confirmed by the experiments described in Rybaltovsky et al (2007), in which a type III sample pre-annealed in vacuum (at 300°C) was treated for a day at room temperature under an atmosphere of dry oxygen at a pressure of 60 Torr. In this case, unlike in the sample treated in air (see Fig. 6.6), there was no significant increase in the concentration of paramagnetic centres. Also there was no increase in the size of the PCs in these samples if after vacuum annealing they were treated only in an atmosphere of water vapour.

In general, prolonged contact with the air atmosphere leads to the formation of an oxide layer in the synthesized powders of nanocrystalline silicon which, according to the experimental results obtained by Radtsig et al (2007) amounts to 7% by weight of the nc-Si particles. This layer is formed by the reactions which include molecular ions O_2^-, the dangling bonds on silicon $\equiv$Si and H_2O molecules. Intermediate steps of these reactions are still unclear, but oxide groups such as $\equiv$Si–O–Si$\equiv$ appear in the end. In addition, $\equiv$SiH, $=$SiH$_2$, $\equiv$SiOH groups form in the surface layer by direct reactions of H_2O molecules with the silicon lattice. As for the groups ($\equiv$SiH), ($=$SiH$_2$), in the samples of type III they can form already during synthesis, as the source material is the monosilane SiH$_4$ (Kononov et al, 2005).

The concentration of paramagnetic centres in powders depends on the conditions of their synthesis, composition and temperature of the surrounding atmosphere. Indeed, the samples of type I, which were synthesized by the plasma-chemical method with the addition of oxygen to argon in the last stage (Rybaltovsky et al, 2006), showed the lowest concentration of the PCs: $(3–5) \cdot 10^{17}$ particles/g. It can be assumed that in this case particles initially have a thicker oxide shell than in the type III and IV samples. With increasing thickness of the oxide shell the processes of formation of the charged pairs with dangling bonds on silicon and O_2^- become more difficult (Szymanski et al, 2001). This factor also explains the marked non-monotonic effect (Kostantinova, 2007) of the change of the concentration of paramagnetic centres in the synthesized samples of nanoporous silicon, when their concentration in the first two months of exposure of the samples in air increases by an order of magnitude, and then falls again. At sequential annealing of the samples in air the efficiency of reactions leading to the disappearance of the PCs and the formation of oxide groups increases, which is partially reflected in the results of experiments by Rybaltovsky et al (2007) and presented in Fig. 6.7.

Upon heating the samples in air to temperatures above 300°C the samples of type III were characterized by the activation of the mechanism of formation of dangling bonds on the silicon atoms as a result of homolytic cleavage of bonds in the groups ($\equiv$Si–H), which corresponds to the annealing temperature. Therefore, the annealing curve for the sample of type III shows an increase in the intensity of the EPR signal to a temperature of

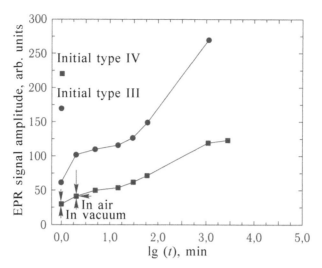

Fig. 6.7. Curves of changes in the intensity of the EPR signal in dependence on the time of exposure to air at room temperature for type III and IV samples, pre-annealed in vacuum at 300°C (Rybaltovsky et al, 2007).

approximately 600°C (Fig. 6.7). For type IV samples this type of effect is not observed, since the (≡SiH) and (≡SiOH) groups form only in the surface layer during the interaction with the molecules of H_2O, arriving from the atmosphere by the reaction (Dai et al, 1992):

$$\equiv Si - Si \equiv + H_2O \rightarrow \equiv Si - H + \equiv Si - OH.$$

The breakdown of the PCs in the samples subjected to treatment at 300°C in vacuum is accompanied by the evolution of hydrogen due to the loss of structural groups ($\equiv SiH_2$) with the formation of centres of two-coordinated silicon. In this case, the possibility of breakdown of the PCs can be viewed from this perspective. First, the transformation of these centres may occur in the same way as for the samples heated in air, because the samples were placed earlier in air for a longer period of time, and in the surface layer the particles will contain O_2 molecules or O_2^- ions. This assumption is supported by the results of experiments by measuring the IR spectra of the samples of porous silicon, where similar conditions (i.e. after heating in vacuum at 300°C) were characterized by the appearance of oxide groups linked to silicon (Hadj Zoubir et al, 1995). Second, a process may take place in which the structural defects formed (=Si:) act as electron donors, injected into the silicon lattice; they cause the reaction:

$$\equiv Si^{\bullet} + \equiv Si^+ + e^- \rightarrow \equiv Si - Si \equiv .$$

The rising concentration of paramagnetic centres after the start of exposure of the silicon nanoparticles in air (pre-treated powders in vacuum) is associated with a recharging mechanism, which involves oxygen molecules and water. As heating of the samples at 700°C corresponds to the decay of ($\equiv$Si–H) groups to form paramagnetic centres ($\equiv$Si), the appearance of the EPR signal of greater intensity after this annealing in vacuum is quite natural. It should be noted that the source of PCs in the form of ($\equiv$Si–H) groups occurs also in the initial samples of type IV due to the reaction of silicon dimers with water molecules from the atmosphere (Dai et al, 1992). Recall that the effect of increasing the concentration of paramagnetic centres after the exposure of these samples in air is much smaller than in the previous case, and completely disappears if the samples are annealed in vacuum at 800°C. It is possible that the observed phenomenon (Rybaltovsky et al, 2007) is due to the transformation of the existing original oxide shell into a denser shell, for example, with its transition from the SiO_2 phase to the SiO_x phase during heat treatment (Sato et al, 2003). The denser shell will prevent the diffusion of oxygen molecules from the air to the surface of the silicon core, which ultimately will lead, on the one hand, to the disappearance of the effects associated with recharging and, on the other hand – to a slowdown of the thermal decomposition processes of the existing PCs that take place in air.

6.3. Effect of carbon impurities on thermal oxidative processes in nanoscale silicon powders

In the practical application of nc-Si powders there are problems with the transformation of their properties depending on the chemical composition of impurities, appearing in the synthesis of these particles and their presence in the air (Delerue et al, 1999; Kovalev & Fujii, 2005; Jia et al, 1997; Terekhov et al, 2008; Kononov et al, 2005; Radtsig et al, 2007; Ischenko and Sviridov, 2006; Rybaltovsky et al, 2006). Given that the surface to volume ratio of the nanoparticle is significantly higher than in the samples of monocrystalline silicon, the effect of impurities falling into the nc-Si during and after synthesis may be decisive in the formation of the properties of nanosilicon. In particular, the optical characteristics and the capacity for oxidative processes in the air. If these questions have frequently been discussed in the literature for single-crystal silicon (See Konstantinova et al, 2005; Cerofolini & Meda, 1997, and references therein), in the case of nc-Si there are only a limited number of studies on this subject (Jia et al, 1997; Kononov et al, 2005; Rybaltovsky et al, 2006; Radtsig et al, 2007).

Besides the main component of the nc-Si powder – silicon atoms, its composition includes oxygen, whose amount can vary widely, as well as hydrogen, nitrogen and carbon, which fall into a particle during or after synthesis during storage in air (Kononov et al, 2005; Rybaltovsky et al, 2006; Radtsig et al 2007; Terekhov et al, 2008). As will be seen from further consideration, the content of carbon impurities in certain kinds of powders of nc-Si can be the third largest after silicon and oxygen (reaching a few percent by weight) and, apparently, has a significant influence on the physico-chemical properties of nanomaterials based on them.

This section is devoted to the study of the spectroscopic manifestations of the impurity carbon in the nc-Si powders and comparative analysis of its impact on the thermochemical processes in the nanosilicon produced by the laser-plasma and plasma-chemical and methods of synthesis annealing in air.

In Rybaltovsky et al, (2009), experiments were carried out on nc-Si powders obtained by the two methods described above – by plasma chemical and laser chemical. Of the currently known methods of synthesis of nc-Si these methods can produce a sufficiently large amount of material per unit time, which is crucial for subsequent use. A detailed description of this method of plasma chemical synthesis is given in (Rybaltovsky et al, 2006). nc-Si powders were synthesized in an argon atmosphere (99.95%). Second, the investigated samples were synthesized by decomposition of silane, SiH_4, in a CO_2-laser. This method is described in detail in (Kononov et al, 2005). The initial materials – crystalline silicon and silane – had a high degree of purity (not less than 99.99% for silicon and 99.95% for SiH_4).

After synthesis all the nc-Si samples were stored for quite a long time at room temperature in contact with the air atmosphere and were then studied by EPR and optical spectroscopy, including measurements of absorption and Raman spectra (see Kononov et al, 2005; Rybaltovsky et al, 2006; Radtsig et al, 2007).

The particle size of powders was monitored by two methods – measuring the specific surface area of the material and the method of Raman scattering (RS) analyzing the position and shape of the band near 520 cm^{-1}. The specific surface area of the powder S (m^2/g) was determined from adsorption isotherms of argon by the samples of silicon at 77 K (the site area for the argon molecule was assumed to be 0.16 nm^2). The obtained isotherms were analyzed by the BET method, which allows to determine the specific surface area of the sample and the parameter (C), which characterizes the heat of adsorption of gas molecules on solid surfaces. The powders, synthesized by different methods, yielded the following values of specific surface: ~70–90 m^2/g for the samples prepared by the plasma-chemical method and ~100–120 m^2/g for the samples prepared by laser chemical decomposition of silane. The value of the constant C is in the range of 30–40, indicating the absence of capillaries and micropores in the synthesized samples. On the assumption that the particles are spherical, the average value of their diameter over the entire ensemble of particles, $d = 6/S\rho$ ($\rho = 2.3$ g/cm^3 – the density of silicon), lies in the range 35–30 nm and 25–20 nm, respectively.

Raman spectra were measured (Rybaltovsky et al, 2006) using the T-64000 monochromator (Jobin Yvon) with the exciting radiation from an argon laser ($\lambda = 514.5$ nm). The measurement procedure for silicon powders was presented in (Rybaltovsky et al 2006). The functions of particle size distribution were measured by transmission electron microscopy. The maximum of the function distribution for the particles produced by the plasma-chemical method is in the range of 10–14 nm, and for the particles obtained by laser chemical synthesis 7–10 nm.

In the pilot study and comparative analysis of the data obtained by EPR spectroscopy on nc-Si powders, the experimental materials were in the form of fine powders of amorphous SiO_2, prepared from aerosil grade A300, the initial value of the specific surface of which is ~300 m^2/g, which corresponds to the average particle diameter of ~8 nm. Carbon was added to the aerosil powder and and its chemical activity was studied by the method of methoxylation of the initial sample surface in the methanol vapour at the temperature 700°C and its vapour pressure of ~100 Torr (Morterra, Low, 1974). Subsequent high-vacuum pyrolysis of such samples is accompanied by incorporation of part of the carbon atoms of the methoxy groups into the silica structure (Radtsig et al, 2007).

Elemental analysis of samples for impurities was performed by laser mass spectrometry in EMAL-2 equipment. The results showed that besides the main component of the silicon powder particles, whose content in the powders obtained by laser-chemical method, ~81 wt.%, in these materials the oxygen content reached about 16 wt.%, and in third place there was the carbon content, ~2 wt.%. In samples of nc-Si, obtained by plasma chemical synthesis the carbon content was ~5 wt.%. A high content of carbon in these samples can be explained if one considers that the synthesis of this powder takes place at a temperature of about 10000°C, and the carbon atoms can enter the reaction volume during evaporation of graphite electrodes.

According to IR spectroscopy data obtained on pressed sheets of nc-Si (Kononov et al, 2005), the carbon can fall onto the surface of the powder particles from the compounds present in ambient air and can then become chemically embedded in it, forming ($\equiv C-H_x$) groups, where $1 \leq x \leq 3$, associated with silicon atoms. These groups appear in the IR absorption spectra in the region 2850–2990 cm^{-1} (Kononov et al, 2005). The content of other impurities, including alkali elements and the elements of the iron group, in the powders, prepared by both plasma chemical and laser-chemical synthesis methods synthesis ranged from $n \cdot 10^{-4}$ to $n \cdot 10^{-3}$ wt.%.

In samples of nc-Si, obtained by laser chemical and plasma chemical synthesis, after preliminary heat treatment in vacuum for about ~30–40 min at 900°C the EPR spectra showed, in addition to the signal from the P_b-centres, another EPR signal (see Fig. 6.8).

It consists of a narrow central line with the g-factor of 2.0020 ± 0.0004, the half-width of which is less than ~1 G, and of two lines of lesser intensity, located symmetrically about the centre line and separated by a distance of 16.1 G.

The total intensity of these satellites is about ~15% of the intensity of the centre line. The intensity of this signal in both samples is comparable and equals about 10^{17} particles/g, which is at least an order of magnitude

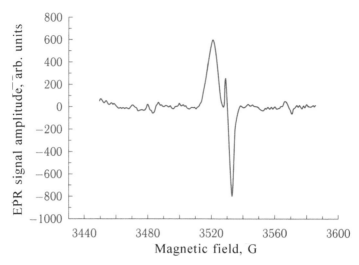

Fig. 6.8. The EPR spectrum of the signal of the pyrolysis products of nc-Si powders (Rybaltovsky et al, 2007).

less than the concentration of P_b-centres. Paramagnetic centres with similar radiospectroscopic values of parameters were recorded for the first time by the EPR method, initially in the oxidation of single-crystal silicon wafers (Kononov et al, 2005), and then in the oxidation of porous silicon (Terekhov et al, 2008). The structure of these paramagnetic centres was deciphered in Radzig (2007), where the use of isotope-enriched samples of silica (^{29}Si, and ^{13}C) showed that it has a structure, $((\equiv Si{-}O{-})_3 Si{-})_3 C^{\cdot}$, i.e. it is noit an intrnsic but an impurity carbon-centred paramagnetic centre not own, and the impurity carbon-centred PMC in fused silica, in which the first coordination sphere of the carbon atom is composed of three silicon atoms.

Note that this signal does not change its intensity after heat treatment in vacuum and subsequent supply of the air atmosphere into the ampoule, but further annealing of the sample in air decreases the concentration of PMC. Figure 6.9 shows the curve of the intensity of the PMC signal for the powder, produced by plasma-chemical synthesis, in dependence on the annealing temperature. Upon annealing, the ampoule with the powder was held for about 5 min at each temperature point. The data presented show that when the temperature reached ~500°C almost all considered PMCs disappear and the effective change in signal intensity started after reaching the temperature of ~300°C.

For the samples obtained by laser-chemical synthesis, changes in the intensity of the ESR signal took place already after annealing at temperatures above 200°C. Therefore, further analysis of this signal was conducted on the samples annealed in air at temperatures of 250–300°C, where the intensity of the signal from the P_b-centres which overlapped this signal is significantly reduced.

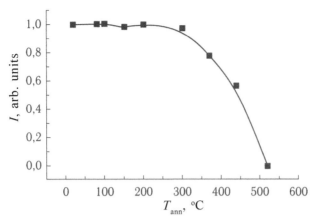

Fig. 6.9. The dependence of the intensity of the EPR signal of carbon radicals on annealing temperature of the samples of nc-Si, pre-treated in vacuum at 900°C (Rybaltovsky, et al, 2009).

One possible mechanism of formation of $(\equiv Si-)_3 C^{\bullet}$ radicals is the thermal decomposition of the C–H bond in the $(\equiv Si-)_3$ C–H group as a result of occurrence of the free-radical reaction (Radzig, 2007):

$$\left(\equiv Si-\right)_3 C - H \rightarrow \left(\equiv Si-\right)_3 C^{\bullet} + H^{\bullet}. \tag{6.4}$$

The activation energy of reaction (6.4), according to the work (Radzig, 2007), is ~100 kcal/mol, and it can occur at an appreciable rate at temperatures above 1000°C.

Thus, in both types of nc-Si powders, in spite of different synthesis methods, vacuum heat treatment formed the $(\equiv Si-)_3 C^{\bullet}$ radicals. If the initial powders of both types are subjected to heat treatment in air for ~30 min at 900–950°C, then subsequent vacuum annealing after this treatment does not lead to the emergence of carbon radicals. However, as our measurements show, the increase in the mass of these samples as a result of heat treatment in air is due to the increase in the mass of the oxide shell due to thermo-oxidative processes by ~20% (for the samples obtained by plasma chemical synthesis), and ~30% (for the samples obtained by the laser chemical method). If it is also considered that the loss of carbon radicals containing carbon powders in the nc-Si occurs when they are warming up in the air (see Fig. 6.9) or in an atmosphere of pure oxygen, it can be assumed that the proposed mechanism of formation of the PMC operates in this case (equation (6.4)).

This conclusion is also confirmed by the above-mentioned result of investigation of (C–H) hydrocarbon groups in powders obtained by the laser-chemical method, after their synthesis and holding in air for several days (Kononov et al, 2005). Incorporation of the carbon from the compounds present in the atmosphere, in the oxide shell of the nc-Si particles is connected apparently with the formation of ($\equiv$Si–CH$_3$) groups and subsequent dehydration when heating the samples. This process can also take place in the nc-Si samples, held in contact with air and obtained by plasma-chemical synthesis.

It is known (Elmor, 1992) that the usual air atmosphere contains up to $n \cdot 10^{-4}$ mol % CH$_4$, and, even in lower concentrations, the molecules of alcohols and formaldehyde. The mechanism of chemical incorporation of these molecules to form ($\equiv$Si–CH$_3$) groups remains unclear. Inclusion of carbon in the composition of nc-Si in this case can also occur during plasma-chemical synthesis during the formation of nc-Si particles. This assumption is indirectly confirmed by the higher concentration of carbon relative to silicon in the samples obtained by the plasma chemical method in comparison with the samples of nc-Si, obtained by laser chemical synthesis.

Given that the solubility of carbon in silicon at temperatures above 1000°C increases with decreasing temperature (Rybaltovsky, et al, 2009), it can be assumed that the carbon concentration in the volume of the groups particles inside the volume of the nc-Si particle changes and near the surface of the nc-Si core the carbon concentration increases. In this case, the method of entry of the carbon atoms into the composition of the silicon particle in this situation will be different from the method of their chemical insertion through the external side surface of the resultant particles in the event of prolonged contact of the particles with the atmosphere. In this case, for example, one can expect the appearance inside the particle of the groups of the ($\equiv$Si–C(–Si$\equiv$)$_3$) type, characteristic of silicon carbide. The different way of entry of carbon into the nc-Si composition can, in principle, affect some thermochemical processes in these particles.

A common property of the ($\equiv$Si–)$_3$C$^\cdot$ radicals, formed in the RSi and nc-Si samples, is that the EPR signal of these centres is saturated at very low levels of microwave power. The presence in the gas phase above the sample of oxygen molecules strongly affects the form of saturation curves. This effect is associated with a decrease in the duration of spin-lattice relaxation of the carbon-centred radical as a result of collisions with paramagnetic oxygen molecules from the gas phase (spin exchange) and indicates that these radicals are located in the surface layer of the particle. Figure 6.10 shows the curves of relative change in the EPR signal intensity of the carbon radicals in the samples of RSi and two types of nc-Si samples, depending on the oxygen pressure above the samples.

Measurements of the signals of the samples were made at the minimum values of the modulation of the microwave field to reduce the effects of the broadening of the absorption lines. The results presented in Fig. 6.10, suggest that the removal of saturation of the EPR signal from the carbon radicals in all cases occurs at an oxygen pressure of ~4 Torr. At the same time, these data support the assumption that the mechanism of removal of saturation for each sample has a number special features: the relative change in signal intensity before and after removal of saturation varies. The greatest

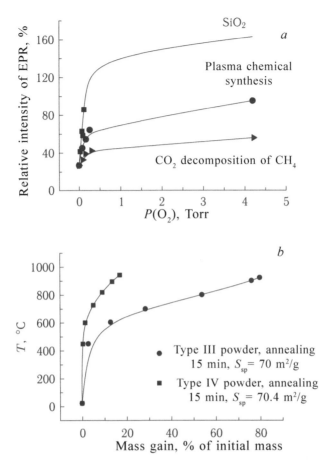

Fig. 6.10. The dependence of the relative EPR intensity on the partial pressure of oxygen (*a*). The curves of mass change in % of the initial mass of the nc–Si samples in sequential annealing in air (*b*). Description of the samples is given in the beginning of Section 6.1 (Rybaltovsky et al, 2009).

change in the signal occurs for the silica sample and the lowest – in the nc-Si sample, obtained by the the laser chemical technique. The mechanism of this phenomenon is not completely clear and requires a separate study.

It is possible to assume that the smaller changes of the signal for the nc-Si samples in comparison with silica are due to the differences in the mechanisms of energy exchange between the spin states of P_b-centres, whose concentration is an order of magnitude greater than the concentration of carbon radicals, and the centres themselves ($\equiv$Si$-$)$_3$C$\cdot$. For this reason, perhaps, the least 'sensitive' to the presence of oxygen are the carbon-containing radicals in the nc-Si samples, obtained by laser chemical decomposition of silane compared with the activated silica samples.

Figure 6.10*b* shows the curves of mass changes in the samples of nc-Si, obtained by plasma-chemical and laser-chemical synthesis in annealing in air. The ordinate is the temperature at which the sample was annealed for about ~15 min. The powder sample was kept in a quartz ampoule and then cooled to room temperature. The abscissa gives the relative magnitude of the mass change of the sample, measured after annealing and is expressed as a percentage. As can be seen from Fig. 6.10*b*, this value increases with increasing annealing temperature, which indicates an increase the mass of the oxide shell of the nc-Si samples. At the same time, this growth is uneven and at a temperature of ~950°C the mass change is equal to ~80% for the sample obtained by laser-chemical synthesis and ~20% for the sample obtained by the plasma chemical method. Increase of the annealing temperature up to about ~1000°C and above leads to almost 100% transition of the sample obtained by the laser-chemical method to the silicon dioxide, and annealing under these conditions of the sample obtained by the plasma method essentially does not change its composition compared with that for the annealing temperature of 950°C.

In the article by Rybaltovsky et al (2009) it is noted that the annealing of the nc-Si powders, obtained by laser decomposition of silane, which was carried at 800°C for an hour, is also accompanied by a complete transformation of the silicon nanoparticles to SiO_2 particles. Increased resistance to thermal oxidative processes in air of the samples obtained by the plasma chemical method, is probably linked to the mechanism of chemical incorporation of the carbon atoms in the composition of the silica particles, as described above, and to the formation of a large number of bonds characteristic of the thermally stable silicon carbide.

Thus, the results of Rybaltovsky et al (2009) show that the samples of nanocrystalline silicon synthesized by two different methods which allow to produce a significant amount of material for further technological development – the plasma chemistry method and laser decomposition of SiH_4 – can contain fairly high concentrations of carbon – to 2–5 wt.%, which can enter the particles either in the process of synthesis and in the following prolonged contact with the atmosphere, as in the first case, or only as a result of prolonged contact with the air atmosphere in the second case.

In these samples, after thermal treatment in vacuum (10^{-3} atm) at 900°C, EPR spectroscopy detected the signal with a spectrum similar to that observed in specially prepared powders of amorphous SiO_2; the signal was identified as a paramagnetic centre of the radical, located on the carbon atom bonded with three silicon atoms. Thus, the observed centre with a *g*-factor equal to 2.0020, and two satellites with a splitting gap of 16 G, belongs to the carbon radicals formed during the thermal reactions in the oxide shell of the nc-Si particles. Their concentration reaches 10^{17} particles/g, which is at least an order of magnitude lower than the concentration of P_b-centres in these samples.

The intensity of the EPR signal from the radicals on carbon is strongly dependent on the power of the probing microwave radiation when the powder particles are held in a vacuum without contact with oxygen molecules. At low pressures of the oxygen atmosphere, about 4 Torr, the intensity of the saturated signal increases without changing its form. These results can be interpreted as the formation of carbon radicals in the surface oxide layers of the particles and as the interaction of oxygen molecules with these radicals. The intensity of the EPR signals of the carbon radicals decreases in annealing of the nc-Si samples in air, and their concentration drops to zero after annealing at temperatures above 500°C in the samples obtained by plasma-chemical synthesis and at 400°C – in the samples obtained laser chemical decomposition of silane. This effect for these radicals is associated with thermal oxidation processes involving oxygen molecules of the ambient atmosphere.

6.4. Effect of water

The mechanism of interaction between water and silicon is the first step in the disproportionation reaction:

$$2Si + 4H_2O \Leftrightarrow SiH_4 + Si(OH)_4.$$

In the second stage, the orthosilicic acid is easily dehydrated:

$$Si(OH)_4 \rightarrow SiO_2 + 2H_2O,$$

and silane readily reacts with water:

$$SiH_4 + 2H_2O \rightarrow SiO_2 + 4H_2.$$

The overall reaction:

$$Si + 2H_2O \rightarrow SiO_2 + 2H_2.$$

It was found that the interaction of water with nanosilicon is influenced by ultraviolet radiation (Bahruji et al, 2009). Water was pre-deaerated, the nc-Si particles had a mean diameter of 75 nm, the reaction was carried out in an aqueous suspension of nc-Si by argon. The thickness of the oxide layer of nc-Si was measured by X-ray spectroscopy on the basis of the ratio of the integrated intensities of the peaks Si $(2p)$ and O $(1s)$. In the initial sample (not heated at all) the layer thickness was equal to 1.3 nm. The surface was measured by the BET method.

In the absence of UV irradiation of the suspension hydrogen was not released at room temperature even during 5 hours, but after irradiation a reliably determined (20.5 ml) amount of hydrogen was released during 5 hours, the first hour of the reaction was the incubation period. The curve of hydrogen evolution was *s*-shaped, characteristic of autocatalytic reactions.

After 5 hours of the reaction the brown nanosilicon powder turned into a gel that could be extracted from the water only by centrifugation. X-ray analysis showed only traces of unreacted silicon.

To study the influence of the thickness of the oxidized surface layer on the interaction of nc-Si with water, the samples were heated in air or argon at temperatures of 100, 200, 300 and 600°C for 5 hours. Preheating of nc-Si on air at a temperature of 37°C for 5 h almost completely interrupted the evolution of hydrogen. The heating in dry argon had the same effect, i.e. the reason for the termination of the reaction of nc-Si with water was not associated with the thickness of the oxidized layer on the surface of nc-Si.

It was found that the thickness of the oxide layer of nc-Si particles, preheated up to 300°C for 5 hours, is equal to the thickness of the oxide layer of the starting material. Pre-annealing at 600°C resulted only in better correspondence of the composition of the oxide layer to the SiO_2 formula as a result of more extensive oxidation of the boundary layer at the Si/SiO_2 interface.

In the initial nc-Si FTIR revealed the (Si–O–Si) and Si–H bonds. After heating up to 600°C the former grew quantitatively, the latter disappeared. The intensity of the bands related to the (Si–OH) bonds remained constant at all temperatures. However, in UV irradiation the number of the OH-groups on the surface of nc-Si greatly increased, indicating the reaction of silicon with water. After the reaction of nc-Si with water for 1 hour (i.e. the incubation period), the surface area of nc-Si increased by an order. This indicated the formation of a porous layer around each nc-Si particle.

In the reaction of silicon with water UV radiation generated electron–hole pairs in the nc-Si and charge carriers were localized in the traps – the surface atoms of silicon, i.e. at the Si/SiO_2 interface. The silicon atom with an electron accepts a proton, forming a Si–H bond, the silicon atom with a hole accepts OH$^-$ forming the (SiOH$^-$) silanol group. Preheating to 37°C eliminates defects – traps of charge carriers, and the nc-Si does not react with water. What are the traps remains unclear.

In the presented experimental results attention should be given to the possibility that the composition of the oxide layer on the nc-Si surface may not correspond completely to the stoichiometry of SiO_2. It is widely believed that there is a transition layer between silicon and silica, in which the atomic ratio Si:O > 1:2, i.e. silicon is not completely oxidized.

The results of modifying the surface of nc-Si with oxygen using water were demonstrated by a group of authors (Gelloz et al, 2006). Nanoporous silicon (P-Si) was obtained by anodic etching of silicon wafers in a mixture of 55 wt.% HF and ethanol (1:1) to a porosity of 68%. The thickness of the porous layer was 0.5–20 μm. The surface of P-Si was covered (i.e. modified) with hydrogen atoms.

The P-Si layer was washed with ethanol, 2 min, and a part of P-Si was oxidized electrochemically in 1 M H_2SO_4 solution until the moment when the potential of the silicon substrate increased abruptly due to oxidation of silicon. Oxidized P-Si (OP-Si) was washed ethanol, 2 min, and dried in a nitrogen flow.

Porous silicon was treated with water vapour: P-Si or OP-Si was placed in an autoclave and water was added in the amount that at 260°C it transformed to vapour. The amount of water was calculated from the equation of an ideal gas. The autoclave was kept for 3 hours at this temperature. Prior to treatment with the vapour there was no PL in P-Si, and in OP-Si it was barely noticeable, but the intensity of the PL of the vapour-treated P-Si increased a thousand times, and for OP-Si 70 times, although the position of the maximum remained at 700 nm. The quantum efficiency of vapour-treated P-Si increased by up to 23%.

Comparison of FTIR spectra before vapour treatment of P-Si and after it shows that all the peaks, corresponding to vibrations of Si–H$_x$ bonds (x = 1–3), disappear after vapour treatment, which means that the surface is covered with oxides of silicon to which the absorption bands of the infrared spectra correspond. The OP-Si oxidized in 1 M H$_2$SO$_4$ solution had Si–H bonds on the surface. Apparently, the thin oxide layer formed during electrochemical oxidation, did not affect the Si–H bonds. After vapour treatment of OP-Si all Si–H bonds disappeared, and the intensity of the peaks, corresponding to the silicon oxides, increased.

The EPR method showed that the density of P_b-defects after vapour treatment of P-Si is much smaller than that in P-Si before vapour treatment. Spectroscopy of PL with time resolution showed that the exciton lifetime in vapour-processed P-Si is 2–3 times longer than before vapour treatment.

Vapour treatment of P-Si increased the resistance P-Si to UV radiation: exposure for 3 h had no effect on the PL intensity, and prior to vapour treatment the PL intensity decreased by half after 15 min of irradiation.

Figure 6.11 shows that vapour treatment dramatically increases the PL quantum efficiency without a significant shift of the peak wavelength of the PL spectrum.

It turned out that as a result of vapour treatment all P-Si were covered with a thin layer of silica and there were no stressed at the P-Si/SiO$_2$ interface. Vapour treatment removes non-radiative defects and thus localizes excitons in the P-Si particles. At the same time the stability in air increases. P-Si is easily oxidized at low temperatures and this accompanied by the formation of surface defects that lead to non-radiative recombination and traps of charge carriers and thus to a deterioration in the quality of nc-Si as a material for various applications. Water vapour treatment improved the electroluminescence (EL) of P-Si by several orders of magnitude compared to macrocrystalline silicon. The quantum efficiency of EL reached 1.1%, and the efficiency with respect to energy 0.4% (Koshida et al, 2007).

It can be assumed that the formation of the SiO$_2$ layer on the surface of nc-Si is a consequence of both oxidation of silicon by atmospheric oxygen and of disproportionation of silicon under the influence of water or its vapour, but in the literature both these processes are called oxidation of

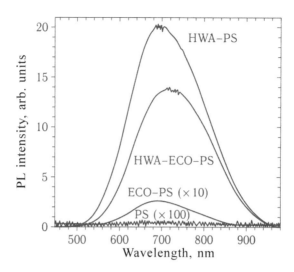

Fig. 6.11. Photoluminescence spectra of porous silicon layers (PS) before and after electrochemical oxidation (ECO-PS). The graph also shows the photoluminescence spectra of these two layers after treatment with water vapor at high temperature (HWA – PS and HWA–ECO-PS) (Gelloz et al, 2006).

silicon. It has been shown and repeatedly confirmed (e.g. Ledoux et al, 2000) that the formation of the SiO_2 layer on nc-Si is a self-limiting process, i.e. oxidation of silicon is terminated after reaching a certain thickness of the SiO_2 layer on the silicon surface. The Deal–Grove one-dimensional model of silicon oxidation, designed for coarse crystalline silicon was extended to spherical geometry (Coffin et al, 2006). The authors have linked the decrease in the rate of oxidation of nc-Si with the tensile stress arising inside the SiO_2 shell, and the radial pressure on the surface of the nc-Si, progressing at the nc-Si/SiO_2 interface during the oxidation of nc-Si.

The two-dimensional model of self-limitation of oxidation of silicon was proposed previously (Liu et al, 1994) for the example of oxidation of silicon nanowires (separated fibers of P-Si). In the oxidation of P-Si, starting at some point, the oxidation is almost completely interrupted. The limiting diameter of P-Si depends on the thickness of the oxide at temperatures below 900°C. It was experimentally shown that a significant inhibition of oxidation of P-Si is due to kinetic factors: the activation energy for diffusion of oxygen from the air to the Si/SiO_2 interface increases with increasing ratio $\gamma = r_o/r_c$, where r_o is the radius of the outer wall of the oxide layer, and r_c is radius of P-Si, i.e. the activation energy of diffusion depends on the thickness of the SiO_2 layer on P-Si. Such a dependence coincides with an increase in the strain energy in the oxide layer, which indicates the dependence of the diffusion activation energy on stresses in the SiO_2 which increase with increase of the value $\gamma = r_o/r_c$.

According to the information presented in (Scheer et al, 2003), oxidation of nc-Si in an oxygen atmosphere is self-limiting at temperatures below 850°C, at which the point of transition of SiO_2 from the elastic to viscous state is not reached, and in the NO atmosphere – at any temperature. In oxidation in oxygen the squeezing effect of the SiO_2 shell on nc-Si inhibits the oxidation reaction. The presence of nitrogen in the oxide shell inhibits the diffusion of oxygen or NO from the outside to the surface of nc-Si through the oxide shell even at a temperature of 1050°C. Modification of the surface of nc-Si by the nitrogen oxide reduces the thickness of the insulating layer on nc-Si.

A group of authors Liao et al. (2006) used the method of tandem differential mobility analysis of solid particles in a gas to determine the mechanism of oxidation of silicon. nc-Si particles were obtained in cold plasma of SiH_4: silane was mixed with a carrier gas – a mixture of argon and helium. The pressure in the chamber was 17 Torr, the kinetics of oxidation of nc-Si was determined at the atmospheric pressure of the gas. The flow of nc-Si, mixed with a carrier gas, passed through a bipolar charging device, after which each particle of nc-Si acquired a charge. The lability of the charged particles was recorded by a differential analyzer, and the nc-Si fell into a tubular furnace at the outlet from which the flow passed through another differential analyzer. At this stage, the carrier gas was nitrogen or a mixture of nitrogen and oxygen. Flows of nitrogen and oxygen were measured carefully.

When using pure nitrogen as a carrier gas, the nc-Si produced after heating in the furnace up to 1100°C had an average diameter of 10.3 nm. TEM showed their crystallinity and the presence of an amorphous shell. However, at the volume fraction of oxygen of 0.045 in the gas mixture after heating up to 1100°C the diameter of nc-Si was 11.10 nm and at the volume fraction of oxygen of 0.73 and greater it was 11.69 nm due to thickening in of the surface layer of SiO_2 or suboxide. The surface layer of silicon oxide acted as a barrier to oxygen diffusion, i.e. the kinetics of the oxidation is limited by diffusion. The first oxide monolayer is formed quickly, because the oxygen reacts directly with silicon and the oxidation reaction then slows down.

The dependence of the properties of nc-Si on modification of their surface with oxygen (and for comparison with hydrogen) was very studied in detail in (Wolkin et al, 1999). Porous silicon was obtained by etching of silicon single crystals with an ethanol solution of hydrofluoric acid (10–25%) under illumination, i.e. the surface of the nc-Si was modified by hydrogen (see Chapter 5, section 'Electrochemical etching'). The samples were washed with ethanol, and some of them kept and measured in argon, and the other (reference) group of samples was left in the air. PL was excited by radiation with wavelengths of 337 and 325 nm. After holding in air for 24 hours, the PL of the samples, which luminesced in argon from blue to orange light, was shifted to the red and its intensity decreased. The magnitude of the red shift increased with increasing porosity. All samples, luminescing under argon by blue and green light, after holding in air started to luminesce with the light with a photon energy of 2.1 eV (590 nm). The samples, luminescing in argon with red light, did not change the PL after exposure to air. The lifetime of the PL in argon and in air was measured in microseconds. In argon, the PL lifetime decreased with increasing porosity from 32 ms (red luminescence) to 0.07 ms (blue luminescence) and after exposure in the air the PL lifetime was 2 ms (for a sample with green luminesce in argon), and then remained constant with increasing porosity of the samples. A large red shift was observed in samples placed in pure oxygen, but there was no red shift in hydrogen or in vacuum.

FTIR showed that in holding in air the composition of the surface of P-Si changes: holding for 3 min results in the formation of Si–O–Si bonds, 100 min – links Si–O–H bonds, and 24 h – the last Si–H bonds disappeared. The red shift of photoluminescence was almost completed after 200 min, and 2/3 of the red shift occurred in the first minutes of exposure to air. The authors assert that both the size of nc-Si and the chemical composition of the surface determine the mechanism of electron–hole recombination. The authors (Wolkin et al, 1999) assume that the observed experimental facts are determined by the Si=O bonds. The electronic structure of silicon nanoclusters in size up to ~5 nm and a single Si=O bond was calculated by quantum mechanics methods. The results were already partly discussed in Chapter 3 and are presented here in Fig. 6.12. These important results deserve further analysis in the context of the current chapter.

According to the chosen model, the nc-Si, passivated by hydrogen, recombines the free exciton for any size of nc-Si, i.e. If there are (Si=O) bonds, the electron is localized on Si and the hole – on oxygen. Three different mechanisms were proposed, depending on the size of nc-Si. Each zone shown in Fig. 6.12 corresponds to different mechanisms. In zone I the free exciton recombines, since the energy levels (Si=O) are outside the band gap. In zone II, the electron trapped by the silicon atom of the Si=O bond and the free hole recombine. The energy of the trapped electron also depends on the size of the nc-Si, but not as much as the energy of the threshold of the conduction band. In zone III both the trapped electron and a hole recombine, and with decreasing size of nc-Si the photon energy in the PL remains constant.

The cover the surface of nc-Si plays an important role especially for particle sizes < 3 nm. nc-Si larger than 3 nm, are not affected by the (Si=O) bonds, and the composition of the surface can be neglected. It is shown that nc-Si oxidizes in the air within seconds.

In the article by Wolkin et al. (1999) PL spectra recorded from the entire set of nc-Si which always contains nc-Si in different sizes and therefore

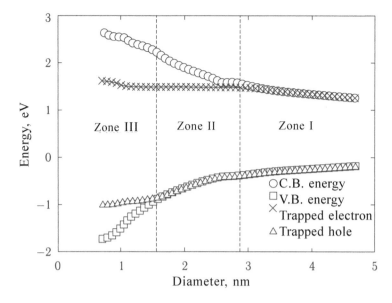

Fig. 6.12. The dependence of the energy of the electronic states of nc-Si on the cluster size and surface passivation. Trapped electron states are *p*-states localized on the silicon atom of the (Si=O) bonds, and the trapped hole states – *p*-states localized on the oxygen atoms (Wolkin et al, 1999).

luminesce by virtue of the size effect in slightly different wavelength intervals of the visible region. All the intervals overlap and a broad, poorly structured PL band is recorded in this case. To avoid this complication it is appropriate to the study of photoluminescence in one single nanocrystal, as carried out in (Martin et al, 2008), with the size of nc-Si not exceeding 3 nm. The particles of nc-Si were obtained by pyrolysis of silane with a CO_2-laser and were covered with native layer of amorphous SiO_2. The PL spectra of two single nc-Si revealed satellites on the red side of the PL peak (zero phonon) indicating strong electron–phonon coupling related to the (Si–O–Si) groups. Experimentally it was found that, starting at the PL quantum energy of 2.1–2.3 eV, the PL energy does not depend on the diameter of nc-Si when the size of nc-Si decreases. The PL quantum energy of 2.1 eV corresponds to electron localization and with decreasing diameter of nc-Si the electrons of the electron–hole pairs do not reach the energy level of the conduction band and remain at the level of energy localization of the electrons. At the photon energy of 2.325 eV the same thing happens in the valence band with the localization of holes. Thus, the surface of nc-Si at the interface nc-Si/SiO_2 is modified by both Si=O and Si–O–Si groups.

Dohnalova et al, 2009, experimentally studied and compared exciton states in the volume of nc-Si and on their partially oxidized nc-Si/H surfaces the band gap of which in the absence of oxygen impurity depends only on the size of the nanocrystals. Nanocrystals were obtained by anodic etching of a Si (100) wafer in a mixture of 50% HF:ethanol:30% H_2O_2 = 1:2.8:0.15. According to HRTEM and Raman spectra, the size of nc-Si is 2–3 nm, but the crystallinity of nanosilicon could not be proved, so we have to rely on the commonly used terms obtained in the synthesis of the results and regard the nanosilicon particles as crystalline. FTIR showed that after oxidation in air the OH-groups dominate over the hydrogen atoms on the surface of the nanoparticles. The photoluminescence of the freshly prepared sample, initially green, changes in one minute to red, and all stages of the evolution of colour through yellow and orange can be seen. Even weak ultraviolet excitation contributed to surface oxidation contributed to nc-Si.

The PL spectra with time resolution showed that there are two PL bands – fast (ns) and slow (hundreds of microseconds). The initial nc-Si/H had a fast PL peak at 525 nm, but after a few minutes this peak shifted to 550 nm and its intensity initially increased slightly but then significantly decreased without changes of the lifetime of the excited state, while the slow component of the PL formed at the same time, increased its intensity, and its peak shifted from 575 to 650 nm, i.e. during the oxidation of the nc-Si surface the slow PL component intensified and the fast one decreased. Some increase in the intensity of the fast component of PL at the beginning of oxidation was explained by the authors by the blocking of P_b-centres by oxygen. As soon as (Si=O) groups appeared on the surface, the interband electron–hole recombination was terminated because the excitons were localized on the (Si=O) surface states. As the number of these bonds increased the slow PL became brighter and the number of impurity levels in the forbidden zone of nc-Si increased. Hence the red shift of the slow became greater.

Concluding remarks

The practical application of nc-Si powders is associated with problems with the transformation of the properties of the material depending on the chemical composition of impurities that appear in the synthesis of nanosilicon and its presence in the atmosphere. Given that in the case of nc-Si powders the surface to volume ratio is significantly higher than in the samples of monocrystalline silicon, the effect of the impurities entering the particle during synthesis and after may be decisive in shaping the properties, in particular, the optical characteristics and the ability to oxidation processes in the air. If these issues have been often discussed in the literature for single-crystal silicon, in the case of nc-Si there are only a limited number of papers devoted to this subject.

Besides the main component of the nc-Si powder – silicon atoms, its composition includes oxygen, whose amount can vary widely, as well as hydrogen, nitrogen and carbon, which fall into a particle during the synthesis or after synthesis during the storage of powder in the air. The content of carbon impurities in certain types of nc-Si powders can be the third largest after silicon and oxygen and has a significant influence on the physico-chemical properties of nanomaterials based on them.

Samples of nanocrystalline silicon synthesized by two different methods that produce a significant amount of material for further technological development: plasma chemistry methods and laser decomposition of SiH_4, can contain fairly high concentrations of carbon – up to 2 wt.%, which can enter the particles either in the process of synthesis followed by prolonged contact with the atmosphere, as in the first case, or only as a result of prolonged contact with the air atmosphere in the second case.

In these samples, after thermal treatment in vacuum (10^{-3} atm) at 900°C, EPR spectroscopy detected a signal similar to the spectrum which was observed in specially prepared powders of amorphous SiO_2, and was identified as a paramagnetic centre of the radical located on the carbon atom, associated with three silicon atoms. Thus, the observed centre with a g-factor equal to 2.0020(3), and two satellites with a splitting of 16 G belong to carbon radicals formed during the thermal reactions in the oxide shell of the nc-Si particles. Their concentration reaches 10^{17} particles/g, which is at least order of magnitude lower than the concentration of P_b-centres in these samples.

The intensity of the EPR signal from radicals on carbon is strongly dependent on the power of the probing microwave radiation in finding the powder particles in vacuum without contact with oxygen molecules. At low oxygen pressures in the atmosphere, about 4 Torr, the intensity of the saturated signal increases without changing its form. But at high oxygen pressures, as shown by measurements on SiO_2 powders (Radtsig, Ischenko, 2011), there was a change in the signal. These results can be interpreted by the formation of carbon radicals in the surface layers of oxide particles and by the interaction of oxygen molecules with these radicals. The intensity of the EPR signals of carbon radicals decreases during annealing of the nc-Si samples in air and their concentration drops to zero after annealing at temperatures above 500°C in the samples obtained by plasma-chemical synthesis, and at 400°C – in the samples obtained by laser chemical decomposition of silane. This effect for these radicals is associated with the processes of thermal oxidation with participation of the oxygen molecules of the surrounding atmosphere.

References for Chapter 6

Browner C. Adsorption of gases and vapors. McGraw-Hill, 1948.

Butyagin P. Y., Streletskii A. N., Berestetskaya I. V., Borunova A. B. Amorphization of silicon during mechanical treatment of powders 3. Gas sorption. Kolloid. Zh. 2001. V. 63, No. 5. P. 699-705.

Ischenko A. A., Sviridov A. A. Sunscreen. II. Inorganic UV filters and their compositions with organic protectors. Izv. VUZ. Ser. Khim. Khim. Tekhnol. 2006. V. 49, No. 12. P. 3-16.

Kononov N. N., Kuzmin G. P., Orlov A. N., Surkov A. A., Tikhonevich O. V. Optical and electrical properties of thin plates made of nanocrystalline powders of silicon. Fiz. Tekh. Poluprovod. 2005. V. 39, No. 7. P. 868-873.

Kostantinova E. A. Photoelectric processes in nanostructured silicon with spin centers. Abstract of PhD. thesis. Moscow, 2007.

Radtsig V. A., Ischenko A. A. Carbon in silica. Kinetika i Kataliz, 2011. V. 52, No. 2. P. 316-329.

Radtsig V. A., et al. Thermo-oxidative processes in nanosized powders of silicon. I. Spectral manifestations. Nanotekhnika. 2007. No, (11). P. 110-116.

Rybaltovsky A. O., et al. The spectral features of water–emulsion composite media containing silica nanoparticles. Optika Spektroskopiya. 2006. V. 101, No. 4. P. 624-631.

Rybaltovsky A. O., et al. Impact carbon impurities on the thermal-oxidative processes in nanoscale powders of silicon. Izv. VUZ. Ser. Khim. Khim. Tekhnol. 2009. No. 10. P. 126-131.

Rybaltovsky A. O., et al. Thermo-oxidative processes in nanoscale powders of silicon. II. Paramagnetic centers. Nanotekhnika. 2007. No. 3 (11). P. 116-121.

Tutorsky I. A., et al. The structure and adsorption properties of nanocrystalline silicon. Kolloid. Zh. 2005. V. 67, No. 4. P. 541-547.

Bahruji H., Bowker M., Davies P. R. Photoactivated reaction of water with silicon nanoparticles. International journal of hydrogen energy. 2009. V. 34. P. 8504–8510.

Cerofolini G. F., Meda L. Mechanisms and kinetics of room-temperature silicon oxidation. J. Non-Cryst. Sol. 1997. V. 216. P. 140–147.

Chang C. S., Lue J. T. Photoluminescence and Raman studies of porous silicon under various temperatures and light illuminations. Thin Solids Films. 1995. V. 259. P. 275–280.

Coffin H., Bonafos C., Schamm S., Cherkashin N., Ben Assayag G., Claverie A., Respaud M., Dimitrakis P., Normand P. Oxidation of Si nanocrystals fabricated by ultralow-energy ion implantation in thin SiO_2 layers. J. Appl. Phys. 2006. V. 99. P. 044302.

Dai D.-X., Zhu F.-R., Luo Y.-C., Davoli J. Dissociative chemisorption of water on the Si (111) 7×7 surface studied at 150 K by x-ray photoelectron spectroscopy and energy loss spectroscopy. J. Phys.: Condens. Matter. 1992. V. 4. P. 5855–5862.

Delerue C., Allan G., Lanoo M. Optical band gap of Si nanoclusters. J. of Luminescence. 1999. V. 80. P. 65–73.

Dohnalová K., Kusová K., Pelant I. Time-resolved photoluminescence spectroscopy of the initial oxidation stage of small silicon nanocrystals. Appl. Phys. Lett. 2009. V. 94. P. 211903.

Elmor T.H. In: Engineered Materials. Handbook. V. 4. Ceramic and glasses, ASM International, 1992. P. 427.

Griskom D. L. Defect structure of glasses. J. Non-Cryst. Sol. 1985. V. 73. P. 51–74.

Hadj Zoubir N., Vergnat M., Delatour T., Burneau A., De Donato Ph., Barres O. Natural oxidation of annealed chemically etched porous silicon. Thin Solid Films. 1995. V. 255. P. 228–230.

Jia J.-H., Wang Y., Chen Z.-X., Zhang L.-D. Influence of oxidation on optical diffuse reflectance spectra in handscale silicon powder. Appl. Phys. A. 1997. V. 65, No.4–5. P. 383–385.

Jolly F., Canfin J. L., Rochet F., Dufour G., von Bardeleben H. J. Temperature effects on the Si/SiO2 interface defects and suboxide distribution. J. Non-Cryst. Sol. 1999. V. 245. P. 140–147.

Konstantinova E.A., Demin V.A., Vorontsov A. S., Ryabchikov Yu.V., Belogorokhov I.A., Osminkina L.A., Forsh P.A., Kashkarov P.K., Timoshenko V. Yu. Electron paramagnetic resonance and photoluminescence study of Si nanocrystals-photosensitizers of singlet oxygen molecules. J. Non-Cryst. Sol. 2006. V. 352. P. 1156–1159.

Konstantinova E.A., Osminkina L.A., Sharov C. S., Timoshenko V. Yu., Kashkarov P.K. Influence of NO2 molecule absorbtion on free charge carries and spin centres in porous silicon. Phys. Stat. Sol. (a). 2005. V. 202. P. 1592–1596.

Koshida N., Gelloz B. Photonic and relateddevice applications of nano-crystalline silicon. Proc. of SPIE. 2007. V. 6775. P. 67750N (9 p.).

Kovalev D., Fujii M. Silicon Nanocrystals: Photosensitizers for Oxygen Molecules. Adv. mater. 2005. V. 17. P. 2531–2544.

Kuz'min G. P., Karasev M. E., Kokhlov E.M., Kononov N.N., Korovin S. B., Plotnichenko V.G., Polyakov S.N., Pustovoy V. I., Tikhonevich O.V. Nanosize silicon powders: the structure and optical properties. Laser Phys. 2000. V. 10. P. 939–945.

Ledoux G., Guillois O., Porterat D., Reynaud C., Huisken F., Kohn B., Paillard V. Photoluminescence properties of silicon nanocrystals as a function of their size. Physical Review B. 2000. V. 62. P. 15942 (10 p.).

Liao Y.-C., Nienow A.M., Roberts J. T. Surface Chemistry of Aerosolized Nanoparticles:Thermal Oxidation of Silicon. J. Phys. Chem. B. 2006. V. 110. P. 6190–6197.

Liu H. I., Biegelsen D.K., Ponce F.A., Johnson N.M., Pease R. F.W. Self-limiting oxidation for fabricating sub-5 nm silicon nanowires. Appl. Phys. Lett. 1994. V. 64. P. 1383–1385.

Mangolini L., Jurbergs D., Rogojina E., Kortshagen U. Plasma synthesis and liquid-phase surface passivation of brightly luminescent Si nanocrystals. Journal of Luminescence. 2006. V. 121. P. 327–334.

Martin J., Cichos F., Huisken F., von Borczyskowski C. Electron-Phonon Coupling and Localization of Excitons in Single Silicon Nanocrystals. Nano Lett. 2008. V. 8, No.2. P. 656–660.

Meier C., Lüttjohann S., Kravets V.G., Nienhaus H., Lorke A., Wiggers H. Raman properties of silicon nanoparticles. Physica E. 2006. V. 32. P. 155–158.

Norton F. J. Permeation of gaseous oxygen through vitreous silics. Nature. 1961. V. 191.P. 701–705.

Prusry S., Mavi H. S., Shukla A.K. Optical nonlinearity in silicon nanoparticles: Effect of size and probing intensity. Phys. Rev. B. 2005. V. 71. P. 113313-1-4.

Radzig V.A. Point defects on the silica surface: structure and reactivity. In: Physico-chemical phenomena in thin films and at solid surfaces. V. 34. Elsevier, 2007. P. 233–345.

Sato K., Izumi T., Iwase M., Show Y., Morisaki H., Yaguchi T., Kamino T. Nuleation and growth of nanocrystalline silicon studied by TEM, XPS and ESR. Appl. Surf. Scie. 2003. V. 216. P. 376–381.

Sato K., Kishimoto N., Oku T., Hirakuri K. Improvement of luminescence degradation in pure water of nanocrystalline silicon particles covered by a hydrogenated amorphous carbon layer. J. Appl. Phys. 2007. V. 102. P. 014302.

Scheer K. C., Rao R.A., Muralidhar R., Bagchi S., Conner J., Lozano L., Perez C., Sadd M., White B. E. Jr. Thermal oxidation of silicon nanocrystals in O_2 and NO ambient. J. Appl. Phys. 2003. V. 93. P. 5637–5642.

Svrček V., Sasaki T., Shimizu Y., Koshizaki N. Aggregation of Silicon Nanocrystals Prepared by Laser Ablation in Deionized Water. J. Laser Micro/Nanoengineering. 2007. V. 2, No.1. P. 15–20.

Szymanski M.A., Stoneham A.M., Schluger A. The different roles of charged and neutral atomic and molecular oxidizing species in silicon oxidation from ab initio calculations. Solid-State Electronics. 2001. V. 45. P. 1233–1240.

Terekhov V.A., Kashkarov V.M., Turishchev S. Yu., Pankov K.N., Volodin V.A., Efremov M.D., Marin D.V., Cherkov A.G., Goryainov S.V., Korchagina A. I., Cherepkov V.V., Lavrukhin A.V., Fadeev S.N., Salimov R.A., Bardakhanov S. P. Structure and optical properties of silicon nanopowders. Mater. Sci. and Eng. B. 2008. V. 147. P. 222–225.

Wolkin M.V., Jorne J., Fauchet P.M., Allan G., Delerue C. Electronic States and Luminescence in Porous Silicon Quantum Dots: The Role of Oxygen. Phys. Rev. Lett. 1999. V. 82. P. 197–200.

7

Thin films deposited from sol of silicon nanoparticles

Thin films of nanocrystalline silicon (nc-Si) are very promising as elements of the solar cells (Shan et al, 2000; Cheng et al, 2004), thin-film transistors (Lee, 2005), gas sensors and devices using the one-electron recharge processes (single electronic devices) (Tiwari, 1996). In the case of solar cells, these prospects are determined by the fact that the fundamental absorption edge of these films can be shifted to photons with energies less than 1 eV, which is not absorbed by monocrystalline silicon (c-Si). In addition, in these films there is no light degradation (Staebler–Wronski effect), observed in films of amorphous hydrogenated silicon (a-Si:H). Applied to thin-film transistors the nc-Si films exhibit high charge carrier mobility and high electrical stability (Lee, 2005; Tiwari, 1996).

Currently, there are several technologies for depositing thin nc-Si (~100 nm) films on the substrate, of which the most frequently used method is deposition in the plasma produced in silane (SiH_4) by the low-pressure radio-frequency discharge (Plasma Enhanced Chemical Vapor Deposition, PECVD).

In this chapter, attention is given to a promising method for the formation of homogeneous thin films (with thicknesses to 30 nm) with size-selective deposition of from sols containing nanocrystalline particles of silicon. Such films (nc-Si films) formed from tightly adjacent to each other crystalline nanoparticles of Si, therefore, their physical characteristics should be similar with the characteristics of films based on porous silicon (P-Si). The optical absorption and the ability to photoluminescence of the P-Si films have been thoroughly studied (see Chapter 3), but insufficient attention has been paid to the transport and dielectric properties of these films in an alternating electric field. A similar situation exists with the nc-Si films, for which there are no details of any studies of conductivity in an alternating electric field (AC conductivity) and dielectric relaxation.

7.1. Synthesis of powders and deposition of films nc-Si

Dorofeev et al (2009) reported a new method for producing thin films of nc-Si, which consists in the fractional deposition of nanoparticles on a substrate by centrifugation of the sol in ethanol. The data obtained in studies the transformation of the particle size distribution in powders of nc-Si, caused by etching in a solution of hydrofluoric acid, HF, and its mixture with nitric acid (HF + HNO_3). The results of a comparative analysis of the absorption spectra of films obtained from the initial powders of nc-Si and powders subjected to etching are also described. The results of spectral studies indicate the presence of significant absorption in the films obtained in comparison with crystalline silicon in the energy region before the edge of the fundamental absorption up to ~0.6 eV. It is shown that with decreasing size of nanoparticles the optical characteristics of the films are strongly affected by the surface conditions that may be related to the coordination of the surface of nanoparticles hydrogen and oxygen atoms.

The results of a comparative analysis of the absorption spectra and Raman scattering spectra of the films obtained from the initial powders and the nc-Si powders subjected to etching (Dorofeev et al, 2009) are presented. It is shown that in films deposited from the sol of the nc-Si powders, subjected to etching, the absorption in the incident photon energy range $1.0 \text{ eV} \leq \hbar\omega \leq 3.0 \text{ eV}$ is mainly determined by the electronic states associated with the surface of the nanoparticles. It was found that in the nc-Si powders and films deposited from these powders, there is an amorphous component, which changes the relative share in the processes of etching of the particles and their deposition on the substrate. From the analysis of Raman spectra it is concluded that this component is due to the oxygen atoms located on the nanoparticle surface and distorting its crystal lattice.

A detailed description of the apparatus and the synthesis conditions of nc-Si powders, which are used in the work of Dorofeev et al (2009), is in the works of Kuz'min et al. (2000), Kononov et al (2005). Briefly, the procedure for the synthesis of silicon nanoparticles is as follows. The reactor chamber filled with a buffer gas (helium or argon) at a pressure of $P \approx 200$ torr included the formation of a thin jet SiH_4, which is heated by the focused radiation of a cw CO_2-laser crossing this jet. As a result of the pyrolysis of silane the molecules of SiH_4 are decomposed and free silicon atoms form. In the processes of collisions with buffer gas atoms and with each other the silicon atoms form particles the average size of which, depending on the buffer gas pressure, may be in the range from 10 to 100 nm. The obtained nc-Si powders were dispersed in ethanol using sonication and centrifuged for 20 min during acceleration of $2000g$. Almost all of the agglomerates nc-Si particles precipitate as a result of this procedure. After initial centrifugation a stable colloidal solution of nc-Si in ethanol remains. Visible changes of the solution, including precipitation, were not observed for 2 years (Dorofeev et al., 2009). An aqueous solution of aluminium dihydrophosphate was added for the subsequent deposition of nanoparticles in the sol.

For the etching of the colloidal solution of nanoparticles nc-Si was added to an aqueous solution of acids (HF + HNO_3) or to an aqueous solution of HF acid and subjected to ultrasonic treatment for different times, depending on the concentration of acids. To stop the etching in the acid solution (HF + HNO_3) 10 times the volume of water was added to the reaction mixture. In the HF solution etching of the particles stopped spontaneously as their surface was passivated. As a result of etching the nanoparticles lost their ability to form a sol. To restore their solubility, they were washed three times with distilled water in an ultrasonic bath with intermediate separation of particles from the washing solutions by centrifugation for 10 min at acceleration of $2000g$. At the same time, nc-Si sols which did not settle in the centrifuge were produced.

To produce films, the substrate with the sol, coagulated by the aluminum hydrophosphate, was subjected to centrifugation, after which the nc-Si film was deposited in the substrate. nc-Si films were deposited in the first, second and third stages of centrifugation of the sols. It was assumed that each subsequent stage of size-selective deposition is characterized by the formation of a film from smaller particles. The thickness of the deposited films of nc-Si in our studies ranged from 40 nm to 2 μm. To increase the mechanical strength, the nc-Si films were annealed in vacuum at a pressure of 10^{-5} torr and temperatures from 423 K to 673 K.

7.2. Methods of analysis

The shape and particle size distribution in the starting nc-Si powders and in the powder subjected were studied in a transmission electron microscope LEO912AB OMEGA (Dorofeev et al, 2009). The size distribution of the nc-Si particles was determined by processing TEM images with program UTH SCSA Image Tool. The thickness of the nc-Si films was determined using the step-profilometer Taly Step (Taylor-Hobbson). The transmission spectra of films in the UV and visible ranges were recorded with a Lambda 900 spectrophotometer (Perkin Elmer). Raman spectra of the films were recorded using a micro-Raman triple spectrograph T-64000 (Jobin Yvon) with the radiation power of the exciting argon laser of 2 mW in backscattering geometry. Typical TEM images of the starting powder and the nc-Si powder subjected to etching, are shown in Fig. 7.1. According to the TEM, the shape of the particles in the original nc-Si powder was nearly spherical, and the particles themselves coagulated in the agglomerates of up to several hundred nanometers.

The size distribution of the nc-Si particles, obtained from computer analysis of the TEM images, is shown in Fig. 7.2. The histograms, shown in the figure, indicated that the maximum size distribution of particles in the initial

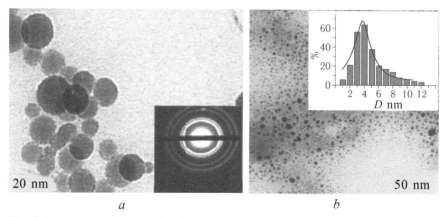

a   *b*

Fig. 7.1. *a* – TEM micrograph of particles of nc-Si, dispersed in ethanol, and the electron diffraction patterns of powders of these particles. *b* – TEM image and particle size distribution of the nc-Si sample, processed in a mixture of acids (HNO$_3$ + HF) (Dorofeev et al, 2009).

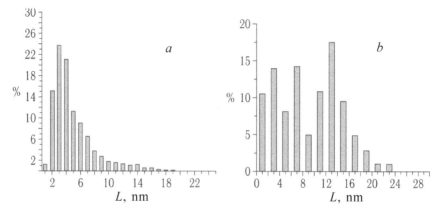

Fig. 7.2. *a* – histogram of the size distribution of particles obtained in the processing of TEM images of nc-Si powder, etched in a mixture of acids (HF + HNO$_3$), see Table 7.1. *b* – histogram of the size distribution of particles obtained in the processing of TEM images of the nc-Si powder etched in HF acid, HF, see Table 7.1 (Dorofeev et al, 2009).

nc-Si powders is in the range of values of the diameter of L(nc-Si)$_{max}$ ~20 nm. Etching the nc-Si for 1 hour in an aqueous solution of acids (5 wt.% HF, 14 wt.% HNO$_3$), shifts the maximum of the distribution into the region L (nc-Si)$_{max}$ ~3 nm, and etching in an aqueous HF solution at a concentration of 12 wt.%, into the region from 7 nm to 13 nm.

A comparison of the histograms in Fig. 7.2*a* and Fig. 7.2*b* shows that in the powder subjected to etching in a mixture of (HF + HNO$_3$), the size distribution of nc-Si has only one pronounced maximum. For the powder subjected to etching in a solution of HF, there is a more complex structure distribution. This feature of the size distribution of nc-Si, exposed to

Table 7.1. Marking of films prepared from sols of starting nc-Si powder (Dorofeev et al, 2009)

Number of film	Impact of acid[*]	The nc-Si film thickness, nm[**]	Stage of fractional-selective deposition
1	None	1600	1
2	5 wt.% HF + 14 wt.% HNO$_3$	215	2
3	12 wt.% HF	180	2
4	12 wt.% HF	45	3

[*] Time of exposure of acids – 1 hour at 293 K. [**] Accuracy of measurements with the step-profilometer Taly Step (Taylor–Hobbson) is 2 nm. However, the roughness of the film the non-uniformity of its tickness results in an increase of the relative measurement error of up to ~15%.

hydrofluoric acid, may be associated with stopping of etching at the moment when the surface of nc-Si particles is saturated with hydrogen. In addition, because the surface of such particles is devoid of the oxide shell, the coagulation rate increases, which leads to the formation of agglomerates of large nanoparticles.

7.2.1. Absorption in UV and visible ranges

To record the transmission spectra, nc-Si particles were deposited from a sol on quartz substrates (Dorofeev et al, 2009). For comparison, the transmission spectra of films deposited from sols of nc-Si in the first, second and third stages of size-selective precipitation were recorded. Marking of the films, prepared from sols of the starting nc-Si powder, is shown in Table 7.1. The transmission spectra were used to calculate the absorption coefficients of the films and their dependences on the energy of the incident photons are shown in Fig. 7.3a. For comparison, the figure also shows the spectrum of absorption of crystalline silicon.

The absorption spectra show that all the films studied (see Table 7.1) show significant absorption in the incident photon energy range 0.6 eV $\leq \hbar\omega \leq 1.0$ eV, which radically distinguishes these spectra from the absorption spectrum of c-Si.

The second important feature of these spectra is that the absorption coefficients of films 2 and 3 are greater than the absorption coefficients of c-Si in the incident photon energy range $1.0 \leq \hbar\omega \leq 3.0$ eV (see Fig. 7.3a). Such a behaviour of the absorption coefficient for the films of microcrystalline silicon was also observed in the works (Richter, Ley, 1981; Jackson, 1983), but the reason for this has not been established. With respect to the films obtained in the present work, the high values of the absorption coefficient in this energy range were quite unexpected. Indeed,

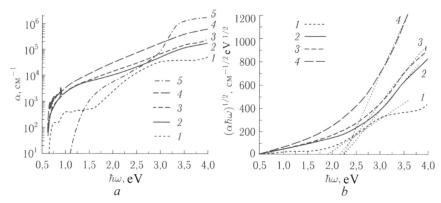

Fig. 7.3. *a* – Absorption coefficients of films 1–4, see Table. 7.1. Line *5* – spectral dependence of the absorption coefficient of monocrystalline silicon (c-Si). *b* – coefficient Absorption coefficients of films 1–4 coordinates $\sqrt{\alpha \cdot \hbar\omega} \sim \hbar\omega$. The thin straight lines of dots dependences correspond to $\sqrt{\alpha \cdot \hbar\omega} = A\left(\hbar\omega - E_g\right)$ for the investigated films (Dorofeev et al, 2009).

if we use the results of TEM studies of nc-Si powders from which the films were deposited, it is necessary to take into account that a significant fraction of the particles in them has the sizes much smaller than 10 nm, and, therefore, in such films on the magnitude of absorption should be affected by the increase of the band gap due to the effect of the quantum constraint. Obviously, such an increase in the band gap should lead to a decrease in absorption for each of the values of the energy of the incident photons.

The influence of the quantum constraint on the optical properties of these films can also be found by analyzing the energy dependence $\sqrt{\alpha \cdot \hbar\omega}$, where $\alpha(\hbar\omega)$ is the absorption coefficient of the film for a photon with energy $\hbar\omega$. The corresponding graphs for the films 1–4 are shown in Fig. 7.3*b*. These graphs show that for film 1 in the incident photon energy range 2.2 eV $\leq \hbar\omega \leq$ 3.0 eV, for films 2, 3 and 4 in the range 3.0 eV$\leq\hbar\omega\leq$ 3.5 eV function $\sqrt{\alpha \cdot \hbar\omega}$ is well approximated by a linear dependence: $\sqrt{\alpha \cdot \hbar\omega} = A\left(\hbar\omega - E_g\right)$. In the case to amorphous silicon, in accordance with the model proposed by Tauc (Tauc et al., 1996; Poruba et al., 2000), the values of band gap E_g, obtained from such dependence, determine the width of the band gap in a-Si:H. A similar procedure was applied also to porous silicon, and a linear extrapolation of $\sqrt{\alpha \cdot \hbar\omega} = A\left(\hbar\omega - E_g\right)$ near the critical points in the absorption corresponding to fundamental shift $\Gamma'_{25} - X_1$ and indirect transition $\Gamma'_{25} - L_1$, gives values E_g equal to 1.80 eV and 2.50 eV, respectively. Since for c-Si at room temperature, $E_g(\Gamma'_{25} - X_1) = 1.12$ eV and $E_g(\Gamma'_{25} - L_1) = 1.80$ eV, the values obtained by the authors (Kovalev et. al., 1996) associated with an increase in the width of the fundamental band gap $\Gamma'_{25}-X_1$ and indirect band $\Gamma'_{25}-L_1$ in porous silicon, P-Si, caused by the quantum constraint.

With respect to films 1, 3 and 2, 4 (see Table 7.1) the quantities E_g equal 1.80 eV, 2.20 eV and 2.25 eV, respectively (see Fig. 7.3b), which is significantly more than the width of the fundamental band gap in c-Si. Since the energy range in which this linear approximation holds, is close to the critical point $\Gamma'_{25}-L_1$, the measured values of E_g are associated with the band gap at this point in the studied films. As already noted, for c-Si the magnitude E_g $(\Gamma'_{25}-L_1) = 1.80$ eV, so it can be argued that for the film 1 the quantum constraint is not observed, whereas in the films 3 and 2, 4 due to width of the quantum limit indirect band $\Gamma'_{25}-L_1$ increases by 0.40 eV and 0.45 eV, respectively.

A comparison of the absorption coefficients of the films 2, 3 and 4 in the incident photon energy region 1.0 eV $\leq \hbar\omega \leq 3.0$ eV shows that the highest absorption is obtained for the thinnest film 4 with a thickness of 45 nm. The type of absorption spectrum of this film is very similar to the absorption spectrum of film 3, which was precipitated from the same nc-Si sol in the preliminary centrifugation stage. Figure 7.2b shows that the particle size distribution of nc-Si, etched in a solution of HF (12 wt.%), has several maxima.

It is natural to assume that the nc-Si particles deposited from the given sol at the preliminary stage of centrifugation are larger than at a later stage. Accordingly, it can be assumed that the film 4 is formed from smaller nc-Si particles than the film 3. Since the average particle sizes in the two films do not exceed 10 nm, then their properties are strongly affected by the electronic states associated with the surface of the particles.

The degree of influence of the surface states increases with decreasing size of the nanoparticles and, as the film 4 is formed from smaller nanoparticles than film 3, the degree of influence of the surface states on the absorption in it should also be higher than in the film 3. Thus, the behaviour of the absorption spectra of the films 3 and 4 gives grounds to assert that in the range of photon energies 1.0 eV $\leq \hbar\omega \leq 3.0$ eV, the absorption is mainly determined by the electronic states associated with the surface of the silicon nanoparticles. The same conclusion applies to the film 2, which is formed from the nc-Si sol exposed to a mixture of (HF + HNO_3). Indeed, as follows from the foregoing analysis, the band gap in this film coincides with the band gap in the film 4 (see also Fig. 7.3b). Therefore, we can assume that the average sizes of the nanoparticles in films of 2 and 4 are similar, and the degree of influence of the surface states on the absorption should also be about the same because of the proximity of the values of their specific surface areas.

However, Fig. 7.3a shows that in the incident photon energy range $\hbar\omega \geq 1.5$ eV the absorption coefficient of the film 4 is about three times greater than the coefficient of the film 3. This difference is probably due to the different conditions of surface passivation of silicon nanoparticles when they are etched in a solution of only hydrofluoric acid, compared with a mixture of acids – (HF + HNO_3). The difference in the values of the absorption coefficients of the films 2 and 4 points to various concentrations of the electronic states associated with the surfaces of the particles of nc-Si, in the films 2 and 4, as well as, perhaps, to the different nature of these states. The difference in the nature of the surface states are mainly determined by the difference in the concentrations of fragments of type ($\equiv$Si–H) and ($\equiv$Si–O–) on the surface of the nanoparticles, depending on the history of its treatment.

7.2.2. Raman scattering

To record Raman spectra, a quartz substrate was deposited with an aluminum film with a thickness of $\approx$300 nm. Subsequently, a nc-Si film from the sol was deposited on the substrate. This was done in order to get rid of the background component, bonded with the quartz substrate, from the scattering spectra.

Dorofeev et al, 2009, analyzed the Raman scattering spectra reported in the original nc-Si powder, as well as in the films deposited in the second stage of centrifugation from the sols of the starting nc-Si powder, etched in a mixture of acids (HF + HNO$_3$) under the etching conditions corresponding to film 2 (see Table 7.1), and in film 2, annealed for 1 hour in a vacuum at a pressure of $P = 10^{-5}$ torr and temperature $T = 673$ K.

The Raman spectra recorded for the films 1–4 are shown in Fig. 7.4.

These spectra are very similar to the Raman spectra, recorded for P-Si in the works (Tsu et al, 1992; Tsang et al, 1992), as well as in clusters of nc-Si (Ehbrecht et al, 1995). Raman spectra of the samples with high accuracy can be approximated by four Lorentzian contours (see Fig. 7.4),

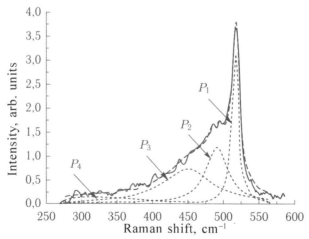

Fig. 7.4. Raman spectrum of film the deposited from a sol containing a powder of nc-Si, pre-etched in a mixture of acids (HF + HNO$_3$), see Table 7.1, for the film 3. The dotted line of the spectrum shows the approximation by Lorentzian contours (peaks P_1, P_2, P_3 and P_4) (Dorofeev et al, 2009).

which will be marked as peaks P_1, P_2, P_3 and P_4. The Raman shift of the maximum of the most intense peak P_1 with respect to the emission frequency of the probe laser is in the range of wave numbers from 515 cm^{-1} to 517 cm^{-1} for all samples studied. For the same maximum peak of c-Si the shift corresponds to the wave number 520.5 cm^{-1}. Thus, for all investigated films, the maximum of the peak P_1 is shifted to lower wave numbers relative to the maximum peak of c-Si (redshift). The presence of peak P_1 in the spectra is determined by Raman scattering of light in the nc-Si particles with longitudinal (LO) and transverse (TO) optical phonons of the central point Γ'_{25} of the Brillouin zone for the silicon crystal lattice.

The magnitude of the red shift of the peak P_1 and its half-width, depending on the nanoparticle size, are well described by a model of the phonon limit (Richter et al, 1981; Campbell et al, 1986). The result of applying this model to the spherical nanoparticles is shown in Fig. 7.5.

The Figure shows that the average particle size of nc-Si in the studied samples is in the range 4–6 nm, regardless of whether the particles of the nc-Si starting powder were subjected to some effect or not. For sols from the nc-Si powders, subjected to etching in a mixture (HF + HNO$_3$), the average particle size, obtained from the model of the phonon limit, agrees well with the particle size corresponding to the maximum particle size distributions obtained by processing the TEM images (Dorofeev et al, 2009). However, for the initial nc-Si powders the average particle sizes determined by these two methods differ by about three times. The difference

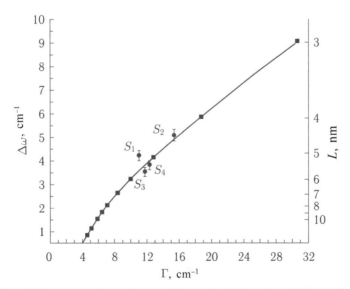

Fig. 7.5. Solid line – the dependence of the half-width and redshift of the peak P_1 on the diameter of spherical nanoparticles of silicon in the phonon limit model. Solid circles S_1–S_4 – the results of the approximation of Lorentzian peak P_1 for samples 1–4 (see Table 7.1). The diagram from the work (Dorofeev et al, 2009).

in the average size of particles, determined from the phonon limit model and processing of TEM images, is determined by two possible reasons.

The first reason stems from the fact that in the model of the phonon limit the nanoparticles are monocrystalline. Therefore, the value of the phonon wave vector q in the nanoparticle can be varied from 0 to $2\pi/d$, where d is the diameter of the particle. If the core of the nanoparticles is polycrystalline, and the average size of the unit crystal lattice in it is L, then the restrictive condition $q \le 2\pi/d$ should be replaced by $q \le 2\pi/L$. Thus, it is possible that the values $L = 4$–6 nm computed from the model of the photon limit particle siz are determined not by the size of the particles in the original nc-Si powder but by the average size of elementary lattices in polycrystalline cores of these particles. From this assumption and the fact that for the nanoparticles subjected to etching the average size, determined by the two methods, are the same, it should also be assumed that in etching of the nanoparticles the remaining cores of nc-Si are monocrystalline. The second possible reason has to do with the well-known fact of the low contrast of the small nanoparticles (particle diameter of 3 nm) in TEM images. Because of the low contrast, the TEM image processing procedure always reduces the relative share of the fine fraction of nanoparticles in the analyzed ensemble.

The Raman shift of the maximum peak P_2 in these samples is in the range from 480 to 495 cm^{-1}. The presence of this peak is determined by Raman scattering of light with TO phonons in a-Si:H. Also, as the peak P_2, the P_3 and P_4 peaks are associated with the amorphous component of silicon in the particles and their appearance is due to scattering with participation of longitudinal optical (LO) and acoustic (LA) phonons.

By comparing the integral intensities of the peaks P_1 (I_c) and P_2 (I_a), we can determine the crystalline volume fraction X_c in the study of silicon particles. In (Dorofeev et al, 2009) the following expression (Voutsas et al, 1995) is used:

$$X_c = \frac{I_c}{I_c + \eta I_a}, \qquad (7.1)$$

where $\eta = \dfrac{\sigma_c}{\sigma_a}$ is the ratio of integral cross sections of backward scattering in crystalline and amorphous fractions (i.e. peaks P_1 and P_2). According to (Kakinuma et al, 1991), the value for silicon $\eta = 0.8$–0.9. In the calculations (Dorofeev et al, 2009) the authors used $\eta = 0.8$. For the samples 1–4 the X_c value is 0.45, 0.35, 0.50 and 0.60, respectively. From these values it follows that almost half of the volume of the particles have a high degree of disorder in the crystal lattice.

A comparison of the values X_c shows that in film 2, deposited in the second stage of centrifugation from the sol with the initial nc-Si powder, X_c is less than in the original powder. The average particle size in this film is less than the average particle size in the starting powder. Accordingly, the ratio of the surface area to the volume of the nanoparticles in it is greater than the corresponding ratio in the initial powder. Therefore, the influence of the surface of the nanoparticles on their general properties in this film should also be greater than in the original powder. Therefore, the decrease in X_c (or increase of the degree of amorphization of the particles) in a film of sample 2 compared to X_c in sample 1 should indicate that the region of disorder is not in the core of the nanoparticles and is on its surface.

However, in the films of the samples 3 and 4 the value of X_c is greater than in the film 2, although the average particle sizes in these films are comparable. This increase in X_c indicates that the degree of disorder of the surface of the particles in the films of the samples 3 and 4 is less than in the film 2.

As the films 3 and 4 were precipitated from the sol of the nc-Si powders, subjected to etching, then this decrease is related to the impact of the acids HF and HNO_3 on the particle surface. One should note the studies (Puzder et al, 2002; Luppi, Ossicini, 2005) which examine the influence of the oxygen atoms on the structure of the clusters of silicon and the degree of ordering of the crystal lattice of Si in the nanoparticles, as well as the studies (Tsang et al, 1992; Zhixun et al, 2000) which report on the modification of the Raman peak, similar to the peak P_2 (see Fig. 7.4), under the influence of oxygen on the surface of porous silicon passivated by hydrogen. The general idea of these studies is marked by the fact that crystal lattice of the nanoparticles, whose surface is completely passivated by hydrogen, showed almost no change compared to the crystalline lattice of silicon. If, however, oxygen atoms which can form (Si–O–Si) and (Si=O) bonds appear on the surface of the nanoparticles, then these atoms distort the lattice at distances up to 0.5 nm. In this region of space the distortion of valence angles bonds (Si–Si) crystal lattice can reach 10°. As a consequence, if the surface of a silicon nanoparticle with a diameter of 3 nm is coated with SiO_2, then the crystalline lattice of such a particle is distorted in an appreciable fraction of the volume. As a consequence, if the surface of P-Si is etched in a solution of HF (see Table 7.1), the Raman spectrum will also contain a peak similar to peak P_1. If P-Si is exposed to the oxygen atmosphere, the spectrum will also show peak P_2 along with peak P_1.

Based on these studies and the analysis of the Raman spectra, discussed in this section, we can state the following. In all the investigated samples, the surface of the nc-Si particles contains appreciable numbers of oxygen atoms, which significantly distort the crystal lattice of the particles and cause the appearance of the peak P_2 in the Raman spectra. Since the average size of the nanoparticles in the film 2 is less than in the film 1, the strength of the effect of these atoms on the lattice structure in sample 2 is greater than in sample 1. This fact is reflected in a reduction of the crystal volume fraction in sample 2.

Etching of nc-Si particles in a solution of acids (HF + HNO_3) also leads to a decrease of the particle size. However, in this case, the total number of oxygen atoms on the surface of the nanoparticles decreases, since some of these atoms are replaced by hydrogen atoms. Therefore, there must be two divergent process in the films 3 and 4. The first is associated with a decrease in the size of nanoparticles and leads to a decrease in X_c, and the other one, associated with a decrease in the number of oxygen atoms on their surface causes an increase of parameter. Since in the experiments the films 3 and 4 showed an increase in X_c in comparison with the film 2, it can be argued that the second process dominates over the first in etching of the nc-Si particles. The increase of X_c in film 4 compared to film 3 indicates that in annealing in vacuum at a pressure of $P = 10^{-5}$ torr and temperature $T = 673$ K an appreciable part of the oxygen atoms leaves the surface of the annealed nanoparticles.

The particle size distributions in the starting powders of nc-Si and the powders, subjected to etching, constructed from the TEM images, indicate that as a result of etching in the HF solution the distribution becomes non-monotonic and is characterized by several maxima, while in etching in a mixture of acids (HF + HNO_3) the particle size distribution, as in the starting powder, has a single maximum (see Fig. 7.2).

The results of spectral studies (Dorofeev et al, 2009) indicate the presence of significant absorption in the films compared to crystalline silicon in the energy range in front of the fundamental absorption edge to ~0.6 eV. The results of processing the absorption spectra were used to calculated the band gap E_g in the studied films. A comparison with the corresponding values of E_g value for c-Si revealed that in the films deposited from the sols of the nc-Si powders, which had previously been treated in solutions of HF and (HF + HNO_3), the width of the indirect band gap $\Gamma'_{25}-L_1$ is increased due to the effect of the quantum constraint. It was found that, other things being equal, a decrease in the average size of the particles which formed the film, leads to an increase in its absorption coefficient. With decreasing size of nanoparticles, the optical characteristics of the films are significantly affected by the surface states which may be associated to the coordination of atoms of hydrogen and oxygen on the surface of the nanoparticles.

The form of the Raman spectra indicates that in the initial nc-Si powders and in the films deposited from sols of these powders, there is an amorphous component, the relative proportion of which varies when etching the nc-Si powder, and also in the deposition of films from the sols with these powders. A comparison of the relative amount of the crystalline fraction in the films deposited from the sols of the starting nc-Si powders and the powders subjected to etching, indicates that the region of the deformed crystal lattice is located close to the surface of the nanoparticles, and the most probable causes of the distortion of the crystal lattice near the surface of the nanoparticles are the oxygen atoms are located on its surface.

7.3. Dielectric and transport properties of thin films deposited from sols containing silica nanoparticles

Currently, there is strong scientific interest in the structures formed from particles of nanocrystalline silicon (nc-Si). This interest is largely determined by the fact that over 10 years it has been possible to develop effective methods of obtaining silicon nanoparticles, capable of vivid, stable photoluminescence in the visible spectrum, which has a high quantum yield (Jurbergs et al, 2006). The main carriers of such nanoparticles are colloidal solutions (sols) based on methanol, chloroform, hexane, etc. Such sols are very promising targets for the development of deposition techniques extremely thin films of nc-Si on different substrates. The use of such films is very promising for the creation of light-emitting elements, which are based on electroluminescence of nc-Si (Anopchenko et al, 2009). In addition, nc-Si films are very promising as elements of the solar cells (De la Torre et al, 2006), thin-film transistors (Min et al, 2002), and devices using the principle of one-electron recharge (single electronic devices) (Tsu, 2000).

In the case where the films consist of nanoparticles with a diameter less than 10 nm, their total characteristics are determined not only by the properties of matter from which the nanoparticles consist, but also by the properties of atoms on the surface of these particles. In other words, such films should usually be regarded as a multi-component environment whose properties are formed by the contribution of both crystalline cores of the nanoparticles and the surface atoms and molecules and air pores that are part of the film.

In the current literature, the greatest number of publications relate to studies of the properties of films consisting of amorphous silicon (a-Si), in which silicon nanocrystals are embedded (Conte et al, 2006; Wang et al, 2003). Such films can be sprayed, for example, in a high-frequency discharge in a mixture of gases SiH_4, Ar or He, and H_2 (PECVD method), followed by high-temperature annealing (Saadame et al, 2003) (see chapter 5 of this monograph).

In the previous sections of this chapter we described a method of forming uniform thin films (thickness to 30 nm) as a result of size-selective deposition from sols containing silicon nanocrystalline particles (Dorofeev et al, 2009). These nc-Si films are formed of closely spaced crystalline nanoparticles of Si, so that their physical characteristics to some level should be similar to those of films based on porous silicon (P-Si). The optical absorption and ability for photoluminescence of the P-Si films have been studied extensively (see, e.g. Kovalev et al, 1996; Brus et al, 1995), but there are only a limited number of studies of the transport and dielectric properties of these films in an alternating electric field. We note here the studies (Ben-Chorin et al, 1995; Axelrod et al, 2002; Urbach et al, 2007), devoted to P-Si. A similar situation exists in relation to the nc-Si films, and there is no mention in the literature of studies of the conductivity in an alternating electric field (AC conductivity) and dielectric relaxation in these films.

This section describes the measurement of dielectric constant of the nc-Si films in the optical range ($5 \cdot 10^{14} \leq$ June $\leq 10^{15}$ Hz) and the frequency range $10 \leq v \leq 10^{6}$ Hz. The AC conductivity (σ_{AC}) of the nc-Si films was also determined in the frequency range $10 \leq v \leq 10^{6}$ Hz as determined by (Kononov et al, 2011). In the optical range of the real ε' and imaginary ε'' components of the complex dielectric constant were determined by the ellipsometric analysis of the light beams incident and reflected from the free surface of the nc-Si film. In the normalized frequency range $10 \leq v \leq 10^{6}$ Hz spectra ε' and ε'' were determined by analyzing the frequency dependence of impedance of the nc-Si films. The optical spectrum of values ε' and ε'' with increasing frequency changed in the range 2.1–1.0 and 0.25–0.8 respectively. Such small values of ε' and ε'' the authors associated with the structure of the nc-Si films. The particles of nc-Si, forming such films, consist of crystalline cores surrounded by the SiO_x shell, $0 \leq x \leq 2$. The SiO_x shell is formed by the interaction of the surface of the Si nanoparticles with the surrounding air.

Kononov et al (2011) used the effective-medium approximation (EMA) developed by Bruggeman (Bruggeman, 1935) to model the structural composition of the nc-Si films. It is shown that good agreement between the frequency dependences of ε' and ε'', derived from the EMA, with the spectra of ε' and ε'' determined from the ellipsometric data, is achieved if the nc-Si films are treated as a two-component medium consisting of SiO and air voids, present in them. A comparison of the absorption spectra obtained for each films from the ellipsometric measurements and as a result of the processing of the transmission spectra, shows that for incident photon energies 2.7–4.0 eV (frequency of $6.6 \cdot 10^{14}$–10^{15} Hz) the spectra are very close to each other. However, in the energy range 1.0–2.7 eV (frequency $2.4 \cdot 10^{14}$– $6.6 \cdot 10^{14}$ Hz) the values of absorption, obtained from the ellipsometric parameters, are significantly higher than the absorption of certain transmission spectra. This result is due to the presence of defects in the SiO_x shell of the nanoparticles, as well as the fact that the method of ellipsometry, which analyzes reflected light with respect to such defects, is more sensitive than transmission analysis.

The analysis of the frequency dependence of the capacitance of the nc-Si films and their impedance spectra in the frequency range 10–10^{6} Hz determined the dispersion of values ε' and ε''. It is found that with increasing frequency the values ε' and ε'' change in the range 6.2–3.2 and 1.8–0.08 respectively. We found that in this frequency range function $\varepsilon'(\omega)$ is well approximated by the Cole–Cole semi-empirical dependence (Cole–Cole dielectric relaxation) (K.S. Cole, R.H. Cole, 1941). At the same time, spectra, the $\varepsilon'(\omega)$ spectra of the nc-Si films are well approximated by the Cole–Cole dependence only in the frequency range greater than $2 \cdot 10^2$ Hz. In the low-frequency region of the spectrum good approximation is attained by combining the Cole–Cole dependence and the term associated with the conductivity of the films. The analysis of approximate dependences were used to define the relaxation times of dipole moments in the nc-Si films, which at room temperature is $6 \cdot 10^{-2}$ s. Conductivity σ_{AC} of the studied nc-Si films in an alternating electric field depends on its frequency by a power law, and in the entire investigated frequency range the exponent is equal to 0.74. Such a behaviour of σ_{AC} indicates that the electrical transport in the films is of the hopping type. From a comparison of the measured frequency dependence $\sigma_{AC}(v)$ with similar dependences following from different models of hopping conductivity, it was found that the behaviour of $\sigma_{AC}(v)$ is most accurately described by the diffusion cluster approximation (DCA) (Dyre, Schrøder, 2000; Schrøder, Dyre, 2002; Schrøder, Dyre, 2008). From the analysis of the dependence of dark conductivity of the films on the humidity of the ambient air, as well as from the temperature dependence of the absorption bands determined by the associated groups (Si–O–H) on the surface of the films, it is concluded that in the frequency range lower than $2 \cdot 10^2$ Hz, the conductivity is related to transport of protons through hydroxyl groups located on the surface of the silicon nanoparticles and bonded by hydrogen bonds.

7.3.1. Impedance spectra and ellipsometry

The process of obtaining nc-Si particles and their deposition on various substrates is described in section 7.1. The samples for the measurement of the impedance spectra were prepared as follows (Kononov et al, 2011). Initially, a glass substrate was sprayed with aluminium electrodes separated from each other by a straight slit 1 mm wide. Then, the substrate prepared by this treatment was coated with nc-Si particles from the sol, and the particles formed a film. The resulting film of nc-Si was then deposited with a third aluminium, resulting in a sandwich-like structure shown in Fig. 7.6.

To achieve the ohmic nature of the conducting contacts, the resultant structure was annealed at 400°C and a pressure of 10^{-5} torr. The impedance spectra were recorded at an amplitude voltage of 10 mV, but the produced films could withstand a voltage up to 25 V without electrical breakdown. In the experiments (Kononov et al, 2011) the ellipsometric angles ψ and Δ

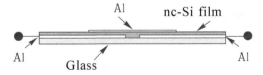

Fig. 7.6. Scheme of a sandwich-like structure for measuring the impedance spectra of nc-Si films (Kononov et al, 2011).

were measured as a function of the wavelength of the light beam incident at an angle Φ_0 on the free flat surface of the nc-Si film. The studied films were deposited on glass and quartz substrates, as well as on quartz substrates with pre-deposited aluminium films. The thickness of the nc-Si films was measured by an atomic force step-profilometer and was in the range of 1–2 μm. When processing the ellipsometric data, the studied nc-Si films were considered as the bulk medium in an air environment. The complex refractive index $N = n - ik$, where n is the refractive index of the film and k – the extinction coefficient, was determined from the expression (R. Azzam, N. Bashara, 1981):

$$N = N_0 \sin \Phi_0 \sqrt{1 + \left(\frac{1-\rho}{1+\rho}\right)^2 \text{tg}^2 \Phi_0}. \qquad (7.2)$$

In equation (7.2) $\rho = \text{tg } \psi \cdot e^{i\Delta}$ and N_0 is the complex refractive index of the environment (air), which in this case is assumed to be equal to unity. It is known that (7.2) gives the exact values of n and k only if the reflection of light from a semi-infinite medium with an atomically clean surface. If there are impurities (or oxide film) at the interface, they make an error in the calculated values. In the study (Tompkins & Irene, 2005) the authors compared the values of n and k for crystalline silicon (c-Si) in the absence and also in the presence of the oxide film on its surface. From this comparison it follows that if the silicon surface is coated with an SiO_2 film with a thickness of up to 2 nm, the values of n in the incident photon energy range 1.0–3.4 eV is almost equal to the refractive index of c-Si. In the range 3.4–5.0 eV it differs from the refractive index of the c-Si by no more than 20%, as well as k. However, in the range 1.0–3.4 eV the value of k (in the presence of the SiO_2 film) is almost two times higher than the extinction of c-Si.

Since the real (ε') and imaginary (ε'') components of the dielectric constant of the medium are related to n and k by the known relations:

$$\varepsilon' = n^2 - k^2, \qquad \varepsilon'' = 2nk, \qquad (7.3)$$

we can expect that the value of ε' calculated from the equation (7.2) for the nc-Si films will be systematically underestimated, and the value the value of ε'' will be overstated.

Nevertheless, the representation of the pseudodielectric functions defined by (7.3) is very convenient and often used in studying the dielectric properties of materials. For example, in (Pickering et al, 1984) the authors used equation (7.3) to study of the dielectric parameters P-Si. With regard to the study carried out in (Kononov et al, 2011), analysis of the spectra obtained using the formula (7.3) was limited to the incident photon energy range in which the films strongly absorbed the probe radiation. In this case, the probe radiation could not reach the substrate surface.

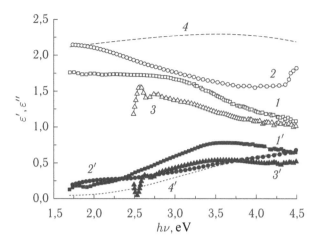

Fig. 7.7. Spectra of real (*1–3*) and imaginary (*1'–3'*) component of the dielectric constant of the nc-Si film, deposited on different substrates: *1, 1'* – a film from the initial (not etched) nanoparticles on a glass substrate *2, 2'* – a film of nanoparticles, pre-etched in a mixture of acids (HF + HNO$_3$), on a quartz substrate, *3, 3'* – nc-Si film from the source of nanoparticles on a glass substrate coated with a preliminary aluminum foil, *4, 4'* – Bruggeman approximation for the value of ε′, ε″ respectively (Kononov et al, 2011).

If the probe radiation is passed to the surface of the substrate on which the film was deposited, the spectra show an interference structure, consisting of alternating maxima and minima. Such a structure in the energy range below 2 eV is clearly visible in Fig. 7.7 (curves *3* and *3'*). Figure 7.7 shows the spectra of the pseudodielectric functions ε′ and ε″ of the nc-Si films, obtained by depositing on a glass substrate initial silicon nanoparticles and also nanoparticles pre-etched in a mixture of acids (HF + HNO$_3$; the weight ratio of 1:3) for 30 min (the quartz substrate). The same figure shows the spectra of ε′ and ε″ of the nc-Si film deposited on a glass substrate which has been previously coated with aluminum foil. The resulting values of ε′ and ε″ are substantially lower than similar values for c-Si.

Figure 7.8 shows the absorption spectra α(E) of the same films, obtained from relations:

$$\alpha(E) = \frac{4\pi v}{c} k = \frac{4\pi E}{ch} k; \qquad (7.4)$$

where $E = hv$ is the incident photon energy and k is the experimentally measured extinction coefficient.

Comparison of the absorption spectra of the nc-Si films produced from unetched nanoparticles, obtained in the ellipsometric measurements and processing of the corresponding spectrum transmission, shows that ellipsometric quantities α are higher than the values calculated from the transmission spectra, and this difference increases with decreasing incident

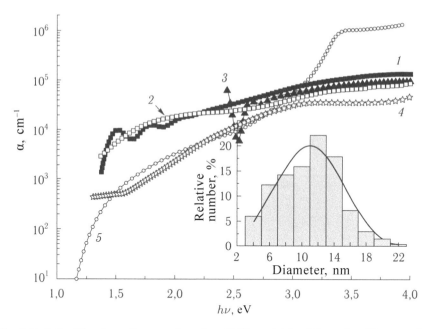

Fig. 7.8. *1–3* – The absorption spectra of nc-Si films obtained from the ellipsometric data (designation of the films is the same as in Fig. 7.1), *4* – absorption spectrum of a film obtained from its transmission spectrum, *5* – absorption spectrum of crystalline silicon (c-Si). The inset shows the dimensional distribution of nanoparticles used for the deposition of films 1 and 3 (Kononov et al, 2011).

photon energy. As noted earlier, this is due to an error of computation of the extinction coefficient using (7.2). However, both spectra show a strong absorption of nc-Si films as compared to c-Si in the energy region below 1.5 eV. Such an increase in absorption in the low-energy photon range is also characteristic of the film deposited from the etched nanoparticles. At the energies greater than 3 eV all the spectra show the absorption smaller than that of the c-Si. The inset in Fig. 7.8 shows that a significant fraction of the ensemble of particles, used to form films, has the external dimensions smaller than 10 nm, so the most likely reason for the decrease of the absorption of photons in the high-energy photon is the increase (due to the quantum constraint) of the width of the band gap in crystalline silicon cores of the nanoparticles forming the film.

The spectra of the dielectric constant of nc-Si films were calculated from the measured frequency dependence of the capacitance of the individual samples, as well as their impedances:

$$Z(v) = Z' - iZ''; \qquad Z(v) = U(v)/I(v), \tag{7.5}$$

where $U(v)$ is the potential difference across the electrodes of the sample and $I(v)$ is the current, flowing through the sample.

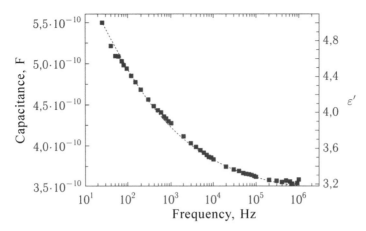

Fig. 7.9. The frequency dependence of capacitance and the real component of dielectric permeability of the nc-Si film. The dotted line shows the approximation of the function (7.11). From (Kononov et al, 2011).

We will analyze the dielectric properties of the system in the form of the Al–nc-Si–Al sandwich, in which a layer of nc-Si was deposited from a sol containing nanoparticles not subjected to etching (Fig. 7.6). The thickness of the film was 2 μm, and the geometric capacitance of the system $C_0 = 1.15 \cdot 10^{-10}$ F. The form of the dispersion of the dielectric constant of this film is typical for other films obtained in this way from similar nc-Si particles.

Figure 7.9 shows the frequency dependence of the capacitance of this system. The capacitance was measured in the parallel connection mode. This figure also shows the spectrum of the active components $\varepsilon'(v)$ of the dielectric permittivity of the film, calculated from the relation:

$$\varepsilon' = \frac{C(v)}{C_0}.$$

(7.6)

The spectra of the real and imaginary permittivities of the film were also determined from the frequency dependence of impedance with the expression:

$$\varepsilon = \varepsilon' - \varepsilon'' = \frac{1}{2\pi v C_0 Z(v)}.$$

(7.7)

Figure 7.10 shows the $\varepsilon'(v)$ and $\varepsilon''(v)$ dependences, calculated by this method for the studied film. Comparison of values $\varepsilon'(v)$ obtained from measurements of $C(v)$ and $Z(v)$ shows good qualitative and quantitative agreement with the values calculated in two different ways, in both cases in the frequency range $10 \le v \le 10^6$ Hz the value $\varepsilon'(v)$ is within 6.0–3.2 and decreases with increasing frequency.

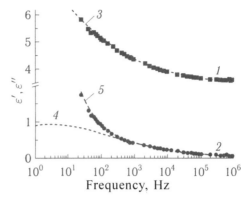

Fig. 7.10. *1, 2* – Frequency dependence of ε' and ε'' respectively, obtained from the impedance spectra. Approximation of the Cole–Cole spectra ε' (*3*), and ε'' without (*4*) and with (*5*) the contribution of free charge carriers taken into account (Kononov et al, 2011).

7.3.2. AC conductivity of nc-Si films

The AC conductivity of the nc-Si films was determined from the known relation:

$$\sigma_{AC}(v) - \sigma(0) = 2\pi v\varepsilon_0\varepsilon''(v), \tag{7.8}$$

where $\sigma(0)$ is the dark conductivity of the films in a dc electric field and $\varepsilon_0 = 8.85 \cdot 10^{-12}$ F/m.

Dielectric constant $\sigma(0)$ of the studied film at room temperature ($T = 297$ K) was $9 \cdot 10^{-10}$ ohm^{-1} m^{-1} and was used to calculate $\sigma_{AC}(v)$. Dependence $\sigma_{AC}(v)$ is shown in Fig. 7.11 on a logarithmic scale. The figure shows that $\sigma_{AC}(v)$ can be well approximated by a power dependence with an exponent of 0.74.

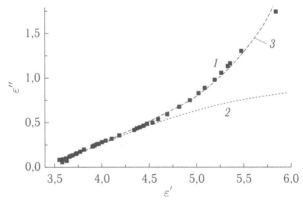

Fig. 7.11. The dependence $\varepsilon'(\varepsilon'')$ of the nc-Si film (*1*); Cole–Cole approximation without (*2*) and taking into account (*3*) the contribution of free charge carriers (Kononov et al, 2011).

7.3.3. The conductivity mechanism

From the analysis of ellipsometric spectra it follows that in the energy range 2.0–4.4 eV and, correspondingly, in the frequency range $5.0 \cdot 10^{14}$–$1.1 \cdot 10^{15}$ Hz of the electromagnetic field, the value of ε' of both studied nc-Si films varies in the range of 2.2–1.6, significantly below the values characteristic in this range for the c-Si. In (Kononov et al, 2011) it is suggested that there are two reasons leading to such low values of ε' and ε''. One of them is that in the process of making the films the nc-Si particles were for some time in contact with atmospheric oxygen, resulting in their surface being covered with a layer ($SiO_x + SiO_2$), $0 \le x < 2$. In (Schuppler et al, 1995) the authors studied the oxidation of silicon nanoparticles and produced thicknesses of the SiO_x layer on their surface depending on the diameter of the nanoparticles. In this section it is shown that in the range of the nanoparticle diameters of 10–3 nm the thickness of the layer ($SiO_x + SiO_2$) was ~1 nm. But this means that the ratio of the volume of crystalline silicon nucleus to the volume of the amorphous SiO_x shell ranges from 100 to 40% respectively. Therefore, the oxidized crystalline silicon nanoparticle with the dimensions smaller than 10 nm should show distinctive amorphous properties. This statement was previously confirmed by analyzing the RS spectra of thin films of nc-Si (Dorofeev et al, 2009). The second reason, which leads to a decrease in the dielectric constant, are air gaps between the nanoparticles, resulting in the formation of films.

In order to evaluate the relationship between the crystalline and amorphous components of the films and their degree of porosity, Kononov et al, (2011) used the Bruggeman EMA model. In the EMA approximation, of the effective dielectric constant ε_e of the inhomogeneous medium consisting of spherical micro-objects with permittivities $\varepsilon_1, \varepsilon_2, \ldots, \varepsilon_{N-1}$, embedded in an environment with ε_N, is determined from the following equation:

$$\sum_{i=1}^{N} f_i \frac{\varepsilon_i - \varepsilon_e}{\varepsilon_i + 2\varepsilon_e} = 0, \quad \sum_{i=1}^{N} f_i = 1, \tag{7.9}$$

where $f_i = V_i \Big/ \sum_{i=1}^{N} V_i$ – is the degree of volume filling of the environment with element with dielectric constant ε_i; V_i – volume occupied by this element.

Initially, to find ε_e of the studied films, Kononov et al, (2011) assumed that the medium is two-phased consisting of nanocrystalline silicon particles and air spaces. In this case, equation (7.9) was reduced to the sum of two terms and it was required, knowing the dispersion law of crystalline silicon, determine the values of f_1 and f_2 so that the approximating dispersion contours coincided with the experimental values $\varepsilon'(v)$ and $\varepsilon''(v)$. However, a satisfactory approximation could not be obtained at any value of f_1 and f_2. Since the degree of oxidation of the nanoparticles is not known, using the two-phase Bruggeman model the authors hypothesized that, on average, each particle does not behave as crystalline silicon and it behaves as the

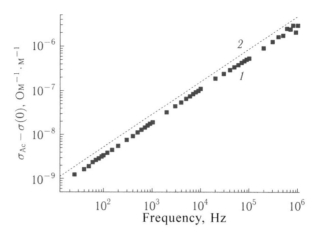

Fig. 7.12. *1* – Frequency dependence of AC conductivity of the film nc-Si; *2* – dependent $\sigma_{AC}(v)$, determined by the DCA model using experimentally measured values of ε_s, ε_∞ and $\sigma(0)$. From (Kononov et al, 2011)

SiO$_x$ environment, where $0 \leq x \leq 2$ was a fitting parameter, as with f_1 and f_2. Spectra $\varepsilon'(v)$ and $\varepsilon''(v)$ for SiO$_x$ in the whole range $0 \leq x \leq 2$ were taken from (Zuter, 1980), in which it was assumed that SiO$_x$ is a mixture of tetrahedra Si–Si$_y$O$_{4-y}$, whose random parameter y takes values from 0 to 4 (random-binding model. Hubner, 1980). Using these spectra, it was possible to get a good approximation of the experimental values $\varepsilon'(v)$ and $\varepsilon''(v)$ at $x = 1$ and $f_1 = f_2 = 0.5$. The approximating EMA spectra for the specified parameters are shown in Fig. 7.12 as dashed lines. Thus, it is shown that, on average, the studied nc-Si films show themselves as a medium consisting of SiO with a porosity of 0.5.

Here we note the already mentioned study (Pickering et al, 1984), in which the recorded spectra $\varepsilon'(v)$ and $\varepsilon''(v)$ were consistent with the data of (Kononov et al, 2011).

Very characteristic are the absorption spectra of nc-Si films, calculated from the ellipsometric data. As can be seen from Fig. 7.8, in the incident photon energy range less than 3 eV, the absorption of the films is greater than the absorption of c-Si, and at higher energies – it is significantly lower. This behaviour of absorption shows that for low-energy photons the main contribution to the absorption comes from the SiO$_x$ shell with a high density of defective states near the interface with the crystalline core, and the photons with energies greater than 3 eV are absorbed in the main crystalline cores of the nanoparticles which have, due to the quantum constraint, a larger band gap than the c-Si.

There are several models interpreting the results of measurements of AC conductivity of the substance. For semiconductors. a good approximation is the model proposed in (A. Goswami, A.P. Goswami, 1973), according to which the conductive material is a composite of the parallel-connected capacitor C_1 and a resistor with a conductivity of G_1 ($G_1 = 1/R$). In order to account for the effect of conducting contacts, a resistor with conductivity G_2 ($G_2 = 1/r$) is connected in series with this group. According to this model, C_1 and G_1 do not depend on the frequency of the applied alternating electric field, but G_1 is dependent on the temperature of the conductive substance.

If we measure the capacitance of the sample in the parallel connection mode, C_p, then the measured value is linked to the quantities C_1, G_1 and G_2 by the following equation:

$$C_p = \frac{C_1 G_2^2}{\left(G_1 + G_2\right)^2 + \left(2\pi v C_1\right)^2}. \tag{7.10}$$

From this equation it is clear that when the conditions of the model (A. Goswami, A.P. Goswami, 1973) are fulfilled, the measured capacitance of nc-Si films must satisfy the condition $C_p \sim v^{-2}$. However, the experimental C_{nc-Si} curve shown in Fig. 7.10 cannot be approximated by the given power-law dependence. It can be assumed that the values of C_1 and G_1 are dependent on frequency. Indeed, in the experiments $G_2 \gg G_1$ and $G_2 \gg v \cdot C_1$, hence $C_p \approx C_1$. Therefore, for the approximation in (Kononov et al, 2011) one can use the following semi-empirical function:

$$C_{nc-si}\left(v\right) = C_\infty + \frac{C}{1 + \left(Av\right)^\beta}. \tag{7.11}$$

From the form of (7.11) it follows that for $v \to \infty$ $C_{nc-Si} \to C_\infty$, and when $v = 0$ $C_{nc-Si} = C_\infty + C \equiv C(0)$. Thus the value of C_∞ in the expression (7.11), is the capacitance of the film at the 'infinitely high frequency', and $C(0) = C_\infty + C$ is its static capacity. Adjustable parameter A in formula (7.11) has the dimension of time and the adjustable parameter β specifies the type of power-law dependence of C_{nc-Si} on the frequency of the applied alternating field. The function (7.11) was a very good approximation of the experimental dependence $C_{nc-Si}(v)$ with the following coefficients: $C_\infty = 3.9 \cdot 10^{-10}$ F, $C = 7.9 \cdot 10^{-10}$ F, $A = 0.5$ and $= \beta$ 0.32. Respectively. $C(0) = 11.8 \cdot 10^{-10}$ F.

The values of the capacitance of the film are related to the real component of its dielectric constant relationship:

$$C_{nc-si}\left(v\right) = \varepsilon'_{nc-si}\left(v\right).C_0,$$

where C_0 is the geometric capacitance. In order to study the film $C_0 = 1.15 \cdot 10^{-10}$ F, and the determined values of capacitances $C(0)$ and C_∞ the corresponding the static and optical dielectric constants are $\varepsilon_s \equiv \varepsilon(0) = 10.3$ and $\varepsilon_\infty = 3.4$. The value of the static dielectric constant of the studied film, which is 10.3, is substantially less than the dielectric constant of crystalline silicon which, as we know, is ~12. Frequency spectra $\varepsilon'(v)$ and $\varepsilon''(v)$ obtained from the impedance of the nc-Si film, are shown in Fig. 7.10.

A good approximation for these spectra was the semi-empirical Cole–Cole relationship (K.S. Cole, R.H. Cole, 1941; Moliton, 2007):

$$\varepsilon = \varepsilon_\infty + \frac{\varepsilon_s - \varepsilon_\infty}{1 + (i\omega\tau)^{1-h}}, \quad 0 \le h \le 1, \tag{7.12}$$

where ε_s and ε_∞ are static and optical dielectric constants, defined above, $\omega = 2\pi v$ is angular frequency, and τ – the time of dipole relaxation. The Cole–Cole dependence is known to be true in the case when the substance has multiple types of dipoles, each with their own relaxation time. Therefore, the quantity τ, appearing in equation (7.12), the relaxation time, averaged over an ensemble of dipole groups represented in the studied nc-Si film.

The approximating Cole–Cole curves are shown in Fig. 7.10 by dashed lines. From this figure it is clear that $\varepsilon'(v)$ is very well approximated in the entire measured frequency range, and for $\varepsilon''(v)$ the Cole–Cole dependence finds good agreement only in the frequency range of $2 \cdot 10^2 \le v \le 10^6$ Hz. The resultant approximation corresponds to the following values: $\varepsilon_s = 10.8$, $\varepsilon_\infty = 3.43$, $\tau = 6 \cdot 10^{-2}$ s and $h = 0.7$. It is important to note the proximity of the value ε_∞ with the values of ε' defined in the optical range of ellipsometry.

By comparing the values of ε_s and ε_∞, corresponding to the Cole–Cole approximation with similar values determined from capacitance measurements, we can see the proximity of their numerical values. The quantity $(1 - h)$ is also very close to the value of the degree β in (7.11). In addition, if we assume that A from (7.11) is the relaxation time is multiplied by 2π, then $\tau = \dfrac{A}{2\pi} = 6.4 \cdot 10^{-2}$, which is also close the average dipole relaxation time corresponding to the Cole–Cole approximation.

The static dielectric permittivity $\varepsilon_s = 10.8$, found from the Cole–Cole relationship, is slightly higher than the value found from the equation (7.11), but it is also less than the value $\varepsilon_s = 12$, characteristic of crystalline silicon. There are two reasons that lead to a decrease in ε_s of the nc-Si film compared to the ε_s value of c-Si. The first reason is related to the presence of air pores in the body of the film. Second – the fact that the size distribution of nanoparticles from which the film is made, there is a large fraction of particles with sizes less than 10 nm (see the inset in Fig. 7.8). In [29] the authors presented the results of calculations of the dielectric constant of silicon nanoparticles, depending on their size. In line with these results, the static dielectric constant decreases with decreasing particle diameter smaller than 10 nm, and for particles with a diameter of 10 nm its value ranges from 11.2 to 10.1, depending on the computational model used.

Figure 7.10 shows that in the frequency range $v \le 2 \cdot 10^2$ Hz there is a marked discrepancy between the approximating Cole–Cole function and experimental dependence $\varepsilon''(v)$. The reason for this discrepancy lies in the fact that the Cole–Cole ratio, which describes the relaxation of the dipole moments of the dielectric, does not take into account the availability of electric charges. The studied nc-Si film does not contain any free charges, as evidenced by a non-zero DC conductivity the value of which, as noted above, at $T = 297$ K is $\sigma(0) = 9 \cdot 10^{-10}$ ohm^{-1} m^{-1}. In accordance with the work (Nakajima, 1972) the frequency corresponding to the maximum dispersion for $\varepsilon''(v)$ is related to $\sigma(0)$ by the relation:

$$\sigma(0) - p(\varepsilon_s - \varepsilon_\infty)\varepsilon_0 \cdot 2\pi v_m,$$

where the numerical coefficient p is approximately equal to unity. Figure 7.10 shows that Cole–Cole approximating function reaches a maximum at a frequency of $v_m = 2.5$ Hz, and this value agrees well with the experimental value $\sigma(0)$ obtained using the Burton–Nakajima–Namikawa equation (Nakajima, 1972).

To account for the conductivity associated with the free electric charges, the relation (7.12) should be written as:

$$\varepsilon = \varepsilon_\infty + \frac{\varepsilon_s - \varepsilon_\infty}{1 + (i\omega\tau)^{1-h}} + \frac{\sigma(0)}{\varepsilon_0\omega}. \tag{7.13}$$

Approximation of the spectrum $\varepsilon''(v)$ of the studied film is shown by a dashed line in Fig. 7.10 (curve 5), which shows that in the whole measured frequency range the function (7.13) gives a very good approximation of the experimental dependence $\varepsilon''(v)$. The effect of free electric charges on the dielectric properties of the nc-Si film is very evident in the Cole–Cole plot on which the value ε'' for each frequency is depicted as a function of ε' (see Fig. 7.11).

From the Cole–Cole approximation (curve 2 in Fig. 7.11) it follows that the function $\varepsilon''(\varepsilon')$ must have the form of a semicircle whose centre lies below the x-axis of ε'. Intersection of this circle with the axis ε' at $\omega = 0$ and $\omega \to \infty$ gives the values of ε_s and ε_∞.

To establish the nature of the transfer of electric charges in the nc-Si films, Kononov et al (2011) investigated the frequency dependence of the specific conduction $\sigma_{AC}(v)$:

$$\sigma_{AC}(v) - \sigma(0) = \varepsilon_0 \cdot 2\pi v \cdot \varepsilon''(v).$$

It is shown that $\sigma_{AC}(v)$ in the whole measured frequency range is well approximated by a power function:

$$\sigma_{AC}(v) = \sigma(0) + A \cdot v^s,$$

with the value $s = 0.74$. This behaviour of $\sigma_{AC}(v)$ means that the electric transport in the film occurs by a hopping mechanism, which in turn is a manifestation of the disorder in the structure of that part of the film in which the charge transfer takes place.

There are several theoretical models describing the hopping conduction in disordered solids. All these models yield a power-law dependence of AC conductivity on the frequency of the alternating electric field: $\sigma(v) \sim v^s$. However, the numerical values of the exponent s are different. Thus, in models (Hunt, 1991) whereby the conductivity is due to tunnelling of electrical charges through the energy barriers separating closely space localized states, the value of s is given by:

$$s = 1 + \left[\frac{q}{\ln\left(v / v_{ph}\right)} \right], \tag{7.14}$$

where $q = 4.5$, depending on the theoretical model, and $v_{ph} \approx 10^{12}$ Hz is the phonon frequency. Equation (7.14) shows that the value of s should decrease with increasing frequency. However, this behaviour is consistent with the experimental data obtained in (Kononov et al, 2011), as well as data of a large number of other experiments (Dyre, Schrøder, 2000).

At the present time it is reliably established (Hunt, 1991; Isichenko, 1992) that an important role in processes of the conductivity of disordered solids is played by percolation processes which result in the electric transport occurring along the trajectories of least resistance (percolation trajectories). The conductive properties of the percolation trajectories are determined by the structure of clusters (percolation clusters) forming a conductive shell of the solid. In strongly disordered solids the percolation trajectories on a small scale show a fractal structure, resulting in their fractal dimension d_f being greater than the topological D, for example, fractal dimension of the trajectory of a Brownian particle $d_f = 2$, at topological $D = 1$ (Isichenko, 1992).

In this regard, we note the theoretical studies (Dyre, Schrøder 2000; Schrøder, Dyre, 2002; Schrøder, Dyre, 2008), which formulated a model of diffusion cluster approximation (DCA). In these studies it is argued that in the percolation regime the greatest contribution to the AC conductivity comes from the so-called diffusive clusters whose fractal dimension is in the range 1.1–1.7. This statement means that the fractal structure of such clusters is simpler than the structure of multiply connected percolation clusters formed in the conductive material above the percolation threshold (backbone – clusters) whose fractal dimension is equal to 1.7 (Isichenko, 1992). At the same time, the structure of the diffusive clusters is more extensive than that of the simply-connected network clusters where breaking of each cluster leads to the disappearance of the current flowing through the cluster (redbonds). The fractal dimension of the redbonds–clusters is 1.1 (Isichenko, 1992). In the mentioned studies the universal dependence of the dimensionless complex conductivity

$$\tilde{\sigma} = \frac{\sigma_{AC}(v) + i\sigma''(v)}{\sigma(0)}$$

on the dimensionless frequency was obtained:

$$\tilde{\omega} = \frac{\varepsilon_0 \left(\varepsilon_s - \varepsilon_\infty \right)}{\sigma(0)} 2\pi v.$$

This dependence is determined by the following expression:

$$\text{In } \tilde{\sigma} = \left(\frac{i\tilde{\omega}}{\tilde{\sigma}} \right)^{d_f/2}. \tag{7.15}$$

The fractal dimension d_f in (7.15) is an adjustable parameter. As a result of processing a large number of experimental plots it was found in the studies (Schrøder, Dyre, 2002; Schrøder, Dyre, 2008) that the best agreement in the frequency range $v > 1$ Hz is achieved at $d_f = 1.35$.

Kononov et al (2011) compared the experimentally measured dependence $\sigma_{AC}(v)$ with the values determined by the formula (7.15). It should be noted that the complex equation (7.15) has no exact analytical solution and must be solved numerically. However, at low frequencies $\omega \rightarrow 0$ and equation (7.15) can be written as

$$\tilde{\sigma} - 1 = \left(i\tilde{\omega} \right)^{d_f/2}.$$

Respectively

$$\sigma_{AC}(v) - \sigma(0) = \sigma(0)^{\left(1 - \frac{d_f}{2} \right)} \cos\left(\frac{\pi d_f}{4} \right) \left(2\pi\varepsilon_0\Delta\varepsilon \right)^{d_f/2} v^{d_f/2}, \tag{7.16}$$

where $\Delta\varepsilon = \varepsilon_s - \varepsilon_\infty$. Substituting into (7.16) the experimentally obtained values of ε_s, ε_∞, $\sigma(0)$ and $s \equiv d_f/2 = 0.74$ gives an approximating dependence for $\sigma_{AC}(v)$, corresponding to the DCA model and shown in Fig. 7.12. The figure shows that the calculated dependence is a sufficiently good approximation of the experimental curve $\sigma_{AC}(v)$ throughout the measured frequency range. However, the calculated dependence gives values of σ_{AC} about 1.5 times larger than the experimental values. This discrepancy is attributed to possible errors in determining numerical values of ε_s, ε_∞ and $\sigma(0)$.

One possible reason, which can lead to the measurement error of $\sigma(0)$ of the nc-Si film is the dependence of $\sigma(0)$ on the ambient humidity. Qualitatively, in (Kononov et al, 2011) the authors established the following pattern: the higher the humidity of laboratory air, the higher (at constant temperature) conductivity $\sigma(0)$ of the nc-Si film. Conversely, if the film is preheated at a temperature of 200°C for more than 15 min, and then cooled to its original temperature, the conductivity of the film decreases by almost two orders of magnitude. Thus, the presence of water in the atmosphere surrounding the film significantly alters its conductive properties. In the study (Nogami, Abe, 1997), a similar phenomenon was observed in the study of ionic conductivity in glasses of fused quartz. In these studies it was shown that the presence on the glass surface glass of the (Si–O–H) bonds, H_2O molecules form complexes with them sustained by hydrogen bonds. These complexes can dissociate to form free H_3O^+ ions and bonded groups (Si–O⁻) according to the scheme:

$$Si - O - H \cdots OH_2 \rightarrow Si - O^- + H^+ : H_2O$$

The points here indicate the hydrogen bond between the atoms of H and O. Proton H^+ can be captured in this case by a neighbouring H_2O molecule:

$$H_2O_{(1)} : H^+ + H_2O_{(2)} \rightarrow H_2O_{(1)} + H^+ + H_2O_{(2)} \rightarrow H_2O_{(1)} + H^+ : H_2O_{(2)}$$

etc.

This scheme allows for proton transport near the surface of the glass (Fig. 7.13).

Returning to the nc-Si films, we note that the particles used for the deposition of the films are hydrogenated nanocrystalline silicon. However, exposure of these particles in the air resulted in the formation of the SiO_x shell on their surface, $0 \leq x \leq 2$. In (Nogami, Abe, 1997; Nogami et al, 1998) the authors calculated the kinetics of the interaction of H_2O molecules with chain-like structures of SiO_2. It is shown that the H_2O molecules in interaction with the SiO_2 surface groups efficiently break the (Si–O–Si) bonds to form the (Si–O–H) groups. The subsequent interaction of H_2O molecules with the (Si–O–H) groups results in the formation of H_3O^+ ions which are characterized by high mobility and can make a noticeable contribution to the proton transport along a SiO_2 chain.

In addition to this process, another possible process is the collective proton conductivity, linked with associated Si–O–H groups, i.e. groups united by hydrogen bonds, as shown in the diagram in Fig. 7.13a (Du et al, 2003). The arrows in Fig. 7.13 show the direction of the transfer of the positive charge. The collective proton conductivity is also possible in the interaction of water molecules with hydroxyl groups in which a surface structure shown in Fig. 7.13b forms.

Since the internuclear distance (O···O) of the (O···H–O) fragment is in the range of 2.5–2.9 Å (Cao et al, 2007), the angle between the H–O–H

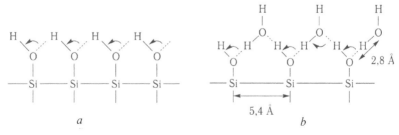

a   b

Fig. 7.13. Diagram illustrating the collective proton conductivity, associated with associate Si–O–H groups (groups, united by hydrogen bonds, as shown in the figure) (Du et al, 2003).

bonds is equal to about ~104°, then there is good spatial coherence of the given surface structure with the lattice constant of crystalline silicon which is known to be 5.4 Å.

With respect to the nc-Si film, analyzed in the article by Kononov et al (2011), there is a direct proof of the existence of such structures. Earlier in the study of the IR spectra of nc-Si powders, similar to those used in this study, it was shown that the spectra there is a broad intense band, the maximum of which is about at 3420 cm^{-1} (Glasser, 1975). In the works (Leite et al, 1998; Kononov et al, 2005), this band is associated with the vibration of (O–H) in the hydrogen-bonded hydroxyl groups. In (Glasser, 1972) it is also shown that when heating the nc-Si particles up to 400°C the intensity of the band near 3420 cm^{-1} markedly decreases and this is accompanied by an increase of the intensity of the narrow band with the maximum near 3750 cm^{-1}, which is identified with the vibration of (O–H) in the isolated (Si–O–H) group. This behaviour of the intensities of the bands 3420 and 3750 cm^{-1} means that the associated (Si–O–H) groups become isolated in heating the nc-Si particles. Accordingly, when heated, the proton conductivity, associated with these groups, should decrease. Thus, the dependence of the conductivity $\sigma(0)$ of the nc-Si films on the humidity of the ambient air together with the temperature behaviour of the absorption bands, associated with (Si–O–H) groups, suggests that the main contribution to the DC dark conductivity of the nc-Si films comes from proton conductivity.

Concluding remarks

Currently, there is strong scientific interest in the structures formed from the particles of nanocrystalline silicon (nc-Si). This interest is largely determined by the fact that over the last 10 years it has been possible to develop effective methods of obtaining silicon nanoparticles, capable of vivid, stable photoluminescence in the visible spectrum, which has a high quantum yield. The main carriers of such nanoparticles are colloidal solutions (sols) based on methanol, chloroform, hexane, etc. Such sols are very promising targets for the development of application techniques on different substrates by extremely thin nc-Si films. The use of such films seems very promising for the creation of light emitting elements, which are based on electroluminescence of nc-Si. In addition, nc-Si films are very promising as elements of the solar cells, thin film transistors and devices that use the principle of one-electron charge (*single electronic devices*).

In the case where the film consists of nanoparticles with a diameter of less than 10 nm, their overall characteristics are determined not only by the properties of the substance from which nanoparticles consist, but also by the properties of atoms on the surface these particles. In other words, such films should be viewed as a multi-component environment whose properties are determined both by crystalline nuclei of nanoparticles and the surface atoms and molecules and air pores forming the films.

References for Chapter 7

Azzam R.M.A., Bashara N.M. Ellipsometry and polarized light.. Amsterdam-NY-Oxford: North-Holland Publishing company, 1977.

Dorofeyev S.G, et al. Optical and structural properties of thin films deposited from sol of Si nanoparticles. Fiz. Tekh. Poluprovod. 2009. V. 43. P. 1460-1467.

Kononov N. N., Dorofeev S.G., Ischenko A.A., Mironov R.A., Plotnichenko V.G., Dianov E. M., Dielectric and transport properties of thin films deposited from sols containing silica nanoparticles. Fiz. Tekh. Poluprovod. 2011 , 45 (8) P. 1068-1078.

Kononov N.N., Kuzmin G. P., Orlov A. N., Tikhonevich O. Optical and electrical properties of thin plates made of nanocrystalline silicon powders. Fiz. Tekh. Poluprovod. 2005. V. 39 (7). P. 868-873.

Anopchenko A., Marconi A., Moser E., Prezioso S., Wang M., Pavesi L., Pucker G., Bellutti P. Low-voltage onset of electroluminescence in nanocrystalline-Si/SiO$_2$ multilayers. J. Appl. Phys. 2009. V. 106. P. 033104 (5 p.).

Aspens D. E. In: Properties of Silicon. London, UK: INSPEC, IEE, 1988. Ch. 2. P. 59–80.

Aspens D. E., Studna A.A. Dielectric functions and optical parameters of Si, Ge, GaP, GaAs, GaSb, InP, InAs, and InSb from 1.5 to 6.0 eV. Phys.Rev. B. 1983. V. 27. P. 985–1009.

Axelrod E., Givant A., Shappir J., Feldman Y., Sa'ar A. Dielectric relaxation and transport in porous silicon. Phys. Rev. B. 2002. V. 65. P. 165429(1–7).

Ben-Chorin M., Averboukh B., Kovalev D., Polisski G., Koch F. Influence of Quantum Confinement on Critical Points of the Band Structure of Si. Phys. Rev. Lett. 1996. V. 77. P. 763–766.

Ben-Chorin M., M"oller F., Koch F., Schirmacher W., Eberhard M. Hopping transport on a fractal: ac conductivity of porous silicon. Phys. Rev. B. 1995. V. 51. P. 2199–2213.

Bruggeman D.A.G. Berechung verschiedener physikalischer konstanten von heterogenen substanzen. Ann. Phys. (Leipzig). 1935. V. 24. P. 636–664.

Brus L. E., Szajowski P. F., Wilson W. L., Harris T.D., Schuppler S., Citrin P.H. Electronic-Spectroscopy and Photophysics of Si Nanocrystals: Relationship to Bulk c-Si and Porous Si. J. Am. Chem. Soc. 1995. V. 117. P. 2915–2922.

Campbell I.H., Faushet P.M. The effect of microcrystalline size and shape on the one phonon Raman spectra of crystalline semiconductors. Solid State Comm. 1986. V. 58. P. 739–742.

Cao C., He Y., Torras J., Deumens E., Trickey S.B., Cheng H.-P. Fracture, water dissociation, and proton conduction in SiO$_2$ nanochains. J. Chem. Phys. 2007. V. 126. P. 211101(1–7).

Cheng J-C., Allen S., Wagner S. Evolution of nanocrystalline silicon thin film transistor channel layers. J. Non-Cryst Solids. 2004. V. 338–340. P. 720–724.

Cole K. S., Cole R.H. Dispersion and Absorption in Dielectrics. J. Chem. Phys. 1941. V. 9. P. 341–351.

Conte G., Feliciangeli M. C., Rossi M. C. Impedance of nanometer sized silicon structures. Appl. Phys. Lett. 2006. V. 89. P. 022118 (1–3).

De la Torre J., Bremond G., Lemiti M., Guillot G., Mur P., Buffet N. Silicon nanostructured layers for improvement of silicon solar cells efficiency: A promising perspective. Materials Science and Engineering. C. 2006. V. 26. P. 427–430.

Du M.-H., Kolchin A., Chenga H.-P. Water–silica surface interactions: A combined quantum-classical molecular dynamic study of energetics and reaction pathways. J. Chem. Phys. 2003. V. 119. P. 6418–6422.

Dyre J. C. Schrø der T.B. Universality of ac conduction in disordered solids. Rev. Mod. Phys. 2000. V. 72, No.3. P. 873–892.

Ehbrecht M., Ferkel H., Huisken F., Holz L., Polivanov Yu.N., Smirnov V.V., Stelmakh O.M., Schmidt R. Deposition and analysis of silicon clusters generated by laser-induced gas phase reaction. J. Appl. Phys. 1995. V. 78. P. 5302–5305.

Forman R.A., Thurber W. R., Aspens D. E. Second indirect band gap in silicon. Solid State Comm. 1974. V. 14. P. 1007–1010.

Glasser L. Proton conduction and injection in solids. Chem. Rev. 1975. V. 75, №1. P. 21–65.

Goswami A., Goswami A. P. Dielectric and optical properties of ZnS films. Thin Solid Films. 1973. V. 16. P. 175–185.

Hubner K. Chemical bond and related properties of SiO_2. Phys. Stat. Sol. A. 1980. V. 61. P. 665–670.

Hunt A. AC hopping conduction: perspective from percolation theory. Phil. Mag. B. 1991. V. 64. P. 579–584.

Isichenko M.B. Percolation, statistical topography, and transport in random media. Rev. Mod. Phys. 1992. V. 64, №4. P. 961–1043.

Jackson W. B., Johnson N.M., Bigelsen D.K. Density of gap states of silicon grain boundaries determined by optical absorption. Appl. Phys. Lett. 1983. V. 43. P. 195–197.

Jurbergs D., Rogojina E., Mongolini L., Kortshagen U. Silicon nanocrystalls with ensemble quantum yields exceeding 60%. Appl. Phys. Lett. 2006. V. 88. P. 233116(3 p.).

Kakinuma H., Mohri M., Sakamoto M., Tsuruoka T. Structural properties of polycrystallline silicon films prepared at low temperature by plasma chemical vapor deposition. J. Appl. Phys. 1991. V. 70. P. 7374–7378.

Kovalev D., Polisski G., Ben-Chorin M., Diener J., Koch F. The temperature dependence of the absorption coefficient of porous silicon. J. Appl. Phys. 1996. V. 80. P. 5978–5983.

Kuz'min G. P., Karasev M. E., Khokhlov E.M., Kononov N.N., Korovin S. B., Plotnichenko V.G., Polyakov S.N., Pustovoy V. I., Tikhonevich O.V. Nanosize silicon powders: the structure and optical properties. Laser Physics. 2000. V. 10. P. 939–945.

Lee C.H., Sazonov A., Nathan A. High-mobility nanocrystalline silicon thin-film transistors fabricated by plasma enhanced chemical vapor deposition. Appl. Phys. Lett. 2005. V. 86. P. 222106(3 p.).

Leite V.B. P., Cavalli A., Oliveira O.N. Jr. Hydrogen-bond control of structure and conductivity of Langmuir films. Phys. Rev. E. 1998. V. 57, No. 6. P. 6835–6839.

Luppi M., Ossicini S. Ab initio study on oxidized silicon clusters and silicon nanocrystals embedded in SiO_2: Beyond the quantum confinement effect. Phys. Rev. B. 2005. V. 71. P. 035340(1–5).

Min R. B., Wagner S. Nanocrystalline silicon thin-film transistors with 50-nm-thick deposited channel layer, 10 cm^2 V^{-1} s^{-1} electron mobility and 108 on/off current ratio. Appl. Phys. A. 2002. V. 74. P. 541–543.

Moliton A. Applied Electromagnetism and Materials. Springer Science and Business Media, 2007. Ch. 1. P. 8. 345 p.

Nakajima T. 1971 Annual report, conference on electric insulation and dielectric phenomena. Washington D. C.: National Academy of Sciences, 1972. 168 p.

Nogami M., Abe Y. Evidence of water-cooperative proton conduction in silica glasses. Phys. Rev. B. 1997. V. 55, №18. P. 12108–12112.

Nogami M., Nagao R., Wong C. Proton Conduction in Porous Silica Glasses with High Water Content. J. Phys. Chem. B. 1998. V. 102. P. 5772–5775.

O'Leary S.K., Lim P.K. On determining the optical gap associated with an amorphous semiconductor: a generalization of the Tauc model. Solid State Comm. 1997. V. 104. P. 17–21.

Pickering C., Beale M. I. J., Robbins D. J., Pearson P. J., Greef R. Optical studies of the structure of porous silicon films formed in p-type degenerate and non-degenerate silicon. J. Phys. C: Solid State Phys. 1984. V. 17. P. 6535–6552.

Poruba A., Feifar A., Remes Z., Springer J., Vanecek M., Kocka J., Meier J., Torres P., Shah A. Optical absorption and light scattering in microcrystalline silicon thin films and solar cells. J. Appl. Phys. 2000. V. 88. P. 148–160.

Puzder A., Williamson A. J., Grossman J. C., Galli G. Surface chemistry of silicon nanoclusters. Phys. Rev. Lett. 2002. V. 88. P. 097401(4).

Richter H., Ley L. Optical properties and transport in microcrystalline silicon prepared at temperatures below 400°C. J. Appl. Phys. 1981. V. 52. P. 7281–7286.

Richter H., Wang Z., Ley L. The one phonon Raman spectrum in microcrystalline silicon. Solid State Comm. 1981. V. 39. P. 625–631.

Saadame O., Lebib S., Kharchenko A.V., Longeaud C., Cabarrocas R.R. Structural, optical and electronic properties of hydrogenated polymorphous silicon films deposited from silane–hydrogen and silane–helium mixtures. J. Appl. Phys. 2003. V. 93. P. 9371–9379.

Schrøder T. B., Dyre J. C. Computer simulations of the random barrier model. Phys. Chem. Chem. Phys. 2002. V. 4. P. 3173–3178.

Schrøder T. B., Dyre J. C. AC hopping conduction at extreme disorder takes place on the percolating cluster. Phys. Rev. Lett. 2008. V. 101. P. 025901 (4 p.).

Schuppler S., Friedman S. L., Marcus M.A., Adler D. L., Xie Y.H., Ross F.M., Chabal Y. L.,

Harris T.D., Brus L. E., Brown W. L., Chaban E. E., Szajowski P. E., Christman S. B., Cit-rin P.H. Size, shape and composition of luminescent species in oxidized Si nanocrystals and H-passivated porous Si. Phys. Rev. B. 1995. V. 52. P. 4910–4925.

Shan A., Vallat-Shauvain E., Torres P., Meier J., Kroll U., Hof C., Droz C., Goerlitzer M.,

Wyrsch N., Vanechek M. Intrinsic microcrystalline silicon (μc-Si:H) deposited by VHF-GD (very high frequency-glow discharge): a new material for photovoltaics and opto-electronics. Mater. Sci. Eng. B. 2000. V. 69–70. P. 219–226.

Stuart B. Infrared Spectroscopy: Fundamentals and Applications. John Wiley & Sons, Ltd. 2004. P. 83.

Tauc J., Grigorovici R., Vancu A. Optical properties and electronic structure of amorphous germanium. Phys. Status Solidi. 1966. V. 15. P. 627–637.

Tiwari S., Rana F., Chan C., Shi L., Hamafi H. Single charge and confinement effect in nanocrystal memories. Appl. Phys. Lett. 1996. V. 69. P. 1232–1234.

Tompkins G.H., Irene E.A. Ed. Handbook of ellipsometry. N.Y.–Heidelberg: William Andrew Publishing, Springer, 2005.

Tsang J. C., Tischler M.A., Collins R. T. Raman scattering from H or O terminated porous Si. Appl. Phys. Lett. 1992. V. 60. P. 2279–2281.

Tsu R. Phenomena in silicon nanostructure devices. Appl. Phys. A. 2000. V. 71. P. 391–402.

Tsu R., Babic D., Ioriatti L. Jr. Simple model for the dielectric constant of nanoscale silicon particle. J. Appl. Phys. 1997. V. 82 (3). P. 1327–1329.

Tsu R., Shen H., Dutta M. Correlation of Raman and photoluminescence spectra of porous silicon. Appl. Phys. Lett. 1992. V. 60. P. 112–114.

Urbach B., Axelrod E., Sa'ar A. Correlation between transport, dielectric, and optical properties of oxidized and nonoxidized porous silicon. Phys. Rev. B. 2007. V. 75. P. 205330 (7 p.).

Voutsas A. T., Hatalis M.K., Boyce J., Chiang A. Raman spectroscopy of amorphous and microcrystalline silicon films deposited by low-pressure chemical vapor deposition.

J. Appl. Phys. 1995. V. 78. P. 6999–7005.

Wang K., Chen H. Shen.W.Z. AC electrical properties of nanocrystalline silicon thin films // Physica B. 2003. V. 336. P. 369–378.

Wovchko E.A., Camp J. C., Glass J.A. Jr. Yates J. T. Jr. Active sites on SiO_2: role in CH_3OH decomposition. Langmuir. 1995. V. 11. P. 2592–2599.

Zhixun M., Xianbo L., Guanglin K., Junhao C. Raman scattering of nanocrystalline silicon embedded in SiO_2. Science in China A. 2000. V. 43. P. 414–420.

Zuter G. Dielectric and Optical Properties of SiO_x. Phys. Stat. Sol. A. 1980. V. 59. P. K109–K113.

Part III

8

Methods for investigating and controlling the structure and properties of nanosilicon

The strong size dependence of the properties of silicon nanoparticles smaller than 5 nm sharply distinguishes them from all other phases of the condensed silicon. If the other phases, for example, a:Si, c-Si or mc-Si, can be characterized by well-defined properties averaged over the volume (see Table 2.1), the properties of even monodisperse nanoparticles vary substantially depending on the size, as demonstrated in previous chapters of this book, and each fraction is as if an entirely new material. There is no possibility of uniquely characterization of the properties of polydisperse systems of nanoparticles, since they strongly depend on the type and width of the size distribution function of particles in the sample volume. Therefore, in contrast to bulk materials, where much of the expected properties of the final product can be said based on a very limited number of characteristics of raw materials and technology parameters, when working with nanomaterials without a careful definition of the majority of physical and chemical parameters of nanoparticles it is often impossible to understand their properties and even more impossible to plan the synthesis of nanomaterials with the desired characteristics.

Unfortunately, from a huge set of currently available experimental analytical methods only a few are effective for the control of nanomaterials with very small particles (Wang, 2000). The difficulty lies in the fact that Si particles of the nanometer size often consist of many phases, for example, core–shell formations with a crystalline core ad an amorphous shell or of a single-phase core of Si, covered with an oxide shell (see Chapter 2 and 5), etc., and the electro-optical properties of the different phases of Si may greatly differ, so each phase should be analyzed selectively. The influence of impurities and defects on the semiconductor properties of nanoparticles is strong and this is combined with the influence of size effects of the particles themselves. It is obvious that the polydisperse system of nanoparticles in this case cannot be characterized by the mean values of the concentration of impurities or defects in the sample material as a defect or an impurity atom in a particle with a diameter of 1 nm and a particle 10 nm in diameter have different concentration, and the impact of this defect on the electronic structure of the particles will be completely different. For example, the introduction of a single atom in a crystal with a diameter of 3 nm ($1.4 \cdot 10^{-20}$ cm^3, 700 atoms) corresponds to concentration of $7.0 \cdot 10^{19}$ atoms/cm^3 in bulk silicon. At this level of doping the bulk silicon is a degenerate semiconductor and behaves like a metal. Hence for the correct performance of systems based on nanosilicon the analytical methods should also have size selectivity.

To directly determine the characteristics of Si nanoparticles or to find out structural features it is necessary to use method used for such commonly for nanostructures, such as transmission electron microscopy (TEM) X-ray photoelectron spectroscopy (XPS), photoluminescent spectroscopy (PL), atomic force microscopy (AFM), scanning tunnelling microscopy (STM), Raman spectroscopy (RS – Raman spectroscopy), Fourier transform infrared spectroscopy (FTIR) and X-ray diffractometry (XRD). Often, the experimental methods of study of nanomaterials are ineffective and then to determine the properties and define the characteristics it is necessary to use rather complex mathematical modelling techniques.

In this chapter we consider the basic methods of structural analysis, suitable to properly study the atomic structure of nanosilicon. Methods of morphological and particle size and also microstructure control of the nanoparticles will be discussed further in Chapters 9 and 10.

8.1. Transmission electron microscopy

One of the main methods of investigation of the morphology and structure of nanosilicon is high-resolution transmission electron microscopy (HRTEM) in various versions. Transmission electron microscopy is partially similar to optical microscopy, since it allows to observe the image of the object with a large magnification and is a direct method of analysis nanostructured systems. The essential difference between these two methods is due to the huge difference used for imaging wavelengths. The wavelength of the electrons in the electron microscope is ten thousand times smaller than the wavelength of visible light in the optical microscope, which allows to minimize the geometrical aberrations and to obtain electron microscope images with the spatial resolution of 0.1 nm. This makes the electron microscopy a powerful tool for studying the smallest details of the structure of materials down to the atomic structure.

There are two types of electron microscopes: 1) transmission electron microscope (TEM), and 2) scanning electron microscope (SEM).

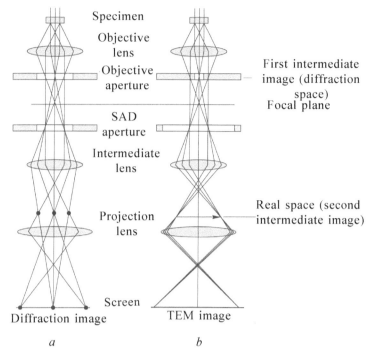

Fig. 8.1. Status of the 'rays' and TEM modes: *a* – microdiffraction mode SAD (selected area diffraction); *b* – mode images (from Yang Leng, 2008; Reimer & Kohl, 2008).

The optical scheme of TEM, like the scheme of conventional optical transmission microscopes, consists of a light source, a condenser lens, the object table for the sample, an objective lens and a projection lens (Fig. 8.1).

The difference between optical and electron microscopes is that the electron microscope uses a beam of electrons instead of visible light, and instead of glass lenses – electromagnetic lenses. Because of the special features of controlling the electron beams in the electron microscope the number of lenses and apertures is greater than in the optical microscope.

The resulting image of the sample in a TEM is usually projected onto a flat fluorescent screen or on a matrix of a mosaic detector (e.g. CCD). The transmission electron microscope can operate in two modes: 1) using a microscopic image, and 2) in the microdiffraction mode. In the first case, the image of the sample is observed when focusing of the optical system is set up at the front focal plane of the objective lens (i.e. the position of the sample)[1]. The diffraction image is observed when the optical system focuses

---

[1]In transmission electron microscopy, as in optical transmission microscopy, the terminology is similar to that used in conventional photography: the sample is photographed object and the screen recording analog film. The front focal point of focus is called, which is closer to the sample, back focus – one that is located on a sample of the objective lens.

on the back focal plane of the objective lens (i.e., on the plane of the first image for the diffracted and transmitted electron beams). The diffraction image is the plane of the reciprocal lattice or, more accurately, is rather the lattice plane of the diffraction image of the crystal. The diffraction mode is designed for measurements by the method of selected-area diffraction (SAD), i.e., microdiffraction.

To use an electron microscope requires a high vacuum, whereas optical microscopes are usually operate in air. The success of the electron-microscopic study on the whole depends on three factors: the right sample preparation, correct data collection appropriate data interpretation. It should be noted in that the complexity of the equipment, techniques and interpretation of experimental measurement results (Shindo, Oikawa, 2006; Flegler et al, 1993; Williams & Carter, 2009; Yang Leng, 2008) electron microscopy greatly exceeds optical microscopy.

In TEM, electrons are accelerated to several hundred thousand volts. The de Broglie wavelength of the electrons in this case is only a few picometres and the spatial resolution of modern TEM reaches 0.1 nm. In addition to high spatial resolution, because of the strong interaction of electrons with matter the volume in which this interaction is very small so that TEM can detect and determine the position of individual rows of atoms, and in principle it is possible to determine the type of atoms on which is electron scattering takes place. For these reasons, the electron microscope can combine in a single tool a large number of methods and means of getting various information about the structure and composition of substances. For example, in addition to obvious imaging, inelastic scattering of electrons can take place in the electron microscope and the energy of these electrons excites the emission of X-rays for which energy dispersive X-ray emission analysis (EDX) can be carried out. Modern TEM are also used for electron energy loss spectroscopy (EELS) and transmission microscopy with energy filtering. The spatial resolution of the order of several nanometers, available for conventional TEM, is insufficient for an accurate study of nanostructures. But when the appropriate method of image processing is used, TEM can reach the resolution of the order of 0.2 nm (HRTEM method).

8.1.1. Image modes

In order to observe a meaningful image of the structure and morphology of the material under study, the image must have sufficient contrast, i.e., the difference of brightness between the adjacent image points of the image. The contrast in optical transmission microscopy is obtained as a result of the difference in the absorption of light in different regions of the sample. In the transmission electron microscope, the contrast is based on the deviation of electrons from their primary direction as they pass through the sample (Reimer & Kohl, 2008). The contrast in the TEM is formed when the number of electrons, scattered from the transmitted beam, changes. There are two mechanisms for obtaining contrast images with the scattering of electrons in a transmission electron microscope: 1) based on the contrast of mass density (thickness); 2) on the basis of the diffraction contrast.

Mass density contrast. Deflection of the electrons passing through the material arises from the interaction of electrons with atoms, including with atomic nuclei. The magnitude of the scattering of an electron at any given point of the sample depends on the mass density (product of density by thickness) at this point. Thus, the difference in the thickness of the specimen will cause changes in the intensity of electron scattering which the TEM screen registers from the electron beams transmitted by the selector objective aperture (Fig. 8.2).

The deviated electrons with the scattering angle greater than α (about 0.01 radians), falling on the aperture of the objective, will be blocked by its ring. Thus, the aperture reduces the intensity of the beam transmitted through the sample. The brightness of the image on the screen is determined by registering the intensity of the electron beam passing through the objective aperture. The intensity of the transmitted beam (I_t) is equal to

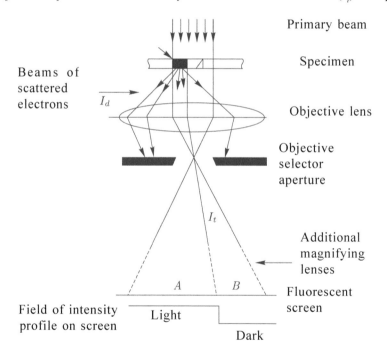

Fig. 8.2. The scheme for obtaining electron microscopic images in the mass density contrast (from Yang Leng, 2008). The oblique arrow, passing through the sample, shows the direct electron beam with Bragg scattering in the crystal, and the scattered beams, passing through the selector diaphragm, may be involved in the formation of dark-field diffraction contrast images.

the primary beam intensity (I_0), reduced by the intensity of beams deflected by the object (I_d) when passing through the sample, i.e.

$$I_t = I_0 - I_d. \tag{8.1}$$

Contrast (C) in TEM expressed by the equation:

$$C = \frac{I_0 - I_t}{I_0}, \tag{8.2}$$

The mass density contrast is found in all materials and is the main mechanism for the imaging of non-crystalline materials, such as amorphous polymers and biological objects. The mass contrast is also often referred to as the absorption contrast, although this name is not accurate, and even erroneous, because the absorption of electrons on the image in mechanism is very small.

Diffraction contrast. The contrast in the transmission electron microscope can also be obtained by diffraction. The diffraction contrast is the primary mechanism for obtaining images of crystalline samples. Diffraction can be considered as a collective deflection of the electrons of the primary beam passing through a crystal object. Electrons can be together scattered by parallel atomic planes, as in Bragg X-ray diffraction and, therefore, Bragg's equation so obtained for x-ray diffraction, is also used to describe the diffraction of electrons.

The terms of Bragg reflection are satisfied for some particular combinations of the angle of orientation of the crystal relative to the direction of electron beams at a given wavelength/energy of the electrons. Under these conditions, the scattered electron beams form a constructive interference when the amplitudes of the electron waves are added up and the observed scattering intensity increases sharply. Constructive interference of the diffracted electrons is usually observed at large angles of deflection, and can be easily cut off by the selector aperture (Fig. 8.3 *a*). Due to the Bragg scattering, which is delayed by the objective aperture ring, the intensity of the electron beam transmitted through the crystal is greatly reduced, and we get an effect similar to the mass density contrast. But there is a big difference between the diffraction contrast and the mass density contrast, because the diffraction contrast is very sensitive to the orientation of the sample relative to the direction of the primary beam electrons in the TEM, whereas the mass density contrast depends only on weight/thickness of the sample.

The diffraction angles (2θ) in TEM, as compared to X-ray diffraction, are very small ($\leq 1°$), so the electron beams, diffracted by the crystallographic plane (*hkl*), can be focused in one spot at the back focal plane of the objective lens. If the primary electron beam is parallel to crystallographic axis, all the diffracted points from the crystallographic zones form a diffraction image on the back focal plane.

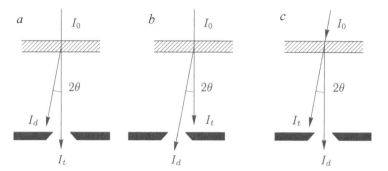

Fig. 8.3. The path of rays through the selector aperture for: bright-field image (*a*); dark-field image (*b*) by moving the selector aperture, dark-field Image (*c*) by deflecting the primary beam without moving the selector aperture. I_0– intensity of the primary electron beam; I_t – the intensity of the direct beam of the electrons transmitted through the sample; I_d – the intensity of the electron beam obtained as a result of Bragg diffraction; 2θ – Bragg angle (from Yang Leng, 2008).

With the help of diffraction contrast bright-field and dark-field images can be obtained in TEM (Figs. 8.2 and 8.3).

The bright-field image is obtained when the selector objective diagram passes only the direct electron beam transmitted through the sample (Fig. 8.3*a*), in the same way as in the formation of images in the mass contrast.

The dark-field image is formed by passing through the selector aperture one of the diffracted beams, as shown in Fig. 8.3*b, c*.

To switch between the bright and dark field modes first it is necessary to select the diffraction mode and also select one of the diffraction spots by placing it into the centre of the selector aperture (Fig. 8.3*b*). Usually this is done in TEM not by displacement of the diaphragm, as shown in Fig. 8.3*b*, but by the deviation of the primary electron beam, which follows the movement of the diffraction pattern (as in Fig. 8.3*c*). In the bright-field mode in a polycrystalline sample darker grains are those for which the Bragg equation is fulfilled and there is a large amount of the energy of the primary beam carried off by diffraction. Tilting the sample, one can change the diffraction angle and place other grains in the Bragg position, and the grains that were dark become lighter and other grains become darker.

In operation in the mass contrast or diffraction contrast mode the selector (objective) aperture, mounted in the back focal plane of the objective lens, cuts the reception angle to $\alpha \leq 2\theta$, i.e. less than the Bragg angles of the coherently scattered electron beams (see Fig. 8.3). The electron beams transmitted in this case form a shadow image on the TEM recording screen that displays the variations in the intensity of electron beams that have passed through different parts of the sample. This image is called bright-field. If the incident beam is tilted so that the Bragg reflection condition is satisfied in scattering by the sample, the selector aperture allows only one of the coherently scattered electron beams to the screen, cutting at the same time the direct beam passing through the sample and all other scattered beams, the screen will show the dark-field image.

The diffraction contrast is very sensitive to the slope of the sample. Even with the slope in the range 1° it is possible to change the image of the grain from bright to dark and vice versa. But on the other hand, the diffraction contrast does not show all grains and only those for which the Bragg condition is satisfied.

Usually, the dark-field image mode is used less frequently than the bright-field image. But in the dark-field image we can detected finer details of the structure of the sample than in the bright-field image, as the background becomes dark, and on the dark background details can be seen better. Because in the dark-field images it is possible to select different diffraction spots (*hkl*), then for one bright-field image we can obtain several slightly different dark-field images, which revealed various details of the structure of the sample. This is another significant difference of TEM in comparison with optical microscopy, in which one bright-field image matches only one dark-field image.

Phase contrast. The contrast of mass density and the diffraction contrast are essentially amplitudes since in both cases the contrast is formed by changing the amplitude of the electron waves passing through the sample. Another way to obtain the image contrast in the TEM can be the use of changes of the phase of the electron wave passing through the material under study. Phase contrast allows to obtain the highest spatial resolution in the study of crystalline materials. For this reason, the method of phase contrast is used to study crystal structures and crystal lattices even more frequently than high-resolution transmission electron microscopy (HRTEM)[2], which we briefly review a little later.

To obtain images with the phase contrast in TEM, is necessary to use at least two frequency coherent electron beams (the passed and diffracted) and stop them interacting with each other. The principle of imaging in TEM in phase contrast is shown schematically in Fig. 8.4.

The phase contrast mode is different from the mass and diffraction contrast modes in that with working with the phase contrast the objective diaphragm is not used. If the selector (objective) diaphragm is completely removed or a diaphragm with a larger reception aperture ($\alpha > 2\theta$) is used, then both the beam Bragg-reflected by the separate crystal of the sample and

---

[2]The preference of TEM images in the phase contrast compared to high-resolution transmission microscopy (HRTEM), which also uses the phase contrast mode, is due to the extremely low representativeness of the images obtained in HRTEM. Because of the extreme localization HRTEM gives the image of only the smallest details in the area of a few square nanometers, which has little common with the average structure of the entire sample. At the same time, using phase contrast, TEM can be used to study the area of several square microns, or even hundreds of square microns whose structure already largely characterizes the studied sample.

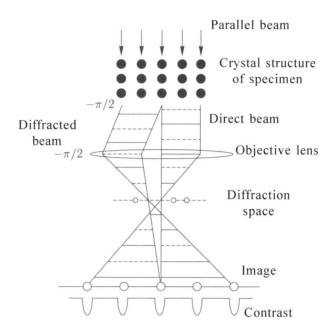

Fig. 8.4. A simplified scheme for obtaining the phase contrast from the crystalline sample in TEM. The objective aperture is either not used at all, or its width α is set greater than the angle of Bragg reflection 2θ by the crystallographic plane (*hkl*) of the observed crystal in the sample so that the beam reflected by Bragg reflection and the direct electron beam travel to the recording screen. The phase difference is formed by diffraction in the crystal and the objective lens. As a result, the phase difference equal to – π is obtained between the direct and diffracted beams. In real cases, the interference pattern can be much more complicated especially when several diffracted beams take part in the formation of the image.

the directly transmitted electron beam will pass through it and interact with one another because of their frequency coherence. The variable path length of different beams to the observation point is expressed in the difference of the phases of the interfering electron waves and creates an interference pattern in the image plane (Fig. 8.4), i.e. the phase contrast appears.

However, TEM imaging with phase contrast is not so simple as images with the mass and even diffraction contrasts. To do this, it is necessary to satisfy several additional conditions: chromatic and spherical aberrations (Shindo and Oikawa, 2006; Brandon & Kaplan, 2008; Reimer & Kohl, 2008) in the microscope should be kept to a minimum, the primary electron beam be monoenergetic (with a minimum spread of the electron energy); the conditions of coherence of the scattering on the sample should be satisfied.

In electron diffraction in the crystalline sample with a periodic lattice the diffracted beam shows the phase shift by $-\pi/2$ compared with the primary beam. When passing through the objective lens the electrons of this beam receive an additional phase shift by further $-\pi/2$. At a meeting

Fig. 8.5. Picture of Si nanowires in phase contrast (from Yang Leng, 2008).

of these electrons with the electrons of the beam transmitted through the sample without diffraction, interferences appear between them taking into account the phase difference and expressed by the alternation of light and dark interference fringes on the screen registering an image (see Fig. 8.4). The images of the fringes indicates the periodicity of the structure of the sample (see example in Fig. 8.5).

The interpretation of the image (Brandon & Kaplan, 2008; Reimer & Kohl, 2008; Shindo and Oikawa, 2006) in the phase contrast is more complicated than the interpretation of images in the mass contrast and phase contrast images are processed widely by mathematical treatment using the Fourier transform methods the consideration which goes far beyond the scope of this chapter.

8.1.2. Studying silicon particles by TEM and HRTEM

High-resolution transmission electron microscopy (HRTEM) is designed to study the structure of materials with atomic resolution and is particularly useful for the study of nanomaterials with periodic and non-periodic structures. In order to ensure this resolution, it is first necessary to get a clear two-dimensional projection of the crystal structure that can be done only at the highest stability of the electron-optical system of the microscope, while minimizing its aberrations, at high mechanical stability of the device itself, using ultra-thin samples. These samples are required due to the fact that the images for HRTEM are usually obtained in the phase contrast mode, and then processed mathematically, taking into account the contrast transfer function, but the theory of phase contrast is well enough satisfied only if the sample thickness is of the order of a small number of atomic layers. In the currently available TEMs with an accelerating voltage of 100 kV or more the images with a spatial resolution better than 0.2 nm are obtained quite easily, but it is much harder to correctly interpret the resulting images to identify the features of the structure material at the atomic level. It is the application of sophisticated methods for processing and interpretation of images that is a feature of HRTEM that distinguishes it from other methods of transmission electron microscopy. To achieve the atomic resolution, i.e. resolution of details of the structure with dimensions less than 0.2 nm, in modem HRTEM the signal transmission function from the sample $f(x, y)$ is 'cleaned' using widely the methods of information theory and mathematical methods of for image processing, the Fourier transform and Fourier filtering methods. A detailed description of the theory and practice methods HRTEM can be found in a number of modern textbooks and monographs, of which we can recommend, for example, Williams & Carter (2009); Spence (2009); Reimer & Kohl (2008), Shindo and Oikawa (2006).

The problem of obtaining TEM images of Si nanoclusters in the matrix, for example, SiO_2 or Si_3N_4, is not simple. Nanoclusters of Si in the Si/SiO_2 and Si/Si_3N_4 systems can be either an amorphous or crystalline phase. The difficulty lies in separating the images of the host matrix from Si nanoclusters, especially when nanoclusters are amorphous, because the difference in the atomic number and atomic density of these two environments is negligible. As a result, the contrast between the clusters of Si and SiO_2 or Si_3N_4 matrix in TEM micrographs is low. In this case, to avoid mistakes, it is better to carry out measurements in two ways: 1) remove the section of the sample in the bright field, and 2) to obtain the dark-field HRTEM image for the same section. True, we can observe Si nanoclusters, if they have crystallinity (i.e. they are Si nanocrystals).

The process of obtaining images depends on the orientation of nanocrystals, so the pictures do shown all the particles. We can observe only those that are oriented in a specific (correct) direction relative to the beam of the electrons incident on the sample. Because of this, the reliability of quantitative estimates of the density (concentration) of the content of Si nanoclusters in matrices such SiO_2/Si_3N_4 is always in doubt. In addition, in the quantitative estimates of TEM images, it is assumed that the sample is homogeneous, and the area of the image is representative of the entire sample. To observe the morphology of the sample in the TEM, bright-field and dark-field image techniques are used.

Images with a resolution comparable to the size of individual atoms, are obtained using the technique of high-resolution transmission electron microscopy resolution (HRTEM). As noted above, HRTEM works on the principle of phase contrast, i.e. a coherent superposition of the incident and elastically scattered electromagnetic waves from Si nanoclusters. The success of this method depends on the correct orientation of the nanoclusters with respect to the primary electron beam and the thickness of the studied sample. Poorly oriented nanoclusters are not visible on the image, and if the thickness of the sample is not comparable with the size of the nanocrystal, the contrast, required for observation of clusters, is generally not obtained. For this reason, the HRTEM method is generally not very effective for the quantitative determination of the density of Si nanoclusters in the matrix.

It should be mentioned that the analysis in the dark field also works only when the clusters are oriented well. However, the thickness of the sample for the dark field method is not so important as for HRTEM. But in the dark field mode a large magnification is not required, as in the case of HRTEM,

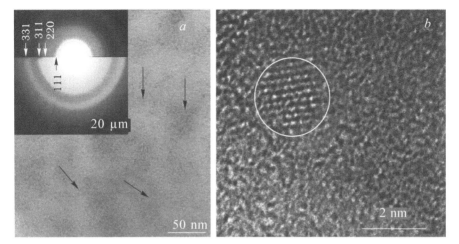

Fig. 8.6. TEM (*a*) and HRTEM (*b*) images of the sample with Si nanoparticles. The inset in (*a*) shows the electron diffraction pattern of the image containing the Si nanoparticles. The image (*a*) was produced in a LEO 912 AB OMEGA electron microscope with an accelerating voltage of 200 kV; HRTEM image (*b*) was obtained in a JEOL-JEM 2100 F/Cs microscope at a voltage of 200 kV. The circle shows the area of a separate Si nanocrystal (Dorofeev et al, 2011).

so the density of particles in the matrix can be clearly seen, and this method can be used to assess the concentration of particles in the matrix. Since the method of obtaining TEM images in the dark field also works only with crystalline clusters and does not identify amorphous clusters, the quantification of the concentration of clusters is usually obtained too low.

An example of studies of compositions based on Si nanoparticles using TEM is shown in Fig. 8.6.

It should be noted that at the huge number of reports on obtaining and studying nc-Si, the editors of the majority of international scientific journals before taking a decision to publish require confirmation by HRTEM that the material belongs to nc-Si, and not to any other forms of silicon.

8.1.3. Transmission electron microscopy with energy loss spectrometry

Energy-loss spectroscopy of electrons (EELS spectroscopy) or electron spectroscopy – EELS (electron energy loss spectroscopy) is carried out in an electron microscope. The principle of electron energy loss spectroscopy is based on the phenomenon which occurs in the interaction of the electron beam with the electrons of the substance in a transmission electron microscope. In interaction each electron of the primary beam can lose the characteristic part of its energy, and the magnitude of this loss depends on the properties of the substance. Thus, EELS is a variety of electronic spectroscopy, based on an analysis of the inelastically scattered electrons that have lost a fixed amount of energy in the process of interaction with the solid.

The term 'spectroscopy of characteristic electron energy losses (SCEEL)' has a double meaning. In general, this term is used to indicate methods for analyzing the electron energy loss throughout the range from 10^{-3} up to 10^4 eV. In spectroscopy and electron microscopy, for ease of review of the processes, the electron energy spectrum is conventionally divided into three parts:

1) low-energy – with energies ranging from fractions of eV to several eV;

2) the area of medium-energy electrons – from a few eV to about a dozen eV;

3) high-energy – from tens of eV to several tens of keV.

In this regard, the term SCEEL is often used in a narrower sense – for designation of research methods of the characteristic losses of only the electrons with energies ranging from several eV to several tens of eV, associated with the excitation of plasmons and electronic interband transitions. The group of the losses in region 2 is the subject of EELS deep level spectroscopy, and the losses in the region 3 are the subject of EELS high-resolution spectroscopy. The most frequent use of the SCEEL method, in the narrow sense of the term, is associated with the solution of such problems as the determination of the density of electrons participating in plasma oscillations, and chemical analysis of samples, including analysis of the distribution of elements in depth.

If the scanning transmission electron microscope is equipped with a device for measuring the EELS, then it can be used to construct maps of the chemical images of nanoclusters and the surrounding matrix. In this method, we cannot receive information about the atomic numbers and the distances between atoms but the chemical phases of the elements present in the sample can be seen. The biggest advantage of this method is the ability to track any type of morphology regardless of crystallinity or amorphous nature of the clusters.

In the SCEEL method, the main characteristic feature of the collective oscillations of valence electrons is the electronic energy loss in the quasi-particles, known as plasmons. For example, knowing the characteristic energy of the plasmons Si and SiO_2 (respectively 17 and 26 eV) and being able to measure the electron energy-loss spectrum with quite a narrow resolution, SCEEL images can easily distinguish between Si particles in the SiO_2 matrix.

Two methods are used for the processing of electron microscopic images obtained in the SCEEL mode: 1) analysis of the area of the sample using the direct image produced in transmission microscopy with energy filtering (EFTEM); 2) analysis of the indirect image, constructed from the measured EEKS spectra, extracting from them the distribution of these signals from the plasmon. This method of obtaining and analyzing images, which was formulated in the article (Jenguillaume & Colliex, 1989), is called the spectral imaging method in a scanning transmission microscope with parallel electron energy loss spectrometry (STEM–PEELS – SPEM–PSCEEL).

In this method, the sample area of interest to researchers is scanned by the electron probe of the microscope producing a superposition of different scanning trajectories. In the study of, for example, Si/SiO_2 these trajectories contain a signal from both Si and from SiO_2, therefore, for clear visualization of the Si particles it is necessary to cuf off the signal of SiO_2. This procedure can be performed either by linear interpolation or by the method of non-negative least squares (Schamm et al, 2008). The experimental data is sufficient to obtain both two-dimensional (2D), and three-dimensional (3D) images of distribution of Si nanoclusters in the SiO_2 matrix if we use a combination of plasmonic losses with electronic tomography produced in energy filtered scanning transmission electron microscopy (Gass et al, 2006). In this method, the 3D distribution of Si nanoclusters is reconstructed on a series of tilted micrographs (Frank, 1992), so the resulting image can be used to identify the form of nanoclusters (see Fig. 8.7).

8.1.4. Energy filtered transmission electron microscopy

Energy filtered transmission electron microscopy with energy filtering (EFTEM) in their spatial resolution (less than 1 nm) has the spatial resolution similar to HRTEM. But unlike HRTEM, the EFTEM method has no difficulty with the visualization of clusters of non-crystalline Si, and it can identify amorphous particles. Moreover, high resolution EFTEM energy gives able to distinguish between the plasmon energy, related to Si and SiO_2.

EFTEM is used to obtain images with a certain range of the electron energies. In studies of Si nanoparticles the filter window of electron energy centred on the energy loss of the Si plasmon equal to 1.6 eV. At this

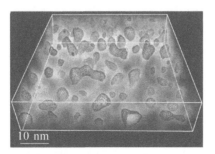

Fig. 8.7. Tomographic reconstruction of 3D images of Si nanoclusters in the SiO_2 matrix. The image shows that the majority of clusters are non-spherical. The figure from Yurtsever et al. (2006).

mode, this method allows to determine the Si nanoclusters regardless of th presence of the crystal structure and crystallographic orientation, but separately, for example, from the SiO_2 shell.

The original images obtained by means of EFTEM are less contrasting compared with, for example, images obtained by STEM–PEELS, but additional image processing can improve the contrast. The quality of images is improved by subtracting the background of the SiO_2 matrix from the original total image. The contrast of the EFTEM images is then comparable with the image of STEM–PEELS (see Fig. 8.8).

A detailed study of a particular sample by STEM–PEELS provides more information because this method allows quantitative analysis, separating the

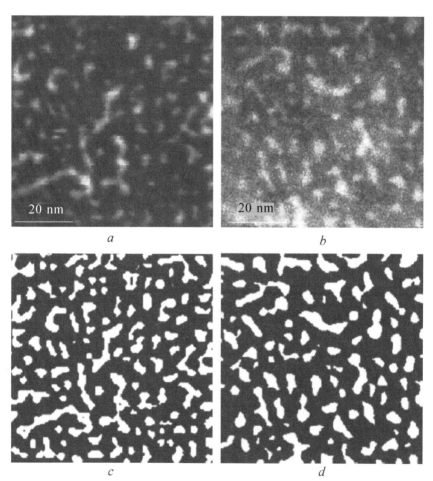

Fig. 8.8. The original images of Si nanoclusters dispersed in the SiO_2 matrix obtained by STEM-PEELS microscopy (*a*) and EFTEM (*b*), and the results of processing the images by reconstruction and intensification, respectively *c* and *d*. Example from Schamm et al. (2008).

individual particles, and even quantitative studies of their morphology in the 3D image. However, the procedure for obtaining images and, especially, their reconstruction of this method are more complicated than EFTEM in the method. At the same time, the process of obtaining and processing the image in EFTEM is much easier and faster, although the results are semi-quantitative. Therefore, EFTEM is more convenient for qualitative analysis of series of samples and for obtaining statistical data.

It should be noted that radiation damage can occur in studies by any methods of transmission electron microscopy (Egerton et al, 2004), which results in changes in the surface morphology and surface states. This problem is particularly acute in work with insulators, such as the study of Si nanoclusters in the insulating SiO_2 or Si_3N_4 matrix. Impact of the electron beam can produce damage such as displacement of atoms, ionization, electrostatic charge, electron beam sputtering of residual impurities of the vacuum on the surface of the sample and radiolysis. At higher exposure times and large doses electron radiation can cause in the sample a local chemical reduction of SiO_2 to Si (Schamm et al, 2008). These morphological changes may also occur at short exposures, if the electron current in the beam is higher.

8.2. Scanning probe microscopy methods

Diagnostics of the structure and shape of nanoparticles is carried out using almost the same experimental methods as for surface diagnostics, for example, in the physics and technology of semiconductors. Methods based on the sensing of surface by some agent acting on it are used mainly for this purpose. Typically, these are fluxes of particles or electromagnetic radiation, which are called primary and which, after interaction with the surface, carry information about its condition. The main requirement for these methods, as in the case of the study of the surface is low penetrating power or high sensitivity to the structure and properties of extremely small amounts of material. These requirements are satisfied in particular by the methods of scanning probe microscopy (SPM), such as scanning tunnelling microscopy and scanning atomic force microscopy, which have become in recent years powerful modern methods for studying the morphology and the local properties of solid surfaces with high spatial resolution.

8.2.1. Scanning tunnelling microscopy

In the early 1980s, a new method was developed for the direct study of the surface structure on the atomic level, known as

(STM). The idea of the measurement scheme used in STM, which includes a sharp metal probe (feeler), located a short distance δ_z above the surface of the object, the X–Y table of the sample holder and the potential difference with a feedback between the end of the probe and the surface sample belongs to Russell Young published in 1966. In 1971, while working at the National Bureau of Standards of the United States, Russell Young created on the basis of this idea a device, called Topografiner (Young et al, 1971; 1972), which is used as a profiler for precise control of optical diffraction gratings, although the resolution achieved in it (30 Å perpendicular to the surface and 4000 Å in its plane), allowed the study of the arrangement of atomic layers. A schematic diagram of the device was published in 1977.

The device consisted of a tungsten field emitter mounted on a piezoelectric platform. Electric current formed when placing the radiator at a distance of about 3 nm from the sample surface between the emitter and the surface which led to the release of the surface electrons and photons. A pair of detectors was used for the registration of these electrons and photons with movement along the emitter surface at a fixed distance between the transducer and the surface, allowing to convert the results of these measurements to the three-dimensional image of the surface. Typically, the vertical resolution was 3 nm, but sometimes Topografiner could reach a resolution of 0.3 nm (which is enough to distinguish on the surface steps with the height of a single atom). However, R. Young failed to achieve atomic resolution in all directions.

The scanning tunnelling microscope that can distinguish individual atoms was created and patented only in 1982. For the development of this method and a demonstration of its capabilities the developers G. Binning and H. Rorer (Binning, Rorer, 1988) were awarded the Nobel Prize in Physics in 1986. The principle of the method is briefly discussed, for example, in the textbook (Kaxiras, 2003, p. 388), in a review by Bakhtizin (2000), in a special monograph by Wang (2000) dealing with methods of control of nanomaterials and nanostructures, or in conjunction with other methods scanning probe microscopy in chapter 3 of the textbook by Di Ventra et al. (2004), as well as in textbooks by Mironov (2004) and Demikhovskii, Filatov (2007). More details methods of scanning probe microscopy, with examples of applications in their present state, are considered in a collective monograph by Bhushan (2010) fully dedicated to these issues.

The operating principle of a scanning tunnelling microscope (STM) is quite simple. The probe (the probe examining a sample surface) is a thin metal edge of the conical shape, mounted on an electromechanical drive (X-, Y-, Z-positioner). When this tip is brought to the surface at a distance $\delta_z < 10$ Å, then the application of a low bias voltage V_s (from 0.01 to 10 V) between the tip and the sample results in a quantum effect of tunnelling of the electron (Wolf, 1985), and tunnelling current I_t of about 10^{-9} A starts to flow through the vacuum gap δ_z. If the electronic states are localized on each atomic site, as often happens, the scanning of the surface of the sample in the X direction and/or Y with simultaneous measurement of the output signal in the circuit Z can produce a picture of the surface structure at the atomic level. This structure can be displayed in two modes: 1) either by the value of the tunnelling current at a constant distance δ_z from the tip to the sample surface, 2) or by change in the distance δ_z at a constant tunnelling current (the second mode is used more often). Using the concept of measurement by Topografiner (Young et al, 1971; 1972), STM is fundamentally different from it by the fact that the first device uses a substantially higher bias voltage V_s between the sample and the tip of the probe, at which an electric current through the gap occurs without the tunnelling effect. That is why P. Young was not able to obtain the atomic resolution with the help of his Topografiner.

It should be noted that the practical realization of this simple idea in the STM is not as simple as it is not easy to analyze raster images. Disregarding the fact that in order to obtain the atomic resolution it is necessary to move the sample or probe in the scanning direction, also with the atomic resolution, there is a problem of creating the angstrom-sized gap between the tip of the probe and the sample surface. In addition, it is obviously necessary to eliminate external vibrations, which influence the positioning of the probe relative to the sample. Therefore, despite the simplicity of the idea, STM is an extremely precise instrument. A rather stern challenge is the correct interpretation of the images which requires purification of initial data from experimental noise by complex mathematical treatment[2]. With the help of STM one can get a map of the surface topography at the atomic level, but only for conductive materials.

If STM was used initially to determine the structure of metal surfaces, now the objects of research are also semi-metals, semiconductors, adsorbed particles, and even biological objects that are no longer need (as in conventional electron microscopes) dehydration and covering with a thin layer of metal. But is should be remembered that in all cases the essential requirement is the presence of a highly conducting surface on which the object under study is placed. Improving the design of STM in the past quarter of a century has led to a truly fantastic resolution which reached 0.2–1.0 Å normal to the surface under study and in the plane of the sample. Equipment itself became simpler and 'slimmer'. The dramatically reduced size allowed this equipment to be easily integrated into other systems, such as scanning electron microscopes. Systems which can scan the surface at the speed of television scanning have been constructed. Thus, in contrast to the electron microscope, STM studies directly the surface, it does not destroy and does not alter the sample, is able to work not only in a vacuum, but also in the gaseous or liquid media (with minor losses in resolution compared with the vacuum) can be combined with other devices in order to obtain devices with qualitatively new characteristics.

As regards the application of STM to study nanosilicon, silicon was the first object and this resulted in rapidly developing interest in scanning

---

[3] The problem of purification of the STM signal is discussed in some detail by M.O. Gally-amov and I. Yaminsky: Scanning probe microscopy: Basic principles, analysis of distorting effects, 2001. Electronic publication is available online http://www.spm.genebee.msu.ru/members/gallyamov/gal_yam/gal_yam1.html.

tunnelling microscopy. This first study, published in a famous article (Binning et al, 1983), was devoted to solving one of the fundamental problems in surface physics – the explanation of the atomic structure of the unit cell of the reconstructed surface Si (111) 7 × 7. The corresponding STM images of the reconstructed surface are shown in Fig. 2.15a, b in Chapter 2 of the book with the notes to the established model of reconstruction.

A very important parameter in the tunnelling microscope is the bias voltage V_s and its polarity. The images obtained at different V_s, are consistent with different energy states, so they look different (Fig. 2.13a, b). Thus, the protrusions observed in the image of the filled states are due to tunnelling into the conduction band of silicon through the dangling bonds by adatoms while the depressions, visible in the image of unfilled states, are determined by the tunnelling of electrons from the valence band or localized states of silicon in the probe tip through the dangling bonds of restatoms atoms located at the corner pits. Thus, the STM is able to display localized electron states, in particular the distribution of the density of states in the direct space and the location of the levels on the energy scale. But this means that STM allows us to observe not the atoms themselves but the distribution in the space around the atoms of the electron density with different energies and gives not only the topography but rather the image of the electronic structure of the surface in the Fermi level energies.

Since the mid-1980s, attempts were made to apply the STM method to study of the electronic states in semiconductor quantum heterostructures. However, the first works to study the local density of states (LDS) in semiconductor heterostructures appeared only at the beginning of the XXI century. The reason for such a long delay were difficulties, first of all, with special requirements for preparing and monitoring the state of STM probes – both the geometry and chemical purity of the surface of the tip of the needle imposed in the study of the electronic structure of low-dimensional semiconductor heterostructures. These requirements were higher than in the study of the atomic structure and spectrum of the LDS of the surface of homogeneous semiconductors and metals (Demikhovskii, Filatov, 2007). Nevertheless, these problems were successfully resolved and it was then possible to visualize the spatial distribution of the LDS, i.e. scanning tunnelling spectroscopy (STS) was developed, which allowed, in effect, to visualize the envelopes of the wave functions of the size-quantized electronic states in quantum wells, quantum dots, quantum point contacts, and other equally interesting quantum-sized structures.

The essence of the scanning tunnelling spectroscopy is as follows. If the STM images are studied at different bias voltages V_s, or the feedback loop is temporarily disabled, we remove the dependence of the tunnelling current I_t on V_s at the constant value of the gap δ_z, then the difference image shows a picture of dangling bonds as well as of other electronic states corresponding to different energies, as in this case the tunnelling electrons will be involved with different energies (from the conduction band, valence band or localized states). In fact, here we measure the dependence $dI_s/dV_s = f(V_s)$, directly related to the local density of states near the Fermi level.

Thus, it has become possible to compare theoretical calculations of the energy spectrum and LDS in low-dimensional structures with the experimentally measured values. However, as a rule, such investigations are carried out at cryogenic temperatures (4.2 K and below). In addition, in the observation by STM/STS of the size-quantized states in semiconductor nanostructures there are special requirements for surface finish. Therefore, such studies are possible only in ultrahigh vacuum (UHV), with residual gas pressure of 10^{-10} Torr or less.

Despite all these advantages, unfortunately, the STM has a fundamental feature, which in most cases does not allow to apply this method to study the atomic structure of the surface of the synthesized Si nanopowders. This feature was noted even by the inventors of STM in their Nobel lecture in 1986 (Binning, Rorer, 1988), and consists in the fact that the atomic structure of the surface of the material can be properly visualized only when it is spotlessly clean. It is the presence of adsorbed impurities and oxides on the surface which hindered for a long time the correct reception of STM images of the reconstructed surface of c-Si (Binning et al, 1983). And as we know from Chapter 5 of this monograph, in most cases to ensure the stability of the synthesized Si nanoparticles their surface must either hydrogenated or hydrosilylated, or protected from degradation by some other coatings.

8.2.2. Atomic force microscopy

In 1986, somewhat later than scanning tunnelling microscopy, Binning et al (Binning et al, 1986) invented a scanning atomic force microscope (AFM). At the heart of AFM is the force interaction between the probe and the surface, which is used to record by special probe in the form of an elastic console with a sharp tip at the end. The force acting on the probe from the surface leads to bending of the console. By recording the amount of bending we can control the strength of the interaction of the probe with the surface. The theory and practice of the AFM method were discussed in detail, for example, in books (Mironov, 2004; Birdi, 2003; Bhushan, 2010)).

In the atomic force microscope, the interaction $A(X, Y, Z)$ is the force interaction between the probe and the sample. The nature of this interaction in the general case is rather complicated (Galyamov Yaminsky, 2001; Mironov, 2004), as it is determined by the properties of the probe, the sample and the environment in which research is conducted. In the studies of uncharged surfaces in the natural atmosphere (air) the main contribution to the force interaction of the probe and the sample is provided by: repulsive forces, caused by mechanical contact of the outer atoms of the probe and the sample atoms, the van der Waals and capillary forces associated with the presence of the adsorbate film (water) on the sample surface. In the case of charged surfaces, a significant contribution is introduced by electrostatic forces. When measuring the response of one or other of these forces, AFM can measure the different characteristics of the analyzed sample.

The main sensor of AFM is a microscopic elastic beam (cantilever), at the free end of which there is fixed a supersharp needle (probe) ending in an almost single atom. The probe glides over the studied surface and due to the interatomic force is attracted to the surface. The attraction force, expressed in the angle of inclination of the microbeam (cantilever) is recorded by the sensor, usually an optical one (laser – photodiode). No electric current is required between the probe and the surface. Maintaining a constant force of gravity by changing the height of the cantilever above the surface, it is possible to construct an atomic topography of the sample. For example, AFM (Erlandsson et al, 1996) was used to obtain the image of the atomic structure of the reconstructed surface of c-Si (111) 7 × 7, as shown in Fig. 2.24.

AFM, through its operating principle, can explore the atomic structure of the surface of any material, not only conductive, so is often used as an alternative method for TEM study of nanoparticles. With the help of this instrument, which is much simpler than the transmission electron microscope, one can also explore the clusters of Si in the SiO_2 matrix but gradual etching of the matrix to show the profile of clusters (Mayandi et al, 2006) is required.

It should be noted that the applicability of AFM to study the shape and external structure of the Si nanoparticles is limited by the same factors, coupled with the purity of the surface, which we discussed at the end of the previous section, dedicated to STM.

8.3. X ray diffraction analysis of nanopowders

X-ray diffraction analysis (XDA) is the primary method of determining the atomic structure of substances and different versions of this method have been used to decode the structure of hundreds of thousands of known compounds to date. The principle of XDA is to establish a mathematical correspondence between the experimentally measurable single crystal or powder sets of X-ray diffraction data and the electron density distribution in crystals of the material under study, and this technique is now well developed and published in many monographs and textbooks (see, e.g., Guinier, 1961; Woolfson, 1997; Will, 2006; Dinnebier & Billinge, 2008). Although the basic and the most accurate results in these methods can be obtained from X-ray diffraction in single crystals, after development by Hugo Rietveld (Rietveld, 1967; Young, 1995) of the method of analysis of powder diffraction patterns, which is known today as the Rietveld method, thousands of crystal structures have already been determined more accurately and even decoded from scratch using the diffraction patterns of powders. It would seem that this method can be most suitable for studying the atomic structure of nc-Si. Unfortunately, there are no similar studies in the scientific literature for the Si particles with sizes below 10 nm in which effects of quantum size restrictions, including photoluminescence in the visible region. For this there are objective reasons, which we briefly review here.

Figure 8.9 shows the X-ray diffraction patterns of Si powders, including the powder obtained by disproportionation of SiO at 200°C by the method described in (Dorofeev et al, 2010) and in section 5.4.2 of this monograph. Diffraction measurements were taken on a high-resolution X-ray diffractometer with the Guinier scheme (transmission focusing geometry): the Huber G670 camera (Gala et al, 2005) with a curved Ge (111) primary beam monochromator, cutting out a line $K\alpha_1$ of the characteristic radiation of the X-ray tube with a copper anode (wavelength $\lambda = 1.5405981$ Å). The diffraction pattern in the range of diffraction angles 2θ from 3° to 100° was registered by a plate bent around the circumference of the chamber (radius 90 mm) with an optical memory (a two-dimensional coordinate IP-detector), and the image from it was read by a special scanner with an angular resolution of 0.005° and converted into a one-dimensional intensity distribution of X-ray scattering depending on the scattering angle, i.e. the distribution I (2θ). Samples for measurement were prepared by the deposition and fixing with a special adhesive of a

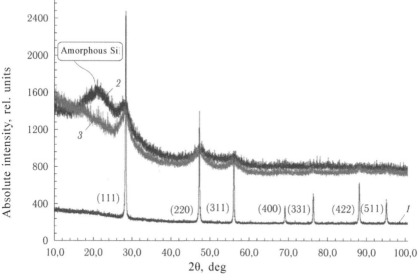

Fig. 8.9. X-ray diffraction patterns of Si powders with different crystallite size: *1* – diffractogram (exposure 30 min) of the reference powder of c-Si (Cline, 2000), in parentheses, in addition to the Bragg peaks there are appropriate interference indices *hkl*; *2* – diffraction pattern of the product (exposure 14 h) containing nc-Si, synthesized from SiO at 200°C (see Dorofeev et al, 2010), *3* – diffractogram (exposure 14 h) of the same product after etching in a solution of hydrofluoric acid.

thin layer of powder on the surface of the mylar film with a thickness of 6 mm.

As seen in Fig. 8.9, the reference powder c-Si (SRM 640 NIST) with an average crystal diameter of about 0.2 mm had high reflectivity and gave narrow Bragg peaks, strictly corresponding to a FCC Bravais lattice with an edge of the unit cell of 0.54311 nm (diamond structure, space group $Fd\bar{3}m$), characteristic of pure crystalline silicon. The reflectivity of the nanosilicon powders was much weaker, so to get from it diffraction patterns with clearly visible Bragg peaks the recording time was increased almost 30 times.

The diffractogram of the reaction product of obtaining nc-Si from SiO at 200°C shows quite clearly only three Bragg peaks and the positions of the maxima of these peaks coincide with the maxima of the peaks of the reference c-Si with the indices *hkl* equal to 111, 220 and 311. The presence of these peaks indicates the presence in the product of the crystalline phase of silicon. The diffraction pattern has a very high level of background and apart from the Bragg peaks of nc-Si with a very large width (greater than 2° with respect to the angle 2θ), it shows clearly a diffuse peak typical for the amorphous phase a-Si. This phase then dissolved and was removed by etching the product with a solution of HF (diffuse peak at the red diffractogram in Fig. 8.10 is absent), which had no noticeable effect on the width and position of the previously observed diffraction peaks, as well as the height of the continuous background, which is due to scattering of X-rays in the mylar film. However, by increasing the concentration of crystal phase in the sample after etching, except for the peaks 111, 220 and 311, the diffraction pattern showed clearly visible traces of the most intense high-angle peaks of crystalline silicon of 331, 433 and 511, which were previously weakened also by the amorphous phase present in the sample.

A very large width of the diffraction peaks indicates either the small size of crystalline blocks of nc-Si or the fantastically large deformations of their crystal lattice. The results of studies of this product by small-angle X-ray scattering, published in the paper (Dorofeev et al, 2010) yielded the maximum of the size distribution function at the diameter of nc-Si particles equal to 2.15 nm which favours the first hypothesis. The responsibility of the small size of the particles for broadening of the diffraction peaks is also confirmed by a simple estimate of the average block size of coherent scattering by the Scherrer formula and the mean relative deformation of the crystal lattice by Wilson's formula with simultaneous clarification with respect to the peaks 111, 220 and 311, taking into account the instrumental function of the diffractometer, calculated by the reference diffraction pattern[3] (see overleaf). As a result, the average diameter of the block of coherent scattering of nc-Si, equal to 2.5 nm at a strain of 0.8%, was obtained. Although the correctness of the values of the estimates, as will

---

[3]Details of the methods of determining the size of crystalline blocks and their deformations to X-ray diffraction data are in, for example, Warren (1969) and Iveronova and Revkevich (1978); Dinnebier & Billinge (2008), or chapter 10 of this monograph.

be shown later in chapter 10 of this monograph, causes a serious and legitimate criticism, the result still confirms that the synthesized material consists of very small silicon nanocrystals. Unfortunately, the resulting diffraction pattern does not identify all the peaks of the crystalline phase and therefore does not allow to clarify the crystal structure by the known classical methods based on the Bragg–Laue theory of x-ray scattering, in which the basic structural information is the intensity of Bragg peaks.

Why is it not possible to obtain good diffraction patterns from the Si nanopowders which could be used to refine by, for example, the Rietveld method the crystal structure of the nanoparticles? The reason can be understood, for example, by examining the results and conclusions of the studies in the works Kaszkur (2000a and 2000b).

8.3.1. Features of X-ray diffraction patterns of nanomaterials

In (Kaszkur, 2000a and 2000b) attention was given to the effect of the size of small metallic clusters on the intensity and shape of the Bragg X-ray diffraction lines. The studies were conducted using mathematical modeling of the atomic structure of the cluster, followed by theoretical calculations of powder diffraction patterns. Atomistic modelling was carried out by the minimization of energy by the method of molecular dynamics (MD) on the basis of many-particle potentials for transition metals with the FCC structure, and the powder diffraction patterns corresponding to the obtained clusters were calculated by the Debye formula (8.5). The comparative ease of packing of atoms in metal clusters allowed the author quite easily simulate the reconstruction[4] of the atomic structures on their surface and to consider changes in the structure reconstructed surface and the closest atomic layers in the adsorption of atmospheric gases. Thus, in these studies the authors considered Pd clusters ranging in size from 2.0 to 6.0 nm.

The validity of the obtained models and the results was confirmed by X-ray diffraction measurements on the powders of palladium nanocrystals (changes in the most intense peak with interference index 111 were monitored). As a result, it was shown that despite the visual similarity of the diffraction patterns of small nanoclusters with the diffraction patterns of polycrystalline powders of the same material, there are significant differences between them. First, with decreasing cluster size the diffraction peaks show different values of the lattice parameter, if they are indexed by the usual method applied to the diffraction patterns of polycrystalline samples. Moreover, since the applicability of a specific particle size crystallographic indexing of small clusters in general begins to lose sense, since the periodicity of the atomic structure is lost.

---

[4] For reconstruction of the surface of crystals see sections 1.1.6 and 2.3.3 of this monograph.

The evolution of the structure of nanocrystals in the process of gradual physical and chemical changes in the surface is manifested in the shifts of the peaks and changes of their form, width and the intensity of powder diffraction. The intensity of the scattering of X-rays by small clusters is approximately proportional to the number of atoms in the cluster, i.e. decreases very rapidly with decreasing particle diameter. In addition, the scattering intensity is strongly affected by the reconstruction of the atomic structure of the surface which results in the scattering of surface atom carrying away part of the intensity of the main Bragg peak to other areas the diffraction pattern corresponding to the reconstructed structure. At very small cluster sizes their structure undergoes marked amorphization. The maximum change in the intensity of the peak due to surface reconstruction of the cluster is approximately equal to the ratio of surface atoms to the total number in the cluster. So in the case of Pd clusters with the size of 2.0–6.0 nm, as discussed in the works (Kaszkur, 2000a and 2000b), reordering of the surface could lead to a decrease of the peak intensity by 20–50%. Computer simulation for Pd shows that the maximum loss of intensity of the peaks due to amorphization ranges from 10% for clusters with the size of 18 Å to 22% for clusters 37 Å in size.

8.3.2. Inapplicability of classical XDA to nanomaterials

An example of an unsuccessful attempt to apply the classical approach of XDA according to Rietveld with the use of the Laue–Bragg theory to clarify the structure of nanocrystals is shown in Fig. 8.10 (Neder & Proffen, 2008, chapter 9).

Diffraction patterns of nanocrystalline and conventional ultrafine powders ZnSe in Fig. 8.10 show a strong broadening of the diffraction lines in the transition to the dimensional range of the nanoparticles (upper figure). Refinement by the Rietveld method using the model of the unit cell of the wurtzite type (space group $P6_3mc$) shows marked differences between the model and the experimental diffraction pattern (bottom figure). Some features of the diffraction patterns are describes unsatisfactorily by this refinement. For example, it is clear that the shape of the profile of an interference peak 113 at an angle of $2\theta = 52°$ and the shape of all peaks at high diffraction angles are not well reproduced. The cause for this difference lies in the fact that the diffraction image of the nanoparticles is strongly affected by stacking faults and local disordering, as well as by deformation defects, which are almost impossible to take into account in the refined crystallographic model.

Another fundamental reason for failure of this attempt is the violation of one of the boundary conditions of the kinematic theory of diffraction which forms the basis of the classic formulas of X-ray diffraction analysis, namely, the conditions of the approximate infinity of the crystal. Proceeding from this condition it can be expected that with a decrease in the crystal size at

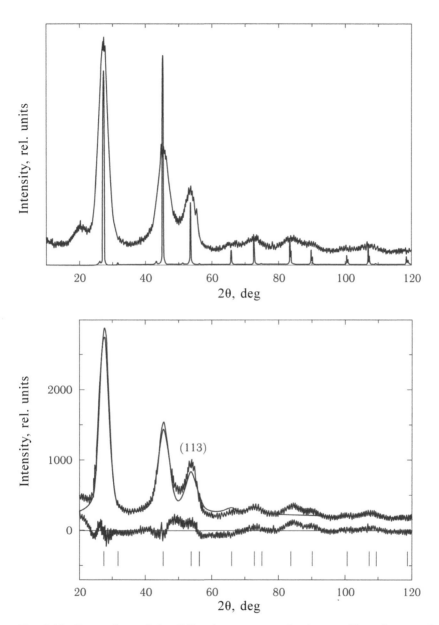

Fig. 8.10. Comparison of the diffraction patterns of polycrystalline (fine powder) and nanocrystalline (particle diameter of about 26 Å or 2.6 nm) samples of ZnSe. Filming took place at the same modes in Bragg–Brentano geometry with CuKα radiation (Figure from Korsunskiy, Neder et al, 2007). The lower figure shows the result of refinement of the structure of the ZnSe nanoparticles by the Rietveld method. The thin line shows the result of refinement, and the bottom curve is the difference between the experiment and the result of refinement (figure from Neder & Proffen, 2008, chapter 9).

some point it will be greatly disturbed, and all the calculations based on the Bragg–Laue theory, will be incorrect.

The results of many other studies (see, e.g. Heap, and Rempel, 2007; Heap, 2009; Tsybulya, 2004; Yatsenko and Tsybulya, 2008; Billinge, 2007; Gianini et al. 2007; Korsunskiy, Neder et al, 2007; Neder and Proffen, 2008) show that the diffraction patterns of nanostructured materials with block sizes less than 5 nm (of the order of 10–15 translations of the unit cell of the normal crystal) can not be described with sufficient accuracy only on the basis of the model of three-dimensional periodic structure and the crystallite size. That is, for nanosized crystals, starting with a certain minimum size, the structural distortions become so important that the unit cells are no longer similar to each other within the Gaussian statistics, and the representation of the intensity $I(\mathbf{H})$ of diffraction of the particle through the structure factor $F(\mathbf{H})$ usually used in X-ray crystallography ceases to be adequate:

$$I(\mathbf{H}) \sim \frac{V_p}{V_{cell}^2} \left| F(\mathbf{H}) \right|^2, \tag{8.3}$$

where V_{cell} is the unit cell volume; V_p is the volume of the diffracting grain; $\mathbf{H}$ is the reciprocal lattice vector defined by integer coordinates h, k, l. Here the structure amplitude is given by

$$F(\mathbf{H}) = \sum_{j=1}^{N_c} f_j \exp\left[2\pi \cdot i \left(\mathbf{H} \cdot \mathbf{r}_j \right) \right] \tag{8.4}$$

and contains all the information about the arrangement of atoms per unit cell, i.e. describes the crystal structure, built a three-dimensional translational multiplication of the unit cell. The summation in this equation is carried out over all N_c atoms in the unit cell, which are indicated by numbers j, and their coordinates in the unit cell are given by the vectors $\mathbf{r}_j$. The expression for the structure factor includes the atomic factors f_j, determining the ability of each scattering atom.

Based on the results of the above studies, it is understandable why there is such large difference of the intensity of Bragg peaks of nanosilicon and standard polycrystalline Si powder (Fig. 8.10), the diameters of the particles in which differ by almost two orders of magnitude. The reason is, first, a very small size of nc-Si, in which the crystal lattice can not even approximately be considered infinite, and the strong effects of surface reconstruction and possible amorphization of the particles that carry a substantial part of the intensity away from the Bragg peaks. This once again confirms that the traditional X-ray diffraction methods, based on the law of the Bragg reflections, are not quite applicable for the analysis of the structure of the material.

The structure of the nanocrystal is non-uniform and the nanoparticle can be considered as a two-phase formation consisting of a core and a shell, in which the positions of much of the atoms are displaced from positions corresponding to the correct lattice. Therefore, in the case of nanoparticles the very concept of the crystal and its underlying assumption of the infinite periodicity are unfair. As the number of surface atoms in the core and the reconstructed surface nanocrystal layer is comparable, and the interatomic distances due to the restructuring of the surface atoms in these two parts of the crystal differ, the structure of the nanocrystal can not be adequately described by a single set of lattice parameters, as is done in the case of microcrystals or microcrystalline powders. Therefore, the structure of nanoparticles should be studied by the methods of analysis not based on theory of Bragg scattering which is mainly used in X-ray diffractometry of the powders in classical X-ray analysis, and should be studied by methods based on other concepts.

8.3.3. X-ray diffraction analysis of total scattering patterns

Current studies of the structure of nanocrystals have been using actively used methods other than X-ray crystallography, which do not involve the spatial periodicity of the atomic structure. Among these methods, the most promising and physically well-grounded are the methods of analysis of total scattering diffraction patterns, such as the direct calculation of the diffraction patterns according to the Debye formula and refinement of the structure obtained by fitting the model to experimental function of total scattering (Debye, 1915; Zanchet et al, 2000; Giannini et al, 2007) as well as methods of pair interatomic correlation functions (Billinge, 2007). A detailed review of theory and practice of these techniques can be found, for example, in the books (Dinnebier & Billinge, 2008; Egami & Billinge, 2003).

With increasing disorder in the crystal to the state when the crystal can hardly be called a crystal (e.g. nanocrystals, or materials with the 'controversial' structure), more and more information is transferred from the Bragg peaks to diffuse scattering. In this case, the structure of the material is incorrectly characterized either by peaks of Bragg scattering or by diffuse scattering in isolation, but it is important to consider the entire scattering pattern as a whole, i.e. the diffraction pattern of total scattering, not dividing it into Bragg and diffuse parts.

In the methods of analysis of total scattering, as in the methods of X-ray analysis by the Bragg–Laue theory, we measure the distribution of intensity $I(2\theta)$ of the radiation scattered by the sample throughout the accessible range of diffraction angles 2θ, but analysis is carried out not only with the peaks of Bragg scattering but with the whole diffraction pattern, including areas where there is mainly coherent diffuse scattering on disordered systems of atoms. Thus, the entire coherently scattered radiation with a total intensity I_{coh} is analyzed without division into Bragg and diffuse.

The analysis of total scattering, which is based on the theory of x-ray scattering, proposed by P. Debye (Debye, 1915), does not assume any the frequency distribution of the atoms in the sample and, therefore, there is no need to search Bragg peaks on the diffraction pattern and it is enough to know the intensity of the elastic scattering at each point of its profile. The only thing that is required for the initial preparation of the experimental diffraction data for analysis is to remove from them the background and inelastic scattering contributions (with a change in wavelength), leaving only intensity $I_{coh}(2\theta)$, relating to coherent scattering which are included in the formulas of the Debye theory. To do this, from the measured diffraction pattern it is necessary to subtract the background effects introduced by the external and instrumental factors, and also inelastic effects, such as, for example, Compton scattering.

Thanks to modern advances of computer technology and computer mathematics, for the analysis of diffraction data from nanostructured materials in the interpretation of diffraction patterns of total scattering it is now possible to use the long-established approach to the interpretation of X-ray diffraction in powder samples, proposed by P. Debye (Debye, 1915, see also James, 1950, Guinier, 1961; Warren, 1969; Iveronova and Revkevich, 1978; Hall, 2000; Cervellino et al, 2006).

XDA of nanomaterials based on the Debye formula
In contrast to the widely used the Laue–Bragg theory of X-ray diffraction, which addresses only the Bragg peaks of the scattering intensity and to which the above formula (8.3) applies, the Debye theory of diffraction does not use the crystallographic constraints, and considers the scattering particle as an object consisting of N atoms, where each j-th atom can occupy any position in the object, which is characterized by the radius vector $\mathbf{r}_j$. Interferences of diffracted X-rays from any pair of atoms are considered. From thermodynamic considerations it is assumed that the probability of the existence of pairs with the same interatomic distances in the scattering medium is high, and the probability of triatomic and higher configurations with the same coordination, which could form the constructive interference, is negligibly small. With this approach to the description of diffraction it is not necessary to use the concepts crystal and the unit cell, as well as symmetry, so this analysis is equally applicable to crystalline, amorphous and even liquid materials, and remains valid for particles of any size.

The rest of the Debye theory (Debye, 1915) uses standard assumptions for kinematic[5] approximation. We consider the flow of parallel flying monoenergetic photons with momentum $|\mathbf{q}_0| = \hbar 2\pi / \lambda$ (ie. the initial flat monochromatic wave with the length λ), which are elastically $\left(|\mathbf{q}_0| = |\mathbf{q}| = \hbar 2\pi / \lambda\right)$ scattered on the atoms of the object the positions of

[5]The boundary conditions of kinematic approximation in the scattering theory of X rays see, for example, (James, 1950; Warren, 1969; Iveronova and Revkevich, 1978; Fetisov, 2007a)

which are given by the vectors r_j. As in the Laue–Bragg theory, only single scattering is taken into account here and the lack of interaction of the incident wave with the diffracted waves is assumed, but there is no need for assumptions about the infinity and the periodicity of the investigated environment. Further, the pairwise addition of complex amplitudes $A_j(Q)$ of radiation waves elastically scattered (diffracted) by atoms is carried out in accordance with relevant phases, and the intensity $I(Q)$ of scattering by the entire N-atomic ensemble is determined in the usual way through the square of the resulting amplitude $A(Q)$, obtained by exhaustive search of all pairs of atoms of the system. For an isotropic sample (for example, a powder with the random orientation of the particles), the result can be averaged over all directions. If we consider the scattering intensity independently of the intensity of the flux of incident photons, i.e. the scattering intensity divided by the intensity I_0 of the primary beam, then after averaging the resultant sum over all orientations we get the function of total coherent scattering I_{coh}, described by the Debye equation (Debye, 1915)

$$I_{coh}(Q) = \sum_j^N f_j^2(Q) + \sum_j^N \sum_{k,k \neq j}^N f_j f_k \frac{\sin(Q.r_{jk})}{Q.r_{jk}}. \tag{8.5}$$

Here, the intensity is normalized to the intensity of the primary beam I_0, j and k label the atoms of the sample in these pairs, $r_{jk} = |\mathbf{r}_j - \mathbf{r}_k|$ – modulus of the vector of the distance between the atoms j and k; $Q = |\mathbf{Q}| = |\mathbf{q} - \mathbf{q}_0| = 2|\mathbf{q}|\sin\theta = \hbar 4\pi(\sin\theta)/\lambda$ if the modulus of the vector of momentum transfer in the scattering of a photon, where 2θ is the diffraction angle, i.e. the angle between the directions of movement of the photon before the collision with an atom and after scattering. The modulus of the vector $\mathbf{Q}$, expressed in units of $\hbar$, i.e.

$$Q = 4\pi\frac{\sin\theta}{\lambda}, \tag{8.6}$$

is commonly referred to as the length of the vector of elastic scattering or the length of the diffraction vector.

It should be noted that diffraction patterns of X-ray scattering are usually obtained in the form of the intensity distribution of scattered X-rays depending on the scattering angle 2θ, which is measured by the diffractometer, so the graphics of the experimental data obtained from one and the same sample in radiation at different wavelengths λ will differ in stretching on the 2θ scale and, in fact, the same scattering functions are difficult to compare. For ease of comparison, diffraction patterns are often recalculated and are shown as depending on the length of the scattering vector Q, which does not depend on wavelength. This means the representation of the scattering intensity as a function on the magnitude of the change of the momentum of a photon in elastic scattering.

Obviously, formula (8.5), under the terms of its derivation, allows us to analyze diffraction on a particle with an arbitrary atomic structure, and the case of translational symmetry it is only one of special cases. This formula does not differences between the Bragg and diffuse scattering, and the same processes diffraction (reverse) the space of all points of measurement, i.e. it describes all points of the diffraction pattern and completely all of the elastic scattering of X-rays of the studied system of atoms. As a result of the calculations based on a set of experimental points of the function of the total coherent scattering using the equation (8.5) it is possible to determine the coordinates of all atoms in the system, which further can be compared either in the case of amorphous substances with a model of the continuous random network (see section 2.4), or with the crystal lattice, if we consider the crystalline material.

It is clear that at a large number of atoms, i.e. for sufficiently large particles, to determine the atomic coordinates using the formula (8.5) we must solve a system of a very large number of equations with a large number of independent parameters. However, due to the advent of modern computers, calculations by this formula for the nanoparticles are quite real, and a number of algorithms for these calculations and analysis of diffraction patterns of nanostructured materials have already been developed. We should note, for example, algorithms and computer programs discussed in the works Neder & Proffen (2008); Cervellino et al. (2006); Yatsenko and Tsybulya (2008); Vorokh (2009). However, it should be mentioned that today the program of structural analysis of particles with the size <100 nm by the Debye formula is only in the development stage and the practical application of this method is so far mostly limited to the exact calculations of X-ray diffraction patterns for materials with a given structure (see, e.g. Neder & Proffen, 2008), where it is used widely and successfully.

The method of paired correlation functions
While the methods of structural analysis by the Debye equation are still being developed, the analysis of the structure of nanomaterials is widely carried out using the well-designed (see Guinier, 1961; Warren, 1969, pp. 22–34; Iveronova and Revkevich, 1978) method of paired distribution functions (PDF) which was previously used for many years in the X-ray analysis of gases, liquids, glasses, amorphous materials, and other disordered systems. This method, like the Debye method, works with the diffraction patterns of total coherent scattering $I_{coh}(Q)$.

The method uses the property of the Fourier transform to establish the relationship between the measured diffraction intensities and the location of pairs of atoms in real space. The PDF considers the distance r between pairs of atoms and can be directly determined in real space using the atomic coordinates. It can also be written in terms of the Fourier transform of the intensity of scattered X-rays. In contrast to the crystallographic methods, in the method PDF, as well as in the above method using the Debye formula, on which the PDF method is actually based, there is no assumptions about the frequency that allows us to study non-periodic structures or aperiodic modifications, and at the same time the crystallographically interpreted materials. This feature is clearly visible, because the peaks in the PDF are obtained directly from pairs of atoms in solids. For example, if a peak on the PD graph shifts toward smaller r, then this directly means that in the given pair of atoms the link is shortened. For this reason, the directly measured PDF gives a very good quantitative 'template', through which we can quickly check intuitive chemical assumptions.

For analysis by the PDF method, the data $I_{coh}(Q)$, derived from experiments, are usually converted by normalization to a dimensionless scattering function $S(Q)$, expressed in units of scattering by a single atom. This representation allows us to compare with each other interpret in the same way not only the data obtained in X-ray diffraction measurements on the radiation with different wavelengths but also the data obtained for the same material using the diffraction of X-rays, neutrons, or electrons, as well as on different devices.

Consider the theoretical foundations of the method of structural analysis using the pair correlation functions in general terms, omitting the derivation and analysis of accurate formulas, which can be found, for example, in the article by Farrow & Billinge (2009) or in chapter 11 of the outstanding book by Guinier (1961). Here, for brevity wee confine ourselves to the case of a single-element material, such as nanosilicon, although this method is also applicable to multi-element materials if we use common formulas discussed in detail in the aforementioned literature.

The experimental basis for analysis by the PDF method is the total scattering function $S(Q)$, obtained from experimentally measured powder diffraction patterns $I_m(Q)$ by the well known methods (Guinier, 1961; Bish & Post, 1989; Egami & Billinge, 2003; Dinnebier & Billinge, 2008). To obtain it from the experimentally measured diffraction patterns, all $I_m(Q)$ should be normalized to the intensity of the primary photon flux I_0, accurately deduct the background from it, correct for polarization and subtract the contribution of inelastic (incoherent) scattering associated mainly with Compton scattering, then the resulting diffraction pattern $I_{coh}(Q)$ must be divided by the total cross section of scattering by the sample, i.e. by the number of scattering atoms N and the average scattering power of per atom, which in the case of X-rays is expressed by the square of the atomic scattering factor $f(Q)$. As a result, if the X-ray diffraction measurements were performed on the radiation with a wavelength sufficiently far from the edges of absorption of the atoms of the material, we can neglect the effects of resonant scattering (see Fetisov, 2007, sections 1.8.2.2 and 4.3.2), the function of total coherent scattering $S(Q)$ is given by

$$S(Q) = \frac{I_{coh}(Q)}{N\langle f(Q)\rangle^2}. \tag{8.7}$$

With this normalization, $S(Q)$ is a dimensionless quantity, and its average over all values of the scattering vector $\langle S(Q) \rangle = 1$. But the scattering function $S(Q)$ in fact continues to be the powder diffraction pattern of elastic scattering of X-rays, free of background, experimental errors and the contribution of incoherent scattering, and the intensity values in this function do not depend on the volume of the sample. In fact, $S(Q)$ is a full structure factor of the sample, with the meaning identical with that of the structure factor F^2 in the formula (8.3), but describes the distribution of atoms not in one elementary cell but in the entire sample.

Taking into account the Debye function (8.5) and the equations (8.7), the reduced function of total scattering changes to

$$F(Q) = Q\left[S(Q) - 1\right], \tag{8.8}$$

which is interesting because it reflects only a particular arrangement of the atoms in the sample for which it is often called the structure function of the sample and, due to the normalization factor Q, does not decay with increasing Q as quickly as function $S(Q)$, which is shown by a real example in Fig. 8.11.

The scattering intensity in the diffraction pattern is determined by the values of atomic factors $f(Q)$, which decrease rapidly with increasing Q, which is expressed in a rapid decrease in the intensity of the diffraction patterns with increasing length of the scattering vector. But the double

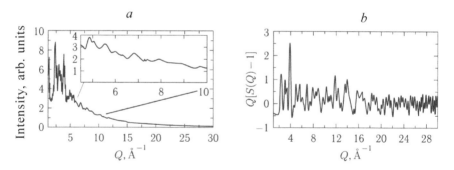

Fig. 8.11. Comparison of experimental total scattering diffractograms $I_{coh}(Q)$ (*a*) with the corresponding reduced function of total scattering (*b*). Measurements of the diffractogram of the exfoliated WS_2 sample were taken at a wavelength $\lambda = 0.202$ Å ($E \sim 60$ keV) from the synchrotron light source CHESS. The inset in the figure *a* on the large scale shows an area of the diffraction pattern from 5 to 10 Å^{-1} to show that it contains characteristic intensity variations, which are not noticeable at less detailed examination of the diffractogram. This figure illustrates the suppression of the intensity and characteristics of its variation at high Q under the influence of the angular dependence of the atomic form factors. The same data after conversion to the reduced scattering function (*b*) clearly reveal significant structural variations due to the intensity of diffuse scattering at large Q, up to 20 Å^{-1}. (Billinge & Kanatzidis, 2004).

normalization (by dividing the scattering intensity by the same small values of $\langle f(Q) \rangle^2$, and multiplication by Q) enhances the high-angle part of the diffraction pattern, and weak features of the coherent scattering at high diffraction angles are amplified and are clearly marked on the graph as functions $F(Q)$, which is evident from a comparison of Fig. 8.11a and 8.11b.

It is known (James, 1950; Guinier, 1961; Vasiliev, 1977; Iveronova and Revkevich, 1978) that the inverse Fourier transform (Fourier inversion) transforms the optical signal (in this case it is contained in the function of total scattering and hence the function (8.8)) in its prototype in the metric space, and the converse is also true, i.e. the direct Fourier transform converts the function of the prototype to the optical signal. Hence the use of the Fourier sine-inversion for the reduced total scattering function (8.8) gives the function (Keen, 2001; Dinnebier & Billinge, 2008):

$$G(r) = \frac{2}{\pi} \int_0^\infty Q\left[S(Q) - 1\right] \sin(Q \cdot r) dQ, \qquad (8.9)$$

describing the variation of atomic density in the metric space, which is called the reduced pair distribution function (RPDF). This function is essentially a function of the probability distribution in the direct space and defines the measure of the probability of finding in the studied material a pair of atoms located at a distance r from each other. Function $G(r)$ has the dimension of Å^{-2}.

The function $S(Q)$ and hence $F(Q)$ cannot be measured from zero to infinity. However, the function $S(Q)$ falls off rapidly, approaching zero as large Q and near zero (see Fig. 18.12a). Therefore, in practice it is measured in the available range from Q_{min} to Q_{max}, which should be the widest possible (not less than 30 Å^{-2} on the high-angle side), and the integral (8.9) is calculated in these limits. With proper choice of the integration limits this method provides sufficiently high accuracy of determination of the function $G(r)$ without the loss of structural information.

An example of conversion of the reduced total scattering function $F(Q)$ to the RPDF $G(r)$ for the case of a crystalline material (nickel powder) is shown in Fig. 8.12.

On the other hand, the same function of the atomic pair distributions, averaged over all orientations of the sample, is expressed in terms of atomic density by the relationship (Guinier, 1961; Warren, 1969; Billinge & Kanatzidis, 2004):

$$G(r) = 4\pi r\left[\rho(r) - \rho_0\right], \qquad (8.10)$$

where ρ_0 denotes the average density of the number of atoms (the number of atoms per unit volume of the material, i.e. the concentration of the atoms); $\rho(r)$ – the density of the number of atomic pairs (concentration of atomic pairs); r – the radial distance between the atoms in this pair.

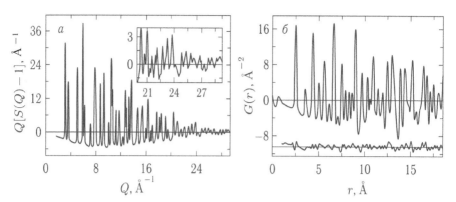

Fig. 8.12. An example of the transformation function of the total scattering in the IDF. a – Reduced structure function $Q[S(Q) - 1]$ for crystalline Ni. The measurements were performed on the radiation from a channel ID-1 of the synchrotron radiation source APS in Argonne National Laboratory. b – RPDF obtained by the Fourier transform of data from a. From an article (Billinge & Kanatzidis, 2004).

Therefore, the function $G(r)$ obtained from the experimentally measured diffraction pattern of total scattering by the transformation (8.9) gives information on the number of atoms in a spherical shell of unit thickness, located at distance r from the reference atom taken as the origin, i.e. gives the radius of the coordination sphere and coordination number. It forms peaks at characteristic distances separating pairs of atoms.

Thus, the PDF method makes it possible without solving complex structural problems, as in the classical Bragg–Laue X-ray analysis or most complex computing as in the method of structural analysis by the Debye formula, to obtain by a relatively simple mathematical operation (Fourier transform) from the experimentally measured diffraction patterns of total elastic scattering information on the radial distribution of the atoms of the material in the usual metric space, which has an obvious physical meaning. The resulting $G(r)$ function can be considered as the experimentally measured function because it is just another form of representation of the experimentally measured diffraction data not in the diffraction but in the usual laboratory space.

In almost visual analysis of $G(r)$ we can immediately obtain the following valuable structural information:

1) interatomic distances (the positions of the peaks of the function);
2) coordination numbers (peak areaa);
3) particle size (including at the values of <2 nm);
4) the mean-square displacements of atoms (the width of the peaks).

In addition, the PDF graphs can be used to estimate deflection of the structures of nanoparticles from the structure of the ideal crystal.

The family of pair correlation functions used in structural analysis
In addition to $G(R)$, the methods of structural analysis in PDF use the whole family of distribution functions (see Keen, 2001; Billinge, 2008), which differ in the way of normalization and are linearly related to each other, but each of them has characteristics that are useful in certain cases.

First, we have already noted the density function of pair distances (DFPD) $\rho(r)$, which determines the density of the number of atom pairs, depending on the interatomic distances and is associated with $G(R)$ by equation (8.10).

Secondly, it is the function $g(r)$, called the pair distribution function. The normalization of this function is such that $g(r) \rightarrow 1$ at $r \rightarrow \infty$, so it has the property that at the lengths r is shorter than the possible distance between adjacent atoms, $g(r)$ vanishes. This function is related to the density function of pair distances $\rho(r)$ by the relation

$$\rho(r) = \rho_0 g(r). \tag{8.11}$$

Therefore, it is clear that $\rho(r)$ should oscillate about the function $g(r)$, asymptotically closer to the average atomic density of the material ρ_0 for large values of r and turning to zero at $r = 0$.

This function of paired distances $G(R)$ can be expressed through these correlation paired functions by the equation

$$G(r) = 4\pi r \left[\rho(r) - \rho_0 \right] = 4\pi r \rho_0 \left[g(r) - 1 \right], \tag{8.12}$$

and, respectively, with the help of this equation, the experimentally measured function $G(R)$ can be transformed into any of these functions.

Finally, we should note the radial distribution function of the atomic density (RDF) well-known in statistical physics and structural chemistry which is expressed by the equation

$$R(r) = 4\pi r^2 \rho(r), \tag{8.13}$$

where ρ is the volumetric concentration of atomic pairs with a characteristic interatomic distance r. The radial distribution function $R(r)$ is interesting since it is associated more than other correlation functions with the physical structure of the material, as $R(r)\,dr$ gives directly the number of atoms in a spherical shell of thickness dr, located at a distance r from the given selected atom. For example, the coordination number N_c, i.e. the number of nearest neighbors, is expressed through it as

$$N_c = \int_{r_1}^{r_2} R(r)\,dr, \tag{8.14}$$

that is defined by the area of the peak corresponding to the considered coordination sphere and located in the range from r_1 to r_2. Among the considered paired correlation functions, the reduced function of paired distances $G(r)$ stands out, first, as it is the direct Fourier transform of the experimental curve of the total scattering. This function oscillates around zero and gradually decreases to zero at high values of r (see Fig. 8.12 b). It also decreases with the approach of r to zero, and the tangent of the slope of this decline is $-4\pi\rho_0$, where ρ_0 is the value of the average atomic density of the material.

From a practical point of view the function $G(r)$ is interesting in that the random errors in the experimental data related to the measurement process, inevitably present in $G(r)$, are constant for all r. That is, the fluctuations of the difference between the calculated and measured values of $G(r)$ have the same value for all r (Billinge, 2008). So, for example, if the fluctuations, observed in the difference curve, decreases with increasing r, then it means that the structural model used in the calculations is not equally good in description of the near- and long-range order and better agrees with experiment at large interatomic distances (it is possible that the chosen model reflects mainly the average crystallographic structure). Such a conclusion can not be made directly on the basis of the difference curve of comparison with other experimental correlation functions (such as $\rho(r)$ and $g(r)$), different from $G(r)$ by normalization.

Another advantage of the function $G(r)$ is that the amplitude of the oscillations gives a direct measure of the structural coherence of the structure of the sample. Thus, in a crystal with an ideal coherence the oscillations of $G(r)$ extend to infinity with a constant peak–peak amplitude. Although in actual measurements the amplitude of the signal $G(r)$ peak-to-peak gradually decreases due to the finite resolution on Q, which is likely to limit the spatial coherence of the measurements and not coherence of the structure itself. The increased resolution with respect to Q enables measurements over a wider range of lengths r. In samples with a degree of structural disorder the signal amplitude $G(r)$ decreases faster than dictated by the magnitude of resolution in Q, and this becomes a useful measure of the structural coherence of the sample. This effect, for example, can be used to measure the diameter of the nanoparticles (Billinge, 2008).

Structural analysis using RPDF. The most detailed method of structural analysis using the distribution function of paired interatomic distances is discussed in the book by Egami & Billinge (2003), which is entirely devoted to this method and its application to crystalline materials in violation of the order and nanomaterials. Structural analysis using correlation functions of the interatomic distances is carried out by calculating the theoretical function $G(r)$ for the pre-selected model of the structure and specification of the model parameters in comparison of theoretical and experimental RPDF by the method of least squares, just as is done in the Rietveld method. In contrast to the usual structural analysis by the Rietveld method of powder diffraction patterns, in this case both the construction of models and analysis are carried out in direct space. Analysis of the data in direct space is more comfortable, clearer and simpler than the interpretation of diffraction patterns, especially in the case of materials with significant structural disorder (as in the case of nanoparticles).

The theoretical reduced function of atomic paired distributions is calculated by the formula

$$G(r) = \frac{1}{r} \sum_{j} \sum_{k} \frac{f_j(0) f_k(0)}{\langle f(0) \rangle^2} \delta(r - r_{jk}) - 4\pi r \rho_0, \tag{8.15}$$

where the sums are computed over all the atoms of the sample. Here ρ_0 is the average atomic density, δ is the Dirac delta function, r_{jk} is the distance between atoms j and k, constants $f(0)$ are the values of the atomic form factors at $Q = 0$, which according to the properties of the atomic scattering factor of X-rays are almost equal to the number of electrons Z in the corresponding atoms; $\langle f(0) \rangle$ is the mean value over the atoms of all elements of the composition of the sample. The experimental function $G(r)$ is determined from the total scattering structure function $F(Q)$ by (8.9) with account of (8.8) with the integration between Q_{min} to Q_{max}.

A number of regression computer programs has been developed for the structural analysis of the function $G(r)$. For example, the program PDFfit (Farrow et al, 2007) uses the method of least squares in the same way as is done in refining structures by the Rietveld method. There are programs that are based on the inverse Monte Carlo method (e.g. program RMCA (2005)[6], as well as a program RMCProfile (Krayzman et al, 2008) for structural analysis when used in conjunction RDF and EXAFS data), where the difference function is minimized using a simulated annealing algorithm, or a certain potential, constructed on the basis of circuit simulation. These approaches are discussed and compared in the book (Egami and Billinge, 2003).

Examples of application of the paired correlation functions to study the structure of silicon are presented in Section 2.4 of this monograph (see Fig. 2.18).

Features of PDF measurements. Although the PDF are actually X-ray powder diffraction patterns, experiments with them for measurements for subsequent correct structural analysis using the Debye equation or by using the paired correlation function is significantly different from the usual collection of diffraction data for X-ray diffraction analysis of microcrystalline powders.

The first major difference stems from the fact that in the RDF all the points of the function contain important structural information, so the scattering intensity in them must be measured with the same reliability, whereas is structural analysis by the Bragg–Laue theory the basic value are Bragg peaks and the intensities between the peaks belong to the background,

---

[6]The instruction is available on the Internet http://www.isis.rl.ac.uk/RMC/downloads/rmca. htm

and the accuracy of their measurement can be donate neglected. As a result, when measuring the RDF there is a paradoxical situation – the lion's share of time is spent on the measurement in interpeak intervals, where the normal X-ray analysis measurements are carried out very quickly.

The second difference is that the correct analysis of the formulas (8.5), (8.9) and (8.15) requires the most complete set of data, otherwise structural information may be lost or distorted by a break in the set, so that the RDF should not be measured in the infinite range, but in the widest possible range of Q. Even if the high-angle region (at large Q) does not contain Bragg peaks, it may show features of coherent scattering (e.g. diffuse) associated with the structure of the sample. For correct structural analysis we should measure $S(Q)$ or $F(Q)$ in the whole range of Q, where structural features of coherent scattering are evident. As is well known (Guinier, 1961; Warren, 1969), these features are damped with increasing Q due to violations of coherence, related to the Debye–Waller factor due to thermal vibrations of atoms and static distortions, contributing a disorder in the arrangement of atoms. Depending on the temperature and the rigidity of the material under study, the structural features no longer appear in the Q values of the order of 30–50 Å^{-1}. For example, the RDF, as shown in Figs. 8.11 and 8.12, these features disappear at Q of the order of 25–30 Å^{-1}. However, this limit is not available for measurements in laboratory diffractometers in which is the hardest radiation is the radiation from the tube with a silver anode, as when working with radiation Ag Kα ($\lambda \approx 0.56$ Å) $Q_{max} \approx 22$ Å^{-1}.

Acceptable results can be obtained on Bragg–Brentano laboratory diffractometers with the characteristic radiation Ag Kα, However, the measurement process will be very long, because measurements should be taken in the range of diffraction angles of about 160 degrees in increments of about 0.01° and with an exposure of at least 100 seconds (see Heap, 2009) at each step (i.e. almost 19 days). Therefore, for correct structural analysis by the pair correlation functions or Debye equation the IDF is usually measured at the synchrotron radiation with energy >45 keV ($\lambda < 0.27$ Å) and to 100 keV ($\lambda = 0.12$ Å). In addition to a wide range of measurements of Q, the advantage of synchrotron radiation is that it is millions of times brighter than the emission of the X-ray tube, which allows to reduce the measurement time a thousand times, and in studying of a cylindrical sample by the Debye–Scherrer method with a coordinate detector, covering a full range of diffraction angles (Fetisov, 2007b), the PDF can be measured even within a second, and in some cases even faster.

8.4. Structural analysis using absorption spectra of X-rays

XAFS spectroscopy is a method of analysis of the atomic structure of materials, based on the study of the absorption spectra of X-rays in the neighbourhood of shock absorption (resonant photoabsorption) of a chemical element (called the reference atom) from the test material. XAFS is the abbreviation of the X-ray absorption fine structure.

This relatively new method of structural analysis required for measurement subtle variations in the wavelength λ (or the corresponding photon energy E) of X-ray emission in a rather wide range, so almost entirely owes its rapid development of the appearance of synchrotron radiation sources, without which its experimental implementation is very problematic. Therefore, the vast majority of the centres of scientific development of XAFS-spectroscopy and its practical application in the world are focused on synchrotron radiation sources or research organizations working closely with these sources and actively use their capabilities. Fundamentals of the theory and practice of the method can be found, for example, in a recently published special monograph (Bunker, 2010) or in a review (Fetisov, 2007b).

8.4.1. Principles and methods of XAFS analysis

Structural analysis of XAFS spectra compares the X-ray absorption spectrum of a monatomic gas, which has no structure, with the absorption spectrum of a condensed medium, which has, at least, a short-range order. The differential spectrum obtained from this comparison carries information about the chemical structure of the substance. This information is reflected in the fluctuations of the absorption spectrum which show a fine structure, which is expressed in the form of waviness of the spectrum near the absorption 'jump' of the selected atom of the substance, as well as in chemical shift of the energy threshold of excitation of the absorption jump. From the analysis of XAFS spectra we can obtained complete information about the atomic structure of the local atom cluster in the vicinity of several coordination shells around the excited atom, including atomic positions, bond lengths, bond angles and parameters of thermal vibrations (Debye–Waller factor). In addition, it is possible to determine the valence of the atoms, the electron density distribution and the band structure of matter.

The carrier of structural information in the XAFS spectra is the coefficient of X-ray absorption, which is easily measured experimentally, and it can equally easily measured for the substances in any aggregate state, unlike X-ray diffraction patterns. Therefore, XAFS-spectroscopy successfully completes the X-ray diffraction analysis and, in some cases, has an absolute advantage over it and replace it where X-ray diffraction analysis is almost 'powerless'. XAFS-spectroscopy is very effective in studying the structure of amorphous materials, liquid and gaseous states, phenomena on solid surfaces, as well as the dynamics of transformations in chemical reactions and external influences on the matter.

The experimental nature of the structural analysis of XAFS spectra is fairly simple, although the theory of the method is very complex and so far not yet complete. In the XAFS experiments, the sample is irradiated

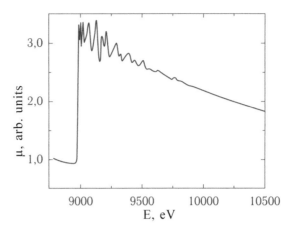

Fig. 8.13. Fine structure of X-ray absorption spectrum near the K-jump of absorption of Cu, measured using synchrotron radiation at a temperature of 10 K.

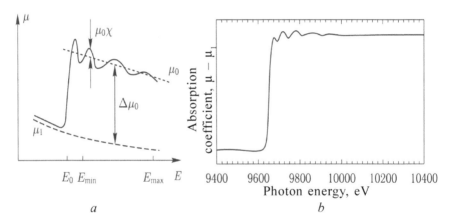

Fig. 8.14. Examples of XAFS spectra. a – schematic representation of the XAFS spectrum for a hypothetical substance composed of atoms of one kind. b – real example of the structure of X-ray absorption spectrum measured near the ionization threshold (absorption edge) of the inner electron shell of an atom of Zn in the semiconductor ZnS, doped with 2% Cd. The graph represents the absorption spectrum after subtracting the background $\mu_1(\varepsilon)$, caused by the shells of Zn atoms with lower ionization energy and the normal absorption by S and Cd atoms.

by a parallel beam of X-rays of known intensity and photon energy, and measurement is taken (directly or indirectly) of the intensity of radiation, attenuated by absorption in the sample. The intensity of the attenuated beam is compared with the intensity of the primary beam to determine the absorption coefficient $\mu(E)$ for a specific wavelength of the absorbed radiation. Measurements are taken in the vicinity of the energy of the absorption jump (resonant photoabsorption) of any of the atoms of the sample. If the photon energy exceeds the energy of the absorption edge, in the measurements with a fine pitch (several eV), starting from the energy of the absorption jump and approximately to 1000–1500 eV in the high-energy side, the dependence of the absorption coefficient on the photon energy show oscillations, as shown in Fig. 8.13. These oscillations are almost periodic sinusoidal and are the fine structure of the XAFS of the absorption spectrum caused by the influence of the structure of matter on the absorption process.

For the quantitative characterization of the fine structure of the absorption spectrum near the absorption jump we use the value of its deviation from the normal absorption. The method for determining this value is shown in Fig. 8.14a depicting the XAFS region of the absorption spectrum on an enlarged scale. The dot and dashed lines in Fig. 8.14a correspond to the normal absorption coefficients calculated in the approximation of the free atom without considering the influence of its environment, and the dotted line represents the total coefficient of normal absorption $\mu_0(E)$, and the dashed line $\mu_1(E)$ – the extrapolation of the contribution to this coefficient from electron shells with the ionization energies smaller than in the inner shell (e.g. L-shell, if μ is considered beyond the threshold of K-ionization). The energy threshold of ionization of the inner shell is marked by E_0, and E_{min} and E_{max} show the boundaries of the interval of the energy of X-ray photons in which EXAFS deviations from the calculated dependence of the total coefficient of normal absorption in the free atoms approximation are detected in the experiments.

The magnitude of the deviations of the measured absorption coefficient observed near the jump on the magnitude of the absorption coefficient of free atoms can be expressed in relative units, as

$$\chi(E) = \frac{\mu(E) - \mu_0(E)}{\Delta\mu_0(E)}. \tag{8.16}$$

The expression (8.16) determines the amplitude of the oscillations in XAFS in parts of the normal absorption coefficient due to the inner electron shell, near the ionization threshold of the measurements. It is seen that the magnitude of $\Delta\mu_0(E) = \mu_0(E) - \mu_1(E)$ in such a narrow energy range as 1000 eV from the K-edge of absorption changes little and, therefore, for the normalization in (8.16) we can use the height of the step of the studied jump absorption, which is often done in practice. The convenience of such a normalization of the amplitude of the oscillations of XAFS is that the quantity $\chi(E)$ is dimensionless and independent of specific forms of representation of μ, and, therefore, it is equally applicable to characteristics of substances in any aggregate state. Function $\chi(E)$ in the practice of EXAFS spectroscopy is called the normalized function EXAFS or EXAFS signal.

Most often, in practice, the background curve of $\mu_1(E)$ is determined by extrapolating the experimental dependence of the total absorption coefficient by outer electronic shells to the high-energy side, as shown schematically by the dashed line in Fig. 8.14*a*, although it is usually required to carry out a further modification of this extrapolation to obtain a satisfactory acceptance of the high-energy data with EXAFS.

Oscillations of the EXAFS signal are caused by scattering of slow photoelectrons emitted by a resonantly absorbing atom on the surrounding atoms and therefore depend on the distance to the scattering atoms and their number in the coordination environment. Function $\chi(E)$ is related to the number of atoms of the coordination environment and the distances to them by the Fourier transform. In the simplest case this relationship is described by Sayers–Stern–Little formula (see, for example, Bunker, 2010; Fetisov, 2007b). With the fitting parameters of this formula to the experimentally measured function $\chi(E)$ and the Fourier transform of the received results we obtain the radial distribution of electron density around the absorbing atom and thus determine the radii of coordination spheres and the number of their population. In fact, the Fourier transform transforms the function $\chi(E)$, defined in the space of momenta k of the photoelectron i.e. the function $\chi(k)$, in the radial distribution of electron density, defined in real space, which is very easy to analyze. Application of the Fourier transform to dependence $\chi^{\exp}(k)$ transforms it into a spatial region, where it has a number of rather sharp peaks, whose position is correlated with the radii of coordination spheres around the central atom, height – with the type and number of atoms in the coordination spheres, and their half-widths – with the amplitude of thermal vibrations. Also, when solving the problem we can

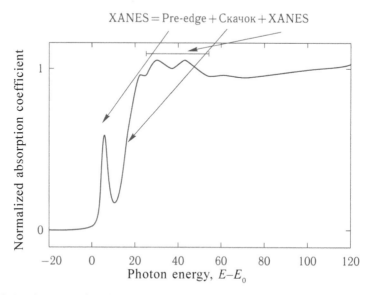

Fig. 8.15. An example of the fine structure of the absorption spectrum near the K-jump of absorption of the Ti atom in a Ba_2TiO_4 sample.

determine the valence angles and bond lengths between the nearest atoms, i.e. to obtain the atomic structure of the substance within a few (usually 2–3) first coordination spheres.

The absorption spectrum, which has a fine structure, can be divided into several sections, each of which is described by its theory. There are the near-edge fine structure XANES (X-ray Absorption Near Edge Structure), located in close proximity to the absorption jump and stretching to an energy of about 30–50 eV above the energy of the jump (shown in Fig. 8.15) and the region of EXAFS, situated above the XANES region extending to energies of the order of 1500–2000 eV. In presenting the material in this monograph, we sometimes will use the generalized name XAFS, which implies immediately all of the mentioned fine structures of absorption without their differences.

Sometimes in the X-ray absorption spectroscopy the pre-edge peak is considered separately from the XANES region (see Fetisov, 2007, section 5.12). The pre-edge peak, which in complex structures and chemical compounds has the form of a structure consisting of a series of peaks, and the position of the step of the absorption jump is very sensitive to chemical bonds and electronic structure the material. Therefore, the spectral region that includes these elements is often considered separately, and in this case, the method is called NEXAFS (Near-Edge X-ray Absorption Fine Structure). However, because of very low energy, slow photoelectrons produced in the absorption of X-ray photons in this field have a very long mean free path and experience multiple scattering, which greatly complicates the construction of a theory for the quantitative interpretation of the spectrum. But even the observation of the qualitative changes in the NEXAFS spectrum can with high reliability detect changes in the electronic structure and nature of chemical bonds in the investigated object. The method is especially effective for studying molecules on surfaces, as measurements of photoelectron yield emissions, used to record NEXAFS spectra, collect information from the thinnest surface layer. The NEXAFS method as well as the XANES method provides information on the composition of the substance, on the chemical bond in the local environment of the absorbing atom and on the atomic structure of the local cluster.

The theory of signal processing techniques $\chi(k)$ has now been sufficiently well developed for the EXAFS range, and for the XANES method (including NEXAFS) these is still complete theory because much more complicated scattering processes occur in it. But this area carried much more chemical information than EXAFS, therefore, it has been used for a long time to solve practical problems by applying approximate theories and complex calculation methods to obtain this information at least semi-quantitatively.

Experimental methods for measuring the XAFS spectra can be divided into two categories: 1) direct measurements in which we measure the weakening of the intensity of the X-ray beams transmitted through the sample, and 2) indirect measurements, in which measurements are taken of the intensity of secondary processes accompanying the absorption of X-rays in the sample. In each category there is a wide variety of techniques (measurement mode), most of which are examined in some detail in a review (Fetisov, 2007b).

In the study of semiconductors indirect measurements of spectra are taken quite often, since they give to the method high flexibility and selectivity as compared with direct measurement of XAFS. The most common are measurement techniques such as the fluorescence measurement mode of the characteristic X-ray emission of the reference element, sometimes known in the literature as FLY or FY (short for Fluorescence Yield); measurement mode of photoelectron emission, which is called TEY (Total Electron Yield); measurement mode of the intensity of the Auger effect, which is called AEY (Auger Electron Yield), or SEXAFS (short for Surface EXAFS) because of the ability to measure only the surface layers of a thickness of several angstroms; the measurement mode of optical luminescence excited by X-rays, known as the XEOL method (X-ray Excited Optical Luminescence) or PLY (Photo-Luminescence Yield). Each of these techniques has unique features that distinguish it from others. For example, the FLY method has the highest sensitivity and for measuring XAFS it is capable of using as reference atoms even doping elements in semiconductors, which are present in very low concentrations. The TEY and AEY methods allow us to investigate the outer layers of nanoparticles, and the XEOL/PLY method can selectively analyze only the structure of luminescent particles. There are high-speed methods producing XAFS spectra within a few picoseconds, which makes it possible to carry out structural studies with the picosecond time resolution. One of the fastest methods is, for example, the dispersion method of measuring the XAFS spectra, called EDEXAFS (Energy Dispersive EXAFS).

8.4.2. Examples of studying nanosilicon by XAFS methods

A striking example of the usefulness and effectiveness of XAFS methods studies of nanosilicon is that using this method Sham et al (1993) shortly after reporting the observation of visible photoluminescence of porous silicon experimentally proved the validity of one of several emerging hypotheses about the causes of this phenomenon. Using the XAFS spectra measured by TEY and XEOL, the authors investigated the luminescence centres in porous silicon and siloxane. The results showed that the luminescence of porous Silicon is not related to siloxane and is explained by the model of quantum-size constraint in the structure of porous silicon.

The authors of the work (Schuppler et al, 1995) conducted a comprehensive study of a large number of different samples of nc-Si with surface oxidation and P-Si passivated by hydrogen, which exhibited photoluminescence in the visible spectrum region. Electron microscopy, infrared spectroscopy, absorption spectroscopy, atomic recoil and photoluminescence spectroscopy were used.

Through a joint analysis of the results obtained by different methods, the authors have drawn conclusions about the structure of the investigated particles and considered some components of the studied systems with luminescence. In the samples of P-Si, having a peak of luminescence in the region < 700 nm, the average morphology was represented by nanoparticles (and not by the fibers) which had dimensions of <15 Å and the crystalline structure. This was confirmed by studies of the short-range order. The results obtained showed that the observed luminescence is not related to the oxidation of the surface of the nc-Si particles, nor to suboxides at the borders of nc-Si, and is completely determined by the structure and the size of the nanocrystals, i.e. quantum-size constraints. Therefore, the photoluminescence properties of nanocrystalline silicon should not depend on the methods of production.

In the paper (Zhang & Bayliss, 1996) NEXAFS and EXAFS spectra were used to investigate samples of porous silicon, possessing photoluminescence in the red, yellow and green bands of the visible spectrum. Luminescence peaks were observed at wavelengths of 690, 580 and 520 nm, and the task of the authors of this paper was to determine how the photoemission peak position is due to the size, the atomic structure and the shape of luminescing P-Si particles. Fourier transform of the EXAFS date was used to determine the coordination numbers for the first three fields of the environment of the Si. Comparison of these results with the energies of the emission peaks of luminescence and the optical band gap measured by photoluminescence spectroscopy on the same samples allowed the authors to conclude that the investigated P-Si samples consisted of two types of quantum nanofibres with a crystal core, and the compliance of each of them with a certain type of luminescence peak was established.

Investigation of the structure of nanoparticles. For example, Fig. 8.16 shows the results of a comprehensive study by TEM and NEXAFS spectroscopy in the TEY and FLY modes obtained by laser ablation of Si nanowires doped with phosphorus (Tang et al, 2002). In these two modes XAFS spectroscopy has a selectivity to different areas of the studied material. In the TEY mode the signal carries information about the structure of the surface. The FLY signal is obtained from the fluorescent characteristic K-radiation of the Si atoms. Because the surface of nanowires may contain chemical compounds where the concentration of Si atoms is less than in pure silicon, the FLY signal is mainly relates to the silicon core of the nanowires.

When comparing the NEXAFS spectra of pure silicon and of the freshly prepared nanowires with the oxide layer, one can notice a distinct peak due to Si oxides. As shown in the absorption spectra in the figure, measured by TEY and FLY, the disappearance of this peak and the appearance of the absorption edge of crystalline Si with increasing thickness of the removed

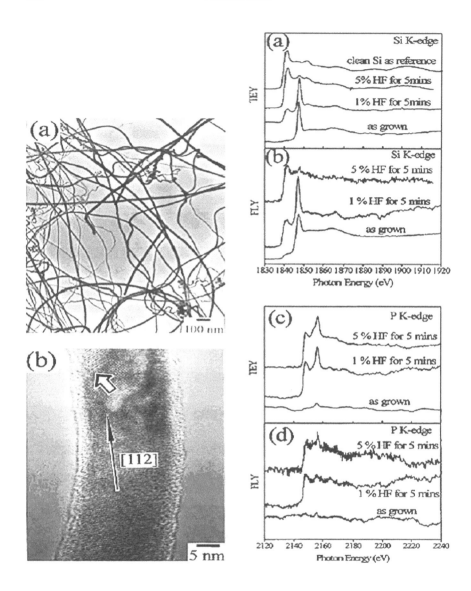

Fig. 8.16. An example of the influence of the depth of etching of Si nanowires, doped with phosphorus, on the NEXAFS spectra, measured by different methods. The concentration of the etchant (solution of HF) and the duration of the etching are in the inscriptions on the graphs of the spectra. On the left is the TEM image of the general morphology of the Si nanowires highly doped with phosphorus (a); HRTEM image showing how the fiber grows along the [112] direction (b). On the right are the NEXAFS spectra, measured in TEY and FLY modes after successive etching of nanofibers in HF: (a) – TEY, as measured by the K-edge of Si, for comparison shows the spectrum of pure silicon; (b) – FLY on the K-edge of Si; (c) – TEY on the K-edge of P; (d) – FLY on the K-edge of P. Figure from (Tang et al, 2002).

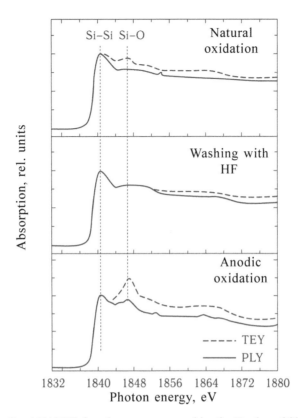

Fig. 8.17. Normalized XANES functions, as measured by the K-edge of Si absorption by the TEY and PLY methods for samples of porous silicon. The results for the aged and naturally oxidized P-Si, for the same sample after washing in HF and for sample subjected to anodic oxidation, are shown. Figure from (Daldosso et al, 2000).

surface layer (with increasing concentration of the etchant solution) is clearly visible. In addition, it is noticeable that the interatomic Si–Si distances in the core of several nanofibres are slightly greater than in bulk crystalline silicon. Even this visual analysis of the NEXAFS spectra suggests that the initial Si nanowires are nanowires with the central part of crystalline silicon doped with phosphorus-coated silica.

Study of the luminescence centres. With the help of EXAFS spectroscopy Sham et al (1993) established a connection between the structure of porous silicon (P-Si) and photoluminescence centres. XANES spectra were measured at transmission of X-rays through the sample. The spectra were registered by both the total electron yield (TEY method) and the yield of optical luminescence (PLY/XEOL) method. Comparison of the data obtained by the TEY and PLY methods (PLY in this case means the same as FLY), showed that the luminescence is due to the nanocrystalline nature of P-Si, and not to the chemicals adsorbed on its surface. The local structure of Si luminescence centres in this material was determined from the XANES spectra, measured by XEOL (Daldosso et al, 2000). Figure 8.17 shows the XANES spectra for P-Si, stored in air and washed in HF before measurements to remove the surface oxide layer, and measured in the TEY and XEOL modes at the K-edge of absorption of Si. The XANES spectra, obtained in the TEY and PLY modes, provide different information. The TEY data, which carry information about the surface layer thickness of a few tenths of a nanometer, showed for the aged samples the presence on the surface of porous silicon of crystal structures containing oxygen atoms. The PLY data selects only the positions emitting light and allow study their structure. As can be seen from the comparison of the TEY and PLY graphs in Fig. 8.18, the PLY spectrum is fully consistent with the NEXAFS spectrum of silicon with the post-threshold peak from Si–Si bonds and does not depend on the presence of the oxide film. The oxygen peak in the PLY spectrum begins to appear only after artificial anodic oxidation, which leads to deeper penetration of oxygen into the nanoparticles. At the same time, the TEY spectra of the oxidized sample show a peak from the Si–O bonds, which disappears after etching the sample in HF. The independence of the form of the lines on the presence of oxygen shows that the emitting positions are not affected by the presence of oxygen.

The publication (Hessel et al, 2008) presents the results of X-ray studies of the formation and luminescent properties of silicon nanocrystals (nc-Si) embedded in an oxide. The particles (nc-Si) were obtained by thermolysis of hydrogen silsesquioxane (HSQ). NEXAFS spectroscopy was used to monitor the evolution of the electronic characteristics of silicon during thermal disproportionation of HSQ from nanocrystalline clusters of elemental Si at 500°C to well expressed nc-Si in the matrix of the a-SiO$_2$ type at 1100°C. When excited by x-ray the nc-Si particles in the oxide luminesce in two bands centred at wavelengths of 780 and 540 nm of the optical spectrum. These photoluminescence bands were used for the measurement of NEXAFS spectra by the XEOL method. Due to the selectivity of the XEOL method to the luminescence centres experiments were carried out to determined the physical and chemical characteristics of these centres. It was found that the emission band of 540 nm is the result of formation of well-defined Si–nc-Si boundary connections with the surrounding oxide matrix. In this paper, using the energy selective luminescence XEOL method, the independently distributed Si nanoparticles with the surface, functionalized with hydrogen were studied. Based on these results, the authors concluded that the cause of the emission from nc-Si particles with sizes smaller than 5 nm are the effects of quantum-size constraint, rather than the surface functional groups and their environment.

In the paper (Daldosso et al, 2003) the XANES method was used for a detailed study of the structure of luminescent Si nanoparticles, embedded in an SiO$_2$ matrix in order to determine their chemical and morphological state of the matrix and also the role of the matrix environment in luminescence.

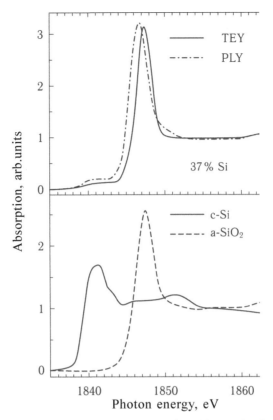

Fig. 8.18. Comparison of the absorption spectra of nc-Si samples in a matrix of SiO_2, obtained by annealing of SiO_x films with the concentration of Si atoms of 37%. Measurements in the recording modes of TEY and PLY. For comparison, the lower figure shows the absorption spectra of the c-Si and a-SiO_2 samples. For comparability the spectra are normalized to a constant height of the band in the range 1850–1860 eV (Daldosso et al, 2003).

The nc-Si particles were obtained by annealing at 1250°C SiO_x films deposited on a mc-Si substrate by vapor deposition in plasma discharge (PECVD method). At annealing of these films excess Si atoms diffuse and form aggregates in the form of nanometer-sized Si clusters, i.e. SiO_x SiO_x is divided into two phases: nc-Si and SiO_2. As in the original film the excess Si atoms are distributed uniformly in thickness, the distribution of nanoparticles after annealing should be also uniform. Experimental verification of this assumption by TEM and Rutherford backscattering spectroscopy (mass spectrometry) showed that this distribution is not affected by the film thickness nor by the interface between the film and the substrate. Depending on the concentration of Si atoms in the original film annealing was accompanied by the formation of nanosilicon particles ranging in size from 0.7 to 2.1 nm and different scatter of the sizes, as described by the Gaussian distribution. The particles emitted bright photoluminescence in the wavelength range 650–950 nm, which is well explained by the nature of the recombination of charge carriers in the nc-Si particles of limited size, i.e. the quantum-size effects. However, there was a quantitative discrepancy between the experimentally measured luminescence energy and the magnitude of energy derived from theoretical calculations.

The samples prepared from SiO_x films with different Si concentrations were measured by the XANES method in the TEY (total electron yield) and PLY (yield of photoluminescence excited by X-ray irradiation) modes. The measurements were carried out at synchrotron radiation of the K-edge of absorption of Si, and TEY and PLY spectra were measured simultaneously. In this case, the TEY measurements were conducted by the compensation current, which was necessary to maintain the neutral charge of the sample exposed to direct X-ray radiation. Naturally, this current must be equal to the current of photoelectrons ejected by the X-rays. The probing depth of the surface layer by this method is equal to the mean free path of electrons and depends on the energy of the photoelectrons. In XANES measurements of SiO_2 layers, according to the authors of this paper, the probing depth was up to 100 nm. The PLY method probes XANES only from the centres of luminescence. The measurement results are shown in Fig. 8.18.

The graph of PLY measurements, Fig. 8.18, clearly distinguishes two jumps of absorption, the first of which begins at an energy of 1839 eV, and the second peak is at an energy of 1847 eV. If the positions of these discontinuities are compared with the positions of the absorption edges in the figure below, the positions of the maxima of the jumps clearly indicate that the first jump of absorption in the PLY spectrum is due to absorption by Si atoms, associated with the Si atoms, while the second absorption edge relates to the Si atoms in the SiO_2 chemical compound.

The peak of the absorption jump is due to the strong contribution of multiple scattering of the photoelectrons, which is very sensitive to the slightest changes in the arrangement of atoms (even without changing the distance to the central reference atom) (Fetisov, 2007b) and to changes in the stoichiometry of the test compound causing changes in the electronic density of states. Comparison of the XANES spectra measured by the TEY and PLY techniques, shows the difference in the position of this peak for the same sample, and the position of the peak in the PLY spectrum also differs from the peak position in pure a-SiO_2.

Figure 8.19 shows the same spectra of all the investigated samples of nc-Si, obtained from the original films with different stoichiometry of SiO_x. The thickness of the films was greater than 200 nm (much larger than the probing depth of the measurement methods), so there was no signal from the substrate in the spectra. If the presented spectra show a low-energy absorption jump, it is clear that its peak intensity depends on the fraction of Si atoms which form the Si nanoclusters as the O atoms are spent on the formation of the a-SiO_2 stoichiometric oxide. The greater proportion

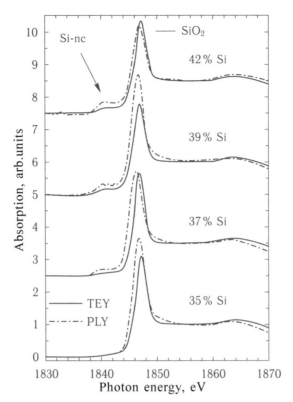

Fig. 8.19. Normalized XANES spectra of different samples of nc-Si in the matrix of a-SiO$_2$, measured simultaneously by TEY and PLY methods. From (Daldosso et al, 2003).

of these atoms, the more low-energy absorption jump becomes evident. For a sufficiently large proportion of these atoms the low-energy jump begins to manifest itself even in the spectra measured by TEY, where the averaging of the signal over the volume is much stronger. A more vivid manifestation of the low-energy absorption jump on the PLY data indicates that the concentration of nc-Si in the luminescence centres is much higher than the average value over the volume of the sample.

By its nature, the luminescence, as measured by PLY, is the result of absorption of X-ray photons in both nc-Si and in SiO$_2$. As can be seen from Fig. 8.20, the PLY XANES signal shows a strong increase in both absorption jumps with increasing concentration of Si atoms in the samples. We also see that profile shape of the peak after the absorption jump, associated with a-SiO$_2$, in the measurements by TEY and PLY is different, as different are the positions of their peaks (the peak in the PLY method is always lower in energy than the TEY peak, approximately by 0.4–0.5 eV). This can be explained by the fact that a-SiO$_2$ in the immediate vicinity of luminescent Si nanoparticles differs in the structure of the short-range order from the

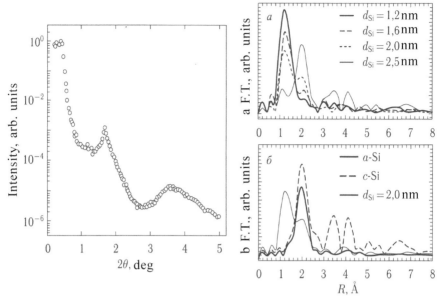

Fig. 8.20. X-ray diffraction measurements and the results of measurements of the EXAFS spectra of silicon nanostructures. Left – X-ray diffraction pattern of a multilayer thin-film Si/CaF$_2$ structure with 100 periods, as measured by the sliding beam incidence method (d'Avitaya et al, 1995). Right: a – the Fourier transform of EXAFS spectra from samples of multilayer thin-film Si/CaF$_2$ structures with the number of periods 50. The thickness of the Si layer in different samples is different; b – Fourier-transform spectra of EXAFS (Bassani et al, 1996) for samples of amorphous (a-Si) and crystalline (c-Si) silicon and the multilayer Si/CaF$_2$ structure with 50 periods. Example from (Ossicini et al, 2003).

secondary structure of the entire volume of the sample, and this border region affects the luminescence process of nc-Si. If it were not so, the shape and position of the peak high-energy shock absorption in the methods of TEY and PLY would have be the same.

Thus, the results of XANES measurements suggest that the nanoparticles obtained by the decomposition of the SiO$_x$ films by high-temperature annealing, are nc-Si nanocrystals, surrounded by a thin layer of SiO$_x$ with variable composition, forming a gap between the nc-Si and a-SiO$_2$. That is, the XANES measurements showed that the interface between the nc-Si and the surrounding SiO$_2$ matrix is not sharp: there is an intermediate region of the amorphous type and variable composition, which connects the nc-Si with stoichiometric a-SiO$_2$, and this region is actively involved in the process of light emission.

The study of these samples by HRTEM with electron energy filtration (EFTEM analysis) confirmed this conclusion. It was expressly found that the Si particles have a crystalline core with distinct atomic planes, and this core was surrounded by a shell, which differs in composition and structure from both the nucleus and the a-SiO$_2$ matrix whose thickness was approximately 1 nm.

Study of multilayer thin-film nanostructures. XAFS spectroscopy is widely used for solving practical problems in the production of silicon nanostructures for photonics (Ossicini et al, 2003), for example, to obtain quantitative information on the local structure of nanoscale layers of Si. So in Bassani et al. (1996) XAFS techniques were used in comprehensive studies of thin films obtained in periodic multilayer thin-film Si/CaF$_2$ structures, whose results are shown in Fig. 8.20. To the right in the figure are the Fourier transformants of the EXAFS spectra, measured by the K-jump of absorption of Si. The upper diagram shows the size distribution of transformants for samples of silicon films of varying thickness, and the chart below shows for comparison similar curves for the bulk samples of amorphous (a-Si) and crystalline (c-Si) silicon.

Comparison of the graphs from samples of amorphous, crystalline silicon and the multilayer thin-film Si/CaF$_2$ shows that in the films there are ordered structures, which manifests the presence of peaks from at least three coordination shells, which coincide in position with the peaks in crystalline silicon. This indicates that the films contain quasi-crystalline regions with dimensions of about 1.5 nm. The peak at 1.3 Å indicates the presence of Si–O bonds which indicates the presence of oxygen in the films that got there by diffusion along grain boundaries. The results presented here indicate that in the production of thin-film structures by molecular beam deposition the Si layers show the formation of crystallites with the dimensions on the order of the thickness of its layer in the multilayer structure.

8.4.3. The DAFS method

Considering the XAFS techniques for the study of the local structure of silicon nanoparticles, it is impossible to ignore the DAFS method closely associated with the former – the study of the diffraction anomalous fine structure (Sorensen et al, 1994). The DAFS method measures the intensity of elastic Bragg scattering of the photon energy near the energy of the absorption jump. This method combines the sensitivity of X-ray diffraction to the long-range order and the crystallographic structure with the sensitivity of absorption spectroscopy techniques to the short range order, and as a result, in addition to the elemental selectivity of the XAFS method, it becomes possible to study the short-range order in an environment of dedicated crystallographic positions. In the region of the extended fine structure DAFS provides the same information about the structure of the short-range order as EXAFS: i.e. the bond lengths, coordination numbers, types of neighbouring atoms and bonds for the disordering of the atoms surrounding the resonantly scattering atoms. In the near-threshold region DAFS allows to obtain the same information as XANES: the valence, the data on unfilled orbitals, and information about relationships for resonant atoms. As a result, DAFS can be used to obtain, as in the EXAFS and XANES methods, information about a specific subsystem of atoms selected by the diffraction condition, and at the same time, DAFS can provide information such as that which is obtained by absorption spectroscopy, i.e. information on the structure of non-equivalent positions of monatomic entities in the unit cell.

The DAFS method measures the intensity of the diffraction Bragg peak depending on the energy of photons at energies close to the threshold of excitation of the absorption jump of one of the elements of matter (central atom). DAFS measurements are taken using a more complex scheme (Fetisov, 2007b) than that used to measure conventional XAFS spectra, since this method requires a search of Bragg reflections and tracking their position in the measurements which require at least a two-circle goniometer.

The intensity of Bragg reflections measured by the DAFS method carries information about the atoms of matter relating to the reflecting crystallographic plane and at the same time it is modulated by the influence of absorption. Thus, modulation of the intensity of the Bragg reflections, depending on the photon energy, actually contains the spectrum of the XAFS signal, but only for parts of atoms defined by the Bragg reflection, i.e. it represents the oscillations of the absorption caused by the local structure (short-range order) of the substance, imposed on the function of background intensity, which is determined by the crystallography of the sample. Thus, the DAFS signal contains a crystallographically selected information on the local structure rather than on its spherically symmetric average value, as it occurs in the XAFS signal.

A striking example of the uniqueness of the analytical capabilities provided by the DAFS method compared with other methods of structural analysis is the study of mixtures of amorphous and crystalline phases or short-range order changes in the transformation of amorphous to crystalline. The complexity of such studies is related to the small size of the resulting clusters (from 100 to 1000 atoms) and the multiphase nature of the medium, in which both the amorphous matrix and several crystalline phases can coexist. XAFS spectra of these materials provide integrated information about the structure, from which it is difficult to separate the contributions of amorphous and crystalline phases. X-ray analysis of this material is virtually impossible because of the small particle size, the heterogeneity of their distribution in the sample and strongly disordered interfaces. But the task is easily solved by combining XAFS and DAFS methods.

The DAFS method allows to select only information about the nanocrystals, completely ignoring the amorphous phase. After obtaining the spectrum of the crystalline phase, it can be separated from the total XAFS spectrum and further analyze the amorphous phase. A. Frenkel and co-workers (Frenkel et al, 2002) for the first time demonstrated the possibility of studying the structure of such a mixture of amorphous and crystalline

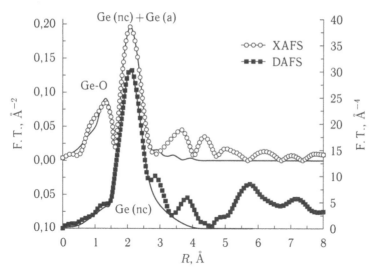

Fig. 8.21. Comparison of the XAFS and DAFS data from a four-phase sample (amorphous and crystalline Ge + Ge oxide in a matrix of amorphous SiO_2), measured near the K-jump of absorption of Ge. At the top there is the module of the Fourier transform (FT) of the XAFS signal (experimental data and their approximation) – a scale for the data is shown on the left, bottom – the module of the Fourier transform of the DAFS signal (scale is shown at right). Graph from (Sorensen et al, 1994).

phases: crystalline germanium nanoparticles, amorphous germanium and a small amount of amorphous germanium oxide in a matrix of amorphous silicon dioxide. The size of nanocrystals was 15–20 nm, and they were evenly distributed in the amorphous matrix.

The spectra, measured on the K-edge of absorption of Ge, are shown in the form of a module of the Fourier transforms in Fig. 8.21. In the XAFS spectrum the peak of the germanium atoms is due to the total contribution of nanocrystals and amorphous phase of Ge, and the peak from the amorphous germanium oxide can also be seen. The peak on the graph the DAFS data is due only to the contribution of Ge nanocrystals. In a joint analysis of these data (detailed analysis and data, see Kolobov et al. (2003)) it was possible to obtain not only the bond lengths in each of the three observed phases, but also the proportion of these phases in the mixture. It was also found that the Ge–Ge distances in the amorphous phase (2.50 ± 0.03 Å) are 3% greater than in nanocrystals (2.44 ± 0.02 Å).

Conclusion

Real nanosilicon powders and composites based on them are polydisperse materials whose properties depend strongly on the size of the Si nanoparticles and statistics of their size distribution, as well as the atomic structure of the particles themselves, which is also often associated with size. In contrast to the material from the bulk Si particles, nanomaterials based on it can be correctly described only by the totality of all properties, particle size and shape, size distribution function of Si nanoparticles, the atomic structure of the particles (both the short- and long-range order), the ratio of shares of a-Si, nc-Si and nanoclusters in the sample.

Unfortunately, methods for efficient and accurate determination of these parameters to determine the correct relationship 'morphology + structure + composition + properties are yet available or are under development. As a rule, the most correct answers can be obtained only through an integrated use of different techniques.

Direct methods of determination of some of these characteristics are microscopic methods where the most complete information can be provided by transmission electron microscopy methods. However, they are able to give only the local characteristics of the sample, which due to the low statistics may be significantly different from the characteristics of the material.

The main methods of structural analysis of nanomaterials based on nanosilicon, as in the case of bulk materials, are indirect methods, such as X-ray diffraction and spectroscopic methods. However, the classical methods of structural analysis based on the Bragg–Laue diffraction theory, which have long been successfully used to decrypt and refine the structures of materials, are practically useless for analyzing the structure of nanoparticles smaller than 5 nm. They are helped by new methods based on a different theoretical framework not limited by the hypotheses of the structural periodicity of the materials, allowing the investigation of both the short- and long-range order. Research of the structure of the particles of this size can also be successfully carried out by methods of analysis of the fine structure of X-ray absorption spectra (XAFS methods). At present a drawback of all these X-ray diffraction and spectroscopic methods is the ambiguity of the results, since there are cases where several different refined models of the structure agree well with the results of measurements. New methods of structural analysis are being developed for a unique solution of these problems, for example, the methods discussed in chapter 10 of this monograph.

References for Chapter 8

Bakhtizin R.Z. Scanning tunneling microscopy – a new method of studying the surface of solids. Soros Educational Journal. 2000. V. 6 (11). P. 1-7.

Billinge S.J.l., Kanatzidis M.G. Beyond crystallography: the study of disorder nanocrystal-linity and crystallographically challenged materials. Chem. Commun. 2004. No. 7. P. 749–760.

Binnig G., Rohrer H.,. Scanning tunneling microscopy – From Birth to Adolescence: Nobel Lecture. Stockholm, December 8, 1986.

Vasiliev D. M. Diffraction methods for studying structures. Moscow, Metallurgiya, 1977.

Vorokh A. S. Disordered atomic structure of cadmium sulfide nanoparticles: PhD dissertation. Ekaterinburg, 2009.

Vorokh A. S., Rempel' A. A. The atomic structure of nanoparticles of cadmium sulfide. Fiz. Tverd. Tela. 2007. V. 49, No. 1. P. 143-148.

Gallyamov M.O., Yaminsky I.V. Scanning probe microscopy: Basic Principles, distortions analysis. 2001. Electronic publishing: http://www.spm.genebee.msu.ru/members/ gallyamov/gal_yam/gal_yam1.html

Guinier A. X-ray diffraction of crystals. Theory and Practice. Moscow. Gos. Izd. Fiz. Mat. Lit., 1961.

Demikhovsky V. Ya., Filatov D. O. The study of electron states in low-dimensional structures using scanning probe microscopy: textbook. Nizhny Novgorod: N. I. Lobachevsky University Press, 2007.

James R. Optical Principles of X-ray diffraction. Moscow: IL, 1950.

Dorofeev S. G., Kononov N. N., Fetisov G. V., Ischenko A. A., Lyao D.-G. Nanocrystalline silicon obtained from SiO. Nanotekhnika. 2010. No. 3 (23). P. 3-12.

Iveronova V. I., Revkevich G. P. The theory of scattering of X-rays. Moscow: Moscow State University, 1978.

Mironov V. L. Fundamentals of scanning probe microscopy: textbook. Nizhny Novgorod. Institute of Physics of Microstructures, RAS, 2004..

Shindo D., Oikawa T. Analytical transmission electron microscopy. M. Tekhnosfera. 2006.

Fetisov G. Something about X-ray analysis, electromagnetic radiation, X-rays, their properties and diffraction. In.: G.V. Fetisov Synchrotron radiation. Methods for studying the structure of matter. Moscow: Fizmatlit, 2007a. Chap. 2. P. 17-90.

Fetisov G. V. X-ray diffraction analysis with synchrotron radiation. In.: G.V. Fetisov Synchrotron radiation. Methods for studying the structure of matter. Moscow: Fizmatlit, 2007b. Chap. 4. P. 338-487.

Fetisov G. V. XAFS spectroscopy for structural analysis. In.: G. V. Fetisov Synchrotron radiation. Methods for studying the structure of matter. Moscow: Fizmatlit, 2007c. Chap. 5. S. 488-579.

Tsybulya S. V. X-ray analysis of nanocrystals: the development of methods and structure of metastable states in the metal oxides of non-stoichiometric composition: PhD Dissertation. Novosibirsk, 2004.

Yatsenko D. A., Tsybulya S. V. Simulation method of the diffraction patterns for nanoscale crystalline systems. Vestnik NGU. Ser. Fizika. 2008. V. 3, No. 4. P. 47-51.

Bassani F., Vervoort L., Mihalcescu I., Vial J. C. Arnaud d'Avitaya F. Fabrication and optical properties of Si/CaF$_2$(111) multi-quantum wells. J. Appl. Phys. 1996. V. 79. P. 4066–4071.

Bhushan B. Microscopy in Nanoscience and Nanotechnology. Springer, 2010.

Billinge S. Local structure from total scattering and Atomic Pair Distribution Function (PDF) analysis. In: Powder diffraction theory and practice. Ed. by R. E. Dinnebier, S. J. L. Billinge. Cambridge, UK: Royal Society of Chemistry, 2008. Ch. 16. P. 464–493.

Billinge S. J. L. Nanostructure studied using the atomic pair distribution function. Z. Kristallogr. Suppl. 2007. V. 26. P. 17–26.

Binnig G., Rohrer H., Gerber Ch., Weibel E. 7×7 Reconstruction on Si (111) Resolved in Real Space. Phys. Rev. Lett. 1983. V. 50, No. 2. P. 120–123.

Binning G., Quate C. F., Gerber C. Atomic force microscopy. Phys. Rev. Lett. 1986. V. 56. P. 930–933.

Birdi K. S. Scanning probe microscopes: Applications in science and technology. CRC Press, LLC, 2003.

Bish D. L., Post J. E. (Eds.) Modern Powder Diffraction (Reviews in Mineralogy. V. 20). Washington, DC: Mineralogical Society of America, 1989.

Brandon D. G., Kaplan W.D. Microstructural characterization of materials. John Wiley & Sons, Ltd., 1999. 424 p. 2 ed., 2008. 550 p.

Bunker G. Introduction to XAFS: A practical guide to X-ray Absorption Fine Structure. Cambridge University Press. 2010. 260 p.

Cline J. P. Use of NIST standard reference materials for characterization of instrument performance. In: Industrial Applications of X-ray Diffraction / Ed. by F.H. Chung, D.K. Smith. Pub. by Marcel Dekker, Inc., 2000. Ch. 40. P. 903–917.

d'Avitaya F.A., Vervoort L., Bassani F., Ossicini S., Fasolino A., Bernardini F. Light emission at room temperature from Si/CaF$_2$ multilayers. Europhys. Lett. 1995. V. 31. P. 25–30.

Daldosso N., Luppi M., Ossicini S., Degoli E., Magri R., Dalba G., Fornasini P., Grisenti R., Rocca F., Pavesi L., Boninelli S., Priolo F., Spinella C., Iacona F. Role of the interface region on the optoelectronic properties of silicon nanocrystals embedded in SiO$_2$. Phys. Rev. B. 2003. V. 68. P. 085327 (8 p.).

Daldosso N., Rocca F., Dalba G., Fornasini P., Grisenti R. New EXAFS Measurements by XEOL and TEY on Porous Silicon. J. Porous Materials. 2000. V. 7. P. 169–172.

Debye P. Zerstreuung von Röntgenstrahlen. Annalen der Physik. 1915. Bd. 46, №1. S. 809–823.

Di Ventra M., Evoy S., Heflin J. R. Introduction to Nanoscale Science and Technology. Springer, 2004. .

Dinnebier R. E., Billinge S. J. L. (Eds.). Powder diffraction theory and practice. Cambridge, UK: Royal Society of Chemistry, 2008. xxii+582 p.

Dorofeev S.G., Ischenko A.A., Kononov N.N., Fetisov G.V. Effect of synthesis temperature on the optical properties of nanosilicon produced from silicon monoxide. Nanotechnology. 2011.

Egami T., Billinge S. J. L. Underneath the Bragg peaks: Structural analysis of complex materials. Oxford, England: Pergamon Press, Elsevier, 2003. 422 p.

Egerton R. F., Li P., Malac M. Radiation damage in the TEM and SEM. Micron. 2004. V. 35. P. 399–409 .

Erlandsson R., Olsson L., Mårtensson P. Inequivalent atoms and imaging mechanisms in ac-mode atomic-force microscopy of Si (111) 7×7. Phys. Rev. B. 1996. V. 54. P. R8309–R8312.

Farrow C. L., Billinge S. J. L. Relationship between the atomic pair distribution function and small angle scattering: implications for modeling of nanoparticles. Acta Crystallogr. 2009. V. A65. P. 232–239.

Farrow C. L., Juhas P., Liu J.W., Bryndin D., Božin E. S., Bloch J., Proffen Th., Billinge S. J. L. PDFfit2 and PDFgui: computer programs for studying nanostructure in crystals. J. Phys. Condens. Matter. 2007. V. 19. P. 335219 (7 p.).

Flegler S. L., Heckman J.W. Jr., Klomparens K. L. Scanning and transmission electron microscopy, an introduction. Oxford University Press, USA, 1993. 240 p.

Frank J. (Ed.). Electron tomography: Three dimensional imaging with the transmission electron microscope, N.Y.: Plenum Press, 1992. 416 p.

Frenkel A. I., Kolobov A.V., Robinson I.K., Cross J.O., Maeda Y. & Bouldin C.E. Direct separation of short range order in intermixed nanocrystalline and amorphous phases. Phys. Rev. Lett. 2002. V. 89. P. 285503.

Gala J., Mogilanski D., Nippus M., Zabicky J., Kimmel G. Fast high-resolution characterization of powders using an imaging plate Guinier camera. Instr., Meth. in physics research. Section A. 2005. V. 551. P. 145–151.

Gass M.H., Koziol K.K., Windle A.H., Midgley P.A. Four-dimensional spectral tomography of carbonaceous nanocomposites. Nano Lett. 2006. V. 6. P. 376–379.

Giannini C., Cervellino A., Guagliardi A., Gozzo F., Zanchet D., Rocha T., Ladisa M. A Debye function based powder diffraction data analysis method. Z. Kristallogr. Suppl. 2007. V. 26. P. 105–110.

Hall B.D. Debye function analysis of structure in diffraction from nanometre-sized particles. J. Appl. Physics. 2000. V. 87, №4. P. 1666–1675.

Hessel C.M., Henderson E. J., Kelly J.A., Cavell R.G., Sham T.-K., Veinot J.G. C. Origin of Luminescence from Silicon Nanocrystals: a Near Edge X-ray Absorption Fine Structure (NEXAFS) and X-ray Excited Optical Luminescence (XEOL) study of oxide-embedded and free-standing systems. J. Phys. Chem. C. 2008. V. 112 (37). P. 14247–14254.

Jenguillaume C., Colliex C. Spectrum-image: The next step in EELS digital acquisition and processing. Ultramicroscopy. 1989. V. 28. P. 252–257.

Kaszkur Z. Nanopowder diffraction analysis beyond the Bragg law applied to palladium. J. Appl. Cryst. 2000a. V. 33. P. 87–94.

Kaszkur Z. Powder diffraction beyond the Bragg law: study of palladium nanocrystals. J. Appl. Cryst. 2000b. V. 33. P. 1262–1270.

Keen D.A. A comparison of various commonly used correlation functions for describing total scattering. J. Appl. Cryst. 2001. V. 34. P. 172–177.

Korsunskiy V. I., Neder R. B., Hofmann A., Dembski S., Graf Ch., R"uhl E. Aspects of the modeling of the radial distribution function for small nanoparticles. J. Appl. Cryst. 2007. V. 40. P. 975–985.

Krayzman V., Levin I., Tucker M.G. Simultaneous reverse Monte Carlo refinements of local structures in perovskite solid solutions using EXAFS and the total scattering pair-distribution function. J. Appl. Cryst. 2008. V. 41. P. 705–714.

Mayandi J., Finstad T. G., Foss S., Thorgesen A., Serincan U., Turan R. Luminescence from silicon nanoparticles in SiO_2: atomic force microscopy and transmission electron microscopy studies. Phys. Scr. 2006. V. T126. P. 77–80.

Neder R. B., Proffen T. Diffuse Scattering and Defect Structure Simulations: A cook book using the program DISCUS. Oxford University Press, 2008. ix+228 p.

Ossicini S., Pavesi L., Priolo F. Light emitting Silicon for microphotonics. Berlin–Heidelberg: Springer-Verlag, 2003. xii+282 p.

Reimer L., Kohl H. Transmission electron microscopy: Physics of image formation. 5 ed. — Springer, 2008. xvi+590 p.

Rietveld H.M. Line profiles of neutron powder-diffraction peaks for structure refinement. Acta Crystallogr. 1967. V. 22. P. 151–152.

Schamm S., Bonafos C., Coffin H., Cherkashin N., Carrada M., Ben Assayag G., Claverie A., Tence M., Colliex C. Imaging Si nanoparticles embedded in SiO_2 layers by (S) TEM-EELS. Ultramicroscopy. 2008. V. 108. P. 346–357.

Schuppler S., Friedman S. L., Marcus M.A., Adler D. L., Xie Y.-H., Ross F.M., Chabal Y. J., Harris T.D., Brus L. E., Brown W. L., Chaban E. E., Szajowski P. F., Christman S.B., Citrin P.H. Size, shape, and composition of luminescent species in oxidized Si nanocrystals and H-passivated porous Si. Phys. Rev. B. 1995. V. 52. P. 4910–4925.

Sham T.K., Jiang D. T., Coulthard I., Lorimer J.W., Feng X.H., Tan K.H., Frigo S. P., Rosenberg R.A., Houghton D. C., Bryskiewicz B. Origin of luminescence from porous silicon deduced by synchrotron-light-induced optical luminescence. Nature. 1993. V. 363. P. 331–334.

Sorensen L.B., Cross J.O., Newville M., Ravel B., Rehr J. J., Stragier H., Bouldin C. E., Woicik J. C. Diffraction anomalous fine structure: unifying x-ray diffraction and x-ray absorption with DAFS. In: Resonant anomalous X-ray scattering: Theory and applications. Ed. by G. Materlik, C. J. Sparks, K. Fischer. North-Holland, 1994. P. 389–420.

Spence J. C.H. High-Resolution Electron Microscopy. 3 ed. Oxford University Press, 2009. 424 p.

Tang Y.H., Sham T.K., J"urgensen A., Hu Y. F., Lee C. S., Lee S. T. Phosphorus-doped silicon nanowires studied by near edge X-ray absorption fine structure spectroscopy. Appl. Phys. Lett. 2002. V. 80. P. 3709–3711.

Wang Z. L. (Ed.). Characterization of Nanophase Materials. Wiley-VCH Verlag GmbH, 2000. 406 p.

Warren B. E. X-ray diffraction. Addison-Wesley, Reading, MS, 1969. 381 p.

Will G. Powder diffraction: The Rietveld method and the two stage method to determine and refine crystal structures from powder diffraction data. Springer-Verlag, 2006. 232 p.

Williams D. B., Carter C.B. Transmission electron microscopy: A textbook for materials science. Springer, 2009. 779 p.

Woolfson M.M. An Introduction to X-ray Crystallography. Cambridge University Press, 1997. 414 p.

Yang Leng. Materials Characterization: Introduction to microscopic and spectroscopic methods. Singapore: John Wiley & Sons, 2008. 384 p.

Young R., Ward J., Scire F. Observation of metal-vacuum-metal tunneling, field emission, and the transition region. Phys. Rev. Lett. 1971. V. 27. P. 922–924.

Young R., Ward J., Scire F. The Topografiner: An instrument for measuring surface microtopography. Rev. Sc. Instr. 1972. V. 43. P. 999 (13 p.).

Young R.A. (Ed.). The Rietveld method. IUCr, Oxford University Press, 1995. 312 p.

Yurtsever A., Weyland M., Muller D.A. Three-dimensional imaging of nonspherical silicon nanoparticles embedded in silicon oxide by plasmon tomography. Appl. Phys. Lett. 2006. V. 89. P. 151920 (3 p.).

Zanchet D., Hall B.D., Ugarte D. X-ray Characterization of Nanoparticles. In: Characterization of Nanophase Materials. Ed. by Zhong Lin Wang. Wiley-VCH Verlag GmbH, 2000. Ch. 2. P. 13–36.

Zhang Q., Bayliss S. C. The correlation of dimensionality with emitted wavelength and ordering of freshly produced porous silicon. J. Appl. Phys. 1996. V. 79. P. 1351–1356.

9

Methods for controlling nanosilicon particle size

In 1988, the effect of the crystallite size on the band gap E_g – the quantum size effect (Furukawa et al, 1988) was observed for the silicon nanocrystals (nc-Si). Then, Lee Kenham (Canham, 1990) observed the effect of visible photoluminescence of porous silicon particles (p-Si) when excited by ultraviolet light, which is absolutely not typical for crystalline silicon (c-Si) – classical indirect-gap semiconductor having $E_g = 1.17$ eV. Subsequent studies (Cuilis et al, 1997; Kumar, 2008; Ledoux et al, 2000; Takeoka et al, 2000) confirmed the dependence of the luminescence spectrum on the size of crystallites of nc-Si. Moreover, it was found that crystals of nc-Si can effectively generate stimulated emission (Pavesi et al, 2000), and they showed a non-linear increase in the intensity of luminescence (Nayfeh et al, 2001), i.e. the laser effect is manifested.

In all studies, nc-Si showed a large width of the luminescence peak (half-width is typically in the range 200–300 meV), which can not be lowered even by measurement at very low temperature (Takeoka et al, 2000). This shows that the width of the luminescence peak in the systems of the nc-Si particles is determined mainly by the inhomogeneity of the material and the spread of the crystallite size. A team of researchers (Sychugov et al, 2005) were the first to measure on a single Si nanocrystal photoluminescence at low temperatures. They have convincingly shown that each Si nanocrystal gives a very narrow photoluminescence peak (half-width of the peak at 35 K is several meV), similar to atomic radiation, and the broadening observed in conventional measurements on ensembles of nc-Si is due to the scatter of the dimensions and differences in the shape of particles.

Synthesis produces most often polydisperse nanosilicon powders, and to link their electron-optical properties with the size it is not enough to know the average size of particles in the powder and it is required to define the size distribution density function (SDDF) with suitable experimental methods (Myatlev et al, 2009). Lee Canham (Canham, 1997) compared the results of several studies and showed that using the average size of particles in polydisperse nanosized systems for determining the size–property relationship, one can easily come to wrong conclusions, for example, the fluorescent properties of porous silicon do not depend on the size, as in (Kanemitsu et al, 1993), or that they vary slowly with changes in size, as in (Schuppler et al, 1995). In this review of the available experimental data it can be understood that in any layer of porous silicon, along with very small photoluminescing particles there is a significant part of the skeleton, which consists of too large particles, in which the effect of the quantum constraint does not occur and in which there is no visible region in luminescence. In such cases, the 'average' particle size of the skeleton has almost nothing to do with the luminescence determined by size, because the efficiency of emission is affected only by one tail of the distribution of sizes (from the smallest structural details).

There are many methods for the particle size control of powders (Allen, 1997; Merkus, 2009), but not many of them can be used for analysis of nanomaterials. These ultrafine systems can be characterized by more sophisticated methods (Kaye, 1999). This is carried out using different spectroscopic techniques, optical laser diffractometry and laser microprobe analysis, or high-resolution transmission electron microscopy (TEM). To control the large quantities of material in a production environment it is also recommended to use the Brunauer–Emmett–Teller (BET) method, which allows to determine the specific surface area of particles S_{BET} by low-temperature adsorption of nitrogen or inert gas, and at a known density material to recalculate this parameter to the mean particle size of certain form (see, e.g Allen, 1997, v. 2, and Merkus, 2009). Each of the methods has its limitations by the type of samples that it can monitor, and limits on the range of available sizes. Different methods also differ in the representativeness of the results of analysis and their metrological characteristics.

The size of the nanoparticles is often determined by measurement of Raman scattering (Raman spectroscopy). Raman spectroscopy was used to perform the first detailed study of particle size distribution in the skeleton of porous silicon (Frohnhoff et al, 1995; Münder et al, 1992). The particle size distributions obtained in these studies were multi-modal (have several distribution maxima). Despite the widespread use of Raman spectroscopy to study the size of nanoparticles, this method has still many difficulties with the correct measurement data and modelling in processing. Nanosilicon, because of its low thermal conductivity, may have a marked shift of the phonon frequencies simply by heating the sample with a laser beam. In addition, the theories for analysis of the line shape of the nc-Si particles in this method are still under development, and the currently available capabilities allow using Raman spectroscopy to correctly evaluate the size of the core of the particles only in the monodisperse samples. Therefore, the particle size distributions obtained by Raman spectroscopy can be trusted if they are confirmed by other methods.

Laser-optical methods, such as static light scattering (laser diffractometry) and dynamic light scattering (photon correlation spectroscopy) allow us to determine not only the size but also the shape of the particles, as well as the size distribution density function. These two types of methods are based on different physical principles and have different ranges of measurement.

High-speed optical laser diffractometry is widely used for the direct determination of particle size and particle size distribution in liquid and gas flows and operates on the principle of measuring the interference of the light diffracted by the particles. Consequently, the minimum resolution limit of this method is physically limited by the diffraction limit and can not be less than half the wavelength of the probe light, i.e. it cannot be used to determine the particle size diameter of less than about ~150 nm – the diffraction limit for the presently existing solid-state lasers.

Optical photon-correlation spectroscopy is an indirect method of determining the particle size by the time dependence of the intensity of light scattering that is due to Brownian motion of particles, which causes local fluctuations of the refractive index of the medium where the particles are distributed. The measured temporary fluctuations in the intensity of scattered light are converted to the diffusion coefficients of particles in a liquid medium with the precisely known viscosity which are then linked with the size of the diffusing particles. In contrast to the laser diffraction method, this method is not limited by the diffraction limit, so the range of particle sizes that can be measured is very wide and extends from ~0.8 nm up to several microns. The method allows to determine the size and size distribution density function and is widely used in many fields of science and technology.

The first significant limitation of this method for the grain size control of the nc-Si composites is the need to prepare suspensions or sols of dispersed particles and the exact knowledge of the viscosity of the medium. Fast agglomeration, inherent to nanoparticles, causes that is not always possible to determined by this method the size of individual particles which are most interesting for the characterization of nc-Si.

Another major disadvantage of laser-optical methods of particle size of any type is that all particles, regardless of their composition, are the same for these methods, so if the system is a mixture of nanoparticles of different compounds, these methods still characterize it as a system consisting of particles of one type. Unfortunately, the amorphous and crystalline particles whose quantum-dimensional properties in nanosilicon significantly differ, are also indistinguishable for these methods.

Free of these shortcomings is high-resolution transmission electron microscopy (HRTEM) in combination with the observation of electron diffraction. This is the most direct method of particle size control of the nanomaterials which allows to determine the size and the particle size distribution density function with the identification of phase and chemical composition.

However, this method has a number of serious obstacles in widening the range of application. First, it is the high price necessary of modern equipment for HRTEM, and as a consequence, its inaccessibility for a wide range of researchers. In addition, the method is local, which complicates its characterization using large amounts of material. Finally, despite the fact that the method allows direct observation of the particles and the direct calculation of the size distribution density function, the results may have significant systematic errors due to the specifics of the selection and preparation of samples for analysis. In addition, image editing, which is almost always necessary before the particle size analysis in order to separate particles superimposed on the images, can lead to uncontrollable errors.

To determine the size of nc-Si, laboratory studies often use methods of X-ray diffraction of powders on the monochromatic radiation. Despite a long history of X-ray diffraction methods for determining the crystal size (Warren, 1969; Iveronova and Revkevich, 1978), the method for identifying the size distribution density function based on analysis of X-ray powder diffraction patterns is relatively new (see chapter 10), since the first real working computer programs to implement it appeared only at the beginning of the XXI century (Leoni et al, 2006; Scardi & Leoni, 2002), so that this method is not yet widely used, but has a very good potential and will be discussed in detail in chapter 10 of this monograph.

It should be noted that other widely used methods for determining the size of the nanoparticles are methods of optical measurements, as small-angle scattering of X-ray (SAXS) and spectroscopy of the fine structure of X-ray absorption (XAFS).

9.1. Methods for determining the size of nanoparticles by dynamic light scattering

The term 'dynamic light scattering' (DLS) refers to the methods for measuring the temporal fluctuations of scattered light intensity caused by the movement particles (in particular Brownian motion) in the liquid or gaseous media. These methods are used to study the characteristics of particles in emulsions and colloidal solutions, and more recently in aerosols. In principle, these methods can be used to determine the size of the particles with diameters from several nanometers to several microns, as well as the particle size distribution function.

Besides the most common name 'dynamic light scattering' (DLS), this method is also referred to as 'quasi-elastic light scattering' (QELS) or 'photon correlation spectroscopy' (PCS). This terminology is somewhat confusing due to the fact that the results of measurement and analysis are processed using these or other special features of the overall process of light scattering in a disperse environment.

The name 'quasi-elastic light scattering' is used because in the scattering of the moving particles due to the Doppler effect, light can change by a small value (from a few Hz to 100 Hz) the frequency in comparison with the from baseline (frequency of visible light used for the study of dynamic scattering is of the order of 10^{14} Hz). In fact, the DLS method in the QELS mode is used very rarely, because it is very difficult to accurately record such small changes of frequency against the huge frequencies of light.

Recently, photon correlation spectroscopy (PCS) has been used quite frequently, because this method is implemented in most cases using devices with the principle of correlation measurements and the autocorrelation signal analysis.

In the PCS method the scattering intensity is measured by an ensemble of particles of monochromatic coherent light is measured under the given angle as a function of time. The rate of change of intensity of scattered light is determined by the Brownian motion of particles in a dispersed system. The measured variations in the intensity with time is characterized by the correlation function, which is converted to the mean translational diffusion coefficient D (or a set of diffusion coefficients). Rapid changes in intensity are manifested in the rapid decay of the correlation function, indicating a high diffusion coefficient. The obtained diffusion coefficient D is converted, using the Stokes–Einstein equation, to the particle size (hydrodynamic diameter x).

In the PCS method, as in the Stokes–Einstein theory of Brownian motion, which is used in the analysis of experimental data, it is assumed that particles are spherical and do not interact with each other.

It should be noted that the colloidal particles in a liquid dispersed system contain a layer of ions and molecules of the environment connected with their surface, which in the Brownian motion also moves with the particles and affects the diffusion coefficient. Therefore, the measured hydrodynamic size of the particle x is somewhat larger than the actual size.

The DLS method was widely used almost only in the last 20 years thanks to the emergence of small lasers with a large coherence length, high-speed photon detectors (with a frequency of the measurement trains of 50 kHz and more), high-performance, high-speed electronic digital correlators and personal computers with a frequency greater than several hundred megahertz. The need for such a method of measuring the dispersion was so great that in the mid-1990s, an international standard was developed for the determination of the average particle size in suspensions by this method (ISO 22412 – Analysis of particle size by photon correlation spectroscopy) and in the process of being adopted now is the ISO standard 22412, combining different methods of size analysis of the particles by dynamic light scattering (ISO, 2008).

Surprisingly, in the preparation of this brief review we could not find Russian-language books, with the exception of very old books (Lebedev et al, 1987, Eskin, 1986), which would discuss at a professional level the current state of the method and its principles, although the apparatus for measuring by this method is widely used in Russia, especially in biology and medicine. Apparently, specialists working with this equipment refer to these publications. Brief articles on this topic in some Russian-language

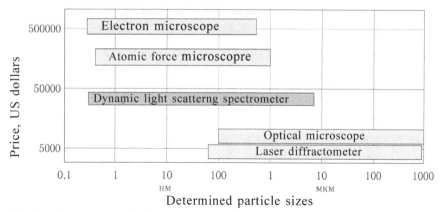

Fig. 9.1. Comparison of prices of equipment and measuring ranges of particle sizes for the analysis methods used in nanotechnology.

books and teaching materials on nanotechnology and nanomaterials, for example in the tutorial (Shabanov et al, 2006), cannot be regarded as a profes- sional consideration of the method, but rather as brief information on the method and its main characteristics, not revealing its details, and, moreover, its development in recent years. In the English-language monographs the current state of dynamic light scattering is widely and deeply discussed. A fairly complete description of the principles of the method of modern advances and equipment can be found in books (Merkus, 2009; Xu, 2002, Gardiner, 1986), as well as a set of review papers on international conferences.

A generalized summary of the principal practical characteristics of the DLS method, according to the book (Merkus, 2009), is given in Table 9.1.

Figure 9.1 shows a comparison of analytical capabilities and cost of DLS equipment with other methods of measuring the size of nanoparticles used in nanotechnology.

9.1.1. The method of photon correlation spectroscopy (PCS)

If a beam of monochromatic light is passed through a suspension of fine particles, such that there is no sedimentation during the measurement time, and a photodetector is used during the time t to continuously monitor the number of photons scattered at an angle θ to the beam direction, we can register the line diagram with fluctuations in intensity, as shown schematically in Fig. 9.2. Just such measurements are made in the PCS method.

The reason for the observed fluctuations in the scattering intensity can be explained from several different perspectives; all of these explanations are valid and can be described by formulas that lead to the same result.

Table 9.1. The main characteristics of dynamic light scattering

Characteristic	Specifications
Principle of the method	Calculations based on the model of dynamic changes of light scattered under a specific angle
The type of the determined diameter of the particle	Hydrodynamic diameter based on scattering intensity
The type of size distribution	The best agreement of the calculations with the experiments in the particle size distribution by the logarithmic-normal law with a small with
The range of diameters permissible for measurement	0.005–1 μm
Calibration of equipment	Not required
Type of sample	Suspensions or emulsions
Mass of sample	From microgramms to milligramms (from 10 μl to 10 ml)
Measurement time	Around 1 min
Reproducibility of mean particle size	Relative error in repeated measurements around 2-5%
Systematic error in determination	Relative error less than 2%
Resolution	Around 30%
Service procedure	Not complicated
Price: price of equipment, running expenses	Depending on specification, 30000-60000 Euro, low
Effect of particle shape	Yes
Standardization	ISO 22412 standard available
Comments	For measuring absolute values of diameter it is necessary to use suspensions and emulsion with very low concentration. At high concentrations, only relative dimensions can be estimated. Many liquids can be used as the mediumn for preparation of samples
Possibilities of measurement in production line	Only if movement of the disperse medium is small. Concentration should not be high
Possible problems and difficulties	At higher concentrations errors may form as a result of interaction of particles and due to the effect of multiple light scattering

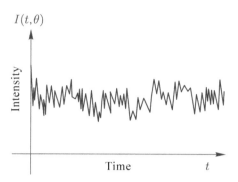

Fig. 9.2. Time dependence of the photodetector signal, measuring $I(t,\theta)$ The intensity the scattering of monochromatic light passing through a suspension of fine particles. Measured by the number of photons scattered at an angle θ to the direction of the primary beam.

1. Fine particles in a suspension are in Brownian motion. In scattering by moving particles, the Doppler effect causes a stochastic change in the frequency of the scattered radiation (quasi-elastic scattering), in the dispersed environment there are local violations of coherence of scattering extinguishing constructive interference and leading to fluctuations in intensity over time.

2. Scattering of light occurs at the boundaries of the sharp change in the refractive index. Such centres are the scattering particles suspended in a liquid (in suspension). The Brownian motion of particles results in stochastic fluctuations in the local refractive index, causing fluctuations in the amplitude (intensity) of light scattering.

3. In a suspension there are always groups of particles (not necessarily located next to each other), which, owing to the successful arrangement, scatter waves with phases that give constructive interference. Due to the constant random displacement of particles in Brownian motion, the phase difference of electromagnetic radiation waves, scattered by different particles, constantly changes resulting in changes of the intensity of constructive interferences and hence the total intensity of scattering by an ensembles of particles fluctuates in time

In all three mechanisms stated above the rate of change of the scattering intensity (sharp fluctuations in Fig. 9.2) is completely dependent on the speed of moving particles, which is determined by their diffusion mobility, i.e. particle size and viscosity of the medium in which they move. And, of course, the effects of temperature, which determines the fluctuations of the solvent molecules causing Brownian motion, is very strong

DLS measurements are commonly taken using the scheme shown in Fig. 9.3.

A device for DLS measurements includes a laser light source, which provides its coherence and vertical polarization, a transparent cell for the

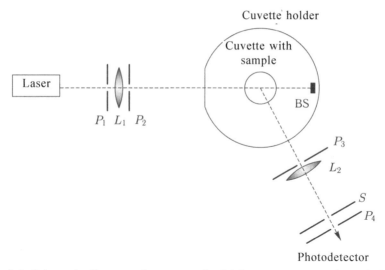

Fig. 9.3. Schematic diagram of apparatus for DLS measurements. The polarization vector of the laser light is perpendicular to the plane of the figure.

sample in the form of dilute suspensions, placed into a cuvette holder, which can have either a normal atmosphere or vacuum, or a filling fluid with a certain refractive index to compensate for refraction in the walls of the cuvette, with a constant temperature maintained, a system of collimating slits, focusing lenses, polarizers, and apertures for cutting the scattered radiation at a certain angle and low-noise photodetector with high temporal resolution.

Illumination of the suspension with a beam of coherent laser radiation leads to Rayleigh scattering of light by particles and due to the coherence of the primary radiation, the beams scattered from different particles are also coherent and can produce constructive or destructive interference. If we make the snapshot (with an exposure of the order of microseconds) of the scattered radiation, we obtain the diffraction pattern (see Fig. 9.4), containing information about the location of particles at the time of measurement.

The diffraction pattern has a scaly appearance, and each speckle (scale) on it is obtained from a group of particles, the location of which leads to constructive interference. If the two-dimensional picture shown in the figure is integrated over the area, we get a one-dimensional instantaneous intensity $I(t)$ at time t. At the next moment the particle due to Brownian motion change position, and depending on the size and shape each particle moves over a distance characteristic of this particle in a random direction, and only pairs of beams from other particles and particle groups will interfere, i.e. the number, intensity and type of speckles in the diffraction pattern changes, and the two patterns measured at different times will give different values of $I(t)$.

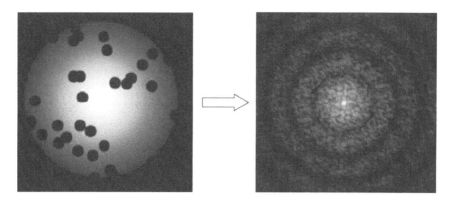

Fig. 9.4. Diagram of the optical conversion of the particle distribution in the speckle diffraction patterns. On the left a system of particles, on the right the corresponding diffraction pattern.

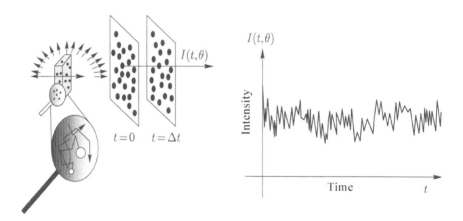

Fig. 9.5. The emergence of the temporal spectrum of DLS intensity from a system of particles in Brownian motion in the fluid. Left due to Brownian motion the particle in the suspension travel, so their distribution at time $t = \Delta t$ differs from that observed at time $t = 0$. On the right, due to changes in the location of the particles in a suspension the scattering intensity of laser light transmitted through the medium, measured under an angle to the primary beam, fluctuates in time t.

The process and results of measurements by DLS are shown schematically in Fig. 9.5.

The fact that the one-dimensional scattering intensity by a system of particles, which are in Brownian motion, must vary (fluctuate) with time, is easy to prove mathematically, describing the process of scattering by an ensemble of particles and the result of observation of intensity. The formalism of describing the scattering of light is almost identical with the description of the scattering of electrons or X-rays. In the formulas for the intensity there is an exponential the exponent of which has the terms $\mathbf{Q} \cdot \mathbf{r}_j$, where the multipliers denote respectively the scattering vector and the coordinate of the j-th particle, which varies due to Brownian motion, leading to a change in intensity.

The observed fluctuations in the scattering intensity can be analyzed either in the time scale with the help of the correlation function (if the analysis is carried out in terms of mechanism 3), or on the frequency scale by means of frequency analysis (If the analysis is performed on the basis of mechanism 1). However, both methods of analysis are linked together by the Fourier transform.

Most often, the size and density of the particle size distribution are determined by DLS using correlation analysis. In correlation analysis we construct the autocorrelation function $G(t)$ of the values of intensity, which is it should be, fades with time, i.e. the correlation decreases increasing the lag between the measurements of intensity. The autocorrelation function characterizes the rate of change in the intensity which depends on the velocity of the particles and therefore is related to the diffusion coefficient. Diffusion coefficient D of Brownian motion, in turn, depends to some extent on the size and density of the particles and the viscosity of the solvent in the suspension. As a result of these relationships, the autocorrelation function of the intensity fluctuations can be converted to the size x and/or particle size distribution $q(x)$. In this transformation it is usually assumed that the particles are spherical and do not interact with each other.

The procedure for determining the particle size and the size distribution density function by DLS can be clearly depicted by the following diagram (Fig. 9.6).

Measurements and data transformation in the PCS method. A typical scheme of the PCS spectrometer is shown in Fig. 9.6. The light source uses a laser generating monochromatic coherent radiation with a vertical polarization. This is often carried out using He–Ne lasers with the output power of 2.5 mW. In the measurements of very small particles it is recommended to use a more powerful laser (about 25 mW), because such particles scatter very little light. Focusing the laser beam on the area of measurement is often carried out using lens L_1.

The sample (a highly diluted suspension of the sample powder in a clear wetting liquid with the precisely known viscosity) is placed in a cubic or circular cylindrical measurement cell (cuvette) made of transparent materials (fused silica or plastic) with a path length of about 1 cm, set in the cuvette holder in which the temperature is known and kept constant during measurements with an accuracy of ±0.3°C. Sometimes the measuring cell is placed the liquid with the desired refractive index to compensate for the refractive index between the cell walls and the environment.

The intensity of the scattered light (number of photons in a given time) is converted into an electrical signal in the photodiode detector or a photomultiplier of a scintillation counter, which is located at a certain angle

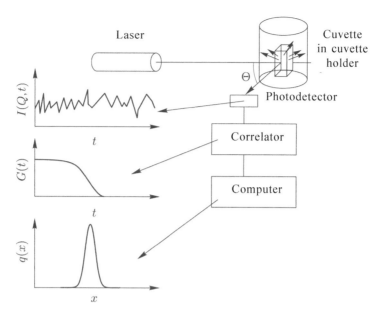

Fig. 9.6. Schematic diagram of the procedure for obtaining the particle size density distribution function by PCS – correlation spectroscopy of light scattering. $I(Q, t)$ – the intensity which depends on the scattering vector Q and the time t; $G(t)$ – autocorrelation function depending on the time t; $q(x)$ – size distribution density function x of the particles in the suspension.

θ to the passing the primary beam of laser radiation behind the lens system and a set of narrow slits or diaphragms (which lately is often replaced by an optical fiber). The most common angle θ, at which the scattering is measured, is fixed at 90°, and sometimes it changes in the 20–160° range using a goniometer or fiber optics to choose the most appropriate length of the scattering vector, which determines the resolution of the method. The measurement zone is determined by the width of the laser beam and the working width of the aperture diaphragm and detector. Typically, this zone is about 10^{-6} cm³ in the measurements at an angle 90°. Such a small size of the measurement zone provides the necessary spatial resolution for adequate detection. The primary laser beam passed through the sample is absorbed by the BS trap (see Fig. 9.3). The signal from the detector is usually analyzed by spectral analysis (in the phase mode), which gives the power spectrum of the scattering or is transmitted into the multi-channel digital correlator (if analysis is carried out by the correlation method).

The measurement procedure is as follows. The cell with the sample is irradiated with a beam laser and the detector continuously measures the number of the scattered light photons falling into its reception window and transmits real-time data in the form of a train of electrical pulses to a digital correlator. In the correlator the time interval generator (clock) divides the resulting time spectrum (pulse train) into equal short intervals Δt (the duration of the intervals can be from nanoseconds to milliseconds). The number of pulses incident on the photodetector in each interval is summed up and loaded into the serial cells of the register of the analyzer memory (analyzer channels) corresponding to specific time lags (time delay) τ measured relative to the first pulse ($\tau = n\Delta t$, $n = 0.1,..$ – number of channels of the analyzer). Then, to obtain the autocorrelation function $G(\tau)$, the correlator multiplies in pairs the contents of the channels of the analyzer and puts the product in their respective channels (channel numbers in the correlator are determined by the delay between the compared data, i.e. by the number N of time intervals Δt between the multiplied values of intensity in the channels of the analyzer). For example, the contents of the first channel of the analyzer is multiplied by itself (delay time $0 \times \Delta t$) and the result is recorded in the first channel of the correlator, is then multiplied by the contents of the second channel of the analyzer (delay time $1 \cdot \Delta t$) and the product is stored in the second channel, then in the channel with a delay of $2 \cdot \Delta t$, etc.

Thus, we compute all the possible products of pairs of data from the entire set, stored in the channels of the analyzer, and the results are recorded in the channels of the correlator with the number equal to the difference between the number of time intervals between the compared pairs of the analyzer channels. The result is the correlation function $G(\tau)$, shown schematically in Fig. 9.6, and an example of a real function is shown in Fig. 9.7.

Extraction characteristics of particles from the autocorrelation function. By definition, for the signal $f(t)$ stochastically changing with time, the autocorrelation function (ACF) is an integral self-convolution

$$\Psi(\tau) = \int f(t) f(t-\tau) dt = f(t) * f(t-\tau), \qquad (9.1)$$

which shows the relationship of the signal, i.e. the function $f(t)$, with its value, shifted in time by the value τ. The * in the shortened form, which is often used in the formulas, indicates that the functions are involved in the operation of the convolution integral.

To analyze the temporal correlations of the light scattering signal (variable $x(t)$, which can be either the intensity $I(t)$, or the amplitude of the scattered field $E(t)$, but more often is simply an electrical pulse from the photodetector) in the DLS measurements the autocorrelation function is denoted by $G(\tau)$ and, according to the definition (9.1), can be written as (Gardiner, 1986)

$$G(\tau) = \lim_{T \to \infty} \left[\frac{1}{T} \int_0^T x(t) x(t+\tau) dt \right] = \langle x(t) \cdot x(t+\tau) \rangle. \qquad (9.2)$$

Here, T is the total time of measurement of the autocorrelation function, τ is

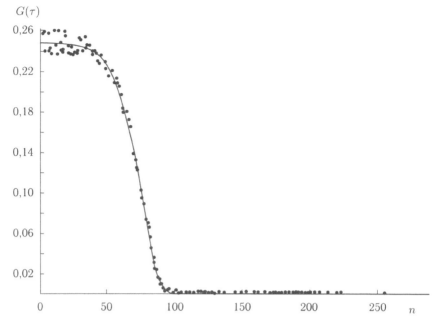

Fig. 9.7. Form of the real measured autocorrelation function (scattering by albumin), n is the number of the correlator channel correlation. The experimental points and the approximating line are shown.

the time delay or time lag between two measurements the signal. From the expression (9.2) it is obvious that $G(\tau)$ is the average over a time interval T of the product of the values of the random function, taken at two different points in time separated by a time period τ. The angle brackets $\langle \cdot \rangle$ in the shortened form denote the averaging over the time interval T.

This function has several important physical meanings and easy to understand properties. If we consider the following two directly following acts of scattering from a system of particles in Brownian motion, then the system has not yet greatly changed, and the information which they carry on the state of the system corresponding to the measured signals $x(t)$ is linked in some way, i.e. the signals $x(t)$ and $x(t + \tau)$ are correlated. If the time τ between acts of scattering is very long, the stochastic system changes beyond recognition, and the two measured signals are virtually independent, i.e. not correlated. These facts can be expressed mathematically as

$$\langle x(t) \cdot x(t+\tau) \rangle \rightarrow \langle [x(t)]^2 \rangle, \ \textit{if } \tau \rightarrow 0.$$

As $\langle [x(t)]^2 \rangle \geq \langle x(t) \rangle^2$ then the value of ACF is maximal at $\tau = 0$, and with increasing time τ either remains constant or decays to the rms value of the signal. The shape and the decay rate depend on the nature of the measured signal $x(t)$.

In the equilibrium stochastic system the autocorrelation function depends strongly on the time interval of measurements τ but does not depend on time t of the beginning of the measurements. Therefore, the time t in the formula (9.2) can be anything, including zero.

Comparison of different ACFs is often made using the normalized autocorrelation function of the intensity of light scattering (NACFI)

$$g^{(2)}(\tau) = \frac{\langle I(0) \cdot I(\tau) \rangle}{\langle I \rangle^2}. \tag{9.3}$$

For the electric field strength E of electromagnetic radiation (the amplitude spectrum of radiation), related in the quadratic manner to the intensity, i.e. $I(t) = E^2(t)$, the correlation is evaluated by the normalized temporal autocorrelation function of the fluctuations of the scattered field, or simply by the normalized autocorrelation function of the field (NACFF), which has the form

$$g^{(1)}(\tau) = \frac{\langle E(0) \cdot E(\tau) \rangle}{\langle I \rangle}. \tag{9.4}$$

The functions (9.3) and (9.4) describing respectively the correlation between the intensity and fields, are linked so-called Siegert relation

$$g^{(2)}(t) = 1 + \beta \left| g^{(1)}(t) \right|^2,$$

where β is the coherence factor, and the superscripts (1) and (2) indicate the normalized autocorrelation function respectively, to the amplitude of the field and the intensity of the scattered light (the autocorrelation functions of the first and second order, since the amplitude and intensity are linked by a quadratic dependence).

One can show mathematically (Kissa, 1999; Merkus, 2009; Xu, 2002, Gardiner, 1986) that the correlations of scattering by the system of particles in Brownian motion, i.e. the NACFI and NACFF, decay exponentially with thr time of measurement. For a system of identical particles (monodisperse system) this attenuation can be expressed by a factor Γ of the rate of decay of correlations and NACFF can be written in the form

$$g^{(1)}(\tau) = \exp(-\Gamma \cdot \tau) \tag{9.5}$$

The rate of decay of the autocorrelation function at Brownian motion, as shown by the solution of the second Fick's equation (diffusion equation), is directly dependent on the coefficient of translational diffusion coefficient D of the particles, and

$$\Gamma = Dq^2. \tag{9.6}$$

Here q denotes the modulus of the vector of scattering of electromagnetic radiation, which is expressed by the usual formula

$$q = \frac{4\pi n}{\lambda_0}\sin(\theta/2),\qquad(9.7)$$

where n is the refraction index of the medium where the particles are situated. It is obvious that under these conditions, the diffusion coefficient D can be determined directly from the known correlation function (9.5), substituting expression (9.6) into it.

At the same time, for identical spherical particles with a diameter of x, that are in Brownian motion in a liquid medium and do not interact with each other, according to the Stokes–Einstein theory the diffusion coefficient of particles is expressed by the formula

$$D(x) = \frac{k_B T}{3\pi\eta x},\qquad(9.8)$$

where k_B is the Boltzmann constant, T is temperature in absolute scale η is the dynamic viscosity of the solvent, x is the apparent hydrodynamic diameter of the particle.

If we substitute (9.8) into (9.6) and then (9.6) into (9.5), then from the resulting equations and from the measured correlation function we can easily determine the size of particle x, if the measurements were performed at constant temperature and the viscosity of the medium is precisely known.

It should be borne in mind that these calculations are made only to explain the principle of determining the particle size obtained by DLS and the formulas derived for real work and accurate determination of the particle size require some modification. In particular, we should take into account the instrumental function and a number of parameters describing the extraneous perturbations introduced in the process of real scattering (see, for example, books (Kissa, 1999; Merkus, 2009; Xu, 2002, Gardiner, 1986)).

In reality, the measured temporal autocorrelation of the intensity of the scattering by the strongly diluted dispersion medium with **monodisperse particles** is given by (Kissa, 1999)

$$G^{(2)}(\tau) = A + B\left[g^{(1)}(\tau)\right]^2 = A + B\exp(-2\Gamma\tau),\qquad(9.9)$$

where the superscript (2) indicates we consider the correlation between the intensities and not the amplitudes, i.e. the autocorrelation function of the second order, A is a term which takes into account the constant background, with the autocorrelation function decreases to the line of this background with infinitely increasing measurement time, and the factor B is an instrumental factor which is determined by the optical measurement.

Thus, taking into account the properties of the autocorrelation function of random variables, in single scattering of light in a monodisperse medium autocorrelation dependence of the intensity can be expressed by the formula

$$G^{(2)}(\tau) = \langle I(0,0) \cdot I(0,\tau) \rangle = \langle I \rangle^2 \left(1 + \exp\left[-2D(x)q^2\tau\right]\right) = \langle I \rangle^2 \left[1 + \exp\left(-\frac{2q^2 k_B T}{3\pi\eta x}\tau\right)\right],$$

(9.10)

where q is the scattering vector. In the formula (9.10) we have used the assumption that the investigated random process is invariant over time, i.e. t can be anything, including and zero. If the viscosity of the solvent η and temperature are known, equation (9.10) allows to estimate the size x of the particles in the suspension directly from the slope of the line of the graph $\ln [G^{(2)}(\tau) - 1] \sim \tau$, as shown in Fig. 9.8.

However, the applicability of this simple method is limited by a system of non-interacting spherical particles and the single scattering of light **in the monodisperse system.**

If the scattering system is polydisperse, the autocorrelation function of the field is a weighted sum of intensity of different autocorrelation functions of particles of different sizes, and the expression (9.5) must have the form

$$g^{(1)}(\tau) = \int_0^\infty C(\Gamma) \exp(-\Gamma \cdot \tau)\, d\Gamma,$$

(9.11)

where $C(\Gamma)$ is a function of the distribution function of the rate of decay of the correlation functions. This equation is the Laplace transform, which is poorly undefined and has no exact solution. The same equation in discrete form is:

$$g^{(1)}(\tau) = \sum_{i=1}^n c_i \exp(-\Gamma_i \cdot \tau).$$

(9.12)

Here, the coefficients c_i are weighted by the intensity contribution to the total rate of decay of scattering correlations of the group of particles, indexed by sequence number i, each of which the particles have one size x_i and are characterized by the diffusion coefficient $D_i = \Gamma_i/q$). To find the diffusion coefficients D_i and the corresponding particle size x_i it is necessary to carry out the inverse transformation of the experimentally measured correlation function. As the summable correlation functions are very close (the difference between them is often on the level of experimental noise), then the inverse transformation is ill-conditioned, so the problem is usually solved by approximate cumulant expansion, if the variation in the size distribution is not very large, or by the regularization methods (Provencher, 1982a,b).

However, the theory of photon autocorrelation spectroscopy is well-established only in the case of single scattering of light. If there are multiple scattering effect, this theory is not valid and gives erroneous results, which become dependent on the concentration and the optical path of light in the sample, i.e. even on the position of the measuring cell. Therefore, the photon autocorrelation spectroscopy method can investigate only very dilute suspensions, in which the effects of multiple scattering are unlikely.

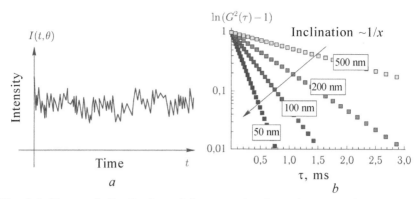

Fig. 9.8. Temporal distribution of the scattering intensity monodisperse system (*a*) and examples of determination of particle size in monodisperse suspensions of several scheduled autocorrelation function (*b*)

Low concentrations of suspensions make this method very sensitive to any impurities in the liquid. Therefore, for ideal measurements it is necessary to use very pure liquid sample preparation and conduct and measurements should be taken in a clean room.

From the condition of the absence of interactions between the particles it is clear that the DLS method can work properly when measurements are taken on very dilute suspensions. Good results with this method, which is most often used in the variant of photon correlation spectroscopy of scattering intensity, can be obtained in measurements on very dilute solutions, but not less than 500 particles in the area of measurement. With increasing concentration above 0.01 vol.% the effect of interaction between the particles starts to operate together with the effect of multiple scattering that distort the results (Merkus, 2009). The achievement of such undesirable concentrations is often easy to spot by the visible turbidity of the solution. At too low concentrations (less than 500 particles in the measurement zone) random concentration fluctuations and the weakness of the scattering signal over background can also distort the result. Turbid suspensions and dyes are often analyzed by the correlation analysis of backscattered light. At the same time, strong absorption of light in poorly transparent systems have leads to a significant decrease in the intensity of scattering, which complicates the detection of scattered light. On the other hand, the strong absorption of light eliminates the effects of multiple scattering of light (the effect of natural filtration), which distort measurements in concentrated solutions.

In the past 10 years special attention has been given to the method of photon cross-correlation spectroscopy (PCCS) of the intensity of light scattering, in which it became possible to screen out by the hardware the contribution of multiple scattering. Also, the method allows granulometric analysis to be carried out even in turbid suspensions. At present, many firms produce commercially available equipment for measurement by this method.

9.1.2. The method of photon cross-correlation spectroscopy(PCCS)

To work with turbid suspensions (i.e. at high particle concentrations) the PCS method was modified to the method that is called photon cross-correlation spectroscopy (PCCS) and in which the effect of multiple scattering is eliminated in the measurements. The basis of this new method is three-dimensional (3D) cross-correlation. With particular geometry of scattering the cross-correlation of scattered light can accurately separate the contributions from the acts of single and multiple scattering. This method can do all that can be done in ACPS but, at the same time, is able to detect the average dynamic particle size and the size distribution function of the particles ranging from 1 nm to several microns, even in turbid suspensions. The principle method is schematically shown in Fig. 9.9.

The primary beam of laser light is divided by a special divisor into two identical beams which are directed by a lens on the same sample volume; two detectors are installed in such a way as to take photons from different beams but with completely identical scattering vectors. Each detector has its own photomultiplier. The signals from the photomultipliers are transmitted for processing in the digital correlation connected with a computer.

In this scheme, it is important to position the detectors to provide the identity of the measurements, from different beams. What this is is shown the following diagram (Fig. 9.10). For a complete suppression of multiple scattering the scattering vectors, received by both detectors, must be identical in magnitude and direction. The scattering volumes should also be identical.

In the cross-correlation method two detectors simultaneously perform independent measurements of one of the analyzed sample volume of the two ACF. These experiments are compared with the cross-correlation function, which is similar to the usual normalized autocorrelation function has the form

$$g^{(2)}(\tau) = \frac{\langle I_A(\tau) \cdot I_B(t+\tau) \rangle}{\langle I_A(\tau) \rangle \cdot \langle I_B(\tau) \rangle}. \tag{9.13}$$

This correlation function is also associated with $g^{(1)}$ and Γ, as the usual autocorrelation correlation function, i.e. by the equations

$$G^{(2)}(\tau) = A\left(1 + B\left[g^{(1)}(\tau)\right]^2\right), \quad g^{(1)}(\tau) = \sum_{i=1}^{n} c_i \exp(-\Gamma_i \cdot \tau).$$

The working formula for the cross-correlation function between signals from two detectors A and B has the form

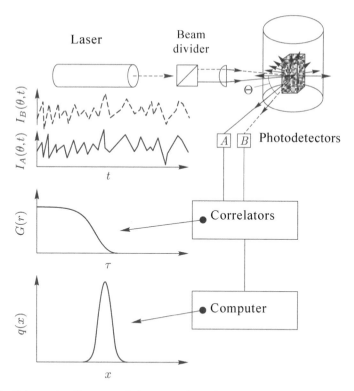

Fig. 9.9. Schematic diagram of the procedure for obtaining the size distribution density function of particles photon cross-correlation spectroscopy.

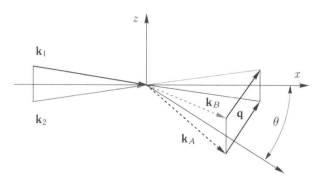

Fig. 9.10. In the FCCS method it is important that the the scattering vectors in measurements by the detectors A and B were identical in magnitude and direction.

$$g^{(2)}(\tau) = \frac{\langle I_A(t) I_B(t+\tau) \rangle}{\langle I_A(t) \rangle \langle I_B(t) \rangle} = \frac{\dfrac{1}{N} \displaystyle\sum_{i=1}^{N} I_A(t_i) I_B(t_i + \tau)}{\dfrac{1}{N^2} \left(\displaystyle\sum_{i=1}^{N} I_A(t_i) \sum_{i=1}^{N} I_B(t_i) \right)}. \qquad (9.14)$$

Calculation by formula (9.14) is carried out in a digital correlator in real time. If only one detector is used, then the formula (9.14) becomes the expression for the autocorrelation function. The numerator in the formula (9.14) is influenced by the temporal correlation of light scattered by particles, while the denominator is proportional to the average intensity of scattered light. The relationship between the intensity correlation function and the function of the size distribution of the particles is given by

$$\sqrt{g^{(2)}(\tau)-1} = g^{(1)}(\tau) \approx \sum_{j=1}^{m}\left[I\left(x_j\right)\right] \cdot \Delta Q_0\left(x_j\right) \cdot \exp\left(-\Gamma\left(x_j\right) \cdot \tau\right), \quad (9.15)$$

where j numbers of the monosized particles, $\Delta Q_0(x_j)$ is the value of the fraction of the particles with a diameter of x_j, i.e. the density distribution for a group of particles with a diameter of x_j. The rest of the terms of this equation have been given earlier.

To determine the size distribution density function of the particles from the experimentally measured cross-correlation function, according to the recommendations of the ISO FDIS 22412 standard (ISO, 2008), equation (9.15) is transformed into a system of n linear equations, in which as the known quantities are the values of the correlation function for n different values of delay time τ.

Determination of values $\Delta Q_0(x_j)$ by the inverse transform of the correlation function (9.15), in fact, is a rather complicated mathematically ill-posed problem and is performed using computer programs based on different approaches (Merkus, 2009). First, the solution must minimize the impact of random noise of measurements, which is usually achieved by the priori assumption about the form of the distribution function and its modalities. Other boundary conditions, simplifying the solution, are also often used. To solve this problem, when the correlation function is reduced to a system of linear equations, often used a computer program CONTIN (Provencher, 1982a,b). This problem and restoration of the parameters of the investigated system of the correlation function of the signal, relating to the class of ill-posed problems, are also solved by a nonlinear least-squares method with the condition of non-negativity and the maximum entropy method (Merkus, 2009).

A program for the analysis of dynamic light scattering Dynals v2.0 was developed by Alango company (Goldin, 2002). This program allows one to process data in three ways: 1) the method of regularization, as the program CONTIN; 2) the method of cumulants, and 3) the method of multiexponential processing. This program is now one of the best programs of its type.

9.1.3. Examples of application of PCS and PCCS

Figure 9.11 shows the comparison of particle sizes in three different systems

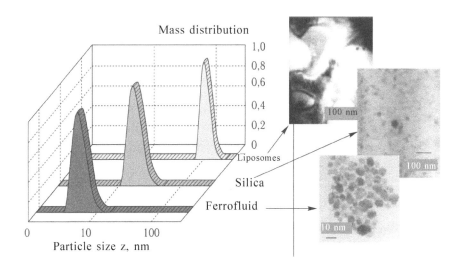

Fig. 9.11. Comparison of the size distributions obtained by PCCS with TEM data for three different materials. Adapted from an article (Witt et al, 2003).

with particles diameters of 3 nm to 200 nm (Witt et al, 2003). The size, determined by cross-correlation analysis of light scattering, are compared with the particle sizes on the TEM images. There is good agreement between the sizes obtained by the two different methods. Unlike TEM, the PCCS method can easily determine the liposome particle size, which when measured in an electron microscope are destroyed by the electron beam.

Figure 9.12 shows the effect of the concentration of particles in suspension on the determination of the particle size by the correlation methods. In this example, the 'monodisperse' latex emulsions in water with a particle size of $x = 107 \pm 10$ nm (according to TEM) were studied. The left figure shows the results of measuring the correlation function of light scattering by a suspension with the transparency of 99.7% and 0.7% (turbid suspension). The measurements were performed in the standard correlation mode (PCS method) and the cross-correlation mode (PCCS method, where the two detectors are used to measure the cross-correlation function). The autocorrelation function of a dilute suspension has the normal form with a linear dependence of the logarithm of the delay time. At the same time, the ACF of the turbid suspension, in which the effects of multiple scattering are strong, strongly deviates from the linear dependence. In the PCCS measurements, multiple scattering is completely eliminated so the cross-correlation function remains linear regardless of the concentration of the suspension.

The right figure shows the mass-weighted particle size distribution. The figure shows that the three-dimensional autocorrelation method (PCCS analysis) gives a unimodal distribution with the expected diameter of about

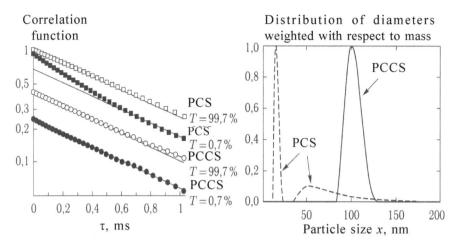

Fig. 9.12. Effect of concentration on the results of the study of the distribution of particles in an aqueous suspension of Latex with a particle size of $x = 107\pm10$ nm (according to TEM). Here: PCS – autocorrelation function of the intensity of light scattering and PCCS – the cross-correlation function. T is the transparency of the suspension. Adapted from an article (Witt et al, 2003).

105 nm even in the case of a very high concentration of particles in the suspension (36.5%, the transparency $T = 0.7$%). At the same time, at this concentration, the PCS method yields a bimodal distribution, which in this case is an erroneous result.

9.1.4. Measuring equipment

The currently available dynamic light scattering spectrometers widely use fiber-optic light guides (Merkus, 2009), thus drastically reducing the size of devices to a desktop or laptop, and increasing their reliability. Below, for example, are photographs of some modern DLS spectrometers (Figs. 9.13–9.15).

9.2. Determination of dispersion by Raman spectra

Raman scattering (Raman effect) is the inelastic scattering of optical radiation on the molecules of matter (solid, liquid or gaseous) accompanied by a noticeable change in its frequency. In contrast to Rayleigh scattering, in the case of Raman scattering the spectrum of scattered radiation contains spectral lines that are not in the spectrum of the primary (exciting) light. The number and location of the emerging lines are determined by the molecular structure of matter. Details of the methods of Raman spectroscopy can be found, for example, in the books (Hollas, 2004; Long, 1997; Grasseli et al, 1984; Smith, 1982; Pentin and Vilkov, 2003).

Fig. 9.13. Industrial device NANOPHOX of Sympatec GmbH – table-top device for controlling nanoparticle dispersion by the PCCS method.

Fig. 9.14. Multiangular dynamic light scattering spectrometer Photocor Complex (Photocor Instruments, Inc.). Designed to measure the size of the nanoparticles, the diffusion coefficients and molecular weight of polymers. The scattering angle from 10° to 150°. The cross-correlation photon counting system for maximum accuracy of measurement of extremely small particles.

Fig. 9.15. Photocor Mini – a device for measuring the size of the nanoparticles. The device meets the international standard for laser measurement of the particle size ISO 22412:2008. Most well-known manufacturers of particle size analyzers by DLS methods are firms: ALV GmbH, Brookhaven Instruments Corp., Malvern Instruments Ltd., Microtrac Inc., Otsuka Electronics Company Ltd., Photocor Instruments, Inc., Sympatec GmbH.

Raman spectroscopy can be described as follows. In the fall of the monochromatic light with a frequency of $\tilde{v}_0$ on the material the light is scattered by the molecules of the material. Scattering can be elastic (without changing the wavelength and wave number) and inelastic (with a change in the wavelength and hence the frequency, resulting in redistribution of energy between the electromagnetic wave and the oscillating molecules). In the frequency spectrum of the scattered light there are frequencies equal to $\tilde{v}_0$ and $\tilde{v}_0 \pm \tilde{v}_M$ where $\tilde{v}_M$ is the vibrational frequency of the studied (scattering) the molecule. In optical spectroscopy the elastic scattering (without changing the frequency) is called Rayleigh scattering and inelastic – Raman scattering (RS). In the course of RS there also occurs changes of the vibrational energy of the scattering molecules, in particular, various modes of excitation of internal energy operate. In the scattering spectrum there are new lines with frequencies that are combinations of the frequency of incident light and the frequency of vibrational and rotational transitions of the scattering molecules. In other words, we can say that Raman scattering is accompanied either by the creation of a phonon (Stokes process) or by the absorption of a phonon (anti-Stokes process). The kinematic and energy diagram of the process of inelastic scattering of light on the atom/molecule is shown in Fig. 9.16.

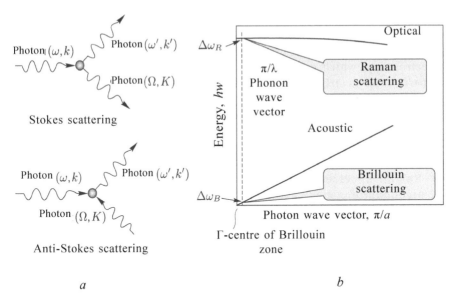

Fig. 9.16. Scheme of the scattering of light on an atom/molecule of the substance with the participation of phonons. a – kinematic scheme of RS of the photon (*in direct space*) with the emission or absorption of a phonon; ω and k denote the cyclic frequency and wave number of the photon, respectively; Ω and K – angular frequency and wave number of phonons. b – the energy (*in the reciprocal space*) scheme of the process of inelastic scattering with optical phonons (Raman scattering) and acoustic phonons (Brillouin scattering).

In Raman spectroscopy the vibrational levels are excited by absorption of a photon with a frequency v_{inc} and re-emission of another photon with a frequency v_{emit}: $E_n = |hv_{inc} - hv_{emit}|$.

There are two cases: 1) $v_{inc} > v_{emit}$ corresponds to Stokes lines and 2) $v_{inc} < v_{emit}$ to anti-Stokes. In practice of Raman spectroscopy measurements are taken of the phonon frequency-determined difference $\Delta v = v_{phon} = |v_{inc} - v_{scat}|$ between the frequencies of the incident v_{inc} and scattered v_{scat} light, where v_{phon} is the frequency of some phonon mode of the optical branch. The value of the scattering frequency is $v_{scat} = v_{inc} \pm v_{phon}$, where the negative sign corresponds to the Stokes lines, and the positive – anti-Stokes.

The scattering of light due to acoustic phonons is called Brillouin scattering, and the scattering due to optical phonons is called Raman scattering. Both of these processes in the reciprocal space are schematically shown in Fig. 9.16b. Since the wave vector of the phonon π/a (where a is the lattice constant, measured in angstroms) is much greater than the wave vector of the photon π/λ (where λ is the wavelength of the laser radiation used in measuring the RS spectra, which typically has a wavelength of 400–700 nm), the scattering of light occurs near the centre of the Brillouin zone (Fig. 9.16b). So the frequency of Brillouin scattering (or Brillouin shift $\Delta\omega_B$) is much less than the frequency (or Raman shift $\Delta\omega_R$) for Raman scattering.

Raman scattering can occur both on the rotational and vibrational oscillations of the molecules and, therefore, the following are considered separately: rotational, vibrational and vibrational–rotational Raman spectroscopy, which differ in the formalism an the theory of analysis of experimental data (Hollas, 2004). However, the rotational transitions in Raman spectroscopy are clearly expressed only in gases, therefore, for the study of solid materials it is most often necessary to use vibrational Raman spectroscopy, and for crystalline materials in which low-frequency librational vibrations are possible, we can also used methods of rotational–vibrational Raman spectroscopy.

Raman spectroscopy is a very sensitive tool for the study of semiconductor nanostructures, including nanosized silicon particles (Marchenko et al, 2000; Meier et al, 2006; Faraci et al, 2006). The vibrational properties of nanostructures – i.e. acoustic and optical phonon modes in bounded systems – and interaction with the photon, and change of the selection rules play a fundamental role in understanding the basic processes of the scattering with a vector with a small length (Faraci et al, 2006). For bulk crystalline silicon the Raman spectrum shows a peak at a frequency of 521 cm^{-1}. Using Raman spectroscopy, for nanocrystals we can obtain experimental information as a function of the size of the crystallites in terms of the shift of the peak from the band belonging to crystalline silicon. In addition, the broadening of the Raman band for nanoparticles provides information on their structural features.

Raman spectroscopy is often used to characterize Si nanocrystals as an express non-destructive method of analysis (Hollas, 2004; Long, 1997; Grasseli et al, 1984; Pentin and Vilkov, 2003). On the basis of the shape and position of the first order peaks in the Raman scattering band, we can:

1. analyze the structure of nanocrystals (carry identification);
2. evaluate the size of the nanocrystals (track size variations);
3. evaluate the strain in the nanocrystals (follow the evolution of the deformation of nanocrystals);
4. evaluate the ratio of the amorphous and crystalline phases (or follow the change of the phase composition).

The sensitivity of Raman scattering spectra to the nanocrystalline Si phase is illustrated in Fig. 9.17, which shows the change in Raman spectra at formation and evolution of Si nanocrystals in a matrix of sapphire (Al_2O_3), depending on the annealing temperature of initial amorphous silicon deposited by ion implantation (Yerci et al, 2006). It is seen in the transformation of amorphous clusters to Si nanocrystals the wide spectral band with a maximum around 480 nm is transformed into a narrow strip with a width of about 3–4 cm^{-1} with a maximum at 521 cm^{-1}, characteristic of the bulk silicon crystal at room temperature. On the other hand, this band for the nanocrystalline Si is broadened and shifted toward lower wave numbers due to the effect of spatial constraints of the phonons. Comparison of the Raman spectra in Fig. 9.17, shows that the formation of nanocrystals in the sample begins at temperatures between 700 and 800 °C for the sample with the implantation dose of $2 \cdot 10^{17}$ Si/cm^2.

9.2.1. The distribution function of the size of nc-Si from Raman spectra

As a result of the law of momentum conservation, the relaxation of excited states in the spectra of light scattering by the crystal structures is accompanied by the formation of narrow Raman lines. This means that the wave vectors of the photons are much smaller than the wave vectors of phonons. Therefore, only phonons with the wave vector $k \approx 0$ can participate in Raman scattering. However, the law of momentum conservation is not applicable to amorphous structures, because they have no long-range order. At the same time, the phonons in the nanocrystals are localized in small crystallites, and their momentum, according to the principle of uncertainty, is poorly defined, which allows phonons with $k \neq 0$ to contribute to the Raman scattering. It is well known that the optical dispersion curves are flat at low values of k and decrease for large k. Thus, in the first order Raman spectra from the nanocrystals we observe asymmetric broadening and red shift.

In the literature there are different models for calculating the size from the shape of Raman lines. For example, the WRL model, developed by Richter and coworkers (Richter et al, 1981) is used widely. This model was modified by other researchers (Campbell & Fauchet, 1986; Paillard et al, 1999; Zi et al, 1996; Islam & Kumar, 2001; Mishra & Jain, 2002). This model is based on the multiplication of the wave function in an infinite crystal in the weighting function W_D (r). The size of the nanosilicon crystals, has been successfully determined using weight functions, such as the Gaussian function and the function Sinc (Paillard et al, 1999; Zi et al, 1996). Paillard and colleagues suggested at a constant weight function, such as Sinc, to regulate the dispersion ratio of the optical phonons, taking into account the anisotropy of the phonon dispersion curves. According to this modified model, the line shape of the Raman spectrum $I(\omega)$ is defined as (Paillard et al, 1999)

$$I(\omega) \alpha A \cdot \left[n(\omega) + 1 \right] \cdot \int \left| C_D(q) \right|^2 \frac{\Gamma_0/\pi}{\left[\omega - \omega(q) \right]^2 + \left(\Gamma_0/2 \right)^2} dq, \qquad (9.16)$$

where $\omega(q)$ is the phonon dispersion in the bulk material, Γ_0 is the natural the Raman line width (RWL) of the bulk material, $C(0,q)$ is the Fourier coefficient of the phonon confinement function, A is a constant, $[n(\omega) + 1]$ is the Bose–Einstein factor. In addition, it is assumed that the dispersion relation $\omega(q)$ of the optical phonons in the nanocrystals must be the same as in bulk Si. The dispersion of phonons is determined either by the formula (Mishra & Jain, 2002):

$$\omega^2(q) = 1,714 \cdot 10^5 + 10^5 \cdot \cos\left(\frac{\pi q}{2} \right), \qquad (9.17)$$

or by the formula (Paillard et al, 1999; Islam & Kumar, 2001):

$$\omega^2(q) = 522^2 - \frac{126,100q^2}{q + 0,53}. \qquad (9.18)$$

Here q is expressed in units of $2\pi/a$, where a is the lattice constant of silicon (0.543 nm), and Γ_0 is approximately equal to 4 cm^{-1}, depending on the system configuration.

Islam and Kumar (Islam & Kumar, 2001) improved the method of determining RWL, having taken into account in the determination of $I(\omega)$ the distribution function of the size σ of the Si nanocrystals:

$$I(\omega) \propto \int_q \left[1 + \frac{(\sigma q)^2}{\alpha} \right]^{-1/2} \frac{\exp\left(\frac{-q^2 L^2}{2\alpha^2} \right)}{\left(\omega - \omega(q) \right)^2 + \left(\Gamma_0/2 \right)^2} d^3q, \qquad (9.19)$$

where a is a constant representing the degree of restriction of the phonon, and L is the average size of nanocrystals.

In a study of Zi et al. (1996) using a microscopic model and taking into account the bond polarizability, the authors determined the correlation

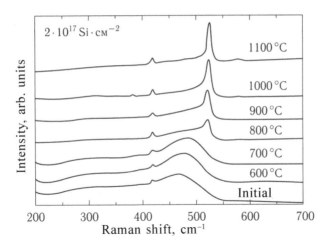

Fig. 9.17. Raman spectra of samples consisting of a sapphire matrix (Al_2O_3) with embedded silicon clusters. Samples were obtained by ion implantation of silicon with a radiation dose of $2 \cdot 10^{17}$ Si/cm² in the matrix and subsequent annealing at various temperatures (Yerci et al, 2006). For comparison, the Raman spectrum of the initial sample without annealing immediately after implantation is shown.

with the Raman shift of the size of Si nanocrystals. Raman shift $\Delta\omega$ was determined defined as in Islam & Kumar (2001), i.e.

$$\Delta\omega = -A \times \left(\frac{\alpha}{L}\right)^{\gamma},$$ (9.20)

where A and γ are fitting refined parameters. The authors found that the latter of these parameters is usually equal to 1.44 for Si nanocrystals in the form of spheres and 1.08 in the approximation of the form of a crystal by a column.

Another modification of the RWL model was proposed in (Faraci et al, 2006), where a particular spatial correlation function was used. Figure 9.18 shows the data from this work. The results of calculation of the Raman frequency red shift due to quantum restrictions for Si nanocrystals with a diameter of 1.2 nm to 100 nm are shown. In the graph, together with calculations using the so-called polarizability model (Zi et al, 1996), there are data according to the theory of Cheng (Cheng & Reng, 2002), calculations using the RWL model (Richter et al, 1981) and by the method proposed by Faraci (Faraci et al, 2006), which are indicated in the figure as 'the present theory'. There are also some literature data. While all the models give the data that are in good agreement with experiment at large sizes, the shift for small sizes obtained using these models is smaller than observed experimentally, which can be attributed at the expense of surface effects in nanocrystals, such as stress (Faraci et al, 2006).

Several models were developed in order to better interpret the experimental: a microscopic force model, a model of bond polarization and a spatial correlation model (Faraci et al, 2006).

However, none of these models gives a satisfactory explanation within the limits of the phonon constraint inside the dot for the simultaneous determination of the shift frequency and line width. Moreover, there is a strong discrepancy between the experimental data and theoretical calculations (Marchenko et al, 2000; Meier et al, 2006; Faraci et al, 2006). In (Meier et al, 2006) experiments were carried out to study the Raman spectra for silicon nanoparticles with a size range between $d = 3.5$–60 nm. Scattering up to the second order was observed. Experimental data were analyzed in the framework of the phonon constraint model. Although this model qualitatively explains the processes of the first order scattering, it is not applicable for the scattering processes of higher order. In the above study the particle size of nanocrystalline silicon, produced the low-temperature plasma chemical method, was also studied by the Brunauer–Emmett–Teller (BET) method.

In bulk crystal, first order Raman scattering exhibits the optical phonon frequency at Γ-point in the Brillouin zone determined by the selection rule $\Delta k = 0$. As the geometrical tensioning of the crystal decreases, this selection rule is removed, and the distribution of scattered light at frequencies is collected in a wide k-interval around Γ. The received RS signal is determined by the weighted integral over k in the first Brillouin zone. Thus, the frequencies near the Γ-point will have the strongest contribution. This is the usual interpretation of the phonon constraint effect. Nevertheless, we can say that with decreasing particle size the optical phonon scattering is manifested as a result of increasing k. These results are shown in Fig. 9.19.

As can be seen from Fig. 9.19, for particle sizes of 60 nm the maximum occurs at 522 cm^{-1}, a value very close to the value for bulk silicon. With decreasing size the shift is shifted toward lower energies. In order to analyze the data, the authors compared the positions of the experimentally determined peaks, the positions of the peaks obtained theoretically based on the model of the phonon constraint, and the Gauss envelope function. All three dependence differ from each other. The results showed a significant shift in frequency for particles with a diameter $d = 27$ nm, obtained by the BET method, whereas in the model construction the expected shift was almost negligible. For smaller particles with $d = 7$–10 nm, the discrepancy between theoretical and experimental data increased and then decreased again (see Fig. 9.20).

As is known, the surface of nanoparticles is very reactive and in the air is always covered with an oxide shell. Therefore, the effective diameter of the particles decreases due to the formation of the oxide shell. Calculations for such particles as shown in Fig. 9.20. Nevertheless, taking into account the oxide layer is insufficient to explain the discrepancy between the experimental and theoretical data. As can be seen from

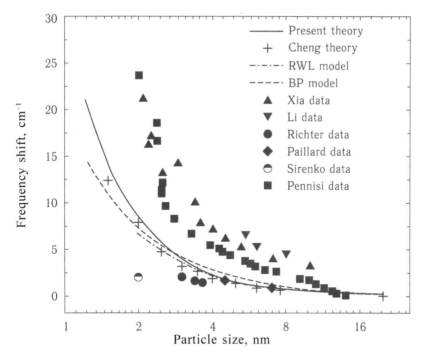

Fig. 9.18. Plots of the red shift of Raman frequencies in relation to the size of quantum dots according to the model (Faraci et al, 2006). For comparison, the BO and RWL models and the experimental points obtained by different authors are also shown,

Fig. 9.20 (with RS shifts), the silicon nanoparticles are characterized by a broad size distribution. However, a large fraction of the particles is significantly smaller than the average particle diameter obtained by the BET method. In this way, we can conclude that the BET measurements give different results compared with Raman spectroscopy experiments.

Soni et al, 1999, presented the Raman spectra for even smaller silicon nanocrystals. With decreasing size from 2.2 nm to 1.4 nm the RS peak shifted from 511 to 502 cm^{-1}. Raman lines are broadened asymmetrically in the direction of decreasing energy. The line broadening arises from the particle size distribution, which leads to strong fluctuations in the wave vectors, in particular for nanocrystals smaller than 2 nm.

A group of Italian researchers (Faraci et al, 2006) presented the results of calculations of the Raman spectra for silicon nanocrystals in the size range 1.2–100 nm. A comparative analysis between the different models and experimental data showed the close correspondence between them for larger particle sizes, while for the small size the experimental data showed a larger shift than calculated. This can be explained by the fact that in RS experiments performed on the deposited clusters one usually obtains an universal layer of known thickness, and it is also difficult to control the particle size distribution.

Amorphous silicon in silicon nanocomposites may appear in the spectrum of Raman scattering as a broadening of the band edge in the 480 cm⁻¹ range. The broad Raman structure near 480 cm⁻¹ corresponds to the transverse optical (TO) mode in amorphous silicon, and the appearance of this band indicates the presence of amorphous components in the samples. In analyzing the Raman spectra we can also determine the relationship between the amorphous and crystalline components in silicon nanocrystals. To this end, the integrated intensities, corresponding to the crystalline and amorphous peaks of S_c and S_a are determined. From the equation $V_c/V_a = \eta \, S_c/S_a$, where η is the relative efficiency of the scattering by crystalline and amorphous silicon, η varies from 1.0 to 1.7.

9.3. Particle size distribution determined by the method of small-angle scattering of X-rays

Small-angle scattering of x-rays yields the diffraction image of the electron density distribution in a multiphase system in which scattering takes place. Under certain initial conditions, this diffraction image can be mathematically linked by the equations with the structure of the scattering system and, solving these equations, we can determine the size and shape of the scattering particles of the multiphase (mostly two-phase) system, if the particles are monodisperse, or the particle size distribution, if the system is polydisperse and has some valid data on the shape of the particles.

The method of small-angle X-ray scattering (SAXS) was developed in the second half of the 1930s (Guinier, 1939) as a means to study the

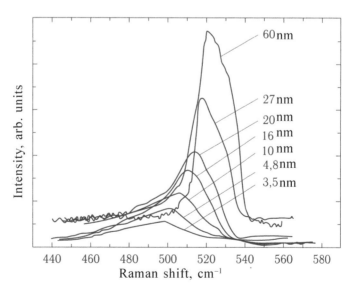

Fig. 9.19. First order Raman spectrum for the Si nanoparticles with diameters ranging from 3.5 to 60 nm determined by BET method. Figure from (Meier et al, 2006).

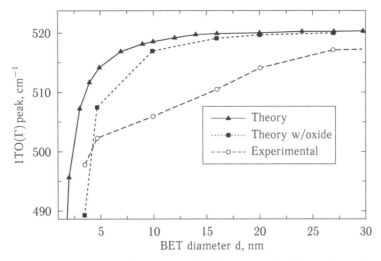

Fig. 9.20. The dependence of the theoretically expected shift of the peak of the phonon constraint model for particles without and with an oxide shell, and the experimental values. Figure from (Meier et al, 2006).

morphology and size of particles of the colloidal size in metals and alloys and has been constantly developed in subsequent years (Guinier & Fournet, 1955; Glatter & Kratky, 1982; Svergun and Feigin, 1986; Glatter & May, 2004). In particular, this method has been used extensively in the past two decades, thanks to the emergence of powerful new sources of radiation, such as X-ray tubes with a rotating anode and sources of synchrotron radiation, as well as through the development of high-performance X-ray optics such as focusing monochromators based on multilayer mirrors (see Fetisov, 2007, p. 306–316) and all kinds of coordinate detectors of X-ray emission (Bergmann, Orthaber et al, 2000), enabling fast and reliable measurement of SAXS. No less significant impact on the development of this method had the appearance of powerful computers and new mathematical methods for processing experimental data, such as methods of indirect Fourier transformation (Glatter, 1980; Bergmann, Fritz et al, 2000; Fritz et al, 2000) and regularization methods (Svergun, 1991, 1992) to solve unstable problems, arising from the transformation of the experimentally observed SAXS diffraction pattern to the information about the microstructure of the sample. At present, SAXS has a solid theoretical base, computer software support and the instrumental basis for solving various problems in the study of the morphology, structure and distribution of particles in greatly different systems and in a variety of conditions.

It should be noted that the SAXS method works in the same range of particle sizes as transmission electron microscopy but unlike TEM it makes it possible to characterize much larger sample volumes (volumes of the order of mm^3 or areas to cm^2) providing a more representative information on the structure of the material, rather than the characteristics of the local volume of the order of μm^3, as in studies by TEM techniques. In addition, measurements by SAXS methods does not require special preparation of samples and the measurements themselves do not require a vacuum and are carried out very quickly (within a few minutes with conventional X-ray sources of the type of unsealed X-ray tubes for X-ray analysis). By virtue of its physical principles this method can examine the matter in solid, liquid and even gaseous states.

For a long time the SAXS method has been mostly used to study the shape and structure of macromolecules, such as biological molecules, in which, thanks to the monodisperse systems, it is possible to determine the external morphology of the molecules, their size, molecular weight, and even the structure (Vachette & Svergun, 2000). But with the development of nanotechnology in the last 15 years, this method has been increasingly involved as a means of the characterization of nanostructured materials, in particular, quantum dots, wires and wells. Many studies have shown (Sasaki, 2005; Britton et al, 2009) that the SAXS method can characterize the dispersion of systems better than the TEM method. In this case, it does not require much effort and expenditure in experiments and can work even in studies *in situ*, providing an opportunity to study the growth of nanoparticles during synthesis, for example, in studies (Mattoussi et al, 1998; Tokumoto et al, 1999), who studied the growth of CdSe and ZnO nanoclusters, or work (Kammler et al, 2004), where with the help ultra small-angle SAXS it was possible to follow the process of formation of silica particles in pyrolysis, or in the work (Diaz et al, 2008), in which SAXS was used to detect the formation of Si nanoclusters in plasma chemical deposition, or in the work of Britton et al. (2009), where the SAXS method was used as a method for controlling the morphology and size distribution of Si nanoparticles, obtained by different methods, for the production of silicon ink.

9.3.1. Basic principles of the method of SAXS

By its nature, SAXS is the X-ray diffraction method of studying the structure of materials. By the type of structural information obtained, the X-ray diffraction methods can be divided into two groups: 1) methods of wide-angle scattering using the intensity of scattering in a wide range of diffraction angles 2θ (from about 3 to $160°$) or, in other words, in the length range of the scattering vector $|\mathbf{q}| = 4\pi\dfrac{\sin\theta}{\lambda}$ from 4 to ~180 nm^{-1}, 2) small-angle scattering methods, using information from a range of scattering angles $2\theta \leq {\sim}3°$ (or $|\mathbf{q}| \leq 4$ nm^{-1}). However, both groups of methods are based on the same physical principles X-ray scattering and very similar measuring circuits (Fig. 9.21).

The SAXS method stands out from other X-ray diffraction methods due to the fact that because of the smallness of the scattering angle and scattering vector, respectively, it has low resolution, so 'it does not see' individual atoms, but efficiently distinguished between the larger particles composed of atoms.

In the method of small-angle X-ray scattering, as its name implies, we measure the scattering of photons going forward in the direction of the beam of monochromatic X-rays as it passes through the sample of the material under study. Typically, measurements are carried out in a cone with an angle to $2\theta = 6°$ relative to the direction of the primary beam of rays. In SAXS we consider only the elastic scattering of x-rays, i.e. the diffraction scattering without any change of wavelength.

From this measurement geometry there are two important consequences: 1) since we consider the forward scattering, the nuclear factors[1] of X-ray scattering almost reach their maximum values equal to the number of electrons in atoms, and the scattering coefficient of any material particle is equal to the number of electrons belonging to atoms contained in it, i.e. the total electron density, 2) as a result of consequence 1, the pattern of scattering intensity will show special features observed only if the scattering sample has areas of markedly different electron density from the rest the sample material. These are the areas of electron density fluctuations in size from tenths of nanometers to several tens of nanometers which are usually measured in the SAXS methods. Interestingly, these circumstances already accurately indicate that the cavities (e.g. nanopores in the material), which differ in the electron density from the rest of the material, also should produce the characteristic features of the diffraction pattern.

Schematic diagrams for the SAXS diffractometers are shown in Fig. 9.22.

SAXS diffractometers are constructed on the basis of two geometrical principles: 1) point collimation with a narrow cylindrical beam of parallel X-rays from a point source, 2) linear collimation with a flat beam of X-rays from a linear light source. The first option is closer to the ideal case of plane-wave scattering and is therefore preferred in terms of interpreting the data, but the devices constructed according to this scheme have a lower luminosity as compared to devices built by the scheme 2. In all SAXS diffractometers a very serious problem is the need to take measurements in the immediate vicinity of the centre of the primary beam of X-rays, which is solved using specific beam traps.

9.3.2. Diffraction patterns of small-angle scattering

SAXS diffractograms are measured after the passage of X-rays through the thin sample, which is in a clear cell, or in powder form with a thin layer bonded to the adhesive tape transparent to X-rays. As shown schematically in Fig. 9.22, in measurements in the geometry with point collimation the diffraction pattern recorded by the

[1]For the atomic scattering factor of X-rays and its main properties, see, for example, Iveronova and Revkevich (1978), or chapter 1 of the book by Fetisov (2007).

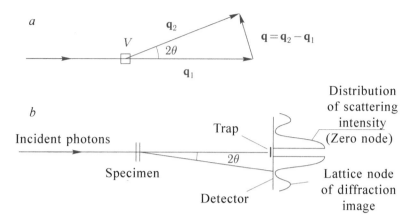

Fig. 9.21. The scheme of the elastic scattering of a photon (or electromagnetic wave) on the electronic density of the sample (plane section of a three-dimensional process). a – the process in the space of pulses (or the inverse diffraction space); $\mathbf{q}_i$ – the momentum vector of a photon (or wave vector), and $|\mathbf{q}_1| = |\mathbf{q}_2|$ – the condition of elastic scattering; $\mathbf{q} = \mathbf{q}_2 - \mathbf{q}_1$ – the scattering vector (the vector of momentum transfer); 2θ – the scattering angle. b – the same process as it looks in the laboratory space (the diffraction pattern is shown for the case of scattering by crystalline substance, which gives the lattice of the diffraction pattern with characteristic nodes). The detector registers the position and intensity of the density of the flux of photons in different directions relative to the direction of the beam of incident photons; in front of the detector along the beam path there is shown a trap, which extinguishes the beam of photons that have passed through the sample without scattering, in order to protect the detector and reduce the background from the direct beam.

detector has a circular symmetry and, in principle, can be reduced to the one-dimensional intensity distribution $I(2\theta)$ by simple summation over the azimuthal angle. The diffraction pattern in Fig. 9.22b does not have the circular symmetry, but can be reduced to an ideal geometry with point collimation by the geometric transformation, when the intensity, corresponding to the scattered rays, emanating not from the centre of the primary beam, are transferred to the new 'correct' positions, corresponding to perfect collimation. This procedure of reducing the measurement results to the ideal point collimation is almost always used in primary processing of the experimental data prior to their structural interpretation using special formulas.

To visualize the location and magnitude of the small-angle scattering signal in the overall picture of the X-ray scattering pattern, Fig. 9.23 shows the real one-dimensional diffraction pattern measured in a wide range of scattering angles 2θ, and Fig. 9.24 shows the result of separation from the diffraction data of the signal of small-angle scattering by the fine-dispersion particle of the suspension.

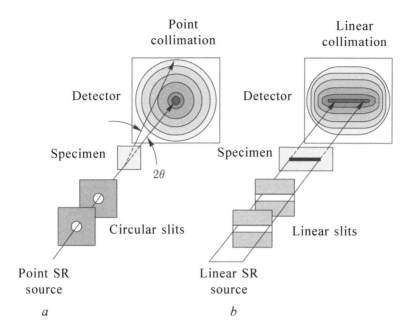

Fig. 9.22. Scheme of measurement of small-angle X-ray scattering. 2θ is the scattering angle. Focusing monochromators and X-ray mirrors used in modern diffractometers, mounted between the source and the collimating slit system, are not shown.

9.3.3. Theoretical basis of microstructure analysis by SAXS

The basics of the theory of the method of small-angle X-ray scattering were laid by French physicist Andre Guinier and Austrian scientists, Otto Kratky and Gunther Porod in the period from late 1930s until the end of the 1940s (see bibliography in the monograph by Guinier & Fournet, 1955; and articles in the book by Glatter & Kratky, 1982) and since then, despite the long road of development and contributions of many scientists, have remained the same, although the technique of analysis has changed significantly. Details of the current theory of the SAXS method can be found, for example, in the book (Svergun and Feigin, 1986) or brochure (Volkov, 2009), or in review articles (Vachette, Svergun, 2000; Glatter & May, 2004), of which we give here only the most basic concepts

Examining the particle of matter with the volume V_p, which is in a vacuum and placed in a parallel beam of X-rays, we determine the intensity of the elastic scattering of the particle. The electric field of electromagnetic waves, incident on the studied material, interacts with the electronic charges of the atoms, causing their dipole fluctuations. The energy of X-rays is large enough to excite the electrons. Oscillating charges due to the acceleration at forced vibrations generate secondary electromagnetic waves with a frequency equal to the frequency of the exciting primary wave, which

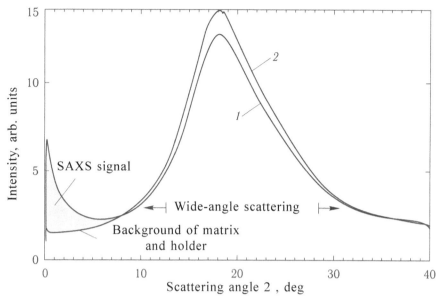

Fig. 9.23. Diffraction patterns of x-ray scattering in the angular range 2θ from 0° to 40°: *1* – a suspension of Si nanoparticles in toluene, located in a quartz capillary; *2* – scattering from the quartz capillary with pure toluene without Si nanoparticles. Measured at a SAXS small angle X-ray diffractometer (radiation Cu Kα, λ = 0.15402 nm). The fill shows the intensity of small-angle scattering by Si nanoparticles on the scattering background from the holder of the sample with the pure solvent.

adding up with each other at large distances, form the total wave with an amplitude

$$A(\mathbf{q}) = \int_{V_p} \rho(\mathbf{r}) \cdot \exp(-i\mathbf{r} \cdot \mathbf{q}) d\mathbf{r}, \tag{9.21}$$

determined by the conditions and the direction of scattering. Here **r** is the vector defining the coordinates of the elements of the scattering volume, and it is assumed that because of the low resolution of the measurement method, we may assume that electron density ρ(**r**) is distributed in the scattering volume continuously. In the formula (9.21), to reduce its bulkiness, we use the shortened form of writing, in fact, it is a triple integral, since **r** is a vector of the three-dimensional space. The integral in (9.21) is by definition a direct Fourier transform[2] as the result of no change can be calculated within the from −∞ to +∞, since the electron density outside the particle is zero. In this case, the transformation converts a function of electron density distribution that exists in the real lab space in its image, obtained by using X-ray scattering in the vector scattering space.

---

[2]Fourier transforms have a number of important properties which we shall use in our arguments, and which can be found, for example, in books (Korn and Korn, 1972; Shostak, 1972; Vasiliev, 1977).

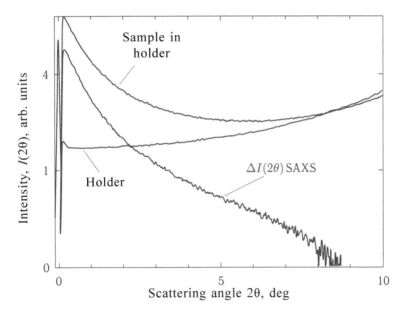

Fig. 9.24. Small-angle part of the diffraction pattern of X-ray scattering, as shown in Fig. 9.22, and the difference curve ΔI, corresponding to the scattering of colloidal particles of Si. Measurements on a diffractometer SAXSess, radiation Cu Kα, $\lambda =$ 0.15402 nm.

Experimentally using X-ray scattering detector can register only as the scattering intensity (i.e. the power of the wave field is proportional to the square of the amplitude) and to evaluate its dependence on the scattering angle 2θ – the angle between the directions of propagation of primary and scattered waves, or that the same as the length of the scattering vector $|\mathbf{q}|$, shown in Fig. 9.21. The scattering intensity I_p $(\mathbf{q})$ the particle is proportional to the square of the amplitude of the scattered wave, which is calculated as the product $A(\mathbf{q}) \cdot A^*(\mathbf{q})$ of complex conjugate amplitudes defined by (9.21). A very important result can be obtained if the scattering intensity is treated by the inverse Fourier transform operation, taking advantage of its properties (Vassiliev, 1977, Korn and Korn, 1972; Shostak 1972), in particular the theorem of multiplication and transformation of the complex conjugate function,

$$\mathrm{Im}^{-1}\left[I_p(\mathbf{q})\right] = \mathrm{Im}^{-1}\left[A(\mathbf{q}) \cdot A^*(\mathbf{q})\right] = \rho(\mathbf{r}) \otimes \rho(-\mathbf{r}) = \int_{V_p} \rho(\mathbf{r})\rho(\mathbf{r}+\mathbf{R})d\mathbf{r} = P(\mathbf{R}),$$

$$(9.22)$$

where $\mathbf{R} = \mathbf{r} - \mathbf{r}'$ is the the current displacement vector with respect to the origin of the coordinates changing at all points within the particle volume, the symbol $\otimes$ denotes the convolution operation of two functions. Expression (9.22) shows that the Fourier transform of the intensity allows for the definition of the convolution function $\rho(\mathbf{r})$ with itself, but an inverted at the origin. The resulting function $P(\mathbf{R})$ is called the autocorrelation function[3]. In X-ray analysis this function is also known as the 'Patterson function' or the 'function of interatomic vectors'. From the expression (9.22) it is clear that the intensity and the autocorrelation function are the reciprocal Fourier transforms, i.e. the possible expression are

$$P(\mathbf{r}) = \int I(\mathbf{q}) \exp(-i\mathbf{q}\cdot\mathbf{r}) d\mathbf{q}, \tag{9.23}$$

$$I(\mathbf{q}) = \int P(\mathbf{r}) \exp(i\mathbf{q}\cdot\mathbf{r}) d\mathbf{r}. \tag{9.24}$$

In SAXS measurements, the particles in the sample are usually oriented randomly or are in Brownian motion, so the measured scattering intensity refers to the average position of the particle and the formula (9.24) can be averaged over all orientations of $\mathbf{r}$.

By averaging over all orientations, the Patterson function (9.23) becomes the correlation function (Debye and Bueche, 1949)

$$\gamma(r) = \langle P(\mathbf{r}) \rangle_\omega = \frac{1}{4\pi} \int_{\omega=0}^{4\pi} P(\mathbf{r}) d\omega, \tag{9.25}$$

which, in contrast to the Patterson autocorrelation function, is the average value of the product of two fluctuations of the electron density at points separated by a distance r. The spherical averaging of the exponential (phase) term gives the expression

$$\langle \exp(i\mathbf{q}\mathbf{r}) \rangle = \frac{\sin(qr)}{qr}$$

– Debye equation (Debye, 1927).

The average intensity

$$I(q) = \langle I(\mathbf{q}) \rangle_\Omega = 4\pi \int_{r=0}^{\infty} \langle p(r) \rangle_\omega \cdot r^2 \cdot \frac{\sin(qr)}{qr} dr,$$

is a direct sine-Fourier transform of the function $p(r) = \gamma(r) \cdot r^2$ – known as a function of the distance distribution of a function of paired distributions.

Thus,

$$I(q) = \langle I(\mathbf{q}) \rangle_\Omega = 4\pi \int_{r=0}^{\infty} \gamma(r) \cdot r^2 \cdot \frac{\sin(qr)}{qr} dr, \tag{9.26}$$

and as the inverse Fourier sine transform

---

[3]In probability theory, the autocorrelation function is a special case of the function of cross-correlation, which is the estimate of the correlation between two continuous functions. The fundamental property of the autocorrelation function is its symmetry: $P(r) = P(-r)$. For continuous functions the autocorrelation is an even function.

$$\gamma(r) = \langle P(\mathbf{r}) \rangle \omega = \frac{1}{2\pi^2} \int_{q=0}^{\infty} I(q) \cdot q^2 \cdot \frac{\sin(qr)}{qr} dq. \qquad (9.27)$$

The obtained formulas provide a tool for determining the size of nanoparticles from the results of measuring the intensity of small-angle X-ray scattering (e.g. through the distribution function of the distance), and are remarkable since they describe the functions of scalar (not vector) arguments.

The SAXS theory often uses an approximate model of the material as a two-phase medium in which it is assumed that the system under study consists of particles with uniform electron density ρ, that are distributed in the matrix (or in the solvent), also having a uniform electron density, but with the value ρ_0. The difference between the average particle density and mean density of the solvent

$$\Delta\rho = \rho - \rho_0$$

is the effective density or contrast of the particles with respect to the solvent.

Neglecting interference effects between the waves scattered by particles and solvent, we can examine the scattering by an ensemble N of particles as the scattering of particles with an effective density $\Delta\rho$ located in a vacuum, if the scattering intensity is assumed to be the difference in the scattering intensity, which is obtained by subtracting the experimentally measured scattering intensity by the solvent from the intensity of scattering by th solution. An example of this subtraction is shown in Fig. 9.24. All these formulas remain valid for the effective density, if the intensity in them is replaced by the intensity difference.

The function $p(r) = \gamma(r) \cdot r^2$, calculated from the experimentally measured intensity distribution of SAXS, is widely used to determine the maximum particle size in monodisperse systems (which are found mainly in biology). However, in materials science, as a rule, we have to deal with polydisperse systems. The function of the intensity $I(q)$ of small-angle X-ray scattering of a polydisperse system depends on the shape of the particles and the distribution of their sizes. Therefore, using this function and on the basis of assumptions about the form of the particles we can determine the particle size distribution. The particles of the polydisperse system can be characterized by some effective size R, and the variation of their size by the size distribution function $D_N(R)$, which corresponds to the number of the particles in the system, the dimensions of which are enclosed between R and $R + dR$. The contribution of such particles to the SAXS intensity function is expressed by the product

$$i_0(q,R) \cdot m^2(R) \cdot D_N(R) \cdot dR,$$

where $i_0(q,R) = \langle F_0^2(q,R) \rangle - i_0(q,R)$ – is the form factor of the particles with the size R, averaged over all orientations and normalized so that $\langle F_0^2 (0, R) \rangle$. The size distribution function of the nanoparticles(R) in polydisperse systems is calculated by the integral equation

$$I(q) = \int_{R=0}^{R\,\max} D_N(R) \cdot \left[\upsilon(R) \cdot \Delta\rho \right]^2 i_0(q,R) \cdot dR. \qquad (9.28)$$

As a function of the size distribution we usually chose distributions such as the Schultz distribution, the Gaussian distribution or the gamma distribution, characterized by two parameters: the average value R_0 and variance ΔR, the definition of which is the aim of the study of polydisperse systems by SAXS.

9.3.4. Analysis of the measurement results

Direct conversion of the experimental curve of the intensity of small-angle scattering to the curve of distribution of the distances $p(r)$ and the size distribution density function $D_N(R)$ is the ill-conditioned problem due to a break of a number of experimental data, leading to strong oscillations of the Fourier transform, and also because of the ambiguity of solutions of convolution equations in the 'clean-up' of the experimental data from the instrumental effects. The first of these difficulties have recently been overcome by using the indirect Fourier transform, proposed by Otto Glatter (Glatter, 1977), and the successful search for the best possible solution can be ensured, for example, through the use of regularization methods proposed by Tikhonov (Svergun et al, 1988) and other methods of finding the optimal model (Bergmann et al, 2000).

Based on these principles, a larger number of computer programs has been developed for the analysis of SAXS data from polydisperse systems and for finding the particle size distribution function. The best known are programs such as GNOM (Svergun, 1992); GIFT (Bergmann, Fritz et al, 2000).

References for Chapter 9

Vasiliev D. M., Diffraction methods for studying structures. Moscow, Metallurgia, 1977. 248.

Volkov V. V., Determination of particle shape according to the small-angle X-ray and neutron scattering. Moscow: IR RAS, 2009. 51. http://nano.msu.ru/files/systems/4_2010/practical/41_full.pdf

Gardiner K. V., Stochastic Methods for Physics. Moscow. Mir, 1986. 528 p.

Grasselli J., Sneyvili M., Balkin B. Application of Raman spectroscopy in chemistry. Academic Press, 1984. 261 p.

Iveronova V. I., Revkevich G. P., The theory of scattering of X-rays. Moscow: Moscow State University, Moscow. 1978. 278.

Korn G., Korn T., Mathematical Handbook for Scientists and Engineers. Moscow. Nauka, 1970. P. 144-155.

Lebedev A. D., Levchuk Yu. N., Lomakin A.V., Noskin V.A., Laser correlation spectroscopy in biology. Kiev: Naukova Dumka. 1987. 256 p.

Myatlev V.D., Panchenko L. A., Riznichenko G. U., Terekhin A. T., Probability theory and mathematical statistics. Mathematical models. Moscow: IC Akademiya. 2009. 320.

Pentin Yu. A., Vilkov L. V., Methods in Physical Chemistry. Academic Press, 2003. 683 p.

Svergun D. I., Feigin L. A., Small-angle X-ray and neutron scattering. Moscow: Nauka, 1986. 280 p.

Smith A., Applied Infrared Spectroscopy. Academic Press, 1982. 328 p.

Fetisov G. Synchrotron radiation. Methods for studying the structure of matter. Moscow: Fizmatlit, 2007. 672.

Shabanova N. A., Popov V. V., Sarkisov P. D., Photon correlation spectroscopy. Chemistry and technology of nanosized oxide. Moscow: Akademkniga, 2006. P. 71-74.

Shostak R. Ya. Operational calculus. M.: Vysshaya Shkola, 1972. 279 p.

Eskin V.E. Light Scattering by Polymer Solutions and properties of macromolecules. Le ingrad: Nauka, 1986. 288.

Allen T. Practicle size measurement. (in 2 volumes). Chapman & Hall. Fifth edition. 1997. V. 1–525 p. and V. 2–251 p.

Bergmann A., Fritz G., Glatter O. Solving the generalized indirect Fourier transform (GIFT) by Boltzmann simplex simulated annealing (BSSA). J. Appl. Cryst. 2000. V. 33. P. 1212–1216.

Bergmann A., Orthaber D., Scherf G., Glatter O. Improvement of SAXS measurements on Kratky slit systems by Göbel mirrors & image plate detectors. J. Appl. Cryst. 2000. V. 33. P. 869–875.

Billinge S. J. L., Kanatzidis M.G. Beyond crystallography: the study of disorder, nanocrystallinity & crystallographically challenged materials with pair distribution functions. Chem. Commun. 2004. No. 7. P. 749–760.

Campbell I.H., Fauchet P.M. The effects of microcrystal size and shape on the one phonon Raman spectra of crystalline semiconductors. Solid State Commun. 1986. V. 58. P. 739–741.

Canham L. T. Skeleton size distribution in porous silicon. In: Properties of porous silicon. Ed. by L. Canham. London: INSPEC, 1997. Part 3.2. P. 106–111. xviii+416 p.

Canham L. T. Silicon quantum wire array fabrication by electrochemical & chemical dissolution of wafers. Appl. Phys. Lett. 1990. V. 57. P. 1046–1048.

Cheng W., Reng S.-F. Calculations on the size effects of Raman intensities of silicon quantum dots. Phys. Rev. B. 2002. V. 65. P. 205305 (9 p.).

Craievich A. F. Synchrotron SAXS studies of nanostructured materials and colloidal solutions. A Review. Materials Research. 2002. V. 5, No. 1. P. 1–11.

Cuilis A.G., Canham L. T., Calcoct P.D. J. The structural and luminescence properties of porous silicon. J. Appl. Phys. 1997. V. 82. P. 909–965.

Deby P., Bueche A.M. Scattering by an inhomogeneous solids. J. Appl. Phys. 1949. V. 20. P. 518–525.

Debye P. Über die Zerstreuung von Röontgenstrahlen an amorphen Körpern. Z. Physik. 1927. V. 28. P. 135–141.

Diaz J.M.A., Kambara M., Yoshida T. Detection of Si nanoclusters by x-ray scattering during silicon film deposition by mesoplasma chemical vapor deposition. J. Appl. Phys. 2008. V. 104. P. 013536 (5 p.).

Egami T., Billinge S. J. L. Underneath the Bragg-peaks: Structural analysis of complex materials. Elsevier Ltd. 2003. 500 p.

Faraci G., Gibilisco S., Russo P., Pennisi A. Modified Raman confinement model for Si nanocrystals. Phys. Rev. B. 2006. V. 73. P. 033307 (4 p.).

Fritz G., Bergmann A., Glatter O. Evaluation of small-angle scattering data of charged particles using the generalized indirect Fourier transformation technique. J. Chem. Phys. 2000. V. 113. P. 9733–9740.

Frohnhoff St., Marso M., Berger M.G., Thöonissen M., Lüth H., M¨under H. An extended quantum model for porous silicon formation. J. Electrochem. Soc. 1995. V. 142, No. 2. P. 615–20.

Furukawa S., Miyasato T. Quantum size effects on the optical band gap of microcrystalline Si:H. Phys. Rev. B. 1988. V. 38. P. 5726–5729.

Glatter O. The interpretation of real-space information from small-angle scattering experiments. J. Appl. Cryst. 1979. V. 12. P. 166–175.

Glatter O., Kratky O. (Eds.) Small angle X-ray scattering. N.Y.: Academic Press, 1982. 515 p.

Glatter O., May R. Small-angle techniques. International Tables for Crystallography. V. C. Kluwer Academic Publishers, 2004. Ch. 2.6. P. 89–112.

Glatter O. A new method for the evaluation of small-angle scattering data. J. Appl. Cryst. 1977. V. 10. P. 415–421.

Glatter O. Determination of particle-size distribution functions from small-angle scattering data by means of the indirect transformation method. J. Appl. Cryst. 1980. V. 13. P. 7–11.

Goldin A.A. Software for particle size distribution analysis in photon correlation spectroscopy.2002.http://www.photocor.ru/downloads/Alango/dynals-manual. htm#references.

Guinier A., Fournet G. Small-angle scattering of X-rays. N.Y.: Wiley, 1955. 268 p.

Guinier A. La diffraction des rayons X aux très faibles angles: Applications à l'etude des phéenomènes ultra-microscopiques. Ann. Phys. (Paris). 1939. No. 12. P. 161–236.

Hollas J.M. Modern spectroscopy. John Wiley & Sons Ltd. 2004. xxvii+452 p.

Islam M.N., Kumar S. Influence of crystallite size distribution on the micro-Raman analysis of porous Si. Appl. Phys. Lett. 2001. V. 78. P. 715 (3 p.).

ISO 13321–1, 1996. Particle size analysis Photon correlation spectroscopy. ISO 22412, 2008. Particle size analysis Dynamic light scattering.

Kammler H.K., Beaucage G., Mueller R., Pratsinis S. E. Structure of flame-made silica nanoparticles by ultra-small-angle X-ray Scattering. Langmuir. 2004. V. 20. P. 1915–1921.

Kanemitsu Y., Uto H., Masumoto Y., Matsumoto T., Futagi T., Mimura H. Microstructure & optical properties of free-standing porous silicon films: Size dependence of absorption spectra in Si nanometer-sized crystallites. Phys. Rev. B. 1993. V. 48. no. 4 P. 2827–30.

Kaye B.H. Characterization of powder and aerosols. WILEY-VCH Verlag GmbH. 1999. 312 p.

Kissa E. Dynamic light scattering (photon correlation spectroscopy, quasielastic light scattering). In: Dispersions: characterization, testing, and measurement. Marcel Dekker, 1999. P. 449–469.

Kumar V. (Ed.). Nanosilicon. Elsevier Ltd., 2008. xiii+368 p.

Ledoux G., Guillois O., Porterat D., Reynaud C., Huisken F., Kohn B., Paillard V. Photoluminescence properties of silicon nanocrystals as a function of their size. Phys. Rev. B. 2000. V. 62. P. 15942–15951.

Leoni M., Confente T., Scardi P. PM2K: a flexible program implementing Whole Powder Pattern Modelling. Z. Kristallogr. Suppl. 2006. V. 23. P. 249–254.

Long D.A. Raman Spectroscopy. McGraw-Hill Inc., 1997. 276 p.

Marchenko V.M., Koltashev V.V., Lavrishchev S.V., Murin D. I., Laser-induced V.G. P. transformation of the microsctucture of SiO_x, $x \approx 1$. Laser Phys. 2000. V. 10. P. 576–582.

Mattoussi H., Cumming A.W., Murray C.B., Bawendi M.G., Ober R. Properties of CdSe nanocrystal dispersions in the dilute regime: Structure and interparticle interactions. Phys. Rev. B. 1998. V. 58. P. 7850–7863.

Meier C., Lüttjohann S., Kravets V.G., Nienhaus H., Lorke A., Wiggers H. Raman properties of silicon nanoparticles. Phisica E. 2006. V. 32. P. 155–158.

Merkus H.G. Dynamic light scattering. In: Particle size measurements. fundamentals, practice, quality. Springer Science+Business Media B.V. 2009. P. 299–317.

Merkus H.G. Particle size measurements. Fundamentals, practice, quality. Springer Science+Business Media B.V. 2009. xii+533 p.

Mishra P., Jain K. P. Raman, photoluminescence and optical absorption studies on nanocrystalline silicon. Mater. Sci. Eng. B. 2002. V. 95. P. 202–213.

M¨under H., Andrzejak C., Berger M.G., Klemradt U., L¨uth H., Herino R., Ligeon M. A detailed Raman study of porous silicon. Thin Solid Films. 1992. V. 221. P. 27–33.

Nayfeh M.H., Barry N., Terrien J., Akcakir O., Gratton E., Belomoin G. Stimulated blue emission in reconstituted films of ultrasmall silicon nanoparticles. Appl. Phys. Lett. 2001. V. 78. P. 1131–1133.

Paillard V., Puech P., Laguna M.A., Carles R., Kohn B., Huisken F. Improved one-phonon confinement model for an accurate size determination of silicon nanocrystals. J. Appl. Phys. 1999. V. 86. P. 1921 (4 p.).

Pavesi L., Dal Negro L., Mazzoleni G., Franzo G., Priolo F. Optical gain in siliconnanocrystals. Nature. 2000. V. 408. P. 440–444.

Provencher S.W. CONTIN: A general purpose constrained regularization program for inverting noisy linear algebraic and integral equations. Comput. Phys. Commun. 1982. V. 27. P. 229–242.

Provencher S.W. A constrained regularization method for inverting data represented by linear algebraic or integral equations. Comput. Phys. Commun. 1982◻. V. 27. P. 213–227.

Richter H., Wang Z. P., Ley L. The one phonon Raman spectrum in microcrystalline silicon. Solid State Commun. 1981. V. 39. P. 625–629.

Sasaki A. Size distribution analysis of nanoparticles using small angle X-ray scattering technique. The Rigaku Journal. 2005. V. 22. no. 1. P. 31–38.

Scardi P., Leoni M. Whole powder pattern modeling. Acta Cryst. 2002. V. A58. P. 190–200.

Schuppler S., Friedman S. L., Marcus M.A., Adler D. L., Xie Y.-H., Ross F.M., Chabal Y. J., Harris T.D., Brus L. E., Brown W. L., Chaban E. E., Szajowski P. F., Christman S. B., Citrin P.H. Size, shape, and composition of luminescent species in oxidized Si nanocrystals & H-passivated porous Si. Phys. Rev. B. 1995. V. 52. P. 4910–4925.

Semenyuk A.V., Svergun D. I. GNOM a program package for small-angle scattering data processing. J. Appl. Cryst. 1991. V. 24. P. 537–540. Skrzypek J. J., Rustichelli F. (editors). Innovative technological materials: Structural properties by neutron scattering, synchrotron radiation & modeling. Springer, 2010. 279 p.

Soni R.K., Fonseca L. F., Resto O., Buzaianu M., Weisz S. Z. Size-dependent optical properties of silicon nanocrystals. J. Lumin. 1999. V. 83–84. P. 187–191.

Stribeck N. X-ray scattering of soft matter. Berlin–Heidelberg: Springer-Verlag, 2007. 258 p.

Svergun D. I. Determination of the regularization parameter in indirect-transform methods using perceptual criteria. J. Appl. Crystallogr. 1992. V. 25. P. 495–503.

Svergun D. I. Mathematical methods in small-angle scattering data analysis. J. Appl. Cryst. 1991. V. 24. P. 485–492.

Svergun D. I., Semenyuk A.V., Feigin L.A. Small-angle-scattering-data treatment by the regularization method. Acta Cryst. 1988. V. A44. P. 244–250.

Sychugov I., Juhasz R., Valenta J., Linnros J. Narrow luminescence linewidth of a silicon quantum dot. Phys. Rev. Lett. 2005. V. 94. P. 087405 (4 p.).

Takeoka S., Fujii M., Hayashi S. Size-dependent photoluminescence from surface-oxidized Si nanocrystals in a weak confinement regime. Phys. Rev. B. 2000. V. 62. P. 16820–16825.

Tokumoto M. S., Pulcinelli S.H., Santilli C.V., Craievich A. F. SAXS study of the kinetics of formation of ZnO colloidal suspensions. J. Non-Cryst. Solids. 1999. V. 247. P. 176–182.

Vachette P., Svergun D. I. Small-angle X-ray scattering by solutions of biological macro-molecules. In: Structure and Dynamics of Biomolecules / Ed. by E. Fanchon, G. Geissler,

J.-L. Hodeau, J.-R. Regnard, P.A. Timmins. N.Y.: Oxford University Press, 2000. Ch. 11. P. 199–237.

Warren B. E. X-ray Diffraction, Addison-Wesley, Reading, MS, 1969. 381 p.

Witt W., Aberle L., Geers H. Measurement of particle size & stability of nanoparticles in opaque suspensions and emulsions with photon cross correlation spectroscopy (PCCS). Particulate Systems Analysis. Harrogate (UK), 2003. 5 p.

Xu R. Scattering intensity fluctuations of particles. P. 83–89, and Photon correlation spectroscopy. P. 223–288. In: Particle characterization: Light scattering methods. Kluwer Academic Publishers, 2002. 398 p.

Yerci S., Serincan U., Dogan I., Tokay S., Genisel M., Aydinli A., Turan R. Formation of silicon nanocrystals in sapphire by ion implantation and the origin of visible photoluminescence. J. Appl. Phys. 2006. V. 100. P. 074301 (5 p.).

Zi J., Buscher H., Falter C., Lugwig W., Zhang K., Xie X. Raman shifts in Si nanocrystals .Appl. Phys. Lett. 1996. V. 69. P. 200 (3 p.).

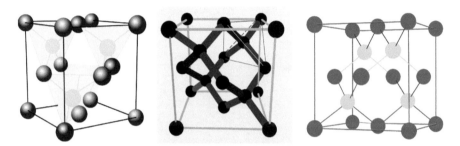

Fig. 2.1. Axonometric images of the elementary cell of crystalline silicon (spheres indicate Si atoms); a) representation of the structure of the elementary cell using coordination tetrahedrons; b) crystallochemical structure of diamond/silicon with indicated covalent bonds (the red lines show the coordination tetrahedron with the Si atom in the centre); c) explanation of the formation of the elementary cell of Si with mutual penetration of two FCC lattices where the grey colour indicates the Si atoms, forming the classic FCC cell, and the green colour the 'additional' Si atoms from the same cell inserted into it with the shift by 1/4 of the length of the spatial diagonal.

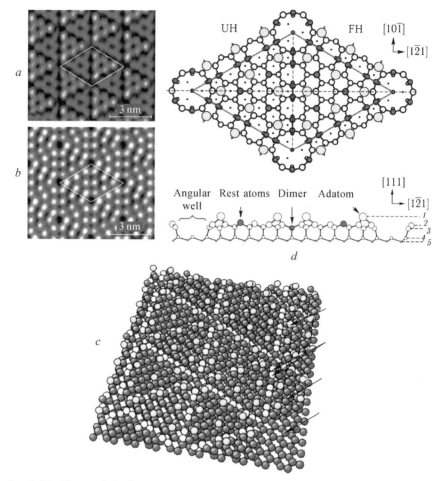

Fig. 2.15. The model of the surface Si (111) 7×7 (DAS model – the model for dimers–adatoms and stacking fault). The atomically clean surface Si (111) with the 7×7 reconstruction. STM images (a) of filled and unfilled (b) electronic states of the surface: c) a schematic representation of the surface (top view and side view) in accordance with the DAS model of Takayanagi; b) axonometric image of the model. The yellow circles in the flat projections indicate the 'additional' atoms (adatoms) of Si, red circles – dimerized Si atoms (in axonometry they are indicated by the white circles), the blue circles – Si atoms (rest atoms) of the second layer, remaining after departure of the adatoms in the upper layer (in axonometry they are indicated by the red circles). The rhombs on the flat projection and the STM images show the two-dimensional surface elementary cell 7×7. Half of the elementary cell, containing the stacking faults, is denoted as FH (faulted half), the half without the stacking forces indicated as UH (unfaulted half). It may be seen that on the STM image of the filled states (a) half of the cell with the stacking fault is brighter. The brightness maximum on the STM image corresponds to the adatoms. From http://thesaurus.rusnano.com/wiki/article14156.

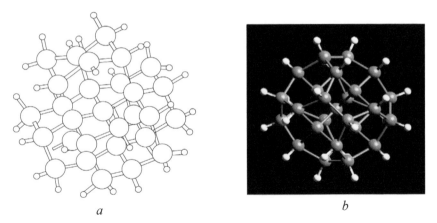

Fig. 4.1. Configuration of Si atoms in the non-reconstructed $Si_{29}H_{36}$ nanoclusters with a diameter of 1.0 nm (a) and computer image of the reconstructed $Si_{29}H_{24}$ (b) nanoclusters (Smith et al., 2005)

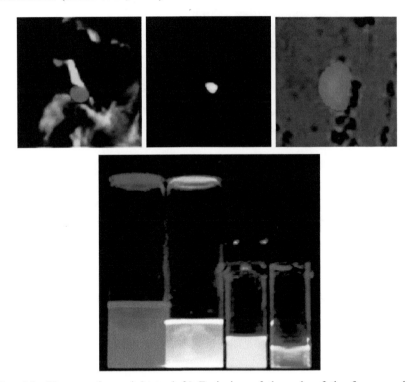

Fig. 4.2. (Bottom, from right to left) Emission of the sols of the four members of the magic family of the silicon nanoclusters with the diameters of 1.0 (Si_{29}) (Si_{123}); 2.15 and 2.9 nm after their separation. Excitation was carried out with a commercial source with a mean wavelength of 365 nm (top view). The emission of blue, green and red colour by the sols with the magic dimensions – 1.0, 1.67 and 2.9 nm (Belomoin et al., 2002).

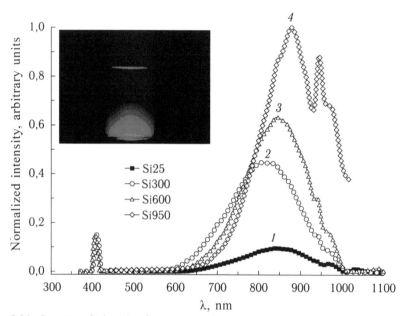

Fig. 5.21. Spectra of photoluminescene excited in Si nanoparticles in the samples Si25, Si300, Si600 and Si950 by radiation with a wavelength of 405 nm. The insert shows the photograph of the glow of nanoparticles in the Si300 sample (Dorofeev et al., 2010).

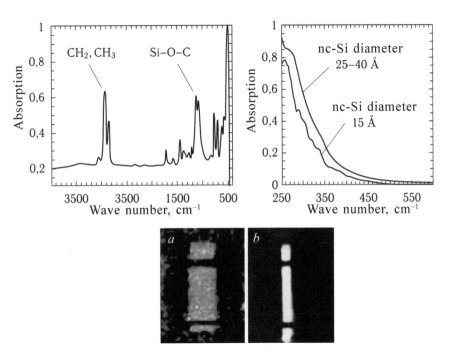

Fig. 5.28. Top left – Fourier IR spectrum of alkoxylated nc-Si. Top right – absorption spectrum of nc-Si. Bottom – photograph of luminescent nc-Si in hexane at excitation at ~320 nm (a); photoluminescence of particles with a diameter of 1.5 nm (b) (Holmes et al, 2001).

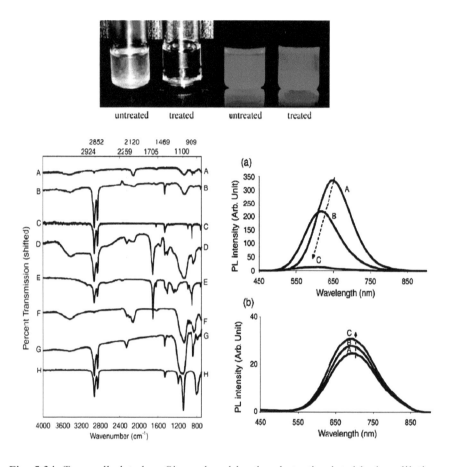

Fig. 5.34. Top – alkylated nc-Si, produced by the photostimulated hydrosylilation reaction of the solution and their photoluminescent properties (two sample on the left – in daylight, two on the right – under UV radiation). Treated – particles were treated with octadecene. Bottom left – Fourier IR spectra of nc-Si/Si–H (A), octadecyl (B), nc-Si/1-octadecene (C), nc-Si/undecylenic acid (D), undecylenic acid (E), nc-Si/Si–OH (F), nc-Si/octadecyl trimethoxysilane (G), octadecyl trimethoxysilane (H). Bottom right – stability of photoluminescence of particles of nc-Si/Si–H (top) and nc-Si/alkyl (Li X., et al., 2004).

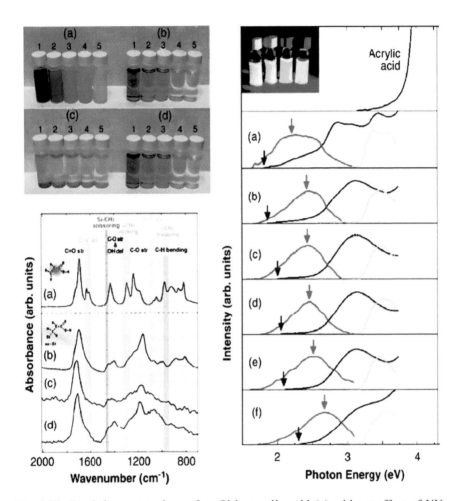

Fig. 5.37. Top left – comparison of nc-Si in acrylic acid (a) without effect of UV radiation, (b) after UV irradiation; (c) and (d) show the samples after storage for 1 week. Bottom left – Fourier IR spectra of propionic acid (a), nc-Si functionalized by propionic acid (b), nc-Si functionalized by propionic acidand isolated by dialysis (c) and resuspended nc-Si, functionalized by propionic acid (d). On the right – luminescence of nc-Si functionalized by propionic acid (Sato, Swihart, 2006).

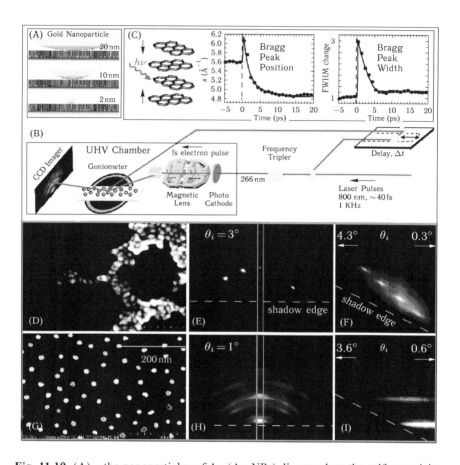

Fig. 11.19. (A) – the nanoparticles of Au (Au-NPs) dispersed on the self-organising molecular surfaces of the interface. (B) – the diagram of the UEnC method. (C) – determination of the diffraction signal as the result of photomechanical response of the multilayers of graphite. (D) – the SEM image of Au-NPs with the size of 20 nm, scattered on the surface of the substrate without buffering. (E) – the diffraction pattern from Au-NPs with the size of 20 nm, showing the areas of Bragg signals of the silicon substrate. (F) – tilting curves, corresponding to the diffraction pattern from Au-NPs with the size of 20 nm (E) and different incidence angles θ_i. (G) – SEM image of Au-NPs with the size of 20 nm with the appropriate buffering. (H) – the diffraction pattern, corresponding to (G), showing the Debye–Scherrer diffraction rings and the spots from the Bragg buffer layer (Si, N, C and a stack of the layers in the self-organised aminosilane, range 2.2 Å, the angle of inclination 31°). (1) – the tilting curve (H). The figure from Raman et al (2008).

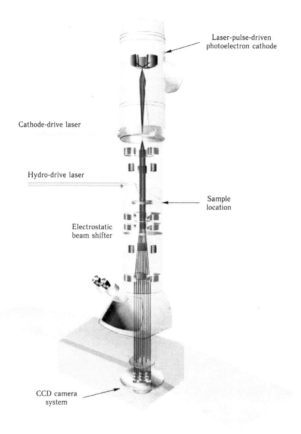

Laser-pulse-driven
photoelectron cathode

Cathode-drive laser

Hydro-drive laser

Sample
location

Electrostatic
beam shifter

CCD camera
system

Fig. 11.20. Equipment for dynamic transmission electron microscopy (DTEM) from (King et al, 2005).

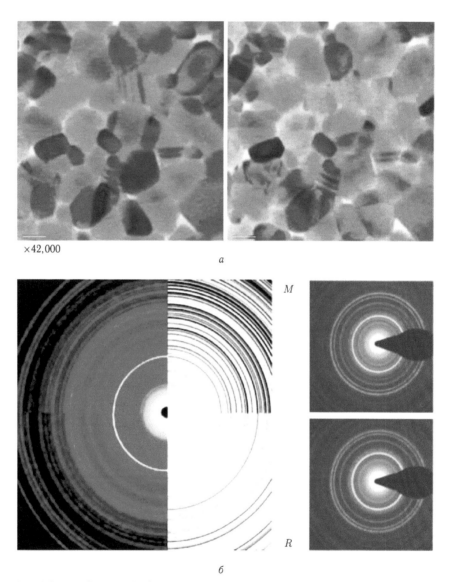

×42,000

a

M

R

б

Fig. 11.21. *a* – images obtained by the UEM method prior to phase transition in the VO$_2$ films (left) and after phase transition (right). Magnification 42 000 (100 nm scale). It should be mentioned that these images will not be seen if the generation of femtosecond pulses of the photoelectrons is blocked. *b* – diffraction patterns, produced by the UEM method prior to phase transition in VO$_2$ (right) and after (left). Diffraction patterns of two phases (monoclinic phase *M* and high-temperature tetragonal phase of rutile *R*), observed in experiments (left part of the figure *b*) and constructed as a result of calculations (right hand part of figure *b*). Analysis was described in (Grinolds et al, 2006).

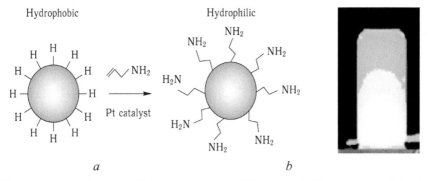

Hydrophobic Hydrophilic

Pt catalyst

a *b*

Fig. 12.1. *a* – schematic of the process of modification of silicon nanoparticles by allylamine described in (Warner et al, 2005). *b* – blue fluorescence of Si nanoparticles modified by allylamine.

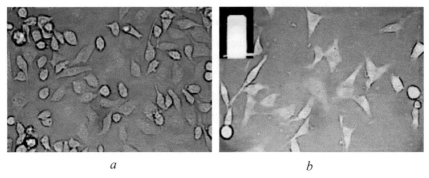

a *b*

Fig. 12.7. Optical image, obtained in fluorescent light, of the cells of the HeLa line (a) and cells of the HeLa line specifically labeled with silicon nanoparticles (b). The inset shows the fluorescent image of silicon nanoparticles hydrophilized with allylamine (Warner et al, 2005).

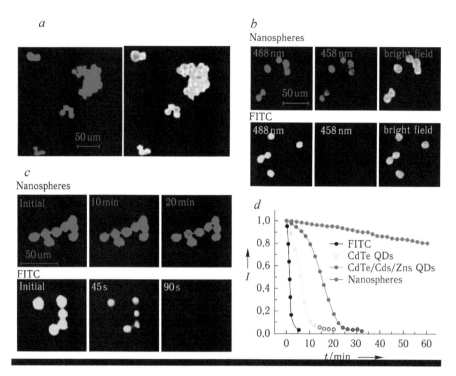

Fig. 12.8. *a* – specific labelling of the cells of HEK2-293T lines by the nanospheres based on silicon particles (left) and the superposition of the optical and fluorescent images (right), excitation of fluorescence at 488 nm; *b* – comparison of the fluorescent signal of the cells of the HEK293T line, specifically labelled with the nanospheres based on silicon nanoparticles (upper) and FITC (lower); *c* – variation of fluorescence in relation to the duration of radiation of the cells of the HEK293T lines, labelled with the silicon nanospheres (upper) and FITC (lower). Fluorescence was excited with an argon laser with a wavelength of 488 nm, a delay time of 8 ms, power 15 mW; *d* – comparison of the photostability of the quantum dots with the composition CdTe and CdTe/CdS/Zns with the core–shell structure and the nanospheres based on silicon nanoparticles – fluorescence of all specimens was excited with a xenon lamp, power 450 W. The data from He et al (2009).

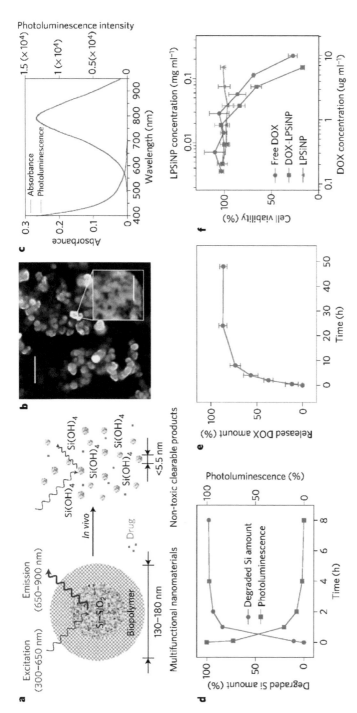

Fig. 12.12. Characteristics of FPSN, from Park et al (2009): a – schematic representation of the dissolution of the FPSN in the organism (*in vivo*); b – the micrograph of FPSN (the size scale 500 nm, in the inset 50 nm); c – the spectra of photoluminescence and absorption of FPSN; d – the dependences of the photoluminescence and the amount of reacted FPSN on time (FPSN were placed in the phosphate buffer saline solution PBS); c – the time dependence of the amount of doxorubicin, extracted from the particles of the carrier of FPN in the PBS solution (this also indicates the complete biodegradation of the silicon nanoparticles); f – investigation of the cytotoxicity of the composite material of the core/shell type DOX/FSPN (DOX – doxorubicin, anticancer drug).

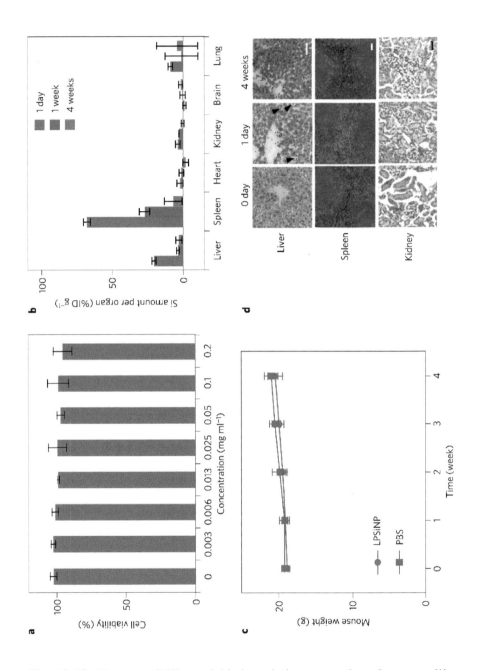

Fig. 12.13. Biocompatibility and biodegradation properties of porous silica nanoparticles: a – *in vitro* cytotoxicity of these nanoparticles; b – *in vivo* distribution and biodegradation of the porous silica nanoparticles in the organism of mice over a period of four weeks; c – change of the mass of the body of the mice after adding the nanoparticles and PBS (for comparison); d – histology of the cells of the lever, spleen and kidneys (the size scale 50 μm). Park et al (2009).

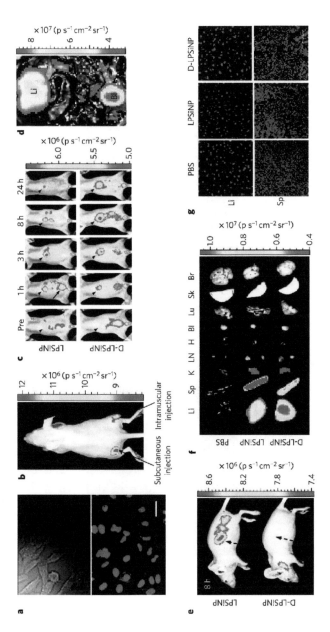

Fig. 12.14. Study FPSN fluorescence of FPSN *in vitro*, *in vivo* and *ex vivo* (Park et al., 2009): *a* – *in vitro* fluorescence of cells of the HeLa line, treated with FPSN (red and blue indicate FPSN and cores of cells respectively, dimensional range 20 μm); *b* – *in vivo* fluorescence of FPSN introduced into the mouse from two sides; *c* – *in vivo* fluorescence FPSN and FPSN further modified with dextran; *d* – *in vivo* images showing removal 1 hour after administration into the bladder (Bl) FPSN (Li – liver); *e* – the same image, as in *c* (line indicates the position of the spleen); *f* – fluorescence showing *ex vivo* biodistribution of FPSN in organs in mice (Li, Sp, K, LN, H, Bl, Lu, Sk and Br correspond to liver, spleen, kidney, lymph nodes, heart, bladder, lung, skin and brain); *g* – histology of the liver and the spleen cells of mice, as shown in Figure *c*, *f* 24 hours after administration (red and blue labelled nanoparticle and cell nuclei, respectively dimensional range 50 μm).

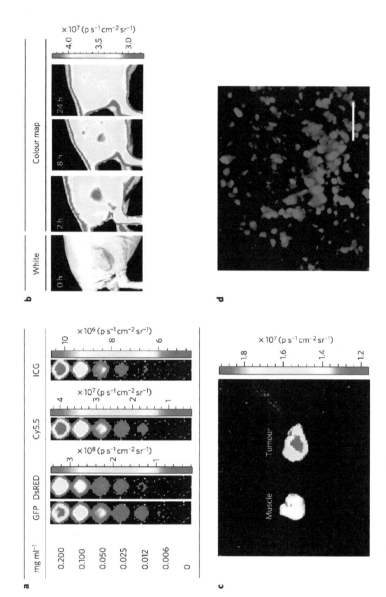

Fig. 12.15. Fluorescence of tumors containing FPSN coated with dextran (Ji-Ho Park et al., 2009): *a* – the intensity of the fluorescence of FPSN depending on the concentration; *b* – fluorescence in tumors MDA-MB-435; *c* – *ex vivo* fluorescence of murine tumors and muscle tissue around the tumor; *d* – fluorescence of murine tumors (red and blue mark FPSN coated with a polymer and cell nuclei, respectively; dimensional range 100 μm).

10

New X-ray diffraction methods for the analysis of the structure and morphology of nanocrystalline powders

Matteo Leoni[1]

The Scherrer formula, the Williamson–Hall plot and the Warren–Averbach method are the most misused tools for the quantitative estimation of the "average size" of a nanocrystalline powder from an X-ray diffraction pattern. In fact, it will be shown that, even if mathematically correct, the result of the traditional analyses is not the one suggested or sought, i.e. the mean of the domain size distribution: a physical interpretation of quoted 'average size' or 'microstrain' values is not always straightforward or possible. Those traditional analysis techniques are based on rather simplified hypotheses and the microstructure information is extracted from partial or pre-processed data (i.e. peak width): the consistency between measurement and results cannot be verified and, in general, is not preserved. Modern alternatives for the analysis of powder diffraction data are here revised, demonstrating the possibility of extracting a physically sound microstructure description, with a level of detail that matches and in most cases surpass that of transmission electron microscopy. The Whole Powder Pattern Modelling (WPPM) and an alternative formalism for cases where faulting is dominant are analysed in detail and some examples of application are provided and commented.

10.1. Introduction

In the last decades, with the boom of nanostructured materials and the discovery of peculiar properties in nanometre-sized systems, there has been an increased need for techniques able to provide reliable microstructural parameters on the nanoscale. Microstructure is also the key for silicon technology, as e.g. the luminescence wavelength depends on the size of the nanoparticles.

[1]Department of Civil, Environmental and Mechanical Engineering, University of Trento, via Mesiano, 38123 Trento (TN), Italy; email: Matteo.Leoni@unitn.it

An obvious choice for microstructure analysis of such systems would be microscopy, as the sub-nm resolution can be easily achieved on last generation transmission electron and scanning probe microscopes (TEMs and SPMs). However, obvious does not always mean better: a large limit of scanning probe microscopes is that just the surface morphology can be measured (no info on the interior of the grains is available). The TEM, on the other hand, provides both the morphology and the internal microstructure information, but the number of grains that can be reasonably measured is certainly too small to guarantee a good statistical significance.

A possible alternative is offered by indirect methods such as X-ray powder diffraction (XRPD). Valuable microstructure information is contained, albeit highly integrated, in a powder diffraction pattern: already at the beginning of last century, in fact, it was observed a close relationship between e.g. the breadth of a diffraction peak and the characteristics of the scattering domains. In the years, several techniques have been proposed for the so-called line profile analysis (LPA), mainly grouped into two broad classes:

   (i) pattern decomposition (PD) i.e. separation of the individual peaks, composing a pattern, fitting with some bell-shaped function and analysis of the parameters of the fit
   (ii) pattern modelling (PM) i.e. fitting of a set of microstructure models to the entire diffraction pattern.

Most of the traditional techniques of LPA, still widely employed by the scientific community, fall in the first category. They are the well known Scherrer formula (SF), Williamson–Hall (WH) plot and Warren–Averbach (WA) method (Scherrer, 1918; Williamson & Hall, 1953; Warren & Averbach 1950, 1952; Warren, 1969; Klug & Alexander, 1974; Langford & Louër, 1996). They all pretend to give some microstructure information under the form of "average domain size" and "microstrain": in most practical cases, however, the true meaning of those terms and the theory behind those methods are highly overlooked.

The Scherrer formula (Scherrer, 1918) relates the Full Width at Half Maximum (FWMH) of a diffraction peak having Bragg position $2\theta_B$ with the so-called "average crystallite size" D:

$$\text{FWHM}(2\theta_B) = \frac{0,94\lambda}{D\cos(\theta_B)}. \tag{10.1}$$

The equation can be simply derived by starting from the intensity of a single crystal of cube shape (edge length $D = Na$ where a is the cell parameter) and evaluating the FWHM:

$$I_P \propto \exp\left[-\pi\left(\frac{Na\Delta(2\theta)\cos(\theta_B)}{\lambda}\right)^2\right] \Rightarrow \frac{1}{2}$$

$$= \exp\left[-\pi\left(\frac{D\cos(\theta_B)FWHM(2\theta)}{\lambda}\right)^2\right] \Rightarrow$$

$$\Rightarrow FWHM(2\theta) = \frac{2\lambda\sqrt{\ln(2)/\pi}}{D\cos(\theta_B)}. \tag{10.2}$$

In order to render the approach independent of the actual shape of the peak, the FWHM is usually replaced with the so-called integral breadth (IB) β, i.e. the ratio between peak area and peak maximum intensity. The IB is the width of the rectangular peak having the same area and intensity as the given one. With this definition, the usual forms of the Scherrer equation are:

$$\beta_{\{hkl\}}(2\theta) = \frac{\lambda K_\beta}{\langle D\rangle_V \cos\theta_B} \quad \text{or} \quad \beta_{\{hkl\}}(d^*) = \frac{K_\beta}{\langle D\rangle_V}, \tag{10.3}$$

where K_β is called the Scherrer constant for the given shape and the second equation is written in reciprocal space d^*. The Scherrer constants appearing in equations (10.1)–(10.3) depend on the choice of the domain shape and they can be recomputed for simple shapes and for a given hkl reflection, as shown e.g. in Langford, Wilson (1978).

The meaning of the average crystallite size obtained from the Scherrer formula is clear for the cube edge, but is not obvious for other directions or shapes. In fact, applying the idea of Bertaut (1949, 1950), the average crystallite size is an average of the length of the columns in which the shape can be subdivided along the scattering direction. Therefore it does not represent the average size we are interested in (i.e. the mean of a side distribution). For non-spherical sizes, the average crystallite size depends on the hkl and the same value can be obtained with different shapes.

In order to compare the broadening corresponding to different directions and to try increasing the level of information that can be extracted, Williamson and Hall (1953) proposed to plot the FWHM or the IB in reciprocal space. For an isotropic domain (size independent of the direction), this plot is expected to be constant in reciprocal space. In most cases, however, a local variation in the integral breadth and an increase with the distance from the origin of the reciprocal space is observed. Williamson and Hall proposed to write the integral breadth in reciprocal space as the combination between the Scherrer formula and the differential of Bragg's law:

$$\beta\left(d^{*}\right)=\frac{K_{\beta}}{\langle D\rangle_{V}}+2\left\langle\varepsilon^{2}\right\rangle^{1/2}d^{*}. \qquad (10.4)$$

The first term is termed 'average crystallite size' and represents the intercept of the line (extrapolation of the integral breadth to the origin of the reciprocal space), whereas the second is the so-called microstrain or root mean strain.

Even if this formalism seems straightforward, there is no physical reason for summing the two contributions up. As we are dealing with integral breadths, the only case where breadths are additive is when peaks are Lorentzian: the Williamson–Hall formalism is therefore strictly valid only for Lorentzian peaks, condition seldom met in practice.

What is to be noted, is that most people employ Scherrer formula and the Williamson–Hall approach without paying much care about the underlying hypotheses (Lorentzian peaks only) and about the true meaning of the results. More, if we consider that in all cases the coherently scattering domains, the true elements visible by XRD, are in general not monodispersed. In the presence of a distribution, the "average crystallite size" obtained with those traditional approaches is not the average size of the domains, i.e. the first moment of the size distribution. Calling the Scherrer results 'average size' contributes to increase the confusion: the most published data quoting an average size of the grains calculated according to the Scherrer equation are wrong.

The microstrain contribution is quite an abstract concept as it provides an evidence for the presence of defects, but it does not identify their source. It should also be borne in mind that the microstrain term accounts for the breadth of the microstrain distribution: should the microstrain be constant or average to a non-zero value, a residual strain (leading to shift in peak position) would be present.

After noticing the Lorentzian peak shape limitation, some authors proposed alternative formalisms to deal with the more general case of a peak being Voigtian. This extends the possible combinations of Lorentz (L) and Gauss (G) functions from LL (Williamson–Hall) to LG, GL and GG. As usual, the choice of either one is subjective.

The Warren–Averbach method is employed as an alternative to the Williamson–Hall plot. The Warren–Averbach method can be considered as the father of all modern Fourier methods of LPA: the Fourier approach will be introduced in the following. Even if it represented a major step forward when it was introduced in the fifties, it does not solve one of the major problems that can be spotted in the Scherrer and Williamson–Hall analyses: there is no direct connection between the measured pattern and the result and the information contained in the whole diffraction pattern is not employed. In fact, all methods analyse just one or a limited number of peak profiles, and there is always an intermediate extraction step in which some pre-processed data (e.g. integral breadth of the peak) are generated.

The analysis is then based on those pre-processed data and not directly on the raw data. The net result is the inconsistency in most cases, between data and result (i.e. the extracted microstructure parameters are incompatible with the original data).

A leap forward in the Line Profile Analysis techniques was the introduction of the Whole Powder Pattern Fitting (WPPF; Scardi, Leoni, Dong, 2000; Scardi, Dong, Leoni, 2001) and Whole Powder Pattern Modelling (WPPM; Scardi & Leoni, 2002, 2004) techniques. In both cases, the full pattern is considered and model parameters are directly refined on the experimental data as e.g. in the Pawley (1981) technique or in the Rietveld method (Toraya, 1991). However, in WPPF some arbitrary shaped functions are used for the description of the peak profiles, limitation removed in WPPM where the profiles are built *ab initio* based on suitable models for the microstructure.

We can consider the Rietveld as the dual of the WPPM method: the first is a structural refinement technique whereas the second is a microstructural refinement method. The major differences between the two techniques are in the generation of the peak profiles: everything is done in reciprocal space in WPPM, whereas angular space is always used in the Rietveld method. Asymmetry due to the passage from one space to the other can be easily missed, especially for broad peaks. Furthermore, most Rietveld refinement codes employ the Williamson–Hall model for the treatment of the broadening, using pseudo-Voigt peaks for the fitting. This is certainly better than the Williamson–Hall method alone that does not simultaneously consider all peaks in the pattern and therefore loses completely the link with the experimental data. The structural constraint on peak intensities can however pose serious problems, as the microstructure extraction is somewhat biased by the structural one: any problem in the description of the structure can in fact have consequences on the microstructure. It would be ideal to see both the structure and the microstructure being simultaneously modelled: but as in the general case this is seldom possible, it is advised to perform the microstructure studies with the less degree of biasing in such a way that the subtle features of each profile are correctly taken into account.

A last note concerns the Multiple Whole Profile Fitting (MWP; Ungár et al. 2001; Ribárik, 2008), the Convolutional Multiple Whole Profile Fitting (CMWP; Ribárik et al. 2004) and extended Convolutional Multiple Whole Profile Fitting (eCMWP; Balogh et al. 2006) methods, independently proposed as possible alternative to traditional LPA methods and to the WPPM. The MWP has serious flaws and is no longer extensively employed. In fact it requires some pre-processing of the data necessary to separate the peaks and to extract their Fourier transform. This transform is then fitted to a model in Fourier space. Again the connection with the data is completely lost. In order to remove this limitation, the CMWP has been proposed, closely resembling a downsized version of the WPPM. A model certain differing between the CMWP and the WPPM concerns the stacking faults. An alternative to the model of Warren for faults, based on the fitting of simulated profiles, has been proposed in the literature (Balogh et al. 2006). This can potentially allow systems with extrinsic and intrinsic types of faults to be considered. Any intermediate or mixed case would have to be independently re-calculated. The application is however limited just to cubic, hexagonal and orthorhombic powders with spherical or ellipsoidal domain shapes and assuming the presence of dislocations and faults. A generalisation and extension would be highly overlapped with the existing literature on WPPM.

The basics of the WPPM method and some relationships with traditional LPA techniques are shortly presented.

10.2. Theoretical basis of the WPPM method

10.2.1. Useful definitions

Some useful definitions are introduced to simplify the formalism provided in the following.

Reciprocal space variable: $d* = \dfrac{2\sin\theta}{\lambda} = \dfrac{1}{d}$.

Bragg peak position $\qquad d^*_{\{hkl\}} = d^*_B = \dfrac{2\sin\theta_{\{hkl\}}}{\lambda} = \dfrac{1}{d_{\{hkl\}}} = \dfrac{1}{d_B}$.

Variable: $\qquad\qquad s = d^* - d^*_{\{hkl\}} = \dfrac{2}{\lambda}\left(\sin\theta - \sin\theta_{\{hkl\}}\right)$.

Unless otherwise noted, each peak profile will be described in reciprocal space with reference to its Bragg position (i.e. when applicable, in terms of the s variable). The Bragg position is the one expected for the reflection family in the absence of any type of defect. Going from the reciprocal space to the diffraction space 2θ involves a trivial change of variables: it is of paramount importance to notice that peaks that are symmetrical in reciprocal space, will forcedly become asymmetrical in 2θ space and viceversa.

10.2.2. Diffracted intensity: Fourier approach

The profile $h(s)$ of the diffraction peak in the powder diffraction pattern of a polycrystalline material results from the folding of an instrumental profile $g(s)$ with sample-related effects $f(s)$ (microstructure) (Klug & Alexander, 1974):

$$h(s) = \int_{-\infty}^{\infty} f(y)g(s-y)dy = f \otimes g(s). \qquad (10.5)$$

Fourier transformation (FT) allows a suitable handling of this expression: the convolution theorem, in fact, states that the Fourier transform of a convolution can be obtained as the product of the FT of the functions to be folded. In synthesis:

$$C(L) = FT\left[h(s)\right] = FT\left[f(s)\right]FT\left[g(s)\right], \tag{10.6}$$

where L is the variable conjugate to s. The properties of the Fourier transform, allow rewriting eq. (10.5) as:

$$h(s) = FT^{-1}\left[C(L)\right] = FT^{-1}\left[FT\left[h(s)\right]\right] = FT^{-1}\left[FT\left[f(s)\right]FT\left[g(s)\right]\right]. \tag{10.7}$$

Equation (10.7) is the basis of all Fourier methods of line profile analysis, but also of the Whole Powder Pattern Modelling: Once the Fourier Transform of each broadening component is known, the global profile can be easily synthesised via product and back transform. Of course the function $h(s)$ is normalised and provides information just on the shape of the profile. The generic profile can be therefore represented (with the usual definition of x) as:

$$I_{hkl}(s) = k(d^*)h(s) = k(d^*)\int_{-\infty}^{\infty} C(L)\exp(2\pi iLs)dL \tag{10.8}$$

where $k(s)$ groups all geometrical and structural terms which are constant or known functions of s (e.g., structure factor, Lorentz polarization factor, etc.), whereas $C(L)$ is the Fourier transform of the peak profile. Any broadening contribution can enter the convolution chain and can therefore be included in the equation: the expressions are known for several cases of practical interest (see e.g. Scardi & Leoni, 2002, 2004, 2005; D'Incau et al. 2007; Leoni & Scardi, 2004; Leineweber & Mittemeijer, 2004; van Berkum, 1994; Cheary & Coehlo, 1992).

Actually, equation (10.8) is not fully general: it implicitly assumes that the broadening contributions act on the entire family of reflection {hkl} and therefore that a multiplicity term can be used. The multiplicity term is included in the $k(d^*)$ function.

Certain types of defects (such as faults, for instance), can act independently on each member of the reflection family. In this case, we should rewrite equation (10.8) more correctly as:

$$I_{\{hkl\}}(s) = k(d^*)\sum_{hkl} w_{hkl}I_{hkl}(s_{hkl}) =$$

$$= k(d^*)\sum_{hkl} w_{hkl}I_{hkl}(s - \delta_{hkl}), \tag{10.9}$$

where w_{hkl} is a weight function,

$$s_{hkl} = d^* - (d^*_{\{hkl\}} + \delta_{hkl}) = s - \delta_{hkl}$$

is the distance, in reciprocal space, from the centroid of the *hkl* component and δ_{hkl} is the shift from the reciprocal-space point $d^*_{\{hkl\}}$ corresponding to the Bragg position in the absence of defects. The sum is over independent profile sub-components, selected on the basis of the specific defects (e.g. 4 for the {111} reflection family in the *fcc* case when faults are present: selection is based on the value of *h+k+l* that can be ±3 or ±1).

The simplest broadening contributions observed in a real material, and thus entering equation and following ones, are certainly those related to the real nature of the instrument, the finite size of the coherently diffracting domain (size effect) and to the presence of defects such as e.g. dislocations and faults (van Berkum, 1994; Scardi & Leoni, 2002). Taking this into account, the $\mathbf{C}(L)$ reads:

$$\mathbf{C}(L) = T^{\mathrm{IP}}(L) A^S_{hkl}(L) \left\langle \exp\left[2\pi i \psi_{hkl}(L)\right]\right\rangle \left\langle \exp\left[2\pi i \phi_{hkl}(L)\right]\right\rangle, \quad (10.10)$$

where $T^{\mathrm{IP}}(L)$ and $A^S_{hkl}(L)$ are the FT of the instrumental profile (IP) and domain size components, respectively, and the terms in brackets ($\langle ... \rangle$) are average phase factors related to lattice distortions (ψ) and faulting (ϕ).

Equation (10.10) can be considered as the core of the WPPM method. A plug and play behaviour is envisaged: new broadening sources can be considered together with the other effects simply by including the corresponding complex Fourier transform in the large product of equation (10.10). A warning should be however given: the approach is valid when the broadening sources can be considered as diluted and independent, i.e. if the correlations between defects can be neglected. If this does not apply, then cross terms need being considered and the whole approach revised.

The main sources of broadening and the corresponding equations are briefly illustrated to show the model parameters and the completely different functional form that allows those microstructure features to be separated.

10.3. Broadening components

10.3.1. Instrument

When dealing with specimen related effects, the primary recommendation is to try limiting the instrument influence. The large the instrumental effects, the larger the error in the evaluation of the microstructure. The instrument starts playing a role when it influences the positions of the peaks (e.g. through the effect of the axial divergence) or when the instrumental broadening is comparable in magnitude with the one due to specimen-related effects.

Two possible ideas can be followed when dealing with the instrumental contribution: modelling the effects using the so-called Fundamental Parameters Approach (FPA; see e.g. Cheary & Coelho, 1992; Kern &

Coelho, 1998) or measuring them by means of a suitable standard. In the Fundamental Parameter approach, the geometry of the instrument and the effect of each optical component are described through analytical raytracing. In this way, the profile is directly synthesised in 2θ. The actual physical parameters (sizes of the optical elements) can be then refined on the pattern of a suitable line profile standard such as the NIST SRM 660b LaB_6 (Cline et al, 2010). The FPA would be the ideal approach if all possible aberrations could be taken into account, as it is of paramount importance that all instrumental features are reproduced when dealing with microstructure effects. However this is usually not possible. Moreover, the FPA provides the peak directly in 2θ space, with an intrinsic difficulty of handling it within the Fourier approach on which the WPPM algorithm is based (a transformation from 2θ into L space would be needed to obtain the $T^{IP}(L)$ function to be inserted in equation (10.10)). Due to the superposition and the nonlinear mapping between the two spaces, handling the peaks in 2θ can cause errors in the analysis of nanostructured materials (Scardi et al, 2011). A possible alternative is to employ an optical setup that limits the aberrations (narrow slits), to measure the diffraction pattern of a line broadening standard, to fit the peaks of the standard using any arbitrary function and to parameterise the parameters of this function versus 2θ. The major problems can be found in the low 2θ region where sample transparency and axial divergence (Klug & Alexander, 1974; Wilson, 1962) can cause asymmetry and shift in the peak positions.

A rather used parameterisation of the instrumental peak profile is the one based on the functions proposed by Caglioti et al. (1958) and by Rietveld (1969). A set of accessible peaks from the standard are fitted using Voigt (or pseudo-Voigt) functions. The Full Width at Half Maximum (FWHM) and the Lorentz (or Gauss) content (mixing parameter η) are then parameterised according to a function in $\tan(\theta)$ and θ, respectively (Caglioti et al, 1958; Leoni et al, 1998; Scardi & Leoni, 1999):

$$FWHM^2 = U\tan^2\theta + V\tan\theta + W,$$
$$\eta = a + b\theta + c\theta^2. \tag{10.11}$$

The parameters of the Fourier transform of a Voigt or pseudo-Voigt are then constrained to those of the Caglioti equations (10.11). Fortunately, the approach is entirely analytical as the Fourier transform of a Voigtian is the combination of a Gaussian (for the Gauss part η) and an exponential (for the Lorentz part). The simplest approach to incorporate the instrumental profile into the WPPM is therefore to use a function such as:

$$T_{pV}^{IP}(L) = (1-k)\exp\left(-\frac{\pi^2\sigma^2 L^2}{\ln 2}\right) + k\exp(-2\pi\sigma L), \tag{10.12}$$

where σ is the half width at half maximum in the reciprocal space, whereas

k is related to the mixing parameter of the pseudo-Voigt η by (Langford, 1992; Scardi & Leoni, 1999):

$$k = \left(1 + \frac{1}{\sqrt{\pi \ln 2}} \frac{(1-\eta)}{\eta}\right)^{-1}. \tag{10.13}$$

10.3.2. Domain size and shape

The main contribution to the line profile broadening is usually that due to the finite size of the scattering domains. For a large part of the last century, a generic 'average crystallite size' was the only information that could be obtained from the broadening of the profile. In the last decade, the possibility was found for extracting information on the shape of the domains and on the distribution of sizes, key information when analysing any nanosized powder.

It is important to remember that grain size and shape are not properties of the material and therefore the use of a symmetry constraint there (such as e.g. using spherical harmonics to describe the shape of the generic scattering object) is not justified. Domain size, in fact, is not a tensor property (Nye, 1987; Scardi & Leoni, 2001).

An easy way to deal with the size broadening contribution comes certainly from the works of Bertaut (1949, 1950) and of Stokes & Wilson (1942). Bertaut proposed to consider the domains as made of columns and to analyse the scattering inside those columns. The column length distribution can be always extracted from the data: more complex modelling involving given shapes or distributions simply modify the way in which the columns are rearranged. Stokes & Wilson introduced the concept of ghost to calculate the Fourier transform for a given shape: the volume common to a domain and its ghost, i.e. a copy of the same domain displaced by a quantity L along the scattering direction $\mathbf{s}$, is proportional to the Fourier transform $A_c^S(L,D)$ for the given object of shape c. The calculation has been already carried out for several simple shapes characterised by a single length parameter (Scardi & Leoni, 2001) and a cubic polynomial in L is always obtained (not necessarily the same over the whole accessible range). In general, therefore:

$$A_c^S(L,D) = \sum_{n=0}^{3} H_n^c (L/D)^n, \tag{10.14}$$

where H_n^c can be found in the literature or calculated by geometrical means. It is somehow possible to relate the Fourier coefficients with the usual sizes obtained with traditional methods. In particular, the area-weighted size $\langle L \rangle_V$ (Warren–Averbach method) and the volume weighted size $\langle L \rangle_S$ (Williamson–Hall method) read:

$$\langle L \rangle_S = -\left[\frac{dA_c^s(L,D)}{dL} \Big|_{L=0} \right]^{-1} = \frac{D}{K_k} = -\frac{D}{H_1}, \tag{10.15}$$

$$\langle L \rangle_V = \left[\beta(s) \right]^{-1} = 2 \int_0^{D/K} A_c^S(L,D)dL = D/K_\beta, \tag{10.16}$$

where $\beta(s)$ is the integral breadth and K_k, K_β are the integral breadth and initial-slope Scherrer constants, respectively (Langford & Wilson, 1978; Scardi & Leoni, 2001).

The results that can be obtained have little physical significance in real cases: the distribution of size (and shape) can play a key role in determining the shape of the profile and the properties of the powder under analysis. The Fourier coefficients for the polydisperse case can be calculated more simply by assuming a functional shape for the distribution. Commonly met distributions are certainly the lognormal and the gamma for which the equations and the corresponding moments read:

$$g_l(D) = \frac{1}{D\sigma_l \sqrt{2\pi}} \exp\left(-\frac{(\ln D - \mu_l)^2}{2\sigma_l^2} \right), \qquad M_{l,n} = \exp\left(n\mu_l + \frac{n^2}{2}\sigma_l^2 \right);$$

$$\tag{10.17}$$

$$g_p(D) = \frac{\sigma_p}{\mu_p \Gamma(\sigma_p)} \left(\frac{\sigma_p D}{\mu_p} \right)^{\sigma_p - 1} e^{-\frac{\sigma_p D}{\mu_p}}, \qquad M_{p,n} = \left(\frac{\mu_p}{\sigma_p} \right)^n \frac{\Gamma(n+\sigma_p)}{\Gamma(\sigma_p)}.$$

$$\tag{10.18}$$

For the given distribution g_i, the scattered intensity reads:

$$I(x) = k(x) \int_0^\infty I_c(d^*,D)g_i(D)V_c(D)dD \Big/ \int_0^\infty g_i(D)V_c(D)dD =$$

$$= k(x)\left[M_{i,3} \right]^{-1} \int_{L=0}^\infty \int_{D=LK}^\infty A_c^S(L,D)g_i(D)D^3 dD e^{2\pi iLx} dL = k(x) \int_{L=0}^\infty A^s(L,D)e^{2\pi iLx} dL,$$

$$\tag{10.19}$$

where:

$$A^S(L) = \left[M_{i,3} \right]^{-1} \int_{LK}^\infty A_c^S(L,D)g_i(D)D^3 dD. \tag{10.20}$$

The equation for the polydispersed case (equation (10.19)) is therefore analogous to that of the monodispersed case, with a suitable definition of the Fourier coefficients for the dispersed powder. Explicit equations can be obtained in some cases. For a lognormal, for instance, the Fourier coefficients can be written as:

$$A_l^S(L) = \sum_{n=0}^{3} \text{Erfc}\left[\frac{\ln(L \cdot K_c) - \mu_l - (3-n)\sigma_l^2}{\sigma_l\sqrt{2}}\right]\frac{M_{l,3-n}}{2M_{l,3}} \cdot H_n^c L^n.$$ (10.21)

Equation (10.21) is clearly not a combination of an exponential and a Gaussian that would be the expected functional form for the Fourier transform of a Voigtian. This is a proof of the inability of traditional LPA methods (based on Voigt functions) to deal with a lognormal distribution of domains, one of the most frequent cases of study.

It is however possible (Scardi & Leoni, 2001) to relate the parameters of the polydispersed system to the size obtained with the traditional methods (Warren–Averbach and Williamson–Hall, respectively):

$$\langle L \rangle_S = \frac{1}{K_k}\frac{M_{i,3}}{M_{i,2}},$$ (10.22)

$$\langle L \rangle_V = \frac{1}{K_\beta}\frac{M_{i,4}}{M_{i,3}}.$$ (10.23)

Using an analytical function for the description of a size distribution can help in stabilising the results (as the size is somehow forced to be zero at very small and very large size values). Some doubts can however raise as to the physical validity of this forcing. This is for instance the case of a multimodal system. A possible alternative has been proposed in the literature, replacing the analytical distribution with an histogram. It has been demonstrated the ability of this model to fit the experimental data (M. Leoni & P. Scardi, 2004). The quality of the measurement and the availability of models describing all contributions to the peak broadening is in most cases the limiting factor for an extensive use of the histogram model: correlations of the small sizes with the background and with features such as e.g. the thermal diffuse scattering (Scardi et al, 2011) can in fact occur. This is the only available solution so far to explore cases where the analytical models are unable to correctly describe the observed broadening. None of the traditional LPA techniques is in any case able to deal with multimodal size distributions.

10.3.3. Lattice distortions

Dislocations are certainly one of the major sources for lattice distortions that can be measured by X-ray powder diffraction. The average phase term due to distortions is, in general, a complex quantity:

$$\left\langle e^{2\pi i \psi_{hkl}(L)} \right\rangle = \left\langle \cos\left(2\pi L d^*_{\{hkl\}} \varepsilon_{\{hkl\}}(L)\right) \right\rangle + i\left\langle \sin\left(2\pi L d^*_{\{hkl\}} \varepsilon_{\{hkl\}}(L)\right) \right\rangle =$$

$$= A^D_{\{hkl\}}(L) + iB^D_{\{hkl\}}(L),$$

$$(10.24)$$

where $\varepsilon_{\{hkl\}}(L)$ is a strain related to the lattice distortion on a coherence distance L. Knowing the actual source of distortion allows the explicit calculation of the various terms. Some simplifying hypotheses are possible for a powder, such as considering the strain being independent of the reflection family subcomponents $\varepsilon_{hkl}(L) \equiv \varepsilon_{\{hkl\}}(L)$.

Traditional LPA methods such as the Warren–Averbach method (Warren & Averbach, 1950, 1952; Warren, 1969), simply take a first order Maclaurin expansion of the equation:

$$A^D_{\{hkl\}}(L) \cong 1 - 2\pi^2 L^2 d^{*2}_{\{hkl\}}\left\langle \varepsilon^2_{\{hkl\}}(L)\right\rangle, \qquad (10.25)$$

$$B^D_{\{hkl\}}(L) \cong -\frac{4}{3}\pi^3 L^3 d^{*3}_{\{hkl\}}\left\langle \varepsilon^3_{\{hkl\}}(L)\right\rangle. \qquad (10.26)$$

to extract the microstrain contribution from the measured data. Just the second order moment of the strain distribution (root mean strain or microstrain) can be obtained this way. The third order is usually neglected and the peaks are therefore considered as symmetric. The strain field moreover, is highly anisotropic especially if line defects (dislocations) are considered. Equation (10.25) should therefore show a dependence on the actual diffraction peak, i.e. anisotropic broadening should occur. The traditional Warren–Averbach method is not able to take this into account and it is therefore not suitable, in its original form, to deal with materials containing dislocations.

The anisotropy in the case of dislocations depends on the relative orientation of the Burgers and diffraction vector with respect to the dislocation line, as well as on the elastic anisotropy of the studied material (Wilkens, 1970a, 1970b). Following the approach of Wilkens based on the early work of Krivoglaz and Ryaboshapka (Krivoglaz & Ryaboshapka, 1963; Krivoglaz, 1969), so far the only available and validated one (Wilkens, 1970a, 1970b; Krivoglaz et al, 1983; Kamminga & Delhez, 2000; Groma et al, 1988; Klimanek & Kuzel, 1988), the distortion Fourier coefficients read:

$$A^D_{\{hkl\}}(L) = \exp\left[-\frac{1}{2}\pi b^2 \overline{C}_{\{hkl\}}\rho d^{*2}_{\{hkl\}}L^2 f\left(L/R'_e\right)\right], \qquad (10.27)$$

where b is the modulus of Burgers vector, $\overline{C}_{\{hkl\}}$ is the so-called average contrast factor, ρ is the dislocation density and R'_e is an effective outer cut-off radius. The term $f\left(L/R'_e\right)$ is a decaying function introduced by

Wilkens to guarantee the convergence of the Fourier coefficients; several formulations have been proposed in the literature (Wilkens 1970, 1970a; van Berkum, 1994; Kaganer & Sabelfeld, 2010), leading to small differences in the final profile.

The anisotropic nature of the broadening due to dislocations is mainly taken into account by the so called contrast or orientation factor C_{hkl} (Wilkens, 1970a, 1970b; Krivoglaz et al, 1983; Kamminga & Delhez, 2000; Groma et al, 1988; Klimanek & Kuzel, 1988; Kuzel & Klimanek, 1989; Wilkens, 1987; Martinez-Garcia et al, 2008, 2009). The average value $\overline{C}_{\{hkl\}}$ is usually employed under the assumption that all slip systems are equally populated. In general, as extensively shown in the recent literature (Popa, 2000; Leoni et al, 2007), the contrast factor is related to the 4th order invariant for the Laue class of the material under study (Leoni et al, 2007)

$$d_{\{hkl\}}^4 \overline{C}_{\{hkl\}} = E_1 h^4 + E_2 k^4 + E_3 l^4 + 2\left(E_4 h^2 k^2 + E_5 k^2 l^2 + E_6 h^2 l^2\right) +$$
$$+4\left(E_7 h^3 k + E_8 h^3 l + E_9 k^3 h + E_{10} k^3 l + E_{11} l^3 h + E_{12} l^3 k\right) +$$
$$+4\left(E_{13} h^2 kl + E_{14} k^2 hl + E_{15} l^2 hk\right). \qquad (10.28)$$

It can be seen that 15 coefficients are needed to describe the anisotropy effects in the general case. For cubic materials, such as silicon, symmetry reduces the number of independent coefficients to two and the average contrast factor reads (Stokes & Wilson, 1944):

$$\overline{C}_{\{hkl\}} = (A + BH) = A + B\frac{h^2 k^2 + h^2 l^2 + k^2 l^2}{(h^2 + k^2 + l^2)^2}, \qquad (10.29)$$

where the two coefficients A and B can be calculated from the elastic constants following the procedures proposed by (Klimanek & Kuzel, 1988; Kuzel & Klimanek, 1989] and generalised in Martinez-Garcia et al, 2008, 2009).

As the calculation of the contrast factor is not an easy task, it is customary to evaluate it for the screw and edge case and to refine an effective dislocation character φ (Ungar et al, 1999):

$$\overline{C}_{\{hkl\}} = \left[\varphi\overline{C}_{E,\{hkl\}} + (1-\varphi)\overline{C}_{S,\{hkl\}}\right] = \left[\varphi A_E + (1-\varphi)A_S\right] + \left[\varphi B_E + (1-\varphi)B_S\right] \cdot H,$$

$$(10.30)$$

where the geometric term H is the same as in equation (10.29). Albeit not completely correct, the approach proposed in equation (10.30) allows dealing with the general case where a mix of dislocations of varying character is acting on equivalent slip systems.

10.3.4. Twin and deformation faults

In many materials the energy for the creation of a fault is quite low. This is the case e.g. of Cu, Ni, Au, for example, but most *fcc* metals more or less fall in this list. The two major types of faults in cubic materials are the so-called deformation and twin faults, whose occurrence probabilities are traditionally denoted as α and β. The analysis of faulting using a Bragg-type method is quite dangerous in the general case as faulting causes the appearance of a diffuse broadening component together with the Bragg one. However, under the hypothesis of diluted presence of faults, the treatment originally proposed by Warren (1969) can be employed to account for the observed modification of the peak position and shape. The approach of Warren presents several flaws that have been corrected, e.g. by Velterop et al. (2000), Leoni et al. (2001), Estevez-Rams et al. (2003, 2008): this allows accounting for the different broadening of the various members of a reflection family and for missing high order terms in the faulting formulae. Considering the limits of the approach, a good information can be obtained up to a few percent of faults on the {111} plane in an *fcc* system (spacegroup *Fm-3m*). The trick is to describe the fcc faulting in hexagonal axes, i.e. as a $\langle 001 \rangle$ stacking problem on the {111} plane. The extension to the bcc case (*Im-3m* spacegroup) is straightforward. More complex is the case of silicon (*Fm-3m*) as the corresponding model has not been developed yet. The differences between the *Fm-3m* and *Fd-3m* cases are limited: the *Fd-3m* is in fact an *Fm-3m* with half of the tetrahedral sites occupied in an ordered way. The rules for broadening are therefore expected to be similar.

In the *fcc* case, referring to equation (10.10), the average phase term due to faulting can be written as:

$$\left\langle \exp\left[2\pi i \phi(L; d^*_{\{hkl\}}, L_0/h_0^2)\right]\right\rangle = A^F_{hkl}(L) + iB^F_{hkl}(L),\tag{10.31}$$

where $L_0 = h + k + l$ and $h_0^2 = h^2 + k^2 + l^2$.

Faulting is one of the typical cases where a complex (sin) term is present, as peak shift and asymmetry in the profiles is expected (unless twin faults are absent). Following the treatment of Warren, a set of recurrence equations can be written for the probability of faulting occurrence. The solution of the recurrence equations is employed to generate the Fourier coefficients for faulting. In particular, by defining:

$$S^2 = 3 - 12\alpha - 6\beta + 12\alpha^2 - \beta^2 + 12\alpha\beta + 24\alpha^2\beta, \qquad Z = \frac{\sqrt{(1-\beta^2)+s^2}}{2},\tag{10.32}$$

and introducing the sign function:

$$
\sigma_{L_0} = \begin{cases} +1 & \text{for } L_0 = 3N+1, \\ 0 & \text{for } L_0 = 3N, \\ -1 & \text{for } L_0 = 3N-1, \end{cases} \qquad N = 0, \pm 1, \pm 2, ..., \tag{10.33}
$$

the Fourier coefficients are obtained as:

$$
A_{hkl}^F(L) = \exp\left(\frac{1}{2}\ln(Z) \left| Ld_{\{hkl\}}^* \sigma_{L_0} \frac{L_0}{h_0^2} \right| \right), \tag{10.34}
$$

$$
B_{hkl}^F(L) = -\sigma_{L_0} \frac{L}{|L|} \frac{L_0}{|L_0|} \frac{\beta}{S} A_{hkl}(L). \tag{10.35}
$$

Besides being asymmetric, each profile subcomponents can also be shifted with respect to the average Bragg position. For the *hkl* subcomponent the shift is:

$$
\delta_{hkl} = -\left(\frac{1}{2\pi} a \, \tan\left(\frac{S}{1-\beta} \right) - \frac{1}{6} \right) d_{\{hkl\}}^* \frac{L_0}{h_0^2} \sigma_{L_0}. \tag{10.36}
$$

It can be seen that in a given reflection family {*hkl*} there can exist subcomponents that are unaffected ($L_0 = 3N$) and affected ($L_0 = 3N \pm 1$) by faulting, contributing to produce peculiar shapes of the corresponding peak profiles.

Again, the simple treatment that can be found in the literature (e.g. the treatment of Warren (1969) or the Warren–Averbach (1950, 1952) method, would be rather erroneous, as it does not take the fine details of the broadening into account.

10.3.5. Assembling the equations into a peak and modelling the data

The various broadening contributions briefly proposed in the previous, can be employed to generate the {*hkl*} peak profile. Starting from the *hkl* component in the simplified case $B_{\{hkl\}}^D \cong 0$:

$$
I_{hkl}(s) = k(d^*) \int_{-\infty}^{\infty} C(L) \exp(2\pi iLs) dL
$$

$$
= k(s) \int_{-\infty}^{\infty} T_{pV}^{IP}(L) A_{hkl}^S(L) \left[A_{hkl}^D(L) \cos(2\pi Ls) + iB_{hkl}^D(L)\sin(2\pi Ls) \right]...
$$

$$
... \left[A_{hkl}^F(L) \cos(2\pi Ls) + iB_{hkl}^F(L)\sin(2\pi Ls) \right] dL \tag{10.37}
$$

we have:

$$
I_{\{hkl\}}(s) = \sum_{hkl} w_{hkl} I_{hkl}(s - \delta_{hkl}). \tag{10.38}
$$

Those equations are employed, via the use of Fast Fourier Transform and suitable space remapping, to generate the peaks present in an observed pattern. Suitable constants are included, possibly through refinable parameters. Intensities are multiplied by the Lorentz and polarization (when needed) terms, and a background is added to the whole pattern.

The parameters of the model are refined within a nonlinear least squares routine based on the Marquardt algorithm, that will match the synthesised pattern directly with the experimental data.

It is worth noting that the number of parameters involved is quite limited, as the shape of each peak is bound to the underlying physical models.

For silicon, for instance, compared to 4 parameters per peak (intensity, width, shape, position) necessary for a Scherrer-type analysis, in WPPM we refine:

- two parameters defining the domain size distribution
- three parameters related to dislocations (ρ, R'_e and φ)
- two parameters related to faulting (α and β)
- one lattice parameter a_0
- a few background parameters (e.g. 4 parameters)

In addition, we refine one parameter per peak (intensity) and we can refine for instance a specimen displacement error for measurements done in Bragg–Brentano geometry. It can be easily seen that the same number of parameters employed to analyse 4 peaks in a typical case of study (16 parameters), allows extracting just some numbers of limited physical significance using traditional LPA methods, or some physical information using the WPPM approach.

10.4. Examples of WPPM Analysis

In the last years the WPPM approach has been employed for the modelling of nanostructured materials both in bulk and powder form. We will present here some applications where the full power of the approach is illustrated. The PM2K software was employed for the analyses.

10.4.1. Nanocrystalline ceria

Diffraction patterns of a cerium oxide specimen obtained by calcinating a gel made with cerium isopropoxide (Leoni et al, 2004; Leoni & Scardi, 2004b) were collected on a Rigaku PMG/VH diffractometer using Cu radiation (40kV, 45mA). The machine has a high resolution setup obtained by employing a narrow set of slits in the primary ($\frac{1}{2}°$ divergence, 2° primary Soller) and on the diffracted (0.15mm antiscatter, $\frac{1}{2}°$ receiving, 2° secondary Soller) beams. A curved graphite analyzer crystal was also employed. The instrumental profile was characterised by modelling the reflections of the NIST SRM 660a standard (LaB_6) using symmetrical pseudo-Voigt functions and parameterising them according to the Caglioti et al. formulae.

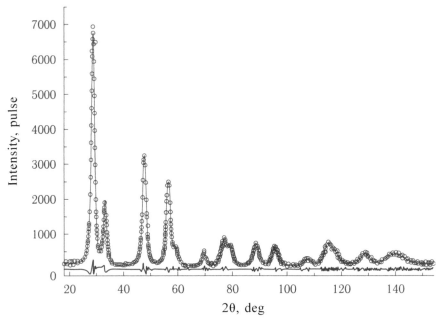

Fig. 10.1. Powder diffraction pattern of the ceria specimen calcined at 400°C. Raw data (dots), WPPM model (line) and difference (line below).

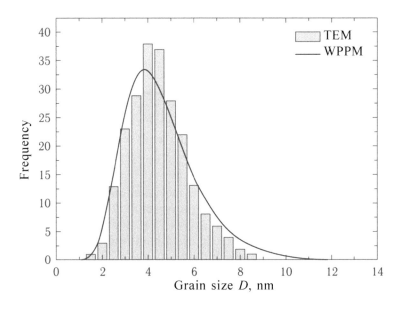

Fig. 10.2. Size distribution obtained with WPPM (line) and from TEM (histogram).

The data shown in Fig. 10.1 (dots) were collected by a scintillation counter in the 18–154° range with a step of 0.05° and a counting time of 60s/step. This type of acquisition is typical for line profile analysis: extremely high signal to noise ratio, very wide acquisition range. The two conditions are necessary in order to collect information on peak tails (important to distinguish between similar models) and to have a larger number of peaks in the diffraction range (important to capture details on the anisotropic broadening).

The data were modelled using the PM2K software implementing the WPPM and the result is shown in Fig. 10.1: the agreement between model and data is clear from the almost flat residual. It is interesting to see that 1800 datapoints were modelled using a very small number of parameters:

- 1 cell parameter
- 6 (fixed) parameters defining the instrumental contribution (5 for the Caglioti plots and 1 for the $K\alpha_{1,2}$ intensity)
- 2 parameters for a lognormal distribution of sizes (μ, σ)
- 3 parameters for dislocations (ρ, R_e, mixing parameter)
- 3 parameters for background
- 1 parameter for specimen displacement
- 16 parameters for the intensity of the peaks

The number of parameters is certainly smaller than that required for a traditional fitting. Instead of the 5 parameters related to the microstructure (size and dislocations), $2 \cdot 16 = 32$ parameters (width and shape for all peaks) would have been employed.

The size distribution extracted by WPPM is shown in Fig. 10.2 as a continuous line. The information is more complete than with traditional methods as, knowing the distribution, the first moment (mean) can be extracted. In this case the mean is about 4.3 nm.

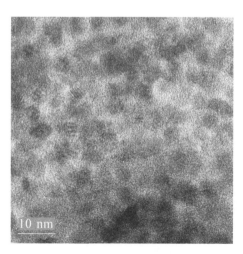

Fig. 10.3. TEM micrograph of the calcined ceria powder.

It is worth comparing the WPPM distribution with the one that can be obtained using transmission electron microscopy (TEM). Figure 10.3 shows a micrograph recorded for the specimen using a 300 kV JEOL 3010 microscope (0.17 nm point-to-point resolution) equipped with a Gatan slow-scan 974 CCD camera (Leoni et al, 2004).

As the particles are well separated, a distribution can be easily obtained from a large set of those micrographs. The operation is quite long and complex and required a certain experience in order to consider only those particles for which an univocal size can be determined (clearly visible, non overlapping with neighbouring particles). A sufficient number of particles need to be considered in order to minimise the errors due to the personal selection operated by the scientist in charge of the size estimation.

The TEM size distribution is shown in Fig. 10.2 as a histogram superimposed to the WPPM distribution. A grand total of 800 domains was considered and their size estimated by measuring two orthogonal directions. Considering the difference between the techniques, the agreement is again excellent. The small residual differences can be due to several factors including issues in the evaluation of the size from the analysis of TEM micrographs, simplified treatment employed in the analysis of the X-ray powder diffraction data via WPPM (e.g. spherical domains, absence of models for surface relaxation).

The agreement can be improved (see e.g. Leoni & Scardi (2004b)), but again, a single diffraction pattern provides the same quality of information as the TEM in term of distribution. The additional information is the one concerning defects, of difficult evaluation under the microscope. A dislocation density of $1.4 \cdot 10^{16}$ m^{-2} was in fact obtained for the specimen analysed here, corresponding to ca. 1 dislocation every couple of grains (Leoni & Scardi, 2004b). With those numbers and considering that most of those dislocations would be impossible to see under the microscope (as they would be oriented in a wrong way with respect to the electron beam), we can understand why dislocations can hardly be seen e.g. in 10.3 and could be easily missed.

10.4.2. Copper oxide

Looking at the results of TEM it can be argued where the true advantage of WPPM is, considering that a visual information on the domain shape, size and distribution are not readily available. A first advantage concerns the higher statistical validity of the WPPM results. This is clear if we consider e.g. that in a TEM analysis only a few hundred grains are (at most) considered whereas at least a few millions are sampled in a powder diffraction pattern. The second advantage certainly lays in the phase sensitivity of diffraction, allowing distinct phases to be analysed simultaneously. Unless each grain is individually sampled and carefully analysed, the assignment of the phase type to a certain size is impossible under the microscope. In a diffraction pattern, conversely, different phases

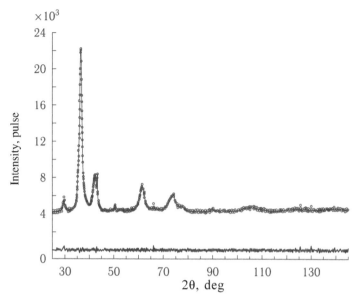

Fig. 10.4. Powder diffraction pattern of the milled cuprite specimen. Initial data (dots), WPPM model (line) and difference (line below).

provide different diffraction peaks: the broadening of the peaks of each phase identified in the system (e.g. through a matching with the data on the known compounds within the ICDD PDF file) can be considered via WPPM.

Figure 10.4 shows, as an example, the diffraction pattern of a ball milled cuprite powder. The powder was obtained by grinding for 5 minutes a commercial (Carlo Erba) Cu_2O powder in a high-energy shutter mill (Fritsch Pulverisette 9). Milling was made in 30 stages (consisting of 10 s milling followed by 120 s cooling at 25°C) to avoid overheating of the powder that would lead, e.g. to recrystallisation or grain growth (Martinez-Garcia et al, 2007).

The diffraction pattern was collected on the same Rigaku PMG-VH diffractometer employed for the analysis of the ceria powder in step scan mode from 18° to 154°, with a step size of 0.1° and a fixed acquisition time of 150 s per point. Again, the acquisition is over a quite wide range and done with the intent to maximise ht signal/noise ration. This time, however, a large set of very broad peaks is present, as well as an intense background. The background is due to part of the fluorescence of copper, not eliminated by the secondary crystal analyser. The broadening of the peaks increasing with the angle, on the other hand, is a clear indication that strain-type effects are quite important, as expected in an extensively milled powder.

The qualitative phase analysis evidenced the presence not just of Cu_2O (cuprite $Pn\bar{3}m$, but also of Cu (*fcc* copper, $Fm\bar{3}m$) and CuO (tenorite, $C\bar{c}$).

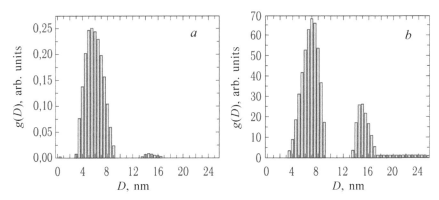

Fig. 10.5. Domain size distribution for cuprite. The extracted distribution and the distribution multiplied by D^3 are shown in (α) and (β), respectively.

The WPPM modelling was thus done considering all three phases and the corresponding result is shown in Fig. 10.4. The size distribution and defect content for all three phases can in this case be extracted. In order to get a more reliable result, the intensity of the peaks for the tenorite (minor and broadened phase) were evaluated and bound to a structural model (in such a way just a scale parameter is refined). Domains were considered as spherical. A good modelling is impossible if all phases are considered as monodispersed. The choice of the distribution makes little difference for the minor phases (copper and tenorite), but has a dramatic effect on cuprite. The peaks, in fact, have a peculiar shape showing a wide base and a rather sharp tip, indication of a possible multimodality in the size distribution. For this reason, copper and tenorite were modelled using a lognormal

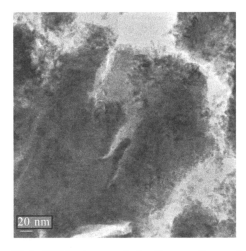

Fig. 10.6. TEM micrograph for the cuprite specimen.

distribution of spheres, whereas a histogram distribution was employed for cuprite (Leoni & Scardi, 2004a).

The resulting distribution is shown in Fig. 10.5a. The presence of a multimodal character is clear. Traditional analysis methods would be unable to correctly identify this distribution: in fact, even if the fraction of larger domains seems small, their weight in the diffraction pattern is large as the volume of the corresponding sphere enters the diffraction peaks. If we multiply the distribution by a volume term D^3 (cf. Fig. 10.5b), we clearly see that the larger grains have more or less the same weight of the small ones.

This issue must always be borne in mind when dealing with nanostructured powders: the smaller the domains and the wider the distribution, the smaller the contribution of the left tail with respect to the right one. This results in large errors associated to the smaller sizes unless an analytical distribution (pinning the small sizes to be finite and small) is employed.

As for the defects, the model based on the Laue invariant for the description of the contrast factor was employed for tenorite (cf. equation (10.28)), whereas the classical description was employed in the cubic case (Martinez-Garcia et al, 2007). The analysis confirms that dislocation densities are quite high here: $\rho = 2.8(5) \cdot 10^{16}$ m^{-2} was obtained, with an effective outer cut-off radius $R_e = 9(3)$ nm and an effective dislocation character $f_E = 0.85(3)$, thus showing a predominance of edge dislocations. Those quite high values are associated to the low shear modulus of cuprite: even if the material is a ceramic, it can be easily deformed and therefore a large quantity of dislocations can be pumped in it.

It is clear that the same detail of information is impossible to be obtained from TEM data: the size information would be impossible to be attributed to each single grain, plus atomic resolution could be hardly reached with this high level of defects (cf. Fig. 10.6).

10.5. Beyond the WPPM

It is clear that the WPPM is a powerful tool for the analysis of the microstructure of nanostructured materials, including silicon. However, WPPM is not an universal tool in that it cannot solve all problems nor deal with all types of specimens. We should in fact not forget that the formalism is valid in the limit of diluted defects i.e. defects that are non-interacting and whose effects can be modelled independently. To overcome the limitations, a few possible paths can be taken:

- employing a bottom up approach: starting from the atomic positions, it is possible to generate the diffraction pattern by using the Debye scattering formula (Debye, 1915). The diffraction pattern can be calculated exactly for a diluted powder of any given object, irrespective of the defect type and content. The disadvantage of this approach lays in speed that even on modern computers (based on CPUs or GPUs) can be prohibitive for very large objects (a few million atoms) (Gelisio et al, 2010).

- changing the point of view and considering a total scattering paradigm: the Pair Distribution Approach (PDF)[2] (Billinge & Thorpe, 2002) is more and more employed for the analysis not just of amorphous, but also of crystalline and nanocrystaline materials. The tools for the analysis of nanopowders are not yet fully developed, and the first application are appearing in the last years. The PDF approach has the advantage of being more flexible than the reciprocal space one for structural studies (interatomic distances can be easily estimated), but it is intrinsically limited for microstructure studies: modelling is not tied to the original dataset and it can suffer of the same problems of consistency observed for the traditional LPA.

- considering diffuse scattering (Welberry, 2004). In order to appreciate diffuse scattering effects, the traditional approach would pass through the identification of a single crystal and the measurement of sections of the reciprocal space for the single crystal. In this case, Bragg as well as diffuse features would clearly appear and the defective system would be possibly described via, e.g. a reverse Monte Carlo approach. This approach is fine for single crystals, but impossible to be realised on powders where the diffuse signal would be highly integrated in the corresponding reciprocal space.

- employing an approach that goes beyond the 3D periodicity on which all traditional methods are based (Leoni et al, 2004, Treacy et al, 1991). This is for instance the case of stacking defects that occur along a given direction. Alternatives to traditional methods have been proposed to describe both Bragg and diffuse scattering, dealing with a 2D periodic approach in which the third direction could show the presence of faulting.

[2]See also section 8.3.3 of this book.

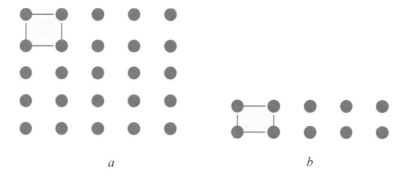

Fig. 10.7. 3D periodic (*a*) and 2D periodic (*b*) approaches.

Fig. 10.8. Two possible objects A and B (e.g. different unit cells) to be stacked.

All those approaches may potentially be employed for a detailed study of nanocrystalline silicon, as the material may present defects that can be hardly modelled in reciprocal space. The last approach proposed above is the only viable solution for the analysis of data collected on laboratory diffractometers operating with Cu radiation.

The 3D periodicity (3D lattice) assumed in the WPPM and in the Rietveld approach is the major limitation towards a full account of lattice stacking defects in real materials. In the 3D periodic case, the unit cell is translated regularly in all three direction in the space in order to regularly cover it (see Fig. 10.7a where the third direction is normal to the picture). Any defect introduced in the structure would result in periodic features, unless a diluted approach (see e.g. the WPPM) is employed.

A possible solution to overcome this limitation is to reduce the dimensionality of the lattice. The unit cell is still taken as building block, but its periodic repetition is limited to two dimensions (see Fig. 10.7b) or, in the more general case, to one. Reducing the dimensionality of the problem allows going form a traditional crystalline material to a system where only short range symmetry is preserved, i.e. a system close to being amorphous. The reduction to two dimensions is already sufficient to deal e.g. with:

- traditional 3D periodic case
- layer disorder (stacking faults limited to one direction only)
- interlayered systems
- modular sets
- ion substitution/ordering
- commensurate modulated structures with modulation along one direction

Alternative approaches exists to deal, at least to a certain extent, with some of those cases (van Smaalen, 2007). Actually, a wide superposition exists between the disordered/interlayered systems and modular materials (Ferraris et al, 2004). The order/disorder theory offers in fact a possible common mathematical background to deal with most of those cases in a consistent way. The study of modular materials, however, is traditionally

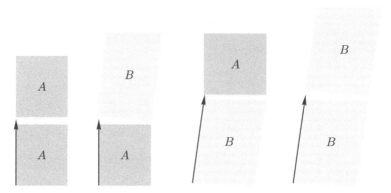

Fig. 10.9. Stacking possibilities involving just $\langle 001 \rangle$ stacking vectors given two objects A and B.

done on single crystals, to easily observe the peculiar features caused by the sequence of modules in the system.

One of the features of a system based on a 2D periodic lattice (corresponding to a lattice plane in the traditional 3D approach), is the need for a description of the sequence of the layers. Periodicity would in fact recover the traditional or modulated cases and therefore a statistical description is needed in all other cases.

To easily illustrate the problem we can consider the case of a system described as a stack of layers created with two different building blocks, A and B, shown in Fig. 10.8.

Sequences involving just one of the two blocks extended periodically in the plane and regularly stacked one on top of the other, lead to traditional crystalline materials with no defects. In the order/disorder theory, those correspond to the so-called MDO (Maximum Degree of Order) polymorphs (Ferraris et al, 2004). Any other stacking sequence involving a regular periodic arrangement of the stacking vectors (those defining the position of the origin of the next layer based on the current one) but also a regular alternation of different blocks would lead to an MDO polymorph as well.

To be possible in reality, the stacking should also preserve the bonding between neighbouring layers: this limits the number of possible

Fig. 10.10. Recursive description of the diffraction from a faulted system. The mean scattered wave from the crystal is given as the sum of the contribution from one layer plus a contribution from a crystal displaced by one layer.

stacking vectors in our model system. If translations or rotations between neighbouring layers are forbidden, then the only remaining stacking vector is $\langle 001 \rangle$ and the possibilities are those illustrated in Fig. 10.9.

When interlayering is possible, the presence of defects cannot be excluded. Most real systems will fall in this category and a statistical description of the layer sequence will be needed.

A probability matrix is usually considered, accounting for all probabilities of occurrence of a given ij sequence with i and j being indices running over all types of objects to be stacked (2 in this case, leading to a 2×2 matrix).

Each individual specimen of our material made out of a statistical alternation of A and B objects may potentially be different from any other one: both the unit cell data and the probability matrix is thus needed to describe a given specimen, but the matrix won't in general be unique for the given material.

Dealing with a specimen that can be described just in statistical terms opens a wide set of problems as for a possible way to calculate its diffraction pattern in a fast and effective way. A trick here is the one proposed by Treacy et al. (1991): the stacking sequence is self similar and therefore it can be described recursively as in a Markov chain. In this way also the corresponding scattering function can be described recursively. The concept is illustrated in Fig. 10.10: the scattering from the crystal domains in the system are considering as the sum of the scattering from one layer plus the scattering of a displaced crystal.

The recursive approach has undoubted advantages in simplifying the mathematics involved in the calculation of the intensity scattered by the entire structure. In fact, the average scattering function from an ensemble of N layers starting with layer i can be written in terms of the layer form factor F_i for the i-th type layer (the equivalent of the structure factor, but calculated for a layer), the transition probability matrix a and the stacking vectors R_{ij} as (Treacy et al, 1991):

$$\phi_i^N(\mathbf{s}) = F_i(\mathbf{s}) + \sum_{j=1}^{N} \alpha_{ij} \exp\left(-2\pi i \mathbf{s} \cdot \mathbf{R}_{ij}\right) \phi_j^{N-1}(\mathbf{s}). \tag{10.39}$$

With the condition:

$$\phi_i^0(\mathbf{s}) = 0. \tag{10.40}$$

i.e. that the recursive function is identically zero for an ensemble of zero layers, the recurrence equations are easily solved to provide the ϕ_i necessary to compute the scattering intensity:

$$I(\mathbf{s}) = \sum_{m=0}^{N-1} \sum_{i=1}^{N} g_i \left(F_i^*(\mathbf{s}) \phi_i^{N-m}(\mathbf{s}) + F_i(\mathbf{s}) \phi_i^{N-m}(\mathbf{s})^* - \left| F_i(\mathbf{s}) \right|^2 \right). \tag{10.41}$$

The term g_i is the fraction of layers of type i, i.e. the probability that the layer of type i exists. The vector g is normalised in such way that:

$$g_i = \sum_{j=1}^{N} g_j \alpha_{ji} \qquad \sum_{j=1}^{N} g_j = 1 \qquad \sum_{j=1}^{N} \alpha_{ij} = 1. \qquad (10.42)$$

As in the traditional approaches by Hendricks and Teller (1942) and Kakinoki & Komura (1965), it is possible to rewrite the whole intensity equation in a more compact matrix form. In fact, by introducing the vectors Φ and $\mathbf{F}$ and the matrix $\mathbf{T}$:

$$\Phi^N(\mathbf{s}) = [\phi_i^N(\mathbf{s})], \quad \mathbf{F}(\mathbf{s}) = \left[F_i(\mathbf{s})\right], \quad \mathbf{T} = \left[\alpha_{ij}\exp\left(-2\pi i\mathbf{s}\cdot\mathbf{R}_{ij}\right)\right], \quad (10.43)$$

the recurrence equation can be reduced to:

$$\Phi^N(\mathbf{s}) = \mathbf{F}(\mathbf{s}) + \mathbf{T}\Phi^{N-1}(\mathbf{s}) = \sum_{n=0}^{N-1} \mathbf{T}^n \mathbf{F}(\mathbf{s}). \qquad (10.44)$$

The intensity for the stack thus reads:

$$I(\mathbf{s}) = \sum_{m=0}^{N-1}\sum_{n=0}^{N-m-1} \left(\mathbf{G}^{*T}\mathbf{T}^n\mathbf{F}(\mathbf{s}) + \mathbf{G}^{T}\mathbf{T}^{*n}\mathbf{F}^{*}(\mathbf{s}) - \mathbf{G}^{*T}\mathbf{F}(\mathbf{s})\right), \qquad (10.45)$$

where the vector $\mathbf{G} = [g_i F_i(\mathbf{s})]$ is employed for convenience. The summations can be done explicitly in the case of a finite number of layers and, by introducing the auxiliary function:

$$\Psi^N(\mathbf{s}) = \frac{1}{N}\sum_{m=0}^{N-1}\sum_{n=0}^{N-m-1} \mathbf{T}^n\mathbf{F}(\mathbf{s}), \qquad (10.46)$$

the intensity finally reads:

$$I(\mathbf{s}) = N\left(\mathbf{G}^{*T}\Psi^N + \mathbf{G}^{T}\Psi^{N*} - \mathbf{G}^{*T}\mathbf{F}\right). \qquad (10.47)$$

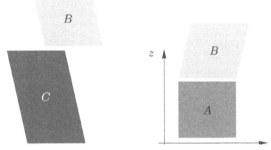

Fig. 10.11. Example of complex stacking of a set of heterogeneous objects: the sequence includes a coherent stacking of A and B followed by an empty layer (e.g. interlayer), a cell C with different parameters and the cell B considered as twin with a stacking fault. Any statistical sequence of any type of objects can be potentially modelled.

Unfortunately, unless e.g. layers are centrosymmetric, the full complex treatment needs being carried out, as a system of equations with complex variables and complex terms needs to be solved. As shown by Treacy et al. (1991), the whole treatment can be further simplified when dealing with a small number of different layer types.

Equation (10.47) allows the diffracted intensity to be calculated in each point in reciprocal space, but does not provide the powder average. A suitable integration is therefore needed in reciprocal space in order to simulate the powder data collection and reconstruct the whole powder diffraction pattern. Any regular or irregular sequence of layers can be employed in the intensity equation, clearly leading to the possibility, illustrated at the beginning of this section, to simulate the pattern of objects showing a symmetry ranging from 3D to 2D. Nothing in the treatment is said about unit cells or layers: the approach can therefore cope with any type of stacking of homogeneous or heterogeneous layers, such as the one presented in Fig. 10.11. It is clear that e.g. extensive interlayering, rototranslation faults, swelling regions can be simultaneously considered on a fully statistical basis.

The average composition of the system can then be reconstructed e.g. from the Perron–Frobenius theorem of Markov chains: the composition are given by the eigenvector with the maximum eigenvalue of the probability matrix. A check of the composition can therefore be easily made, based just on statistical considerations.

It is not yet possible, as in the WPPM, to extract the information relative just to a single point in reciprocal space and apply extra broadening models to it: the directionality of the stacking and of the faults present in it cause the intensity to be distributed along rods in reciprocal space. This limits the possibility of considering other broadening models simultaneously. Both an instrumental broadening and an effective size-broadening term (considering platelet-type domains) have been introduced in (Treacy et al, 1991) in order to consider the largest quota of broadening effects often experimentally observed.

The simulation approach is implemented in the DIFFaX software (Treacy et al, 1991) and has been used for long time for the simulation of the most diverse types of faulting problems. Simulation, however, limits the result to the sole qualitative or semi-quantitative level. In order to extract quantitatively valid results, an approach intermediate between the Rietveld and the WPPM one was proposed by Leoni et al. (2004) and implemented in the DIFFaX+ software. The intensity is calculated using the recursive approach and the pattern is then completed with instrumental and size broadening and with aberrations terms. The model parameters (i.e. structure and microstructure, considering also the transition vectors and the layer transition probability matrix) are refined using a conditioned nonlinear least squares algorithm allowing the constraints on the transitions to be fully preserved. Any type of diffraction data (laboratory X-rays and synchrotron

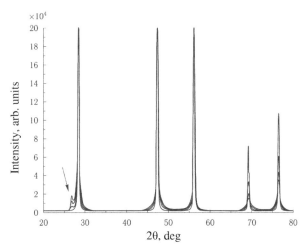

Fig. 10.12. Simulation of the diffraction pattern of silicon with increasing quantity of random defects. Two types of layers were employed, one providing the correct stacking sequence and one for the twinned one. Simulations correspond, respectively, to 0, 5, 10, 15% probabilities of twin faulting. The extra feature appearing when twins are present is marked with an arrow.

radiation, neutrons, electrons) can be modelled as in the Rietveld approach (Leoni et al, 2004). However, any type of defect along one direction, i.e. the cases illustrated above, can be handled. The application to complex systems such as clays (Gualtieri et al, 2008; Argüelles et al, 2010), intermetallics (Leineweber & Leoni, 2009), LDH (Johnsen & Norby, 2009), perovskites, superconductors, pharmaceuticals, multiferroics (Fuess et al, 2007) etc. is ongoing and new fields of applications are under current investigation.

To witness the ability to reproduce features observed in the diffraction pattern of silicon, Fig. 10.12 shows the results of a simulation done using DIFFaX+. The patterns were simulated in Bragg–Brentano geometry using monochromatised CuK$_\alpha$ radiation (graphite monochromator) and regard silicon with increasing quantity of defects. It is clear that a small defect content do not significantly alter the diffraction pattern.

The features become even fainter if peaks are extremely broad as in nanosized powders. This explains the ability of traditional models, based on diluted defects conditions, to extract quantitative information from the data. Increasing the faulting probability increase the incidence of constructive interference from atoms situated on the two sides of the fault and not positioned on a common lattice: this causes the appearance of diffuse features in the pattern, commonly observed, e.g. in highly deformed powders and often misinterpreted (when e.g. the Rietveld method is used) as impurity phase or experimental errors. The appearance of a diffraction band, resembling the peak of an impurity, is clear in Fig. 10.12. This feature is a common indicator for the occurrence of faults and the recursive approach is a fast and reliable method to account for it.

Conclusions

Both traditional and advanced methods are available for the analysis of powder diffraction pattern of nanocrystalline and defective materials. Traditional methods represent a fast way to qualitatively assess some trend of behaviour: advanced analysis models are, however, the key to extract valuable and quantitative physical information on the material under study. The PM2K and DIFFaqX+ software implementing the methods discussed here, can be obtained from the author (Matteo.Leoni@unitn.it).

Acknowledgements

The author wants to acknowledge the continuous scientific and technical support of P. Scardi and M.D'Incau (University of Trento). The Italian government is also acknowledged for financial support through the project FIRB Futuro in Ricerca RBFR10CWDA,

References for Chapter 10

1. Scherrer P. Bestimmung der Grösse und der inneren Struktur von Kolloidteilchen mittels Röntgenstrahlen. Nachr. Ges. Wiss. Göttingen. 26 September 1918. V. 26. P. 98–100.
2. Williamson G.K., Hall W.H. X-ray line broadening from filed aluminium and wolfram. Acta Metall. 1953. V. 1. P. 22–31.
3. Warren B. E., Averbach B. L. The effect of cold-work distortion on X-ray patterns. J. Appl. Phys. 1950. V. 21. P. 595 (5 p.); Warren B. E., Averbach B. L. The separation cold-work distortion and particle size broadening in x-ray patterns. J. Appl. Phys. 1952. V. 23. P. 497–512.
4. Warren B. E. X-ray Diffraction, Addison-Wesley, Reading, MS, 1969. 381 p.
5. Klug H. P., Alexander L. E. X-ray Diffraction Procedures for Polycrystalline and Amorphous Materials. 2 ed. N.Y.: Wiley, 1974.
6. Langford J. I., Louër D. Powder diffraction. Rep. Prog. Phys. 1996. V. 59. P. 131–234.
7. Langford J. I., Wilson A. J. C. Scherrer after sixty years: A survey and some new results in the determination of crystallite size. J. Appl. Cryst. 1978. V. 11. P. 102–113.
8. Bertaut E. F. X-ray study of the distribution of crystallite dimensions in a crystalline powder (in French). Comptes Rendus Acad. Sci Paris. 1949. V. 228. P. 492–494; Bertaut E. F. Raies de Debye–Scherrer et repartition des dimensions des domaines de Bragg dans les poudres polycristallines. Acta Cryst. 1950. V. 3. P. 14–18.
9. Scardi P., Leoni M., Dong Y.H. Whole diffraction pattern-fitting of polycrystalline fcc materials based on microstructure. Eur. Phys. J. B. 2000. V. 18. P. 23–30.
10. Scardi P., Dong Y.H., Leoni M. Line profile analysis in the Rietveld method and whole-powder-pattern fitting. Mat. Sci. Forum. 378–381. 2001. P. 132–141.
11. Scardi P., Leoni M. Whole powder pattern modeling. Acta Cryst. 2002. V. A58. P. 190–200.
12. Scardi P., Leoni M. Whole Powder Pattern Modelling: Theory and application. In: Diffraction Analysis of the Microstructure of Materials / Ed. by E. J. Mittemeijer, P. Scardi. Berlin: Springer-Verlag, 2004. P. 51–91.
13. Pawley G. S. Unit-cell refinement from powder diffraction scans. J. Appl. Cryst. 1981.

V. 14. P. 357–361.

14. Toraya H. Position-constrained and unconstrained powder-pattern-decomposition methods. In: The Rietveld Method / Ed. by R.A. Young. Oxford: Oxford University Press, 1993. P. 254–275.

15. Ungar T., Gubicza J., Ribarik G., Borbely A. Crystallite size distribution and dislocation structure determined by diffraction profile analysis: principles and practical application to cubic and hexagonal crystals. J. Appl. Cryst. 2001. V. 34. P. 298–310.

16. Ribarik G. Modeling of diffraction patterns properties: PhD thesis. Budapest: Eötvös University, 2008.

17. Ribarik G., Gubicza J., Ungar T. Correlation between strength and microstructure of ball-milled Al–Mg alloys determined by X-ray diffraction. Mat. Sci. Eng. A. 2004. V. 387–389. P. 343–347.

18. Balogh L., Ribarik G., Ungar T. Stacking faults and twin boundaries in fcc crystals determined by x-ray diffraction profile analysis. J. Appl. Phys. 2006. V. 100. P. 023512 (10 p.).

19. D'Incau M., Leoni M., Scardi P. High-energy grinding of FeMo powders. J. Mat. Research. 2007. V. 22. P. 1744–1753.

20. Leoni M., Scardi P. Nanocrystalline domain size distributions from powder diffraction data. J. Appl. Cryst. 2004. V. 37. P. 629–634.

21. Scardi P., Leoni M. Diffraction whole-pattern modelling study of anti-phase domains in Cu_3Au. Acta Materialia. 2005. V. 53. P. 5229–5239.

22. Leineweber A., Mittemeijer E. J. Diffraction line broadening due to lattice-parameter variations caused by a spatially varying scalar variable: its orientation dependence caused by locally varying nitrogen content in ε-$FeN_{0.433}$. J. Appl. Cryst. 2004. V. 37. P. 123–135.

23. van Berkum J.G.M. Strain Fields in Crystalline Materials, PhD Thesis, Technische Universiteit Delft, Delft, The Netherlands, 1994.

24. Cheary R.W., Coelho A.A. A fundamental parameters approach to X-ray line-profile fitting. J. Appl. Cryst. 1992. V. 25. P. 109–121.

25. Kern A.A., Coelho A.A. A New Fundamental Parameters Approach in Profile Analysis of Powder Data. Allied Publishers Ltd. 1998.

26. Cline J. P., Deslattes R.D., Staudenmann J.-L., Kessler E.G., Hudson L. T., Henins A., Cheary R.W. Certificate SRM 660a. NIST Gaithersburg, MD, 2000.

27. Wilson A. J. C. X-ray Optics. 2 ed. London: Methuen, 1962.

28. Caglioti G., Paoletti A., Ricci F. P. Choice of collimator for a crystal spectrometer for neutron diffraction. Nucl. Instrum. 1958. V. 3. P. 223–228.

29. Rietveld H.M. A profile refinement method for nuclear and magnetic structures. J. Appl. Cryst. 1969. V. 2. P. 65–71.

30. Leoni M., Scardi P., Langford J. I. Characterization of standard reference materials for obtaining instrumental line profiles. Powder Diffr. 1998. V. 13. P. 210–215.

31. Scardi P., Leoni M. Fourier modelling of the anisotropic line broadening of X-ray diffraction profiles due to line and plane lattice defects. J. Appl. Cryst. 1999. V. 32. P. 671–682.

32. Nye J. F. Physical Properties of Crystals: Their Representation by Tensors and Matrices. Reprint edition. Oxford: Oxford Univ. Press, 1987.

33. Scardi P., Leoni M. Diffraction line profiles from polydisperse crystalline systems. Acta Cryst. 2001. V. A57. P. 604–613.

34. Stokes A.R., Wilson A. J. C. A method of calculating the integral breadths of Debye-Scherrer lines. Mathematical Proceedings of the Cambridge Philosophical Society. 1942. V. 38. P. 313–322.

35. Wilkens M. The determination of density and distribution of dislocations in deformed single crystals from broadened X-ray diffraction profiles. Phys. Stat. Sol. (a). 1970. V. 2. P. 359–370.

36. Wilkens M. Fundamental Aspects of Dislocation Theory. V. II / Ed. by J.A. Simmons, R. de Wit, R. Bullough. Nat. Bur. Stand. (US) Spec. Publ. №317. Washington, DC, USA, 1970. P. 1195–1221.

37. Krivoglaz M.A., Martynenko O.V., Ryaboshapka K. P.. Phys. Met. Metall. 1983. V. 55. P. 1;

38. Kamminga J.-D., Delhez R. Calculation of diffraction line profiles from specimens with dislocations. A comparison of analytical models with computer simulations,. J. Appl. Cryst. 2000. V. 33. P. 1122–1127.

39. Groma I., Ungar T., Wilkens M. Asymmetric X-ray line broadening of plastically deformed crystals. I. Theory. J. Appl. Cryst. 1988. V. 21. P. 47–54.

40. Klimanek P., Kuzel R. Jr. X-ray diffraction line broadening due to dislocations in non-cubic materials. I. General considerations and the case of elastic isotropy applied to hexagonal crystals. J. Appl. Cryst. 1988. V. 21. P. 59–66.

41. Kuzel R. Jr., Klimanek P. X-ray diffraction line broadening due to dislocations in non-cubic crystalline materials. III. Experimental results for plastically deformed zirconium. J. Appl. Cryst. 1989. V. 22. P. 299–307.

42. Wilkens M. X-ray line broadening and mean square strains of straight dislocations in elastically anisotropic crystals of cubic symmetry. Phys. stat. sol. (a). 1987. V. 104. P. K1–K6.

43. Martinez-Garcia J., Leoni M., Scardi P. Analytical contrast factor of dislocations along orthogonal diad axes. Phil. Mag. Lett. 2008. V. 88, №6. P. 443–451.

44. Martinez-Garcia J., Leoni M., Scardi P. A general approach for determining the diffraction contrast factor of straight-line dislocations. Acta Crystallogr. A. 2009. V. 65. P. 109–119.

45. Popa N. C. Diffraction-line shift caused by residual stress in polycrystal for all Laue groups in classical approximations. J. Appl. Cryst. 2000. V. 33. P. 103–107.

46. Leoni M., Martinez-Garcia J., Scardi P. Dislocation effects in powder diffraction. J. Appl. Cryst. 2007. V. 40. P. 719–724.

47. Stokes A. R., Wilson A. J. C. The diffraction of X-rays by distorted crystal aggregates I. Proc. Phys. Soc. Lond. 1944. V. 56. P. 174–181.

48. Ungar T., Dragomir I., Revesz A., Borbely A. The contrast factors of dislocations in cubic crystals: the dislocation model of strain anisotropy in practice. J. Appl. Cryst. 1999. V. 32. P. 992–1002.

49. Velterop L., Delhez R., de Keijser Th.H., Mittemeijer E. J., Reefman D. X-ray diffraction analysis of stacking and twin faults in f.c.c. metals: a revision and allowance for texture and non-uniform fault probabilities. J. Appl. Cryst. 2000. V. 33. P. 296–306.

50. Leoni M., Scardi P. Some notes on Warren's theory of stacking faults for fcc crystals. In: Proceedings of the Size-Strain III conference / Ed. by P. Scardi, M. Leoni, E. J. Mittemeijer. Trento–Italy, 2–5/12/2001. P. 104.

51. Estevez-Rams E., Leoni M., Scardi P., Aragon-Fernandez B., Fuess H. On the powder diffraction pattern of crystals with stacking faults. Phil. Mag. 2003. V. 83, №36. P. 4045–4057.

52. Estevez-Rams E., Welzel U., Penton Madrigal A., Mittemeijer E. J. Stacking and twin faults in close-packed crystal structures: exact description of random faulting statistics for the full range of faulting probabilities. Acta Crystallogr. 2008. V. A64. P. 537–548.

53. Leoni M., Di Maggio R., Polizzi S., Scardi P. X-ray diffraction methodology for the microstructural analysis of nanocrystalline powders: Application to cerium oxide. J. Am.

Ceram. Soc. 2004. V. 87, №6. P. 1133–1140.

54. Leoni M., Scardi P. Grain surface relaxation effects in powder diffraction. In: Diffraction Analysis of the Microstructure of Materials / Ed. E. J. Mittemeijer, P. Scardi. Berlin: Springer-Verlag, 2004. P. 413–454.

55. Martinez-Garcia J., Leoni M., Scardi P. Analytical expression for the dislocation contrast factor of the <001>{100} cubic slip-system: Application to Cu_2O. Phys. Rev. B. 2007. V. 76, No. 17. P. 174117 (8 p.).

56. Debye P. Zerstreuung von Röntgenstrahlen. Annalen der Physik. 1915. V. 351, No. 6. P. 809–823.

57. Gelisio L., Azanza-Ricardo C., Leoni M., Scardi P. Real-space calculation of powder diffraction patterns on graphics processing units. J. Appl. Cryst. 2010. V. 43. P. 647–653.

58. Billinge S. J. L., Thorpe M. F. Local structure from diffraction. N.Y.: Kluwer, 2002.

59. Welberry T. R. Diffuse X-ray scattering and models of disorder. Oxford: Oxford University Press, 2004.

60. Leoni M., Gualtieri A. F., Roveri N. Simultaneous refinement of structure and microstructure of layered materials. J. Appl. Cryst. 2004. V. 37. P. 166–173.

61. Treacy M.M. J., Newsam J.M., Deem M.W. A general recursion method for calculating diffracted intensities from crystals containing planar faults. Proc. Roy. Soc. Lond. 1991. V. A433. P. 499–520.

62. van Smaalen S. Incommensurate crystallography. Oxford: Oxford University Press, 2007.

63. Ferraris G., Mackovicky E., Merlino S. Crystallography of modular materials, Oxford University Press (Oxford), 2004.

64. Hendricks S., Teller E. X-ray interference in partially ordered layer lattices. J. Chem. Phys. 1942. V. 10. 21 p.

65. Kakinoki J., Komura Y. Diffraction by a one-dimensionally disordered crystal. I. The intensity equation. Acta Cryst. 1965. V. 19. P. 137–147.

66. Gualtieri A. F., Ferrari S., Leoni M., Grathoff G., Hugo R., Shatnawi M., Paglia G., Billinge S. Structural characterization of the clay mineral illite-1M. J. Appl. Cryst. 2008. V. 41. P. 402–415.

67. Argüelles A., Leoni M., Pons C.H., De la Calle C., Blanco J.A., Marcos C. Semi-ordered
crystalline structure of the Santa Olalla vermiculite inferred from X-ray powder diffraction. American Mineralogist. 2010. V. 95. P. 126–134.

68. Leineweber A., Leoni M. Refinement of layer-faulting in Nb_2Co_7 intermetallic compound using DIFFaX+. Z. Kristallogr. Suppl. 2009. V. 30. P. 423–428.

69. Johnsen R. E., Norby P. A. Structural study of stacking disorder in the decomposition oxide of MgAl layered double hydroxide: A DIFFaX+analysis. J. Phys. Chem. 2009. V. C113. P. 19061–19066.

70. Fuess H., Schoenau K.A., Schmitt L.A., Knapp M., Leoni M., Maglione M. Structural reaction of PZT under in situ conditions using synchrotron powder diffraction- influence and stability of nanostructures. Acta Cryst. 2007. V. A63. P. S. 43.

71. Krivoglaz M.A., Ryaboshapka K. P.. Fiz. Met. Metalloved. 1963. V. 15. P. 18.

72. Krivoglaz M.A. Theory of X-ray and thermal neutron scattering by real crystals. N.Y.: Plenum Press, 1969.

73. Kaganer V.M., Sabelfeld K.K. X-ray diffraction peaks from correlated dislocations: Monte Carlo study of dislocation screening. Acta Cryst. 2010. V. A666. P. 703–71.

11

Methods of femtosecond spectroscopy and time-resolved electron diffraction

Methods of laser spectroscopy, which have been especially intensively developed since the early 70s of last century (Letokhov, 1987; Demtreder, 1985) allow to investigate the matter at the atomic and molecular levels with high sensitivity, selectivity, spectral and temporal resolution. Depending on the type of interaction of light with the test substance, these methods are divided into linear methods, methods based on single-quantum interaction, and non-linear method based on the non-linear single-quantum and multiquantum interaction.

The spectral instruments use tunable lasers – from the far infrared to the vacuum UV, which provides excitation of almost any quantum transitions of atoms and molecules. Tunable lasers with a narrow radiation band, in particular, injection lasers in the infrared lasers, and dye laser in the visible region (and in conjunction with a non-linear frequency conversion – in the near UV and near IR regions) make it possible to measure the true shape of the absorption spectrum of the sample without affecting the spectral instrument. Using tunable lasers increases the sensitivity of all known spectroscopic methods (such as absorption, fluorescence) for both atoms and molecules. These lasers have been used to develop principally new highly sensitive methods: intracavity laser spectroscopy, coherent anti-Stokes Raman scattering (CARS), resonant photoionization laser spectroscopy. The latter method is based on the resonance excitation of the particle by laser pulses, whose frequency is exactly tuned to the frequency of the resonance transition, and subsequent ionization of the excited particles by absorption of one or more photons from the additional laser pulse.

With sufficient intensity of laser pulses the resonant photoionization efficiency is close to 100%, such as the detection efficiency of the ion by the electron multiplier. It provides high sensitivity of the method and the ability to detect trace elements in samples at 10^{-10}–10^{-12}% in conventional experiments, and in the special – at the level of single particles (Laser Analytical Spectroscopy/Ed. V.S. Letokhov, Moscow, 1986).

The high intensity allows for the non-linear interaction of light with atoms and molecules, through which a large proportion of particles can be translated into an excited state, and forbidden single-photon resonance and multiphoton transitions between the levels of atoms and molecules, unobserved at low light intensity, are also likely to take place. The short (controlled) duration of radiation can excite high-lying energy levels in a time shorter than the relaxation time of any quantum state. Lasers with ultrashort (picosecond and femtosecond) pulses were used in the development of spectroscopy methods with time resolution of 10 femtoseconds. These methods are used to study the primary photophysical and photochemical processes involving excited molecules and to study short-lived particles (radicals, complexes, nanoparticles).

The high monochromaticity of laser light is utilized for measurement of spectra with almost any desired spectral resolution and, moreover, enables to selectively excite atoms and molecules of the same species in a mixture, leaving unexcited molecules of other species, which is especially important for analytical applications.

Taking into account the diversity of ways of energy relaxation of excited particles and, consequently, different ways to detect, the following methods of laser spectroscopy are available (Letokhov, 1987):
- absorption and transmission, based on measuring the transmission spectrum of the sample (insensitive to the 'fate' of the excited particles);
- opto-calorimetric (for example, opto-thermal, opto-acoustic), based on direct measurement of absorbed energy in the sample;
- fluorescence, based on the measurement of the fluorescence intensity as a function of the wavelength of the exciting laser;
- opto-galvanic, in which the excitation of particles is recorded by the change of conductivity;
- photoionization – the appearance of charged particles.

The devices used in laser spectroscopy, are fundamentally different from conventional spectroscopic instruments. In devices that use lasers with a tunable frequency, there is no need for expansion of the radiation in the spectrum with dispersive elements (prisms and diffraction gratings), which are the major parts of the normal spectral instruments. Sometimes laser spectroscopy uses apparatus in which light can be decomposed into a spectrum by means of non-linear crystals. Multistage laser excitation spectra are more selective than the conventional absorption spectra and can be efficiently combined, for example, chromatography and mass spectrometry.

Methods of time-resolved electron diffraction and coherent transmission electron microscopy have been recently applied to study nanoparticles and nanomaterials. The literature to date contains no reports on studies of nanosilicon using this technique. However, in order to attract the attention of researchers in this field, in section 11.5 of this chapter we briefly describe the possibilities of electronic crystallography (UEC) and electronic nanocrystallography (UEnC) and dynamic transmission electron microscopy for the study of nano-objects.

11.1. Femtosecond spectroscopy

To observe the motion of the atoms, it is necessary to have a research tool in the femtosecond time scale, because to travel a distance of ~1 Å the atom requires about 150 femtoseconds. The first experiment in the femtosecond time scale was held in 1986 at the California Institute of Technology, USA (Zewail, 1999). In 1999 A. Zewail received the Nobel Prize for the 'study of the transition states of chemical reactions using femtosecond spectroscopy'. A revolutionary breakthrough in experimental methods and technologies based on the use of femtosecond light pulses duration led to the emergence of new areas of research: femtochemistry and femtobiology.

11.1.1. Experimental procedure

Currently, the main ideology of the experimental study of ultrafast dynamics of molecules and condensed state of the matter lies in the 'pump–probe' technique, which uses two light pulse (Fig. 11.1). First, the exciting pulse (also called a pump pulse) 'starts' the process under study, and the second probe pulse 'reads' the information on the change of the system, passing through the sample with an adjustable time delay.

With the help of the delay line we can change the optical path which the probe pulse travels, and, consequently, the time of his arrival in the sample. A change of the distance by 1 mm leads to a delay time of 3.3 fs. So one can record the response of the molecular system to excitation with a step of 3.3 fs, getting more information 'per frame' about the course of the process under investigation, caused by the excitation pulse. Zero time is the time when the exciting and probing pulses arrive at the sample almost simultaneously. The response is represented by recording of, for example, photoinduced absorption, fluorescence, rotation of the polarization plane. Probing pulses can also be ultrashort pulses of electrons and X-ray radiation (see section 11.6 of this chapter).

In addition to ultrashort pulse duration (on the scale of the processes taking place), which allows us to study the evolution of the system with high temporal resolution, the femtosecond pulse has a number of features

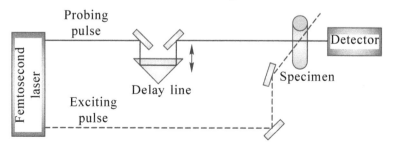

Fig. 11.1. Scheme of the 0 From Sarkisov and Umansky (2001).

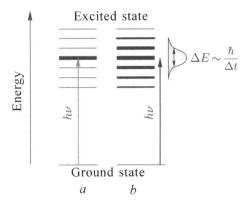

Fig. 11.2. *a* – steady state, formed under the action of continuous laser radiation, *b* – unsteady state (wave packet), formed under the influence of pulsed laser radiation. Thicker lines indicate the population of the state (Sarkisov and Umansky, 2001).

(Sarkisov and Umansky, 2001). The width of a femtosecond pulse is related to its duration by the uncertainty relation for energy (Fig. 11.2). The shorter the pulse, the larger range of energies of the excited state it can reach. Femtosecond femtosecond pulses can coherently excite several vibrational states. This type of excited state is a coherent vibrational wave packet, which is a non-steady state of the investigated system.

Over the past decade, it became possible to conduct research not only in the gas but also in the condensed phase (solutions, interphase boundaries, polymers, etc.), as well as in the mesoscopic phase (e.g. clusters, nanoparticles, nanotubes). Modern experimental methods allow us to change the amplitude and phase characteristics of femtosecond pulse excitation. The pulse of a simple form is decomposed into separate frequencies, and then certain frequencies increase, while others weaken or are removed thus producing the momentum with specified characteristics. The resulting frequencies can be used to select the parameters of the exciting pulse.

11.1.2. The method of femtosecond absorption spectroscopy

Figure 11.3 shows a diagram of the installation of method of femtosecond absorption spectroscopy at the Laboratory of Bio- and Nanophotonics of the N.N. Semenov Institute of Chemical Physics of the Russian Academy of Sciences (RAS) (Sarkisov et al, 2006). A similar system was also developed at the Institute of Spectroscopy, RAS. Depending on the task, it is possible quite easily to rebuild the system for different schemes of the spectral experiments with femtosecond resolution.

Pulses with a duration of ~80 fs and an energy of ~10 nJ at a wavelength of 805 nm and generated by a solid-state femtosecond laser Tsunami (Spectra-Physics), using a sapphire titanate crystal as the active medium. The pulse repetition rate is 80 MHz. The femtosecond laser is pumped

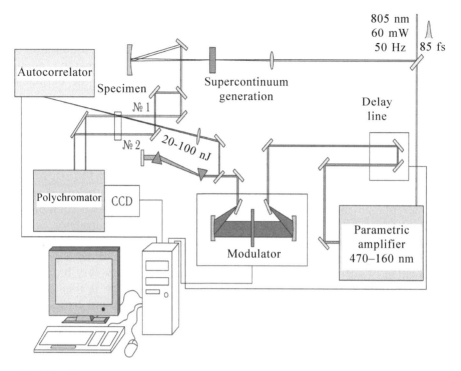

Fig. 11.3. The experimental setup of the method of femtosecond absorption spectroscopy (N.N. Semenov Institute of Chemical Physics, RAS; Sarkisov et al, 2006).

using a Millennia continuous solid-state diode-pumped solid state laser (Spectra-Physics). The pumping wavelength was 530 nm, the power 5 watts. Femtosecond laser radiation is directed to the regenerative amplifier Spitfire (Spectra-Physics), pumped by a solid-state pulsed laser with diode pumping Evolution X (Spectra Physics), while the pumping power is 9 W, pulse repetition frequency 1 kHz, and the wavelength 527 nm.

After amplification, the pulses have an energy of ~1 mJ, duration ~80 fs, repetition frequency of 50 Hz and a carrier wavelength of 805 nm. At the exit from the amplifier there is the pump–probe circuit, i.e. the beam is divided into two parts. The beam average power of 400 mW is directed to the preparation of the exciting pulse. With the help of a parametric light amplifier (NOPA) the beam at a wavelength of 805 nm is transformed to radiation at a wavelength in the range 470–1600 nm. Then, at the output of a parametric amplifier, the beam passes through the line delay, consisting of a precision hollow retroreflector mounted on a moving platform controlled by a step motor. The motor is controlled by computer and provides a minimal movement (step) of 0.5 mm, corresponding to a total delay of 3.4 fs. The delay line determines the time delay between the excitation pulse and the probing pulse during the experiment.

To change the amplitude and phase characteristics of the exciting pulse the radiation is passed through an amplitude-phase modulator based on an LCD matrix with 128 cells. After passing through the modulator to compensate for the incident chirp, optical elements and dispersive media are used to direct the beam to a quartz prism compressor, the output of which passes through the attenuator.

The exciting pulse prepared in this matter with a given wavelength, energy, amplitude and phase characteristics is focused on the sample, where it intersects with the probing beam No. 1 at a slight angle. The diameter of the waist at the focus of the pump is about ~200 µm, and the excitation energy, which is directed onto the sample, is as a rule 20–100 nJ. At the same time the exciting beam can be overlapped by a controlled mechanical shutter placed in front of the cuvette. The excitation pulse (pumping), passing through the cuvette, fall into the monopulse autocorrelator, which measures the duration of the exciting pulse. The second (probe) beam with a power of 10 mW is focused on the cell with water and, as a result of generation of supercontinuum, the pulses become spectrally very broad. The spectrum ranges from 400 nm to 900 nm. Then the beam is then divided through a semitransparent mirror into two beams (No. 1 and No. 2), with approximately equal energies. Both beams are focused on the specimen (diameter of the waist 120 µm) with a spacing in the cell relative to each of 3–4 mm. The beam No. 2 is used to get rid of the signal of the instrumental function during recording, while the other beam (No. 1) is probing and intersects with the exciting beam in the sample. On passage through the cell both beams (No. 1 and 2) enter the polychromator. Their spectra are recorded in a CCD-camera SPEC–10 (Roper Scientific), with which the data is sent to a computer for initial processing.

To obtain the differential absorption spectra at each time delay it is necessary to record spectra of the probing radiation in the presence and absence of the exciting radiation. The analyzed differential absorption spectra (ΔA) are calculated from the formula:

$$\Delta A = \log\left(\frac{A_1}{A_2}\right)^* - \log\left(\frac{A_1}{A_2}\right), \qquad (11.1)$$

where A_1 and A_2 are the spectra of the probing pulse No. 1 and No. 2, measured in the open (*) and closed gates of the exciting beam. 50 differential spectra are accumulated for each value of the time delay between exciting and probing pulses. This technique of recording the spectra ensures the sensitivity in respect of the spectrum, equal to $5 \cdot 10^{-4}$ rel. units of optical density. The zero time delay is taken as the moment of maximum overlap of the pumping pulse and probe pulses at a given wavelength. Experimentally, the zero time is defined as the middle of the non-resonant electronic response of the sample with the buffer at the time of the overlap of the pump and probe pulses. This allows us to construct a curve of zero delay over the entire range of 400–900 nm. The accuracy determining the of zero delay time is 3–6 fs.

11.1.3. Photoluminescence properties of silicon quantum dots

The photoluminescence properties of silicon quantum dots (Si-QDs) can be adapted to the surface states of nanoparticles through the effective relationship of these states and the bulk resonant states of the central nucleus. Therefore, in the last decade, the methods for the synthesis of silicon quantum points with controlled surface properties and size distribution function of the central Si nucleus have been intensively developed. Although the band gap of nanosilicon naturally increases with decreasing particle size, resonant electronic states of the surface can be changed by changing the structure and chemical composition of the n-Si coating. The energy of interaction of surface and bulk states defines the time of both the photoinduced electron transfer from the excited surface states in the conduction band and the capture of the conduction band electrons by surface states. While the origin of the PL and the influence of the size distribution of quantum dots can be studied by steady-state luminescence spectroscopy, using optical spectroscopy with temporal resolution, such as femtosecond spectroscopy of the transition state, allows to control the photoinduced electron transfer processes between the surface and bulk states and the capture of charge carriers in the subpicosecond time intervals.

11.2. Surface states as a means to control the photoluminescence of silicon quantum dots

One of the main goals of nanophotonics and optoelectronics based on silicon (Hersam et al, 2000; Wilcoxon & Samara, 1999) is the creation of Si-QDs with optically or electrically controlled luminescence properties. Luminescence of silicon quantum dots having a size smaller than the Bohr radius for free excitons (4.3 nm) is associated with the radiative recombination of charge carriers limited by the size of the Si-NPs core (Ledoux et al, 2002; Kanemitsu, 1995). The effect of the quantum constraint allows to change the wavelength of the luminescence in the visible range with the change of the size of the quantum dot (Canham, 1990; Wolkin et al, 1999; Zhou et al, 2003; Meier et al, 2007; Martin et al, 2007; Kuntermann et al, 2008). On the other hand, the effect of the quantum constraint is associated with a large value of the amplitude of the wave function of the charge carriers at the surface of the particles which was originally naturally passivated by oxidation. In natural passivation the shell is a layer of amorphous SiO_x, which has a wide range of structural defects, mainly optically inactive. Defective structures can effectively capture the charge carriers and thus reduce the quantum yield of luminescence (Brus, 1996; Allan et al, 1996; Kimura, 1999). Therefore, the surface of Si-QDs need to be modified by saturating the dangling bonds and eliminating structural defects.

One of the central problems in the manufacture of silicon quantum dots, especially for nano-optoelectronics, is the functionalization of their surface by binding to molecules that have an electronic structure that allows injection of charge carriers in the volume and at the same time optical control of charge separation and of the fluorescent properties of the particles. The successful functionalization of the surface of the organic compounds assumes knowledge of the quantitative characteristics of the dynamics of the photoexcitation and identification of different pathways of the transfer of photoinduced charge carriers between the electronic states of the surface and the core of the particles. Understanding ultrafast dynamics of charge carriers in silicon quantum dots with nanoscale spatial resolution in the subpicosecond time scale can be achieved by using femtosecond laser spectroscopy (Kuntermann et al, 2008; Klimov, 2007; Trojanek et al, 2006).

The use of ultrashort pulses to excite the Si-QDs can make a choice between different ways of decay of carriers: in the localized surface states and in the volume of the particle. The bulk states experience spatial and temporal evolution due to relaxation of the momentum and energy due to scattering of charge carriers; carrier trapping and radiative recombination. As a result, the photoexcited surface states can interact with the states of the conduction band or by phase transfer of charge carriers or transfer of excitation energy.

The interaction between the electronic states of the surface states and the states of the conduction band of Si-QDs with the modified surface can be studied by direct observation of the dynamics of photoexcited charge carriers by the methods of femtosecond spectroscopy.

11.3. Si-QDs with the modified surface and their luminescent properties

The PL spectrum of Si-QDs in the radiative recombination of spatially constrained excitons can be changed in the visible region from the red to blue with decreasing average particle size from 4 to 2 nm (Wolkin et al, 1999; Meier et al, 2007; Martin et al, 2007). The quantum efficiency of photoluminescence increases with decreasing particle size and depends critically on the chemical composition of the surface. In particular, the oxide shell SiO_x, which formed by natural oxidation of silicon nanoparticles, has dangling bonds that can efficiently capture electrons or excitons of the conductivity band (Wolkin et al, 1999). In contrast to the dangling bonds on which the non-radiative recombination of electron–hole pairs can occur, the excited states of the Si=O bonds, defects in Si–SiO_x interphase centres or oxygen vacancies in the amorphous SiO_x shell responsible for the luminescence in the blue region, as shown in (Wolkin et al, 1999; Kanemitsu et al, 1996; Duong et al, 2003).

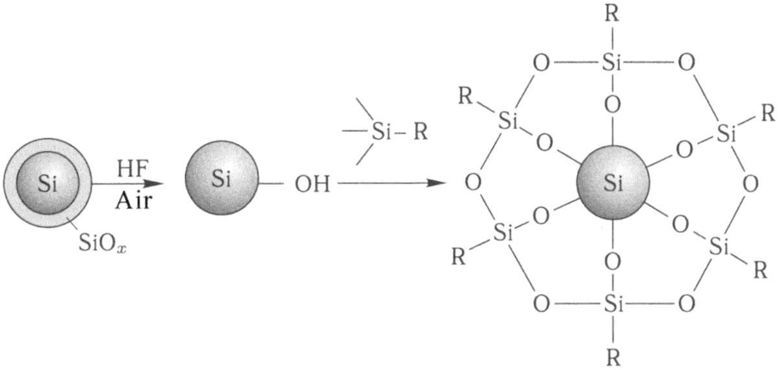

Fig. 11.4. Scheme of the silanization reaction of the surface of a silicon nanoparticle (Fojtik & Henglein, 2006).

To increase the quantum efficiency of photoluminescence of nanosilicon, the surface of Si-QDs needs to be modified, at least, to saturate the dangling bonds. This can be achieved either by complete oxidation of the surface or the formation of (Si–C) bonds by the use of organic compounds. Full passivation of the surface results in the chemical stability to the effect of oxygen and moisture and creates conditions for work in the solvents. In recent years significant progress with the use of unsaturated organic compounds to form covalent bonds at the surface of silicon nanoparticles was achieved using the mechanochemical synthesis method (Andrew et al, 2007) or various hydrosilylation reactions (Buriak, 2002; Buriak et al, 2001; Hua et al, 2006; Rogozhina et al, 2006; Hua and Ruckenstein, 2005; Lie et al, 2002; Nelles et al, 2007). Stable oxide layers were also produced on the surface by controlled oxidation with subsequent silanization (Fojtik & Henglein, 2006). To remove the oxide layer from the surface of nanoparticles produced during the decomposition of silane (Fojtik and Henglein, 2006), samples were etched in hydrofluoric acid. The surface was covered with (Si–H) bonds, which were oxidized to (Si–OH) under controlled conditions. For this purpose oxidation is carried out on the surface the phases propanol-2:HF and cyclohexane saturated with atmospheric oxygen (Fojtik and Henglein, 2006). In this case, we can not exclude the subsequent oxidation of the surface containing OH-groups and, at least, partial formation of (SiO_x) structures. In the future, Si-QDs can be stabilized by interaction with $R_nSiX_{(4-n)}$ molecules, where X – alkoxy, acyloxy, amino group or chlorine, which are subjected to hydrolysis with the formation of terminal OH-groups (see Fig. 11.4).

The PL spectrum of Si-QDs, silanized with chloro-dimethyl-vinylsilane and dispersed in cyclohexane with excitation with radiation with a wavelength 260 nm is shown in Fig. 11.5.

The asymmetric emission band of 370 nm corresponds to the average energy distributions of surface states, which, as it can be assumed, are

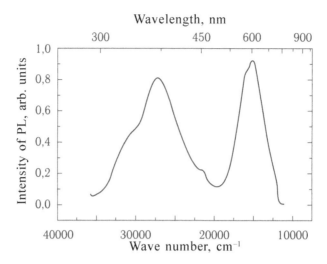

Fig. 11.5. The fluorescence spectrum of Si-QDs, silanized with chloro-dimethyl-vinylsilanes and dispersed in cyclohexane when excited by radiation with a wavelength of 260 nm (Cimpean et al, 2009).

localized on the defects or Si=O bonds at the Si–SiO$_x$ interface (Wolkin et al, 1999) of the structurally disordered SiO$_x$ shell (Kuntermann et al, 2008). On the other hand, the band at 660 nm, which has nearly the Gaussian form, can be attributed to the radiative recombination of spatially constrained excitons and, therefore, corresponds to the Gaussian distribution of Si-QDs, having an average size of 3.3 nm according to the Gaussian distribution of the band gap. In addition, the size distribution of Si-QDs determines both the relative intensity of the exciton photoluminescence and photoluminescence of surface states (Delerue et al, 1993), as the initially photoexcited electrons of the conduction band can be captured by the lower excited surface states of particles that are smaller than 2.5 nm (Wolkin et al, 1999), and the transfer of electrons to the conduction band of Si-QDs larger than 3 nm can be associated initially with the photoexcited overlying surface states (Kuntermann et al, 2008).

Buriak et al (1999, 2002) developed methods of surface protection of Si-NPs using hydrosilylation reactions with various organic compounds. The developed methods of synthesis have been successfully used to protect surface of the Si-QDs due to formation of stable covalent bonds (Si–C) in the reactions activated by heat, the action of a catalyst or UV light (Hua et al, 2006; Rogozhina et al, 2006; Hua and Ruckenstein, 2005; Lie et al, 2002; Nelles et al, 2007). Cimpean et al (2009) developed a method of hydrosilylation using ethinyl derivatives when heated in m-xylene and nitrogen to prevent oxidation of the surface of nanosilicon (Fig. 11.6).

Effective electron interaction between the ethinyl derivatives and silica core, bonded with the surface of Si-QDs, can be achieved using π-conjugated

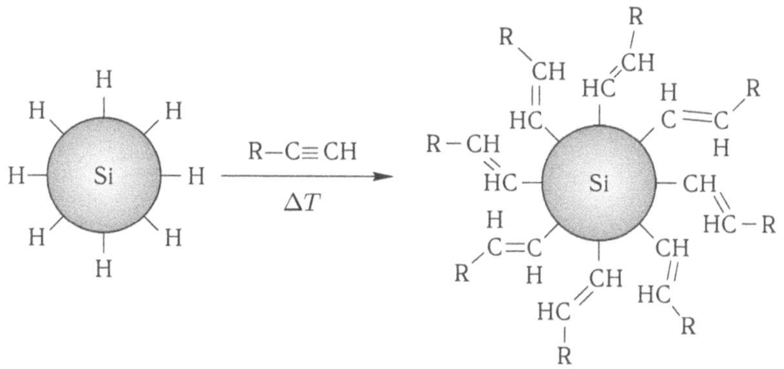

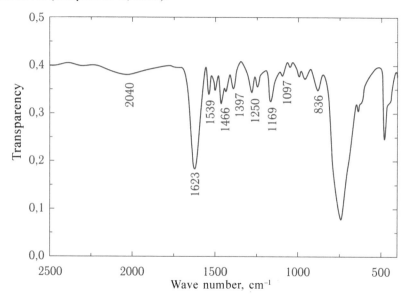

Fig. 11.6. The reaction scheme of thermally activated hydrosilylation using etinilnyh derivatives (Cimpean et al, 2009).

Fig. 11.7. FTIR spectra of Si-QDs after hydrosilylation with 2-ethinyl pyridine (Cimpean et al, 2009).

compounds: 2-ethinyl pyridine, 4-ethinyl pyridine and 3-ethinyl thiophene. Quantitative confirmation of the completeness of the process of hydrosilylation of the surface of the Si-QDs can be obtained in the analysis of IR spectra of, for example, 2-ethinyl pyridine (Fig. 11.7).

The absence of absorption in the region 1000–1100 cm^{-1}, in which the stretching ν(Si–O) vibration frequencies appear clearly indicates the complete removal of the SiO$_x$ shell. In addition, a very strong absorption at 740 cm^{-1} and a weak peak of 1250 cm^{-1}, which relate to deformation vibrations δ(Si–C) and the valence vibrations ν(Si–C), respectively,

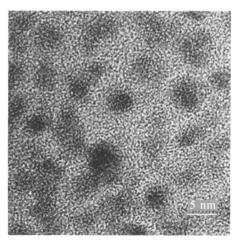

Fig. 11.8. TEM micrograph of Si-QDs, hydrosilylated with 2-ethinyl pyridine (Cimpean et al, 2009).

confirm the formation of covalent Si–C bond. The peaks of 1539 cm^{-1} and 1466 cm^{-1} refer to $v(C = C)$ and $v(C–N)$ fluctuations, identifying pyridine. FTIR spectra obtained after 4 and 6 months after the synthesis of the samples are identical, which confirms the chemical stability of the surface coverage of Si-QDs. The shape and size of hydrosilylated Si-QDs were investigated by TEM (see Fig. 11.8). The spherical shape and crystallinity are obviously supported by the etching process and the thermally activated hydrosilylation procedure; the average size of Si-QDs was 2–3 nm.

In accordance with theoretical predictions, based on the quantum-effect, Si-QDs, having a size of 2 to 3 nm, show PL in the visible region between 450 and 550 nm. PL spectra of Si-QDs, hydrosilylated with 2-ethinyl pyridine, 4-ethinyl pyridine and 3-ethinyl thiophene, dispersed in ethanol, are shown in Fig. 11.9.

It should be emphasized that neither the reagents used for thermal hydrosilylation (2-ethinyl pyridine, 4-ethinyl pyridine and 3-ethinyl thiophene) or their derivatives, obtained after binding to the surface of the Si-QDs, have fluorescence spectra at wavelengths above 400 nm. This means that the PL spectra can be attributed to the luminescence due to radiative recombination of spatially constrained excitons. Therefore, differences in the spectral positions of the PL maxima of the photoluminescence due to slightly different average sizes of Si-QDs.

The Gaussian distribution with a maximum of 2.2 nm adequately describes the silicon nanoparticles hydrosilylated with 4-ethinyl pyridine. Accordingly, the fluorescence spectrum in this case has a shape close to Gaussian. Wider and asymmetric fluorescence band for the Si-QDs, hydrosilylated with 2-ethinyl pyridine and 3-ethinyl thiophene can be explained by the non-Gaussian size distribution the studied nanosilicon particles. Using the comparative method with the fluorescence of rhodamine

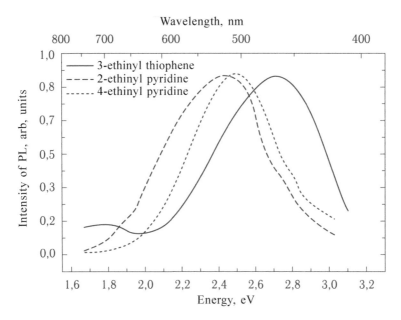

Fig. 11.9. The PL spectra of dispersions in ethanol of Si-QDs, hydrosilylated with 2-ethinyl pyridine, 4-ethinyl pyridine and 3-ethinyl thiophene. The spectra obtained by excitation at a wavelength of 390 nm (Cimpean et al, 2009).

6G, Williams et al, (1983) defined the quantum yield of luminescence, which was found to be ~30%. High values of the quantum yield demonstrate the manifestation of the quantum constraint effect, which leads to an increase in the probability of the optical phonon-less and phonon-induced radiative transitions. This is confirmed by a sharp decrease of the Stokes shift, which clearly illustrates the range of Si-QDs, hydrosilylated with 3-ethinyl thiophene (see Fig. 11.10).

11.4. Ultrafast relaxation dynamics of excitons in silicon quantum dots

As described in the previous section, the steady-state photoluminescence spectroscopy provides important information about the surface and bulk states of oxidized and hydrosilylated Si-QDs. The efficiency of photoluminescence and the PL spectrum of the silicon core of the nanoparticles are critically dependent on the energy difference of the excited surface states and conduction band states. Since the photoinduced electron transfer (PET) from the initially excited surface states in the conduction band states and the capture of conductivity band electrons by the low-lying surface states occur in the picosecond time interval, information about the interactions of surface and bulk electronic states of Si-QDs can be obtained only by using methods of ultrafast spectroscopy. Investigation of

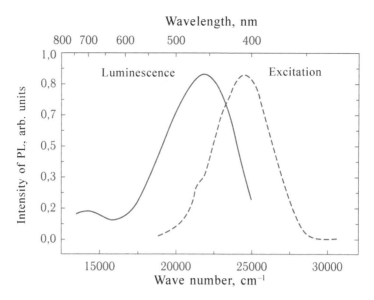

Fig. 11.10. Luminescence and excitation spectra of Si-QDs, hydrosilylated with 3-ethinyl thiophene (Cimpean et al, 2009).

ultrafast dynamics of the excitons of Si-QDs using femtosecond absorption spectroscopy were performed in colloidal solutions (Kuntermann et al, 2008), films of porous silicon (Owrutsky et al, 1995; Matsumoto et al, 1993) and films of amorphous dioxide silicon (Klimov et al, 2003).

Matsumoto et al. (1993) observed two different in time dynamic processes of decay with time constants of 5 ps and 15 ps, which have been interpreted as the transfer of excitation energy of the bulk states to luminescent surface states and their subsequent decay due to photoluminescence, respectively. Observation of the dynamics of decay of the exciton in the sub-10 ps time interval was confirmed in the work (Klimov, 2007) and attributed to the effective capture of the exciton by the surface states. In Owrutsky et al. (1995) in the study of thin films of porous silicon the authors observed a decrease in the intensity of fluorescence which is also describes well by the biexponential function. However, the fastest component (0.8 ± 0.2 ps) was attributed to the process of thermalization of charge carriers, and the slower one (30 ps) to the path of the luminescence. A recent study by femtosecond spectroscopy of the transition state of the exciton dynamics of Si-QDs in colloidal solutions has allowed to resolve the contradiction in the interpretation of the data for the interim interval between 0.6 ps and 4 ps. It was found that the temporal difference in the mechanisms of exciton relaxation and excitation is determined by the size of Si-QDs (Kuntermann et al, 2008).

Trojanek et al. (2006) used the method of femtosecond upconversion luminescence to study particles of nanoporous silicon obtained by sol-gel technology and dispersed in a SiO_x matrix. The PL of Si nanostructures was excited pump pulses with a duration of 80 fs at a wavelength of 400 nm, and the time evolution was monitored with time delays in the picosecond scale. The PL spectrum consists of a broad line between 600 and 800 nm and was interpreted as the radiative decay of an exciton in a closed structure, similar to Si-QD. The transitional state, recorded at 630 nm, was determined by the instantaneous rise and biexponential decline in intensity with time constants of 400 fs and 16 ps. The origin of subpicosecond PL was attributed to the radiative recombination of excitons inside Si-QD, while the slower PL was associated with the processes of capture in surface states. The relative contribution of these two processes to the biexponential decay of luminescence, as was established in Trojanek et al. (2006), depends on the size of the Si nanocrystal as well as on the structure and thickness of the oxide shell. These studies clearly show that the ultrafast dynamics of exciton relaxation in Si-QDs and similar structures is critically dependent on the structure and chemical composition of surface layers and the particle size distribution, as well as on the surrounding of the solvent molecules and the amorphous matrix.

Compared with the femtosecond spectroscopy of the transition state, the method of upconversion luminescence provides additional information on the ultrafast relaxation dynamics of the excited surface and bulk states of Si-QDs. It is due to the fact that the temporal evolution of the PL spectrum can detect the spectral changes that may be associated with ultrafast charge transfer passing through the fluorescent state. On the other hand, the spectral diffusion of the photoinduced absorption lines, which is regulated in the femtosecond absorption spectroscopy of the transition state, generates s signal from both dark and fluorescent states.

In the experiments performed by Kuntermann et al. (2008) by absorption spectroscopy of the transition state with a time resolution of ~270 fs, Si-QDs in a non-polar solvent (e.g. cyclohexane or m-xylene) were excited by pulses of a duration of 150 fs at a wavelength of 390 nm. Temporal evolution of the electronic absorption spectra is associated with the initial excited surface states or states of the conduction band, which are controlled by pulses of white light with a time delay. Figure 11.11 shows the time evolution of the absorption spectra (circles), recorded from 715 nm to 430 nm using diagnostic pulses of white light with a time delay of 0.2 ps to 2770 ps.

During the first 0.7 ps in the 620 nm range a broad band is formed by photoinduced absorption ($\Delta A > 0$) – see equation (11.1). At the time of 1.8 ps this absorption band begins to weaken, and there are two new absorption bands in the 680 nm and 450 nm. With increasing time to 2.8 ns, the weakening of photoinduced absorption bands in the spectrum is increasingly applied to photoinduced bleaching ($\Delta A < 0$) in the absorption of the ground state and the emergence of a distinct minimum in the region of 450 nm. Photoinduced illumination should be observed initially, because the

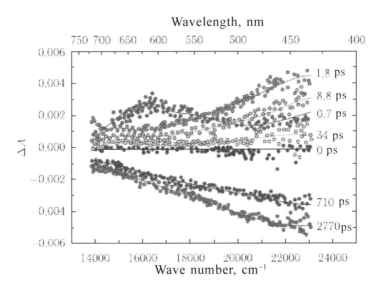

Fig. 11.11. Temporal evolution of the absorption spectra of the transition state of silanized quantum dots of silicon with different delay times between 0.2 ps and 2.3 ns. The value of ΔA is defined by equation (11.1). Figure from (Kuntermann et al, 2008),

excitation of electrons of the valence band to the conduction band is due to the simultaneous reduction of populations of the valence band. It is obvious that in the early stages, up to 10 ps, the bleaching spectrum enlightenment is completely hidden under the photoinduced bands of absorption due to transfer of hot electrons to states with higher level of excitation. These electrons can reach an equilibrium state with the lattice (i.e. thermalized) on the subpicosecond time scale by emission of optical phonons, and subsequently, within hundreds of picoseconds can be scattered by acoustic phonons in one of the valleys of the conduction band (intraband scattering), while full recovery of the population density of the valence band can be expected on the microsecond scale corresponding to the exciton radiation lifetime (Meier et al, 2007; Fojtik & Henglein, 2006).

The transient absorption spectrum, recorded at 2770 ps, accurately reflects the spectrum at time delays greater than 1200 ps and, therefore, it has been suggested that it corresponds to the initial bleaching spectrum. Therefore, by subtracting the illumination spectrum from the transient absorption spectra for time delays from 0.7 ps to 1200 ps, we can get a complete set of photoinduced absorption spectra.

Subsequent approximation of each spectrum by the Gaussian function leads to seven Gaussian bands centred at 685 nm, 625 nm, 581 nm, 549 nm, 505 nm, 455 nm and 417 nm. The position of the maximum of the spectrum and its width were fixed, and the amplitude was varied for each spectrum to achieve the best possible coordination of each spectrum

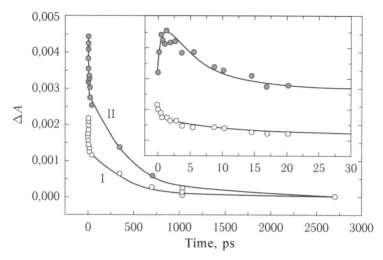

Fig. 11.12. Temporal evolution of Gaussian amplitudes, averaged for the bands of 685 nm, 455 nm and 416 nm (solid circles), as well as bands of 625 nm, 581 nm, 549 nm and 505 nm (empty circles); solid line – fit with bi- and triexponential dependences for the function of luminescence decay. The behaviour of these curves for a small time interval is shown in the inset picture (Kuntermann et al, 2008).

(see Fig. 11.12, approximating lines). Since the amplitude of the three lines (685 nm, 455 nm and 416 nm, curve I, empty circles), and four lines (625 nm, 581 nm, 549 nm and 505 nm, curve II, solid circles) showed very similar time dependences, the mean value was calculated for each set of amplitudes, which is shown as a function of time in Fig. 11.12. The behaviour of these curves for a small time interval illustrated in the inset in Fig. 11.12.

Photoinduced absorption (curve II) appears within the time response of the installation (270 fs). This curve is approximated by three exponential functions with time constants: τ_{el-ph} = 0.6 ps, τ_{trap} = 9 ps and τ_{ph-ph} = 350 ps. The fastest process corresponds to the electron–phonon scattering τ_{el-ph} = 0.6 ps, involving LO phonons with energies of 53 meV (Kuntermann et al, 2008). Capture of the exciton by low-lying surface states is responsible for the slower process, τ_{trap} = 9 ps (Wolkin et al, 1999; Kuntermann et al, 2008; Klimov, 2007; Trojanek et al, 2006). It is assumed that the slowest process with a time constant τ_{ph-ph} = 350 ps corresponds to the scattering of hot electrons by acoustic phonons in one of the valleys of the conduction band (intraband scattering). The latter process is due to the interaction of conduction band electrons with acoustic phonons, which leads to emission. On the other hand, spectral diffusion, observed in the time evolution of the absorption spectra (see Fig. 11.12), is clearly reflected in the rise of the curve I in the range 500 fs. Explanation of the ultrafast dynamics is associated with the transition of an electron from the originally populated highly-excited state to the surface resonance state of the conduction band.

The corresponding transitions in absorption appears to originate from
the **k**-states of the conduction band far from the centre of the Brillouin
zone. For comparison, the instantaneously formed band of photoinduced
absorption at 620 nm (curve I) refers to the transitions of hot electrons near
k = 0, which can decay through the electron–phonon scattering (0.6 ps),
followed by intraband scattering (350 ps) and the capture of an exciton (9 ps).
In accordance with the three-band model, proposed in (Wolkin et al, 1999),
surface states determined by the Si=O bonds in the SiO$_x$ shell, can be
assumed to lie above the conduction band of Si-QDs, if the quantum dot
is larger than 3 nm. Therefore, the observed electronic transition from the
initially photoexcited surface states to the state of the core of the particle
belongs to the Si-QDs having a size of >3 nm. According to the work of
Martin et al. (2007), in the photoluminescence of individual Si quantum
dots for particle size <2.7 nm was studied in detail, the band gap width of
the Si-QDs is greater than the energy of surface states which may in this
case capture carriers. These authors suggest that the localization of electrons
and holes in the Si=O bond of the Si–SiO$_x$ interface occurs stepwise and is
caused by the electron–phonon interaction. As a consequence, we can assume
that the pumping momentum at 390 nm for Si-QDs initially generates
small quantum-constrained excitons followed by hot electrons near **k** = 0,
which can lose energy by ultrafast electron–LO phonon scattering and
trapped surface states. Also, electronic interactions between surface and
bulk states of 2-ethinyl pyridine, 4-ethinyl pyridine or 3-ethinyl pyridine

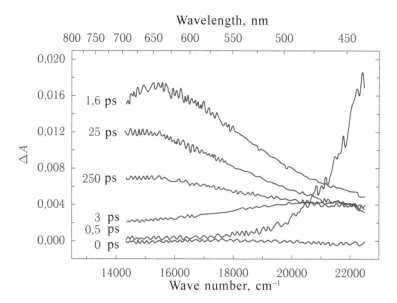

Fig. 11.13. Temporal evolution of the spectra of sols in cyclohexane Si-QDs,
hydrosilylated with 3-ethyl tiophene. The spectra were obtained using femtosecond
absorption spectroscopy for different values of delay time (Cimpean et al, 2009).

covering the Si-QDs, depend on the size of the quantum dot. Therefore, Cimpean et al. (2009) obtained Si-NPs ranging in size from 2 to 3 nm and by the method of femtosecond absorption spectroscopy they studied ultrafast electronic transitions between the excited states of the quantum dot surface points covered with ethinyl derivatives, and the states of the conduction band.

As an example, Fig. 11.13 shows the time evolution of the spectra, obtained by femtosecond absorption spectroscopy of Si-QDs, hydrosilylated with 3-ethinyl tiophene and dispersed in m-xylene. Originally there is an absorption band with a maximum at 450 nm, which decays during 1.6 ps; at the same time, a new absorption band forms in the region 650 nm. This spectral diffusion can be explained by photoinduced electronic transition from the initially excited state of 3-ethinyl tiophene to the quasi-resonant state of the conduction band of the particle core. Apparently, the initially observed photoinduced absorption at 450 nm is due to the state S1, associated with the surface of 3-ethinyl tiophene, while the subsqent appearance of absorption bands in the 650 nm range refers to the conduction band electrons. To illustrate the dynamics of increase of absorption intensity, due to the conduction band electrons, two transitions, recorded at 650 nm and 450 nm, are shown in Fig. 11.14.

Obviously, the photoinduced absorption line at 450 nm is formed instantaneously and, thus, its appearance is determined by the resolution of the experimental setup (270 fs). In contrast to this process, the absorption band of 650 nm reflects the relatively slow rise of absorption intensity of

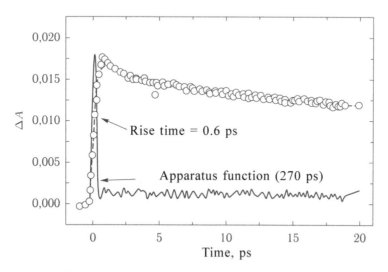

Fig. 11.14. Evolution of the transitions in the 450 nm (lower curve) and 650 nm (curves with circles of experimental points) ranges for short time intervals (Cimpean et al, 2009).

which is connected with the dynamics of the population of the conduction band through the electronic transition from 3-ethinyl thiophene.

Analysis of the dynamics of rise of absorption was performed with a combination of deconvolution and approximation by the biexponential dependence. The resulting value of the rise time of 0.6 ps refers to the photoinduced electron transition, and the decay times of 0.5 ps and 20 ps are associated with the processes of electron–LO phonon scattering and capture, respectively. Unlike silanized silicon quantum dots, for the Si-QDs, hydrosilylated with 3-ethinyl thiophene, pumping by an excitation pulse at 390 nm leads mainly to the initial population of the state S1 of the molecules covering the surface of the particle.

This can be explained by the fact that for a sufficiently dense covering of the surface the 3-ethinyl thiophene molecules act as an antenna for the light pump pulse.

Femtosecond transient absorption spectroscopy and photoluminescence spectroscopy of silanized and hydrosilylated silicon quantum dots, which are in a colloidal solution, provides unique information on the dynamics processes of the exciton and allows to study its relationship with the energetics of the surface states and the size of Si-QDs. The main condition for obtaining reliable information about the effectiveness of charge transfer processes and their temporal scales, as well as for understanding the nature and intensity of interactions of the surface and bulk states by using the described method is the reproducible synthesis of Si-QDs with a narrow Gaussian size distribution and protection of the surface of the particles with stable molecules of a precisely defined chemical composition (Cimpean et al, 2009).

11.5. Methods for studying coherent 4D structural dynamics

To understand the dynamic features of molecular systems with a complex landscape of potential energy surfaces, research is needed in the associated 4D space–time continuum. The introduction of time sweep to diffraction methods and the development of the principles of studying the coherent processes have opened up new approaches to the study of the dynamics of wave packets intermediate products and transition states of the reaction centre, short-lived compounds in gaseous and condensed media.

The use of picosecond and femtosecond pulses of electronic diagnostic pulses, synchronized with the pulses of the exciting laser radiation, predetermined the development of ultrafast electron crystallography, X-ray diffraction with time resolution and dynamic transmission electron microscopy. One of the promising applications of the developed diffraction methods is their use for the characterization and visualization of the processes occurring upon photoexcitation of free molecules and biological objects and analysis of the surface and thin films.

The set of spectral and diffraction methods, using different physical principles, complementing each other, allowing to carry out photoexcitation and diagnostics of the dynamics of nuclei and electrons at the time sequences of ultrashort duration, opens up new possibilities of research, providing the required integration of the Structure–Dynamics–Function triad in chemistry and biology and materials science.

Currently, the method of ultrafast diffraction is being intensively developed. Great opportunities to study the 4D structural dynamics are offered by ultrafast electron crystallography and electron microscopy with time resolution from micro- to femtoseconds.

11.5.1. Ultrafast electron crystallography (UEC)

The method of UEC (Ultrafast Electron Crystallography) allows us to study transient non-equilibrium structures that give decisive information for understanding phase transitions and coherent dynamics of the nuclei in the solid state of the surface of macromolecular systems. In recent years, intensive development of the application of this method to study the dynamics of nano-objects that are in the field of laser radiation (Dwyer et al, 2006; Ruan et al, 2004; 2007) has taken place,

The possibility of combining high spatial resolution (to two decimal places or thousandths of an angstrom), and high time resolution (femtosecond area) allows us to study the processes of restructuring and redistribution of energy in real time. The experimental setup for studies by ultrafast electron crystallography is shown in Fig. 11.15.

The method of ultrafast electron crystallography (UEC) makes it possible to obtain information about the coherent dynamics of the structure in the photoinduced phase transitions in nanoparticles and macromolecules, on solid surfaces, in thin films and interfacial areas. It allows to explore the dynamic processes at the level of constituent elements of the system.

The general scheme of the UEC is shown in Fig. 11.16. A crystal containing adsorbed atoms or molecules is impacted at an angle of $\theta < 5°$ by an electron bunch with an energy of 30 keV (wavelength ~0.07 Å), designated as the wave vector $\mathbf{k}_i$. The resulting diffraction pattern gives information about the structure of the surface, determined by the substrate and the adsorbed particles. In (Ruan et al, 2004), the substrate temperature control was performed using a pulsed IR-laser (typically 120 fs duration and a wavelength of 800 nm); UV radiation, 266 nm, was also used. The reference point of the time when exciting and diagnosing pulses come to the surface of the crystal at the same time was defined as the zero time $t_{z0} = 0$ (see the inset in the upper left part of Fig. 11.16). A vacuum of 10^{-10} torr was maintained in the system (Fig. 1.15).

The observed diffraction patterns, depending on the time delay between diagnosing and exciting pulses, $\Delta I(\theta_i, s; t_{ref}, t)$, are difference curves related to changes in the structure of the transition state:

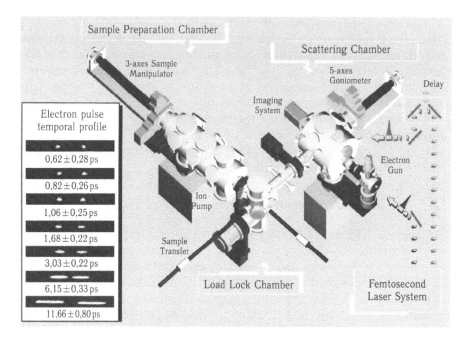

Fig. 11.15. Equipment for studies by ultrafast electron crystallography. The insert on the left side of the picture – the image of electron bunches of varying duration, obtained in operation of equipment in the chamber mode with linear scan (Lobastov et al, 2003).

$$\Delta I\left(\theta_i, s; t_{\text{ref}}, t\right) = I\left(\theta_i, s; t_{\text{ref}}\right) - I\left(\theta_i, s; t\right), \tag{11.2}$$

because the reference time t_{ref} can be selected both before and after the arrival of the exciting pulse.

The diffraction pattern in the given geometry of the experiment reflects the structure in the reciprocal space (Fourier transform). For a 2D monolayer of atoms in the reciprocal space there are diffraction 'rods' separated by distances a and b (Fig. 11.16) in the reciprocal lattice space. The rods correspond to the constructive coherent interference of waves. As the monolayer is introduced into the crystalline substrate, the rods in the diffraction pattern are undergoing changes due to modulation of the interplanar distances (Fig. 11.16). For electrons, the Ewald spheres, defined by the vector $\mathbf{k}_i$, are large, and the diffraction pattern, depending on the size θ_i, has both bands at small scattering angles and Bragg spots at large angles in Laue zones. From these reflexes we can get information about the static structure of the surface and the lattice in scattering of high-energy electrons.

In the presence of sufficiently high temporal resolution it is possible possibility to take additional measurement for the following reasons. First, it is possible to diagnose the structural changes of the surface layer (and its restructuring) directly in real time. Second, there is a significant separation

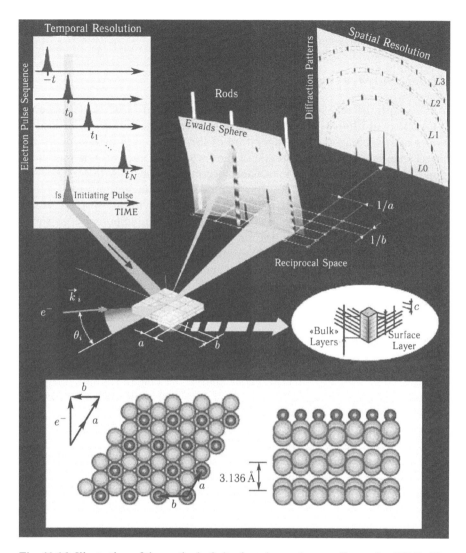

Fig. 11.16. Illustration of the method of ultrafast electronic crystallography (UEC). The electron bunch is directed to the surface of the crystal (Si 111), at an angle $\theta < 5°$. Ewald spheres and Laue zones (L_0, L_1, ...) are shown, see text. The bottom of the figure shows the structure and the distance between the bilayers (3.136 Å). The surface layer shows adsorbed atoms or molecules (Ruan et al, 2004).

of time scales for the processes occurring in the surface layer and in the perpendicular direction to this layer. In this way, it is possible to separate and diagnose the initial non-equilibrium structures, rather than the structures that form during the propagation of the exciting pulse. Third, if the surface is used as a matrix, it is possible to explore the strengthening of the mutual influence of the substratum and the supporting material, and display the structural dynamics of the process.

11.5.2. Surfaces and crystals

An example of research reflecting the potential of UEC is described in Vigliotti et al (2004). The structural dynamics of the surface of crystalline GaAs after increasing the temperature of the crystal was determined. From the change in Bragg diffraction (shift, line width and intensity), it was shown that the 'compression' and 'expansion' occurs from –0.01 Å to +0.02 Å and that the 'transition temperature' (the exact definition in Vigliotti et al. (2004) is not given) reaches the maximum value (1565 K) within 7 ps (Fig. 11.17). The onset of changes in the structure lags behind the temperature, as shown by the evolution of non-equilibrium structures.

These results in (Zewail, 2006) were compared with the results of non-thermal femtosecond optical sensing (Sundaram, Mazur, 2002). The GaAs surface was functionalized by a monolayer of chemically bonded chlorine atoms. The ultrashort time intervals showed compression followed by expansion, which is due to an increase in the phonon temperature. At longer periods of time there was restructuring and evolution to the equilibrium state. The observed structural dynamics can be divided into three modes: changes, which include electronic redistribution without motion of nuclei (from femtoseconds to several picoseconds); the coherent non-equilibrium lattice expansion (increase in time from 7 ps); restructuring and diffusion of heat (from 50 ps to nanoseconds).

Similar studies were carried out for silicon crystals in the presence of and in the absence of adsorbates. The choice of the ground state as a system frame shows the changes in the structure caused by the initial momentum, from the sample in the ground state with the 'negative' time delay to the observed changes in positive time (Fig. 11.17). Changes in the structure were manifested by the shift of the rocking curve at the time of the in-phase Bragg peak, whereas the increase in the amplitude of the oscillations was reflected in the broadening of the peaks. As in the case of GaAs, the movement of surface and bulk atoms was observed. The femtosecond increase of electronic temperature was followed by an increase in the population of optical levels which, after the picosecond delay, causes acoustic waves (expansion and contraction of the lattice) and, finally, the heating of the lattice. With the use of UEC it has become possible to observe the ultrafast surface and volume dynamics and track the rearrangement of the structure and diffusion over long periods of time.

11.5.3. Structural dynamics of the surface

To date, it has been shown that the UEC method can be used to study the structural changes of the surface, the effect of adsorbed molecules

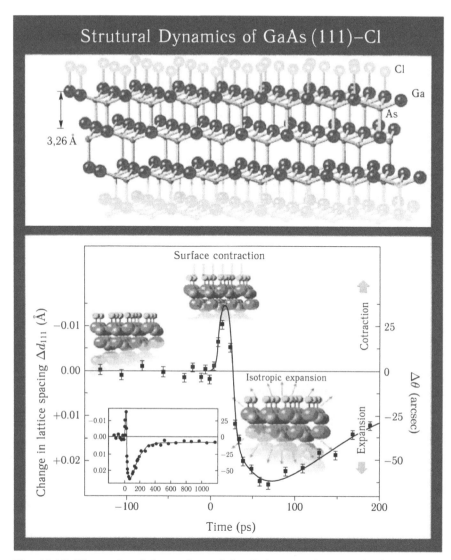

Fig. 11.17. The study of the surface of the GaAs crystal, covered with chlorine atoms. Only changes from the equilibrium positions of the lattice constants (see text) are shown. From Zewail (2006).

and the transition from the crystal to the liquid phase (Zewail, 2006). Reports were published on a study of the silicon surface with a variety of adsorbed molecules: hydrogen, chlorine, and trifluoroiodomethane (CF_3I). The action of laser radiation is coherent restructuring of the surface layers with subangström mixing of the atoms after the ultrafast laser pulse heating the surface. The non-equilibrium dynamics of the surface structure with a step of 2 ps with a total time of up to 10 ps on the basis of the change of interference lines, Bragg spots and diffraction rings was observed.

In the transition from hydrogen to chlorine the oscillation amplitude is reduced to 0.1 Å. The response time for the system with chlorine molecules is similar to the system with hydrogen atoms, however, it is characterized by slow dynamics. In this connection, it is necessary to take into account the significant increase in mass and electronegativity of chlorine compared to hydrogen. Capture of electrons by the surface leads to change of the potential and, consequently, to an increase of the amplitude of motion of the nuclei of the adsorbed molecule. The adsorption of molecules trifluoroiodomethane with a lower effective electronegativity, leads to a decrease in the energy of adsorption and reduction of time of the dynamic response of the system to external excitation (Dwayer, 2006; Zewail, 2006).

First results were obtained to demonstrate the possibility of direct observation of the structural dynamics of adsorbed layers offering new opportunities for the analysis of the state of matter on the surface in 4D space-time continuum of repeating events subfemtosecond temporal resolution.

11.5.4. Time-resolved electron nanocrystallography (UEnC)

The UED, UEC, UEM and DTEM methods (Dynamical Transmission Electron Microscopy) (King et al, 2005) give direct information on structural changes occurring in the test object in real time, and are used at present for quantitative studies of dynamic phenomena occurring in the nanosized objects in the time interval from pico to femtoseconds (see review articles (Dwyer et al, 2006; Ruan et al, 2009). In recent years, there was a significant decrease in the length of the electron bunch and a significant increase of the accelerating voltage, which allowed to obtain electronic pulses of femtosecond duration. The technique of ponderomotive acceleration of the wave front was proposed to reduce the mismatch between the velocity of light and electron pulses and compress the electron bunch (Baum, Zewail, 2006). These advances have opened new opportunities for research of coherent structural dynamics of nanomaterials with a femtosecond time resolution.

One of the important stages of nanoparticle research by the UEnC method is the preparation of the matrix surface on which the studied nanoparticles are deposited (Fig. 11.18). This stage was described in detail in a review article (Dwyer et al, 2006).

The effectiveness of the UEnC method is shown in several studies in the investigation of the transition of graphite into diamond (Raman et al. 2008), homogeneous photoinduced structural transition of gold nanocrystals (Ruan et al, 2007), the dynamics of interfacial charge (Murdick et al, 2007) and in molecular electronics (Wang et al, 2005). Thus, the high sensitivity and resolution of the UEnC method open the possibility of studying processes such as surface melting of nanoparticles, the non-equilibrium structural

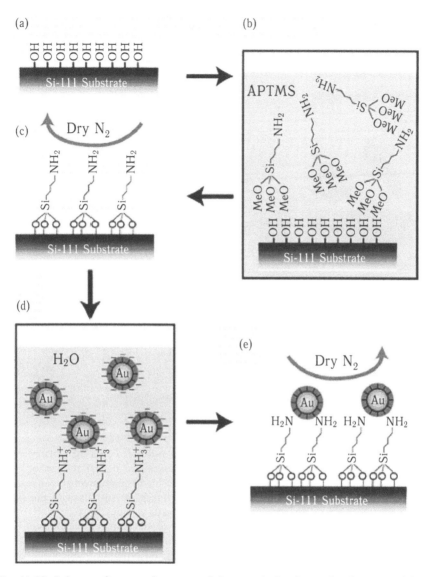

Fig. 11.18. Scheme of preparation steps of the sample for the study of nanoparticles by UEnC: (a) – cleaning the surface of a silicon substrate; (b), (c) – surface functionalization; (d), (e) – the distribution of nanoparticles on the substrate surface. Figure from (Ruan et al, 2009).

dynamics of phase transitions and the response of adsorbed molecules on the non-equilibrium structural changes of the surface.

Ruan et al (2007) studied Au nanoparticles by the UEnC method, with the dimensions of the particles distributed from 2 to 20 nm (Fig. 11.19). Processes of reversible melting of the surface and recrystallization in the

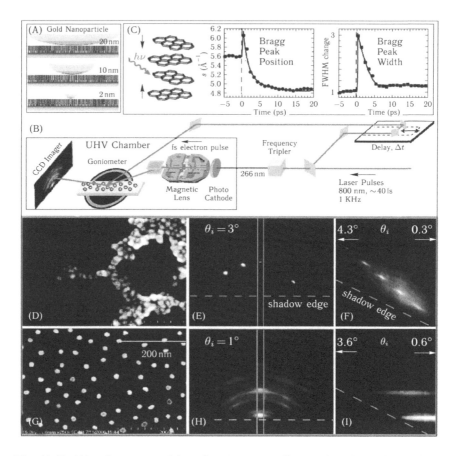

Fig. 11.19. (A) – the nanoparticles of Au (Au-NPs) dispersed on the self-organising molecular surfaces of the interface. (B) – the diagram of the UEnC method. (C) – determination of the diffraction signal as the result of photomechanical response of the multilayers of graphite. (D) – the SEM image of Au-NPs with the size of 20 nm, scattered on the surface of the substrate without buffering. (E) – the diffraction pattern from Au-NPs with the size of 20 nm, showing the areas of Bragg signals of the silicon substrate. (F) – tilting curves, corresponding to the diffraction pattern from Au-NPs with the size of 20 nm (E) and different incidence angles θ_i. (G) – SEM image of Au-NPs with the size of 20 nm with the appropriate buffering. (H) – the diffraction pattern, corresponding to (G), showing the Debye–Scherrer diffraction rings and the spots from the Bragg buffer layer (Si, N, C and a stack of the layers in the self-organised aminosilane, range 2.2 Å, the angle of inclination 31°). (1) – the tilting curve (H). The figure from Raman et al (2008).

subpicosecond time scale and spatial resolution up to units picometres were studied. In ultrafast photoinduced melting processes of nanoparticles, which were carried out under non-equilibrium conditions, they determined the initial phases of lattice deformation, the non-equilibrium electron–phonon interaction and, in melting, the formation of collective bonds and poor coordination of atoms, transforming nanocrystallites to nanofluids. Structural excitation during premelting and the coherent transformation from crystal to liquid, with coexistence of phases at photomelting, differ from the recrystallization process in which 'hot forms' of the lattice and the liquid phase coexist as a consequence of thermal contacts. The extent of structural changes and thermodynamics of melting are dependent on the size of nanoparticles (Raman et al, 2008).

The possibilities of application of the method, not only to the study of structural changes, but also the redistribution of the charge and energy at the interphase boundaries were shown [see review article (Ruan et al, 2009)]. The UEnC method currently allows us to study such low surface densities of ~6 particles/μm^2, in fact, demonstrating the possibility of studying an isolated nanoparticle. Upon reaching the submicron dimensions of the diagnostic electron bunch we should expect fundamentally new results of the study of 4D dynamics of nanostructures with a combination of diffraction and spectroscopic methods with ultrahigh temporal resolution (Baum & Zewail, 2007).

11.6. Dynamic transmission electron microscopy

Transmission electron microscopy with a wide range of tools has long been a powerful method in many fields of study, allowing to achieve the resolution of fractions of a nanometer but not providing ultrashort resolution in time. Optical microscopy using fluorescent probes of, for example, green fluorescent protein, provided the opportunity to visualize phenomena occurring *in vitro* (Zewail & Thomas, 2010). However, despite the possible time resolution, as regards the spatial resolution the best methods are usually limited to the use of the optical wavelength range 200–800 nm.

In a review article (King et al, 2005) and the recently published monograph (Zewail & Thomas, 2010) the authors describe the development of methods for 4D DTEM and UEM, respectively. Images and diffraction patterns in (Lobastov et al, 2005) were obtained at an accelerating voltage of 120 keV for materials (single gold crystals, amorphous carbon and polycrystalline aluminum) and biological cells of the intestine of rats. The gated beam contains an average of one electron per pulse with a reference dose of a few electrons in $Å^2$, but the pulses are completely controlled in space and time.

Conceptually, these works are based on methods used in UED, UEC, and DTEM, but with some differences – namely, the implementation of synchronized pulses of single electrons to form an image in the UEM. The experimental setup for 4D DTEM is shown in Fig. 11.20.

The proposed approach differs from the approach used in the work (Bostanjoglo, 2002) which an enormous single to ~10^8 electrons and the electron pulse duration ~20 ns was used. These pulses, used to study laser-induced melting of metals, contained a large number of electrons which is

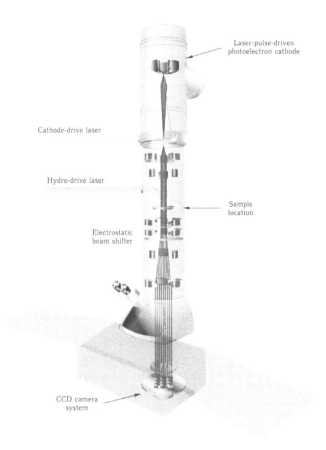

Fig. 11.20. Equipment for dynamic transmission electron microscopy (DTEM) from (King et al, 2005).

not desirable for achieving the formation of images of ultrashort pulses. Moreover, as noted in (Bostanjoglo, 2002), due to the fact that the time window for imaging in these experiments is expressed in nanoseconds, the uncertainty in spatial resolution due to the statistical noise is of the order of micrometers. It turned out that the use of amplified femtosecond optical pulses (with a much higher peak power density of $\sim 10^{12}$ V/cm^2) resulted, first, in the formation of an electron beam, which was insensitive to the focusing of the optical system of the microscope, and, secondly, to a significant lengthening of the pulses due to electron–electron repulsion.

Vanadium dioxide VO_2 undergoes a phase transition of the first kind from the low-temperature monoclinic phase (M) to high-temperature tetragonal phase of rutile (R) at $\sim 67°C$. Since its discovery (almost half a century ago), this phase transition has been the subject of intensive research. Grinolds et al (2006) first obtained the results using the 4D UEM with the use of single electrons. In this case, there was no effect of lengthening of the electron momentum due to their Coulomb (space) repulsion. The possibility of obtaining a sequence of images ('movies') on the atomic scale of spatial resolution with an ultrashort time resolution was indicated. In particular, it was shown that it is possible to investigate the metal–dielectric ultrafast phase transition in vanadium dioxide VO_2. The diffraction patterns (atomic scale) and UEM images (nanometer scale) show the structural phase transition in VO_2 nanoparticles with a characteristic hysteresis with a time resolution of 100 fs (Fig. 11.21).

The 4D UEM method was used to obtain images of polycrystalline materials and single crystals (see also the monograph by Zewail & Thomas, 2010). To calibrate, the images were also obtained when locking the femtosecond pulses fed into the electron microscope. In this case, the pattern was not observed; this confirms that the electrons, generated in the electron microscope, were indeed obtained optically and that the thermal electrons, generated by the cathode, can be neglected. The electron microscope can operate in the UEM and DTEM modes. Studies with the resolution of atomic dimensions were carried out using UEM in the diffraction mode by setting the intermediate lens to select the back focal plane of the lens as the object. The 4D UEM method is being rapidly developed further (Zewail & Thomas, 2010).

Conclusion

Motion of the atoms can be observed using the TRED (UED) methods in real time. Rupture of chemical bonds, their formation and change in the geometry of the molecule occur at a rate close to 1000 m/s. Consequently, the registration of the dynamics at the atomic scale, i.e. at distances of the order of angström requires (on average) the temporal resolution of 100 fs. Regardless of the fact that the molecule is isolated or enters the composition of any phase, ultrafast transformations in it are a dynamic process involving the concerted rearrangement of the electron and nuclear subsystems of the reacting molecules.

According to the transition state theory (Evans & Polanyi, 1935, 1938) for the rate of unimolecular reactions, we have a frequency factor of kT/h – in fact this is the frequency of transition to the final products through the energy barrier of chemical reactions. At room temperature, its value is $\sim 6 \cdot 10^{12}$ s^{-1}, which corresponds to the time of ~ 150 fs. Typical intramolecular fluctuations occur over a time interval of hundreds of femtoseconds. That is why the use of the term 'ultrafast diffraction' is justified to some extent. In this case it is clear that the diffraction process is determined for electrons by the accelerating voltage and a single act of scattering occurs in the range of units of attoseconds for electrons with energies of ~ 100 keV.

In 1927, Davisson and Germer and independently Thomson and Reid discovered the phenomenon of electron diffraction in the crystal. This finding pertained to 'static' diffraction. After the first experiments,

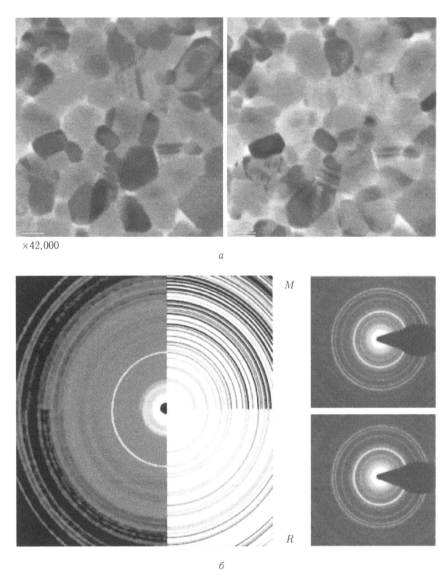

×42,000

a

M

R

b

Fig. 11.21. *a* – images obtained by the UEM method prior to phase transition in the VO$_2$ films (left) and after phase transition (right). Magnification 42 000 (100 nm scale). It should be mentioned that these images will not be seen if the generation of femtosecond pulses of the photoelectrons is blocked. *b* – diffraction patterns, produced by the UEM method prior to phase transition in VO$_2$ (right) and after (left). Diffraction patterns of two phases (monoclinic phase *M* and high-temperature tetragonal phase of rutile *R*), observed in experiments (left part of the figure *b*) and constructed as a result of calculations (right hand part of figure *b*). Analysis was described in (Grinolds et al, 2006).

performed by Mark and Virl' in 1930 (see the monograph: Mark and Virl', 1933), the electron diffraction method conceptually remained unchanged until the early 80's. Only 50 years later it became possible to introduce in the fourth dimension in electron diffraction – time, introducing the concepts of structural dynamics, and research in the 4D space–time continuum.

Electron diffraction with time resolution opened the possibility of direct observation of processes occurring in the transition state of the substance. The time resolution of about 100 fs was achieved for this purpose, which corresponds to the transition time of the system through the energy barrier of the potential surface that describes a chemical reaction – the process of breaking and the formation of new bonds of the interacting substances. Thus, it became possible to study the coherent nuclear dynamics of molecular systems and condensed matter.

In the past two decades, it became possible to observe the motion of the nuclei in the time interval corresponding to the period of oscillations of the nuclei. The observed coherent changes in the nuclear subsystem in these time intervals determine the fundamental shift from the standard kinetics to the dynamics of the phase trajectory of the single molecule, the tomography of the molecular quantum state (Ischenko et al, 1998).

Currently, the method of ultrafast diffraction is being developed intensively. Great opportunities to study the 4D structural dynamics are offered by the methods ultrafast electron crystallography and electron microscopy with time resolution from micro- to femtoseconds.

References for Chapter 11

Buchachenko A. L. Chemistry at the turn of the century: accomplishments and forecasts. Usp. Khimii. 1999. T. 68. P. 99-118.

Demtreder W. Laser spectroscopy. Basic principles and experimental techniques, trans. from English. Moscow: Nauka, 1985. 608 p.

Zharov V. P., Letokhov V. S., Laser optico-acoustic spectroscopy. Moscow: Nauka, 1984. 320.

Letokhov V. S. (Editor). Laser analytical spectroscopy. Moscow: Nauka, 1986. 318 p.

Letokhov V. S. Laser photoionization spectroscopy. Moscow: Nauka, 1987. 320.

Mark G., Wirl' R. Electron diffraction. Leningrad–Moscow: Gos. Tekh. Teor. Izd. 1933. 191 p.

Sarkisov O.M., Umansky S. Ya. Femtochemistry, Usp. Khim. 2001. V. 70, No. 6. S. 515-538.

Allan G., Delerue C., Lannoo M. Nature of luminescent surface states of semiconductor nanocrystallites. Phys. Rev. Lett. 1996. V. 76. P. 2961–2964.

Baum P., Zewail A.H. Breaking resolution limits in ultrafast electron diffraction and microscopy. Proc. Nat. Acad. Sci. U. S.A. 2006. V. 103. P. 16105–16110.

Baum P., Zewail A.H. Attosecond Electron Pulses for 4D Diffraction and Microscopy. Proc. Natl. Acad. Sci. U.S.A. 2007. V. 104. P. 18409–18414.

Bergmann K., Shore B.W. Molecular dynamics and spectroscopy by stimulated emission pumping. Singapore: World Scientific, 1995. 387 p.

Bostanjoglo O. High-speed electron microscopy. Adv. Imaging Electron Phys. 2002. V. 121. P. 1–48.

Buriak J.M. Organometallic chemistry on silicon and germanium surfaces. Chem. Rev. 2002. V. 102. P. 1271–1308.

Buriak J.M., Stewart M. P., Geders T.W., Allen M. J., Choi H. C., Smith J., Raftery D., Canham L. T. Lewis acid mediated hydrosilylation on porous silicon surface. J. Am. Chem. Soc. 1999. V. 121. P. 11491–11502.

Canham L. T. Silicon quantum wire array fabrication by electrochemical and chemical dissolution of wafers. Appl. Phys. Lett. 1990. V. 57. P. 1046–1048.

Cimpean C., Groenewegen V., Kunterman V., Sommer A., Kryschi C. Ultrafast exciton relaxation dynamics in silicon quantum dots. Laser & Photon. Rev. 2009. V. 3, No. 1–2. P. 138–145.

Delerue C., Allan G., Lannoo M. Theoretical aspects of the luminescence of porous silicon // Phys. Rev. B. 1993. V. 48. P. 11024–11036.

Duong P.H., Lavallard P., Oliver A., Itoh T. Temperature dependence of photoluminescence from localized states of silicon nanocrystals in silicon-implanted quartz. Phys. Stat. Sol. C. 2003. No. 4. P. 1271–1274.

Dwyer J. R., Hebeisenl C. T., Ernstorfer R., Harb M., Deyirmenjian V.B., Jordan R. E., Miller R. J.D. Femtosecond electron diffraction: making the molecular movie. Phil. Trans. R. Soc. A. 2006. V. 364. P. 741–778.

Evans M. S., Polanyi M. Inertia and driving force of chemical reactions. Trans. Faraday Soc. 1938. V. 34. P. 11–24.

Evans M. S., Polanyi M. Some applications of the transition state method to the calculation of reaction velocities especially in solutions. Trans. Faraday Soc. 1935. V. 31. P. 875–894.

Fojtik A., Henglein A. Surface Chemistry of Luminescent Colloidal Silicon Nanoparticles. J. Phys. Chem. B. 2006a. V. 110. P. 1994–1998.

Germanenko I.N., Li S. T., El-Shall M. S. Decay Dynamics and Quenching of Photoluminescence from Silicon Nanocrystals by Aromatic Nitro Compounds. J. Phys. Chem. B. 2001. V. 105. P. 59–66.

Grinolds M. S., Lobastov V.A., Weissenrieder J., Zewail A.H. Four-dimensional ultrafast electron microscopy of phase transitions. Proc. Nat. Acad. Sci. U.S.A. 2006. V. 103. P. 18427–18431.

Heinz A. S., Fink M. J., Mitchell B. S. Mechanochemical synthesis of blue luminescent alkyl/alkenyl-passivated silicon nanoparticles. Adv. Mat. 2007. V. 19. P. 3984–3988.

Hersam M. C., Guisinger N. P., Lyding J.W. Silicon-based molecular nanotechnology. Nanotechnology. 2000. V. 11. P. 70–76.

Hua F., Erogbogbo F., Swihart M. T., Ruckenstein E. Organically capped silicon nanoparticles with blue photoluminescence prepared by hydrosilylation followed by oxidation. Langmuir. 2006. V. 22. P. 4363–4370.

Hua F., Swihart M. T., Ruckenstein E. Efficient surface grafting of luminescent silicon quantum dots by photoinitiated hydrosilylation. Langmuir. 2005. V. 21. P. 6054–6062.

Ischenko A.A. Schafer L & Ewbank J.D. Tomography of the molecular quantum state by time-resolved electron diffraction. Proceedings SPIE. 1999. V. 3516. P. 580–587.

Ischenko A.A., Schafer L., Ewbank J. Time-resolved electron diffraction. Ed. by J.R. Helliwell, P.M. Rentzepis. — Oxford University Press. 1997. Ch. 13. — 442 p.

Kanemitsu Y. Light emission from porous silicon and related materials (Review Article). Phys. Rep. 1995. V. 263. P. 1–91.

Kanemitsu Y. Light-emitting silicon materials. J. Luminescence. 1996. V. 70. P. 333–342.

Kimura K. Blue luminescence from siliconnanoparticles suspended in organic liquids. J.

Cluster Sci. 1999. V. 10. P. 359–380.

King W. E., Campbell G.H., Frank A., Reed B., Schmerge J. F., Siwick B. J., Stuart B. C., Weber P.M. Ultrafast electron microscopy in materials science, biology, and chemistry. J. Appl. Phys. 2005. V. 97. P. 111101–111127.

Klimov V. I. Spectral and dynamical properties of multiexcitons in semiconductor nanocrystals. Annu. Rev. Phys. Chem. 2007. V. 58. P. 635–673.

Knipping J., Wiggers H., Rellinghaus R., Roth P., Konjhodzic D., Meier C. Synthesis of high purity silicon nanoparticles in a low pressure microwave reactor. J. Nanosci. Nanotech. 2004. V. 4. P. 1039–1044.

Kuntermann V., Cimpean C., Brehm G., Sauer G., Kryschi C., Wiggers H. Femtosecond transient absorption spectroscopy of silanized silicon quantum dots. Phys. Rev. B. 2008. V. 77. P. 115343 (8 p.).

Ledoux G., Guillois O., Porterat D., Reynaud C., Huisken F., Kohn B., Paillard V. Photoluminescence properties of silicon nanocrystals as a function of their size. Phys. Rev. B. 2000. V. 62 (23). P. 15942–15951.

Lie L.H., Duerdin M., Tuite E.M., Houlton A., Horrocks B. R. Preparation and characterisation of luminescent alkylated-silicon quantum dots. J. Electroanalytical Chemistry. 2002. V. 538–539. P. 183–190.

Lobastov V.A., Srinivasan R., Zewail A.H. Four-dimensional ultrafast electron microscopy. Proc. Nat. Acad. Sci. U.S.A. 2005. V. 102. P. 7069–7073.

Lobastov V.A., Srinivasan R., Vigliotti F., Ruan C. Y., Feenstra J., Chen S., Park S. T., Xu S., Zewail A.H.. In: UltraFast Optics IV, Springer Series in Optical Sciences / Ed. F. Krausz, G. Korn, P. Corkum, I. Walmsley. Berlin: Springer, 2003. P. 413.

Macklin J. J., Trautman J.K., Harris T.D., Brus L. E. Imaging and time-resolved spectroscopy of single molecules at an interface. Science. 1996. V. 272. P. 255–258.

Martin J., Cichos F., Huisken F., von Borczyskowski C. Electron-phonon coupling and localization of excitons in single silicon nanocrystals. Nano Lett. 2008. V. 8. P. 656–660.

Matsumoto T., Wright O. B., Futagi T., Minura H., Kanemitsu Y. Ultrafast electronic relaxation processes in porous silicon. J. Non-Cryst. Solids. 1993. V. 164–166. Part 2. P. 953–956.

Meier C., Gondorf A., Lüttjohann S., Lorke A., Wiggers H. Silicon nanoparticles: Absorption, emission, and the nature of the electronic band gap. J. Appl. Phys. 2007. V. 101. P. 103112 (8 p.).

Murdick R.A., Raman R.K., Murooka Y., Ruan C.-Y. Photovoltage dynamics of the hydroxylated Si (111) surface investigated by ultrafast electron diffraction. Phys. Rev. B. 2007. V. 77. P. P. 245329–245336.

Nelles J., Sendor D., Ebbers A., Petrat F.M., Wiggers H., Schulz C., Simon U. Functionalization of silicon nanoparticles via hydrosilylation with 1-alkenes. Colloid Polym Sci. 2007. V. 285. P. 729–736.

Nelles J., Sendor D., Petrat F.-M., Simon U. Electrical properties of surface functionalized silicon nanoparticles. J. Nanopart. Res. 2010. V. 12. P. 1367–1375.

Owrutsky J. C., Rice J.K., Guha S., Steiner P., Lang W. Ultrafast absorption in free standing porous silicon films. Appl. Phys. Lett. 1995. V. 67. P. 114755 (3 p.).

Raman R.K., Murooka Y., Ruan C.-Y., Yang T., Berger S., Tomanek D. Direct observation of optically induced transient structures in graphite using ultrafast electron crystallography. Phys. Rev. Lett. 2008. V. 101. P. 077401 (7 p.).

Rentzepis P.M. (Ed.). Time-Resolved Electron and X-ray Diffraction. Proceedings SPIE. V. 2521. Bellingham, WA, 1995.

Rogozhina E.V., Eckhoff D.A., Gratton E., Braun P.V. Carboxyl functionalization of ultrasmall luminescent silicon nanoparticles through thermal hydrosilylation. J. Mater. Chem. 2006. V. 16. P. 1421–1430.

Ruan C.-Y., Murooka Y., Raman R.K., Murdick R.A. Dynamics of size-selected gold nanoparticles studied by ultrafast electron nanocrystallography. Nano Lett. 2007. V. 7. P. 1290–1296.

Ruan C.-Y., Murooka Y., Raman R.K., Murdick R.A., Worhatch R. J., Pell A.R. The development and applications of ultrafast electron nanocrystallography. Microscopy and Microanalysis. 2009. V. 15. P. 323–337.

Ruan C.-Y., Vigliotti F., Lobastov V.A., Chen S., Zewail A.H. Ultrafast Crystallography: Transient Structures of Molecules, Surfaces and Phase Transitions. Proc. Nat. Acad. Sci. U.S.A. 2004. V. 101. P. 1123–1128.

Sarkisov O.M., Gostev F. E., Shelaev I.V., Novoderezhkin V. I., Gopta O.A., Mamedov M.D., Semenov A. Yu., Nadtochenko V.A. Long-lived coherent oscillations of the femtosecond transients in cyanobacterial photosystem I. Phys. Chem. Chem. Phys. 2006. V. 8. P. 1–8.

Trojanek F., Neudert K., Maly P., Dohnalova K., Pelant I. Ultrafast photoluminescence in silicon nanocrystals studied by femtosecond up-conversion technique. J. Appl. Phys. 2006. V. 99. P. 116108 (3 p.).

Vigliotti F., Chen S., Ruan C.-Y., Lobastov V.A., Zewail A.H. Ultrafast electron crystallography of surface structural dynamics with atomic-scale resolution,. Angew. Chem. Int. Ed. 2004. V. 43. P. 2705–2710.

Wang D., Dong H., Chen K., Huang R., Xu J., Li W., Ma Z. Low turn-on and high efficient oxidized amorphous silicon nitride light-emitting devices induced by high density amorphous silicon nanoparticles. Thin Solid Films. 2010. V. 518. P. 3938–3941.

Wilcoxon J. P., Samara G.A., Provencio P.N. Optical and electronic properties of Si nanoclusters synthesized in inverse micelles. Phys. Rev. B. 1999. V. 60. P. 2704–2714.

Wilkinson A.R., Elliman R.G. Maximizing light emission from silicon nanocrystals — The role of hydrogen. Nuclear Instruments and Methods in Physics Research, B. 2006. V. 242. P. 303–306.

Williams A. T. R., Winfield S.A., Miller J.N. Relative fluorescence quantum yields using a computer-controlled luminescence spectrometer. Analyst. 1983. V. 108. P. 1067–1071.

Wolkin M.V., Jorne J., Fauchet P.M., Allan G., Delerue C. Electronic states and luminescence in porous silicon quantum dots: The role of oxygen. Phys. Rev. Lett. 1999. V. 82. P. 197–200.

Zewail A.H. Femtochemistry: Atomic-scale dynamics of the chemical bond using ultrafast lasers (Nobel Lecture). December 8, 1999.

Zewail A.H., Thomas J.M. 4D electron microscopy. Imaging in space and time. Imperial College Press, 2010. 341 p.

Zewail A.H. 4D Ultrafast Electron Diffraction, Crystallography, and Microscopy. Annu. Rev. Phys. Chem. 2006. V. 57. P. 65–103.

Zhou Z., et al. Electronic structure and luminescence of 1.1- and 1.4-nm silicon nanocrystals: Oxide shell versus hydrogen passivation. Nano Lett. 2003. V. 3. P. 163–167.

Part IV

12

Fluorescent labels based on nanosilicon

S. Sizova[1], A.A. Ischenko

The intensive development of a broad class of analytical methods based on the use of different fluorescent labels, has made them one of the most important experimental techniques in many scientific disciplines. In particular, their application in biotechnology and medicine has led to the emergence and development of methods facilitating the study of living cells and cellular structures, the fundamental cell processes and methods of recording biospecific interactions, finding applications in medical diagnostics and a variety of biological analyses.

In recent years, the approaches to the visualization of processes at the level of cells, tissues and whole organisms, based on the introduction of specialized fluorescent labels, such as semiconductor nanocrystals, nanoshells, nanorods, metal nanospheres, carbon nanotubes, were rapidly developed.

Today special interest is attracted by silicon nanoparticles and composites based on them. These nanomaterials usually do not cause allergic reactions, can be potentially be split in the body and removed from it, and in the pores one can place medications or killers of cancer cells. The main advantage of such porous nanoparticles in comparison with optically active materials (gold nanoparticles, semiconductor nanocrystals CdSe, CdTe, PbSe, PbS, carbon nanotubes) for use in *in vivo* applications is that these silicon nanoparticles are biodegradable and do not show toxic properties.

Park et al (2009) described a method for producing nanoparticles of porous silicon, which not only has a sufficiently high luminescence, but also exhibits bioactive properties. Porous silicon nanoparticles were obtained by electrochemical etching of a silicon substrate in an alcohol solution of HF, followed by ultrafiltration and coating of isolated nanoparticles

[1]The Shemyakin–Ovchinnikov Institute of Bioorganic Chemistry, Russian Academy of Sciences, Moscow.

with a polymer membrane for a prolonged dissolution in the body. Past studies of biocompatibility and biodegradation properties of porous silicon nanoparticles have confirmed the safety of using this kind of material in *in vivo* applications.

A more detailed description of the above nanoparticles and their potential application in biomedical applications will be the subject of this review.

12.1. Producing nanosilicon for biological applications

Nanocrystalline silicon and composite materials based on it also are an alternative to organic fluorescent labels, traditionally used in biological research. Such materials are widely in demand in various fields of chemistry, physics and materials science, as evidenced by the output in recent years of a large number of publications devoted to the study of these substances.

Features of the spectral-structural properties of silicon nanocomposites, namely the dependence of the UV spectrum of the size distribution function of the particles, chemical composition of the membrane, and the degree of crystallinity of the central core made it possible to control the spectral characteristics of the materials obtained with the use of such nanocomposites. The use of high-productivity plasma-chemical technology related to the evaporation of crystalline silicon in the plasma discharge, as well as the technology of monosilane decomposition in the laser field allows us to adjust not only the size of the synthesized nanocrystalline silicon particles, but also the chemical composition of the surface layer, which offers the additional possibility to control the optical properties of materials obtained on the basis of nc-Si. For nanoparticles of 10 nm and above (containing $>10^4$ silicon atoms), the absorption features in the UV and visible wavelengths are largely determined by the properties of conventional crystalline or amorphous silicon. Silicon nanocomposites having a diameter of the central core of 5 nm or less, exhibit the quantum size effect which has a significant impact on their optical properties.

In general, the methods of producing silicon nanoparticles are more complex than the well-known protocols for obtaining nanoparticles based on the elements of the groups II–VI, group, the so-called quantum dots. The previously published methods for obtaining silicon nanoparticles, including chemical synthesis (Bley & Kauzlarich, 1996), recovery of SiO_2 with carbon in electric arc furnaces (Heath, 1992), the method of laser pyrolysis of silane, electrochemical etching of silicon substrates in a concentrated hydrofluoric acid (Wang et al. in 2004; Yamani, 1998) are based on technologies developed for the synthesis of porous silicon nanoparticles (Jung, 1991). Production of silicon nanoparticles by electrochemical etching traditionally produces a wide particle size distribution and various forms of and, consequently, different physico–chemical properties, and it is well know known that most applications require particles of uniform size.

Thus, the methods of unification of the size of silicon nanoparticles,, such as centrifugation (Belomoin et al, 2002), selective deposition (Wilson & Szajowski, 1993), exclusion chromatography and capillary electrophoresis (Rogozina et al, 2006), also require careful consideration.

The main physico–chemical properties of silicon to facilitate the active involvement of nanomaterials based on nc-Si for application in biomedicine and biotechnology technology are:

- prevalence (the proportion of silicon in the Earth's crust is about 27%);
- biocompatibility (the body of a healthy person weighing 50–70 kg contains 0.5–1.0 g of silicon, making it the third trace element content after iron and zinc);
- biodegradability (silicon dissolves in the form of nanoparticles in the human body at a rate of 1 nm in an acidic medium and up to 1000 nm in an alkaline medium per day with the formation of orthosilicic acid);
- proven technology for producing nanoporous forms of silicon can control the size of the granules and the degree of their porosity.

However, the silicon nanoparticles produced by synthesis are hydrophobic and dispersible only in organic solvents, therefore, their use in the bioanalysis requires a change in the surface properties of nc-Si, and the hydrophilization and functionalization of the surface of the nanoparticles.

Etching of crystalline nanosilicon immediately after synthesis is the inevitable procedure carried out to reduce the crystalline core of nanoparticles for the generation of luminescence and for crushing pyrolytic powders. In fact, crystalline nanosilicon, obtained by laser pyrolysis, represents agglomerates of a few tens of nanometers, which are formed due to coagulation of particles and merger during growth (Borsella et al, 1997; Herlin-Boime et al, 2004; Swihart, 2003). Etching of silicon dioxide can be conducted in the presence of hydrofluoric acid and by alkaline etching, while the crystalline nanosilicon can be treated only by alkaline etching.

Nanoparticles of crystalline silicon after etching show unstable fluorescence, are not dispersed in most solvents, and are easily oxidized in the air. Therefore, nanocrystalline silicon particles with terminal SiH groups are modify by silanization of Si–O bonds or hydrosilylation through the Si–C bond.

12.1.1. Surface modification of nanocrystalline silicon by silanization

The method of modification by organosilicon reagents with different functional groups results in the formation of nanoparticles containing surface groups, mostly amino groups, which (as the amino groups of other carriers) can be converted into aldehyde groups by condensation with glutaraldehyde or in arylamine groups with nitrobenzoyl chloride or carboxyl groups by condensation with succinic anhydride. Li et al. (2004) modified the crystalline silicon nanoparticles after pretreatment with a solution

'Piranha' (a mixture of 25% of hydrogen peroxide and 98% sulphuric acid in a ratio of 1:2) or nitric acid by a solution octadecyltrimethoxysilane in toluene, and obtained fine powders of nanocrystalline silicon which had a bright fluorescence. The study of the fluorescent properties showed that this modification leads to stabilization of the fluorescent properties of nanocrystalline silicon. For example, nanocrystalline silicon particles with terminal Si–Cl groups with blue fluorescent were synthesized by Zou et al. (2004) by reduction with sodium naphthalide followed by silanization. The nanocrystalline silicon particles, modified by this procedure, showed stable photochemical properties in non-polar organic solvents. Li et al. (2004) produced luminescent nanocrystalline silicon by surface modification with organosilane and polyaniline for later use as optical emitters. This method of modification led to significant improvements in the fluorescent properties and photostabilization of the nanoparticles. Later, Kang et al. (2009) synthesized aqueous dispersions of nanoparticles of crystalline silicon with the Si/SiO_xH_y core–shell structure, with a particle diameter of 3 nm; luminescence of the nanoparticles was activated by oxidation using a mixture of ethanol and hydrogen peroxide. The results suggest that the surface modification of the nanoparticles of crystalline silicon with the Si–O bonds has virtually no effect on the luminescence of the crystalline core of nanosilicon.

12.1.2. Surface modification of nanocrystalline silicon by hydrosilylation

Surface Si–H groups of crystalline silicon nanoparticles can be converted into radicals bonded to the silicon atoms, by exposure to high temperature or by photoactivation. High-activity silicon radicals can be attached to the compounds with carbon–carbon bonds with the formation of Si–C bonds; this procedure, as already mentioned several times in this book, is called hydrosilylation.

Hua et al. (2005) first synthesized the nanoparticles of crystalline silicon whose surface was passivated predominantly by hydrogen by increasing the HF/HNO_3 ratio in the etching process and then performing hydrosilylation under an argon atmosphere. The resulting nanoparticles were characterized by stable luminescence in the range from green to red.

Since most alkenes are commercially available, the method of modifying the nanosilicon surface by hydrosilylation is suitable for further bioconjugation with biological molecules. Wang and colleagues (Wang et al, 2004) have demonstrated the possibility of joining of DNA molecules to silicon nanoparticles. Choi et al. (2008) obtained conjugates of luminescent crystalline silicon nanoparticles with the streptavidin protein by hydrosilylation.

In addition to alkenes, alkynes can possibly be used. The possibility of bonding the alkynes to the silicon dimers derived from silicon nanoparticles,

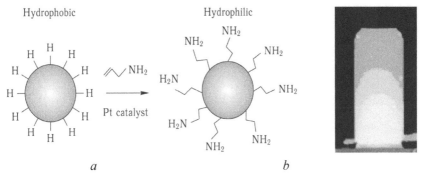

Fig. 12.1. *a* – schematic of the process of modification of silicon nanoparticles by allylamine described in (Warner et al, 2005). *b* – blue fluorescence of Si nanoparticles modified by allylamine. (For colour images please see the colour section in the middle of the book).

luminescent in the blue region, by the method of high-energy mechanical grinding was demonstrated in Hua et al. (2006).

A method of producing hydrophilic silicon nanoparticles was first described in (Li & Ruckenstein, 2004), where acrylic acid was attached to the silicon nanoparticles by hydrosilylation. Aqueous dispersions of nanocrystals based on the crystalline silicon were also produced (Warner et al, 2005) by covalent attachment of allylamine to the silicon nanoparticles by hydrosilylation initiated by a platinum catalyst (Fig. 12.1). The resulting hydrophilized crystalline silicon nanoparticles were used to tag the cells of the HeLa line. Amino groups on the surface of the nanoparticles are suitable for subsequent binding to the compounds containing carboxyl, epoxy groups, ketones, acyl chlorides, aldehydes. However, this method can not be used for the modification of silica nanoparticles that emit in the long-wave range, because the presence of amino groups on the surface leads to quenching of fluorescence.

Stable aqueous dispersions of fluorescent silicon nanoparticles that emit fluorescence in a wide range of wavelengths were produced in (Sato & Swihart, 2006) by hydrosilylation of acrylic acid in the presence of HF. The diameter of the nanoparticles of crystalline silicon and, consequently, the colour of fluorescence – from yellow to green – can be controlled by varying the etching time.

Hua et al. (2006) prepared silicon nanoparticles, fluorescent in the blue region, by hydrosilylation followed by oxidation of the nanoparticles with yellow fluorescence either by UV irradiation in solution or by heating in the air. It was found that the cause of the blue fluorescence is not the quantum size effect and incomplete oxidation of silicon but some other not yet fully understood mechanism.

The authors of the study (Kang, 2009; He et al, 2009) produced aqueous dispersions of composite materials consisting of silicon nanocrystals and polyacrylic acid. The dispersions were prepared by hydrosilylation of acrylic

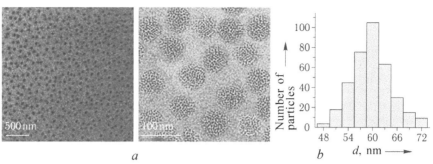

Fig. 12.2. *a* – TEM images of nanocomposites based on nanocrystalline silicon and polyacrylic acid with a particle diameter of ~60 nm, *b* – particle size distribution (He et al, 2009).

acid on the surface of silicon nanoparticles with subsequent UV radiation-induced photopolymerization of acrylic acid. TEM images and the size distribution function of the silicon nanoparticles in these dispersions are shown in Fig. 12.2. Despite the good dispersibility in water, satisfactory colloidal stability in biological fluids could not be reached.

Erogbogbo et al. (2008) synthesized biocompatible fluorescent nanocontainers based on crystalline silicon nanoparticles to visualize cancer cells. Fluorescent silicon nanoparticles, obtained by etching with a mixture of nitric and hydrofluoric acids, were modified with compounds such as styrene and octadecyl. The resulting nanoparticles were easily dispersible in most non-polar organic solvents. The modified nanoparticles were then mixed with phospholipids and functionalized with polyethylene glycol (PEG) in chloroform. After evaporation of chloroform the authors obtained nanocontainers, easily dispersible in water, with a diameter in the range 50–120 nm. Aqueous dispersions retained colloidal stability for at least 2 months when stored at 4°C. The fluorescence intensity decreased slightly with increasing temperature from 25 to 40°C and remained constant for transfer of nanoparticles in to the phosphate salt buffer solution.

12.2. Applications of nanosilicon in biomedicine and biotechnology

As noted above, the crystalline silicon nanoparticles are dispersed only in non-polar organic solvents, which are toxic to living organisms. However, in recent years, interest in nanomaterials such as nanocrystalline silicon and composites based on it has been increasing, and several works devoted to the use of nanocrystalline silicon in biological applications have already been published (Choi, Wang et al, 2008; He et al, 2009; Erogbogbo et al, 2008; Jin et al, 2007; Choi, Zhang et al, 2008). The unique optical properties of nanocrystalline silicon are very promising for the use of silicon nanoparticles in bioimaging, cell labeling, as carriers of nucleotides for gene therapy, applications in immunoassay, etc. Nanoparticles of crystalline silicon can be divided into two types on the basis of the colour of the emitted fluorescence – namely, silicon nanoparticles that emit in the orange spectral region with a long fluorescence decay (a few tens of microseconds), which are traditionally produced by oxidation of the surface (Borsella et al, 1997; Huisken et al, 2002; Herlin-Boime et al, 2004; Li et al, 2003), and silicon nanoparticles that emit in the blue spectral region with a fast fluorescence decay time (tens of nanoseconds) (Akcakir et al, 2000; Yamani et al, 1998; Belomoin et al, 2000).

'Blue' nanoparticles may contain on their surface various chemical groups suitable for conjugation of biomolecules. 'Orange' nanoparticles with a long fluorescence decay are promising for the study of spatial and temporal distribution of biomolecules, imaging of biological objects (viruses, cellular organelles, cells, tissues) *in vitro* and *in vivo*. It is also expected that the sensitivity of the infrared sensors based on these nanoparticles exceeds by 2–3 orders of magnitude the sensitivity of the sensors fluorescing in the visible region of the spectrum due to the effect of transparency of tissue in the near infrared range (for example, absorption by biological molecules (hemoglobin, water etc.) is minimal in the near infrared), which will record the events not only at the level of the body and tissues, but also at the cellular and molecular levels (Klostranec et al, 2006), Fig. 12.3.

Nanoparticles of crystalline silicon containing a surface protective film of silicon oxide with a particle diameter of 2.5–8 nm emit in the wavelength range 560–850 nm (Huisken et al, 2002). In addition, the two-photon excited

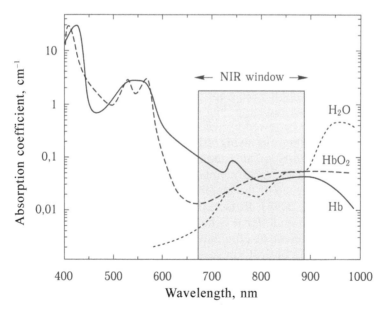

Fig. 12.3. The IR spectrum of water and hemoglobin in the *in vivo* mode (Klostranec et al, 2006).

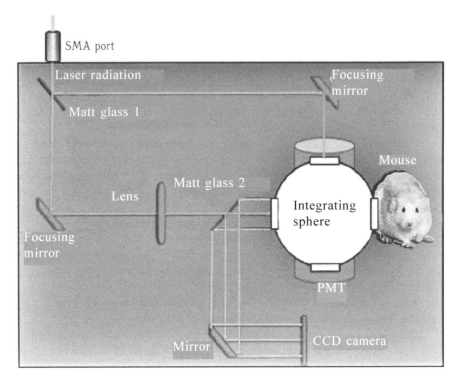

Fig. 12.4. The experimental setup for bioimaging with an integrating sphere. To eliminate the autofluorescence of biological tissues, a time delay must be introduced between laser flashes for excitation of fluorescence and fluorescence signal reception. When using 'red' silicon nanoparticles, the delay time between laser flashes and detection of the fluorescent signal is 10 ns, the signal acquisition time – 1 ms (Stratonnikov et al, 2006).

fluorescence, covering the same spectral range, was observed after excitation by femtosecond laser radiation (900 nm) of silicon nanoparticles synthesized by laser pyrolysis, followed by wet oxidation (Falconieri et al, 2008).

Another advantage of using silica nanoparticles with a large decay time is the possibility of measurements with time resolution to separate the effects of fast (several nanoseconds) autofluorescence of the skin from the luminescence of the majority of biomaterials (Dahan et al, 2001; Stratonnikov et al, 2006). In fact, the delay of a few tens of nanoseconds between excitation and signal detection (see the measurement diagram in Fig. 12.4) is sufficient to eliminate the contribution of autofluorescence emitted in the same optical range.

The fluorescence signal obtained in the continuous mode (1 min) after injection of a colloidal solution of silicon nanoparticles produced by laser pyrolysis (2 mg/ml) into the tail vein of mice is shown in Fig. 12.5.

The manifestation of acute toxicity was not observed. However, the signal intensity was insufficient for examining the spatial localization

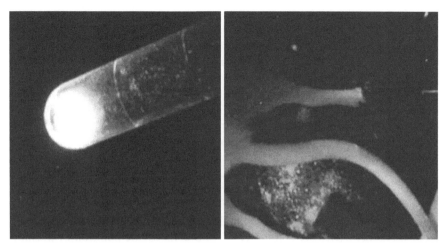

Fig. 12.5. Photoluminescence, recorded 1 min after injection into the tail vein of a mouse of 0.5 ml of an aqueous dispersion of silicon nanoparticles (2 mg/ml) (Ryabova et al, 2006).

of fluorescent silicon nanoparticles inside the organs. Despite the weak intensity of fluorescence, the silicon nanoparticles can serve as a good alternative to coatings of optical probes, due to the lack of appropriate compensation of the autofluorescence of the timeline.

Nanoparticles of silicon with a fast decay time of fluorescence ('blue' nanoparticles) are also promising for use in a number of biological applications. For example, Wang et al. (2004) received conjugates of 'blue' nanoparticles (with a particle diameter of 1–2 nm) with 50-member oligonucleotides. The silicon nanoparticles, whose surface was covered with hydrogen, obtained by electrochemical etching. The conjugation was carried out by two successive photoinduced reactions with the subsequent formation of carboxyamide groups. The resulting nanoparticle conjugates with oligonucleotides were characterized by good dispersibility in water and emitted fluorescence in the blue region of the spectrum for a minimum of 7 days with a quantum yield of about 8%. Monodisperse silica nanoparticles containing amino groups to bind DNA can be used to create non-viral vectors for *in vivo* and *in vitro* gene therapy.

Later, Choi, Wang et al. (2008) reported receiving conjugates of 'blue' silicon nanoparticles with molecules of streptavidin using a bifunctional crosslinking agent. The resulting conjugates were characterized by fluorescence in the blue region of the spectrum, i.e. they retained the fluorescent properties of the initial silicon nanoparticles. By capillary electrophoresis it was revealed that 4–5 silicon nanoparticles are attached to a streptavidin molecule, while the binding affinity of streptavidin to biotin remained unchanged. Silicon nanoparticles, covalently bonded to streptavidin molecules, are promising for use in different types of bioanalytical tests based on specific high-affinity interaction between the detected ligand and a reagent containing an antiligand.

A method of producing aqueous dispersions of 'blue' silicon nanoparticles, covalently bonded to the allylamines, is described in Warner et. al. (2005). The resulting dispersion of the nanoparticles hydrophilized with allylmine, were used to tag the cells of the HeLa line. Cells were incubated for 12 h in the above dispersion of nanoparticles (0.2 nm) and then washed twice with a phosphate-buffered saline to remove non-specifically bound nanoparticles. Fluorescent images were captured using a digital camera mounted on an optical microscope with a wide field of view in the excitation of fluorescence by the radiation with a wavelength of 365 nm, which indicates that the fluorescence emission by silicon nanoparticles in the cytosol. In addition, the 'blue' hydrophilized silicon nanoparticles have been used to label Vero cell lines (Tilley & Yamamoto, 2006). Nanoparticles were transfected into cells by the addition of plasmids enclosed in liposomes. After incubation and washing, the cells were fixed, fluorescent images were obtained by a CCD camera, installed on a fluorescence microscope, using a multipole filter and a mercury lamp as a source of fluorescence excitation.

In Choi (2008) nc-Si was prepared by direct electrochemical reduction of octyltrichlorosilane in an anhydrous electrolyte at room temperature and normal pressure. Hydrophilization and subsequent covalent binding to the streptavilin was carried out using the bifunctional linker 4-azido-2,3,5,6-tetrafluorobenzoic acid (ATFBA) (photoactivated reagent) and succinimide ether (a schematic representation of modification, hydrophilization and conjugation is shown in Fig. 12.6). The activity of these conjugates nc-Si-streptavidin was tested in the reaction with biotinylated polystyrene microspheres.

An example of the use of silica nanoparticles modified with allylamine, as a fluorescent marker to label the cells, is shown in Fig. 12.7 (Warner et al, 2005). Sudeep et al (2008) described the results of studies of the photostability of silicon nanoparticles hydrophilized by amphiphilic polymers based on PEG (polyethylene glycol). The above particles were dissolved in both organic solvents and aqueous solutions, and the absorption spectrum of the modified nanoparticles were consistent with the spectrum of the initial silicon nanoparticles.

He et al. (2009) used fluorescent nanospheres (diameter of the particles in the range of 60–200 nm) formed from polyacrylic acid, with the included silicon nanoparticles for labeling kidney cells of the HEK293T line. Using laser scanning confocal microscopy, they obtained contrast fluorescent images. It should be noted that the fluorescence of silicon nanoparticles may be excited by laser radiation with a wavelength of 458 and 488 nm, whereas for the fluorescent images of cells labeled with FITC, only a laser with a wavelength of 488 nm can be used.

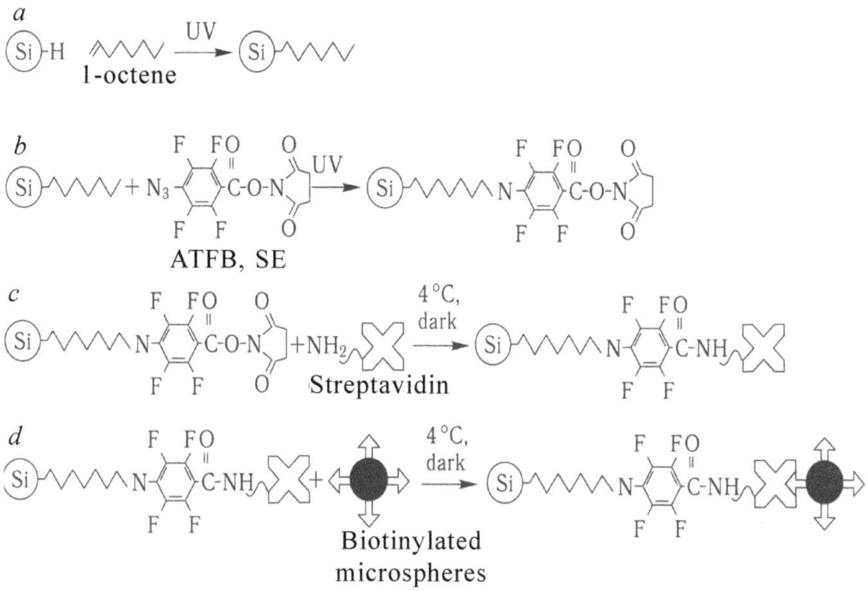

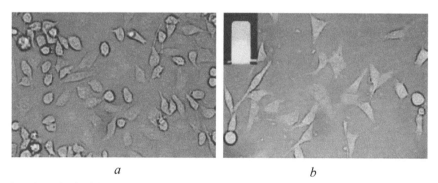

Fig. 12.6. *a* – surface passivation of nc-Si with 1-octadecene; *b* – interaction ATFBA and the succinimide ester of octadecenoic groups on the surface of nc-Si; *c* – formation of an amide bond between streptavidin and activated groups of the linker, *d* – complex formation between the conjugate of nc-Si with streptavidin and biotin, immobilized on polystyrene microspheres (Choi, 2008).

Fig. 12.7. Optical image, obtained in fluorescent light, of the cells of the HeLa line (a) and cells of the HeLa line specifically labeled with silicon nanoparticles (b). The inset shows the fluorescent image of silicon nanoparticles hydrophilized with allylamine (Warner et al, 2005). (For colour images please see the colour section in the middle of the book).

Comparative tests of cells labelled with FITC and silicon nanospheres showed that the nanospheres on the basis of silicon nanoparticles are characterized by greater photostability, which is a necessary requirement to labels for research in real time (Fig. 12.8).

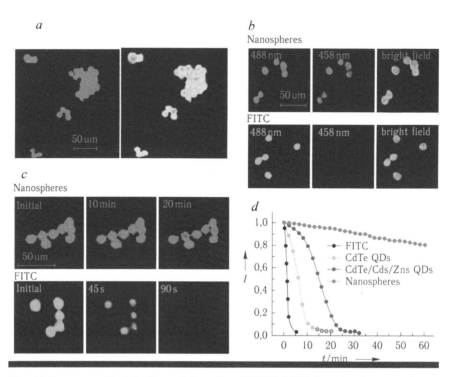

Fig. 12.8. *a* – specific labelling of the cells of HEK2-293T lines by the nanospheres based on silicon particles (left) and the superposition of the optical and fluorescent images (right), excitation of fluorescence at 488 nm; *b* – comparison of the fluorescent signal of the cells of the HEK293T line, specifically labelled with the nanospheres based on silicon nanoparticles (upper) and FITC (lower); *c* – variation of fluorescence in relation to the duration of radiation of the cells of the HEK293T lines, labelled with the silicon nanospheres (upper) and FITC (lower). Fluorescence was excited with an argon laser with a wavelength of 488 nm, a delay time of 8 ms, power 15 mW; *d* – comparison of the photostability of the quantum dots with the composition CdTe and CdTe/CdS/Zns with the core–shell structure and the nanospheres based on silicon nanoparticles – fluorescence of all specimens was excited with a xenon lamp, power 450 W. The data from He et al (2009). (For colour images please see the colour section in the middle of the book).

In the paper (Erogbogbo et al, 2008) the authors presented the results of labelling cancer cells with silicon nanoparticles incorporated in phospholipid micelles with end PEG groups (Panc-1), by laser confocal microscopy. The micelles with incorporated silicon nanoparticles, functionalized with amino- or transferrin groups, at a concentration of 8 mg/ml were added to the cells and incubated for 2 h at 37°C. After a three-fold washing with a phosphate buffer solution to remove unbound micelles with the included silicon nanoparticles, the cells were fixed and fluorescence images obtained after irradiation by a laser with a wavelength of 405 nm. Control experiments using PEG-micelles with included silicon nanoparticles not functionalized with amino or transferrin groups, showed no fluorescent signal from the cells. These results showed that the silicon nanoparticles can be a valuable tool in biomedical optical diagnostics.

Thus, it can be claimed that the silicon nanoparticles are promising materials for use as fluorescent labels because of low price, ease of functionalization, biocompatibility and biodegradability.

12.3. Biodegradable porous silicon nanoparticles for *in vivo* applications

Fluorescent porous silicon nanoparticles (FPSN), obtained by electrochemical etching, are promising for use in biological applications *in vivo* due to biocompatibility (Bayliss et al, 1999) and biodegradability (Canham, 1995), the possibility of using in multiplex analysis (Cunin et al, 2002) and the presence of a controlled porous nanostructure for inclusion and directional drug delivery (Salonen et al, 2008).

It should be noted that the degree of biodegradability of nanosilicon depends on the particle size and the level of acidity of the medium. In the case of nanoparticles with a diameter in the range 1–10 nm, the dissolution in the human organism can take from hours to several months. Ultimately nanosilicon is completely removed from the body, which distinguishes it from other types of nanoparticles. In addition, since the discovery of photoluminescence of porous silicon in 1990 (Canham, 1990), a number of methods have been developed to produce FSPN (Heinrich et al, 1992; Wilson, 1993; Mangolini & Kortshagen, 2007; Wang et al, 2004; Li & Ruckenstein, 2004), suitable for producing silicon nanoparticles for biological applications.

For example, in Park et al. (2009), fluorescent nanoparticles of porous silicon were obtained by electrochemical etching of a silicon substrate in an alcohol solution of hydrofluoric acid, with removal of the porous silicon film, followed by sonication and separation of porous silicon nanoparticles by filtration through a membrane with a pore size of 0.22 mm (Fig. 12.9).

To activate fluorescence, the obtained porous nanoparticles were incubated in water for two weeks. Immediately after the formation of a porous layer the silicon dangling bonds on the pore surface were passivated mainly by hydrogen which was eventually replaced by oxygen. During the activation phase, the pores etched with hydrogen showed the build-up of silica. Such structures exhibit a strong luminescence caused by defects localized at the silicon–silicon dioxide interface (Heinrich et al, 1992; Wilson, 1993; Godefroo et al, 2008). Chemical destruction of the coating on the silicon surface leads to the formation of a layer of silicon dioxide emitting fluorescence upon irradiation with ultraviolet light (Fig. 12.10 and 12.11).

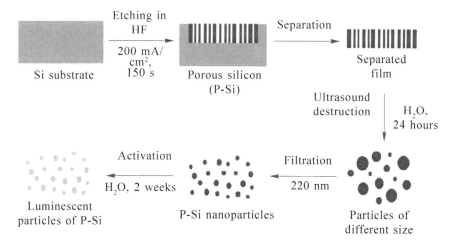

Fig. 12.9. The scheme for obtaining fluorescent porous silicon nanoparticles (FPSN) by electrochemical etching (Park et al, 2009).

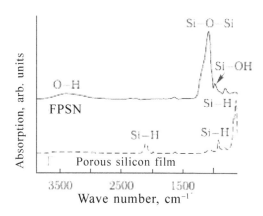

Fig. 12.10. The IR spectra of porous silicon films and FPSN. The surface of the porous silicon film, passivated with hydrogen, shows build-up of silica in the pores. A layer of silicon oxide also passivates the surface of the nanoparticles, which leads to the formation of oxides in the interfacial region and gives a strong contribution to the IR spectrum (Godefroo et al, 2008).

Preparation conditions of FPSN nanoparticles were optimized in order to obtain the pore volume and surface area that are optimum for including therapeutic agents and prolonged circulation *in vivo*, while maintaining satisfactory rate of degradation. Studies were carried out using spherical FPSN, average diameter of 126 nm and a developed micro- and mesoporous structure (pore diameter 5–10 nm).

Upon excitation of fluorescence of FPSN with an UV lamp the fluorescence was recorded in the range 650–900 nm, i.e. in the spectral

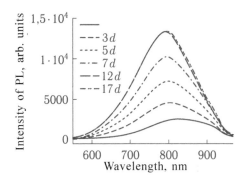

Fig. 12.11. Dynamics of fluorescence intensity of FPSN during activation in deionized water at room temperature (d = 24 h after placing FPSN in water). In the process of activation over time there was an increase in fluorescence intensity and slight shift of the fluorescence band toward shorter wavelengths (Godefroo et al, 2008).

region where the biological tissue is most transparent (Weissleder, 2001). The quantum yield of FPSN in ethanol was 10.2% (based on standard – Rhodamine 101). FPSN were characterized by significantly greater photostability compared with fluorophores, as cyanine or phycoerythrin Cy5, Cy7 (Wang et al, 2004; Li & Ruckenstein, 2004). When placing FPSN in a phosphate-buffered saline, pH 7.4, with a temperature of 37°C and the mass concentration below the solubility limit of orthosilicic acid (0.1–0.2 mg/ml SiO_2), Piryutko (1959) observed quenching of fluorescence and complete dissolution of the nanoparticles (Figure 12.12c, d,).

To study the potential use of FPSN in medicine for diagnosis and therapy of cancer, the anti-cancer drug doxorubicin (DOX) was included FPSN (DOX-FPSN, 43.8 mg DOX per 1 mg of FPSN). Positively charged molecules of doxorubicin were bonded with the negatively charged SiO_2 surface of the pores by electrostatic interactions. The relatively slow exit of the drug, observed at physiological pH and temperature, reached the optimum level after 8 h (Fig. 12.12e).

The FPSN-DOX complex showed a slightly greater cytotoxicity than the DOX drug itself, whereas the original FPSN did not show any cytotoxic properties (Fig. 12.12f). It is important to note that the content of DOX in the FPSN-DOX complex remained virtually unchanged during the circulation in the bloodstream, and the FPSN-DOX complex was eventually delivered to the liver and the spleen.

The biocompatibility and biodegradation of FPSN in conditions *in vitro* and *in vivo* were also studied. Studies (Park et al, 2009), carried out on HeLa line cells showed no apparent toxicity of FPSN in the range of the studied concentrations (Fig. 12.13a).

In order to study the conditions *in vivo*, Park et al. (2009) injected FPSN into the bloodstream of a mouse. After some time FPSN left the bloodstream and accumulated in the liver and the spleen (Fig. 12.13 b), after

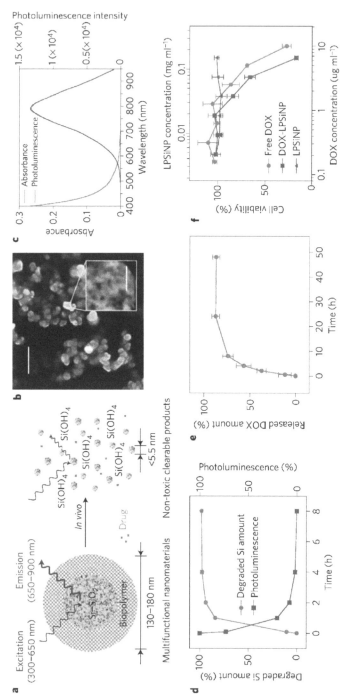

Fig. 12.12. Characteristics of FPSN, from Park et al (2009): *a* – schematic representation of the dissolution of the FPSN in the organism (*in vivo*); *b* – the micrograph of FPSN (the size scale 500 nm, in the inset 50 nm); *c* – the spectra of photoluminescence and absorption of FPSN; *d* – the dependences of the photoluminescence and the amount of reacted FPSN on time (FPSN were placed in the phosphate buffer saline solution PBS); *c* – the time dependence of the amount of doxorubicin, extracted from the particles of the carrier of FPN in the PBS solution (this also indicates the complete biodegradation of the silicon nanoparticles); *f* – investigation of the cytotoxicity of the composite material of the core/shell type DOX/FSPN (DOX – doxorubicin, anticancer drug). (For colour images please see the colour section in the middle of the book).

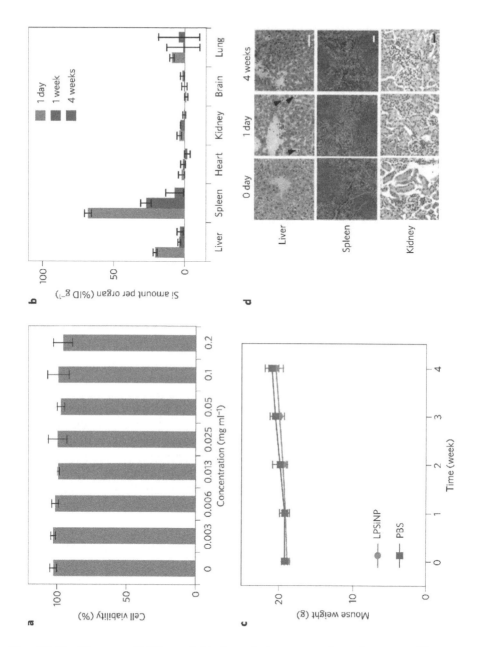

Fig. 12.13. Biocompatibility and biodegradation properties of porous silica nanoparticles: a – *in vitro* cytotoxicity of these nanoparticles; b – *in vivo* distribution and biodegradation of the porous silica nanoparticles in the organism of mice over a period of four weeks; c – change of the mass of the body of the mice after adding the nanoparticles and PBS (for comparison); d – histology of the cells of the lever, spleen and kidneys (the size scale 50 μm). Park et al (2009). (For colour images please see the colour section in the middle of the book).

which is completely transformed into orthosilicic acid and was excreted in the urine within 4 weeks. Within 4 weeks the body weight of the mice in the experimental and control groups increased by about the same amount, indicating that FPSN do not exhibit toxic properties (Fig. 12.13c).

To eliminate a possible manifestation of the toxic action of FPSN, the kidneys, liver and tissue of the mice of experimental group after 1 day and 4 weeks after injection FPSN were examined. Histological studies showed no signs of toxic lesions of the internal organs of mice (Fig. 12.13d).

Also investigated was the ability to visualize cells (*in vitro*) and deep located tissues and organs (*in vivo*) using FPSN (Fig. 12.14). 2 h after treatment of the cells of the HeLa line with the FPSN nanoparticles there was a strong fluorescence when excited by radiation with wavelengths of 370, 488 and 750 nm (two-photon excitation, Fig. 12.14a). For *in vivo* studies, the bloodstream of the mouse was injected subcutaneously and intravenously with a dispersion of FPSN and with FPSN and additionally modified with a biocompatible polymer dextran (D-FPSN); the observed fluorescence was then much greater than the level of tissue autofluorescence (Figure 12.14c, e). Histological studies of tissues showed that the initial biodegradation of FPSN is faster than that of D-FPSN (Fig. 12.14c–g). These results indicate the possibility of using FPSN for non-invasive visualization of the distribution of nanoparticles and their biodegradation in living organisms and to study the localization of nanoparticles in the organs by microscopy methods.

For visualization of deep-seated tumors in the *in vivo* conditions, the fluorescence excitation wavelength of FPSN must be chosen in the near-red region of the spectrum in order to obtain maximum absorption of the tissue and minimal absorption of chromophoric proteins such as, for example, hemoglobin. FPSN emit fluorescence in the range 810–875 nm at the excitation of fluorescence in spectrum ranges of 615–665 and 710–760 nm or two-photon excitation by the light of the near-infrared region (Fig. 12.15a). Like the nc-Si, emitting fluorescence in the near infrared region, the quantum yield of FPSN decreases with increasing wavelength of excitation (Kim et al, 2003), the intensity of fluorescence is sufficient to visualize the FPSN in the internal organs with the usual fluorescence microscope.

The FPSN solution, coated with dextran (the rate of 20 mg per 1 kg of body weight), was administered to mice with implanted MDA-MB-435 tumors, with FPSN passively accumulated in tumors (Fig. 12.15b). Visualization with light with a smaller wavelength (blue or green filter) led to a decrease in staining of the tumor. Studies of fluorescence *ex vivo* and histological studies confirmed the presence of FPSN coated with dextran, in the tumor.

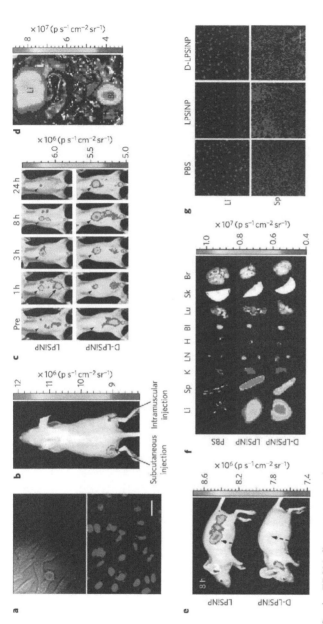

Fig. 12.14. Study FPSN fluorescence of FPSN *in vitro*, *in vivo* and *ex vivo* (Park et al., 2009): *a* – *in vitro* fluorescence of cells of the HeLa line, treated with FPSN (red and blue indicate FPSN and cores of cells respectively, dimensional range 20 μm); *b* – *in vivo* fluorescence of FPSN introduced into the mouse from two sides; *c* – *in vivo* fluorescence FPSN and FPSN further modified with dextran; *d* – *in vivo* images showing removal 1 hour after administration into the bladder (BI) FPSN (Li – liver); *e* – the same image, as in *c* (line indicates the position of the spleen); *f* – fluorescence showing *ex vivo* biodistribution of FPSN in organs in mice (Li, Sp, K, LN, H, BI, Lu, Sk and Br correspond to liver, spleen, kidney, lymph nodes, heart, bladder, lung, skin and brain); *g* – histology of the liver and the spleen cells of mice, as shown in Figure *c*, *f* 24 hours after administration (red and blue labelled nanoparticle and cell nuclei, respectively dimensional range 50 μm). (For colour images please see the colour section in the middle of the book).

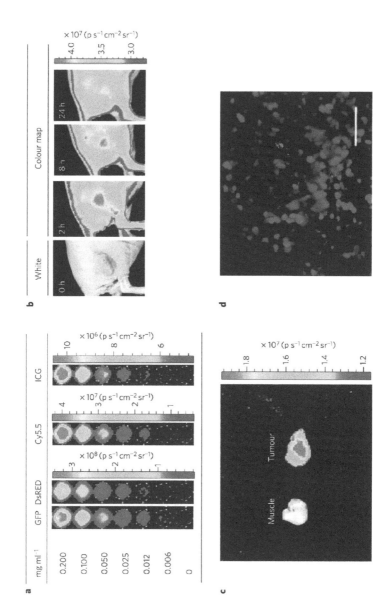

Fig. 12.15. Fluorescence of tumors containing FPSN coated with dextran (Ji-Ho Park et al., 2009): *a* – the intensity of the fluorescence of FPSN depending on the concentration; *b* – fluorescence in tumors MDA-MB-435; *c* – *ex vivo* fluorescence of murine tumors and muscle tissue around the tumor; *d* – fluorescence of murine tumors (red and blue mark FPSN coated with a polymer and cell nuclei, respectively; dimensional range 100 μm). (For colour images please see the colour section in the middle of the book).

Concluding remarks

At present, there are different tomographic methods to study the functions of living organisms. Most tomographic techniques allow to study organisms at the level of physiology, whereas for the successful use of recent advances of proteomics and genomics we require tomographic systems to allow registration molecular and cellular events. High sensitivity, allowing recording of many regulatory processes in cells, is typical of optical methods in the visible and near infrared region, among them the primacy belongs to the fluorescence methods. However, all living organisms, on the one hand, absorb light in this spectral range, on the other hand, they effectively dissipate it. Therefore, the main challenges in developing methods for optical tomography is the correct selection of the spectral properties of contrast agents for molecular events. Currently, one of the most common approaches to this problem is to use semiconductor quantum dots, based on CdSe(Te). The wavelength of maximum fluorescence depends on both the size of the nucleus and the nature of the semiconductor.

This requirement is satisfied by the following semiconductors: CdSe, CdTe, PbSe, PbS, InP. Being slightly inferior to the best fluorescent labels in the quantum yield of fluorescence (~70% at room temperature), the nanocrystals are superior to them by several orders of magnitude in the values of the absorption cross section of the exciting light. As a result, the brightness of the nanocrystals is so high that it allows detection of individual objects using a conventional fluorescence microscope. Semiconductor nanocrystals have two major advantages that distinguish them from organic fluorophores: 1) a wide range of narrow bands fluorescence, whose position depends on the diameter of the nanoparticle and is the controllable parameter in the possibility of excitation by radiation with a single wavelength; the excitation wavelength can be varied to obtain the maximum signal/noise ratio, taking into account the particular object under study; 2) the high photostability of the nanocrystals, 100–4000 times higher than the photostability of the best organic fluorophores. Recently methods were developed for the synthesis of photostable nanocrystals of InP/ZnS, but the low brightness of the fluorescence and the broad distribution of particle size limit their use in biochemistry. A disadvantage is also the toxicity of the ions included in the quantum dots, which requires the build-up of special protective shells.

Cautious attitude to the use of quantum dots *in vivo* was recently suggested by several authors: ICP MS was used to demonstrate the accumulation of quantum dots in the animal body after administration of intravenous injection. In addition, the nanometer size of the crystals can lead to their passive or active transport and accumulation of cell organelles, leading to unpredictable delaying effects. Therefore, the quantum dots based nanosilicon are of special interest. As we know, silicon is not toxic and, moreover, is one of the elements of life as part of a series of enzymes and proteins.

The results obtained in recent years show that, depending on the method of preparation, silicon nanoparticles may have a bright luminescence in the near infrared region (Ischenko et al, 2010). Nanocrystals of the nc-Si/protective shell composition with a diameter of 1.5–4.5 nm, which are an alternative to organic fluorescent labels, are traditionally used in biological studies due to their high brightness and the possibility of obtaining fluorescence throughout the optical range and availability. One of the main problems of using such fluorescent NC is the difficulty of obtaining biocompatible fluorescent complexes based on them and easily conjugate with biological molecules, which limits their application in biotechnology and biological studies as effective fluorescent biomarkers and sensors. To solve this problem, methods are being developed of modification and functionalization of the surface of the nc-Si to create highly efficient and specific fluorescent biomarkers.

The obtained preliminary results for the biodegradability and biocompatibility of fluorescent silicon nanoparticles show promise of development of the direction to create fluorescent labels on their basis for the application in medical and biotechnological applications.

References for Chapter 12

Ischenko A. A., Dorofeyev S. G., Kononov N. N., Ol'khov A. A. A method for producing nanocrystalline silicon, which has a bright sustainable photoluminescence. The patent of the Russian Federation from 01.2011 on an application for a patent on 11.12.2009. Registration number 2009145916.

Ryabov A. V., Stratonikov A. A., Loschenov V. B. Laser spectroscopy technique for estimating effectiveness of photosensitizers in biological media. Kvant. Elektron. 2006. V. 36 (6). P. 562-568.

Akcakir O., Therrien J., Belomoin G., Barry N., Muller J.D., Gratton E., Nayfeh M. Detection of luminescent single ultrasmall silicon nanoparticles using fluctuation correlation spectroscopy. Appl. Phys. Lett. 2000. V. 76. P. 1857–1859.

Bayliss S. C., Heald R., Fletcher D. I., Buckberry L.D. The culture of mammalian cells on nanostructured silicon. Adv. Mater. 1999. V. 11. P. 318–321.

Belomoin G., Therrien J., Nayfeh M. Oxide and hydrogen capped ultrasmall blue luminescent Si nanoparticles. Appl. Phys. Lett. 2000. V. 77. P. 779–781.

Bley R.A., Kauzlarich S.M. A Low-temperature solution phase route for the synthesis of silicon nanoclusters. J. Am. Chem. Soc. 1996. V. 118. P. 12461–12462.

Borsella E., Botti S., Cremona M., Martelli S., Montereali R.M., Nesterenko A. Photoluminescence from oxidised Si nanoparticles produced by CW CO_2 laser synthesis in a continuous-flow reactor. J. Mater. Sci. Lett. 1997. V. 16. P. 221–223.

Canham L. T. Bioactive silicon structure fabrication through nanoetching techniques. Adv. Mater. 1995. V. 7. P. 1033–1037.

Canham L. T. Silicon quantum wire array fabrication by electrochemical and chemical dissolution of wafers. Appl. Phys. Lett. 1990. V. 57. P. 1046–1048.

Choi J. Silicon Nanocrystals: biocompatible fluorescent nanolabel: PhD thesis. — University of Maryland, USA. 2008.

Choi J., Wang N. S., Reipa V. Conjugation of the photoluminescent silicon nanoparticles to streptavidin. Bioconj. Chem. 2008. V. 19. P. 680–685.

Choi J., Zhang Q, Reipa V., Wang N. S., Stratmeyer M. E., Hitchins. V.M., Goering P. L. Comparison of cytotoxic and inflammatory responses of photoluminescent silicon nanoparticles with silicon micron-sized particles in RAW264.7 macrophages. J. Appl. Toxicology. 2008. V. 29. P. 52–60.

Cunin F., Schmedakes T.A., Link J. R., Li Y. Y., Koh J., Bhatia S.N., Sailor M. J. Biomolecular screening with encoded porous-silicon photonic crystals. Nature Mater. 2002. V. 1. P. 39–41.

Dahan M., Laurence T., Pinaud F., Chemla D. S., Alivisatos A. P., Sauer M., Weiss S. Time-gated biological imaging by use of colloidal quantum dots. Opt. Lett. 2001. V. 26. P. 825–827.

Erogbogbo F., Yong K.-T., Roy I., Xu G., Prasad P.N., Swihart M. T. Biocompatible luminescent silicon quantum dots for imaging of cancer cells. ACS Nano. 2008. V. 2. P. 873–878.

Falconieri M., D'Amato R., Fabbri F., Carpanese M., Borsella E. Two-photon excitation of luminescence in pyrolytic silicon nanocrystals. Physica E. 2008. V. 41, №6. P. 951–954.

Godefroo S., Hayne M., Jivanescu M., Stesmans A., Zacharias M., Lebedev O. L., Van Tendeloo G., Moshchalkov V.V. Classification and control of the origin of photoluminescence from Si nanocrystals. Nature Nanotech. 2008. V. 3. P. 174–178.

He Y., Kang Z.-H., Li Q.-S., Tsang C.H.A., Fan C.-H., Lee S.-T. Ultrastable, highly fluorescent, and waterdispersed silicon-based nanospheres as cellular probes. Angew. Chem., Int. Ed. 2009. V. 121. P. 134–138.

Heath J. R. A liquid-solution-phase synthesis of crystalline silicon. Science. 1992. V. 258. P. 1131–1133.

Heinrich J. L., Curtis C. L., Credo G.M., Kavanagh K. L., Sailor M. J. Luminescent colloidal silicon suspensions from porous silicon. Science. 1992. V. 255. P. 66–68.

Herlin-Boime N., Jursikova K., Trave E., Borsella E., Guillois O., Fabbri F., Vicens J., Reynaud C. Laser-grown silicon nanoparticles and photoluminescence properties. Mater. Res. Soc. Symp. Proc. 2004. P. 818.-M13.4.1.

Hua F., Erogbogbo F., Swihart M. T., Ruckenstein E. Organically capped silicon nanoparticles with blue photoluminescence prepared by hydrosylilation followed by oxidation. Langmuir. 2006. V. 22. P. 4363–4370.

Hua F., Swihart M. T., Ruckenstein E. Efficient surface grafting of luminescent silicon quantum dots by photoinitiated hydrosilylation. Langmuir. 2005. V. 21. P. 6054–6062.

Huisken F., Ledoux G., Guillois O., Reynaud C. Light emitting silicon nanocrystals from laser pyrolysis. Adv. Mater. 2002. V. 14. P. 1861–1864.

Jin Y., Kannan S., Wu M., Zhao J.X. Toxicity of luminescent silica nanoparticles to living cells. Chem. Res.Toxicol. 2007. V. 20. P. 1126–1133.

Jung K.H., Shih S., Hsieh T. Y., Kwong D. L., Lin T. L. Intense photoluminescence from laterally anodized porous Si. Appl. Phys. Lett. 1991. V. 59. P. 3264–3266.

Kang Z., Liu Y., Tsang C.H.A., Ma D.D.D., Fan X., Wong N.-B., Lee S.-T. Water soluble silicon quantum dots with wavelength-tunable photoluminescence. Adv. Mater. 2009. V. 21. P. 661–664.

Kim S., Bawendi M.G. Oligomeric Ligands for luminescent and stable nanocrystal quantum dots. J. Am. Chem. Soc. 2003. V. 125. P. 146–52.

Kim S., Lim Y. T., Soltesz E.G., De Grand A.M., Lee J., Nakayama A., Parker J.A., Mihaljevic T., Laurence R.G., Dor D.M., Cohn L.H., Bawendi M.G., Frangioni J. V. Near-infrared fluorescent type II quantum dots for sentinel lymph node mapping. Nature

Biotech. 2003. V. 22. P. 93–97.

Klostranec J.K., Chen W. C. Quantum dots in biological and biomedical research: recent progress and present challenges. Adv. Mater. 2006. V. 18. P. 1953–1964.

Li X., He Y., Talukdar S. S., Swihart M. T. Process for preparing macroscopic quantities of brightly photoluminescent silicon nanoparticles with emission spanning the visible spectrum. Langmuir. 2003. V. 19. P. 8490–8496.

Li X., He Y., Swihart M. T. Surface functionalization of silicon nanoparticles produced by laser-driven pyrolysis of silane followed by HF_HNO3 etching. Langmuir. 2004. V. 20. P. 4720–4727.

Li Z. F., Ruckenstein E. Water-soluble poly(acrylic acid) grafted luminescent silicon nanoparticles and their use as fluorescent biological staining labels. Nano Lett. 2004. V. 4. P. 1463–1467.

Li Z. F., Swihart M. T., Ruckenstein E. Luminescent silicon nanoparticles capped by conductive polyaniline through the self-assembly method. Langmuir. 2004. V. 20. P. 1963–1971.

Mangolini L., Kortshagen U. Plasma-assisted synthesis of silicon nanocrystal inks. Adv. Mater. 2007. V. 19. P. 2513–2519.

Park J.-H., Gu L., von Maltzahn G., Ruoslahti E., Bhatia S.N., Sailor M. J. Biodegradable luminescent porous silicon nanoparticles for in vivo applications. Nature materials. 2009. V. 8, No. 4. P. 331–336.

Piryutko M.M. The solubility of silicic acid in salt solutions. Russ. Chem. Bull. 1959. V. 8. P. 355–360.

Rogozina E.V., Eckhoff D.A., Gratton E., Braun P.V. Carboxyl functionalization of ultrasmall luminescent silicon nanoparticles through thermal hydrosilylation. J. Mater. Chem. 2006. V. 16. P. 1421–1431.

Salonen J., Kaukonen A.M., Hirvonen J., Lehto V.-P. Mesoporous silicon in drug delivery applications. J. Pharm. Sci. 2008. V. 97. P. 632–653.

Sato S., Swihart M. T. Propionic-acid-terminated silicon nanoparticles: synthesis and optical characterization. Chem. Mater. 2006. V. 18. P. 4083–4088.

Stratonnikov A.A., Meerovich G.A., Ryabova A.V., Saveleva T.A., Loshchenov V. B. Application of backward diffuse reflection spectroscopy for monitoring the state of tissues in photodynamic therapy. Quantum Electron. 2006. V. 36 (12). P. 1103–1110.

Sudeep P.K., Page Z., Emrick T. PEGylated silicon nanoparticles: synthesis and characterization. Chem. Commun. 2008. Iss. 46. P. 6126–6127.

Tilley R.D., Yamamoto K. The microemulsion synthesis of hydrophobic and hydrophilic silicon nanocrystals. Adv. Mater. 2006. V. 18. P. 2053–2056.

Wang L., Reipa V., Blastic J. Silicon nanoparticles as luminescent label to DNA. Bioconjugate Chem. 2004. V. 15. P. 409–412.

Wang L., Reipa V., Blasic J. Silicon nanoparticles as a luminescent label to DNA. Bioconjugate Chem. 2004. V. 15. P. 409–412.

Warner J.H., Hoshino A., Yamamoto K., Tilley R.D. Water soluble photoluminescent silicon quantum dots. Angew. Chem. (Int. Ed.). 2005. V. 44. P. 4550–4554.

Weissleder R. A clearer vision for in vivo imaging. Nature Biotech. 2001. V. 19. P. 316–317.

Wilson W. L., Szajowski P. E. Quantum Confinement in Size-Selected, Surface-Oxidized silicon nanocrystals. Science. 1993. V. 262. P. 1242–1244.

Yamani Z., Ashhabab S., Nayfeh A., Thompson W.H., Nayfeh M. Red to green rainbow photoluminescence from unoxidized silicon nanocrystallites. J. Appl. Phys. 1998. V. 83. P. 3929–3931.

Zou J., et al. Solution synthesis of ultrastable luminescent siloxane-coated silicon nanoparticles. Nano Lett. 2004. V. 4. P. 1181–1186.

13

Porous silicon as a photosensitizer of generation of singlet oxygen

V.Yu. Timoshenko[1]

Porous silicon (P-Si), composed of silicon nanocrystals (nc-Si) with characteristic dimensions of 1–5 nm, has the property of photosensitization of singlet oxygen – an excited state of molecular oxygen, involved in many important biochemical processes. Photosensitivity of generation is observed both in the atmosphere of gaseous oxygen and in various liquid media saturated with molecular oxygen. The mechanism of photosensitivity is a direct electron exchange between the photoexcited nc-Si and the oxygen molecules adsorbed on the surface of nc-Si. The quantitative characteristics and features of this process are studied using spectroscopy, photoluminescence and electron paramagnetic resonance. The results obtained and also the results of experiments *in vitro* indicate the possibility of using P-Si in the photodynamic therapy of cancer.

13.1. Introduction

It is known that singlet oxygen ($^1O_2$), which is an electronically excited state of molecular oxygen, is involved in many important biochemical processes in living nature, including those in the reactions with aromatic compounds, fats, proteins, nucleic acids, vitamins, etc. (Arnold et al, 1968). As examples of such reactions one can mention photohemolysis and photodynamic therapy (PDT). The latter is a combination of processes of accumulation of drug–photosensitizer $^1O_2$ and its photoexcitation for the destruction of tumors, including cancers (Moser 1998).

In accordance with the laws of quantum physics, the excitation of oxygen molecules in the singlet state and the process of reversed relaxation to the ground (triplet) state are forbidden by the selection rules on the back, so that $^1O_2$ is a long-lived excited state of oxygen molecules (Turro, 1991). This creates favorable conditions for energy transfer from $^1O_2$ molecules to other molecules, their complexes or solids in the process of relaxation

[1]Department of Physics, M.V. Lomonosov Moscow State University.

in the ground state. The energy transmitted in this case can be used for the activation of numerous chemical and biochemical reactions, which is used in various applications of singlet oxygen (Kearns 1971).

One of the most important biomedical applications of $^1O_2$ is PDT which uses the phenomenon of photosensitized generation (PSG) of this active form of oxygen. PSG requires for its occurrence a special substance – photosensitizer (PS), effectively absorbing light and capable of transmitting photoexcitation energy to oxygen molecules. Examples of PS include organic dyes, namely, porphyrins, chlorins, methylene blue and others (Moser 1998).

Recently it was found that silicon nanocrystals (nc-Si) in layers or powders of the so-called porous silicon (P-Si), formed by electrochemical etching of crystalline silicon, can act as efficient PS of generation of $^1O_2$ (Kovalev et al, 2002). The observed effect was explained by the resonant process of energy transfer of excitons in nc-Si with the size of 1–5 nm to the molecules of O_2, adsorbed on the surface of the nanocrystals (Gross et al, 2003). Such a process is effective because of the large specific surface area (10^2–10^3 m^2/g) of small nanocrystals (Bisi et al, 2000), and long enough lifetimes of excitons $\sim 10^{-5}$–10^{-3} s. (Kovalev et al, 1999). The process of energy transfer manifests itself as a strong quenching of exciton photoluminescence (PL) of P-Si, which can be used to analyze the process of PSG of $^1O_2$ (Gross et al, 2003). At the same time, the process of excitonic PL quenching is, in essence, indirect indication of the generation of $^1O_2$ and often does not allow for quantitative analysis of the effectiveness of the process. A direct method for diagnosing the appearance of $^1O_2$ molecules is the registration of the luminescence (phosphorescence) at the wavelength of 1270 nm, which corresponds to a photon energy of 0.98 eV (Kearns 1971). The above fluorescent diagnosis was successfully applied to study quantum the effectiveness of FSH in the dye solutions $^1O_2$ (Krasnovsky 1998) and powders and suspensions of nc-Si, obtained from P-Si (Fujii et al, 2004). Despite the fact that quenching of exciton PL of nc-Si indicates a high efficiency of the PSG (Kovalev & Fujii 2005), comparison of data on the generation of molecular oxygen PS (porphyrins), introduced in layers of P-Si, indicates a strong deactivation (quenching) of $^1O_2$ in a developed porous structure of the material (Chirvony et al, 2006), which significantly reduces the concentration of molecules of the formed $^1O_2$ and limits the possibility of using this material as a PS (Gongal'skii et al, 2010).

Another physical method for measuring the concentration of $^1O_2$ is the electronic paramagnetic resonance (EPR), which uses the paramagnetism of $^1O_2$ molecules due to the contribution of the orbital motion of the atoms (Kearns 1971). The photosensitized generation of P-Si is studied by EPR in the so-called X-band ($\sim$9.5 GHz), which are clearly visible paramagnetic centres such as unsaturated ('dangling') bonds of silicon atoms interacting with $^1O_2$ molecules (Konstantinova et al, 2006, Konstantinova et al, 2008). The EPR spectroscopy of the Q-band (33–34 GHz) allows us to directly observe the absorption lines associated with transitions in the $^1O_2$ molecule which can be used to measure the concentration of the latter in the gas phase in PSG in P-Si (Konstantinova et al, 2008).

The quenching of the exciton PL of nc-Si obtained by mechanical dispersing P-Si and associated with PSG of $^1O_2$ was observed in various organic liquids (Fujii et al, 2004) and in water (Fujii et al, 2005), which indicates the possibility of biomedical applications. Indeed, *in vitro* experiments demonstrate significant suppression of the fission process (proliferation) of cancer cells in the presence of photoexcited nc-Si (Timoshenko et al 2006).

In the following we discuss the fundamental aspects of physical and chemical properties of molecular oxygen and silicon nanocrystals, necessary to understand the photosensitized generation of singlet oxygen, and also the practical use of this phenomenon.

13.2. The electronic configuration of the oxygen molecule

Oxygen is one of the key chemical elements on the planet with 21% in the atmosphere, 89% in sea water (by weight) and 47% in the earth's shell. Almost all living organisms on the Earth use oxygen for chemical reactions that take place with the release of energy, or for breathing. In 1840, Michael Faraday discovered that oxygen is magnetic. Almost a century later, Robert Millikan explained the reason for the magnetic properties of oxygen, using the advanced at the quantum approach. It was found that the oxygen molecule in its ground triplet ($^3\Sigma$) energy state has two unpaired electrons. The inner magnetism of the electron was predicted Uhlenbeck and Goudsmit one year earlier. The presence of unpaired valence electrons in the molecule in its ground state is a rare property that determines the characteristics of its chemical activity, since the beginning of the reaction with triplet oxygen usually requires additional expenditure of energy (heat). Consequently, in the triplet state the oxygen molecule has paramagnetic properties and has a relatively weak chemical activity. However, the oxygen molecule may be in an excited – singlet state, characterized by valence electrons with antiparallel spins belonging to different ($^1\Sigma$) or one ($^1\Delta$) atoms.

The possible existence of singlet oxygen was first demonstrated experimentally in 1931 by Hans Kautsky, an assistant professor at the Heidelberg University. He conducted experiments with trypaflavine dye adsorbed on silica gel. Kautsky showed that the presence of oxygen quenches and saturates the fluorescence of trypaflavine. In the same system, a colorless dye of malachite Green, also adsorbed on the surface of silica gel, restored its bright green colour. Kautsky assumed that energy is transferred from the optically excited trypaflavine to oxygen molecules, which in its turn colours the colourless dye to bright green color.

The electronic configuration of the oxygen atom is as follows:

$$O \Rightarrow (1s)\uparrow\downarrow (2s)\uparrow\downarrow (2p_x)\uparrow (2p_y)\uparrow (2p_z)\uparrow\downarrow .$$

The LCAO model (linear combination of atomic orbitals) provides the following picture of the filling of molecular orbitals (MOs) in the ground state:

$$O_2 \Rightarrow (1\sigma_g)^2(1\sigma_u)^2(2\sigma_g)^2(2\sigma_u)^2(3\sigma_g)^2(1\pi_u)^4(1\pi_g)^2.$$

The electronic properties of oxygen are determined by six electrons in π MO. The configuration of the spins in the ground state is as follows:

$$O_2 \Rightarrow (\text{core})(\pi_x)\uparrow\downarrow(\pi_y)\uparrow\downarrow(\pi*_x)\uparrow(\pi*_y)\uparrow.$$

The spectroscopic term for this condition is $^3\Sigma_g^-$. Unpaired electrons belonging to two different MOs are responsible for the paramagnetism of oxygen molecules. The electronic filling of the ground state and first excited state is shown in Fig. 13.1.

In the excited state, which is a singlet, the two paired electrons occupy the same π_g MO. O_2 $(^1\Delta_g^+)$ – the lowest excited state, which is chemically known as $^1O_2$. The transition energy in this state is equal to 0.98 eV (22.5 kcal/mol) and the lifetime is 45 min at low pressures of oxygen. However, collisions with other molecules cut this time to 14 min at a pressure of 760 Torr. The upper excited state is $^1\Sigma_g^+$, in which two paired electrons occupy two different π_g MO. The energy of the transition to this state is 1.63 eV (37.5 kcal/mol) and time life is 7 seconds. States $(^1\Delta_g^+)$ and $^1\Sigma_g^+$ can be observed in the gas phase, only in small amounts. This is possible due to the fact that the oxygen molecule can pass from one state to another. The scheme of this process is shown below:

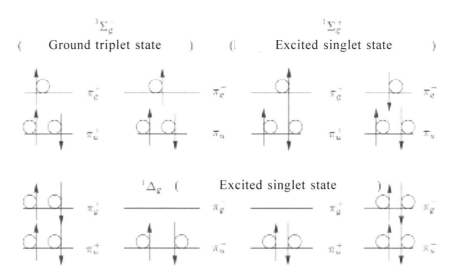

Fig. 13.1. Energy levels of the oxygen molecule.

$$O_2(^1\Delta_g^*) + O_2(^1\Delta_g^*) \Rightarrow O_2(^1\Sigma_g^+) + O_2(^3\Sigma_g^+);$$
$$O_2(^1\Sigma_g^+) + Q + M \Rightarrow O_2(^1\Delta_g^*) + Q + M,$$

where Q is the donor molecule, and M is the third body, introduced for the implementation of the law of conservation of energy and momentum.

13.3. Photoluminescent diagnosis of singlet oxygen with its generation in porous silicon

As noted in the introduction, characteristic changes were recently observed in the PL spectra when illuminating P-Si in an oxygen atmosphere (Kovalev et al, 2002). It was found that in oxygen there is a strong quenching of photoluminescence of P-Si with hydrogen coating of the surface. The degree of quenching was dependent on temperature and partial pressure of oxygen. The observed effect was attributed to photosensitization of singlet oxygen in excitation energy transfer from excitons in the nc-Si to oxygen molecules.

The essence of the observed effect is seen from Fig. 13.2, which presents the results of measurements at low temperatures. Curve *1* shows the PL spectrum of the layer of P-Si, placed in a vacuum at 50 K. The large difference in the shape and size of silicon nanocrystals causes a considerable width of the PL band. Letting molecular oxygen into the cryostat leads to a significant suppression of the intensity of PL (curve *2*). The strongest suppression of the spectrum is observed at a photon energy of 1.63 eV

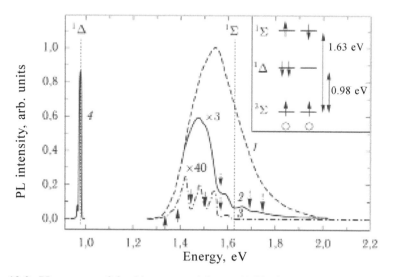

Fig. 13.2. PL spectra of freshly prepared layers P-Si: *1* – *T* = 50 K, the sample in a vacuum; *2* – *T* = 50 K, oxygen pressure 10^{-2} mbar, *3* – *T* = 5 K, the oxygen pressure 10^{-4} mbar; *4* – the emission corresponding to $^1\Sigma$–$^3\Sigma$ transition. Insert – the energy level scheme of molecular oxygen (Kovalev et al, 2002).

(indicated by the dotted vertical line), which coincides with the energy excitation of one $^1\Sigma$ state. The local minima in the spectrum above this energy (marked by arrows) are located at equal distances on the energy scale, they are removed from the $^1\Sigma$ states at integer intervals approximately equal to 62 meV. Further reduction in temperature to 5 K (curve 3) leads to the complete quenching of PL at energies above ~1.62 eV and the appearance of additional extrema in the fine structure of the spectrum indicated by the arrows. They are observed even when at the exciting photon energies below 1.63 eV and are separated from the $^1\Delta$ state by the integer number of intervals, each of which is ~62 meV. Thus, the supply of oxygen on the surface of the sample is characterized by quenching of the PL spectrum with two specific values of the photon energy inherent in the process of activation of oxygen (0.98 eV and 1.63 eV).

All the features of the PL spectrum favour the generation of singlet oxygen on the surface of porous silicon nanocrystals. Indeed, direct evidence is the detection of light emitted in the process of relaxation of singlet oxygen in the triplet (ground) state $^3\Sigma$. In view of the short lifetime of the state $^1\Sigma$ the direct experimental observation of $^1\Sigma - {}^3\Sigma$ transition is not possible. However, it was found that the quenching of PL is always accompanied by the appearance of a narrow line at 0.98 eV (curve 4 in Fig. 13.2), which is the result of $^1\Delta-{}^3\Sigma$ radiative transitions in a molecule of oxygen. Consequently, there is an unambiguous proof of the formation of singlet oxygen.

Extra dissipation of energy is essential for non-resonant energy transfer. In silicon, this process is accompanied by the emission of transverse optical phonons, which is most effective when the energy gap of the silicon nanocrystal is equal to the excitation energy of oxygen plus the energy of a number of transverse optical phonons. This results in the presence of local lows in the PL spectrum. In addition, for the radiative recombination of excitons it is necessary to compensate the quasi-momentum associated with the indirect-band structure of bulk silicon. Compensation is achieved by emission of phonons with an energy of ~56 meV (the local minima are marked by arrows below the energy of the $^1\Sigma$ state).

Figure 13.3 shows the spectral dependence of the PL signal quenching in an oxygen environment. These dependences determine the efficiency of energy transfer from excitons to oxygen molecules. To exclude the component associated with the size of silicon nanocrystals, the differential spectral dependence, which were obtained by dividing the PL spectra obtained in a vacuum by the PL spectra, obtained in an oxygen atmosphere, were analyzed. At low temperatures the spectral features (arrows), which relate to the contributions of TO phonons of silicon to the energy transfer process, become evident at all temperatures, the strongest PL quenching is observed at energies coinciding with the energy of the $^1\Sigma$ state. Quenching can be very strong at temperatures below 100 K, even at low oxygen pressures. At these temperatures the oxygen completely covers the surface

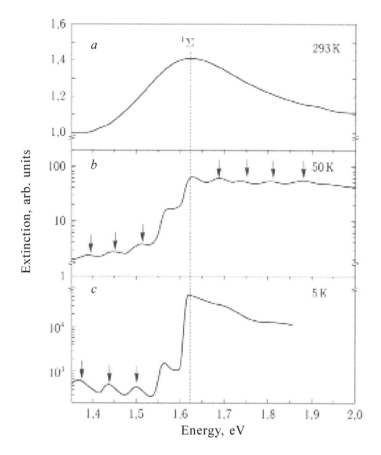

Fig. 13.3. The spectral dependence of the PL quenching for hydrogen-passivated samples: $a - T = 293$ K, oxygen pressure of 1 bar, and $b - T = 50$ K, oxygen pressure 10^{-2} mbar; $c - T = 5$ K, oxygen pressure 10^{-4} mbar. The contribution from the transverse optical (TO) phonons shown by arrows (Kovalev et al, 2002).

of the nanocrystals, whereas at room temperature the interaction of silicon and oxygen occurs only as a result of random collisions.

Gross et al, (2003) showed experimentally that the energy transfer from the exciton to molecular oxygen is carried out by a direct mechanism of electron exchange. Simultaneous photoexcited electron hopping to the oxygen molecule and the compensation of the hole remaining in the nanocrystal, i.e. annihilation of a triplet exciton and the excitation of the molecule with a change in the spin are allowed processes. For brevity this process will be called energy transfer. For electronic exchange the distance between the exciton and the physisorbed oxygen molecule must be small enough so that the probability of electron tunnelling is sufficient for this process. Experiments were performed on the PL spectra for freshly prepared samples of P-Si with the Si–H coating and for strongly oxidized samples with a surface monolayer consisting of Si–O bonds (length of the Si–O bond in 3 times more Si–H bonds). It was found that for strongly oxidized samples PL quenching is about three orders of magnitude smaller than for the freshly prepared samples. This can be explained by the fact that the electron tunnelling probability depends exponentially on the distance between the interacting objects in accordance with the mechanism of direct electronic exchange. Thus, the spectroscopic evidence of efficient energy transfer from excitons in Si nanocrystals to molecular oxygen, adsorbed on their surface, was obtained.

13.4. Photosensitization of singlet oxygen in powders of porous silicon and aqueous suspensions

13.4.1. Photosensitization of singlet oxygen in the gas phase

In this section we consider the possibility of using P-Si as a photosensitizer to generate singlet oxygen. Figure 13.4 shows the measured (at room temperature) PL spectra of samples of P-Si in a vacuum and in an oxygen atmosphere (Ryabchikov et al, 2007).

After the inlet oxygen at a pressure of 760 Torr, a decrease of PL intensity in comparison with its value in vacuum (curves *1* and *2*) was observed, and the integral difference in the PL spectra was 70%. The maximum PL quenching is observed at the same time in the wavelength range 760 nm (1.63 eV photon energy), which corresponds to energy of the $^3\Sigma \rightarrow {}^1\Sigma$ transition in the molecule of O_2. The PL spectrum of P-Si after oxygen pumping to a pressure of 10^{-4} Torr is represented by curve *3*. One can see a partial recovery of the PL signal up to 60% of its initial value,

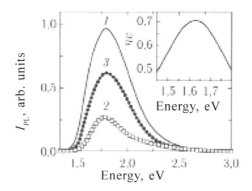

Fig. 13.4. The PL spectra of P-Si in a vacuum (*1*), in an oxygen atmosphere at a pressure of 760 Torr (*2*) and subsequent evacuation (*3*). The inset shows the spectral function of the effectiveness of energy transfer from excitons in nc-Si to oxygen molecules (Ryabchikov et al, 2007).

which can be associated with the formation of non-radiative recombination centres at partial oxidation of the samples. Therefore, the decrease in the intensity of the PL signal can be explained by the competition of two mechanisms: non-radiative recombination at surface defects, which in most cases owe their existence to silicon dangling bonds, and the transfer of the energy of the excitons to the molecules of O_2, adsorbed on the surface of nc-Si.

The inset of Fig. 13.4 shows the spectral function of the efficiency of energy transfer from excitons in nc-Si to the oxygen molecules adsorbed on their surface. This function is given by (Ryabchikov et al, 2007):

$$\eta_r = 1 - \frac{I_{PL}^{ox}}{I_{PL}^{vac}},\tag{13.1}$$

where I_{PL}^{ox}/I_{PL}^{vac} is the ratio of PL intensity of P-Si in an oxygen atmosphere to its value in vacuum. The maximum value of this function can be used to determine what proportion of energy is transferred to the oxygen molecules.

The maximum spectral function of efficiency of energy transfer is located at the wavelength of 760 nm, which is equal to the energy difference between the ground and excited states of O_2 molecules (1.63 eV). Accounting for the contribution of non-radiative recombination at the surface of silicon nanocrystals can be made by comparing PL spectra of P-Si in vacuum prior to generation of singlet oxygen and after evacuation of oxygen. As our analysis shows, in the range 730–885 nm (1.4–1.7 eV) the ratio of curves 1 and 3 corresponds to 1.5–1.6, but in addition increases at lower and higher photon energies. The latter fact suggests an additional non-radiative recombination in small and large nanocrystals. In the region of energy transfer to the oxygen molecule (1.6 eV) that is of interest to us, we can estimate the contribution of the process of non-radiative recombination at defects to the quenching of the PL as 30–40%.

The change of PL with the suply of oxygen molecules can also be obtained from the analysis of PL kinetics. An important characteristic of excitons is the value of their lifetime, which can be obtained by approximating the experimental curves with the formula:

$$I = I_0 \cdot \exp\left(-\frac{t}{\tau}\right),\tag{13.2}$$

where I is the PL intensity at time t, I_0 is the intensity of the initial point in time. τ is the lifetime of the excitons. Moreover, the lifetime of the excitons in silicon nanocrystals increases with their size (Delerue et al, 1999). The increase in the number of defects during the adsorption of oxygen molecules leads to a decrease in the lifetime of the excitons, as these defects can act as centres of non-radiative recombination either by the Shockley–Read mechanism or by energy transfer by the Förster–Dexter mechanism.

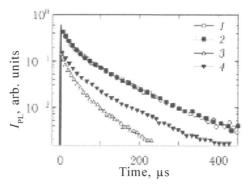

Fig. 13.5. PL kinetics of powder P-Si: *1*– in an atmosphere of oxygen P_{O_2} = 1 Torr, *2* – after the evacuation P_{O_2} = 1 Torr, *3* – in oxygen P_{O_2} = 200 Torr, *4* – vacuum after P_{O_2} = 200 Torr (Ryabchikov et al, 2007).

Figure 13.5 shows typical PL kinetics of freshly prepared powder P-Si, taken at room temperature at a wavelength of 760 nm (the energy 1.63 eV photons) at a pressure of oxygen in the cell with the sample P_{O_2} = 1 and 200 Torr, curves *1* and *3*, respectively. Curves *2* and *4* related to the evacuated cells with P-Si after adsorption of oxygen at pressures P_{O_2} = 1 Torr, and P_{O_2} = 200 Torr, respectively. By analyzing changes in the kinetics of the PL, we can estimate the lifetime of excitons in nc-Si.

As can be seen from the curves shown in Fig. 13.5 and 13.6, the admission of oxygen with P_{O_2} = 760 Torr in a cell with the sample decreases the lifetime of excitons 1.3–2 times. This may be due to both the process of energy transfer of oxygen molecules, and the fact that in the excitation non-equilibrium charge carriers in the samples of P-Si various processes of their recombination can take place. To determine the nature of the observed effect, the lifetime of the excitons at several wavelengths: 730, 760, 880 nm, was estimated

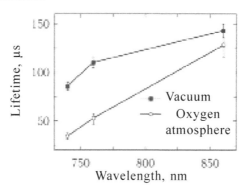

Fig. 13.6. The dependence of the exciton lifetime on the wavelength for II-Si in vacuum and in an oxygen atmosphere (oxygen pressure in the cell with the sample 760 Torr).

These results suggest that the lifetimes of excitons decrease with the adsorption of oxygen molecules. Excitons with less energies recombine at a lower rate, which agrees well with the data obtained from the kinetics of the PL in vacuum. A significant decrease in the lifetime of excitons is observed at a wavelength of 760 nm. The energy of excitons at the given wavelength coincides with the energy necessary for the transfer of the oxygen molecule to the excited state (singlet) state.

The formation of singlet oxygen may be accompanied by the formation of silicon–oxygen bonds on the surface of silicon nanocrystals which results in a deterioration in the photoluminescence properties of the samples. Confirmation of validity of this assumption can be obtained by comparing the PL spectra in the vacuum and immediately after the evacuation of oxygen. This comparison suggests that part of the oxygen molecules can form oxide complexes of silicon nanocrystals not on the surface.

Data on changes in time, including their not fully reversible change after the evacuation of oxygen, are consistent with the analysis of the PL spectra (Kovalev et al, 2004). They show both the effective transfer energy of oxygen and the appearance of non-radiative recombination centres at the surface of the silicon nanocrystals, associated with the oxidation process. At short exposure time oxide complexes may be island-shaped which does not lead to a substantial change in the characteristics of the PL. The spectral dependence of the efficiency of energy transfer can also be used to conclude that in the process of generation of singlet oxygen may be accompanied by the formation complex oxygen complexes such as superoxide O_2^- and dimers. The latter are a combination of two oxygen molecules that are in the singlet state, related to each other through van der Waals forces. Formation of these structures is observed mainly at pressures over atmospheric oxygen (Gross et al, 2003). Superoxide O_2^- is also capable of receiving energy from the excitons and this process can be detected using photoluminescence. However, due to transmission of energy to the superoxide, the maximum of the spectral dependence of the quenching function will be shifted towards higher energies, and the peak will appear at the wavelength range 590–620 nm, which corresponds to the range of photon energies 2–2.1 eV.

13.4.2. EPR diagnostics of photosensitization of singlet oxygen

In addition to the influence of oxygen on the PL of P-Si, we investigated the influence of oxygen on EPR spectra of P-Si. Figure 13.7 shows typical EPR spectra of microporous silicon in vacuum and oxygen at incident power on a sample of microwave radiation of 200 mW. For the chosen value of the intensity of the microwave radiation the EPR signal from the P_b-centres in the vacuum is in the saturation regime (Konstantinova et al, 2006, 2007, 2008). Note that the spin relaxation of dangling bonds of

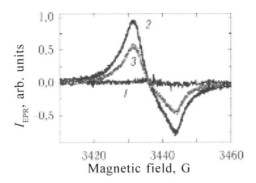

Fig. 13.7. The EPR spectrum of P-Si: *1* – vacuum 10^{-6} Torr, *2* – in oxygen with no lighting, *3* – in an oxygen atmosphere under illumination (Konstantinova et al, 2006).

silicon from the excited state to the ground state when the samples of P-Si are in vacuum, takes place by a low-intensity electron–phonon mechanism through the transfer of excitation energy to the lattice phonons by the spin-orbital interaction. Direct spin–spin interaction of P_b-centres has almost no place, since one P_b-paramagnetic centre has on average an estimated 100 nanocrystals of microporous silicon.

As is known from the literature data, triplet oxygen molecules physisorbed on a solid surface, can effectively dissipate the energy of the excited paramagnetic centres, causing their rapid relaxation to the ground state (Kiselev and Krylov, 1978). This process is realized through magnetic dipole–dipole interaction between the spins of silicon dangling bond ($S_e = 1/2$) and triplet oxygen molecules $^3O_2$ ($S_{TO} = 1$). As a result, the relaxation time of the P_b-centres in the presence of paramagnetic molecules $^3O_2$ decreases compared to vacuum. Thus, the process of absorption of microwave energy in the admission of oxygen, as can be seen from Fig. 13.7, is more intense: the EPR signal amplitude of P-Si in oxygen is 1.5 times greater than in a vacuum.

At low incident microwave radiation power on the sample the saturation effect does not hold. Indeed, the probability W of the resonance induced transition per unit time between the Zeeman energy levels is so low, that even with 'slow' electron–phonon relaxation of the P_b-centre its characteristic lifetime in the excited state is less than the characteristic time of absorption of microwave quanta. Thus, there is no appreciable change in the amplitude of the EPR signal of P-Si during pumping of oxygen.

Illumination of P-Si layers with the light with a photon energy equal to or exceeding the energy of the band gap of nc-Si, forming the sample, is accompanied by the formation of excitons with the binding energy greater than the thermal energy kT (Kovalev et al, 1999). Part of the excitons with the annihilation energy near 1.63 eV recombines with the resonance transfer of energy to $^3O_2$ molecules through direct electronic exchange, with the result that they transfer to the singlet state (Kovalev et al, 1999). Thus, the concentration of $^3O_2$ molecules is reduced, and the process of dipole–dipole relaxation of the spin centres is less efficient (increased characteristic relaxation times of the P_b-centres). As a consequence, the EPR signal is saturated and its amplitude decreases (Konstantinova et al, 2006, 2008). At a low power of the microwave radiation incident on the sample the saturation effect is absent, which leads to the coincidence of spectra for micro-P-Si in oxygen with and without illumination in vacuum.

To determine the selectivity of ensembles of silicon nanocrystals to the type of environment, EPR spectra were recorded for P-Si, placed in a nitrogen atmosphere; the nitrogen molecules are diamagnetic. It was found that the signals in the absence and presence of light are the same, and in the parameters the lines resemble the spectra of samples of microporous silicon in a vacuum at the same values of microwave radiation power. Indeed, the nitrogen molecule in its ground state has the full electron spin equal zero, and thus the magnetic interaction with the spin defects in P-Si is absent, and the outer channel of relaxation is suppressed. In this case, the electron–phonon relaxation mechanism is dominant and this is equivalent to the sample being in a vacuum. Thus, any other factors affecting the decrease in the amplitude of the EPR spectra in oxygen under illumination are excluded, i.e. the presence of triplet oxygen molecules is a crucial paramagnetic relaxation factor.

Note that the effect of photosensitization of singlet oxygen in the layers of P-Si is partially reversible if defect formation under illumination is kept to a minimum. Using data from the EPR spectra of the micro-P-Si in an oxygen atmosphere with and without illumination, we can estimate the concentration of the formed singlet oxygen. The initial concentration of triplet oxygen is calculated as the ratio of Avogadro's number to the molar volume, and at a pressure of 760 Torr is 10^{19} cm^{-3}. From the EPR spectra of micro-P-Si in an oxygen atmosphere it can be seen that illumination decreases the intensity of the ESR signal by about 40% compared with the case in the dark. Thus, considering the initial concentration of triplet oxygen, we find that the concentration of singlet oxygen at a pressure of 760 Torr is $4 \cdot 10^{18}$ cm^{-3}.

13.4.3. Quantitative analysis of the photosensitization of singlet oxygen

The efficiency of photosensitization of singlet oxygen also depends on the properties of silicon nanocrystals. These properties may be the chemical composition and surface morphology, the number of defects in the bulk and on the surface of the samples. However, the most important parameter of P-Si is the porosity. This parameter, in contrast to the composition and morphology of the surface, can be easily varied in the preparation of the samples. In connection with this, Kovalev et al (1999) evaluated the energy transfer time and the efficiency of generation of singlet oxygen, depending on the porosity of the samples. The energy transfer time was estimated excluding non-radiative recombination at defects. This is due to the fact that the observed lifetimes of PL samples in vacuum are close to the natural radiative lifetimes (50–100 ms) of excitons in nc-Si. At the same time, as follows from the analysis of time and PL spectra of the samples (see above), the maximum contribution to photoluminescence quenching of the excitons due to non-radiative recombination of excitons at the surface defects can be 30–40%. Therefore, our assessments of time and efficiency of photosensitivity are approximate. The energy transfer time in this approximation has been calculated by the following formula:

$$\frac{1}{\tau_{tr}} = \frac{1}{\tau} - \frac{1}{\tau_0}, \tag{13.3}$$

where τ_0 is the exciton lifetime obtained from the analysis of the kinetics of the PL signal of P-Si in a vacuum, τ is the lifetime of the excitons, calculated from the kinetics of the PL of P-Si in an oxygen atmosphere. The efficiency of energy transfer from excitons in silicon nanocrystals to oxygen molecules was assessed in relation to the area under the kinetics of PL of P-Si in a vacuum and under the kinetics of the PL of P-Si in an oxygen atmosphere as follows:

$$\eta_J = 1 - \frac{\int I_{PL}^{ox} dt}{\int I_{PL}^{vac} dt}, \tag{13.4}$$

where I_{PL}^{vac} is the PL signal intensity in a vacuum; I_{PL}^{ox} is the PL intensity in an oxygen atmosphere. The results of calculating the transfer efficiency and energy transfer time can be seen in Fig. 13.8.

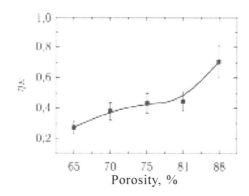

Fig. 13.8. The efficiency of generation of singlet oxygen, depending on the porosity of P-Si samples (Ryabchikov et al, 2007).

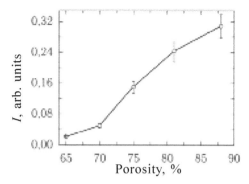

Fig. 13.9. The dependence of the intensity of the PL signal on the porosity of the samples of P-Si, used for the photosensitized generation of singlet oxygen (Ryabchikov et al, 2007).

Figure 13.8 shows that the effectiveness of photosensitization increases with increasing porosity of the sample. This can be explained by an increase in the quantum yield of PL of P-Si with increasing porosity. This is confirmed by the fact that in the highly porous samples the PL intensity increases (Fig. 13.9). The number of adsorbed oxygen molecules can also increase due to the larger specific surface area of the sample. The estimates of the energy transfer time confirm the high efficiency of singlet oxygen in highly porous samples of P-Si.

The efficiency of generation of singlet oxygen may be affected by factors such as the number of defects on the surface of silicon nanocrystals because the defects formed in the as-prepared samples can cause non-radiative recombination of excitons, or may play the role of trapping centres for non-equilibrium electrons and holes. To study this factor, investigations were carried out on P-Si samples which, after preparation, were exposed to the natural process of oxidation in air. The process of forming a natural oxide should lead to a decrease in the number of defects and dangling bonds on the surface. As a result of this process, the probability of non-radiative recombination of excitons in such systems should decrease, which leads to an increase in the signal intensity of the PL of P-Si and an increase of the efficiency of generation of singlet oxygen.

As already stated, the condition of the surface of silicon nanocrystal ensembles and, as a consequence, the efficiency of generation of singlet oxygen are strongly dependent on the oxygen pressure. To clarify the features of the adsorption mechanism, the dependence of the quenching function on the oxygen pressure in the cell with the sample was studied.

Figure 13.10 shows the experimental dependence of the efficiency of generation of singlet oxygen on the pressure of the molecules obtained for the freshly prepared P-Si powder. The presented dependence makes it possible to estimate the intensity of energy transfer from excitons to oxygen

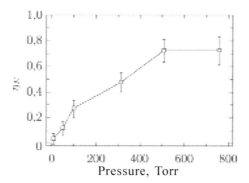

Fig. 13.10. The dependence of the efficiency of generation of singlet oxygen molecules on the pressure oxygen for powder P-Si. (Ryabchikov et al, 2010).

molecules. As can be seen from Fig. 13.10, the energy transfer process is most effective at high oxygen pressures.

The intensity of the PL signal when the pressure changes from 10 to 760 Torr according to a non-linear law, decreasing by a factor of 4. This is because with increasing pressure in the cell with the sample the concentration of oxygen molecules, adsorbed on the surface of the silicon nanocrystals, also increases. It can be assumed that the degree of suppression, which characterizes the intensity of the generation of singlet oxygen, is proportional to the number of molecules adsorbed on the surface of the sample. However, as shown in Fig. 13.10, this dependence saturates at an oxygen pressure of 500 Torr. This could mean that at a given pressure the entire surface is covered with a monolayer of oxygen molecules.

Using the data on oxygen adsorption on graphite, Zarifyants, et al (1962) obtained the dependence of the concentration of adsorbed oxygen on the number of molecules impinging on the sample surface. The number of

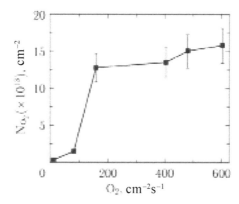

Fig. 13.11. The dependence of the number of singlet oxygen molecules, adsorbed on the surface of silicon nanocrystals, on the number of molecules that strike the surface of the sample.

molecules interacting with the surface of the sample was calculated from the equation (Kiselev, Krylov, 1978):

$$\pi = P_{O_2} \cdot N_A \cdot (2\pi MRT)^{-1/2}, \tag{13.5}$$

where M is the molecular weight of O_2, R is the universal gas constant, T is temperature, N_A is Avogadro's number. The results obtained (see Fig. 13.11) showed that at the number of oxygen molecules interacting with the silicon surface equal to $n = 160$ cm$^{-2} \cdot$ s^{-1}, the concentration of singlet oxygen molecules, adsorbed on the surface, reaches $N_{Ad} = 10^{14}$ cm^{-2}.

A further increase in the number of oxygen molecules that interact with the surface does not lead to a substantial increase in the concentration of adsorbed molecules. Based on these results, we can conclude that at $N_{Ad} = 1.3 \cdot 10^{14}$ cm^{-2} the surface of silicon nanocrystals is completely covered by a monolayer of molecules. However, a rigorous approach to saturation is not observed. With a further increase in the oxygen concentration in the cell with the sample the number of adsorbed oxygen molecules continues to increase. For example, for $n = 600$ cm^{-2} s^{-1}, the concentration of adsorbed molecules of singlet oxygen is $N_{Ad} = 1.6 \cdot 10^{14}$ cm^{-2}. This could mean that at the oxygen pressures close to atmospheric, the process of formation of the monolayer coating can be disrupted, and the process of monomolecular adsorption smoothly changes to multimolecular that may be accompanied by the formation of superoxide radicals O_2^- and dimers on the surface of the Si nanocrystals.

The above-mentioned high efficiency of the generation rate of singlet oxygen on the surface of P-Si does not necessarily mean its greater concentration in the ambient gas phase, as it will also depend on the rate of non-radiative deactivation of the molecules. The latter may significantly increase on the surface of nc-Si due to a complex fractal structure of P-Si (Gongalsky et al, 2010).

13.4.4. Photosensitization of singlet oxygen in liquids

A number of studies demonstrated the possibility of photosensitized generation of singlet oxygen with P-Si, dispersed in various organic and inorganic liquids. The standard approach for this is the registration of changes of optical transmittance of the molecules–labels under the action of singlet oxygen. In (Fujii, Usui et al, 2004) the authors found changes in the absorption spectra of 1,3-diphenylisobenzofuran (DPBF), dissolved in benzene, as a result of interaction with the photoexcited particles of powder P-Si with a concentration of 2 mg/ml (Fig. 13.12).

Fujii, Minobe et al (2004) were able to observe the generation of $^1O_2$ by its luminescence with a photon energy of 0.975 eV on the background of a broad PL band of P-Si dispersed in C_6F_6 (Fig. 13.13). This was possible because of the long lifetime of singlet oxygen ($\tau_\Delta = 25$ ms) in this substance.

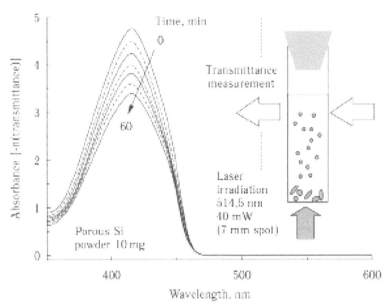

Fig. 13.12. The absorption spectra of DPBF, dissolved in benzene, together with the dispersed particles ofP-Si, at different times of photoexcitation of the latter. The wavelength and excitation power were respectively 514.5 nm and 40 mW at a spot diameter of 7 mm. The inset shows a schematic of the experiment (Fujii, Usui et al, 2004).

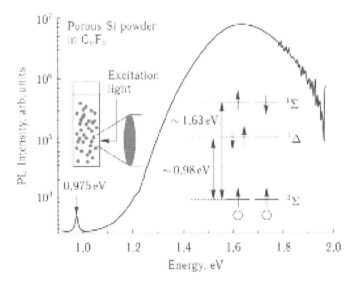

Fig. 13.13. The luminescence spectra of P-Si, dispersed in C_6F_6, at room temperature. The line with the energy of 0.975 eV corresponds to luminescence of singlet oxygen. Inserts show the experimental scheme and the scheme of electronic levels of O_2 molecules (Fujii, Minobe et al, 2004).

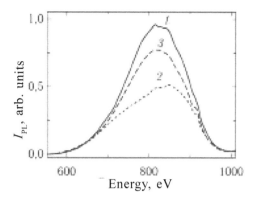

Fig. 13.14. PL spectra of aqueous suspensions based on silicon nanocrystals saturated with molecular oxygen: *1* – pressure in the cell 10 Torr, *2* – in an oxygen atmosphere P_{O_2} = 760 Torr, *3* – subsequent pumping oxygen to a pressure of 10 Torr.

For practical applications of the process of generation of singlet oxygen, it is advisable to use aqueous suspensions of nc-Si. Therefore, the study of the photosensitization of generation of singlet oxygen was carried out in aqueous suspensions, which had been prepared on the basis of porous silicon. Suspensions were prepared by mixing the P-Si powder in distilled water in the ratio: 1 ml of water per 5 mg of powder. An ultrasonic bath was used to obtain homogeneous suspensions based on the P-Si powders.

Figure 13.14 shows the resultant PL spectra of the P-Si suspension prepared with the above parameters. Despite a marked decrease in the intensity of the PL signal of aqueous suspensions of silicon nanocrystals in comparison with the case of the powder P-Si, the generation of singlet oxygen in suspensions was also observed. The PL spectra were used to calculate the spectral function of the efficiency of energy transfer from excitons in nc-Si to oxygen molecules in the case of aqueous suspensions of silicon nanocrystals, which is represented in Fig. 13.15. The comparative analysis of the intensity ratio of PL with no dissolved oxygen and oxygen-saturated suspensions showed that the efficiency of energy transfer to oxygen molecules is about 40%. Thus, in aqueous suspensions it is also possible to generate singlet oxygen, but with less efficiency.

The estimation of the lifetime of excitons in porous silicon–water systems, obtained from the PL kinetics, showed that the admission of oxygen into the cell with the suspension results in a decrease in the lifetime of excitons by a factor of 1.5. The process of generation of singlet oxygen in the P-Si–water systems also leads to the beginning of irreversible adsorption processes at the semiconductor surface. This is confirmed by the difference between the PL spectra, recorded with no dissolved oxygen (Fig. 13.14, curve *1*) and after the evacuation of oxygen (curve *3*). The discrepancy between the spectra may be indicative of surface oxidation of

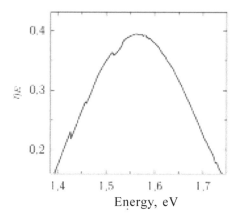

Fig. 13.15. The spectral function of the efficiency of energy transfer from excitons in nc-Si to molecules of oxygen to the aqueous suspension of P-Si.

silicon nanocrystals. The intensity of the generation of singlet oxygen in a colloidal solution also depends on the concentration of the powder P-Si.

Thus, the ratio of the intensities of the PL signal for the powder P-Si in the absence of oxygen and in the oxygen atmosphere at increasing silicon concentration to 10 mg per 1 ml changes from 1.6 to 1.9. This fact confirms that these systems may be used in future in producing singlet oxygen for biomedical purposes.

13.5. Biomedical aspects of photosensitized generation of singlet oxygen in porous silicon

Photoexcited silicon nanocrystals have been used to suppress the multiplication of cancer cells *in vitro* (Timoshenko et al, 2006). Figure 13.16 shows the dependence of the number of fibroblast cells of mice on the concentration of nc-Si, normalized to the value in the control series where no nc-Si was added. It can be seen that at the concentration of nc-Si in the nutrient solution of ~0.5 g/l and above there is a significant reduction in the number of cancer cells after irradiation. At a concentration of 2.5 g/l the death of around 80% of the cells was recorded. At the same time, in the dark nc-Si had virtually no effect in the entire concentration range. Consequently, we can conclude that the suppression of reproduction of cancer cells is caused by the impact of singlet oxygen produced at photo-excitation of nc-Si. The analysis of DNA of cells exposed to the effect of photoexcited nc-Si showed (Kudryavtsev et al 2006) that after illumination in the presence of nc-Si to a concentration of more than 0.1–0.5 g/l they 'die' by the apoptosis mechanism, i.e. 'programmed death' (Steller, 1995) and the concentration dependence of the number of remaining cells is close to that shown in Fig. 13.16. Note that to use the observed effect in practice

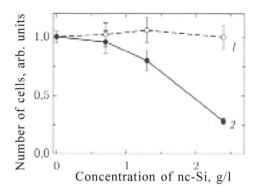

Fig. 13.16. Dependence of the number of cells on the concentration of nc-Si after holding for 1 h in the dark (*1*) or under illumination (*2*) defined by changes of the optical density and normalized to values in the control group, where nc-Si was not added (Timoshenko et al, 2006).

it is necessary to carry out *in vivo* experiments, which can give more precise information about the effectiveness of P-Si as a photosensitizer for PDT.

Thus, the results obtained in the photosensitization of generation of singlet oxygen in the powders and suspensions of nc-Si show the high efficiency of this process and the possibility of its use for suppression of the reproduction of cancer cells. The latter can obviously be used in photodynamic therapy of cancer.

References for Chapter 13

Arnold S. J., Kubo M., Ogryzlo E.A. Oxidation of organic compounds. Advan. Chem. Ser. 1968. V. 77. P. 133–142.

Belyakov L.V., Vainshtein Yu. S., Goryachev D.N., Sreseli O.M. The cruical role of singlet oxygen in the formation of photoluminescence from nanoporous silicon. Semiconductors. 2009. V. 43 (10). P. 1347–1350.

Bisi O., Ossicini S., Pavesi L. Porous silicon: a quantum sponge structure for silicon based optoelectronics. Surf. Sci. Rep. 2000. V. 38. P. 1–126.

Bruggeman D.A.G. Berechnung verschiedener physikalisher Konstanten von heterogen Substanzen. Annalen der Physik. 1935. V. 24. P. 636–664.

Calcott P.D. J., Nash K. J., Canham L. T. Kane M.J.& Brumhead D. Identification of radiative transitions in highly porous silicon. J. Phys., Condens. Matter. 1993. V. 5. P. L91–L98.

Canham L. T. Nanoscale semiconducting silicon as a nutritional food additive. Nanotechnology. 2007. V. 18. P. 85704 (6 p.).

Cantin J. L., Schoisswohl M., von Bardeleben H. J., Zoubir N.H., Vergnat M. Electron-paramagnetic-resonance study of the microscopic structure of the Si(001)-SiO₂ interface. Phys. Rev. B. 1995. V. 52. P. R11599–R11602.

Chirvony V., Bolotin V., Matveeva E., Parkhutik V. Fluorescence and $_1O_2$ generation properties of porphyrin molecules immobilized in oxidized nano-porous silicon matrix. J. Photochem., Photobilogy: A Chemistry. 2006. V. 181. P. 106–113.

Chirvony V., Chyrvonaya A., Ovejero J., Matveeva E., Goller B., Kovalev D., Huygens A., de Witte P. Surfactant-modified hydrophilic nanostructured porous silicon for the photosensitized formation of singlet oxygen in water. Advanced Materials. 2007. V. 19. P. 2967–2972.

Cullis A.G., Canham L. T., Calcott P.D. J. The structural and luminescence properties of porous silicon. J. Appl. Phys. 1997. V. 82. P. 909–965.

Delerue C., Allan G., Lannoo M. Optical band gap of Si nanoclusters. J. Luminescence. 1999. V. 80. P. 65–73.

Delerue C., Lannoo M., Allan G., Martin E., Mihalcescu I., Vial J. C., Romestain R., M¨uller F., Bsiesy A. Auger and coulomb charging effects in semiconductor nanocrystallites. Phys. Rev. Lett. 1995. V. 75. P. 2228–2231.

Delley B., Steigmeier E. F. Size dependence of band gaps in silicon nanostructures. Appl. Phys. Lett. 1995. V. 67. P. 2370–2372.

Dexter D. L. A theory of sensitized luminescence in solids. J. Chem. Phys. 1953. V. 21. P. 836–850.

Dolmans D. E. J.G. J., Fukumura D., Jain R.K. Photodynamic therapy for cancer. Nature Reviews. Cancer. 2003. V. 3. P. 380–386.

Egorov S. Yu., Kamalov V. F., Koroteev N. I., Krasnovsky A.A. Jr., Toleutaev B.N., Zinukov S.V. Rise and decay kinetics of photosensitized singlet oxygen luminescence in water. Measurements with nanosecond time-correlated single photon counting technique. Chem. Phys. Lett. 1989. V. 163, №4–5. P. 421–424.

F¨orster T. Zwischenmolekulare Energiewanderung und Fluoreszenz. Ann. der Phys. (Leipzig). 1948. V. 2. P. 55–75. (In German).

Fujii M., Kovalev D., Goller B., Minobe S., Hayashi Sh., Timoshenko V. Yu. Time-resolved photoluminescence studies of the energy transfer from excitons confined in Si nanocrystals to oxygen molecules. Phys. Rev. B. 2005. V. 72. P. 165321 (8 p.).

Fujii M., Minobe S., Usui M., Hayashi Sh., Gross E., Diener J., Kovalev D. Singlet oxygen formation by porous Si in solution. Phys. Rev. B. 2004. V. 70. P. 085311.

Fujii M., Usui M., Hayashi Sh., Gross E., Kovalev D., Knzner N., Diener J., Timoshenko V. Yu. Chemical reaction mediated by excited states of Si nanocrystals – singlet oxygen formation in solution. J. Appl. Phys. 2004. V. 95. P. 3689–3693.

Fujii M., Usui M., Hayashi Sh., Gross E., Kovalev D., Künzner N., Diener J., Timoshenko V. Yu. Singlet oxygen formation by porous Si in solution. Phys. Stat. Sol. (a). 2005. V. 202. P. 1385–1389.

Gongal'skii M.B., Konstantinova E.A., Osminkina L.A., Timoshenko V. Yu. Detection of singletoxygen in photoexcited porous silicon nanocrystals by photoluminescence measurements. Semiconductors. 2010. V. 44, No. 1. P. 1–4.

Gross E., Kovalev D., Künzner N., Koch F., Timoshenko V. Yu., Fujii M. Spectrally resolved electronic energy transfer from silicon nanocrystals to molecular oxygen mediated by direct electron exchange. Phys. Rev. B. 2003. V. 68. P. 115405-1-11.

Grossweiner L. I., Fernandez J.M., Bilgin M.D. Photosensitisation of red blood cell haemolysis by photodynamic agents. Lasers Med. Sci. 1998. V. 13. P. 42–54.

Halliwell B., John M. C. Free Radical in Biology and Medicine. 2 ed. — Oxford: Clarwndon Press, 1982.

Harper J., Sailor M. J. Photoluminescence quenching and the photochemical oxidation of porous silicon by molecular oxygen. Langmuir. 1997. V. 13. P. 4652–4658.

John G. C., Singh V.A. Porous silicon: theoretical studies. Phys. Rep. 1995. V. 263. P. 93–

151.

Kautsky H. Quenching of luminescence by oxygen. Trans. Faraday Soc. 1939. V. 35. P. 216–219.

Kearns D. R. Physical and chemical properties of singlet molecular oxygen. Chem. Rev. 1971. V. 71. P. 395–427.

Konstantinova E.A., Demin V.A., Timoshenko V. Yu. Investigation of the generation of singlet oxygen in ensembles of photoexcited silicon nanocrystals by electron paramagnetic resonance spectroscopy. J. Exper. Theoret. Physics. 2008. V. 107. P. 473–481;

Konstantinova E.A., Demin V.A., Vorontzov A. S., Ryabchikov Yu.V., Belogorokhov I.A., Osminkina L.A., Forsh P.A., Kashkarov P.K., Timoshenko V. Yu. Electron-paramagnetic resonance and photoluminescence study of Si nanocrystals-photosensitizers of singlet oxygen molecules. J. Non-Cryst. Solids. 2006. V. 352. P. 1156–1159.

Kovalev D., Fujii M. Silicon nanocrystals: photosensitizers for oxygen molecules. Advanced Materials. 2005. V. 17 (21). P. 2531–2544.

Kovalev D., Gross E., Diener J., Timoshenko V. Yu., Fujii M. Photodegradation of porous silicon induced by photogenerated singlet oxygen molecules. Appl. Phys. Lett. 2004. V. 85. P. 3590–3593.

Kovalev D., Gross E., K¨unzner N., Koch F., Timoshenko V. Yu., Fujii M. Resonant electronic energy transfer from excitons confined in silicon nanocrystals to oxygen molecules. Phys. Rev. Lett. 2002. V. 89. P. 137401 (4 p.).

Kovalev D., Heckler H., Polisski G., Koch F. Optical properties of Si nanocrystals. Phys. Stat. Sol. (b). 1999. V. 215. P. 871–932.

Krasnovsky A.A. Jr. Singlet molecular oxygen in photobiochemical systems: IR phosphorescence studies. Membr. Cell Biol. 1998. V. 12, №5. P. 665–690.

Krasnovsky A.A. Jr., Egorov S. Yu., Nasarova O.V., Yartsev B. I., Ponamarev G.V. Photosensitized formation of singlet molecular oxygen in solutions of water soluble phorphyrins. Direct luminescent measurements. Studia Biophysica. 1988. V. 124, No. 2–3. P. 123–142.

Krasnovsky A.A. Jr., Roumbal Ya. V., Ivanov A.V., Ambartzumian R.V. Solvent dependence of the steady-state rate of $_1O_2$generation upon excitation of dissolved oxygen by cw 1267 nm laser radiation in air-saturated solutions: Estimates of the absorbance and molar absorption coefficients of oxygen at the excitation wavelength. Chem. Phys. Lett. 2006. V. 430. P. 260–264.

Krasnovsky A.A. Jr., Drozdova N.N., Roumbal Ya.V., Ivanov A.,V., Ambartzumian R.V. Biophotonics of molecular oxygen: activation efficiencies upon direct and photosensitized excitation. Chinese Optics Lett. 2005. V. 3. P. S1–S4.

Kueng W., Silber E., Eppenberger U. Quantification of cells cultured on 96-well plates. Anal. Biochem. 1989. V. 182. P. 16–19.

Langmuir I. The constitution and fundamental properties of solids and liquids. Part I. Solids. J. Am. Chem. Soc. 1916. V. 38. P. 2221–2295.

Langmuir I. J. The constitution and fundamental properties of solids and liquids. Part I. Solids. J. Am. Chem. Soc. 1916. V. 38. P. 2221–2295.

Lepine D. J. Spin-dependent recombination on silicon surface. Phys. Rev. B. 1972. V. 6. P. 436–441.

Moser J.G. Photodynamic tumor therapy: 2nd and 3rd Generation Photosensitizers. Amsterdam: Harwood Academic Publishers, 1998.

Palenov D.A., Zhigunov D.M., Shalygina O.A., Kashkarov P.K., Timoshenko V. Yu. Specific features of dissipation of electronic excitations energy in coupled molecular-solid systems based on silicon nanocrystals on intense optical pumping. Semiconductors. 2006. V. 41. P. 1351–1355.

Park J.H., Gu L., von Maltzahn G., Ruoslahti E., Bhatia S.N., Sailor M. J. Biodegradable luminescent porous silicon nanoparticles for in vivo applications. Nature Materials. 2009. V. 8. P. 331–336.

Parkhutik V., Chirvony V., Matveeva E. Optical properties of porphyrin molecules immobilized in nano-porous silicon. Biomolecular Engineering. 2007. V. 24. P. 71–73.

Pastor E., Balaguer M., Bychto L., Salonen J., Lehto V.-P., Matveeva E., Chirvony V. Porous silicon for photosensitized formation of singlet oxygen in water and in simulated body fluid: Two methods of modifucation by undecylenic acid. J. Nanosci. Nanotechnology. 2008. V. 8. No. 12. P. 1–7.

Rioux D., Laferriere M., Douplik A., Shah D., Lilge L., Kabashin A., Meunier M.M. Silicon nanoparticles produced by femtosecond laser ablation in water as novel contamination-free photosensitizers. J. Biomedical Optics. 2009. V. 14, №2. P. 021010-1-4.

Ryabchikov Yu.V., Belogorokhov I.A., Vorontsov A. S., Osminkina L.A., Timoshenko V. Yu., Kashkarov P.K. Dependence of the singlet oxygen photosensitization efficiency on morphology of porous silicon. Phys. Stat. Sol. (a). 2007. V. 204. P. 1271–1275.

Sanders G.D., Chang Y.-C. Theory of optical properties of quantum wires in porous silicon // Phys. Rev. B. 1992. V. 45. P. 9202–9213.

Schmidt R., Brauer H.-D. Radiationless deactivation of singlet oxygen (Delta) by solvent molecules. J. Am. Chem. Soc. 1987. V. 109. P. 6976–6981.

Snyder J.W., Skovsen E., Lambert J.D. C., Ogilby P.R. Subcellular, time-resolved studies of singlet oxygen in single cells. J. Am. Chem. Soc. 2005. V. 127. P. 14558–14559.

Steller H. Mechanisms and genes of cellular suicide. Science. 1995. V. 267. P. 1445–1449.

Timoshenko V. Yu., Osminkina L.A., Vorontzov A. S., Ryabchikov Yu.V., Gongalsky M.B., Efimova A. I., Konstantinova E.A., Bazylenko T. Yu., Kashkarov P.K., Kudryavtsev A.A. Silicon nanocrystals as efficient photosensitizers of singlet oxygen for biomedical applications. Proc. of SPIE. 2007. V. 6606. P. 66061E-1-4.

Turro N. J. Modern Molecular Photochemistry. University Science, Sausalito, CA, 1991.

Uhlir A. Electrolytic shaping of germanium and silicon. Bell Syst. Tech. 1956. V. 35. P. 333–347.

Wasserman H.H. Introductory remarks. In: Singlet Oxygen. V. 40 / Ed. by H.H. Wasserman. Academic Press, 1979.

Young R.H., Wehrly K., Martin R. L. Solvent effects in dye sensitized photooxidation reactions. J. Am. Chem. Soc. 1971. V. 93. P. 5774–5779.

Zarifyants Yu. A., Kiselev V. F. Fedorov G.G. Differential heat of adsorption of oxygen and water vapor on the surface of graphite. Dokl. Akad Nauk SSSR. 1962. V. 144. P. 151.

Kiselev V. F., Krylov O. V. Adsorption processes on the surface of semiconductors and dielectrics. Moscow: Nauka, 1978.

Konstantinova E. A., Demin V. A., Timoshenko V. Yu., Kashkarov P. K. EPR diagnostics of photosensitized generation of singlet oxygen on the surface of the Si nanocrystals. Pis'ma Zh. Eksp. Teor. Fiz. 2007. V. 85 (1). P. 65-68.

Kudryavtsev A. A., Lavrovskaya V. P., Osminkina L. A., Vorontsova A. S., Timoshenko V. Yu. Phototoxity of silicon nanocrystals. Fiz. Meditsina. 2006. V. 16, No. 2. P. 4-8.

Timoshenko V. Yu, et al. Silicon nanocrystals as photosensitizers of active oxygen for biomedical applications. Pis'ma Zh. Eksp. Teor. Fiz. V. 83, No. 9. P. 492-495.

14

Sunscreens

A.A. Ischenko, AA Krutikova[1]

In modern society, tan is a sign of health and attractiveness, but in terms of medicine the sunburn is a defensive reaction on the skin to the damage caused by ultraviolet rays. The main objective of the sunburn is to prevent further damage that can lead to dangerous changes in the skin. Acute effects of ultraviolet radiation are manifested in the form of skin redness (solar erythema) and/or burn. Chronic exposure may lead to weakening of the immune system (immunosuppressive effect), skin cancer (Urbach, 1993; Glanz, 2007), benign abnormalities of melanocytes (freckles, melanocytic nevi, solar and senile lentigo), photoaging is the result of chronic exposure of keratinocytes, blood vessels and fibrous tissue (Diffey, 2002). Sunburn is a risk factor for both melanoma skin cancer and for non-melanoma skin cancer (Green & Battistutta, 1990; Marks & Whiteman, 1994). To prevent the harmful effects of solar radiation it is necessary to avoid being under the sun from 10:00 to 16:00 of the day, wear clothing to protect from the sun and use sunscreen cosmetics products. According to the research we can say that sunscreen creams reduce the number of non-melanoma skin cancers provoked by UV radiation on the skin of animals (Gurish et al, 1981), and reduce burns, caused by UV radiation on the skin of people (Pathak et al, 1985).

Sunscreens are complex compositions of substances, which together provide effective protection of the human skin from solar radiation. The main components of sunscreen use organic and inorganic sunscreens or a combination thereof. Organic sunscreens absorb ultraviolet radiation due to the presence of chromophore groups in their structure, and the action of inorganic sunscreen filters is based on the physical mechanism of protection: on the phenomena of scattering, absorption and reflection of ultraviolet radiation by inorganic particles. Nanocrystalline silicon attracts interest as a promising new sunscreen agent. Its feature is the high coefficient of absorption in the ultraviolet region with transparency in the visible region. This combination of physical properties of the silicon nanoparticles with

---

[1]M.V. Lomonosov Moscow State University of Fine Chemical Technologies (MITKhT).

their biological compatibility allows to develop highly effective sunscreens that preserve the aesthetics when applied to light skin.

The efficiency and quality of sunscreens is dependent on many factors. A variety of known sunscreens filter allows us to find the optimal composition to create the most effective sunscreen. In this chapter we discuss the prospects of nc-Si to create effective sunscreens.

14.1. Solar radiation and its effect on the skin

Approximately 50% of solar energy reaching the Earth's surface in the form of visible light, 40% in the form of infrared radiation, and only a fraction of 10% – in the form of ultraviolet radiation (see Fig. 14.1a). The intensity of sunlight is maximum in the green light range (wavelength ≈500 nm) and decreases rapidly with decreasing wavelength in the ultraviolet range. The ultraviolet range is conventionally divided into three areas by the degree of impact on the human skin (Mckinlay & Deffey, 1987):
- rays with wavelengths of 200–290 nm (UV-C);
- rays with wavelengths of 290–320 nm (UV-B);
- rays with wavelengths of 320–400 nm (UV-A).

The Earth's atmosphere tends to absorb almost 100% of UV-C, 90% of UV-B and approximately 10% of UV-A radiation (Moloney et al, 2002).

Various changes can take place in the human body under the influence of solar radiation: redness of the skin, age spots and damage to DNA, which triggers the development of cancer.

Figure 14.1b shows the spectra of the damaging effect of light on the skin and DNA cells. It is evident that DNA is most severely damaged

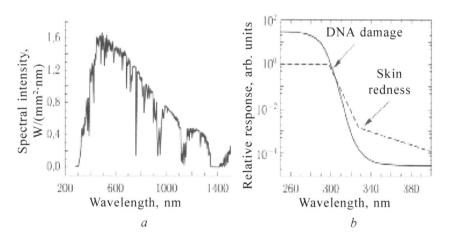

Fig. 14.1. The emission spectrum of the sun on the Earth's surface (Gueymard et al, 2002) (a) and spectrum of the action of light on skin cells and DNA (Roelandts, 1998; Setlow, 1974) (b).

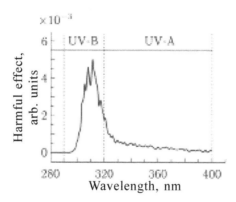

Fig. 14.2. The spectrum of the damaging effect of sunlight on the skin.

when exposed to sunlight light with a wavelength of 260–320 nm. Thus, under the action of ultraviolet light pyrimidine dimers (Sutherland, 1985) and 8-hydroxy-2′-deoxyguanosine form in the cells, and mutations in the gene p53, is responsible for resistance to the growth of malignant tumors, vimentin (Applegate et al, 1997; Bernerd et al, 2000; Liardet et al, 2001; Gelis & Al, 2003) take place.

In the 400 nm range solar radiation mainly causes reddening of the skin. Multiplying the spectra shown in Fig. 14.1*a* and 14.1*b*, we obtain an integral relationship to the damaging effect of each wavelength (Fig. 14.2). As can be seen from Fig. 14.2, the most dangerous zone occurs in the region 290–320 nm (UV-B region). So sunscreens should protect the skin mainly from the rays in this range of the spectrum.

The human skin can be divided into three layers: the stratum corneum, epidermis and dermis. Figure 14.3 schematically shows the amount of UV radiation that passes through each layer of the skin (Marzulli & Maibach, 1983).

UV-C (UVC) rays with the shortest wavelengths have the highest energy and are most dangerous. However, almost all the solar rays in this range are arrested in the atmosphere.

UV-B (UVB) rays reach the Earth's surface. In human skin, they penetrate into the epidermis, but do not reach the dermis. UV-B radiation is most active during the summer. The rays of this region have a strong damaging effect and cause erythema, edema, pigment changes, hyperplasia of the keratinocytes. UV-B radiation in the wavelength region of the spectrum leads to photocarcinogenesis, photoaging, and immunosuppression (Young, 1990). The effect of UV-B radiation leads to sunburn cells and mutations p53.

UV-A (UVA) rays have the lowest energy, but at the same time have the highest penetrating power. In human skin, UV-A rays pass through the epidermis and reach the middle layers of the dermis (Fig. 14.3). The dermis contains melanocytes which produce the pigment melanin, responsible for

Wavelength, nm

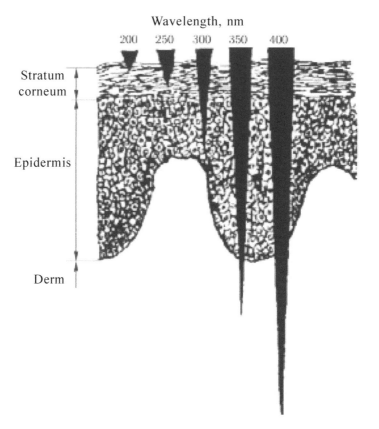

Fig. 14.3. Schematic representation of the penetration of UV radiation in the skin.

skin colour. Due to the formation of melanin tan appears in human skin to protect the skin from sunburn. UV-A radiation can cause photoallergy, phototoxicity and to a lesser extent photoaging and photocarciogenesis (Young, 1990; Diffey, 1991). UV-A radiation leads to an increase in vimentin (Stanfield et al, 1989) and increased pigmentation of the skin of types 3 and 4 (on a scale of phenotypes proposed by Fitzpatrick) (Diffey & Farr, 1991).

In order to assess the impact of solar radiation, the term minimal erythermal dose (MED) was introduced. It is the energy exposure of solar radiation, causing a subtle erythema of the previously non-irradiated skin (Buka, 2004). The radiation dose required for the development of erythema is very individual. For light skin 1 MED equals 200–300 J/m^2. However, it still depends on the type and individual sensitivity of human skin. The maximum dose that a person receives for a day theoretically is 15 MED.

Features of spectral-structural properties of nanocomposites – dependence of the UV spectrum of the particle size distribution function, chemical composition of the membrane, the degree of crystallinity of the central core – provide opportunities to control the spectral characteristics of the materials obtained with the use of these nanocomposites.

The use of high-productivity plasma chemical technology related to the evaporation of crystalline silicon in a plasma discharge, as well as the technology of monosilane decomposition in the laser field allows us to adjust not only the particle size of synthesized nanocrystalline silicon, but also the chemical composition of the surface layer (see chapter 5), which provides an additional opportunity to control the optical properties of materials derived from nc-Si (Sennikov et al, 2009).

For nanoparticles of 10 nm and above (containing $>10^4$ silicon atoms), absorption features in the UV and visible wavelengths are much determined by the properties of conventional crystalline or amorphous silicon. Silicon nanocomposites silicon, having a diameter of the central core of 5 nm or less, show the quantum size effect (see section 2.7 and chapter 4), providing a significant influence on their optical properties. The use of this effect allows management of the absorption and luminescent properties of different materials, For example, emulsion composite materials with sunscreen properties.

14.2. Using nc-Si as a sunscreen component

Among the objects with modified optical properties nanocrystalline Silicon is the most attractive, due to biocompatibility of this material (see chapter 12). In the works (Beckman, 2003; review article Ischenko, Sviridov, 2006) the authors first proposed the use of an emulsion composite based on nanocrystalline silicon as an effective UV protection cosmetic product in the range 200–400 nm. It was suggested to produce sunscreen composites utilizing the effect of photon absorption of the UV range without re-mission of a photon with other performance characteristics. This effect can be simply implemented in the nc-Si particles, since the change of their size and surface modification enables control of their optical properties. Knowledge of the relationship of the structure of nanoscale particles with their optical properties, in turn, can purposefully create a new technology for sunscreen composites (Ischenko, Sviridova, 2006).

Reducing the size of silicon nanoparticles entails a radical change in the band parameters of the material (Brus, 1986), which in turn leads to change its optical properties (Delerue et al, 1999; Knief & Niessen, 1999, Heitmann et al, 2005, Weng-Ge D., 2011).

For silicon particles smaller than 5 nm in the determination of their properties there are significant quantum size effects, which allow to control the fluorescent (Li et al, 2003, Efremov et al, 2004) and absorption characteristics of materials based on them. Thanks to the shift of the edge of the main absorption of nanocrystalline silicon to the visible and ultraviolet

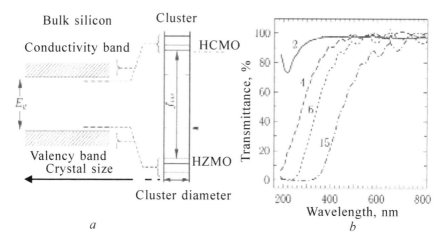

Fig. 14.4. *a* – Schematic diagram of the correlation that relates the band gap of bulk silicon with the electronic states in the nanocrystal, *b* – the transmission spectrum of different segments of the Si/SiO$_2$ films annealed at 1373 K (Soni et al, 1999).

region of the spectrum due to the increase of the band gap, the material of the nc-Si can efficiently absorb the ultraviolet light rays.

Figure 14.4*a* shows a schematic coordinating diagram of electronic states of transition from bulk silicon to nanoclusters. Figure 14.4*b* shows the transmission spectrum of different segments of the Si/SiO$_2$ film annealed at 1373 K. The numbers 2 and 15 belong to the segments, rich in SiO$_2$ and Si, respectively (Soni et al, 1999). With decreasing particle size of nanocrystalline silicon the absorption edge shifts toward lower wavelengths.

From the data presented in (Ranjan et al, 2002), we can estimate the values of the size of silicon nanocrystallites resulting in effective absorption of UV radiation: 2.0 nm – good (absorbance at a wavelength of

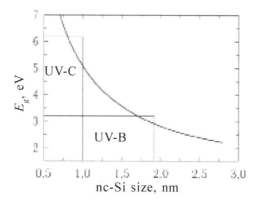

Fig. 14.5. The calculated energy dependence of the electronic transition in nc-Si on the particle size of the nanocrystalline cluster (Ranjan et al, 2002).

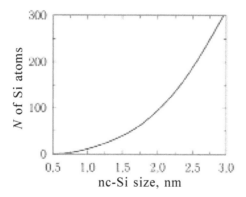

Fig. 14.6. The number of Si atoms in the nanocrystal core, depending on the diameter of nc-Si (Ranjan et al, 2002).

approximately 400 nm, edge of the UV region), 1.8 nm – better, 1.6 – the best size for absorbing the UV-B band of the solar spectrum (see Fig. 14.5).

Thus, selecting the function of the particle size distribution, it is possible to absorb radiation by these particles in the entire UV range. Ranjan et al, (2002) also calculated the number of silicon atoms in the core of the nanocrystal as a function of particle size (Fig. 14.6).

When the number of silicon atoms in the core of the nanocrystal is 165–175, the size of silicon nanoparticles is approximately equal to 2.4 nm, and at 125–135 atoms it is about 2.2 nm. The optimal number of silicon atoms in the nanocrystal core for efficient absorption of UV radiation is 95–100, which corresponds to the size of the nanocrystals of about 2.0 nm.

Ogut et al (1997), theoretically examined the band gap for Si quantum dots depending on the 1–3 mm size range. Later, van Buuren et al (1998) presented 'state-of-the-art' measurements of the edges of the passbands of Si quantum dots with an average diameter from 1 to 5 nm. Summarizing the measured shifts of the conduction band, the valence band and the band gap of bulk silicon, the authors obtained the width the band gap of quantum dots. It turned out that the experimentally determined band gap is much larger compared with the results of theoretical calculations described by Altman et al, (2001). Thus, the available theoretical calculations (van Buuren et al (1998) and experimental observations do not agree. This discrepancy increases with decreasing size of the quantum dots.

For nanoparticles of 10 nm and above (containing about ~10^4 Si atoms) the absorption characteristics in the UV and visible wavelengths are largely determined by the properties of conventional crystalline or amorphous silicon. These characteristics depend on a number of reasons: on the availability of structural defects and impurities, on the phase state and certain other conditions (Knief & Niessen, 1999). By changing the size distribution function of nanoparticles, their concentration in the emulsion, and appropriately modifying the state of their surface, it is possible to target change the spectral characteristics of the nanocomposite material as a whole. It is necessary to seek by changes in the characteristics of the silicon nanocomposite to get the best effect in reducing the transmission at wavelengths shorter than 400 nm (see Altman et al, 2001), which determines the properties of the produced sunscreen creams.

14.3. Modern sunscreens

Ultraviolet radiation is non-uniform and is divided into four regions depending on the effects on human skin: UV-C – 200–290 nm; UV-B – 290–320 nm; UV-A I – 320–380 nm; UV-A II – 380–420 nm. UV filters are sunscreens which are divided into two classes: organic and inorganic (Ischenko, Sviridova, 2006, Sviridova, Ischenko, 2006; Pathak, 1982; Moloney et al, 2002). Depending on the absorption spectrum of the UV – the filters are divided into UV-A (UV-A) and UV-B (UV-B) filters (Liardet & Scaletta, 2001). Principles for the protection of organic and inorganic sunscreens are different. Organic sunscreens are substances containing the chromophore groups to effectively absorb ultraviolet light. Usually there are some vegetable oils, derivatives of benzene, phenol, hinilin derivatives, cinnamates, salicylates, benzophenones, dibenzoylmethanes (Schauder, 1997; Lowe, 1997; Diffey, 1997; Roelandts, 1998; Buka, 2004). All these molecules contain the carbonyl group C=O. The electronic transition $n - \pi^*$ in the carbonyl compounds corresponds to the wavelength, located in the 270 nm range for ketones and close to 290 nm for aldehydes. These substances absorb the harmful short-wave UV radiation (250–340 nm) and convert the remaining energy to safe long-wave radiation (above 380 nm) (Otto, 2003).

Sunscreen materials containing inorganic particles can scatter, absorb or reflect UV radiation. At present, the inorganic sunscreens use fine titanium dioxide particles or zinc oxide (Popov et al, 2005; Pinnell et al, 2000). Previously, it was assumed that the effect of fine particles of TiO_2 and ZnO is based on the pure scattering of UV light. Currently, however, it is believed that the weakening of the UV radiation is the result of combined scattering and absorption of both products. Zinc oxide effectively absorbs almost all UV (380 nm and shorter). Titanium dioxide is mostly an effective UV-B absorber and UV-A scatterer. At the wavelength of 360 nm, approximately 90% of weakening for TiO_2 is due to scattering while for ZnO 90% is due to absorption.

The main physical parameter which characterizes the absorption and scattering of light by materials is the refractive index n. For TiO_2 in the visible region $n = 2.6$. Most cosmetic products are made from components with $n < 1.5$. Therefore it is very difficult to 'hide' TiO_2 in the cosmetic product. Zinc oxide has a lower refractive index, $n = 1.9$. In this regard, it is more attractive for use in sunscreens (Ischenko, Sviridov, 2006).

International Agency for Research on Cancer (IARC) classifies titanium dioxide as a possible human carcinogen – group 2B (Johannes et al, 2006).

In order to effectively use a sunscreen, it is necessary to ensure maximum absorption and scattering in the UV region with minimum absorption and scattering in the visible region. In this case the sun cream will protect against UV light, while remaining transparent on the skin, which is important for its consumer qualities. The optimum particle size taking into account the refractive index of TiO_2 and the UV wavelength range is from 20 to 150 nm (Ischenko, Sviridova, 2006). Preservation of this particle size determines the effectiveness of the material. In natural agglomeration of the particles the material is pigmented, losing efficiency in the UV field. The titanium dioxide pigment has a particle size from 150 to 300 nm with the standard geometric deviation (SGD) to 50 nm. SGD 1.35 gives ~5% for particles <100 nm. In Popov et al. (2005), the authors, using the Mie theory, determined the optimal size of titanium dioxide TiO_2, equal to 62 nm, which is most effective in protecting skin from UV radiation in the wavelength range 307–311 nm. This result is due to absorption, while the scattering does not make a significant contribution. Many studies have shown that when using inorganic filters with the size not less than 20 nm, they do not penetrate the stratum corneum (Durand, 2009; Gontier et al, 2009; Newman, 2009). Particle smaller than 15 nm can already pass through the stratum corneum (Baroli, 2007; Menzel, 2004).

The effectiveness of the sunscreen of titanium dioxide and zinc oxide is also affected by their concentration in the product. By measuring the spectra of diffuse reflection of microcrystalline zinc oxide and microcrystalline titanium dioxide with the concentration of 2% and 6%, it is shown that in comparison with microcrystalline titanium dioxide, zinc oxide absorbs UV radiation in a longer wavelength region of the spectrum (Pinnell et al, 2000; Gelis, 2003).

The sunscreens may be in the form of an emulsion, gel, mousse, spray, hard pencil. But the effectiveness of each of them with the same sunscreen agents will be different. A common parameter for assessing the effectiveness of sunscreens is the sun protection factor SPF (Schulz & Hohenberg, 2002; Neylor & Kevin, 1997) – the ratio of minimal erythemal dose (MED) of radiation, which causes redness, covered with the sunscreen material, to the MED, causing redness of the skin, not treated with sunscreen:

$$SPF - \frac{MED \text{ with UV filter}}{MED \text{ without UV filter}}. \tag{14.1}$$

In determining the SPF, the sunscreen is deposited with a density of 2 mg/cm^2 (Sternberg & Larco, 1985). However, the actual amount of sunscreen that people use in real life is much less – from 0.5 to 1.3 mg/cm^2 (Sternberg & Larco, 1985; Bech-Thomson & Wulf H., 1993). In practice, therefore, consumers use only a quarter or half the thickness of the sunscreen material, which is used in determining the SPF in the laboratory. This means that

Table 14.1. The effectiveness of sunscreens based on TiO$_2$ and ZnO (Tarras-Wahlberg et al, 1999)

TiO$_2$, wt.%	ZnO, wt.%	SPF	SPF ratio UVA/UVB
7.5	0	20	0.59–0.60
0	6	6	0.90
2.5	2.5	12	0.63
2.5	7.5	20	0.81

the consumer skin is protected much less efficiently than indicated on the cream label (Tarras-Wahlberg et al, 1999).

The use of titanium dioxide and zinc oxide with a concentration of 2.5 wt.% each in the mixture leads to SPF = 9, with a ratio of UF-A/UF-B equal to 0.68, which significantly higher than the UF-A/UF-B ratio for pure TiO$_2$, equal to 0.59–0.60.

Table 14.1 shows the effectiveness of sunscreens based on TiO$_2$ and ZnO.

The highest SPF, equal to 18.5 and the UF-A/UF-B ratio of 0.81 with protection in a wide range can be achieved using a mixture of 7.5 wt.% ZnO and 2.5 wt.% TiO$_2$. In addition, to increase the SPF and ensure protection of the skin in a wide UV spectral range, the titanium dioxide and zinc oxide are used in combination with conventional organic UV filters. Usually, the SPF of the composition containing organic and inorganic filter compounds is higher than the arithmetic sum of the SPFs of individual UV filter substances (Lademann, 2005). Consequently, there is a synergic effect between organic and inorganic UV filters (Lademann, 2005).

In the European, American and Japanese industries for sunscreens when evaluating their effectiveness attempts were made to standardize the methodology used, comprising a light source, the method of application, reference samples and the number of trial participants (Neylor & Kevin, 1997). There is a method for determining the SPF *in vitro*, as described in the papers (Diffey, 1991; Lavker, Gerberick & Veres, 1995). The *in vitro* methods are based on measurement of the reduction of the energy of the UV spectrum after passing through a thin film of a sunscreen product applied on the corresponding surface. The measurements are carried out on a special spectrophotometric equipment. An important condition for these methods is that the surface on which the sample is deposited should be similar to the skin surface as much as possible. The basis of calculating the SPF in this group of methods are mathematical techniques that allow relate the true SPF, determined by the *in vivo* method, with the characteristics of the absorption spectrum of the substance.

Another developing area of determining the optical properties of photoprotection means is the fluorescence spectroscopy of the skin (Sinichkin, etc. 2001; Utz et al, 1996; Utz et al, 1996; Sinichkin et al, 2003). There are two groups of methods of studying the skin using fluorescence spectroscopy and related to the *in vivo* methods. In one of them, an exogenous fluorescent dye (dansyl chloride, acridine orange, etc.) is introduced into the skin where it plays the role of a 'source of intradermal radiation' with the known range of excitation and fluorescence. The sunscreen preparation, deposited on the skin, absorbs the exciting radiation, changing the spectrum and intensity of the injected fluorescence dye. For more accurate measurement the analysis should be carried out quickly and efficiently.

Another *in vivo* method is based on measuring the intensity of the autofluorescence of the skin and its reflection coefficient before and after application of the sunscreen composition. There are factors that limit the applicability of this method: its spectral dependence of the quantum yield of fluorescence of the collagen, spectral dependence of absorption of UVA exciting radiation by chromophores of the epidermis and the presence of fluorophores in the stratum corneum, contributing an additional contribution to the overall autofluorescence of the skin.

14.4. Preparation of specimens of emulsion compositions with nc-Si

The UV-active compositions include emulsion media with the distributed encapsulated silicon nanoparticles. The proposed cosmetic product is a promising sunscreen (Beckman et al, 2003).

Water–oil emulsions containing different amounts of nc-Si were prepared. The preparation procedure for water–oil emulsions is presented in Table 14.2.

Table 14.2. Formulation of water–oil emulsion (Beckman et al, 2003)

Ingredients	Manufacturer	Mass concentration, %
Propylene glycol	(TU 06/09/2434)	3
Water, distilled		adds up to 100
Beeswax	PRIGNIZER, CH	3
Alkaline solution of 1% or triethanolamine	GOST 6484–96	1
Nanocrystalline silicon	SSC RF GNIIChTEOS and Prokhorov General Physics Institute of RAS	0.1–2
Preservative Caton C6	ROM – HAAS	0.05

The preparation process. The water–oil cream foundation (Beckman et al, 2003) is biphasic and is an emulsion of the type of water in oil. Each phase is separately heated to 358 K, mixed and homogenized. Then, while stirring 0.25, 0.30, 0.50, 1.00, 1.50 and 2.00 wt.% of nanocrystalline silicon produced by the above techniques is added. Dispersion and homogenization of the system is performed using an UZDN-2 ultrasonic generator with a frequency of 22 Hz, the maximum intensity of 30 W/cm^2 for 5 min. The specified processing time (5 min) is the result of preliminary experiments to optimize the parameters of ultrasound exposure.

The dispersing effect of ultrasonic fields is based on the phenomenon of cavitation, which occurs in the medium during the passage of acoustic waves of high intensity and is manifested in the emergence and collapse of gas bubbles due to a local pressure drop. This leads to a secondary shock wave, which promotes efficient 'deagglomeration' of the particles of the dispersed phase. In the process of processing the original clusters are reconstructed by homogenization.

14.4.1. Measurements of transmission spectra in the UV and visible region

The transmission spectra of composite materials based on nanocrystalline silicon in the range 200–850 nm were measured in (Tutorsky et al, 2005; Ischenko, Sviridova, 2006) in a SPECORD-M40 two-channel spectrophotometer (Carl Zeiss, Jena). The SPECORD-M40 spectrophotometer is a microprocessor-controlled dual-beam instrument for measuring the absorption and transmission as a function of the wave number or wavelength. The spectrophotometer has a double-diffraction monochromator in the ultraviolet region of the spectrum, consisting of pre- and main monochromator, and a diffraction monochromator with a filter for the visible spectrum. Continuous measurements of results are carried out in the instrument with a modulator and a rotating mirror creating a signal with a periodic sequence of the values: measurement–dark–comparative–dark. This signal is received by a photomultiplier and amplified and then fed to the microprocessor. The processed signal is fed to a PC via a special interface, which is not included in the basic equipment unit, where the signal is finally formed into a digital file. Measurements of the transmission spectra of emulsions with nanocrystalline silicon were carried out using two methods.

The first method consisted in measuring the collimated transmittance in the emulsion layer of a fixed thickness in the direction parallel to the probe radiation incident on the sample, which corresponded to changes in the optical density at the observation wavelength. In this case, the photodetector collected the light, passing through a given emulsion.

The use of media with high optical density such as water–oil emulsions based on nc-Si, requires a certain sensitivity in carrying out measurements.

For this purpose special cells were designed and then made from two plane-parallel quartz plates around 0.8 mm thick, 15 mm in diameter.

To fix a definite gap between the plates, an annular diaphragm of the metal foil 20 μm thick was placed on one of the plates. The emulsion layer was uniformly deposited on the surface of the plate inside the annular diaphragm and pressed against the second plate. The whole structure was clamped between the two annular flanges, allowing fixing of the assemble cell cuvette. The cell was then placed in the measuring channel of the spectrophotometer.

The proposed method allowed the control of the specific thickness of the emulsion layer (d) in different dimensions: $d = 20 \pm 2$ μm. To obtain reproducible results, a series of measurements was carried out in each experiment. Error values obtained for the measurement bandwidth were up to 10% in each experiment.

When applying the emulsion layer on a quartz plate, it was important to ensure the absence of air bubbles in the emulsion layer and flowing of the emulsion to the surface of the annular diaphragm. The presence of bubbles in the layer can strongly distort the value of transmission by increasing it, and the leaking of emulsion on the surface of the foil changes the thickness of the layer.

In order to eliminate distortions in the measurement of losses from the emulsion layer associated with the optical losses and reflections of the quartz plates themselves, an empty cell was placed in the comparison channel or an empty cell was placed in the working channel before starting the measurements of the emulsion layer, for which the losses in this wavelength range were also recorded.

As an example, demonstrating the ability of this method of measurement, Fig. 14.7 shows the transmission spectra of water–oil emulsions with a different content of nanocomposites nc-Si I and II (Ischenko et al, 2005)[2]. The curves for the concentrations of C to 0.5 wt.% vary quite strongly depending on the transmittance, and for concentrations of 1 wt.% and higher the transmittance varies little with wavelength, namely, the maximum transmittance measured for the 20 μm for $\lambda = 850$ nm in both cases is less than 1%. It should be noted that for the same concentration of nc-Si (with $C < 0.5$ wt.%) In the first case the transmittance increases much more rapidly with increasing wavelength than in the second case.

It is seen that with increasing concentration of silicon powder the transmittance of the given emulsions decreases in the entire range of wavelength. Moreover, for samples of type-II transmittance is much lower than in the type I samples, with the same mass concentrations of the powder (especially in the area shorter than 600 nm). Given the similar sizes of nc-Si particles in both cases these changes in the transmission spectra can be attributed to differences in the chemical composition of the surface layers

---

[2]Nanocomposite nc-Si of type I is a nanocomposite with the oxide shell; nanocomposite nc-Si of type II is a nanocomposite with an oxynitride shell around the nc-Si core.

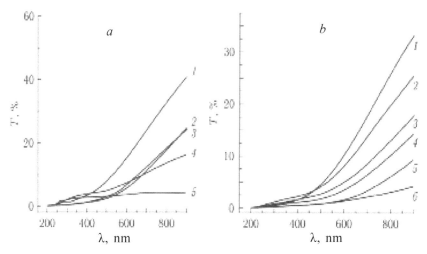

Fig. 14.7. The transmission spectra of a sample of type I – nc-Si in the oxide shell (*a*) and type II nc-Si in an oxynitride shell (*b*) corresponding to changes in optical density, *a* – concentration C_{nc-Si} = 0.10 (*1*), 0.25 (*2*), 0.50 (*3*), 1 (*4*), 1.50 wt.% (*5*), *b* – C_{nc-Si} = 0.10 (*1*) 0.25 (*2*), 0.50 (*3*), 1.0 (*4*), 1.5 (*5*), 2.0 wt.% (*6*) (Ischenko et al, 2005).

of these nc-Si particles, resulting in their synthesis: the oxide shell in the case of I (*a*) and the oxynitride shell in case II (*b*).

14.4.2. Measurement of transmission spectra in the integrating sphere

To take into account the scattering of the radiation passing through the sample, experiments were carried out using a technique for measuring the transmittance spectra in an integrating sphere. Such measurements give more information about changes in the nature of the transmittance of the given composite, taking into account the diffuse scattering forward in the solid angle 2π.

Measurements were taken using the standard sphere of Carl Zeiss, which is included in the SPECORD-M40 spectrophotometer. The spectral range of the device at work with the sphere is limited by the wavelength from 200 nm in the UV spectral region. The standard sphere of the Carl Zeiss system is adapted to measure the diffuse reflection. To measure the diffuse transmittance, special cuvettes, quartz glass for them and other devices were developed. The schematic diagram of measurements of diffuse transmittance through the SPECORD-M40 two-channel spectrophotometer is shown in Fig. 14.8. In the sphere there are four windows, two input and two output. The cuvettes are mounted in front of the input window. When measuring the

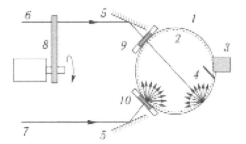

Fig. 14.8. Schematic diagram of the measurements of diffuse transmittance using the two-channel spectrophotometer SPECORD-M40, and an integrating sphere: *1* – integrating sphere, *2* – diffusely reflecting coating *3* – photomultiplier, *4* – valves, *5* – the optical system of the spectrophotometer, *6* – beam of the comparison channel, *7* – beam of the measurement channel, *8* – ray switch, *9* – cell without the sample, *10* – cell with the sample (Rybaltovsky et al, 2007).

diffuse transmittance the output windows were closed with inserts coated with a reflective coating of barium sulphate, $BaSO_4$.

Before measuring the spectra of the samples the spectrophotometer was calibrated. A cell with a quartz window was positioned in the place of the cell with the sample. It was thus possible to take into account the Fresnel reflection of the external surfaces of the glass at subsequent measurements of the spectra of the sample. Fresnel reflection at the sample–glass interface was neglected because of the small difference in the refractive indices. Calibration was performed according to the standard procedure for the SPE-DORD-M40 system. During calibration of the zero signal the beam of the measuring channel was closed with a screen; at 100% calibration, the beam of the measuring channel was open.

The emulsion with nanocrystalline silicon was deposited on quartz glass 1 mm thick, 10 mm in diameter, clamped with a similar quartz glass so that the layer thickness was $d = 20$ mm. The layer thickness was regulated, as in the first method of measuring the spectra, with a foil lining between the plates. The actual thickness of the emulsion layer was $d = 20 \pm 2$ μm.

When measuring the diffuse transmittance spectra of the emulsion composites with nanocrystalline silicon all the rays passing through the sample uniformly filled the integrating sphere, and their density was fixed by means of photomultiplier. The results were displayed on a printer as well as written to a file for further analysis.

The spectra shown in Fig. 14.9 indicates that for the type I samples there is a sufficiently large signal bandwidth over the entire spectral range, including the UV region of 230–400 nm. The level of transmittance in this case is weakly dependent on the concentration of nc-Si in the emulsion. Highs in the region of 230, 280 and 400 nm are likely to be associated with the absorption band from the pure base and bands corresponding

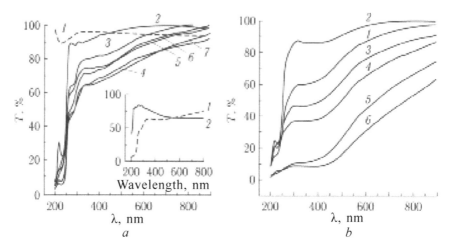

Fig. 14.9. The transmission spectra of the sample within the scope of the type I and type II, and the sample used. The inset to the figure a shows the transmission spectra of the initial sample with 1% nc-Si (*1*) and after heat treatment at 800 °C (*2*): *a* – *1* – clear emulsion concentration, C_{nc-Si} = 0.10 (*2*), 0.25 (*3*), 0.50 (*4*), 1.00 (*5*), 1.50 (*6*), 2.00 wt.% (*7*), *b* – C_{nc-Si} = 0,10 (*1*), 0.25 (*2*), 0.50 (*3*), 1.00 (*4*), 1.50 (*5*), 2.00 wt.% (*6*) (Rybaltovsky et al, 2007).

to the oxide shell and the silicon core. For samples of type II there is a marked dependence of the signal bandwidth of the concentration of nc-Si: at increasing concentration from 0.1 to 2.0 wt.% transmittance decreases sharply and falls below 7% in the 200–450 nm.

Comparing the results presented in Figs. 14.7 and 14.9, we can conclude that for the nc-Si I particles the relative contribution of scattering effects to the total value of the transmittance at 200–450 nm in the measurements using the integrating sphere is larger than for the particles of type II. At the same time, for the nc-Si type II particles in this range the controlling contribution comes from the absorption of light photons by these particles. This is especially noticeable in the 200–450 nm range at high concentrations of particles in the emulsion. If we consider that in both cases the nc-Si particles, according to Raman measurements, have the silicon cores of similar sizes, all the mentioned differences in the transmittance spectra are apparently connected with the presence of outer shells of different chemical composition in these particles and, correspondingly, with their different optical properties. As we know, silicon oxynitrides have much greater losses in the UV region compared with the silicon oxides (Sahu et al, 1990). Therefore, the total losses through absorption for the samples of type II may be higher than in the samples of type I, if we assume that the thickness of the shell is a large value compared with the total diameter o the particle. At the same time, however, the effects of scattering for the type I particles due to the presence of the oxide shell can be controlling when measuring the transmittance spectra in the integrating sphere. This assumption is validated by the following observation. If the silicon nanocomposite is subjected to preliminary heat treatment in air for 1 hour at a temperature of 873–1073 K, followed by a comparative measurement of the transmittance spectra using the integrating sphere in the emulsions containing the initial and heat-treated powders, it can be seen that the magnitude of the transmittance in the UV wavelength range is higher than for the heat-treated powder. As evidenced by the results of IR and Raman measurements, heat treatment of these powders increases the thickness of the oxide layer by reducing the size of the silicon core.

When considering the mechanisms of formation of spectral losses due to introduction of the silicon nanocomposite into the emulsion, it must be taken into account that the Fresnel reflectivity coefficient R for the oxide layer can be several times smaller than for the oxynitride layer because the associated refractive index n changes from 1.46 (for SiO_2) to 2.0 (for Si). This fact should be taken into account to obtain more information about the formation of loss spectra of samples of this type, involving measurements of reflection spectra.

Measurements were also taken of the transmittance spectra in emulsions using the integrating sphere for type IV samples, synthesized by the plasma chemical method in an argon atmosphere without the addition of oxygen and nitrogen. The transmission spectra are shown in Fig. 14.10.

In contrast to the samples of types I and II, 0.50 wt.% of nanocrystalline silicon was added to the type IV samples which resulted in a significant weakening of the transmittance of electromagnetic radiation in the UV region – up to 10%.

We now consider theoretically the changes of the value of the transmittance taking into account the scattering and reflection for pure silicon nanoparticles as a function of their size for certain wavelengths. Popov et al (2006) considered a medium with a thickness of 20 μm with

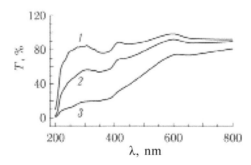

Fig. 14.10. Transmittance spectrum in the integrating sphere of the nc-Si sample, synthesized by the plasma-chemical method in argon without the addition of oxygen and nitrogen (samples type IV); $C_{nc-Si} = 0.10$ (*1*), 0.25 (*2*), 0.50 (*3*) wt.% (Rybaltovsky et al, 2007).

Table 14.3. The refractive index of silicon for the considered wavelengths (Handbook of Optical Constants of Solids, San Diego, Academic Press, 1998, vol. I, pp. 561–565, vol. II, pp. 575–579)

Wavelength, nm	Refractive index	Wavelength, nm	Refractive index
290	$4.525 - i \cdot 5.158$	602	$3.943 - i \cdot 0.025$
350	$5.442 - i \cdot 2.989$	697	$3.787 - i \cdot 0.013$
400	$5.570 - i \cdot 0.387$	795	$3.607 - i \cdot 0.007$
500	$4.298 - i \cdot 0.073$		

a refractive index $n = 1.4$, with uniformly distributed spherical silicon nanoparticles with a concentration of 0.50% by volume, which roughly corresponds to 1.2 wt.%. The light reflected and transmitted through a layer is detected in the solid angle 2π. Calculations were carried out according to the methodology used in such cases for the titanium oxide nanoparticles using the Mie scattering theory and the Monte Carlo method.

Calculations were performed for the nanoparticles with the following parameters: diameter 10–200 nm, the concentration 0.50 vol.% or 1.00 vol.%, which corresponds to 1.2 wt.% and 2.3 wt.%, respectively (at a density of 2.33 g/cm³ and the water–emulsion medium surrounding the particles of 1.00 g/cm³). In the calculations it was assumed that the particles are suspended in a 20 μm infinitely wide layer of a transparent (non-absorbing) medium with a refractive index $n = 1.4$ in the whole wave length range (290–800 nm). The refractive index of the material of the particles is shown in Table 14.3. The results are presented in Fig. 14.11.

From these data it follows that the minimum value of transmittance is obtained for the size of nanoparticles at which the reflection is maximum. The increase of reflection compensates the small 'dip' (4–10% of incident radiation) in the high level of absorption (80%). With the decrease of the wavelength of radiation this particle size range is shifted toward smaller sizes of 60–100 nm for $\lambda = 400$ nm, 40–100 nm for $\lambda = 350$ nm and 30–100 nm for $\lambda = 290$ nm. In addition, the calculations show that for particles of the same size (in our case for the particle size of 10 nm) the value of transmittance in the UV region decreases quite rapidly with decreasing wavelength of observation, from 85% for $\lambda = 400$ nm to 25% at $\lambda = 300$ nm (see Fig. 14.11*d*).

In addition, the calculations show that for particles of the same size (in our case for the particle size of 10 nm) the value of transmittance in the UV region decreases rapidly with decreasing wavelength of observation. Thus, the theoretical analysis of the dependence of the transmittance on the size of silicon nanoparticles in emulsions with the optical properties similar to

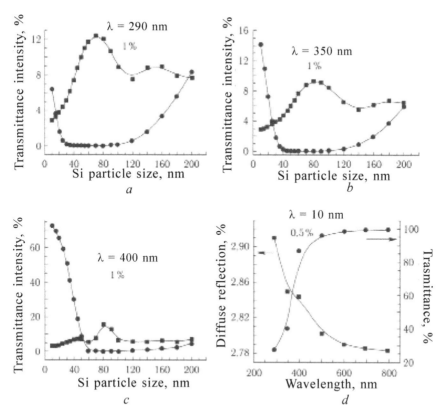

Fig. 14.11. Depending of diffuse reflection (■) and transmittance into the sphere (●) of the suspension of silicon nanoparticles (volume fraction 1.0%, *a–c*) and 0.5%, *d*) in 20 μm layer of water–emulsion medium for incident radiation with a wavelength of 290 nm (*a*), 350 nm (*b*) and 400 nm (*c*) of the average diameter of the silicon nanoparticles, *d* – for different wavelengths of incident radiation at a particle diameter of 10 nm (Popov et al, 2007).

water showed that for particles size of 10–30 nm, the transmittance must be quite large (greater than 50%) at exposure to UV radiation λ = 400 nm. This value is then determined, in general, mainly by the effects of scattering on nanoparticles in the emulsion (absorption is small). As the particle size increases the contribution from scattering decreases, and the value of transmittance of the emulsion in this case is determined by the dominant effect of absorption and also diffuse reflection.

The results also show that the boundary of the minimum values of transmittance is strongly dependent on particle size and shifted to larger particle sizes for large radiation wavelengths. For particles 10 nm in size, the transmittance is greatly reduced during the transition to the UV region of the spectrum (300–400 nm).

In a real situation, as follows from the pilot study, the spectral characteristics of silicon-containing emulsions are strongly affected by effects of the shells of silicon particles having a different chemical composition in comparison with the core.

14.4.3. Measurements of the Raman spectra

Raman spectra (RS) of light were measured (Rybaltovsky et al, 2007) with a T-64000 RS spectrograph (Jobin Yvon). The radiation source was an argon laser, Ar^+: Stabilite® 2017 Ion Laser (Spectra-Physics), at a wavelength of 514.5 nm. In addition the powerful basic line, the laser emits hundreds of other lines (ghosts), whose intensity is several orders of magnitude less than basic, but is comparable with the intensity of the Raman signal. These bands can be completely eliminated by using a prism monochromator placed behind the laser. Replacement lenses, placed on a special cartridge, allow focus the laser beam to a spot diameter of 1 to 20 μm. The scattered radiation is focused by the lens L to the entrance slit of a triple monochromator. The accuracy of measurement of the Raman spectra is 1 cm^{-1} near the wavelength of 514.5 nm (19435 cm^{-1} on an absolute scale) in the given configuration of the spectrograph.

Measurements are taken using either samples of silicon-containing emulsions in the form of droplets on a glass plate or specially prepared samples based on a silicate glue, mixed in a weight ratio of 20:1 with the test sample of nc-Si. In the latter case, the sample can be reused in different measurements. Initially Raman spectra of pure water–oil emulsions and silicate glue are recorded. This procedure is carried out in order to ensure the absence of lines in the range of interest of 450–550 cm^{-1}, where, according to the work (Marchenko et al, 2000), there are the main Raman bands characterizing various states of silicon. In addition, in the registration of each Raman spectrum additional measurements are performed to control the heating of the nc-Si sample as a result of absorption of the exciting laser radiation. Heating of the samples results in broadening and shift to shorter wavelengths of the spectrum of the Raman bands (Marchenko et al, 2000).

Figure 14.12 presents the Raman spectra for the type I, dispersed in the water–oil emulsion with silicate glue. The reference spectrum is the Raman spectrum of polycrystalline silicon, the main band at 521 cm^{-1} with a half-width of 3.5 cm^{-1}. The figure shows that the Raman band of the type I sample is shifted to lower frequencies by 2.5 cm^{-1} and broadened by 25% relative to the reference. The results are identical for the same sample dispersed in the emulsion with silicate glue.

Comparison of peak shapes of the investigated and standard samples indicates that the synthesized nanomaterial is close to the crystalline phase by its nature, whereas the width of the amorphous silicon Raman band should be about 100 cm^{-1} with a maximum in the region 480–490 cm^{-1} (Marchenko et. al, 2000). The broadening of the RS peak in the samples is probably due to the small particle size (the smaller the particle, the larger

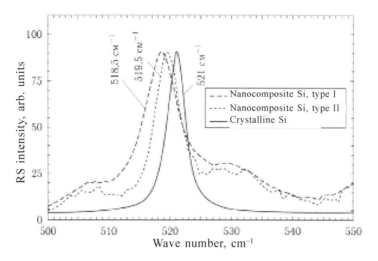

Fig. 14.12. Raman spectrum for sample types I and II (Rybaltovsky et al, 2007).

the number of surface bonds and accordingly, the greater the imperfection of the crystal).

Analysis of the Raman spectra of the composites containing powdered silicon allows not only to establish the presence of amorphous and crystalline phases in the synthesized samples, but also to evaluate the average size of nc-Si nanoclusters. The particle sizes in the first approximation can be calculated using the model of the spatial constraint of phonons in a spherical crystallite with diameter L and the scattering of light on them (Faraci et al, 2006). The shift of the Raman band to lower frequencies compared to the polycrystalline silicon is associated with the particle size decreasing to the nanometer range (Meier et al, 2006). Guided by the approximate relation presented there that binds the value of the shift in inverse centimeters with the particle size in nanometers, we obtain:

$$\Delta \gamma = \frac{S^2}{4.2 \gamma_0 L},$$ (14.2)

where S – the speed of sound in the crystal, c – the speed of light in the crystal, γ_0 – the position of the peak in the single crystal, L – the size of nc-Si. This ratio can be used when the shape of Raman bands for the sample and the polycrystalline silicon is not very different from each other. Using this relation, one may obtain the value of the average particle diameter of nc-Si. In samples of type I, it was equal to 10 ± 2 nm, and in samples of type II 15 ± 5 nm.

To conclude this section, it should be mentioned that the proposed method of estimating the size of nc-Si does not have a high degree of accuracy, since the average speed of sound in the crystal and the average value of the refractive index n in the visible spectrum are used. Of course, the accuracy of estimates also decreases with increasing size of nc-Si, i.e. when approaching the maxima of the RS peaks of the studied powder and crystalline silicon plate. However, as seen from the experiments, such assessments is quite justified. In a similar situation in the study of the spectral properties of fine-grained films of nc-Si, the authors (Soni, 1999; Tripathy, 2001) obtained estimates of the size of the crystallites from the Raman spectra which are in quite good agreement with measurements data obtained from transmission spectra in these films. For more accurate analysis of the RS results we should consider not only the shift of the RS line CD but also its width. It should be noted that the electron microscopic analysis of synthesized powders of type I, presented in Ischenko et al (2005), gives the size of nc-Si in agreement with the RS data, indicating the existence of particles with a diameter of 10 nm. In the case of type II nanocomposites X-ray phase analysis (Tutorsky et al, 2005) allowed us to determine the particle size the order of 20 nm, which is also close to the value obtained from the relevant Raman spectra measurements. It should also be noted that the electron microscope analysis of the synthesized samples agrees with the data of the RS values the size of nc-Si, encapsulated in a silica shell.

The experimental results were used to develop 'The method of experimental investigation of the spectral characteristics of water–emulsion composite media containing silicon nanoparticles of silicon' (Rybaltovsky et al, 2007), which includes measurements of the optical transmittance spectra of the emulsion media taking into account the effects of scattering and Raman spectra measurements, which can be used to determine the size of the central core of the nanocomposite. This technique allows to conduct a rapid analysis of the characteristics of the transmittance in the UV and visible range taking into account scattering and the size of the particles. This technique can be applied to other emulsion materials based on the emulsion of nanoparticles of titanium oxide and zinc which are also regarded as protectors against UV radiation.

14.5. nc-Si composites with natural biologically active compounds

The spectrum of the protective effect of the composition that is based on nanocrystalline silicon can be improved with the introduction of physiologically active additives that prevent oxidative processes associated with the effect on the skin of free radicals and active oxygen (Isaev et al, 2007). The task is achieved by ensuring that the cosmetic product to protect from UV radiation, including the water–oil base and nanocrystalline silicon, additionally contains vitamin E, squalene and Eykonol.

The latter two compounds are natural seafood, separated from fat and tissue of cold-water fish. Eykonol contains fatty polysaturated acids of class ω-3, contributing to increased metabolism in human cells. Squalene is a substance extracted from the liver deep-water sharks – participates in the synthesis of high-density lipoprotein, increases the immunity of healthy cells:

$$(CH_3)_3C-CH(CH_3)_2C(CH_3) \quad CH(CH_3)_2C(CH_3) \quad CH(CH_3)CH \quad C(CH_3)(CH_3)_2CH$$

$$=CH(CH_3)(CH_2CH \quad CHC(CH_3)_2$$

In combination with vitamin E, squalene is an effective antioxidant. All these products are a source of protein, B and PP vitamins, natural trace elements and mineral salts (Isaev et al, 2007).

In 2010 a study was published that demonstrates the use of nanoparticles of English ivy (*Hedera helix*) as an alternative sunscreen (Xia et al, 2010).

Concluding remarks

According to numerous studies, ultraviolet radiation is dangerous to the humans and a factor leading to the development of various diseases, including cancers of the skin. The entire range of UV radiation (100–400 nm) is usually divided into three regions: UV-A (315–400 nm), UV-B (280–315 nm) and UV-C (100–280 nm) (Diffey, 1991).

Despite the fact that the Earth's ozone layer blocks ~98.7% of UV radiation from penetration through the atmosphere, a relatively small part of the total intensity of solar radiation, including UV-A and some parts of the UV-B, reaches the surface of our planet, and may have adverse effects on humans (Pathak, Mason, 2002). UV-C, as a rule, does not reach the Earth's surface, but precisely because of its ability to cause DNA damage this region of the spectral range is often used in modelling the impact of UV in the laboratory.

Depending on the time spent in the sun, the harmful effects UF-A/UF-B radiation includes mechanisms of immediate disorders in the human body – sunburn, and long-term (pending) issues, such as skin cancer, melanoma, cataracts and weakened immune systems (Hockberger, 2002; Longstreth et al, 1998). The main mechanism for UV damage involves oxidative stress and protein denaturation, while the shortwave UV-B radiation causes DNA damage, mainly in the form of formation of pyrimidine dimers and some hazardous products of photostimulated reactions (Friedberg, 2003).

Induced mutation of DNA when exposed to UV radiation is one of the leading causes of skin cancer. Currently more than one million cases and over 12,000 deaths per year are diagnosed in the USA (American Cancer Society[3]. The demand for protection of skin from the harmful effects of UV radiation from the sun is becoming increasingly important due to exhaustion of the ozone layer (van der Leun, de Gruijl, 1993; Smith et al, 1992). Sunscreen tools that are designed to protect human skin usually use the synergic effect of UV-protective properties of organic and inorganic

---

[3]http://www.cancer.org/docroot/ped/content/ped_7_1_what_you_need_to_know_about_skin_cancer.asp

components on the basis of reflection, scattering or absorption of UV radiation, should by definition provide significant protection from damage by solar ultraviolet radiation.

Previously, the development of sunscreens was associated with inorganic UV filters. For example, particles of titanium dioxide (TiO_2) and zinc oxide (ZnO). These sunscreen materials are often not transparent and give the skin a white tinge, which makes them unattractive to consumers (Wolf et al, 2001). However, concern about the need of increased protection from UV radiation has meant that, in recent years, thousands of tons of nanomaterials on the basis of nanosized particles of metal oxides were introduced into cosmetic products and thousands of tons of nanomaterials are currently used by hundreds of millions of people each year (Nohynek et al, 2007). However, as shown by statistical studies, the result of the use of these means is not a positive one. The question naturally arises and is being actively discussed in the literature: "Why increasing use of sunscreens does not lead to a decrease number of diseases that are associated with exposure to UV radiation?". Moreover, there are individual studies that show that the increase in the number of disease almost symbatically changes with increasing consumption of sunscreen cosmetics. This issue requires immediate objective study.

Many studies show that when applied to the skin for about ~8 hours or less, inorganic nanoparticles – filters do not penetrate through the cornea layer of the skin (Durand et al, 2009; Gontier et al, 2009; Newman et al, 2009; Oberdorster, In 2004). However, in this case usually the subject are nanoparticles having the average size of 20 nm. At the same time, samples of healthy skin are always used. As noted in many sections of this monograph the nanomaterial is polydispersed which, in general, is the rule. The average size gives rather weak response, but in this case, neglecting the influence of the fine fraction of the size distribution function of particles may have a decisive influence on the human body. Studies of ultrafine nanoparticles materials found that TiO_2, ZnO and iron nanoparticles with the average size smaller than 15 nm, can penetrate through the human skin (Baroli et al, 2007; Menzel et al, 2004). Other studies have also demonstrated the penetration of particles of 4 nm and 60 nm for TiO_2 through the healthy skin of nude mice after long exposure – from 30–60 days (Wu et al, 2009). This penetration leads to an increase in skin aging, pathological effects in the liver and the accumulation of particles in the brain.

It is known that the properties of nanomaterials, as man-made objects, integrable in human activity, significantly differ from the natural properties, in an environment where humanity has survived for thousands of years. Cosmetics manufacturers often ignore this factor, even neglecting the proper compliance with federal regulations of nanomaterials (Buzea et al, 2007). Numerous examples presented in this monograph show that with decreasing particle size the nano-objects change many of the physical and chemical properties. This includes, for example, colour, solubility, strength of material, electrical conductivity, magnetic properties, mobility (including environment and human), chemical and biological activity (Oberdorster et al, 2005).

The increase in the surface area to volume ratio also increases the chemical activity, increasing the speed of processes, which in the case of bulk materials can be very slow. For example, the effect of the oxygen of the air atmosphere on a nanoparticle can lead to the formation of singlet oxygen and other active forms of oxygen-containing particles (Reactive Oxygen Species, ROS (Nel et al, 2006), see also chapter 13 of this monograph). ROS, detected in the metal nanoparticles of the oxides TiO_2 and ZnO, carbon nanotubes and fullerenes, is the active component of oxidative stress, inflammation, with consequent damage to DNA, proteins and membranes (Nel et al, 2006). In addition, the effect of ROS opens the possibility of occurrence of some reaction channels involving free radicals (Dunford et al, 1997). In the case of small TiO_2 nanoparticles the DNA damage was considerably greater than for the large particles (Donaldson et al, 1996). Although microparticles of TiO_2 (~500 nm) are also capable of causing DNA breaks, mid-sized nanoparticles (~20 nm) cause the complete destruction of the DNA helix at low doses of UV radiation or even without its influence.

No less important is the problem with applying sunscreen cosmetics products, associated with inhalation products released by materials, food and penetration through the skin. Proceeding into the bloodstream, nanomaterials can be distributed within the body and several organs and tissues such as brain, liver, spleen, kidneys, heart, bone marrow and nervous system, which are indefinitely exposed to nanoparticles for an unknown period of time (Oberdorster et al, In 2005). Depending on their stability in the human body (the biodegradation rate), damage can be associated with the presence of ROS on their surface, and also with the accumulation of nanoparticles in human organs and tissues which can disrupt their normal functioning. Hussain et al, (2005) showed significant cytotoxicity of nanoparticles of TiO_2, Fe_3O_4, Al, and MoO_3 at a concentration of 100–250 µg/ml (presumably due to the influence of ROS on BRL 3A liver cells of rats). Studies of carbon nanotubes have also shown inhibition of cell growth and toxicity of the material tested for kidney disease in animals (Oberdorster et al, 2005). The stability of nanomaterials in the environment also leads to brain damage and increased mortality of several species of animals living in water (Luo, 2007; Zhu et al, 2006).

Because of the potential toxicity of metal-containing nanoparticles, to replace them in sunscreen media it is essential to find alternative ingredients materials that are not toxic and, at the same time, are effective in blocking ultraviolet radiation (Xia et al, 2010). As shown in several studies (see chapter 12), nanosilicon, as a biocompatible and biodegradable material, can become one of the most promising materials for use in sunscreens.

References for Chapter 14

Beckman D., Belogorokhov A. I., Guseinov Sh. L., Ischenko A. A., Storozenko P. A., Tutorsky I.A. Cosmetic for UV protection. 05.06.2003.

Gusev A. I., Rempel A. A. Nanocrystalline materials. Moscow: Fizmatlit, 2000. 224.

Efremov M.D., et al. Visible photoluminescence of silicon nanopowders created by evaporation of silicon with a powerful electron beam. Pis'ma Zh. Eksp. Teor. Fiz. 2004. V. 80, No. 8. P. 619–622.

Isaev V. A., Belogorokhov A. I., Es'kova E. V., Ischenko A. A., Storozenko P. A., Tutorsky I. A. Cosmetic agent for protection against ultraviolet radiation. Application for patent from 09.12.2004. Incoming number 039165. Join patent number 2004136008. Russia.

Ischenko A. A., Sviridov A. A. Sunscreen. II. Inorganic UV filters and their compositions with organic protectors. Izv. VUZ. Ser. Khimiya Khim. Tekhnol. 2006. V. 49. No. 12. P. 3-16.

Kononov N. N., Kuzmin G. P., Orlov A. A., Surkov A. A., Optical and electrical properties of thin plates made of nanocrystalline powders of silicon. Fiz. Tekh. Poluprovod. 2005. V. 39, No. 7. P. 868-873.

Nakamoto K., Infrared Spectra of Inorganic and Coordination Compounds. Moscow: Mir. 1966. P. 147.

Otto M. Modern methods of analytical chemistry. Moscow. Tekhnosfera. 2003.

Popov A. P., Priezzhev A. V. The method of computing the effectiveness of the protective properties of nanoparticles in irradiation of materials and tissues in the UV-A and UV-B bands. MR GSSSD 120 Dep. Federal State Unitary Enterprise 'Standartinform.' 03.03.2006. 36.

Rybaltovsky A. O., Koltashev V. V., Sviridov A. A., Popov A. P., Ischenko A. A. Technique for experimental study of the spectral characteristics of water-emulsion composite media containing nanoparticles of silicon. Cert. GSSSD ME 131-2007. Dep. Federal State Unitary Enterprise Standartinform. 12. 07. 2007. 26.

Rybaltovsky A. O., Radtsig V. A., Sviridova A. A., Ischenko A.A. Thermo-oxidative processes in nanoscale powders of silicon II. Paramagnetic centers. Nanotekhnika. 2007. V. 3. No. 11. P. 116-121.

Sviridova A. A., Ischenko A.A. Sunscreens. I. Classification and mechanism of action of organic UV filters. Izv. VUZ. Ser. Khimiya Khim. Tekhnol. 2006. V. 49. No 11. P. 3-14.

Sennikov P.G., et al., Producing soy of nanocrystalline silicon by plasma chemical vapor deposition of silicon tetrafluoride. Fiz. Tekh. Poluprovod. 2009. V. 43, No. 7. P. 1002-1006.

Sinichkin Yu. P., Dolotov L. E., Zimnyakov D. A., Tuchin V. V., Umts S. R. Special workshop on optical biophysics. In vivo reflectance and fluorescence spectroscopy of human skin: textbook. Saratov: Saratov University Press, 2003. 159.

Sinichkin Yu. P., Utts S. R. In vivo reflectance and fluorescence spectroscopy of human skin. Saratov: Saratov University Press, 2001. 92.

Tutorsky I. A., AI Belogorokhov, Ischenko AA, Storozenko PA The structure and adsorption properties of nanocrystalline silicon. Kolloid. Zh. 2005. V. 67. No 4. P. 541-547.

Utts S. R., Knushke P., Sinichkin Yu. P. Rating photoprotective agents using in vivo fluorescence spectroscopy. Vestn. Dermatol. 1996. No. 2. P. 15-21.

A reference action spectrum for ultraviolet induced erythema in human skin / Ed. by A. F. McKinley, B. L. Diffey. CIE J. 1987. V. 6. P. 17–22.

Altman I. S., Lee D., Chung J.D., Song J., Choi M. Light absorption of silica nanoparticles. Phys. Rev. B. 2001. V. 63. P. 161402.

American Cancer Society. http://www.cancer.org/docroot/ped/content/ped_7_1_what_you_need_to_know_about_skin_cancer.asp.

Applegate L.A., Scalletta C., Fourtanier A., Mascotto R., All E. Expression of DNA damage and stress proteins by UVA irradiation of human skin in vivo. Eur. J. Dermatol. 1997. V. 7. P. 215–219.

Bai G. R., Qi M.W., Xie L.M., Shi T. S. The isotope study of the Si−H absorption peaks in the FZ-Si grown in hydrogen atmosphere. Sol. Stat. Comm. 1985. V. 56, No. 3. P. 277–281.

Baroli B., Ennas M.G., Loffredo F., Isola M., Pinna R., Lopez-Quintela M.A. Penetration of metallic nanoparticles in human full-thickness skin. J. Invest. Dermatol. 2007. V. 127. P. 1701–1712.

Bech-Thomson N., Wulf H. Sunbathers'application of sunscreen is probably inadequate to obtain the sun protection factor assigned to the preparation. Photodermatol Photoimmunol Photomed. 1993. V. 9. P. 242–244.

Bernerd F., Vioux C., Asselineau D. Evaluation of the protective effect of sunscreens on in vivo reconstructed human skin exposed to UVB or UVA irradiation. Photochem Photobiol. 2000. 71(3): P. 314–320.

Bisi O., Ossicini S., Pavesi L. Porous silicon: a quantum stronge structure for silicon based optoelectronics. Surface Science Report. 2000. V. 38. P. 1–126.

Borghei A., Sassella A., Pivac B., Pavesi L. Characterization of porous silicon inhomogeneties by high spatial resolution infrared spectroscopy. Sol. St. Comm. 1993. V. 87, No. 1. P. 1–4.

Brus L. Electronic wave functions in semiconductor clusters: Experiment and theory. J. Phys. Chem. 1986. V. 90. P. 2555–2560.

Buka R. L. Sunscreens and insect repellents. Current Opinion in Pediatrics. 2004. V. 16. P. 378–384.

Buka R. L. Sunscreens and insect repellents. Current Opinion in Pediatrics, 2004. V. 16. P. 378–384.

Buzea C., Pacheco I. I., Robbie K. Nanomaterials and nanoparticles: sources and toxicity // Biointerphases. 2007. V. 2. P.MR17–MR71.

Canham L. T., Groszek A. J. Characterization of microporous silicon by flow calorimetry: comparison with a hydrophobic SiO_2 molecular sieve. J. Appl. Phys. 1992. V. 72, No. 4. P. 1558–1565.

Delerue C., Allan G., Lannoo M. Optical band gap of Si nanoclusters. J. Lumin. 1999. V. 80. P. 65–73. Diffey B. L. Solar ultraviolet radiation effects on biological systems. Phys. Med. Biol. 1991. V. 36. P. 299–328.

Diffey B. L., Farr P.M. Sunscreen protection against UVB, UVA and blue light: an in vivo and in vitro comparison. Br. J. Dermatol. 1991. V. 124. P. 258–263.

Diffey B. L. What is light?. Photodermatol. Photoimmunol. Photomed. 2002. V. 18. P. 68–74.

Donaldson K., Beswick P.H., Gilmour P. S. Free radical activity associated with the surface of particles: a unifying factor in determining biological activity?. Toxicol. Lett. 1996. V. 88. P. 293–298.

Dunford R., Salinaro A., Cai L., Serpone N., Horikoshi S., Hidaka H., Knowland J. Chemical oxidation and DNA damage catalysed by inorganic sunscreen ingredients. FEBS Lett. 1997. V. 418. P. 87–90.

Durand L., Habran N., Henschel V., Amighi K. In vitro evaluation of the cutaneous penetration of sprayable sunscreen emulsions with high concentrations of UV filters. Int. J. Cosmet. Sci. 2009. V. 31. P. 279–292.

Faraci G., Gibilisco S., Russo P., Pennisi A. R., Rosa S. Modified Raman confinement model for Si nanocrystals. Phys. Rev. B. 2006. V. 73. P. 033307.

Friedberg E. C. DNA damage and repair. Nature. 2003. V. 421. P. 436–440.

Gelis C., Al E. Assesment of the skin photoprotective capacities of an organo-mineral broad-spectrum sunblock on two ex vivo skin models. Photodermatol Photoimmunol Photomad. 2003. V. 19. P. 242–253.

Gelis C. et al. Assesment of the skin photoprotective capacities of an organo-mineral broad-spectrum sunblock on two ex vivo skin models. Photodermatol Photoimmunol Photomed. 2003. V. 19. P. 242–253.

Giudice M., Bruno F., Cicnelli T., Valli M. Structural and optical properties of silicon oxynitride on silicon planar waveguides. Appl. Optics. 1990. V. 29, No. 24. P. 3489–3496.

Glanz K., Buller D. B., Saraiya M. Reducing ultraviolet radiation exposure among outdoor workers: State of the evidence and recommendations. Environmental health. 2007. V. 6, No. 22. 11 p.

Glinka Y.D. Size effect in self-trapped exciton photoluminescence from SiO_2-based nanoscale materials. Physical Review B. 2001. V. 64. P. 085421.

Gontier E., Ynsa M.-D., Biro T., Hunyadi J., Kiss B., Gaspar K., Pinheiro T., Silva J.-N.,

Filipe P., Stachura J. Is there penetration of titania nanoparticles in sunscreens through skin? A comparative electron and ion microscopy study. Nanotoxicology. 2009. V. 2. P. 218–231.

Green A., Battistutta D. Incidence and determinants of skin cancer in a high risk Australian population. Int. J. Cancer. 1990. V. 46. P. 356–361.

Gueymard C.A., Myers D., Emery K. Proposed reference irradiance for solar energy testing. Solar Energy, 2002. V. 73. P. 443–467.

Gurish M. F., Roberts L.K., Krueger G.G., Daynes R.A. The effect of various sunscreen agents on skin damage and the induction of tumour susceptibility in mice subjected to ultraviolet irradiation. J. Invest. Dermatol, 1981. V. 76. P. 246–251.

Handbook of optical Constants of Solids. Proc. San Diego: SPIE Acad. Press, 1998. Part I. P. 561–565; Part II. P. 575–579.

Heitmann J., M¨uller F., Zacharias M., G¨osele U. Silicon nanocrystals: Size Materrs. Adv. Mater. 2005. V. 17, No. 7. P. 795–803.

Hockberger P. E. A history of ultraviolet photobiology for humans, animals and microorganisms. Photochem. Photobiol. 2002. V. 76. P. 561–579.

Huang F.-Ch., Lee J.-F., Ch.-K. L., Chao H.-P. Effects of cation exchange on the pore and surface structure and adsorption characteristics of montmorillonite. Colloids Surfaces A. Physicochem. Eng. Aspects. 2004. V. 239. P. 41–47.

Hussain S.M., Hess K. L., Gearhart J.M., Geiss K. T., Schlager J. J. In vitro toxicity of nanoparticles in BRL 3A rat liver cells. Toxicol. In Vitro. 2005. V. 19. P. 975–983.

IARC, 2006. IARC monograph on the evalution of carcinogenic risk to humans, carbon black and titanium dioxide. (93) Lyon, International Agency for Research on Cancer.

Ischenko A.A., Sviridova A.A., Zaitseva K.V., Rybaltovsky A.O., Bagratashvili V.N., Belogorokhov A. I., Koltashev V.V., Plotnichenko V.G., Tutosrky I.A. Spectral properties of siliceous nanocomposite materials. Proc. SPIE. 2005. V. 6. P. 6146.

Jacobs J. F. Ibo van de Poel, Patricia Osseweijer. Sunscreens with Titanium Dioxide (TiO_2) Nano-Particles: A Societal Experiment. Nanoethics.-2010.-V.4.-P. 103–113.

Knief S., Niessen W. Disorder, defects, and optical absorption in a-Si and a-Si:H. Phys. Rev. B. 1999. V. 59, No. 20. P. 12940–12946.

Lademann J. et al. Synergy effects between organic and inorganic UV filters in sunscreens. J. of Biomed. Optics. 2005. V. 10. P. 014008.

Lavker R.M., Gerberick G. F., Veres K. et al. Cumulative effects from repeated exposures to suberythemal doses of UVB and UVA in human skin. J. Am. Acad. Dermatol. 1995. V. 32. P. 53–62.

Li X., He S., Talukdar S. S., Swihart M. T. Process for preparing macroscopic qauntities of brightly photoluminescent silicon nanoparticles with emission spanning the visible spectrum. Langmuir. 2003. V. 19. P. 8490–8496.

Liardet S., Scaletta C. Protection against pyrimidine dimers, p53, and 8-hydroxy-2'-deoxyguanosine expression in Ultraviolet-Irradiated human skin by sunscreens: difference between UVB+UVA and UVB alone sunscreens. J. Invest. Dermatol. 2001. V. 117. P. 1437–1441.

Liardet S., Scaletta C., Al E. Protection against pyrimidine dimers. P. 53. and 8-hydroxy-2'-deoxyguanosine expression in ultraviolet-irradiated human skin by sunscreens: difference between UVB+UVA and UVB alone sunscreens. J. Invest. Dermatol. 2001. 117(6): P. 1437–1441.

Longstreth J., de Gruijl F.R., Kripke M. L., Abseck S., Arnold F., Slaper H. I., Velders G., Takizawa Y., van der Leun J. C. Health risks. J. Photochem. Photobiol. B. 1998. V. 46. P. 20–39.

Lowe N. UVA photoprotection, Sunscreens. N.Y.: Marcel Dekker, 1997. 256 p.

Luo J. Toxicity and bioaccumulation of nanomaterial in aquatic species. Journal of the US SJWP. 2007. V. 2. P. 1–16.

Marchenko V.M., Koltashev V.V., Lavrishchev S.V., Murin D. I., Laser-induced V.G. P. transformation of the microsctucture of SiOₓ, x ≈ 1. Laser Phys. 2000. V. 10, No. 2. P. 576–582.

Marks R., Whiteman D. Sunburn and melanoma: How strong is the evidence?. Br. Med. J. 1994. V. 308. P. 75–76.

Marzulli F., Maibach H. Dermatoxicity. 2 ed. N.Y.: McGraw-Hill, 1983. 327 p.

Mckinlay A. F., Deffey B. L. A reference action spectrum for ultraviolet induced erythema in human skin. Human Exposure to Ultraviolet Radiation: Risks and Regulations. Amsterdam: Elsevier. 1987. P. 83–87.

Meier C., Lüttjohann S., Kravets V.G., Nienhaus H., Lorke A., Wiggers H. Raman properties of silicon nanoparticles. Phisica E. 2006. V. 32. P. 155–158.

Menzel F., Reinert T., Vogt J., Butz T. Investigations of percutaneous uptake of ultrafine TiO_2 particles at the high energy ion nanoprobe LIPSION. Nucl. Instrum. Methods Phys. Res. Sect. B. 2004. V. 219–220. P. 82–86.

Moloney F. J., Collins S., Murphy G.M. Sunscreens. safety, efficacy and appropriate use. Am. J. Clin. Dermatol. 2002. V. 3. P. 185–191.

Moloney F. J., Collins S., Murphy G.M. Sunscreens. Safety, efficacy and appropriate use. Am. J. Clin. Dermatol. 2002. V. 3, No. 3. P. 185–191.

Nel A., Xia T., Madler L., Li N. Toxic potential of materials at the nanolevel. Science. 2006. V. 311. P. 622–627. Newman M.D., Stotland M., Ellis J. I. The safety of nanosized particles in titanium dioxide- and zinc oxide-based sunscreens. J. Am. Acad. Dermatol. 2009. V. 61. P. 685–692.

Neylor M., Kevin C. The case of sunscreens. A review of their use in preventing actinic damage and neoplasia. Arch. Dermatol. 1997. V. 133. P. 1146–1154.

Nohynek G. J., Lademann J., Ribaud C., Roberts M. S. Grey goo on the skin? Nanotechnology, cosmetic and sunscreen safety. Crit. Rev. Toxicol. 2007. V. 37. P. 251–277.

Oberdorster E. Manufactured nanomaterials (fullerenes. C. 60) induce oxidative stress in the brain of juvenile largemouth bass. Environ. Health Perspect. 2004. V. 112. P. 1058–1062.

Oberdorster G., Maynard A., Donaldson K., Castranova V., Fitzpatrick J., Ausman K., Carter J., Karn B., Kreyling W., Lai D. Principles for characterizing the potential human health effects from exposure to nanomaterials: elements of a screening strategy. Part. Fibre Toxicol. 2005. V. 2. P. 8.

Oberdorster G., Oberdorster E., Oberdorster J. Nanotoxicology: an emerging discipline evolving from studies of ultrafine particles. Environ. Health Perspect. 2005. V. 113. P. 823–839.

Öğüt S., Chelikowsky J., Louie S. Quantum confinement and optical gaps in Si nanocrystals. Phys. Rev. Lett. 1997. V. 79. P. 1770–1773.

Pathak M.A. Sunscreens: Topical and systematic approaches for protection of human skin against harmful effects of solar radiation. J. Am. Acad. Dermatol. 1982. V. 7. P. 285–312.

Pathak M.A., Fitzpatrick T. B., Greiter F. J., Kraus E.W. Principles of photoprotection in sunburn and suntanning, and topical and systemic photoprotection in health and diseases. J. Dermatol. Surg. Oncol. 1985. V. 11. P. 575–579.

Pathak S.K., Mason N. J. Our shrinking ozone layer. Resonance. 2002. V. 7. P. 71–80.

Pinnell S. R., Faerhurst D., Gillies R., Mitchnick M.A., Kollias N. Microfine Zinc Oxide is a superior sunscreen ingredient to microfine titanium dioxide. Dermaol surg. 2000. V. 26. P. 309–314.

Popov A. P., Lademann J., Priezzhev A.V., Myllyla R. Effect of size of TiO_2 nanoparticles embedded into stratum corneum on ultraviolet-A and ultraviolet-B sun-blocking properties of the skin. J. Biomed. Opt. 2005. V. 10. P. 064037 (5 p.).

Popov A. P., Priezzhev A.V., Lademann J., Myllyla R. TiO_2 nanoparticles as an effective UV-B radiation skin-protective compound in sunscreens. J. Phys. D: Appl. Phys. 2005. V. 38. P. 2564–2570.

Ranjan V., Kapoor M., Singh V.A. The band gap in silicon nanocrystallites. J. Phys.: Condens. Matter-2002. V. 14. P. 6647–6655.

Roca P., Cabarrocas I., Hammea S., Sharma S.N., Viera G., Bertran E., Costa J. Nanoparticle formation in low-pressure silane plasmas: bridging the gap between a-Si:H and μc-Si films. J. Non-Cryst. Sol. 1998. V. 227–230. P. 871–875.

Roelandts R.N. Shedding light on sunscreens. Clin Exp Dermatol. 1998. V. 23. P. 147–157.

Sahu B. S., Agnihotri O. P., Jain S. C., Mertens R., Kato I.. Appl. Opt. 1990. V. 29. P. 3189–3496.

Schauder S. I. I. Contact and photocontact sensitivity to sunscreens. Contact Dermatitis. 1997. V. 37. P. 221–232.

Schulz J., Hohenberg H. Distribution of sunscreens on skin. Advanced Drug Delivery Reviews. 2002. V. 54. P. S157–S163.

Setlow R.B. The wavelengths in sunlight effective in producing skin cancer: a theoretical analysis. Proc. Natl Acad. Sci. USA. 1974. V. 71. P. 3363–3366.

Shirai H., Arai T., Nakamura T. Control of the initial stage of nanocrystallite silicon growth monitored by in-situ spectroscopic ellipsometry. Appl. Surf. Scie. 1997. V. 113–114. P. 111–115.

Smith R. C., Prezelin B. B., Baker K. S., Bidigare R. R., Boucher N. P., Coley T., Karentz D., MacIntyre S., Matlick H.A., Menzies D. Ozone depletion: ultraviolet radiation and phytoplankton biology in antarctic waters. Science. 1992. V. 255. P. 952–959.

Soni R.K., Fonseca L. F., Resto O., Buzaianu M., Weisz S. Z. Size-dependent optical properties of silicon nanocrystals. J. Lumin. 1999. V. 83–84. P. 187–191.

Stanfield J.W., Fildt P.A., Csortan J. S., Kromal L. Ultraviolet A sunscreen evaluation in normal subjects. J. Am. Acad. Dermatol. 1989. V. 20. P. 744–748.

Sternberg C., Larco O. Sunscreen application and its importance for the protection factor.

Arch Dermatol. 1985. V. 121. P. 1400–1226.

Sutherland B.M., Cimino J. S., Delihas N., Shih A.G., Oliver R. P. Ultraviolet light-induced transformation of human cells to anchorage-dependent growth. Canser Res. 1985. V. 40. P. 2409–2411.

Tarras-Wahlberg N., Stenhagen G., Larko O., Rosen A., Wennberg A., Wennerstom O. Changes in ultraviolet absorption of sunscreens after ultraviolet irradiation. J. Invest. Dermatol. 1999. V. 113. P. 547–553.

Theis W. Optical properties of porous silicon. Surf. Science Rep. 1997. V. 29. P. 91–192.

Tripathy S., Soni R.K., Ghoshal S.K., Jain K. P. Optical properties of nano-silicon. Bulletin of Materials Science. 2001. V. 24, No. 3. P. 285–289.

Tsai C., Li K.H., Campbell J. C., Hance B. V., White J.M. Laser-induced degradation of the photoluminescence intensity of porous silicon. J. Electr. Mater. 1992. V. 21, No. 10. P. 589–591.

Urbach F. Enviromental risk factors for skin cancer. Skin Carcinogenesis in Man and in Experimental Models. In: Recent Results in Cancer Reserch Series 128. Ed. by E. Hecker,

E.G. Jung, F. Marks, W. Tilgen. Berlin: Springer-Verlag, 1993. P. 243–262.

Utz S.R., Knushke P., Sinichkin Yu. P. In vivo evaluation of sunscreens by spectroscopic methods. Skin Res. Technol. 1996. V. 2, No. 3. P. 114–121.

van Buuren T., et al. Changes in the electronic properties of Si nanocrystals as a function of particle size. Phys. Rev. Lett. 1998. V. 80. P. 3803–3806.

van der Leun J. C., de Gruijl F.R. UV-B radiation and ozone depletion: effects on humans, animals, plants, microorganisms, and materials. In: Influences of ozone depletion on human and animal health. Ed. by M. Tevini. Ann Arbor: Lewis Publishers, 1993. P. 95–123.

Wen-Ge D., Jing Y., Ling-Hi M., Shu-Jie W., Wei Y., Guang-Sheng F. Dependence of opical absorption in silicon nanostructure on size of silicon nanoparticles. Commun Theor. Phys. 2011. V. 55. P. 688–692.

Wolf R., Wolf D., Morganti P., Ruocco V. Sunscreens. Clin. Dermatol. 2001. V. 19. P. 452–459.

Wu J., Liu W., Xue C., Zhou S., Lan F., Bi L., Xu H., Yang X., Zeng F.-D. Toxicity and penetration of TiO_2 nanoparticles in hairless mice and porcine skin after subchronic dermal exposure. Toxicol. Lett. 2009. V. 191. P. 1–8.

Xia L., Lenaghan S. C., Zhang M., Zhang Zh., Li Q. RNaturally occurring nanoparticles from English ivy: an alternative to metal-based nanoparticles for UV protection. Journal of Nanobiotechnology. 2010. V. 8, No. 12. http://www.jnanobiotechnology.com/content/8/1/12.

Xie Y.H., et al. Luminescence and structural study of porous silicon films. J. Appl. Phys. 1992. V. 71, No. 5. P. 2403–2407.

Young A.R. Senescence and sunscreens. British J. of Dermatol. 1990. V. 122, No. 35. P. 111–114.

Zhao B., Uchikawa K. McCormick JR, Ni CY, Chen JG, Wang H Ultrafine anatase TiO_2 nanoparticles produced in premixed ethylene stagnation flame at 1 atm. Proc Combust Inst. 2005. V. 30. P. 2569–2576.

Zhu S., Oberdorster E., Haasch M. L. Toxicity of an engineered nanoparticle (fullerene. C. 60) in two aquatic species. Daphnia and fathead minnow. Mar. Environ. Res. 2006. V. 62 (Suppl.). P. S5–S9.

Zhu Y., Wang H., Ong P. P. Preparation and thermal stability of silicon nanoparticles. Appl. Surf. Scie. 2001. V. 171. P. 44–48.

15

Polymer nanocomposites based on nanosilicon

To create new polymeric materials with useful properties based on nc-Si: sunscreen films and coatings, photoluminescent and electroluminescent composites (Kumar, 2008; Olkhov et al, 2010; Bagratashvili et al, 2005; Koch, 2009), light-fast dyes (Ischenko et al, 2010), an important synthetic task is to embed these nanoparticles in the polymer matrix. In the simplest case, the role of the polymer matrix is reduced to ensure that the necessary physical and mechanical properties of the composite (mechanical strength, flexibility, adhesion to the substrate, protection against aggressive media, etc.) are achieved. In addition, the matrix should prevent the agglutination of the nanoparticles and should create suitable conditions for the uniform distribution of nc-Si in the volume of the sample or its fixation on the surface. Finally, if the matrix itself has special optical, conductive and photovoltaic properties, the addition of the functional properties of nc-Si particles will produce a material with a complex of new features.

The first section of this chapter describes a new method for obtaining organic–inorganic hybrid hydrogels that are promising materials for applications in biotechnology, medicine, cosmetics. These gels are systems in which the continuous phase is an organic polymer, and the discrete phase – inorganic nanoparticles. The conditions of formation (the influence of molecular weight, polymer concentration, the polymer–precursor ratio, as well as the start time of gel formation) of the hybrid hydrogels have not been studied sufficiently, which makes it difficult to produce such materials with the desired set of properties.

The second part describes the film material with near-zero spectral transmittance of medium wave UV radiation (290–330 nm), which reduces the yield of vegetable crops because of the energy costs of plant growth to be protected, reduces the shelf life of food products and affects the performance characteristics of components in the electronic industry. The material has a long service life due to increased resistance to photodegradation of the polymer under the action of UV radiation and a significant (~20%) increase in the strength properties.

The third section describes a simple and effective approach to the production of polymer nanocomposites – a method of introducing by diffusion of nanoparticles into the polymer matrix with the use of supercritical fluids (SCF), in particular, supercritical CO_2 as a medium for transport of the nanoparticles in the polymer matrix. This method is quite general and can be used to introduce other nanoparticles (structure-forming, magnetic, etc.) into the polymer matrix to produce functional composite systems with specified properties. The resulting polymer nanocomposites mc-PTFE/nc-Si, having stable bright photoluminescence, can be used, for example, for creating decorative elements, in advertising, for the production of thin polymer films and fibres that are used to protect against forgery of various documents and securities.

One of the promising applications of nanosilicon is to produce electroluminescent materials based on nc-Si. They are composites, in which the nc-Si particles are in a matrix of a conductive polymer forming an interlayer of a certain thickness between the particles. Another intensively developing area of use is associated with the deposition of silicon nanoparticles on flexible polymer substrates to produce solar cells. These issues are described in detail in several monographs (see, e.g. Kumar, 2008; Koch, 2009; Pavesi, Turan, 2010).

15.1. Polymeric hydrogels

One of the types of polymer matrix can be polymeric hydrogels, which have been widely used for more than 50 years for solving scientific and applied problems (Ruben, 1978; Okano, 1998; Galaev, 1995). The known methods for producing hydrogels and polymer gels, in general, are based on the formation of a three-dimensional grid due to the formation of covalent or ionic bonds between the macromolecules. Crosslinking of the polymer chains to each other in the space frame is possible with involvement of hydrogen or coordination bonds, van der Waals forces, and hydrophobic interactions. Macromolecular crosslinking agents may also be nanoparticles of different chemical nature (Averochkina et al, 1993; Lozinsku et al, 2007; Bakeeva et al, 2008; Loos et al, 2003; Kirilina et al, 2009; Bakeeva et al, 2010). Systems where the continuous phase is an organic polymer and the discrete nanophase is represented by inorganic inclusions, and the components interact on the molecular level, are called hybrid materials (nanocomposites), and the corresponding hydrogels – organic–inorganic hybrid hydrogels (OIHH).

Zubov et al (2011) developed approaches to obtaining OIHH based on polyvinylpyrrolidone (PVP) and the products of hydrolytic polycondensation of tetramethoxysilane (TMOS) as structure formers, with the inclusion of nc-Si particles. The spectral properties and potential areas of use of such materials have been studied.

The synthesis of nc-Si was performed in an argon plasma in a closed gas loop. The plasma reactor was an evaporator–condenser, working in the low-frequency arc discharge. The raw material was a powder of silicon which was fed to the reactor in a gas stream from a dosing device. In the reactor the powder evaporated at a temperature of 8000–10000°C. At the exit from the high-temperature plasma zone the resultant gas–vapour mixture was subjected to rapid cooling by gas jets, thus condensing the silicon vapour with the formation of an aerosol. The finished powder was collected by a fabric filter. From the filter the powder was discharged under an inert atmosphere into a box in a sealed container or moved in the microencapsulation system, where the surface of the powder particles was deposited with an inert protective layer, protecting the powder from inclement weather (Tutorsky et al, 2005; Koch, 2009).

Initial testing of the properties of nc-Si was carried out using complementary techniques (Rybaltovsky et al, 2007). The visual picture of the nc-Si powder particles was produced by electron microscopy. Electron micrographs of samples of nc-Si (0.1 wt.%) were obtained with an LEO912AB OMEGA electron microscope. Visual image analysis revealed that the particles of nc-Si have a spherical shape with a fuzzy outline; perhaps this is he manifestation of the shell of the particles. High-resolution TEM micrographs, obtained by varying the magnification, allowed by using the UTHSCSA Image Tool software to classify the particle size, to construct the distribution function for all the samples and to determine the average nc-Si particle diameter, which was equal to 27±3 nm.

IR spectra were recorded using a Nicolet Impact 410 FTIR spectrometer operating in the wavelength range of 2.5–25 μm (4000–400 cm^{-1}), with the diffuse reflectance attachment for measuring the spectra of the powdered materials. For these studies, a thin layer of powder with a thickness of several microns was placed in a specially prepared cell with windows made of polished silicon plates. When measuring the IR spectra of nc-Si, the spectral dependences of the absorption coefficient showed intense absorption bands located at frequencies of 461, 799, 978, 1072 and 1097 cm^{-1}. The appearance of intense bands indicated the formation of SiO_2 or SiO_x ($x = 1.5$–2.0).

Raman spectra (RS) of light were measured with a T-64000 monochromator (Jobin Yvon) with the exciting radiation from an argon laser ($\lambda = 514.5$ nm). The RS band of the nc-Si powder was shifted to lower frequencies by 1.5–2.5 cm^{-1} relative to the single-crystal silicon. In addition, the RS peak for the powdered silicon was on average ~25% wider than the peak of single crystal silicon, whose width was ~4 cm^{-1}. To calculate the average diameter of nc-Si, we can use the model of the spatial constraint of phonons and scattering of light on them in a spherical crystallite. Using the procedure described by Rybaltovsky et al (2007), one can determine the size of the core of the nc-Si particles, which was found to be 17±2 nm.

To study the transmittance spectra of these composite materials in the range 200–850 nm, Zubov et al (2011) used a Specord-M 40 spectrophotometer (Carl Zeiss, Jena). The measurements were carried out using two methods. The first technique allowed to measure the collimated transmittance in the direction parallel to the probe radiation incident on the sample, which corresponded to the change in optical density at the wavelength of observation. The results are shown in Fig. 15.1a.

In the second method, measurements of the transmittance spectra in the work of Zubov et al (2011) were taken using an integrating sphere, which was placed under the sample to be studied (a technique described in section 14.4.2 of this monograph). In this case, the photodetector collected not only the light directly transmitted through the layer of the sample but also the radiation scattered into the solid angle 2π. A detailed description of the methodology of spectral analysis of nc-Si is presented in Rybaltovsky et al (2007).

X-ray powder diffraction measurements of the produced nc-Si were carried out in a diffractometer with the Guinier geometry (diffraction of the focused X-ray beam transmitted through the powder sample): G670 camera (Huber) with a curved Ge(111) monochromator of the primary beam, selecting the line $K\alpha_1$ (wavelength $\lambda = 1.540598$ Å) of the characteristic radiation of the X-ray tube with a copper anode. The diffraction pattern in the range of diffraction angles 2θ from $3°$ to $100°$ was registered with a flexible imaging plate (IP) detector bent according to the camera radius of

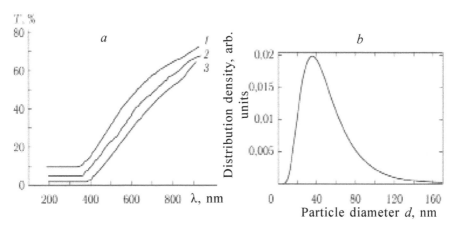

Fig. 15.1. *a* – absorption spectra of the PVP–nc-Si/SiO$_2$ film: *1* – concentration of nanosilicon 0.5 wt.%; *2* – 1.0 wt.%; *3* – 1.5 wt.%. *b* – the function of the size distribution density of the crystallites of nc-Si, obtained on the basis of simulation of the complete profile of the X-ray diffractograms and refinement of the model parameters by fitting to experimental diffractogram of the powder (the model formalism includes the instrumental function, the coherent size of spherical particles *d*, the distribution function Log $N(d)$, the dislocation broadening according to the Krivoglaz–Wilkens theory and the effect of stacking faults in the theory of B. Warren). From an article by Zubov et al, 2011.

90 mm. The phase composition of the samples was evaluated by comparing the diffraction patterns with each other and with the reference diffraction patterns. The particle size distribution was determined by modelling the microstructure of the particles (Scardi, 2008)[1], building a complete profile of the diffraction patterns taking into account the instrumental function of the diffractometer and the lognormal distribution of particle sizes. The model parameters were refined by fitting the model by the non-linear least-squares method to the complete profile of the experimental diffraction pattern (Figure 15.1b) using a PM2K computer program (Leoni & Confente, 2006).

It has been shown that OIHH forms on the basis of linear water-soluble polymers (poly-N-vinylpyrrolidone (PVP), poly-N-vinylcaprolactam) and inorganic nanoparticles – the products of hydrolytic polycondensation of tetramethoxysilane (TMOS) (Loos et al, 2003; Kirilina, 2009; Bakeeva et al, 2010). A gel network in these systems formed at the expense of sustainable hydrogen bonds between the surface HO-groups of silica and O = C-groups located in the lateral substituents of chain links of poly-N-vinylamides. Since the surface layers of nc-Si particles may contain HO-groups as a result of the oxidation of silicon, we can expect that these particles will also participate in the formation of bonds with hydrophilic polymer chains and become nodes in OIHH.

The experiments carried by Zubov et al (2011) confirmed that by varying the concentration of nc-Si particles in the range of 2–5 wt.% in a water solution of PVP (average molar mass, $\overline{M}_w = 1.3 \cdot 10^6 D$) we can produce OIHH, which is a very dark-coloured material that does not have satisfactory physical and mechanical properties and fully absorbs light in the UV range. A method of producing mixed-type OIHH was proposed for the manufacture of homogeneous transparent films with the required physico-mechanical and optical characteristics. In this case, base-forming particles are silica particles derived from sol–gel transitions in a solution of TMOS in a PVP solution, and 'fellow participants' in the gelation process are the nc-Si particles.

The initial mixture is an aqueous solution of PVP, to which added different amounts of an aqueous dispersion of particles of nc-Si (0.5–1.5 wt.%) and TMOS were added. In the course of gelation there was no aggregation of nc-Si particles and no separation of components. The properties of uniform/homogeneous OIHH (Table 15.1, Fig. 15.2) depend on both the polymer concentration and the amount of injected TMOS, whereas the given amounts of nc-Si particles practically do not change them. The physical and mechanical properties of OIHH were measured as in deformation by uniaxial tension and uniaxial compression. The calculation of the elastic modulus (E) was performed by the standard equation (Kuleznev, Shershnev, 2007).

The measurement results are presented in Fig. 15.2 and show that the elastic modulus of gels (E) increases with increasing concentration of PVP and TMOS, and the equilibrium degree of swelling decreases, which indicates an increase in the effective crosslink density of the gel.

---

[1]See also chapter 10 of this book.

Table 15.1. The influence of the PVP–water–TMOS mixture on the elastic modulus of the OIHH gel (E) in uniaxial tension (Zubov et al, 2011)

The initial concentration of aqueous solution of PVP, wt.%	The concentration of TMOS in the reaction mixture, wt.%	E, kPa
6	3.9	0.7
	5.4	1.1
8	4.2	1.3
	7.2	1.7
10	3.9	2.3
	6.4	4.0
	9.0	10.0

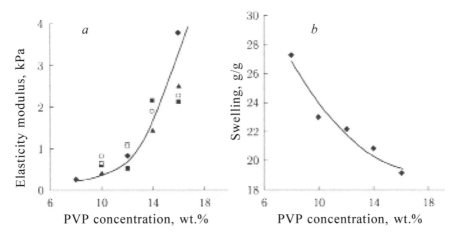

Fig. 15.2. a – The dependence of the elastic modulus of samples of OIHH PVP–nc-Si/SiO$_2$ on the concentration of PVP at a constant concentration in the reaction mixture of TMOS with 6.2 wt.% (uniaxial compression). b – dependence of the equilibrium degree of swelling of OIHH PVP–nc-Si/SiO$_2$ on the concentration of PVP at a constant concentration in the reaction mixture of TMOS 6.2 wt.% (Zubov et al, 2011).

The optical properties of OIHH are very different after adding the nc-Si particles to their composition. If OIHH without nc-Si particles were optically transparent in the wavelength range of 200–1000 nm, after adding the nc-Si particles the material obtained had uniform yellow–brown colour, the intensity and colour of which are dependent on the nc-Si content. Figure 15.1a shows the transmittance spectra of OIHH obtained using the nc-Si of different concentrations.

The nc-Si particles have a strong influence on the absorption of light, which increases with increasing nc-Si particle concentration. To study the optical characteristics, OIHH were formed on quartz glass plates. After evaporation of water the produced films had good adhesion to the substrate and good mechanical strength.

Thus, in the work of Zubov et al (2011) the authors proposed a new approach to the inclusion of nc-Si in the matrix of mixed OIHH based on linear water-soluble polymers in which the role of the structure builder is played by silica nanoparticles, and the optical properties are provided by the presence of a given number of the nc-Si particles. Obviously, this approach can be used to include other nanoparticles (fluorescent, magnetic, etc.) in the polymer matrix in order to produce functional composite systems with desired properties.

15.2. Nanocomposite films with UV-protective properties based on polyethylene

It is known that UV radiation in the spectral range 290–330 nm considerably reduces the yield of vegetable crops due to the energy cost of growth of plants to be protected from it (Briston, Katan, 1993). UV radiation in the 200–420 nm range reduces the shelf life of food products (Russian Federation patent No. 2 000 130 235) and impairs the performance of electronic industry components (Russian Federation patent No. 2 001 121 489). It was also found that the freshness of the food, sealed in clear food packaging, in the presence of oxygen absorbers can suddenly improve, if the transparent packaging material contains a UV absorber (Russian Federation patent No. 2 222 560). There are various ways to protect against UV radiation. Typically, the polymers are injected with substances containing chromophoric groups that absorb ultraviolet light, or transform it to infrared radiation.

Introduction to polymers of nanoparticles of metal oxides and non-metals, with the screening effect of UV radiation, offers extensive opportunities to create nanopolymer thermoplastic composites based on thermoplastics with different chemical structure with a high level of performance. Such materials in the form of films will find applications in agriculture, packaging industry and medicine.

Ol'khova et al (2010) obtained film materials with UV-protective properties, based on low-density polyethylene and nanocrystalline silicon synthesized by the plasma-chemical method.

The starting materials for the preparation of thin-film nanocomposites were low-density polyethylene (LDPE), grade 10803–020, and ultrafine crystalline silicon. Silicon powders were prepared by plasma chemical recondensation of coarse crystalline silicon powder (nc-Si). The synthesis of nc-Si was carried out in an argon plasma in a closed gas cycle in a plasma evaporator–condenser, running in an arc low-frequency discharge. After synthesis, the nc-Si particles were subjected to microencapsulation, which created on their surfaces a protective shell of SiO_2, protecting the powder from the inclement weather and making it stable in the storage. In

Ol'khov et al (2010) silicon powders of two batches were used: nc-Si-36 with the specific surface of particles of ~36 m^2/g, and nc-Si-97 with the specific surface of ~97 m^2/g, according to the BET method.

Pre-mixing of the polyethylene powder with the nc-Si powder was performed using a closed cam chamber of the company Branberder (Germany) at a temperature of 135±5°C, for 10 min, the rotor speed was 100 min^{-1}. Two compositions of LDPE+nc-Si were prepared: 1) the composition PE+0.5% nc-Si-97 based on nc-Si-97, containing 0.5 wt.% silicon, 2) composition PE+1% nc-Si-36 based on nc-Si-36, containing 1.0 wt.% silicon.

Films 85±5 mm thick were produced in a semi-industrial extrusion unit for tubular films ARP-20–150 (Russia) at temperatures of 120–190°C in the zones of the extruder and the extrusion head and the speed of the screw of 120 min^{-1}. The technological parameters of nanocomposites were selected on the basis of the conditions of thermal stability and the characteristic melt viscosity of the polymer recommended for treatment.

Ol'khov et al (2010) investigated the mechanical properties and optical transparency of the polymer films, their phase composition and crystallinity, as well as the relationship of the mechanical and optical properties with the microstructure of polyethylene and the grain size of the nc-Si modifying powder.

The physical and mechanical properties of the films under tension (in the extrusion direction) were measured using an EZ-40 universal tensile machine (Germany) according to the GOST 14236–71 standard. Tests were carried out on rectangular samples 10 mm wide with the working section of 50 mm. The speed of the clamp was 240 mm/min. Tests were conducted on a set of five parallel samples.

The optical transparency of the films was evaluated by absorption spectra. The absorption spectra of the films were measured using an SF-104 spectrophotometer (Russia) in the wavelength range 200–800 nm. To obtain absorption spectra of the polyethylene films and composite films of PE+0.5% nc-Si-97 and PE + 1% nc-Si-36, the samples of size 3×3 cm, secured in a special holder for ensuring uniform tensioning of the film, were studied.

To investigate the phase composition of materials, the degree of crystallinity of the polymer matrix, the size of single-crystal blocks in the nc-Si powders and in the polymeric matrix, as well as the size density distribution function (SDDF) of the crystallites in the starting nc-Si powders, the authors used X-ray analysis on the basis of the data of wide-angle scattering of monochromatic X-rays.

X-ray diffraction measurements were carried out in a Guinier chamber G670 Huber, described earlier in this chapter. The measurements were performed on the initial powders of nc-Si-36 and nc-Si-97, on a pure LDPE film, labelled here as PE, and the composite films PE+0.5% nc-Si-97 and PE+1.0% nc-Si-36. To eliminate the effect of instrumental distortion, measurements were also taken on the standard diffraction pattern SRM660a NIST of the crystalline powder LaB_6, certified by the National Standard Institute of the USA, which were subsequently used as the instrumental function of the diffractometer.

Samples of the initial powders of nc-Si-36 and nc-Si-97 for X-ray diffraction were prepared by applying a thin layer of a powder on a substrate made of a special film with a thickness of 6 mm (MYLAR, Chemplex Industries Inc., Cat. No: 250, Lot No: 011671). Film samples of LDPE and its composites were placed in the holder of the diffractometer without any substrate, but to minimize the effect of the texture two layers of the film, oriented by the extrusion lines perpendicular to each other, were used.

The X-ray diffraction data for the phase and particle size analysis were interpreted (Fetisov et al, 2010) using two different methods of full-profile analysis: 1) the approximation of the diffraction profile by analytic functions, polynomials and splines with the decomposition of diffraction patterns into its component parts, and 2) a method for modelling diffraction patterns based on the physical principles of x-ray scattering. To approximate and decompose the profile of the diffraction patterns, the authors used the software package WinX^{POW} ver. 2.02 (Stoe, Germany) (WINXPOW Version 1.06, 1999), and the simulation of diffraction patterns in the analysis of the particle size distribution was carried out using the PM2K software (Leoni & Confente, 2006, version 2009). The results of mechanical tests of the material prepared in Ol'khov et al (2010) are presented in Table 15.2, which shows that the addition of nc-Si particles improved the mechanical properties of polyethylene.

The results presented in Table 15.2 show that the addition of silica powders increases the mechanical properties of the films, and the effect of improving the mechanical properties is more pronounced in the case of the PE+0.5% nc-Si-97 composite, where the relative elongation at fracture significantly increased in comparison with pure polyethylene. The transmittance spectra of the investigated films are shown in Fig. 15.3.

It is seen that the addition of the nc-Si powders reduces the transparency of the film in the entire investigated wavelength range, but a particularly strong decline in transmittance (almost 20 fold) was observed in the wavelength range 220–400 nm, i.e. in the UV region.

The phase composition of the materials and their components in Ol'khov et al (2010) was investigated using the data obtained in wide-angle X-ray

Table 15.2. Mechanical characteristics of nanocomposite films based on LDPE and nc-Si (Ol'khov, etc. 2010)

Sample	Tensile strength, kg/cm^2	Elongation at rupture, %
PE	100 ± 12	200–450
PE + 1% nc-Si-36	122 ± 12	250–390
PE + 0.5% nc-Si-97	118 ± 12	380–500

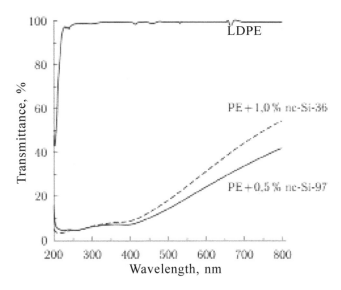

Fig. 15.3. Transmittance spectra of LDPE films and nanocomposite films PE+0.5% nc-Si-97 and PE+1.0% nc-Si-36 (Ol'khov, et al, 2010).

scattering. The measured X-ray diffraction patterns of the starting powders nc-Si-36 and nc-Si-97 as regards the intensity and position of the Bragg peaks are completely consistent with the phase of pure crystalline silicon (the cubic unit cell of diamond-type, space group $Fd\bar{3}m$, cell parameter a_{Si} = 0.5435 nm). The size distribution density function (SDDF) of the crystallites in a powder was restored by X-ray powder diffraction patterns using the computer program PM2K (Leoni & Confente, 2006), which uses the method of modelling the complete profile of the diffraction pattern based on the theory of the physical processes of diffraction of X-ray rays (Scardi, 2008, chapter 10 of this monograph). Simulation was carried out assuming the spherical form of the crystallites and the lognormal distribution of their size, and the deformation effects from linear and planar defects in the crystal lattice were taken into account. The SDDFs obtained for the initial nc-Si powders are shown graphically in Fig. 15.4, with the caption giving the statistical parameters of the distributions. These distributions are characterized by important parameters such as Mo(d) – the position of maximum SDDF (distribution mode); $\langle d \rangle_V$ is the average crystallite size of the sample volume (mean arithmetic size), and Me(d) is the median of the distribution, which determines the size d, indicating that the particles with diameters smaller than this size are half the powder volume.

The results shown in Fig. 15.4 show that the initial nc-Si powders contain particles smaller than 10 nm, which particularly efficiently absorb UV radiation. The SDDF modes of both powders are very close, but the median SDDF of the nc-Si-36 powder is much larger than that of the nc-

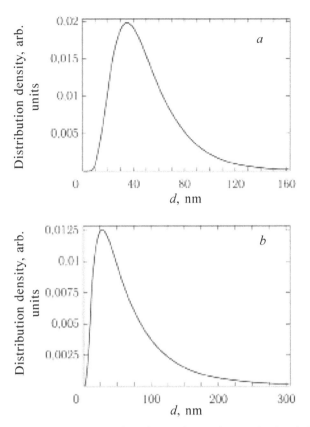

Fig. 15.4. SDDF of the crystals of nc-Si powders, obtained from X-ray diffraction patterns using PM2K software(see Ol'khov et al, 2010): a – nc-Si-97, $Mo(d) = 35$ nm, $Me(d) = 45$ nm, $\langle d \rangle_V = 51$ nm, b – nc-Si-36, $Mo(d) = 30$ nm, $Me(d) = 54$ nm, $\langle d \rangle_V = 76$ nm.

Si-97 powder. This suggests that the number of crystallites with diameters smaller than 10 nm per unit volume of the nc-Si-36 powder is much smaller than in the unit volume of the nc-Si-97 powder. In the composition of the nc-Si-36 powder there is quite a lot of particles with a diameter greater than 100 nm and even larger particles with 300 nm, whereas the size of the particles in the nc-Si-97 powder is not greater than 150 nm, while the bulk of the crystallites have a diameter less than 100 nm.

The phase composition of the films was evaluated on the wide-angle diffraction patterns of X-ray scattering only qualitatively, because due to the complexity of the scattering diffraction patterns and the presence of the texture the quantitative phase analysis of polymer films is practically impossible (Strbeck, 2007). In the phase analysis of polymers it is often necessary to settle for comparative qualitative analysis, which allows one to monitor the evolution of the structure depending on the parameters of the production technology.

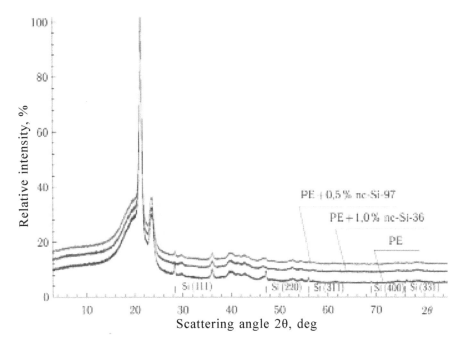

Fig. 15.5. Diffraction patterns of the investigated composite films in comparison with the diffraction pattern of pure polyethylene. At the bottom the vertical lines show the reference position of the diffraction lines of silicon with their interference indices (*hkl*) (Ol'khov et al, 2010).

The measured diffraction patterns of wide-angle X-ray scattering by the investigated films are shown in Fig. 15.5. The diffraction patterns are typical of the type of polymer. As a rule, polymers are two-phase systems consisting of an amorphous phase and regions with the long-range order, conventionally called crystals. Their diffraction patterns are the superposition of the intensity of scattering by the amorphous phase, having the form of a broad halo in the low-angle region (in this case, in the region 2θ between 10° and 30°), and the peaks of the intensity of the Bragg scattering of the crystalline phase.

The data in Fig. 15.5 are presented in the scale of relative intensities (the intensity of the highest peak of the diffraction pattern is taken as 100%). For ease of review the curves are shown with an offset on the ordinate. If the scattering graphs are depicted without the offset, the profiles of the diffraction patterns of composite films are completely consistent with the diffractogram of the pure LDPE film, with the exception of the peaks of crystalline silicon which are not on the PE diffraction pattern. This indicates that the addition of the nc-Si powder has not changed the crystal structure of the polymer. The diffractograms of the films with silicon show clearly distinguishable peaks of crystalline silicon (at the bottom the primes show their reference position with the corresponding indices of interference). The heights of the identical silicon peaks (i.e. peaks with the same index) on the diffraction patterns of the composite films PE + 0.5% nc-Si-97 and PE + 1.0% nc-Si-36 differ by about a half, which corresponds to the ratio of mass concentrations of Si, defined in their production.

The degree of crystallinity of the polymer films (the volume fraction of crystalline ordered regions in the material) in Ol'khov et al (2010) was determined on the diffraction patterns in Fig. 15.5 for a series of samples only semi-quantitatively (more/less). The essence of the method for determining the crystallinity is analytical separation of the diffraction pattern profile to Bragg peaks from crystalline regions and the diffuse peak of the amorphous phase (Strbeck, 2007), as shown in Fig. 15.6.

The profiles of the peaks, including the peak of the amorphous phase, are approximated by a pseudo-Voigt function, the background – by the Chebyshev polynomials of the fourth order. The non-linear method of the least-squares was used to minimize the difference between the intensity of pixels of the approximating and experimental curves. At the same time, the

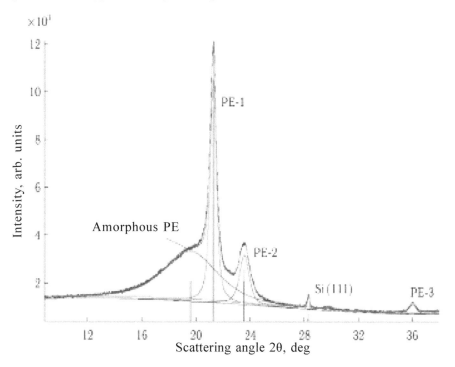

Fig. 15.6. Decomposition the diffraction pattern into individual diffraction peaks and the background using the approximation of the complete profile by analytic functions on the example of the data from a PE + 1% nc-Si-36 sample. PE-n denotes the Bragg peaks of crystalline polyethylene with order numbers n from left to right. Si(111) is the Bragg peak of silicon nc-Si-36. Vertical dashes indicate the positions of the maxima of the peaks (Ol'khov et al 2010).

Table 15.3. The characteristics of ordered (crystalline) regions in polyethylene and its composites with nc-Si (Ol'khov et al, 2010)

Samples	PE			PE + 1% nc-Si-36			PE + 0.5% nc-Si-97		
Crystallinity	46%			47.5%			48%		
2θ,°	d, Å	ε	2θ, °	d, Å	ε	2θ, °	d, Å	ε	
21.274	276	8.9	21.285	229	7.7	21.282	220	7.9	
23.566	151	12.8	23.582	128	11.2	23.567	123	11.6	
36.038	191	6.8	36.035	165	5.8	36.038	162	5.8	
Average values	206	$9.5 \cdot 10^{-3}$		174	$8.2 \cdot 10^{-3}$		168	$8.4 \cdot 10^{-3}$	

width and the height of the approximating functions, the position of their maxima and integral areas, as well as the parameters of the background were clarified. The ratio of the integrated intensity profile of scattering of the amorphous phase to the total integrated intensity of the scattering of all phases except for the crystalline silicon particles gives the fraction of the amorphous state of the sample, and the degree of crystallinity is obtained as the difference between unity and the amorphous fraction.

It was assumed that, thanks to a production technology, the films have the identical texture, which is confirmed by the coincidence of the relative intensities of all the diffraction peaks in Fig. 15.5, and the samples consist of only the crystalline and amorphous phases of the same chemical composition. Therefore, the obtained values of the degree of crystallinity must correctly reflect the trend of its change in modification of LDPE with the nc-Si powders, although due to the texture of the films they can quantitatively differ significantly from the actual concentration of crystalline regions in the material. The obtained values of the degree of crystallinity are shown in Table. 15.3.

Another important characteristic of the crystallinity of the polymer is the size d of the ordered regions in it. To determine the crystallite size and their maximum strain ε, X-ray analysis is often carried out using the width of the Bragg peaks at half maximum intensity (Iveronova, Revkevich, 1978), the so-called half-width of the Bragg line. In this study, the dimensions of the crystallites in the polyethylene matrix were calculated for three well-pronounced peaks in the diffraction patterns of the films in Fig. 15.5. The polyethylene peaks are located at angles 2θ approximately equal to 21.28°, 23.57° and 36.03° (see the peaks of PE-1, PE-2 and PE-3 in Fig. 15.6). The size d of the ordered regions and the maximum relative deformation ε of the lattice are calculated by joint solutions of Scherrer and Wilson equations (Iveronova, Revkevich, 1978) using the half-widths of the peaks identified by the approximation by analytical functions and taking into account the experimentally measured instrumental function of the diffractometer. The calculations were performed using WinX^{POW} size/strain programme. The resultant values of d and ε as well as their average values for the investigated films are presented in Table 15.3. The table lists the refined positions of the maximum peaks considered in the calculations.

Ol'khov et al (2010) proposed a technology to obtain LDPE films and composite films LDPE + 1% nc-Si-36 and LDPE + 0.5% nc-Si-97 of the same thickness (85 μm). The concentration of modifying agents nc-Si in the composite films is consistent with a given composition, as confirmed by X-ray phase analysis. Direct measurements revealed that the addition of nc-Si powders reduced the transparency of the polyethylene over the entire wavelength range, but a particularly strong decrease in bandwidth (up to 20 times) was observed in the wavelength range 220–400 nm, i.e. in the UV region. The suppression effect of UV radiation is especially strong in the LDPE+0.5% nc-Si-97 film, although the concentration of the silicon additives in this material is smaller. This result can be explained by the fact that, according to the experimentally obtained SDDF the number of particles with dimensions smaller than 10 nm per unit volume/mass of the nc-Si powder is greater than in the nc-Si-36 powder.

Direct measurements in Ol'khov et al (2010) were used to determine the mechanical characteristics of the films – tensile strength and relative elongation at rupture (Table 15.2). The results show that the addition of silicon powders increases the strength of the films by about 20% compared with pure polyethylene. The composite films compared to pure polyethylene also have a higher elongation at rupture; this improvement is especially pronounced in the PE + 0.5% nc-Si-97 composite. The observed improvement in the mechanical properties correlates with the degree of crystallinity of the films and the average size of the crystalline blocks in them (Table 15.3). According to the results of X-ray analysis, the highest crystallinity was found in the LDPE + 0.5% nc-Si-97 film, which also contained the smallest crystalline ordered regions, which should help to increase strength and ductility.

15.3. Immobilization of luminescent nanosilicon in the matrix of microdispersed polytetrafluoroethylene with supercritical carbon dioxide

One of the simplest and most effective new approaches to the creation of polymer nanocomposites is the method of implementation of the diffusion of nanoparticles in the polymer matrix using supercritical fluids (SCF), in particular, supercritical CO_2 as a medium for transport of nanoparticles into the polymer matrix. This approach, proposed and implemented in the works by Bagratashvili, et al (2010a), Nikitin, et al (2008) includes: 1) the effect of swelling of the polymer in SCF, leading to an increase in the internal free volume of the polymer matrix, 2) effective delivery (transport) of nanoparticles into the internal free volume of the polymer due to extremely low values of surface tension and viscosity in the SCF conditions, and 3) compression (shrinkage) of the polymer with the nanoparticles introduced into its free volume, promoting matrix immobilization (matrix isolation) of the nanoparticles in the volume of the polymer and preventing their subsequent agglutination.

The approach based on the introduction by diffusion of the nanoparticles into polymers with the SCF has certain advantages over other methods of production of luminescent nanocomposites. They include the relative ease of and ecological purity of the process, effective solution of the problem of agglomeration of nanoparticles, the lack of the negative influence of the diffusion medium (e.g. organic solvents) on the properties of the polymer composite. The composite then may become more homogeneous and optically transparent due wash-out of its low molecular weight fractions (Bagratashvili et al, 2010a; Glagoleva et al, 2010).

In Bagratashvili et al (2010b) the method of immobilization of the matrix with the SCF was used to produce photoluminescent polymer nanocomposites based on photoluminescent nc-Si particles and microparticles of polytetrafluoroethylene (n-PTFE). The structural characteristics and spectral properties of the original nc-Si and the polymer nanocomposite – mc-PTFE/nc-Si – were studied. The fine PTFE powder with a particle size of 1–2 μm was obtained by heating industrial fluoroplastic, which transforms a block polymer to the fine-dispersed state with a fraction of the amorphous phase of 90–95 wt.%. The fine powder can be formed into film samples by pressing.

Synthesis of nanocrystalline silicon is based on the reaction of disproportionation of silicon monoxide when heated:

$$2SiO \rightarrow SiO_2 + Si \tag{15.1}$$

The reaction product was subjected to etching in concentrated hydrofluoric acid (HF, 49 wt.%) to remove the SiO_2, formed by heating SiO. Upon completion of the etching process, nc-Si was washed with methanol and distilled water to stop the etching process and remove HF. The detailed description of the synthesis technique is given in Dorofeev et al (2012).

Protection of the surface of silicon nanoparticles. The procedure of thermally-activated hydrosilylation of the surface of the nc-Si particles (Dorofeev et al, 2012), filled with fragments (Si–H) under the scheme:

was used.

The organic compound containing a terminal double bond, used in the hydrosilylation reaction was 1-octadecene. The process of hydrosilylation of the nc-Si surface leads to a significant increase in the intensity of photoluminescence of nanosilicon. The photoluminescence of nc-Si in the solution remains stable for 7 months (maximum observation time) in relation to the time of hydrosilylation of the surface of the nanoparticles.

Thermogravimetric analysis. The nc-Si samples, pre-dried at $120\,^{\circ}C$, whose surface was covered with 1-octadecene, were studied by the methods of differential scanning calorimetry (DSC) and thermogravimetric analysis (TGA) at a heating rate of 5 deg/min in air from $20\,^{\circ}C$ to $400\,^{\circ}C$ (NETZSCH STA 409 PC/PG). The mass of the sample, placed in a corundum crucible, was 6 mg.

Synthesis of nanocomposites. Experiments with the diffusion SCF introduction of nc-Si in the microdispersed polymer powders were conducted in the following sequence (Bagratashvili et al, 2010b). A high-pressure chamber made of stainless steel was loaded with polymeric microparticles, mc-PTFE (in the amount of 200 mg), dispersed in nc-Si sol on the basis of hexane (~0.5 ml). The resulting suspension was placed in a container made of polystyrene, closed on one side with filter paper with microorifices for free access of supercritical CO_2. The chamber with the container was closed, sealed, and then was filled with gaseous carbon dioxide and the pressure was increased to 10 MPa at room temperature. The heater was then switched on and the temperature was raised to $60\,^{\circ}C$. Upon reaching the desired temperature the pressure of CO_2 was increased up to 20 MPa and the system was under these conditions for 1 h. Thereafter, the chamber pressure was gradually lowered to atmospheric pressure, the system in the reactor was allowed to cool down to room temperature, and the samples were then removed from the reactor. A detailed description of the methodology is contained in Bagratashvili et al (2010a) and Glagoleva et al (2010).

Structural and spectral studies. The nc-Si samples obtained in Bagratashvili et al (2010b) were studied by transmission electron microscopy (LEO 912 AB OMEGA, accelerating voltage 200 keV). Radiographs of small-angle scattering were measured in a SAXSess diffractometer (Anton Paar, Austria): monochromatization and focusing of the primary beam from the linear focus of the X-ray tube by a special synthetic multilayer thin-film elliptical mirror, the wavelength of the radiation λ (Cu K_α) = 1.5418 Å; radiation was registered with a curved X-ray-sensitive plate with optical memory (IP-detector), the image from which was read in the digital form with a special laser scanner and converted into the one-dimensional 'intensity– diffraction angle' distribution by the program SAXSquant which is part of the diffractometer.

The Raman spectra (RS) of nc-Si particles and the powders of the resulting polymer nanocomposite mc-PTFE/nc-Si were registered in the equipment the basis of which was a triple monochromator T-64000 (Jobin Yvon, France) with the exciting radiation from an argon laser (λ = 514.5 nm). A single-crystal silicon plate was used as a standard. Measurements of RS were taken at the minimum laser power to eliminate effects due to heating of the sample when exposed to radiation.

The spectra of photoluminescence (PL) of nc-Si were studied using diode lasers at excitation wavelengths of 407 nm and 532 nm and also

LED emission (~405 nm at the maximum intensity). In this case, the PL spectra were recorded on a Avantes – AVA-Spec-2048 spectrophotometer. In addition, the PL spectra of nc-Si particles and mc-PTFE/nc-Si powders were excited by the same argon laser; however, the exciting radiation was the line $\lambda = 458$ nm (power 20 mW). Transmittance spectra were recorded in the 200–1100 nm range on the two devices (Avantes-AVA-Spec-2048 and SF-104, Interphotophysics).

Initial testing of the properties of the synthesized nc-Si particles was carried out using complementary techniques (Bagratashvili et al, 2010b). Analysis of the images of TEM microphotographs (Fig. 15.7a) revealed that the maximum size of the synthesized nc-Si particles in all the samples did not exceed 5 nm. The diffraction rings were identified with atomic planes of silicon by comparing their diameter with the diameter of the circles in the electronic diffraction structure of gold, registered on the same microscope. From this comparison, the distance between the atomic planes in the studied nc-Si and the Miller indices corresponding to these distances was determined. The analysis results show that the diffraction rings closest to the centre correspond to the atomic planes of silicon (111), (220), (311) and (331). The radiographs of nc-Si sols in hexane was recorded in a quartz capillary tube in the transmission mode in the range of diffraction angles 2θ from $0°$ to $30°$. Before analysis the experimental data were processed by: 1) the mathematical transformation of the intensity–the scattering angle data sets, measured on the linear focus of the X-ray tube, to the form which they should have at point collimation of the primary beam of X-rays, and 2) the allocation of the signal of the small-angle scattering by subtracting the background of the capillary and the solvent.

The distribution density function (DDF) of the size of nc-Si particles was determined by Fourier analysis of the received small-angle scattering signal using computer program GIFT (Bergmann et al, 2000). In analysis, the intensity of small-angle scattering was transformed, using indirect Fourier transform, from the momentum space to the direct space of the dimensions and the non-linear method of least squares was used to search the size distribution of the nanoparticles that best fits the experimental data (Bergmann et al, 2000a; Bergmann et al, 2000b).

Thermogravimetry data obtained in air on the solid nc-Si sample, covered with octadecene, are shown in Fig. 15.8. A small weight loss in the temperature interval 170–220°C is connected, apparently, with the loss of the residual solvent (hexane, from sol, in which the nanoparticles were precipitated) in the shell of 1-octadecene. At temperatures above 220°C the rate of the weight loss was quite high and accompanied by a significant endothermic effect due to the likely expansion of the shell. Thus, it can be concluded that the investigated nanoparticles are quite stable in air at temperatures below 220°C.

Transmittance spectra of nc-Si, dispersed in hexane, were measured in the range from 200 nm to 1100 nm. Taking into account the

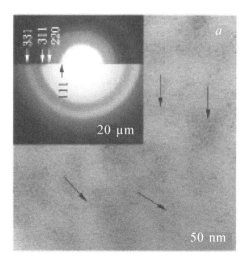

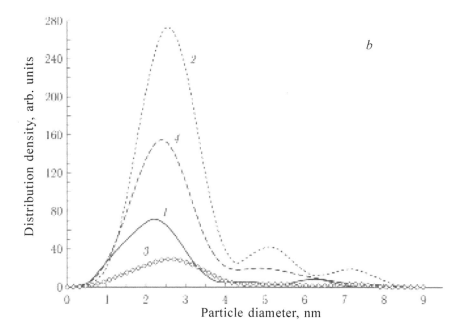

Fig. 15.7. *a* – TEM photomicrograph of nc-Si. The arrows indicate the larger particles size in the range of 5–7 nm. The inset shows a part of the diffraction pattern obtained with increased exposure time. *b* – functions of the size distribution density of the particles, determined from the analysis of small-angle X-ray scattering. *1–4*) the numbers denote the samples obtained at temperatures of 200, 300, 450 and 950°C respectively (Bagratashvili and others, 2010*b*).

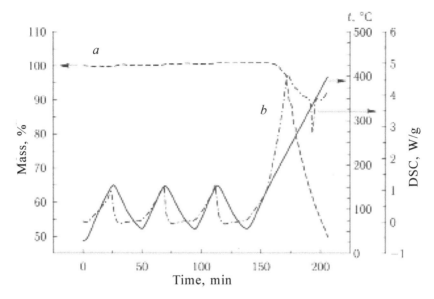

Fig. 15.8. The results of thermogravimetric analysis (*a*) and differential scanning calorimetry (*b*) of the sample of nc-Si, coated with a shell of 1-octadecene (Bagratashvili, etc. 2010*b*).

transmittance of the cell with pure hexane, the recorded transmittance spectra were used to calculate the corresponding absorption coefficients of nc-Si (Dorofeev et al, 2011). The mass concentration of nc-Si in all the studied solutions was similar and amounted to $C = 75$ mg/ml. The effective absorption length of nc-Si in the solutions was calculated from the results of structural investigations by TEM (Fig. 15.7*a*) and the size distribution density functions of the particle was determined from the analysis of small-angle X-ray scattering (Bergmann et al, 2000a), Fig. 15.7*b*, according to which the maximum size of nc-Si does not exceed 5 nm and 7 nm, respectively. At a mass concentration of 75 µg/ml, the concentration of nc-Si in the solutions is not lower than $5.5 \cdot 10^{14}$ cm^{-3}. Accordingly, at this concentration the average distance between the adjacent particles of nc-Si does not exceed 120 nm. Therefore, in determining the absorption coefficient in the wavelength range from 200 nm to 1100 nm, the nc-Si sol with discretely distributed nanoparticles in it can be seen as a continuous medium with an effective thickness, determined by the formula:

$$d = \frac{C \cdot L}{\rho},$$
(15.2)

where ρ is the density of single-crystal silicon and L is the thickness of the cell in the direction parallel to the direction of propagation of the probing radiation. The value of L in our experiments was equal to 1 cm in accordance with the calculations performed within the framework of the described approach, and at the boundaries of the absorption edge of nc-Si there was a blue shift with respect to single-crystal silicon, the magnitude of which was ~0.8 eV.

All the PL spectra registered by the Avantis Ava–Spec-2048 spectrometer were corrected for the spectral function of the hardware device. To do so, the radiation generated by a tungsten filament, heated to a temperature of 2400°C, was recorded. The resulting spectrum was then compared with the spectrum of blackbody radiation with the same temperature, corrected for the grayness factor of tungsten. The spectral function of the spectrometer was obtained by dividing the theoretical emission spectrum of tungsten by the spectrum registered by the instrument.

Polymer nanocomposites mc-PTFE/nc-Si, obtained after the diffusion introduction of nc-Si particles into the mc-PTFE microparticles using supercritical CO_2, had a pinkish hue. They were washed in ethanol, dried, and then were studied using the spectroscopic methods mentioned above.

Figure 15.9 shows the spectra of photoluminescence of the original sol of nc-Si particles in hexane and the nanocomposite powder mc-PTFE/nc-Si in excitation by the argon laser. As can be seen from Fig. 15.9, the solution contains a rather broad (~80 nm) PL band. Note that the PL spectra of nc-Si with excitation at wavelengths $\lambda_1 = 407$ nm and $\lambda_2 = 532$ nm were identical. The band gap, calculated from the experimental absorption spectra for

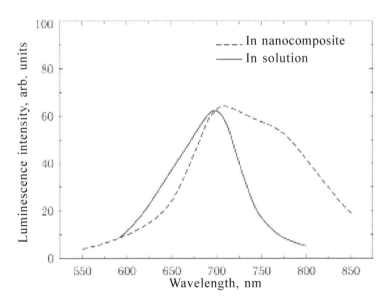

Fig. 15.9. The PL spectrum of solution of nc-Si in hexane and polymer nanocomposite – mc-PTFE/nc-Si excited by an argon laser ($\lambda = 458$ nm; $P = 20$ mW). The solvent – hexane, and the original powder mc-PTFE were tested in advance for the absence of fluorescent impurities in the range 450–900 nm (Bagratashvili et al, 2010b).

the studied nc-Si particles is 2.48 eV (Dorofeev et al, 2011). In this case, the direct exciton transitions would have resulted in a more short-wave photoluminescence with a peak at ~550 nm. Efremov et al (2004) published the theoretically calculated dependence, which allows to connect the average size of nc-Si and the band gap of nanocrystalline silicon. Our results from the analysis of the Raman spectra are consistent with calculations made in the work of Efremov et al (2004), based on which the particles with a parameter of the band gap of 2.5 eV correspond to the size of ~3 nm. We observed red luminescence relating to the intraband transitions for these nanoparticles.

In the synthesized nanocomposite mc-PTFE/nc-Si the maximum of the main band of photoluminescence (related to nc-Si) is shifted by ~15 nm to longer wavelengths with respect to the maximum of the nc-Si band in the solution, but the band itself is strongly structured. A comparison of signal intensities of photoluminescence of the sol of nc-Si and the powder mc-PTFE/nc-Si allowed to make an assessment of the characteristic number of nc-Si particles, embedded in the conditions of our experiments in a single particle of mc-PTFE (size 1–2 μm) as ~10^3–10^4.

Figure 15.10 shows the RS spectrum of the resulting nanocomposite mk-PTFE/nc-Si. It consists of lines with a peak at 510 cm^{-1} and the long-wavelength wing extending into the region to 400 cm^{-1}. For estimates of the mean diameter of the core of nc-Si, D, in the approximation of spherical nanoparticles we can use a simplified relation between the magnitude of the shift $\Delta\gamma$ with the average particle diameter of nc-Si (Rybaltovsky et al, 2006):

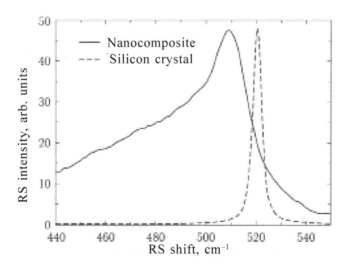

Fig. 15.10. Raman spectrum of mc-PTFE/nc-Si and crystalline silicon plate (Bagratashvili and others, 2010b).

$$\Delta y = \frac{S^2}{4c^2 \gamma_0 D},$$

(15.3)

where S is the speed of sound in the crystal, c is the speed of light in the crystal, γ_0 is the wavenumber corresponding to a single crystal. These estimates give a value $D = 2,5 \pm 0.5$ nm.

DDF maxima of particle sizes (Fig. 15.7b) correspond to values of D (nm): 2.15 (1–200°C), 2.50 (2–300°C), 2.60 (3–450°C) and 2.35 (4–950°C). The resulting values agree well with both the data obtained from the Raman spectrum and from values of the average diameter of silicon nanoparticles, calculated on the basis of the model of the quantum constraint, and the experimentally determined values of the width of the band gap E_g (Dorofeev et al, 2011). Note also that the Raman spectra, obtained in a wide range of wave numbers (from 20 cm^{-1} to 2000 cm^{-1}) for the pure mc-PTFE powder and the synthesized polymer nanocomposite mc-PTFE/nc-Si, contain virtually the same lines, associated with fluctuations of the polymer chains of the matrix, and show no bands that could be attributed to the appearance of vibrations of (Si–F) groups. This indicates a lack of chemical interaction of the polymer surrounding the nanoparticles and silicon nanocrystals.

The approach used in the works by Bagratashvili et al (2010b) and Nikitin et al (2008) is quite general and can be used to add other nanoparticles (structure-forming, magnetic, etc.) to the polymer matrix to produce functional composite systems with specified properties. The resulting polymer nanocomposites mc-PTFE/nc-Si having stable bright photoluminescence can be used, for example, for creating decorative elements, in advertising, for the production of thin polymer films and fibres that are used to protect against forgery of various documents and securities.

Concluding remarks

The polymeric materials in which nanosilicon is immobilized are of practical interest, such as in the development of new UV radiation protectors, as well as films and coatings with stable fluorescence. Among the advantages of such nanocomposite materials compared with other known UV protectors are their environmental purity, thermal stability and the absence of biologically harmful compounds in degradation under the effect of UV radiation. In addition, changing the size distribution function of nanoparticles, their concentration in the polymer matrix and accordingly modifying the state of their surface, the spectral characteristics of the nanocomposite as a whole can be changed in the required direction.

The method of spectral analysis of the properties of nanocomposite polymer materials has specific features related to, *inter alia*, the need to separate the absorption in the core of the silicon nanoparticles, the spectral characteristics of the surface layer, as well as the need to account for the absorption of the polymer matrix.

The study of polymer nanocomposites with immobilized silicon is carried out using a set of measurements of the spectral transmittance characteristics, including the measurement of the optical density of the polymer material, the transmittance spectra in an integrating sphere and Raman scattering (RS). The RS method is used for the systematic control of the set of samples of nanocomposites of the simple sample preparation and execution, as well as efficiency in comparison with, for example, X-ray or TEM. The spectra allow to obtain both the spectral characteristics of the nanomaterial and the average size of an ensemble of particles of the core of the nanocomposite. This in turn opens control over the electromagnetic radiation passing through the composite material containing silica nanoparticles.

References for Chapter 15

Averochkina I. A., Papisov I. M., Matvienko V. N. Pattern formation in aqueous solutionsof sols of polysilicic acid and some polymers. Vysokomol. Soed. A. 1993. V. 35, No. 12. P. 1986-1990.

Bagratashvili V. N. , et al. Perpekt. Mater. 2010a. No 2. P. 39-45.

Bagratashvili V. N., Dorofeyev S. G., Ischenko A. A., Koltashev V. V., Kononov N. N., Krutikova A. A., Rybaltovsky A. O., Fetisov G. V. Immobilization of luminescent nanosilicon microdispersed in the matrix of polytetrafluoroethylene with supercritical carbon dioxide. Supercritical Fluids. Theory and Practice. 2010b. V. 5,No. 2. P. 79-91.

Bagratashvili V. N., Belogorokhov A. I., Ischenko A. A., Storozenko P. A., Tutorsky I.A. Management of the spectral characteristics of ultra-multiphase systems based on nanocrystalline silicon in the UV wavelength range. Dokl. RAN. Fiz. Khim. 2005. V. 405. P. 360-363.

Bakeeva I. V., Kolesnikova Yu. A., Kataeva N. A., Zaustinskaya K. S., Gubin S.N., Zubov V. P. Gold nanoparticles as crosslinking agents in the formation of hybrid nanocomposites. Izv. Akad. Nauk. Ser. Khim. 2008. No. 2. P. 329-336.

Bakeeva I. V., Ozerin L. A., Ozerin A. N., Zubov V. P. The structure and properties of organo-inorganic hybrid hydrogel of poly-N-vinylcaprolactam-SiO₂. Vysokomol. Soed. A. 2010. V. 52, No. 5. P. 776-786.

Briston, G. H., Katan L. L. Polymer films. M.: Khimiya. 1993. 384 p.

Galaev I. Yu. 'Smart' polymers in biotechnology and medicine. Usp. Kmih. 1995. V. 64, No. 5. P. 505-524.

Glagoleva N. N., et al. N. Long-lived excited state spiroantrooxsazine after its matrix isolation in halogenated polyolefins by supercritical fluid impregnation. Supercritical fluids. Theory and Practice. 2010. V. 5, No. 1. P. 73-79.

Efremov M. D., et al. Visible photoluminescence of silicon nanopowder created by evaporation of silicon by the powerful electron beam. Pis'ma Zr. Eks. Teor. Fiz. 2004. V. 80. P. 619-622.

Zubov V. P., Ischenko A. A. , Storozenko P. A., Bakeeva I. V., Kirillina Yu. O., Fetisov G.V. Organic-inorganic hybrid hydrogels containing nanocrystalline silicon. Dokl. RAN. , 2011. V. 437, No. 6. P. 768-771.

Iveronova V. I., Revkevich G. P. The theory of scattering of X-rays. Moscow: Moscow State University. 1978. 278.

Ischenko A. A., Dorofeyev S. G., Kononov N. N., Eskova E. V., Ol'khov A. A. Request for the grant of the patent, 16.12.2009. Registration number 2009146715.

Kirilina Yu. O., Bakeeva I. V., Bulichev N. A., Zubov V. P., Organic-inorganic hybrid hydrogels based on linear poly-N-vinylpyrrolidone, and hydrolytic polycondensation products of tetramethoxysilane. Vysokomolek. Soed. B. 2009. V. 51, No. 4. P. 705-713.

Kuleznev V. N., Shershnev V. A. Chemistry and physics of polymers. M.: KolosS, 2007. 129 p.

Nikitin L. N., Gallyamov M. O., Said-Galiyev E. E., Khokhlov A. R., Bouznik V. M. The supercritical carbon dioxide as the active medium for chemical processes involving fluoropolymers. Ros. Khim. Zh.. 2008. T. LII, No. 3. P. 56-65.

Ol'khov A. A., Goldshtrah M. A., Ischenko A. A.. Request for the grant of the patent, 04.12.2009, Registration number 2009145013.

Ol'khov A. A., et al. Nanocomposite films with UV-protective properties based on polyethylene with ultrafine silicon. Plast. Massy. 2010. No. 9. P. 40-46.

Rybaltovsky A. O., Ischenko A. A., Koltashev V. V., Popov, A. P., Sviridova A. A. Methods of experimental study of the spectral characteristics of water–emulsion composite media containing silica nanoparticles. GSSSD certificate from DOE 131-2007. July 12, 2007.

Rybaltovsky A.O. et al. The spectral features of water–emulsion composite media containing nanoparticles of silicon. Optika Spektroskopiya. 2006. V. 101. P. 626-633.

Tutorsky I. A., Belogorokhov A. I., Ischenko A. A., Storozenko P. A. The structure and adsorption properties of nanocrystalline silicon. Kolloid. Zh. 2005. V. 67, No. 4. P. 541-547.

Fetisov G. V. X-ray phase analysis. Analytical chemistry and physico-chemical methods of analysis. T. 2/ Ed. A. A. Ischenko. Moscow: IC Academy, 2010. Chap. 11. P. 153-184. 416 p.

Samyuels S.-B. Methods and compositions for the protection of polymers from UV radiation. Patent RU No. 2000130235. IPC C08L101/00, C08J5/18, C08K5/00, C08K5/3492, C08K5/3435.

Pfendner R., Hoffman, K., Meier, F., Rotzinger B. Synthetic polymers containing mixtures - additives for enhanced action. Patent RU No. 2001121489. IPC 7 C08L101/00, C08L23/00.

Samyuels S.-B. Methods and compositions for the protection of polymers against UV radiation. Patent RU No. 2222560. IPC 7 C08L101/00, C08J5/18, C08K5/3492, C08K5/3435.

Bergmann A., Fritz G., Glatter O. Solving the generalized indirect Fourier transform (GIFT) by Boltzmann simplex simulated annealing (BSSA). J. Appl. Cryst. 2000. V. 33. P. 1212.

Bergmann A., Orthaber D., Scherf G., Glatter O. Improvement of SAXS measurements on Kratky slit systems by Göbel mirrors and image plate detectors. J. Appl. Cryst. 2000. V. 33. P. 869–875.

Dorofeev S.G., Kononov N.N., Ischenko A.A. Synthesis temperature effect on optical properties luminescent nc-Si, gained from SiO. Adv. Mater. 2011 (to be published).

Gallyamov M.O., Chaschin I. S., Gamzazade A. I., Khokhlov A.R. Chitosan Molecules Deposited from Supercritical Carbon Dioxide on a Substrate: Visualization and Conformational Analysis. Macromol. Chem. Phys. 2008. V. 209, No. 21. P. 2204–2212.

Gallyamov M.O., Qin S., Matyjaszewski K., Khokhlov A., Mölle M. Motion of single wandering diblock-macromolecules directed by a PTFE nano-fence: real time SFM observations. Phys. Chem. Chem. Phys. 2009. V. 11, No. 27. P. 5591–5597.

Glatter O. Determination of particle-size distribution functions from small-angle scattering data by means of the indirect transformation method. J. Appl. Cryst. 1980. V. 13. P. 7–11.

Koch C. C. (Ed.). Nanostructured Materials. Processing, Properties, and Applications. — N.Y.: William Andrew Publishing, 2009. — 752 p.

Kumar V. Nanosilicon / Ed. by V. Kumar. Elsevier Ltd., 2008. — 368 p.

Ledoux G., Porterat D., Reynand C., Huisken F., Kohn B., Paillard V.. Phys. Rev. B. 2000. V. 62. P. 15942.

Leoni M., Confente T., Scardi P. PM2K: a flexible program implementing Whole Powder Pattern Modelling. Z. Kristallogr. Suppl. 2006. V. 23. P. 249–254.

Loos W., Verbrugge S., Du Prez F. E., Bakeeva I.V., Zubov V. P. Thermo-Responsive Organic/Inorganic Hybrid Hydrogels based on Poly(N-vinylcaprolactam). Macromol. Chem. and Phys. 2003. V. 204, No. 1. P. 98–103.

Lozinsku V. I., Bakeeva I.V., Presnyak E. P., Damshkaln L.G., Zubov V. P. Cryostructuring of polymer systems. XXVI. Heterophase organic–inorganic cryogels prepared via freezing–thawing of aqueous solutions of poly(vinyl alcohol) with added tetramethoxysilane. J. Appl. Polym. Sci. 2007. V. 105. P. 2689–2702.

Okano T. (Ed.). Biorelated Polymers and Gels. San Diego: Acad. Press, 1998. 568 p.

Pavesi L., Turan R. (Eds.) Silicon Nanocrystals. Fundamental, Synthesis and Applications. Wiley-VCH Verlag GmbH & Co., 2010. 627 p.

Ruben M. (Ed.). Soft Contact Lenses: Clinical and Applied Technology. N.Y.: Wiley, 1978. xiii+496 p.

Scardi P., Leoni M. Line profile analysis: pattern modelling versus profile fitting. J. Appl. Cryst. 2006. V. 39. P. 24–31.

Scardi P. Recent advancements in whole powder pattern modeling. Z. Kristallogr. Suppl. 2008. V. 27. P. 101–111.

St°ahl K. The Huber G670 imaging-plate Guinier camera tested on beamline I711 at the MAX II synchrotron. J. Appl. Cryst. 2000. V. 33. P. 394–396.

Strbeck N. X-ray scattering of soft matter. Berlin–Heidelberg: Springer-Verlag, 2007. xx+238 p.

Svrcek V., Fujiwara H., Kondo M. Improved transport and photostability of poly(methoxyethylexyloxy-phenylenevinilene) polymer thin films by boron doped freestanding silicon nanocrystals. Appl. Phys. Lett. 2008. V. 92. P. 143301–(3).

WINXPow Version 1.06. STOE & CIE GmbH Darmstadt, Germany, 1999.

Wolkin M.V., Jorne J., Fauchet P.M., Allan G., Delerue C. Electronic states and luminescence in porous silicon quantum dots: The role of oxygen. Phys.Rev. Lett. 1999. V. 82. P. 197–200.

16

Applications of nanosilicon in solar energy

On the global commercial market nearly 85 trillion ($8.5 \cdot 10^{13}$) kWh of energy per year is bought and sold, including the U.S. where the energy consumption is about 25 trillion ($2.5 \cdot 10^{13}$) kilowatt hours per year, which corresponds to more than 260 kWh per person per day. At the same time, the Sun delivers to Earth 10 000 times more free energy than is actually used throughout the world. If only 1% of the US territory was used for the installation of solar equipment (photovoltaic batteries or solar systems for hot water) operating with an efficiency of 10%, then this country would be fully provided with energy. The amount of solar energy falling on the roofs and walls greatly exceeds the annual energy consumption of the residents of these homes in most countries of the world. The energy of sunlight and heat can be used either through solar collectors or photovoltaic cells. The effectiveness of the latter method can be improved by using nanosilicon.

The action of photovoltaic (PV) elements, which are the basis of converters of solar energy into electricity, is based on the phenomenon of the internal photoelectric effect in semiconductors. The photoelectric converters of solar energy used most commonly silicon, with the addition of other elements, forming a structure with the p–n junction. The thickness of the semiconductor does not exceed 0.2–0.3 mm. Photovoltaic cells are produced using also other semiconductor materials, and the elements can consist of a single p–n junction, as well as many junctions, including metal-semiconductor contacts (Schottky barriers). A good overview of the design of solar photovoltaic energy elements and materials for their production can be found, for example, in the article (Meytin, 2000).

Modern production of solar cells is almost entirely based on silicon. About 80% of all modules are made using polycrystalline or single crystal (c-Si) silicon, and the remaining 20% use hydrogenated amorphous silicon (a-Si: H). Crystalline solar cells are most used. Photovoltaic cells of single-crystalline silicon have the best efficiency (14%), but they are significantly more expensive than solar cells made from polycrystalline silicon, the efficiency of which on average is still only 11%.

In advanced development and production there are several types of thin-film PV alternatives, which may in the future win the market. Photocells of the following materials have been developed most efficiently:
- amorphous silicon (a-Si: H),
- telluride/cadmium sulphide (CTS),
- copper–indium or copper– gallium diselenide (CIS or CIGS),
- thin-film crystalline silicon (c-Si film),
- nanocrystalline dye-sensitized electrochemical solar cells (nc-dye).

16.1. Silicon solar cells

Solar power cells are not produced from separate photocells but from solar modules or panels. The solar module is a battery of interrelated solar cells (SC) assembled under a glass cover. The more intense the light falling on the photocells, and the larger their area, the more electricity is produced and the greater the current.

As regards design, the simplest solar cell based on single-crystal silicon is a semiconductor diode, consisting of semiconductors in metallurgical contact with different types of conductivity (electronic n and hole p), as shown in Fig. 16.1.

When the front side of the photoelectric cell is illuminated (Fig. 16.2) with light Φ with photon energy hv higher than the band gap of the semiconductor, the absorbed photons generate non-equilibrium electron-hole pairs.

The electrons generated in the p-layer near the p–n junction, by the contact difference of the transition potentials are carried into the n-region. Similarly, the excess holes if they are in the n-layer, are partially transferred by the contact potential difference to the p-layer. As a result, the n-layer acquires an excess negative charge, and the p-layer – positive. The initial contact difference potential between p- and n-layers of a semiconductor decreases, and a voltage appears in the external circuit, as shown in Fig. 16.2. The negative pole of the power source corresponds to n-layer and

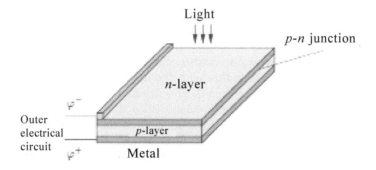

Fig. 16.1. The scheme of construction of solar photoelectric panels.

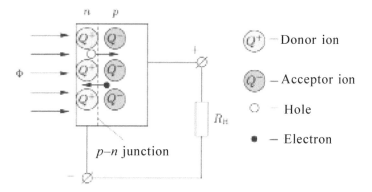

Fig. 16.2. Photovoltaic cell with a connected load R_H.

p-layer – positive. If the external circuit with the load closed, the current generated by the light in the photovoltaic cell will flow through it, and the load will match the power of this current to be used in various applications.

Production of semiconductor structures based on single-crystal silicon is a technically complex and expensive process. Currently, the market is experiencing a shortage of silicon, and the price of single-crystal silicon ingots of 'solar' quality[1] is about $100/kg. Therefore, the attention of engineers and scientists have been drawn to the cheaper materials such as amorphous silicon (a-Si: H) and polycrystalline semiconductors including polycrystalline silicon. Theoretically, the efficiency of solar cells made from these materials is lower than in their single crystal counterparts, but their production is much cheaper.

Based on the price of 1 kilowatt of produced electricity, hydrogenated amorphous silicon (a-Si:H) and polycrystalline silicon (pc-Si) are at present one of the most promising solar energy materials. Solar panels from α-Si:H are produced industrially, and have a number of advantages: higher open-circuit voltage, V_{oc} to 0.8–1.1 V, is achieved through a greater band gap, $E_g \sim 1.6$ eV; relative ease of use on large areas, use of a variety of substrates of easily available materials (glass, stainless steel, polyamide), low cost. Theoretically, their efficiency can reach 16%. At present, the maximum efficiency of the experimental laboratory cells based on a-Si:H is only about 12%, which is somewhat lower than that of monocrystalline solar cells. In reality, this value for industrial samples of a-Si:H cells is about 10%. However, it is possible that with the development of technology the efficiency of the cells based on a-Si:H will reach the theoretical ceiling.

The optical absorption of amorphous silicon is 20 times higher than that of crystalline silicon. Therefore, for a substantial absorption of visible light

---

[1]Impurities and lattice defects in single-crystal silicon do not just reduce the efficiency of conversion of sunlight into electricity, but also lead to rapid degradation of solar cells. The-refore, monocrystalline solar batteries are produced from defect-freeSi single crystals with an impurity content of $\sim 10^{-8}$%.

it is sufficient to use a a-Si:H film thickness of 0.5–1.0 mm instead of the expensive 300 µm single-crystal silicon wafers. In addition, with existing technologies of preparation of thin films of amorphous silicon with a large area it is not required to carry out cutting, grinding and polishing operations required in the manufacture of solar cells based on single-crystal silicon. Compared with polycrystalline elements, silicon products based on a-Si:H are produced at a low temperatures (~300°C), so we can use cheap, for example, glass substrates, which will reduce the consumption of silicon 20 times.

The simplest design of the solar cell produced from a-Si:H is the metal–semiconductor structure (Schottky diode). However, the open circuit voltage (no external load), generated by a simple solar cell based on a-Si:H with a Schottky barrier, is typically less than 0.6, so they are not used widely in solar energy. To increase the photoelectric conversion efficiency of solar cells based on amorphous silicon, multilayer thin film structures with $p–n$ junctions are used, because it is much easier to produce such structures from amorphous silicon than from single crystal. Compared with Schottky diodes, higher efficiency is obtained, for example, in solar cells based on amorphous silicon with a three-layer structure of the $p–i–n$ type (Si of p-type–Si of i-type (insulator) – Si of n-type), shown in Fig. 16.3.

EMF in the photovoltaic cell of this type is generated in the area of semiconductor transitions, as in the SC on the basis of monocrystalline silicon. The 'thick' layer of undoped a-Si:H in this structure absorbs a significant share of the light and gives an additional number of carriers. Such structures can be made both on the metal and the glass substrate.

As seen from the illustrated structures of amorphous silicon solar cells, their production relates to the field of thin-film technology. Therefore,

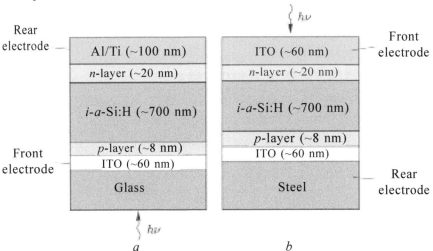

Fig. 16.3. Photovoltaic structure of $p–i–n$ amorphous silicon on glass (a) and steel (b) substrate.

they are often called thin-film solar cells (TFSC). Advantages of TFSC and solar cells from these modules are obvious – low weight, the ability to produce flexible solar cells on flexible substrates (e.g. metal foils or polymer films); easy insertion into virtually any building design; ease of transportation and installation of high speed, and a considerably lower price compared to monocrystalline solar panels. Therefore, the despite the fact that the thin-film silicon solar cells still have much lower efficiency than monocrystalline, consumer interest in them is growing rapidly, their production is increasing and more resources are invested in their production and development of new technologies and designs.

The problem with all of the above types of solar cells, greatly limiting their efficiency, is the limited bandwidth of the solar spectrum (Fig. 16.4), that they use.

The solar spectrum consists of:
- UV wavelength 0.28–0.38 μm invisible to the human eyes and making up approximately 2% of the solar spectrum;
- light waves in the visible range of 0.38–0.78 μm, representing approximately 49% of the spectrum;
- infrared wavelength 0.78–3.0 μm, which accounts for a large of the remaining 49% of the solar spectrum.

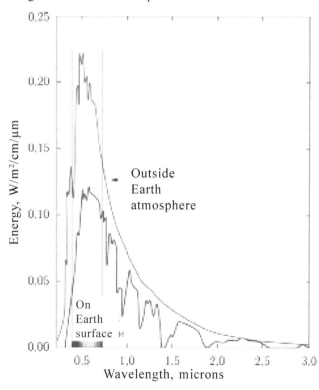

Fig. 16.4. The spectrum of solar radiation incident on the Earth.

The working range of the spectrum for the solar cell is determined by the energy of light quanta and the energy band gap of the semiconductor. This region starts from the red edge (the fundamental absorption edge) of the measured wavelength λ_{rb}. Excitons in a semiconductor can be generated by all the photons with a length smaller than the wavelength λ_{rb}, which is defined by

$$\lambda_{rb} = \frac{c}{v} = \frac{ch}{hv} = \frac{1.24}{hv} = \frac{1.24\{\mu m \cdot eV\}}{\Delta E_g\{eV\}}.$$

For germanium, silicon and gallium arsenide at 300 K λ_{rb} is 1.88, 1.11, 0.873 μm, respectively.

Photons with energies above ΔE_g, but not more than $(2-3)\Delta E_g$, give to the generation of the exciton only part of their energy equal to ΔE_g. The rest of the photon energy is converted into heat. Low-energy photons with a wavelength $\lambda > \lambda_{rb}$, i.e. with the photon energy $hv < \Delta E_g$, cannot form electron–hole pairs. For them, the semiconductor is transparent, but by interacting with the phonons, they, nevertheless, also heat the semiconductor.

Thus, in a bulk semiconductor a single photon can generate only one exciton (electron–hole pair), even if the energy of the absorbed photon is much higher than the band gap. The remaining energy of the photon will be spent on heating the material of the SC which is worsened by the heat input from the low-energy photons. This greatly impairs the technical specifications of SC[2]. In conventional solar cells such 'mismanagement' wastes up to 70% of the energy of the incident light.

In modern manufacturing technology of silicon TPSC the band of the solar spectrum is expanded and the efficiency of solar cells increased using tandem heterostructures that combine solar cells made of amorphous and microcrystalline silicon. The bottom line is that the band gap a-Si:H is ~1.9 eV, whereas for crystalline Si it is ~1.12 eV (Bresler et al, 2004). Therefore, the upper layer of amorphous silicon effectively uses short-wave radiation, passing almost without absorption the red and infrared radiation, which is effectively absorbed by the next layer of polycrystalline silicon. As a result, the conversion efficiency of solar light is increased by 50% compared to conventional single-layer solar cells made of amorphous silicon. The device of this structure is shown in Fig. 16.5.

16.2. Improving the efficiency of solar cells with nc-Si

Due to the recently experimentally observed intense photoluminescence and the many-electron internal photoeffect in nanocrystalline silicon, interest in this material by the manufacturers of photovoltaic generators has been rapidly increasing. These properties have the potential to greatly improve

---

[2]The heating of a silicon solar cell by one degree above 25°C causes it to loose 0.002 V in voltage, i.e. 0.4%/degree. On a bright sunny day, the element heats up 60–70°C and loses due to this are 0.07–0.09 V of voltage.

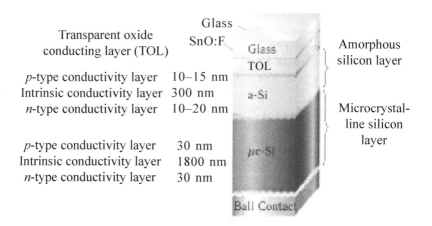

Transparent oxide
conducting layer (TOL)

Glass
SnO:F

p-type conductivity layer   10–15 nm
Intrinsic conductivity layer 300 nm
n-type conductivity layer   10–20 nm

p-type conductivity layer   30 nm
Intrinsic conductivity layer 1800 nm
n-type conductivity layer   30 nm

Amorphous
silicon layer

Microcrystal-
line silicon
layer

Fig. 16.5. The scheme of the modern silicon tandem thin film solar module (cross section).

the efficiency of solar cells. But the solar cells are not yet produced commercially from nc-Si because nc-Si is not produced in sufficient quantities, especially semiconductor purity.

There are many methods proven in the laboratory as well as theoretical developments to produce new solar cells with improved performance based on the nc-Si, as well as the use of this material to enhance the effectiveness of commercially available solar cells.

It is possible to significantly improve the efficiency of conventional silicon (both amorphous and single crystalline) photovoltaic solar cells with application to the surface of thin coatings of silicon nanoparticles. This coating has the property of photoluminescence that converts ultraviolet radiation into visible light that can be effectively transformed to electricity in a photocell.

Ultrathin films of monodisperse Si nanoparticles are easily applied to a layer of polycrystalline silicon in the solar cell structure, as carried out in Stupca et al. (2007). Measurements of the open circuit voltage of the solar cell modified with a film of particles with a diameter of 1 nm with blue luminescence, or a film of particles 2.85 nm in red luminescence showed that the voltage of these solar cells is significantly higher than that of conventional solar cells made from polycrystalline silicon, the and characteristic of converting light into electricity in the UV/blue regions in the modified SCs was higher by 60%. At the same time, the increase of efficiency in the visible light was about 10% and 3% respectively in the case of modification with films with red and blue luminescence. In addition to increasing the power of the SCs, the film of Si nanoparticles reduced the heating of the element by the 'uselessly' absorbed part the solar spectrum. These results indicate that in the modification of the SC of pc-Si with a thin layer of nc-Si the role of the effects of resonance transport across the nanofilm and the current conversion by the Schottky barrier mechanism at the nanoparticles–metal interface becomes important. A patent was granted for this method of increasing the capacity of solar cells (Nayfeh et al, 2009).

The possibility of such a power increase of the SC with the inclusion of Si nanofilms in the structure was tested in a number of other studies (Nayfeh & Mitas, 2008) using a commercially produced solar photovoltaic cell battery made of polycrystalline silicon. The experiments were carried out using a cell of the size 12 × 12 cm produced by BP Solarex, made of p-type silicon with a thin layer of n-type at the top to which was affixed a thin film of Si_3N_4 which was supposed to protect the cells from heating, filtering out infrared radiation. Structurally, the collection of photoelectricity over the entire surface of the cells was carried out by a grid of 50 lines (front contacts), made of a silver–copper alloy. The thickness of the silver/ copper grid lines was larger than that of the nitride coating, therefore the grid wires protruded from the nitride coating. As an experiment, the surface of the photovoltaic elements was deposited with an ultrathin film of Si nanoparticles, luminescent in the visible light range under the effect of UV radiation. The deposited film densely covered the surface of the element and, at the same time, through the projecting conductors they were in electrical contact with the circuit of the photovoltaic cell.

In the experiments it was found that a coating of silicon nanoparticles significantly increased the efficiency of the photovoltaic element. The nanosilicon coating improves the performance of industrial solar cells made of polycrystalline silicon in the ultraviolet region by 60–70%. Photoelectric conversion efficiency of the cells increased at the expense of, first, the transformation of harmful UV radiation, which only destroys a cell in the visible light and in electron–hole pairs. The film of the Si particle with the size of 2.85 nm, with red light luminescence, also resulted in an increase of 7–13% in the visible region as a result of the expansion of the band gap to 2.15 eV because of the size confinement quantum effect. It has been shown that a film of silicon nanoparticles had a long-time stability under ultraviolet light and atmospheric effects, because in the process of applying the nanoparticles were spontaneously assembled in close-packed structure.

A significantly greater increase in the efficiency of solar cells can be achieved using nanocrystalline silicon (in the form of quantum dots or short nanowires) directly for the creation of SC structures rather than amorphous or crystalline silica in the usual way, as shown in Fig. 16.6.

In this case, we can use immediately all of the interesting features of the properties of the silicon quantum dots, such as photoluminescence in the visible region; blue shift of the absorption edge, which can be adjusted by resizing the particles, and the ability to multiexciton output of the internal photoeffect.

According to the model of quantization of the conduction band of a semiconductor crystal with a decrease in its size to quantum dots (Al. L.

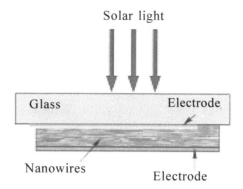

Fig. 16.6. The structure of the solar cell with Si nanoparticles.

Efros, A.L. Efros, 1982) the band gap of the semiconductor can greatly increase with the reduction of the nanocrystal size (see chapter 1 of this monograph, Fig. 1.6). As a result, the absorption edge can be shifted quite considerably to shorter wavelengths. As we have seen, the multi-layered heterogeneous structures with different width of the forbidden zones are already used in the manufacture of thin-film solar cells for more efficient use of the solar spectrum. At present, far the most efficient are thin film solar cells based on a three-layer a-Si/a-SiGe/a-SiGe structure, where each layer has its own width of the optical gap. The middle layer of i-type with a slit width of 1.6 eV absorbs light in the green area and the lower with a high Ge content a long-wavelength light. Tunnelling transitions for between layers, and the entire structure absorbs light from 350 to 900 nm by the photoeffect. However, this band covers only 2/3 of the energy of the incident solar radiation on the solar cell. At the same time, based on nanoparticles, including only silicon ones, but with the correctly selected sizes, it is possible to a fundamentally new SE, which will effectively work with the entire solar spectrum without exception.

For quite a long time it has been shown theoretically that in a quantum dot (nanocrystal) of a semiconductor a single photon with sufficient energy could generate more than one exciton. The essence of this phenomenon lies in the quantum structure of the conduction band of a semiconductor nanocrystal (see, e.g. Ellingson et al, 2005, 2006).

In a bulk semiconductor a single photon in the energy region of the solar spectrum can usually produce in the valence band only one electron capable of moving into the conduction band, i.e. creates only one hole pair (exciton) in a semiconductor. Theoretically, the two-electron photoelectric effect, called impact ionization in the theory of semiconductors, can exist when a photon with very high energy which photons from the solar spectrum do not have, can communicate to an electron from the valence band, the energy which is so high that this hot electron in a collision with a neighboring atom ionizes it itself, having lost some energy, following the secondary electron, goes into the conduction band. But the likelihood of such events for the solar cell is negligible because for impact ionization the photon must have the energy equal to the energy of photons of soft X-ray, whose number in the solar radiation on Earth is very small, plus the probability of the process itself is very low. In contrast to the bulk semiconductor nanocrystals smaller than 10 nm have a multilevel conduction band. Therefore, a photon with an energy of only half greater than the band gap can eject two electrons right out of the valence band, which will settle at different quantum levels of the conduction band of nanocrystal. If the photon energy is three times the width of the forbidden zone, three electrons are ejected into the conduction band, etc. Thus, the multiexciton generation mechanism (MEG) operates in nanocrystalline semiconductors. The MEG effect increases the photovoltaic current and basically allows us to create SC with high efficiency and power (Ellingson et al, 2006).

In 2004, the MEG effect was first observed experimentally in nanocrystals of PbSe (Ellingson et al, 2005). This effect has been studied systematically over many years in the National Renewable Energy Laboratory (NREL), in Golden, Co, on a number of semiconductors, largely with a pragmatic view for transfer to industry. The colloidal systems of nanocrystals of PbS, PbSe, PbTe have already been studied. Thus 7-exciton generation was obtained in PbSe nanocrystals (Schaller et al, 2006). In 2007, the effect of multiexciton generation was also found in nanocrystalline silicon particles, which were in a colloidal solution (Beard et al, 2007). Computer simulation shows that due to the effect that the many-electron internal photoemission in the high-energy part of the solar spectrum, which is most intense, even single-layer thin-film solar cells from nanocrystalline silicon can have a theoretical limit of energy conversion of sunlight into electricity more than 40%.

The research results obtained for multiexciton photoelectron emission of nanocrystalline semiconductors, conducted by NRL, were published in numerous articles, reports, and monographs (e.g. Ellingson et al, 2005; Ellingson et al, 2006; Shabaev et al. 2006; Beard et al, 2007) and summarized in the reviews of Beard and Ellingson (2008); Nozik (2008). The use of the MEG effect in solar technology based on nanocrystalline materials discussed in the book Nozik (2006), as well as articles by Luque et al. (2007); Luther (2008).

16.3. Photovoltaic windows

The American corporation Octillion Corp.[3] is developing a new technology to produce photovoltaic cells from silicon nanocrystal (nc-Si), which can turn ordinary domestic windows into solar cells, capable of producing

---

[3]Since the beginning of 2009 Octillion Corp. became known as the New Energy Technologies, Inc. Internet site: http://www.newenergytechnologiesinc.com/index

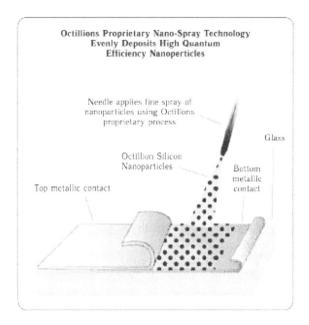

Fig. 16.7. The scheme covers the application of photovoltaic electric spray-lf-Si on window glass.

electricity from passing through them of sunlight, with no appreciable loss of its own transparency, and does not require significant changes in the infrastructure of the building[4].

This work of Octillion Corp. is in agreement with the University of Illinois (the University of Illinois in Urbana-Champaign), using the patented technology for manufacture of photovoltaic cells from nanosilicon. New photovoltaic cells are manufactured using an electrochemical process using ultrasound, in which silicon particles are obtained with a diameter of 1 to 4 nm with very high luminescence providing a high coefficient of conversion of photons with short wavelengths (50 to 60%). The scheme of the specially developed process of applying electric photovoltaic coatings by electric spraying on window glass is shown in Fig. 16.7.

In electrojet deposition in this technology of silicon nanoparticles the high efficiency of conversion of UV radiation of sunlight in the visible light remains unchanged, and the produced film has the required thickness and provides the theoretically predicted (for the film of these particles) coefficient of photoelectric conversion.

When a thin film of silicon nanoparticles is applied (sprayed) on the window glass, the ultraviolet light, which falls on the window, starts to be absorbed and converted into electrical current. At the appropriate combination the thin film acts as a nanosilicon photovoltaic solar cell,

---

[4] http://www.octillioncorp.com/OCTL_20070207.html

having the ability to convert sunlight into electrical energy. The use of such generators of gratuitous electricity does not require any changes in construction technology. Photovoltaic windows can be simply installed in existing homes and apartments.

The first sample of such a laboratory glass window (under the name NanoPower Window) was produced by Octillion Corp. in the middle of 2007[5]. Tests showed that the prototype provides the planned degree of photoelectric conversion, and the manufactured cell has no electrical shunting which could lead to failure of the solar cell. After successful testing of laboratory samples the corporation is committed to quickly bringing its development to the mass production of nanopower windows, and, as stated in March 2008, is engaged in optimization and reducing the cost of the coating process, the selection of an optimal transparent conductive material of the upper and lower contact layers and optimizing the photoconversion of the spectral composition of sunlight.

Currently, the company New Energy Technologies, Inc. manufactures commercially windows with the SolarWindow™ sprayed coating according to its unique patented technology.

References for Chapter 16

Bresler M. S., Gusev O. B., Terukov E. I., Froizheim A. Edge electroluminscence of silicon: amorphous silicon–crystalline silicon heterostructure. Fiz. Tverd. Tela. 2004. V. 46(1). P. 18–20.

Meitin M. Photovoltaics: materials, technology, prospects, Elektronika: Science, Technology, Business. 2000. No. 6. P. 40–46.

Efros Al.L., Efros A.L. Interband absorption of light in a semiconductor sphere. Fiz. Tekhnol. Poluprovod. 1982. V. 16. No. 7. 1209–1214.

Beard M. C., Ellingson R. J. Multiple exciton generation in semiconductor nanocrystals: Toward efficient solar energy conversion. Laser & Photonics Reviews. 2008. V. 2, No. 5. P. 377–399.

Beard M. C., Knutsen K.K., Yu P., Luther J., Song Q., Ellingson R. J., Nozik A. J. Multiple exciton generation in colloidal silicon nanocrystals. Nano Lett. 2007. V. 7. P. 2506–2512.

Ellingson R., Beard M., Johnson J., Murphy J., Knutsen K., Gerth K., Luther J., Hanna M., Micic O., Shabaev A., Efros A. L., Nozik A. J. Nanocrystals generating >1 electron per photon may lead to increased solar cell efficiency. SPIE Newsroom. 2006. Article No. 10.1117/2.1200606.0229 (4 p.).

Ellingson R. J., Beard M. C., Johnson J. C., Yu P., Micic O. I., Nozik A. J., Shabaev A., Efros A. L. Highly efficient multiple exciton generation in colloidal PbSe and PbS quantum dots. Nano Lett. 2005. V. 5. P. 865–871.

Luque A., Martí A., Nozik A. J. Solar cells based on quantum dots: multiple exciton generation and intermediate bands. MRS Bull. 2007. V. 32. P. 236–241.

Luther J.M., Law M., Beard M. C., Song Q., Reese M.O., Ellingson R. J., Nozik A. J.

---

[5]Next energy news, 8.1.2007, http://www.nextenergynews.com/news/next-energy-news1.8a. html.

Schottky solar cells based on colloidal nanocrystal films. Nano Letters. 2008. V. 8. P. 3488–3492.

Nayfeh M.H., Mitas L. Silicon Nanoparticles: New photonic and electronic material at the transition between solid and molecule. In: Nanosilicon. Ed. by V. Kumar. Elsevier Ltd., 2008. Ch. 1. P. 1–78.

Nayfeh M.H., Stupka M. Turki Al Saud T., Alsalhi M. Patent application title: SiliconNanoparticle Photovoltaic Devices. IPC8 Class: AH01L3100FI. USPC Class: 136256. Chicago, IL US, (12-17-2009). http://www.faqs.org/patents/app/20090308441

Nozik A. J. Quantum structured solar cells. In: Nanostructured Materials for Solar Energy Conversion/ Ed. dy T. Soga. Elsevier B.V., 2006. Ch. 15. P. 485–516.

Nozik A. J. Multiple exciton generation in semiconductor quantum dots. Chem. Phys. Letters, Frontiers in Chemistry. 2008. V. 457. P. 3–11.

Schaller R.D., Sykora M., Pietryga J.M., Klimov V. I. Seven excitons at the cost of one: redefining limits for conversion efficiency of photons into charge carriers. Nano Lett. 2006. V. 6. P. 424–429.

Shabaev A., Efros L., Nozik A. J. Multi-exciton generation by a single photon in nanocrystals. Nano Lett. 2006. V. 6. P. 2856–2863.

Stupca M., Alsalhi M., Al Saud T., Almuhanna A., Nayfeh M.H. Enhancement of polycrystalline silicon solar cells using ultrathin films of silicon nanoparticle. Appl. Phys. Lett. 2007. V. 91. P. 063107 (3 p.).

Index

T - #0291 - 071024 - C16 - 234/156/33 - PB - 9780367378516 - Gloss Lamination